Lecture Notes in Computer Science 5441

Commenced Publication in 1973
Founding and Former Series Editors:
Gerhard Goos, Juris Hartmanis, and Jan van Leeuwen

Tülay Adali   Christian Jutten
João Marcos Travassos Romano
Allan Kardec Barros (Eds.)

Independent Component Analysis and Signal Separation

8th International Conference, ICA 2009
Paraty, Brazil, March 15-18, 2009
Proceedings

 Springer

Volume Editors

Tülay Adali
University of Maryland, Baltimore County
Department of Computer Science and Electrical Engineering
ITE 324, 1000 Hilltop Circle, Baltimore, MD 21250, USA
E-mail: adali@umbc.edu

Christian Jutten
Domaine Universitaire, GIPSA-lab
BP 46, 38402 Saint Martin d'Hères Cedex, France
E-mail: Christian.Jutten@gipsa-lab.inpg.fr

João Marcos Travassos Romano
FEEC / Unicamp, Departamento de Microonda e Óptica (DMO)
Avenida Albert Einstein 400, 13083-852 Campinas, Sao Paulo, Brazil
E-mail: romano@dmo.fee.unicamp.br

Allan Kardec Barros
Universidade Federal do Maranhão
Centro Tecnológico, Curso de Engenharia Elétrica
Avenida dos Portugueses, s/n, Bacanga, 65080-040 São Luís, MA,Brazil,
E-mail: allan@ufma.br

Library of Congress Control Number: 2009921803

CR Subject Classification (1998): C.3, F.1.1, E.4, F.2.1, G.3, H.1.1, H.2.8, H.5.1, I.2.7

LNCS Sublibrary: SL 1 – Theoretical Computer Science and General Issues

ISSN   0302-9743
ISBN-10   3-642-00598-5 Springer Berlin Heidelberg New York
ISBN-13   978-3-642-00598-5 Springer Berlin Heidelberg New York

springer.com

© Springer-Verlag Berlin Heidelberg 2009
Printed in Germany

Typesetting: Camera-ready by author, data conversion by Scientific Publishing Services, Chennai, India
Printed on acid-free paper   SPIN: 12626304   06/3180   5 4 3 2 1 0

Preface

This volume contains the papers presented at the 8th International Conference on Independent Component Analysis (ICA) and Source Separation held in Paraty, Brazil, March 15–18, 2009. This year's event resulted from scientific collaborations between a team of researchers from five different Brazilian universities and received the support of the Brazilian Telecommunications Society (SBrT) as well as the financial sponsorship of CNPq, CAPES and FAPERJ.

Independent component analysis and signal separation is one of the most exciting current areas of research in statistical signal processing and unsupervised machine learning. The area has received attention from several research communities including machine learning, neural networks, statistical signal processing and Bayesian modeling. Independent component analysis and signal separation has applications at the intersection of many science and engineering disciplines concerned with understanding and extracting useful information from data as diverse as neuronal activity and brain images, bioinformatics, communications, the World Wide Web, audio, video, sensor signals, and time series.

Evidence of the continued interest in the field was demonstrated by the healthy number of submissions received, and of the 137 papers submitted 97 were accepted for presentation. The proceedings have been organized into seven sections: theory, algorithms and architectures, biomedical applications, image processing, speech and audio processing, other applications, and the special session on evaluation. Within each section, papers have been organized alphabetically by the first author's last name. However, the strong interaction between theory, method, and application inevitably implies that many papers could have equally been placed in alternative categories.

The conference featured a lecture session on semi-blind methods, emphasizing the importance of the small modification introduced to the conference title last year by removing the word blind. Indeed, semi-blind techniques, and other factorization methods, most important of which are the non-negative decompositions, continued to be emphasized. We also received a significant number of papers that concentrated on time-frequency domain approaches and in particular sparse decompositions and representations. The application areas were dominated by submissions focusing on biomedical and speech and audio applications as well as image processing. We would like to thank Emmanuel Vincent, Pau Bofill, and Shoko Araki, organizers of the Special Session on evaluation. Our warmest thanks go to our plenary speakers, Andrzej Cichocki from RIKEN, Japan, David J. Field from Cornell University, USA, Simon Haykin from McMaster University, Canada, and Mark D. Plumbley from Queen Mary University of London, UK.

There are many people that should be thanked for their hard work, which helped produce the high-quality scientific program. First and foremost we would like to thank all the authors who have contributed to this volume. Without them there would be no proceedings. In addition, we thank the members of the Organizing Committee and the reviewers for their efforts in commissioning the reviews, and for their help in selecting the very best papers for inclusion in this volume and for providing valuable feedback that helped improve the papers. We thank in particular Romis Attux , Charles Cavalcante, and Ricardo Suyama for their hard work and constant support throughout the process. Their close attention to every detail has been the most important factor in the achievement of this volume. Gratitude is due to Ursula Barth and other Springer staff for their solicitude and diligence in the preparation of this volume and the CD-ROM Proceedings. Thanks also go to the members of the ICA International Steering Committee for their continued advice and ongoing support for the ICA conference series.

Last, but not least, our deepest thanks go to our Local Organizing Committee and to Roberta Cavalcante from Ikone. Thanks to their initiative our conference came to the Southern Hemisphere for the first time. We express our gratitude for this unique opportunity of attending such a high-level conference in Paraty and enjoying all the beautiful sites around this wonderful village.

January 2009

Tülay Adali
Christian Jutten
Allan Kardec Barros
João Marcos Travassos Romano

Organization

Executive Committee

General Chairs	Allan Kardec Barros (Federal University of Maranhão, Brazil)
	João Marcos Travassos Romano (University of Campinas, Brazil)
Technical Program Chairs	Tülay Adali (University of Maryland, Baltimore County, USA)
	Christian Jutten (Grenoble Institute of Technology, France)
Evaluation Chairs	Shoko Araki (NTT Communication Science Laboratories, Japan)
	Pau Bofill (Polytechnic University of Catalonia, Spain)
	Emmanuel Vincent (IRISA - INRIA, France)
European Liaison	Pierre Comon (University of Nice, France)
Far East Liaison	Andrzej Cichocki (RIKEN Brain Science Institute, Japan)
North American Liaison	José Carlos Príncipe (University of Florida, USA)

Organizing Committee

Romis Attux	University of Campinas, Brazil
Charles C. Cavalcante	Federal University of Ceará, Brazil
Mariane R. Petraglia	Federal University of Rio de Janeiro, Brazil
Ricardo Suyama	University of Campinas, Brazil
Hani Camille Yehia	Federal University of Minas Gerais, Brazil

Technical Program Committee

Luis Almeida	University of Lisbon, Portugal
Romis Attux	University of Campinas, Brazil
Massoud Babaie-Zadeh	Sharif University of Technology, Iran
Charles Cavalcante	Federal University of Ceará, Brazil
Jonathon Chambers	King's College London, UK
Seungjin Choi	POSTECH, Korea
Pierre Comon	University of Nice, France
Sergio Cruces	University of Seville, Spain
Adriana Dapena	University of La Coruña, Spain
Justin Dauwels	Massachusetts Institute of Technology, USA
Yannick Deville	Paul Sabatier University, France

Konstantinos Diamantaras	TEI of Thessaloniki, Greece
Zhi Ding	University of California, Davis, USA
Deniz Erdogmus	Northeastern University, USA
Simone Fiori	University of Perugia, Italy
Pando Gr. Georgiev	University of Cincinnati, USA
Mark Girolami	University of Glasgow, UK
Rémi Gribonval	INRIA, France
Zhaoshui He	RIKEN Brain Science Institute, Japan
Kenneth Hild	University of California, San Francisco, USA
Aapo Hyvärinen	Helsinki University of Technology, Finland
Eleftherios Kofidis	University of Piraeus, Greece
Juha Karhunen	Helsinki University of Technology, Finland
Ali Mansour	Curtin University of Technology, Australia
Ali Mohammad-Djafari	CNRS, Orsay, France
Eric Moreau	Université de Toulon et du Var, France
Saïd Moussaoui	Nantes University, France
Klaus-R. Mueller	Fraunhofer FIRST, Germany
Esa Ollila	Helsinki University of Technology, Finland
Noboru Ohnishi	Nagoya University, Japan
Mariane Petraglia	Federal University of Rio de Janeiro, Brazil
Dinh Tuan Pham	CNRS, Grenoble, France
Mark Plumbley	Queen Mary, University of London, UK
Luis Castedo Ribas	University of La Coruña, Spain
Justinian Rosca	Siemens, USA
Hiroshi Sawada	NTT CS Labs, Japan
Ignacio Santamaría	University of Cantabria, Spain
Ricardo Suyama	University of Campinas, Brazil
Fabian J. Theis	University of Regensburg, Germany
DeLiang Wang	Ohio State University, USA
Lei Xu	The Chinese University of Hong Kong, China
Arie Yeredor	Tel-Aviv University, Israel
Vicente Zarzoso	University of Nice, France
Andreas Ziehe	Fraunhofer Institute, Germany

Reviewers

Luis Almeida	Jonathon Chambers	Iván Durán
Shoko Araki	Seungjin Choi	Deniz Erdogmus
Romis Attux	Pierre Comon	Simone Fiori
Massoud Babaie-Zadeh	Sergio Cruces	Mark Girolami
Paulo B. Batalheiro	Adriana Dapena	Rémi Gribonval
Florian Blöchl	Justin Dauwels	Peter Gruber
Pau Bofill	Yannick Deville	Harold Gutch
Luis Castedo	Konstantinos	Diego Haddad
Charles C. Cavalcante	Diamantaras	Zhaoshui He
A. Taylan Cemgil	Zhi Ding	Kenneth Hild

Aapo Hyvärinen
Juha Karhunen
Eleftherios Kofidis
David Luengo
Ali Mansour
Ali Mohammad-Djafari
Morten Mørup
Eric Moreau
Saïd Moussaoui
Klaus-Robert Mueller
Andrew Nesbit
Yasunori Nishimori

Noboru Ohnishi
Esa Ollila
Mariane Petraglia
Ronald Phlypo
Dinh-Tuan Pham
Mark Plumbley
David Ramirez
Justinian Rosca
Ignacio Santamaría
Auxiliadora Sarmiento
Hiroshi Sawada
Mikkel Schmidt

Ricardo Suyama
Fabian Theis
Steven Van Vaerenbergh
Javier Via
Emmanuel Vincent
DeLiang Wang
Lei Xu
Arie Yeredor
Vicente Zarzoso
Andreas Ziehe

Sponsoring Institutions

University of Campinas (UNICAMP), Brazil
Federal University of Maranhão (UFMA), Brazil
Federal University of Rio de Janeiro (UFRJ), Brazil
Federal University of Ceará (UFC), Brazil
Federal University of Minas Gerais (UFMG), Brazil
Brazilian Telecommunications Society (SBrT), Brazil
Coordination for Improvement of Higher Education Personnel (CAPES), Brazil
National Council of Technological and Scientific Development (CNPq), Brazil
Research Foundation of the State of Rio de Janeiro (FAPERJ), Brazil

International ICA Steering Committee

Luis Almeida   University of Lisbon, Portugal
Shun-ichi Amari   RIKEN Brain Science Institute, Japan
Jean-François Cardoso   ENST, Paris, France
Andrzej Cichocki   RIKEN Brain Science Institute, Japan
Scott Douglas   Southern Methodist University, USA
Rémi Gribonval   INRIA, France
Simon Haykin   McMaster University, Canada
Christian Jutten   Grenoble Institute of Technology, France
Te-Won Lee   University of California, San Diego, USA
Shoji Makino   NTT Communication Science Laboratories,
                               Japan
Klaus-Robert Müller   Fraunhofer FIRST, Germany
Noboru Murata   Waseda University, Japan
Erkki Oja   Helsinki University of Technology, Finland
Mark Plumbley   Queen Mary, University of London, UK
Paris Smaragdis   Adobe Systems, USA
Fabian J. Theis   University of Regensburg, Germany
Ricardo Vigário   Helsinki University of Technology, Finland

Table of Contents

Theory

Algorithms and Architectures

Biomedical Applications

Image Processing

Speech and Audio Processing

Other Applications

Special Session on Evaluation

Stationary Subspace Analysis

Paul von Bünau, Frank C. Meinecke, and Klaus-Robert Müller

Machine Learning Group, CS Dept., TU Berlin, Germany
{buenau,meinecke,krm}@cs.tu-berlin.de

Abstract. Non-stationarities are an ubiquitous phenomenon in time-series data, yet they pose a challenge to standard methodology: classification models and ICA components, for example, cannot be estimated reliably under distribution changes because the classic assumption of a stationary data generating process is violated. Conversely, understanding the nature of observed non-stationary behaviour often lies at the heart of a scientific question. To this end, we propose a novel unsupervised technique: Stationary Subspace Analysis (SSA). SSA decomposes a multivariate time-series into a stationary and a non-stationary subspace. This factorization is a universal tool for furthering the understanding of non-stationary data. Moreover, we can robustify other methods by restricting them to the stationary subspace. We demonstrate the performance of our novel concept in simulations and present a real world application from Brain Computer Interfacing.

Keywords: Non-Stationarities, Source Separation, BSS, Dimensionality Reduction, Covariate Shift, Brain-Computer-Interface, BCI.

1 Introduction

The assumption that the distribution of the observed data is stationary is a cornerstone of mainstream statistical modelling: ordinary regression and classification models, for instance, rely on the fact that we can generalize from a sample to the population. Even if we are primarily interested in making accurate predictions as in Machine Learning, differences in training and test distributions can cause severe drops in performance [9], because the paradigm of minimizing the expected loss approximated by the training sample is no longer consistent. The same holds true for unsupervised approaches such as PCA and ICA.

As many real data sources are inherently non-stationary, researchers have long tried to account for that. In inferential statistics, the celebrated Heckman [4] bias correction model attempts at obtaining unbiased parameter estimates under sample selection bias; cointegration methods [2] aim at discovering stable relationships between non-stationary time-series. In order to improve predictive performance, several approaches have been developed: explicit modeling of the non-stationary process; tracking rsp. adapting to non-stationarity via online learning [7]; constructing features that are invariant under non-stationarities

T. Adali et al. (Eds.): ICA 2009, LNCS 5441, pp. 1–8, 2009.

[1] and correcting for biased empirical risk estimates under covariate shift by reweighting [10].

In this paper, we propose a novel decomposition paradigm that, to the best of our knowledge, has not been explored so far. The premise is similar in spirit to the ICA setting [5], in that we assume that a multivariate time-series was generated as a mixture of sources. However, instead of assuming independent sources, we suppose that some of the sources are stationary while others are not. The proposed Stationary Subspace Analysis (SSA) algorithm then finds a factorization into a stationary and a non-stationary component from observed data. Apart from shedding light on the nature of the non-stationarities, this decomposition can be used to build robust learning systems by confining them to the stationary subspace as we exemplify in an application to Brain-Computer-Interfacing.

Previous work along similar lines has addressed the question of whether two samples come from the same distribution [3], without trying to find subspaces where the distribution stays the same. While ICA finds independent sources, SSA divides the input space into a stationary and a non-stationary component regardless of independence within or between the subspaces.

2   Problem Formalization

We assume that the non-stationary behaviour of the data generating process is confined to a linear subspace of the D-dimensional data space, i.e. there exists a d-dimensional subspace that is undisturbed by the nonstationarity. Formally, we assume that the analyzed system generates d stationary source signals $s^{\mathfrak{s}}(t) = [s_1(t), s_2(t), \ldots, s_d(t)]^\top$ (also referred to as $\mathfrak{s}$-*sources*) and $D - d$ non-stationary source signals $s^{\mathfrak{n}}(t) = [s_{d+1}(t), s_{d+2}(t), \ldots, s_D(t)]^\top$ (also $\mathfrak{n}$-*sources*).

The observed signals $x(t)$ are then modeled as linear superpositions of these sources

$$x(t) = As(t) = \begin{bmatrix} A^{\mathfrak{s}} & A^{\mathfrak{n}} \end{bmatrix} \begin{bmatrix} s^{\mathfrak{s}}(t) \\ s^{\mathfrak{n}}(t) \end{bmatrix} \tag{1}$$

where A is an invertible matrix.[1] The columns of $A^{\mathfrak{s}}$ span the subspace that the $\mathfrak{s}$-sources live in, we will refer to this space as $\mathfrak{s}$-*subspace*. Similarly, the span of $A^{\mathfrak{n}}$ will be called $\mathfrak{n}$-*subspace*. The goal is now to estimate a linear transformation B from the observed data $x(t)$ that separates the $\mathfrak{s}$-sources from the $\mathfrak{n}$-sources, i.e. factorizes the observed signals according to eq. (1).

Figure 1 shows an example of partly non-stationary two-dimensional time series, i.e. the upper time course is non-stationary and the lower one is stationary. Due to the non-stationarity, the scatter plots of different parts of the data are quite different. Note, that PCA or ICA would not be able to perform the separation task due to the strong correlations between the two signals.

---

[1] For simplicity we assume A to be a square matrix. Note that we do *not* assume the sources to be statistically independent or even uncorrelated.

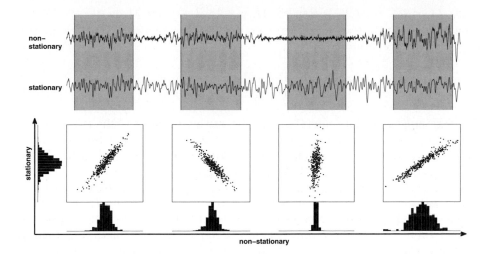

Fig. 1. A two-dimensional time series consisting of a one dimensional stationary and a one-dimensional non-stationary part. The marginal distributions of different sections of the time series are different in the n-subspace (horizontal histograms), but constant in the s-subspace (vertical histogram on the left side). Note, that the distributions in these subspaces are in general not independent.

3 Identifiability and Uniqueness of the Solution

Given the mixing model, the first question is whether it is in principle possible to invert it given only the observed (i.e. mixed) signals. In other words: can the mixing matrix A be uniquely identified or are there symmetries that give rise to multiple solutions?[2] Obviously, the basis *within* each of the two subspaces (stationary or not) can only be identified up to arbitrary linear transformations in these subspaces. However, the answer to the question whether or not the two subspaces themselves are uniquely identifiable is less obvious. Let us write the estimated mixing and demixing matrices as

$$\hat{A} = \begin{bmatrix} \hat{A}^{\mathfrak{s}} & \hat{A}^{\mathfrak{n}} \end{bmatrix} \quad \text{and} \quad \hat{B} = \hat{A}^{-1} = \begin{bmatrix} \hat{B}^{\mathfrak{s}} \\ \hat{B}^{\mathfrak{n}} \end{bmatrix}. \tag{2}$$

The matrices $\hat{B}^{\mathfrak{s}} \in \mathbb{R}^{d \times D}$ and $\hat{B}^{\mathfrak{n}} \in \mathbb{R}^{(D-d) \times D}$ denote the projections from the observed data into the estimated s- and n-subspaces. If we express the true s- and n-subspaces as linear combinations of the respective estimated subspaces

$$A^{\mathfrak{s}} = \hat{A}^{\mathfrak{s}} M_1 + \hat{A}^{\mathfrak{n}} M_2$$
$$A^{\mathfrak{n}} = \hat{A}^{\mathfrak{s}} M_3 + \hat{A}^{\mathfrak{n}} M_4 \tag{3}$$

---

[2] And if so, how do these solutions differ?

(with $M_1 \in \mathbb{R}^{d \times d}$, $M_2 \in \mathbb{R}^{(D-d) \times d}$, $M_3 \in \mathbb{R}^{d \times (D-d)}$, and $M_4 \in \mathbb{R}^{(D-d) \times (D-d)}$) then the composite transformation (the true mixing followed by the estimated demixing matrix) reads

$$\hat{B} A = \begin{bmatrix} \hat{B}^{\mathfrak{s}} A^{\mathfrak{s}} & \hat{B}^{\mathfrak{s}} A^{\mathfrak{n}} \\ \hat{B}^{\mathfrak{n}} A^{\mathfrak{s}} & \hat{B}^{\mathfrak{n}} A^{\mathfrak{n}} \end{bmatrix} = \begin{bmatrix} M_1 & M_3 \\ M_2 & M_4 \end{bmatrix} \tag{4}$$

The estimated $\mathfrak{s}$- and $\mathfrak{n}$-sources can now be written in terms of the true sources:

$$\hat{s}^{\mathfrak{s}} = M_1 s^{\mathfrak{s}} + M_3 s^{\mathfrak{n}}$$
$$\hat{s}^{\mathfrak{n}} = M_2 s^{\mathfrak{s}} + M_4 s^{\mathfrak{n}} \tag{5}$$

Given that the estimated transformation separates stationary from non-stationary signals, M_3 must vanish since estimated signals that contain non-stationary source signals will be non-stationary as well. The estimated $\mathfrak{n}$-sources, on the other hand, might contain contributions from the true $\mathfrak{s}$-sources. So the matrices M_1, M_2, M_4 remain unconstrained[3].

From equation (3) and $M_3 = 0$ we see, that the estimated $\mathfrak{n}$-subspace is identical to the true $\mathfrak{n}$-subspace while the estimated $\mathfrak{s}$-subspace is a linear combination of true $\mathfrak{s}$- and $\mathfrak{n}$-subspaces[4]. Since this linear combination is arbitrary, the estimated $\mathfrak{s}$-subspace can always be chosen such that it is orthogonal to the $\mathfrak{n}$-subspace. If we additionally chose orthogonal bases *within* each of these estimated subspaces, we have effectively restricted ourselves to the estimation of an *orthogonal* mixing matrix.

This means that we can restrict our search for the mixing matrix to the space of orthogonal matrices even if the model allows general (non-orthogonal) mixing matrices. As a result, the estimated $\mathfrak{s}$-sources will be linear combinations of the true $\mathfrak{s}$-sources but well separated from the $\mathfrak{n}$-sources.

4   Estimating the Stationary Subspace

We will now formulate an optimization criterion that allows us to estimate a demixing matrix $\hat{B}$ given N sets of data $\mathcal{X}_1, \ldots, \mathcal{X}_N \subset \mathbb{R}^D$. These datasets may for example correspond to epochs of a time series as in Figure 1. More precisely, we want to find an orthogonal transformation $\hat{B}$ such that the first d components of the transformed data $\hat{s}(t) = \hat{B} x(t)$ are as stationary as possible. Since these components are completely determined by the sub-matrix $\hat{B}^{\mathfrak{s}}$, the cost function can only depend on this sub-matrix, even if the optimization takes place in the full space of orthogonal matrices. For the latter calculations it is convenient to express $\hat{B}^{\mathfrak{s}}$ in terms of the complete demixing matrix as $\hat{B}^{\mathfrak{s}} = I^d \hat{B}$, where $I^d \in \mathbb{R}^{d \times D}$ is the identity matrix truncated to the first d rows.

---

[3] Apart from the mere technical assumption of invertibility of $\hat{A}$.

[4] Note that for the estimated sources this is the other way round: while the estimated $\mathfrak{s}$-sources are mixtures of the true $\mathfrak{s}$-sources only, the estimated $\mathfrak{n}$-sources are mixtures of both the true $\mathfrak{n}$- and $\mathfrak{s}$-sources.

We will consider a set of d estimated sources as stationary, if the joint distribution of these sources stays the same over all sets of samples. Therefore, our objective function is based on minimizing the pairwise distance between the distributions of the projected data which we measure using the Kullback-Leibler divergence. For technical reasons, we only consider differences in the first two moments, i.e. two distributions are assumed to be the same if they have the same mean and covariance matrix. Consequently, our method is ignorant of any non-stationary process that does not change the first two moments. Even though the rationale for this restriction is purely technical, its practical consequences should be limited.[5] In order to compute the KL divergence, we need to estimate both densities. Since we have restricted ourselves to distributions whose sufficient statistics consist of the mean and covariance matrix, a natural choice is to use a Gaussian approximation according to the maximum entropy principle [6].

Let $(\hat{\mu}_i, \hat{\Sigma}_i)$ be the estimate of mean and covariance for dataset $\mathcal{X}_i$. Then we find the demixing matrix as the solution of the optimization problem

$$\hat{B} = \underset{BB^\top = I}{\operatorname{argmin}} \sum_{i<j} \operatorname{KL}\left[\mathcal{N}(I^d B \hat{\mu}_i, I^d B \hat{\Sigma}_i (I^d B)^\top) \,\|\, \mathcal{N}(I^d B \hat{\mu}_j, I^d B \hat{\Sigma}_j (I^d B)^\top) \right].$$

To stay on the manifold of orthogonal matrices (or, more formally speaking: the special orthogonal group SO(N)), we will employ a multiplicative update scheme: starting with $B = I$, we multiply B in each iteration by an orthogonal matrix, $B^{\mathrm{new}} \leftarrow RB$. At each step, the estimated projection to the $\mathfrak{s}$-subspace is then given by $I^d RB$ and we can write the projected mean and covariance matrix of data set $\mathcal{X}_i$ as

$$\hat{\mu}_i^{\mathfrak{s}} = I^d RB \hat{\mu}_i \quad \text{and} \quad \hat{\Sigma}_i^{\mathfrak{s}} = I^d RB \hat{\Sigma}_i (I^d RB)^\top,$$

and the loss function as

$$L_B(R) = \sum_{i<j} \log \frac{\det(\hat{\Sigma}_j^{\mathfrak{s}})}{\det(\hat{\Sigma}_i^{\mathfrak{s}})} + \operatorname{tr}\left((\Sigma_j^{\mathfrak{s}})^{-1} \Sigma_i^{\mathfrak{s}} \right) + (\mu_j^{\mathfrak{s}} - \mu_i^{\mathfrak{s}})^\top (\Sigma_j^{\mathfrak{s}})^{-1} (\mu_j^{\mathfrak{s}} - \mu_i^{\mathfrak{s}}).$$

The rotation R can be parametrized as the matrix exponential of an antisymmetric matrix, $R = e^M$ with $M = -M^\top$, where each element M_{ij} can be interpreted as generalized rotation angle (i.e. rotating axis i towards axis j). Using this, we can express the gradient of the function $L_B(R) = L_B(e^M)$ w.r.t. M in terms of the corresponding gradient w.r.t. R (see e.g. [8]) as

$$\partial L_B / \partial M |_{M=0} = (\partial L_B / \partial R) R^\top - R (\partial L_B / \partial R)^\top. \tag{6}$$

The gradient of the loss function with respect to R is

$$\frac{\partial L_B}{\partial R} = I^{d\top} I^d R \sum_{i<j} B \left(\hat{\Sigma}_i Q \hat{\Sigma}_j^{-1} + \hat{\Sigma}_j^{-1} Q \hat{\Sigma}_i + \hat{\Sigma}_j Q \hat{\Sigma}_i^{-1} + \hat{\Sigma}_i^{-1} Q \hat{\Sigma}_j \right.$$

$$\left. + \left(\hat{\Sigma}_j^{-1} + \hat{\Sigma}_i^{-1} \right) Q D_\mu + D_\mu Q \left(\hat{\Sigma}_j^{-1} + \hat{\Sigma}_i^{-1} \right) \right) B^\top \tag{7}$$

---

[5] Real world nonstationary processes will hardly go unnoticed by the first two moments. Moreover, estimates of higher-order moments are less stable themselves.

with $Q = B^\top R^\top I^{d\top} I^d RB$ and $D_\mu = (\hat{\mu}_i - \hat{\mu}_j)(\hat{\mu}_i - \hat{\mu}_j)^\top$. The factor of $I^{d\top} I^d$ from the left ensures that the lower $D - d$ rows of this matrix gradient vanish. This means, that in the gradient w.r.t. M the lower right block has to be zero. Furthermore, since every summand in the sum is symmetric, the skew-symmetry of eq. (6) makes the upper left block in $\partial L_B / \partial M$ vanish as well. Thus the gradient has the shape

$$\frac{\partial L_B}{\partial M}\Big|_{M=0} = \begin{bmatrix} 0 & Z \\ -Z^\top & 0 \end{bmatrix}$$

where the non-zero part $Z \in \mathbb{R}^{d \times (D-d)}$ corresponds to the rotations between coordinates of the s- and n-space. Note that the derivative w.r.t. the rotations within the two spaces must vanish, because they do not change the solution. Thus we can reduce the number of variables to $d(D - d)$. The optimization is then carried out using a standard conjugate gradient procedure in angle space with multiplicative updates.

5   Simulations

We investigate the performance of SSA under different scenarios based on simulated data. In order to make the analysis more concise, we first consider the hypothetical case where the true mean and covariance matrix of the data generating process are known and then examine the impact of the estimation error in a second set of experiments. The experimental setup is as follows: we randomly

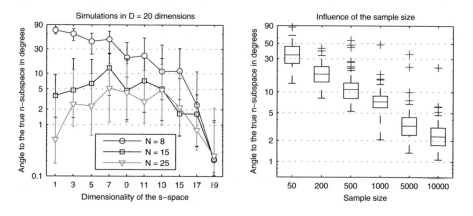

Fig. 2. The left plot shows the results of the simulations for a 20 dimensional input space. The performance of the method is measured in terms of the median angle to the true n-subspace (vertical axis) for several dimensionalities (horizontal axis) and number of datasets $N = 8, 15, 25$ (red, blue and green curve) over 100 random trials. The error bars stretch from the 25% to the 75% quantile. The right plot shows the performance under varying numbers of samples in each dataset for fixed $d = 5$ and $N = 10$ in $D = 10$ input dimensions.

sample N covariance matrices and means such that the s-subspace is spanned by the first d coordinates. Moreover, we randomly sample a well-conditioned mixing matrix A that is then applied to each mean and covariance matrix. Note that ICA cannot in principle separate the s- from the n-space because we have allowed for arbitrary correlations between them. For each trial, in order to avoid local minima, we restart the optimization procedure five times and then select the solution with the lowest objective function value. The accuracy is measured as angle between the estimated n-subspace and the ground truth.

From the results shown in the left plot of Figure 2, we can see that the likelihood that the true s-sources can be found grows with the number of available datasets and scales with the degrees of freedom $d(D - d)$. In order to analyze the influence of the number of samples within each dataset $n = |\mathcal{X}_i|, 1 \leq i \leq N$, we fix the other parameters and vary n. The right plot in Figure 2 shows the result: the performance clearly increases with the number of available samples whereas the method remains applicable even in the small sample case.

6   Application to Brain-Computer-Interfacing

We demonstrate that SSA can be used for robust learning on EEG dat (49 channels) from Brain-Computer-Interfacing (BCI) where the task is to distinguish imagined movements of the right/left hand. Imagined movements of the hand are known to produce characteristic spatio-temporal signal patterns on the respective contralateral hemispheres. However, the frequency content of these signals over the motor cortex (μ-rhythm) is in the same range as the occipital α-rhythm. The strength of the α-rhythm is task unrelated and strongly correlated to the tiredness or the exposure to visual stimulation of a subject.

In our experiment, we induce changes in the strength of the α-rhythm by first extracting it from a separate artefact measurement session (using ICA) and

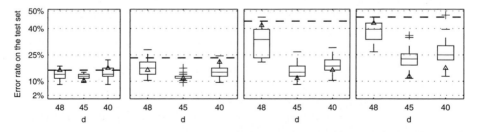

Fig. 3. Results on the BCI data for varying levels of change in the strengths of the α component between test and training data (increasing from left to right). For each scenario, the test error rate of the CSP baseline method is indicated by the dashed black line. The boxplots show the distribution of the test error rates on the s-subspace for varying dimensionalities $d = 48, 45, 40$ over 100 runs of the optimization. The blue triangle indicates the error rate on the s-subspace that attained the minimum objective function value over all runs.

then superimpose it on the data in varying strengths in order to induce realistic yet controlled non-stationarities. See [1] for a full exposition of the experimental setup. The data is divided into three parts: a training data set that contains strong α-activity which corresponds to a wakeful resting brain state and a test data set, where we change the strength of the α-components. From the test portion, we set aside the first 30 trials for adaptation. Then we estimate the s-space over the training and adaptation part and apply the standard CSP algorithm within the s-space. The performance is measured as the misclassification rate on the test set. The experimental results presented in Figure 3 show that the classification accuracy can be retained even under very strong non-stationarities if the learning algorithm is restricted to a 45 dimensional s-subspace.

7   Conclusion and Future Work

We have presented the first algorithm for decomposing a multivariate time-series into a stationary and a non-stationary component. A number of interesting questions remain: first of all, how can we choose the dimensionality of the s-space from data? Secondly, can we extend the algorithm to measure non-stationarities in higher order moments? Finally, we will pursue further applications beyond the neurosciences.

References

1. Blankertz, B., Kawanabe, M., Tomioka, R., Hohlefeld, F., Nikulin, V., Müller, K.-R.: Invariant common spatial patterns: Alleviating nonstationarities in brain-computer interfacing. In: Platt, J.C., Koller, D., Singer, Y., Roweis, S. (eds.) Advances in Neural Information Processing Systems, vol. 20, pp. 113–120. MIT Press, Cambridge (2008)
2. Engle, R.F., Granger, C.W.J.: Co-integration and error correction: Representation, estimation, and testing. Econometrica 55(2), 251–276 (1987)
3. Friedman, J., Rafsky, L.: Multivariate generalizations of the Wald-Wolfowitz and Smirnov two-sample tests. The Annals of Statistics 7(4), 697–717 (1979)
4. Heckman, J.J.: Sample selection bias as a specification error. Econometrica 47(1), 153–162 (1979)
5. Hyvärinen, A., Karhunen, J., Oja, E.: Independent Component Analysis. Wiley, New York (2001)
6. Jaynes, E.T.: Information theory and statistical mechanics. Physical Review 160, 620–630 (1957)
7. Murata, N., Kawanabe, M., Ziehe, A., Müller, K.-R., Amari, S.: On-line learning in changing environments with applications in supervised and unsupervised learning. Neural Networks 15(4-6), 743–760 (2002)
8. Plumbley, M.D.: Geometrical methods for non-negative ica: Manifolds, lie groups and toral subalgebras. Neurocomputing 67, 161–197 (2005)
9. Quiñonero-Candela, J., Sugiyama, M., Schwaighofer, A., Lawrence, N. (eds.): Dataset Shift in Machine Learning. MIT Press, Cambridge (2008)
10. Shimodaira, H.: Improving predictive inference under covariate shift by weighting the log-likelihood function. Journal of Statistical Planning and Inference 90(2), 227–244 (2000)

Reference Based Contrast Functions in a Semi-blind Context

Marc Castella[1] and Eric Moreau[2]

[1] Institut TELECOM; TELECOM & Management SudParis; UMR-CNRS 5157
9 rue Charles Fourier, 91011 Evry CEDEX, France
marc.castella@it-sutparis.eu
[2] University of Sud Toulon Var, ISITV, LSEET UMR-CNRS 6017
Av. G. Pompidou, BP56, 83162 La Valette du Var CEDEX, France
moreau@univ-tln.fr

Abstract. We deal with blind signal extraction in the framework of a convolutive mixture of independent sources. Considering so-called reference signals, we generalize former identifiability conditions. Based on this result, we propose to incorporate some a priori information in the references. We show the validity of reference based contrast functions in two semi-blind situations. The results are confirmed by computer simulations.

1 Introduction

This paper deals with the problem of source extraction in the general case of a convolutive mixture. This issue finds applications in many contexts such as blind equalization for digital communication signals,... Moreover, this problem is a key component in blind source separation methods based on a multi-stage (or iterative) approach (often called deflation [4,5]).

One interesting way to tackle this problem consists in optimizing an adapted criterion called contrast function. However, finding identifiability conditions which state the precise context where source signals can be extracted remain important. This paper considers the above two points.

In this context, reference based methods have recently been introduced [1,2]. They consider a reference signal which provides a supplementary degree of freedom and which can be used in different ways. Here we propose to incorporate some a priori information in the reference signal and we show its effectiveness. New identifiability results which generalize former conditions on the reference are first proposed. Then, we show the validity of reference based contrast functions in two semi-blind interesting situations. Finally, computer simulations illustrate our results.

2 Problem Statement

2.1 Mixing Model

We consider a Q-dimensional ($Q \in \mathbb{N}, Q \geq 2$) discrete-time signal which is called vector of *observations* and denoted by $\mathbf{x}(n)$ (in the whole paper, n stands for

T. Adali et al. (Eds.): ICA 2009, LNCS 5441, pp. 9–16, 2009.
© Springer-Verlag Berlin Heidelberg 2009

any integer: $n \in \mathbb{Z}$). It results from a linear time invariant (LTI) multichannel system $\{\mathbf{M}\}$ described by the input-output relation:

$$\mathbf{x}(n) = \sum_{k \in \mathbb{Z}} \mathbf{M}(k)\mathbf{s}(n - k) \triangleq \{\mathbf{M}\}\mathbf{s}(n), \tag{1}$$

where $\mathbf{M}(n)$ is the sequence of (Q, N) impulse response matrices and $\mathbf{s}(n)$ is a N-dimensional ($N \in \mathbb{N}^*$) unknown and unobserved column vector, which is referred to as the vector of *sources*. The multichannel deconvolution problem consists in estimating a multivariate LTI system $\{\mathbf{W}\}$ operating on the observations, such that the vector $\mathbf{y}(n) = \sum_{k \in \mathbb{Z}} \mathbf{W}(k)\mathbf{x}(n - k) \triangleq \{\mathbf{W}\}\mathbf{x}(n)$ restores the N input sources. Different situation can occur depending on context: when no information is available on the mixing system and when the sources cannot be observed, the problem is referred to as the *blind source separation* (BSS) problem. On the contrary, when some information is assumed to be known on the sources or on the mixing system, the problem is sometimes referred to as *semi-blind source separation* (S-BSS).

The matrix transfer function $\mathbf{M}[z_1]$ of the mixing channel is given by the following z-transform (where z_1 is used instead of z for reasons that appear later)

$$\mathbf{M}[z_1] = \sum_{n \in \mathbb{Z}} \mathbf{M}(n)z_1^{-n} . \tag{2}$$

A similar definition holds for the matrix transfer function $\mathbf{W}[z_1]$ of the separator $\{\mathbf{W}\}$. We define the combined mixing-separating (N, N) LTI filter $\{\mathbf{G}\}$ by its impulse response $\mathbf{G}(n) = \sum_{k \in \mathbb{Z}} \mathbf{W}(k)\mathbf{M}(n - k)$.

In an iterative approach, the sources are extracted one by one. We consider one row of the separator $\{\mathbf{W}\}$, which is a $(1, Q)$ LTI row filter $\{\mathbf{w}\}$ with output $y(n)$ given by:

$$y(n) = \sum_{k \in \mathbb{Z}} \mathbf{w}(k)\mathbf{x}(n - k) \triangleq \{\mathbf{w}\}\mathbf{x}(n). \tag{3}$$

Similarly, $\{\mathbf{g}\}$ denotes the $(1, N)$ row filter given by the row of $\{\mathbf{G}\}$ which corresponds to $\{\mathbf{w}\}$. We have then:

$$y(n) = \sum_{k \in \mathbb{Z}} \mathbf{g}(k)\mathbf{s}(n - k) \triangleq \{\mathbf{g}\}\mathbf{s}(n). \tag{4}$$

We say that the separation is achieved when only one component of $\{\mathbf{g}\}$ is non zero, say the i_0th:

$$\{\mathbf{g}\} = \left(0, \ldots, 0, \{g_{i_0}\}, 0, \ldots, 0\right). \tag{5}$$

When the source signals are assumed to be temporally i.i.d. (independent and identically distributed) signals, a more restrictive separation condition is considered. In this case, $\{\mathbf{g}\}$ should satisfy the following condition in addition to (5):

$$\exists l \in \mathbb{N} : \quad g_{i_0}(k) = 0 \text{ if } k \neq l \text{ and } g_{i_0}(l) = \alpha \in \mathbb{C}^* \tag{6}$$

The parameters i_0, l, and α correspond respectively to the well-known indeterminacies of BSS (or S-BSS): permutation of the sources, delay and amplitude factor. In the case of non i.i.d. sources, a filtering ambiguity is included by specifying no constraint on $\{g_{i_0}\}$.

2.2 Reference Based Separation

In order to be able to solve the BSS or S-BSS problem, we introduce the following assumption on the sources:

A1. The source vector components $s_i(n), i \in \{1, \ldots, N\}$ are *mutually independent*, stationary and zero-mean processes with unit variance.

In addition, we assume that there exist an additional signal $r(n)$, which is referred to as a "reference" signal. This reference signal is used in order to facilitate the source separation and the following assumption is made:

A2. The signals $s_i(n), i \in \{1, \ldots, N\}$ and $r(n)$ are jointly stationary up to the fourth order.

It has been shown that $r(n)$ can generally be constructed from the observations. The corresponding constraints on $r(n)$ being quite weak, an efficient BSS procedure has been proposed based on the maximization of a reference based contrast function [1,2]. In the following Section, we give general conditions on $r(n)$ so that the extraction of a particular source is allowed. Furthermore, it is rather natural to use the reference signal in a S-BSS context: we indeed propose in Section 4 to include in $r(n)$ some a priori information on the source which is being extracted. The validity of the corresponding separation criteria is proved.

3 Generalized Identifiability Conditions

Higher Order Statistics. We consider higher order cumulants and define the following fourth order cross-cumulant:

$$C^r_{s_i s_j}(\mathbf{n}) = \mathrm{Cum}\left\{s_i(n), s_j(n - n_1)^*, r(n - n_2), r(n - n_2)^*\right\}. \qquad (7)$$

Note that it exists and depends on $\mathbf{n} = (n_1, n_2)$ only according to assumption A2. We consider a multidimensional z-transform of order two w.r.t. the variables $\mathbf{n} = (n_1, n_2)$. It is defined by:

$$C^r_{s_i s_j}[\mathbf{z}] = \sum_{\mathbf{n} \in \mathbb{Z}^2} C^r_{s_i s_j}(\mathbf{n}) z_1^{-n_1} z_2^{-n_2} \qquad (8)$$

where $\mathbf{z} = (z_1, z_2) \in (\mathbb{C}^*)^2$. For the signal $\mathbf{x}(n)$, the cumulants $C^r_{x_i x_j}(\mathbf{n})$ and their respective z-transforms $C^r_{x_i x_j}[\mathbf{z}]$ are defined similarly to the definitions given by (7) and (8). Let us introduce the matrices $\mathbf{C}^r_{\mathbf{x}}[\mathbf{z}]$ and $\mathbf{C}^r_{\mathbf{s}}[\mathbf{z}]$ whose (i,j) components are $C^r_{x_i x_j}[\mathbf{z}]$ and $C^r_{s_i s_j}[\mathbf{z}]$ respectively.

Cumulant Decomposition. Using the multilinearity property of cumulants, it can be verified that the following decomposition holds:

$$\mathbf{C}_{\mathbf{x}}^r[\mathbf{z}] = \mathbf{M}[z_1 z_2]\mathbf{C}_{\mathbf{s}}^r[\mathbf{z}]\mathbf{M}[z_1^{-1}]^{\mathrm{H}}. \tag{9}$$

Second order statistics appear formally as a particular case of the preceding results. When there is no reference signal, (7) corresponds to the correlation for which we adopt the specific notation:

$$\Gamma_{s_i s_j}(n_1) \triangleq \mathrm{Cum}\left\{s_i(n), s_j(n-n_1)^*\right\} \qquad \mathbf{\Gamma}_{\mathbf{s}}(n_1) = \left(\Gamma_{s_i s_j}(n_1)\right)_{(i,j)\in\{1,\dots,N\}^2}.$$

Corresponding to (8), the power spectral matrix of the sources is defined by:

$$\mathbf{\Gamma}_{\mathbf{s}}[z_1] \triangleq \sum_{n_1 \in \mathbb{Z}} \mathbf{\Gamma}_{\mathbf{s}}(n_1)z_1^{-n_1}. \tag{10}$$

Note that contrary to (9), z_2 does not appear here. Similar notations hold for the observations $\mathbf{x}(n)$ ($\mathbf{\Gamma}_{\mathbf{x}}(n_1)$ and $\mathbf{\Gamma}_{\mathbf{x}}[z_1]$ respectively). Similarly to (9), we have the well-known relation:

$$\mathbf{\Gamma}_{\mathbf{x}}[z_1] = \mathbf{M}[z_1]\mathbf{\Gamma}_{\mathbf{s}}[z_1]\mathbf{M}[z_1^{-1}]^{\mathrm{H}}. \tag{11}$$

Identifiability Condition. Based on (9) and (11), the following general identifiability result can be proved. The proof proceeds along similar lines as the one in [3] and it is not provided because of lack of space.

Proposition 1. *Assume that*

A3. the matrix $\mathbf{C}_{\mathbf{s}}^r[\mathbf{z}]$ is diagonal, and the first diagonal element of $\mathbf{C}_{\mathbf{s}}^r[\mathbf{z}]$ is a function distinct from all other diagonal elements.

Then, the first source of the mixture is identifiable in the following sense: for any $N \times Q$ matrix $\mathbf{W}[z_1]$ such that the $N \times N$ combined channel-equalizer z-transform matrix $\mathbf{G}[z_1] = \mathbf{W}[z_1]\mathbf{M}[z_1]$ is irreducible, if $\mathbf{W}[z_1]\mathbf{\Gamma}_{\mathbf{x}}[z_1]\mathbf{W}[z_1^{-1}]^{\mathrm{H}}$ and $\mathbf{W}[z_1 z_2]\mathbf{C}_{\mathbf{x}}^r[\mathbf{z}]\mathbf{W}[z_1^{-1}]^{\mathrm{H}}$ are both diagonal, then:

$$\mathbf{G}[z_1] = \mathbf{P} \begin{pmatrix} \alpha & 0 \dots 0 \\ 0 & \\ \vdots & \tilde{\mathbf{Q}} \\ 0 & \end{pmatrix} \tag{12}$$

where $\alpha \in \mathbb{C}, \alpha \neq 0$ and $\mathbf{P}$ is a permutation matrix. In addition, $\tilde{\mathbf{Q}}$ is a unitary matrix.

The above result corresponds to the possibility of extracting the first source. Indeed, if $\mathbf{P}$ is the identity matrix, the first row of the above $\mathbf{G}[z_1]$ is such that the corresponding row LTI filter satisfies both (5) and (6).

4   Contrast Functions

4.1   BSS Contrast Functions

A commonly used approach in BSS consist in finding criteria, which yield separation when they reach their maximum value: such criteria are by definition called *contrast functions*. The modulus of the fourth-order auto-cumulant is one of the most popular contrast [4,5]. In [1], a more general expression of the following contrast has been introduced:

$$C_r\{y(n)\} \triangleq |\kappa_r\{y(n)\}| \text{ where: } \kappa_r\{y(n)\} \triangleq \text{Cum}\{y(n), y(n)^*, r(n), r(n)^*\} \quad (13)$$

with $r(n)$ the reference signal. It has been proved that (13) is a valid contrast under the constraint $\mathbb{E}\{|y(n)|^2\} = 1$ and additional constraint on the reference signal $r(n)$.

Consider the specific situation where the reference $r(n)$ depends only on the source signal which is extracted (say $s_1(n)$) and is independent on the other source signals. One can then see that only the first element of the matrix $\mathbf{C}_\mathbf{s}^r[\mathbf{z}]$ is non zero and condition **A3** of Proposition 1 is satisfied. This situation should thus be very favorable in order to extract $s_1(n)$ and we illustrate two such situations of interest. Of course, in such a situation, the original permutation ambiguity is reduced and necessarily, the first source is extracted, that is $i_0 = 1$ in (5)-(6).

4.2   S-BSS Contrast Function: Information on the Signal Phase

We consider in this paragraph the situation where the reference signal reads:

$$\begin{cases} r(n) = \epsilon(n)\frac{s_1(n)}{|s_1(n)|} \text{ where:} \\ |\epsilon(n)| = 1 \text{ and } \epsilon(n) \text{ is independent of the source signals.} \end{cases} \quad (14)$$

This situation corresponds to the case where the reference contains some information on the phase of $s_1(n)$. This information is corrupted by the perturbation process $\epsilon(n)$ which randomly changes the phase.

Lemma 1. *In the case of i.i.d. source signals, if $r(n)$ is given by (14), we have (where ϵ and s_1 denote $\epsilon(n)$ and $s_1(n)$ for readability):*

$$C_r\{s_j(n-k)\} = \begin{cases} 0 & \text{if } (j,k) \neq (1,0), \\ |\mathbb{E}\{\epsilon\}|^2 \left[\left|\mathbb{E}\left\{\frac{s_1^2}{|s_1|}\right\}\right|^2 + |\mathbb{E}\{|s_1|\}|^2\right] & \text{if } (j,k) = (1,0). \end{cases} \quad (15)$$

Proof. For $(j,k) \neq (1,0)$, the equality $\kappa_r\{s_j(n-k)\} = 0$ follows from independence and the vanishing property of the cumulants. Now, if $(j,k) = (1,0)$, by developing the cumulant in term of moments, we obtain:

$$\kappa_r\{s_1(n)\} = \mathbb{E}\{|s_1|^2|\epsilon|^2\} - \mathbb{E}\{|s_1|^2\}\mathbb{E}\{|\epsilon|^2\} - \left|\mathbb{E}\left\{\epsilon\frac{s_1^2}{|s_1|}\right\}\right|^2 - |\mathbb{E}\{\epsilon|s_1|\}|^2$$

The result then follows from the independence of $\epsilon(n)$ and $s_1(n)$ and after simplification.   □

We can now state the following result:

Proposition 2. *Considering the reference signal given by (14), the criterion C_r in (13) is a contrast function for i.i.d. sources under the condition $\mathbb{E}\{\epsilon(n)\} \neq 0$. In addition, the source identified by C_r is $s_1(n)$.*

Proof. Note that, except in the degenerate case where $s_1(n) = 0$ almost surely, $C_r\{s_1(n)\} > 0$ when $\mathbb{E}\{\epsilon(n)\} \neq 0$. The proposition then follows from Lemma 1 and a straightforward application of [1, Prop. 1].   □

4.3   S-BSS: Information on the Signal Modulus

This paragraph is concerned with the case where the reference signal has the same modulus as $s_1(n)$ but a random phase. All the results and comments in the previous section are adapted hereunder. More precisely, the reference signal reads:

$$\begin{cases} r(n) = \epsilon(n)|s_1(n)| \\ |\epsilon(n)| = 1, \mathbb{E}\{\epsilon(n)\} = 0 \text{ and } \epsilon(n) \text{ is independent of the sources.} \end{cases} \quad (16)$$

Lemma 2. *In the case of i.i.d. source signals, if $r(n)$ is given by (16), we have (where ϵ and s_1 denote $\epsilon(n)$ and $s_1(n)$ again for readability):*

$$C_r\{s_j(n-k)\} = \begin{cases} 0 & \text{if } (j,k) \neq (1,0), \\ \mathbb{E}\{|s_1(n)|^4\} - \mathbb{E}\{|s_1(n)|^2\}^2 & \text{if } (j,k) = (1,0). \end{cases} \quad (17)$$

Proof. For $(j,k) \neq (1,0)$, the equality $\kappa_r\{s_j(n-k)\} = 0$ follows from independence and the vanishing property of the cumulants. Now, if $(j,k) = (1,0)$, by developing the cumulant in term of moments, we obtain:

$$\kappa_r\{s_1(n)\} = \mathbb{E}\{|s_1|^4|\epsilon|^2\} - \mathbb{E}\{|s_1|^2\}\mathbb{E}\{|\epsilon s_1|^2\} - |\mathbb{E}\{\epsilon s_1|s_1|\}|^2 - \mathbb{E}\{\epsilon s_1^*|s_1|\}|^2$$

The result then follows after simplification, using the independence of $\epsilon(n)$ and $s_1(n)$ and the assumption $\mathbb{E}\{\epsilon(n)\} = 0$.   □

Proposition 3. *Considering the reference signal given by (16), the criterion C_r in (13) is a contrast function for i.i.d. sources if $s_1(n)$ has non constant modulus. In addition, the source identified by C_r is $s_1(n)$.*

Proof. Note that, if $s_1(n)$ does not have constant modulus, $C_r\{s_1(n)\} > 0$ since it is the variance of $|s_1(n)|$. The proposition then follows from Lemma 1 and a straightforward application of [1, Prop. 1].   □

4.4   Generalization

As mentioned, the reference signal $r(n)$ defined by (14) or (16) satisfy the conditions of Proposition 1 because it depends on $s_1(n)$ only. Obviously, if $\tilde{r}(n)$ is a scalar filtering of $r(n)$, it also depends on $s_1(n)$ only. More precisely, one can see that any scalar filtering of a reference given by (14) or (16) remains a valid reference signal under the same conditions as the above given ones. Finally, we would like to stress that the results in Section 4 remain valid in the case of non i.i.d. sources, although not presented here due to the lack of space.

5   Simulations

We considered three different kind of sources:

- real valued, uniformly distributed with mean zero and unit variance.
- complex valued, QAM4 and QAM16, that is the real and imaginary parts are independent taking their values with equal probability in $\{\pm 1/\sqrt{2}\}$ for QAM4 sources and in $\{\pm 1/\sqrt{5}, \pm 3/\sqrt{5}\}$ for QAM16 sources.

For each of the above choices, a set of $N = 3$ mutually independent and temporally i.i.d. sources have been generated. They have been mixed by a $Q \times N$ randomly chosen finite impulse response (FIR) filter of length 3 and with $Q = 4$ sensors. We then used the contrast $\mathcal{C}_r$ in (13) and the associated algorithm proposed in [1] to test the effectiveness of our result. The different choices for $r(n)$ are detailed next. All presented results correspond to averaged values over 1000 Monte-Carlo realizations of the mean square error (MSE) on the estimated source.

5.1   Information on the Signal Phase

The original reference $r(n)$ is given by (14), where:

- $\epsilon(n)$ is a binary i.i.d. Bernoulli process with $P(\epsilon(n) = 1) = p$ and $P(\epsilon(n) = -1) = 1 - p$ in the case of real-valued sources.
- $\epsilon(n) = e^{\imath \theta(n)}$ and $\theta(n)$ is uniformly distributed on an interval $[-\vartheta, \vartheta]$ in the case of complex-valued sources.

Table 1. Average (1000 realizations) MSE on the reconstructed source (real-valued sources with uniform distribution). $r(n)$ is given by (14).

Number of samples	Reference: $r(n)$				Reference: $\widetilde{r}(n) \leftrightarrow$ "$r(n)$+filt."			
	$p = 1$	$p = 0.9$	$p = 0.7$	$p = 0.5$	$p = 1$	$p = 0.9$	$p = 0.7$	$p = 0.5$
1000	0.0078	0.0249	0.1548	1.0780	0.0379	0.0945	0.7079	1.1340
5000	0.0015	0.0050	0.0331	1.0799	0.0130	0.0748	0.2470	1.1303
10000	7.57e-4	0.0025	0.0168	1.0762	0.0073	0.0142	0.1383	1.1335

Table 2. Average (1000 realizations) MSE on the reconstructed source (complex valued sources with QAM4 or QAM16 distribution). $r(n)$ is given by (14).

Sources	Number of samples	Reference: $r(n)$			
		$\vartheta = 0$	$\vartheta = \frac{\pi}{4}$	$\vartheta = \frac{\pi}{2}$	$\vartheta = \pi$
QAM4	1000	5.84e-5	0.0055	0.0334	1.2114
	5000	2.11e-6	0.0011	0.0067	1.2160
	10000	4.98e-7	5.40e-4	0.0034	1.2119
QAM16	1000	0.0027	0.0087	0.0395	1.2116
	5000	5.23e-4	0.0017	0.0080	1.2133
	10000	2.63e-4	8.65e-4	0.0040	1.2126

According to Section 4.4, we consider also $\widetilde{r}(n)$ as a reference, where $\widetilde{r}(n)$ has been obtained by a FIR scalar filtering of $r(n)$. The filter have three taps with randomly driven coefficients. Different values of the parameters p, ϑ and different sample sizes are considered. The results are reported in Table 1 for real-valued sources and in Table 2 for complex-valued sources.

One can observe the effectiveness of the separation when $p \neq 0.5$ and $\vartheta \neq \pi$: it should indeed be no surprise that no separation is obtained for $p = 0.5$ or $\vartheta = \pi$ since in this case, $r(n)$ is independent of $s_1(n)$.

5.2   Information on the Signal Modulus

Now the original reference $r(n)$ is given by (16), where:

- $\epsilon(n)$ is a binary i.i.d. process with $P(\epsilon(n) = 1) = P(\epsilon(n) = -1) = 1/2$ in the case of real-valued sources.
- $\epsilon(n) = e^{i\theta(n)}$ and $\theta(n)$ is uniformly distributed on the interval $[-\pi, \pi]$ in the case of complex-valued sources.

Similarly to the previous paragraph, we considered $r(n)$ and $\widetilde{r}(n)$ as reference signals, where $\widetilde{r}(n)$ is a scalar filtering of $r(n)$. The results are showed in Table 3. The reader can see the effectiveness of our method for uniformly distributed and QAM16 sources. On the contrary, QAM4 sources have constant modulus and thus do not satisfy the conditions of Proposition 3.

Table 3. Average (1000 realizations) MSE on the reconstructed source. $r(n)$ is given by (16).

Sources		Uniform			QAM16			QAM4
Number of samples		1000	5000	10000	1000	5000	10000	10000
Reference	$r(n)$	0.0238	0.0045	0.0022	0.0548	0.0100	0.0050	1.2080
	$\widetilde{r}(n) \leftrightarrow$ "$r(n)$+filt."	0.2433	0.0651	0.0408	0.6247	0.1930	0.1190	1.2690

References

1. Castella, M., Rhioui, S., Moreau, E., Pesquet, J.C.: Quadratic higher-order criteria for iterative blind separation of a MIMO convolutive mixture of sources. IEEE Trans. Signal Processing 55(1), 218–232 (2007)
2. Kawamoto, M., Kohno, K., Inouye, Y.: Eigenvector algorithms incorporated with reference systems for solving blind deconvolution of MIMO-IIR linear systems. IEEE Signal Processing Letters 14(12), 996–999 (2007)
3. Hua, Y., Tugnait, J.K.: Blind identifiability of FIR-MIMO systems with colored input using second order statistics. IEEE Signal Processing Letters 7(12), 348–350 (2000)
4. Simon, C., Loubaton, P., Jutten, C.: Separation of a class of convolutive mixtures: a contrast function approach. Signal Processing (81), 883–887 (2001)
5. Tugnait, J.K.: Identification and deconvolution of multichannel linear non-gaussian processes using higher order statistics and inverse filter criteria. IEEE Trans. Signal Processing 45(3), 658–672 (1997)

On the Relationships between MMSE and Information-Theoretic-Based Blind Criterion for Minimum BER Filtering

Charles C. Cavalcante[1] and João M.T. Romano[2]

[1] GTEL, DETI, Federal University of Ceará, Brazil
charles@gtel.ufc.br
[2] DSPCom, FEEC, University of Campinas, Brazil
romano@dmo.fee.unicamp.br

Abstract. In this paper we present a relationship between supervised (MMSE) and unsupervised criteria for minimum bit error rate (BER) filtering. A criterion based on the probability density function (pdf) estimation which has an information theoretical approach is used to link the MMSE criterion and the maximum *a posteriori* one in order to obtain a linear filter that minimizes the BER. An analytical relationship among those three criteria is presented and analyzed showing the limits imposed to achieve minimum BER without training sequences when the pdf estimation-based criterion is considered.

1 Introduction

A classical strategy for the optimization of an adaptive equalizer is the use of a training sequence. Using this approach, the optimum solution is provided by the minimum mean square error (MMSE) supervised criteria [1].

Whenever such sequence is not available, an unsupervised, or *blind*, processing is employed in the optimization procedure, based in some known statistical characteristics of the involved signals [2]. Even that most blind algorithm have higher computational complexity they have a lower information complexity since they require less information about the signal than supervised strategies.

Despite its frequent use in supervised strategies, the MMSE is not the optimum solution in practical systems [3]. The minimization of the bit error rate (BER) is more useful due to the importance of such measure in practice. Further, it is known that MMSE does not achieve minimum BER when the equalizer does not have an appropriate length [4].

Some works have considered an optimization criterion based on minimum BER [4,5]. However, they rely on supervised optimization criteria. So the following question arises: *when a training sequence is not available or desired is it possible to perform minimum BER filtering?* This paper aims to provide an answer to this question.

A probability density function (pdf) estimation-based blind criterion was proposed in [6]. Using a parametric model that matches the statistical characteristics of the transmitted signal, the equalizer is designed to minimize the divergence

T. Adali et al. (Eds.): ICA 2009, LNCS 5441, pp. 17–24, 2009.
© Springer-Verlag Berlin Heidelberg 2009

between the pdf of the equalized signal and such parametric model. This idea is based on an information-theoretic approach that aims to maximize the relative entropy of the system [7].

In this paper, we present a relationship that shows that it is not possible to achieve minimum BER using the proposal in [6]. Using the maximum *a posteriori* (MAP) criterion, which minimizes the BER, we derive a relationship between the MMSE, MAP and the blind criteria proposed in [6]. This result shows an important property of the blind filtering approach when minimum BER is required.

The rest of the paper is organized is follows. Section 2 describes the system model. The blind criterion is revisited in Section 3 and the relationship between the blind criterion and minimum BER approach is presented in Section 4. Finally, our conclusions are stated in Section 5.

2   System Model

The considered base-band system model is depicted in Figure 1.

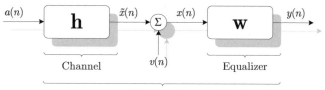

Fig. 1. Base-band system model

The transmitted sequence is represented by $\mathbf{a}(n)=[a(n) \cdots a(n-N-M+1)]^T$, where N and M are, respectively, the channel and equalizer lengths. It is assumed independent and identically distributed (i.i.d) symbols $a(n) \in \mathcal{A}$, and the set $\mathcal{A}$ has cardinality S. The channel is represented by a FIR filter given by $\mathbf{h} = \begin{bmatrix} h_0 \cdots h_{N-1} \end{bmatrix}^T$.

The additive noise denoted in vectorial way by $\mathbf{v}(n)=\begin{bmatrix} v(n) \cdots v(n-M+1) \end{bmatrix}^T$ is white, Gaussian, uncorrelated from the transmitted sequence and has variance σ_v^2.

The equalizer, which has finite impulse response (FIR) denoted by $\mathbf{w}(n) = \begin{bmatrix} w(n) \cdots w(n-M+1) \end{bmatrix}^T$, is fed by the channels outputs $x(n) = \tilde{x}(n) + v(n)$ where $\tilde{x}(n) = \sum_{i=0}^{N-1} h_i a(n-i)$ are the noiseless channel outputs.

The equalizer output is denoted in vectorial representation by $y(n) = \mathbf{w}^T(n)\mathbf{x}(n)$, where $\mathbf{x}(n) = \begin{bmatrix} x(n) \cdots x(n-M+1) \end{bmatrix}$. This model will be used in the rest of this paper.

3   Blind Criterion for PDF Estimation: A Review

Let $\mathbf{w}_{\text{ideal}}$ be an ideal linear equalizer, its output can be written as

$$y(n) = \mathbf{w}_{\text{ideal}}^{T}\mathbf{x}(n), \tag{1}$$

where

$$\mathbf{x}(n) = \mathbf{H}\mathbf{a}(n) + \mathbf{v}(n) \tag{2}$$

and $\mathbf{H}$ is the $M \times (N + M - 1)$ *convolution matrix* of the channel [8].
   Then, using Eq. (2) in (1), it is possible to write:

$$\begin{aligned}
y(n) &= (\mathbf{H}\mathbf{a}(n) + \mathbf{v}(n))^{T}\,\mathbf{w}_{\text{ideal}} \\
&= \mathbf{a}^{T}(n)\mathbf{H}^{T}\mathbf{w}_{\text{ideal}} + \mathbf{v}^{T}(n)\mathbf{w}_{\text{ideal}} \\
&= \mathbf{a}^{T}(n)\underbrace{\mathbf{H}^{T}\mathbf{w}_{\text{ideal}}}_{\mathbf{g}_{\text{ideal}}} + \mathbf{v}^{T}(n)\mathbf{w}_{\text{ideal}} \\
&= \mathbf{a}^{T}(n)\mathbf{g}_{\text{ideal}} + \vartheta(n) \\
&= a(n - \delta) + \vartheta(n),
\end{aligned} \tag{3}$$

where $\mathbf{g}_{\text{ideal}}$ is the ideal system response, δ is a delay and $\vartheta(n)$ is a random variable (r.v.) assumed with zero mean independent Gaussian samples with variance σ_{ϑ}^{2} [9].
   Eq. (3) states that the pdf of the signal on the output of the equalizer is a mixture of Gaussians given by:

$$p_{Y}(y) = \frac{1}{\sqrt{2\pi\sigma_{\vartheta}^{2}}} \cdot \sum_{i=1}^{S} \exp\left[-\frac{|y(n) - a_{i}|^{2}}{2\sigma_{\vartheta}^{2}}\right] \cdot \Pr(a_{i}), \tag{4}$$

where the a_{i} are the possible values of $a(n - \delta)$ which are also symbols of the transmitted alphabet $\mathcal{A}$, and $\Pr(a_{i})$ denotes the probability of a discrete-valued symbol taking the value a_{i}.
   Since the pdf of the desired equalized signal is known, we wish to design a criterion that forces the adaptive filter to produce signals with the same (or similar) pdf than the ideal one. It is then interesting to use the well known measure of similarities between strictly positive functions (such as the pdfs), the *Kullback-Leibler Divergence* (KLD) [8].
   In order to use the KLD to provide pdf estimation, a parametric model which is a function of the filter parameters is constructed [9]. A natural choice is the same model of mixture of Gaussians, as the pdf of the data recovered by an ideal equalizer, like the one in Eq. (4). Then

$$\Phi(y, \sigma_{r}^{2}) = A \cdot \sum_{i=1}^{S} \exp\left(-\frac{\left|\mathbf{w}^{T}(n)\mathbf{x}(n) - a_{i}\right|^{2}}{2\sigma_{r}^{2}}\right) \cdot \Pr(a_{i}), \tag{5}$$

is the chosen parametric model, where σ_{r}^{2} is the variance of each Gaussian in the model and where $A = \frac{1}{\sqrt{2\pi\sigma_{r}^{2}}}$. In pattern classification field these kind of parametric functions, which are used to measure similarities against other functions, are called *target functions* [10].

Then, applying KLD to compare Equations (4) and (5) yields:

$$
D_{p(y)||\Phi(y,\sigma_r^2)} = \int_{-\infty}^{\infty} p(y) \cdot \ln\left(\frac{p(y)}{\Phi\left(y,\sigma_r^2\right)}\right) dy
$$

$$
= \int_{-\infty}^{\infty} p(y) \cdot \ln\left(p(y)\right) dy - \int_{-\infty}^{\infty} p(y) \cdot \ln\left(\Phi(y,\sigma_r^2)\right) dy.
$$

(6)

Minimizing (6) is equivalent to minimizing only the $\Phi\left(y,\sigma_r^2\right)$-dependent term, since the remaining term provides the entropy of the variable y, that is:

$$
J_{\text{FPC}}\left(\mathbf{w}\right) = -E\left\{\ln\left[\Phi\left(y,\sigma_r^2\right)\right]\right\}
$$

$$
= -E\left\{\ln\left[A\cdot\sum_{i=1}^{S}\exp\left(-\frac{|y(n)-a_i|^2}{2\sigma_r^2}\right)\cdot\Pr(a_i)\right]\right\}.
$$

(7)

The **Fitting pdf (FP)** criterion corresponds to minimizing $J_{\text{FPC}}\left(\mathbf{w}\right)$. Furthermore, it is known that minimizing Eq. (7) corresponds to finding the entropy of y if $\Phi\left(y,\sigma_r^2\right) = p_Y(y)$ [10, p. 59].

A stochastic algorithm for filter adaptation is given by:

$$
\mathbf{w}(n+1) = \mathbf{w}(n) - \mu_w \nabla J_{\text{FPC}}\left(\mathbf{w}(n)\right)
$$

$$
\nabla J_{\text{FPC}}\left(\mathbf{w}(n)\right) = \frac{\sum_{i=1}^{S}\exp\left(-\frac{|y(n)-a_i|^2}{2\sigma_r^2}\right)\left(y(n)-a_i^*\right)}{\sigma_r^2\cdot\sum_{i=1}^{S}\exp\left(-\frac{|y(n)-a_i|^2}{2\sigma_r^2}\right)}\mathbf{x}(n),
$$

(8)

where μ_w is the step size. The choice of the step size is done according the condition of convergence of adaptive algorithms [1]. Typically, μ_w has the same value than the step size for other blind algorithms, e.g. the CMA one. For the σ_r^2 the optimum value would be the variance of the Gaussian noise, since the model in Eq. (5) should be the best representation of the pdf of the recovered signals without intersymbol interference. However, we do not assume the knowledge about the SNR at the receiver. Instead, we select an arbitrary value for σ_r^2 that allows a proper recovering of the symbols.

The adaptive algorithm shown in Eq. (8) shows an important property of the algorithm: it takes into account the phase of the transmitted symbols.

The computational complexity of this algorithm is proportional to the computation of S exponentials which are required by Eq. (8). Thus, its complexity is a little higher than other LMS-like algorithms.

Due to the nature of the approach used to derive the criterion FP it turns out that it is very similar to the results from entropy estimators that we can find in the literature of blind source separation, see for instance [11]. However, the rationale of our approach is very different from those. They reach a similar model making use of a Gaussian kernel to implement a nonparametric pdf estimator. In our case, we assume the knowledge about the noise pdf form in order to construct the criterion.

● The parameter σ_r^2:

As shown in the previous section, the parametric model used to update the filter coefficients is also σ_r^2-dependent. This parameter plays an important role since it is the variance of each Gaussian in the parametric model.

Moreover, σ_r^2 is also important for the convergence rate because it modifies the effective step size, that is, $\mu_{\text{eff}} = \frac{\mu_w}{\sigma_r^2}$. In the classification field this parameter is similar to the *temperature* one in annealing processes [7].

A numerical problem that arises with the use of the FPA is the nonconvergence for very small values of σ_r^2. This is due to the Gaussians being very sharp and much more difficult to fit the data on them. This model has also been considered in [12], where the ideal pdf of the received signal is assumed to be a mixture of impulses and later a Gaussian mixture model is considered in order to make the assumption more realistic and feasible.

4   Minimum BER: Supervised and Blind Criteria

In order to allow the analysis of a minimum BER criterion, we consider the MAP one.

The MAP criterion aims to maximize the probability of recovering a symbol a_i given that $y(n)$ has been observed at the equalizer output. Then, the MAP criterion can be written as [13]:

$$J_{\text{MAP}}(\mathbf{w}) = E\left\{\ln\left[p\left(a_i \,|y(n)\,\right)\right]\right\}, \tag{9}$$

where we are considering the logarithm in Eq. (9) in order to simplify computations [13].

Let us write the *a posteriori* probability density functions using the Bayes' rule as [13]:

$$p\left(a_i \,|y(n)\,\right) = \frac{p\left(y(n) \,|a_i\,\right) \cdot \Pr\left(a_i\right)}{\sum\limits_{j=1}^{S} p\left(y(n) \,|a_j\,\right) \cdot \Pr\left(a_j\right)}. \tag{10}$$

Here, we have to change the way we use to present the FPC. The parametric model $\Phi(y)$ represents the sum of probabilities of a possible transmitted signal a_i given that $y(n)$ has been observed. Since there is no knowledge about the transmitted symbol itself, $\Phi(y)$ is then the sum of all conditional probabilities of the received signal $y(n)$ given the transmission of symbol a_i. In other words, we can write:

$$\Phi(y) = \sum_{i=1}^{S} p(y(n)|a_i) \cdot \Pr(a_i). \tag{11}$$

Of course, if we assume that the signal is corrupted by AWGN, we obtain Eq. (5). Besides, we can use the gradient of J_{FPC} without the stochastic approximation in the form

$$\nabla J_{\text{FPC}}(\mathbf{w}) = -E_Y\left\{\frac{E_A\left\{A \cdot \exp\left(-\frac{|y(n)-a(n)|^2}{\sigma_r^2}\right)[y(n) - a^*(n)]\right\}}{\sigma_r^2 \cdot E_A\left\{A \cdot \exp\left(-\frac{|y(n)-a(n)|^2}{\sigma_r^2}\right)\right\}}\mathbf{x}\right\}, \tag{12}$$

where E_Y and $E_{\mathcal{A}}$ stand for expectation with respect to the variables Y and $\mathcal{A}$, respectively and $A = 1/\sqrt{2\pi\sigma_r^2}$.

As in [14], we can define an auxiliary function given by

$$\psi(y,a) = \frac{A \cdot \exp\left(-\frac{|y(n)-a(n)|^2}{\sigma_r^2}\right)}{E_{\mathcal{A}}\left\{A \cdot \exp\left(-\frac{|y(n)-a(n)|^2}{\sigma_r^2}\right)\right\}}, \tag{13}$$

that measures "how sure" is the decision of symbol $a(n)$ since only $y(n)$ has been observed and the signal has a conditional pdf given as a Gaussian.

Then, comparing Equations (10) and (13), we can observe that considering the Gaussian model for the conditional pdf we have the same measure [9].

Using such consideration and supposing that σ_r^2 is chosen appropriately, the MAP criterion can be written as [9]:

$$E\left\{\ln\left[p\left(a_i\left|y(n)\right.\right)\right]\right\} = E\left\{\ln\left[\frac{p\left(y(n)\left|a_i\right.\right)\cdot\Pr\left(a_i\right)}{\Phi(y)}\right]\right\}$$
$$J_{\mathrm{MAP}} = E\left\{\ln\left[p\left(y(n)\left|a_i\right.\right)\cdot\Pr\left(a_i\right)\right]\right\} \underbrace{-E\left\{\ln\left[\Phi(y)\right]\right\}}_{J_{\mathrm{FPC}}}. \tag{14}$$

It is worth mentioning that the conditional probability $p\left(y(n)\left|a_i\right.\right)$ concerns the assumed model to the signal at the output of the equalizer and we are also assuming an ideally equalized signal in presence of additive Gaussian noise. We then have:

$$p\left(y(n)\left|a_i\right.\right)\cdot\Pr\left(a_i\right) = \frac{1}{\sqrt{2\pi\sigma_\vartheta^2}}\exp\left(-\frac{|y(n)-a_i|^2}{2\sigma_\vartheta^2}\right)\cdot\Pr\left(a_i\right). \tag{15}$$

Therefore, we can rewrite Eq. (14) as

$$J_{\mathrm{MAP}} = -\frac{1}{2\sigma_{\vartheta^2}}E\left\{|y(n)-a_i|^2\right\} + \ln\left[\frac{\Pr(a_i)}{\sqrt{2\pi\sigma_\vartheta^2}}\right] + J_{\mathrm{FPC}}$$
$$J_{\mathrm{FPC}} - J_{\mathrm{MAP}} = \frac{1}{2\sigma_{\vartheta^2}}E\left\{|y(n)-a_i|^2\right\} - \ln\left[\frac{\Pr(a_i)}{\sqrt{2\pi\sigma_\vartheta^2}}\right]. \tag{16}$$

Now, we need to explore the right side of Eq. (16) in order to provide an appropriate relationship.

Eq. (11), as we stated before, corresponds to the sum of all probabilities of symbol a_i given the observation $y(n)$.

This is due to the blind processing, when there is no information about the transmitted symbol in a given time instant. In the case of supervised processing, the transmitted symbol is known at each time instant and there is no need of computing the contribution of all symbols from alphabet $\mathcal{A}$. Thus, the parametric model for the supervised case is given by [9]:

$$\Phi(y) = p(y(n)|a(n)). \tag{17}$$

Considering Eq. (17), and also assuming the ideally recovered signal immersed in AWGN, that is, the conditional probability as given by Eq. (15), we can write $J_{\mathrm{FPC}}(\mathbf{w})$ in Eq. (7) for the supervised case as:

$$J_{\text{FPC}}(\mathbf{w}) = -E\left\{\ln\left[A \cdot \exp\left(-\frac{|y(n) - a(n)|^2}{2\sigma_r^2}\right) \cdot \Pr(a_i)\right]\right\}$$

$$= \frac{1}{2\sigma_r^2}E\left\{|y(n) - a(n)|^2\right\} - \ln[A \cdot \Pr(a_i)]. \tag{18}$$

Clearly, the cost function in Eq. (18) is the MMSE cost function up to scaling and translation effects. However, the optimization of Eq. (18) with respect to $\mathbf{w}$ provides the same solution than the classical MMSE cost function given as [1]:

$$J_{\text{MMSE}}(\mathbf{w}) = E\left\{|y(n) - a(n)|^2\right\}. \tag{19}$$

Therefore, the cost function in Eq. (18) will be denoted J_{MMSE} as it stands for the MMSE in the supervised case.

Observing Eq. (16) and comparing the right side with Eq. (18) we can see that it is the same. Thus, the following relationship can be given considering $\sigma_r^2 = \sigma_v^2$ [9]:

$$J_{\text{MAP}} = J_{\text{FPC}} - J_{\text{MMSE}}. \tag{20}$$

Eq. (20) provides an important issue about relationships of blind and supervised criteria for minimum BER filtering using the FPC criterion. It shows that when there is no knowledge about the transmitted signal, the FPC does not achieve minimum BER. So, it is not possible to perform minimum BER filtering with this criterion without the knowledge of the transmitted sequence. Further, since the criteria are defined as positive functions we can also write the following inequality for the FPC and MAP:

$$J_{\text{FPC}} \geq J_{\text{MAP}}, \tag{21}$$

showing that achieving the minimum for J_{FPC} does not necessarily imply achieve minimum BER.

On the other hand, we may start the analysis from the MMSE criterion in order to quantify the achieved fraction of minimum BER.

According to the MMSE criterion, given by Eq. (19), one can see that the expectation is taken w.r.t. the possible values of the r.v. $a(n)$. This makes necessary the knowledge of the pdf of the transmitted symbols.

Performing minimization of BER, in terms of maximum *a posteriori* probabilities, also requires the minimization of the joint pdfs $p(y, a_i)$. For this sake, it is required the estimation of the pdf from the equalizer output and the transmitted sequence.

When we deal with finite length equalizers, the estimate of the joint pdfs is easier to be achieved since the number of probabilities is much higher to be taken into account. Otherwise, we reach a suboptimum solution that does not provides minimum BER.

5   Conclusions and Perspectives

In this paper we have presented a relationship between supervised and unsupervised criteria aiming a minimum bit error rate filtering.

The unsupervised criteria is based on estimating the probability density function of the signal on the equalizer output using a parametric model. The Kullback-Leibler divergence is used to minimize the divergence between the equalizer output pdf and the parametric model.

This criterion presents some interesting properties that are based on the evaluation of the conditional probabilities of received signal over all possible transmitted ones. As a result, the presented approach allows to achieve a relationship between the supervised and unsupervised criteria.

The obtained relationship indicates that minimum BER is not attained with the blind criteria because it requires the instantaneous knowledge of the transmitted signal. Thus, such relationship is given in terms of the minimum mean square error and FPC criteria for achieving maximum *a posteriori* probabilities.

The main perspective for future works is the investigation and proposal of a semi-blind criterion that can possibly minimize BER and some other metric (e.g. Kullback-Leibler one) with a good compromise of computational and information complexity.

References

1. Haykin, S.: Adaptive Filter Theory, 3rd edn. Prentice-Hall, Englewood Cliffs (1996)
2. Ding, Z., Li, Y.G.: Blind Equalization and Identification. Marcel Dekker, New York (2001)
3. Brossier, J.M.: Signal et Communication Numérique: égalisation et synchronisation. HERMES (1997)
4. Yeh, C.C., Barry, J.R.: Adaptive Minimum Bit-Error Rate Equalization for Binary Signaling. IEEE Trans. on Comm. 48(7), 1226–1235 (2000)
5. Phillips, K.A.: Probability Density Function Estimation Applied to Minimum Bit Error Rate Adaptive Filtering. Master's thesis, Virginia Polytechnic Institut and State University (May 1999)
6. Cavalcante, C.C., Cavalcanti, F.R.P., Mota, J.C.M.: A PDF Estimation-Based Blind Criterion for Adaptive Equalization. In: Proceedings of IEEE Int. Symposium on Telecommunications (ITS 2002), Natal, Brazil (September 2002)
7. Haykin, S.: Neural Networks: A Comprehensive Foundation, 2nd edn. Prentice Hall, Englewood Cliffs (1998)
8. Haykin, S. (ed.): Unsupervised Adaptive Filtering, vol. II. John Wiley & Sons, Chichester (2000)
9. Cavalcante, C.C.: On Blind Source Separation: Proposals and Analysis of Multi-User Processing Strategies. PhD thesis, State University of Campinas (UNICAMP) - Brazil (2004)
10. Bishop, C.M.: Neural Networks for Pattern Recognition. Oxford University Press, UK (1995)
11. Erdogmus, D., Hild II, K.E., Príncipe, J.C.: Online Entropy Manipulation: Stochasticc Information Gradient. IEEE Sig. Proc. Letters 10(8), 242–245 (2003)
12. Amara, R.: Égalisation de Canaux Linéaires et Non Linéaires, Approche Bayésienne. PhD thesis, Université Paris XI, France (2001)
13. Proakis, J.G.: Digital Communications, 3rd edn. McGraw-Hill, New York (1995)
14. Sala-Alvarez, J., Vázquez-Grau, G.: Statistical Reference Criteria for Adaptive Signal Processing in Digital Communications. IEEE Trans. on Sig. Proc. 45(1), 14–31 (1997)

Blind Separation of Cyclostationary Signals

Nhat Anh Cheviet[1], Mohamed El Badaoui[1],
Adel Belouchrani[2], and Francois Guillet[1]

[1] LASPI EA3059, University of Saint Etienne, Jean Monnet, France
{nhat.anh.che.viet,mohamed.elbadaoui,guillet}@univ-st-etienne.fr
[2] Elec. Eng. Dept., Ecole Nationale Polytechnique, Algiers, Algeria
adel.belouchrani@enp.edu.dz

Abstract. In this paper, we propose a new method for the blind source separation with assuming that the source signals are cyclostationarity. The proposed method exploits the characteristics of cyclostationary signals in the Fraction-of-Time probability framework in order to simultaneously separate all sources without restricting the distribution or the number of cycle frequencies of each source. Furthermore, a new identifiability condition is also provided to show which kind of cyclostationary source can be separated by the second-order cyclostationarity statistics. Numerical simulations are presented to demonstrate the effectiveness of the proposed approach.

Keywords: Blind source separation, cyclostationary signals, Fraction-of-Time, identifiability condition.

1 Introduction

Blind source separation (BSS) aims to recover a set of source signals and/or a linear combination (mixing) from data that are recorded from a set of multiple sensors. The common mixture model for BSS is the instantaneous mixing. The source signals are often assumed to be zero mean, *stationary, ergodic* and statistically independent while the additive noise is Gaussian with the same variance. To our knowledge, there are many different methods that have been proposed to solve this problem since the eighties. These methods can be classified into several major approaches: non-gaussianity, maximum likelihood, minimum mutual information, neural network modelling and algebraic. Moreover, there are several famous algorithms which are based on the algebraic approach such as AMUSE [1], FOBI [1], SOBI [2], JADE [3]. All these above algorithms are designed to apply to the stationary signal but can be used for cyclostationary signal in some particular case [2].

In this paper, the source signals are assumed to be cyclostationary. Cyclostationary signals (CS) include virtually all man-made signals encountered in telecommunications, in mechanics, in econometrics, etc. In addition, the theory of CS has been studied in term of two alternative approaches: the orthodox approach based on stochastic processes framework where all statistical parameters are defined in term of a probability space (an ensemble of signal together

T. Adali et al. (Eds.): ICA 2009, LNCS 5441, pp. 25–33, 2009.

with a probability rule) and the more recently developed approach based on the Fraction-of-Time (FOT) probability framework where all statistical parameters are defined through infinite-time averages of a single time-series rather ensemble averages of a stochastic process [4]-[8]. On the other hand, for an only observed sample, the question of *ergodicity* (or *cycloergodicity*) is the primary importance (see [5], [6]). Therefore, the theory of the second order and higher order cyclostationary (SOCS, HOCS, respectively) using stochastic process requires a thorough understanding of cycloergodicity and complicates the development of a single data record estimators of the parameters of the theory, since the parameters are based on an ensemble, whereas the estimators operate on a single sample path [5].

Recently, several algorithms have been proposed to take advantage of the cyclostationarity property to solve the above problem such as methods in [9]-[14]. Although these algorithms are successful under several assumed conditions such as: the CS is assumed to be cycloergodic, the source shares only one cycle frequency, the cyclic autocorrelation function of each source at its cycle frequency at zero lag is positive ($R_{s_i}^{\alpha_i}(0) > 0$) or the CS has to satisfy the condition of identification of source [11], etc. But, in general, such requirements can not be always satisfied, e.g. absence of cycloergodicity; each CS exhibits more than one cycle frequency; shares one or more cycle frequencies with the other CS; the condition of source in [11] can not be satisfied, $R_{s_i}^{\alpha_i}(0)$ is negative or complex, etc. Particularly, most of these above algorithms (except [14]) are proposed using the stochastic process framework.

Thus, in order to overcome the aforementioned limitations, the statistical analysis framework used throughout this paper is FOT probability framework, which obviates the concept of cycloergodicity and avoids some difficulties related to the estimation of the SOCS and HOCS parameters. In addition, by resorting to the FOT probability framework that has been developed by W.A. Gardner, C.M. Spooner, A. Napolitano [4]-[8] for cyclostationary signals, we propose in this paper an identifiability condition, an algorithm for the blind separation of cyclostionary signals. The proposed algorithm is an extension of the SOBI algorithm [2]. The remainder of this paper is organized as follows. In Section 2, the relevance theory of cyclostationarity signal in the FOT probability framework is firstly reviewed. Next, in Section 3, the signal model and the assumption for BSS are briefly presented. In Section 4, the new identifiability condition and the proposed algorithm are provided. Furthermore, in Section 5, several different types of simulation experiments are presented to demonstrate the efficiency of the proposed algorithm. Finally, this paper is concluded in Section 6.

2   Cyclostationary Signal

In the FOT probability framework, signals are modelled as single function of time (called time-series). Briefly, a time-series $s_1(t)$ is said to exhibit SOCS with cycle frequency $\alpha \neq 0$ if there exists at least one of two functions $R_{s_1^*}^\alpha(\tau)$, $R_{s_1}^\alpha(\tau)$:

$$R_{s_1^\varepsilon}^\alpha(\tau) \triangleq \left\langle s_1(t+\tau_1)s_1^\varepsilon(t+\tau_2)e^{-j2\pi\alpha t}\right\rangle_t \tag{1}$$

is not identically zero for some values $\tau \overset{\Delta}{=} [\tau_1, \tau_2]^T \in \mathbb{R}^2$ [4]-[6], $\varepsilon = \pm 1$ where $s_1^1(t) \overset{\Delta}{=} s_1^\star(t)$ and $s_1^{-1}(t) \overset{\Delta}{=} s_1(t)$. In (1), $R_{s_1^\star}^\alpha(\tau)$, $R_{s_1}^\alpha(\tau)$, the operator $\langle \cdot \rangle_t$, the superscript $\star$ and ^{T} denote the cyclic autocorrelation function, the cyclic conjugate autocorrelation function, the time-averaging operation, the complex conjugate and the transpose of a vector, respectively.

For a real time-series, two functions $R_{s_1^\star}^\alpha(\tau)$, $R_{s_1}^\alpha(\tau)$ are equivalent. Thus, $s_1(t)$ is said to be purely stationary if $R_{s_1}^\alpha(\tau)$ is not identically zero for only $\alpha = 0$. On the other hand, if $R_{s_1}^\alpha(\tau)$ is not identically zero for only $\alpha = n/T_0$, for some integers n including $n = 0$, then $s_1(t)$ is said to be purely cyclostationary with period T_0. Otherwise, if $R_{s_1}^\alpha(\tau) \neq 0$ for some values of α that are not all integer multiples of one non-zero value, then $s_1(t)$ is said to be almost cyclostationary with multiple incommensurate periods [4].

For complex-valued time-series, two functions $R_{s_1^\star}^\alpha(\tau)$, $R_{s_1}^\alpha(\tau)$ are distinct. For certain complex-valued signal type, the cycle frequency sets $\{\alpha\}$, $\{\eta\}$ such that $R_{s_1^\star}^\alpha(\tau) \neq 0$, $R_{s_1}^\eta(\tau) \neq 0$ are disjoint, as illustrated in [5], [6]. For many complex-valued communication signal models, the cycle frequency sets depend on the modulation type: e.g. the sets $\{\eta\}$ are related to the carrier frequency (or carrier offset), whereas the sets $\{\alpha\}$ are related to the pulse rate, etc.

Let I_k, I be the cycle frequency sets of time-series $s_k(t), k = 1, \ldots, N$ and the cycle frequency sets of all source signals, thus, one can be expressed as:

$$I_k \overset{\Delta}{=} \underset{\tau \in \mathbb{R}^2}{\cup} \{\alpha, \eta \in \mathbb{R}^* : R_{s_k^\star}^\alpha(\tau) \text{ or } R_{s_k}^\eta(\tau) \neq 0\} \tag{2}$$

$$I \overset{\Delta}{=} \overset{N}{\underset{k=1}{\cup}} I_k - \overset{N}{\underset{\substack{k,l=1 \\ k \neq l}}{\cup}} (I_k \cap I_l) \tag{3}$$

Equation (3) takes into account that some signals can be shared the same cycle frequencies. Furthermore, if $s_1(t)$ exhibits SOCS, then $L_{s_1^\varepsilon}(t, \tau)$ i.e. the second-order lag product of $s_1(t)$, contains a poly-periodic or just periodic component $p_{s_1^\varepsilon}(t, \tau)$ and an aperiodic residual component $m_{s_1^\varepsilon}(t, \tau)$:

$$L_{s_1^\varepsilon}(t, \tau) \overset{\Delta}{=} s_1(t + \tau_1)s_1^\varepsilon(t + \tau_2) = p_{s_1^\varepsilon}(t, \tau) + m_{s_1^\varepsilon}(t, \tau) \tag{4}$$

where $\langle m_{s_1^\varepsilon}(t, \tau)e^{-j2\pi\alpha t}\rangle_t = 0, \forall \alpha \neq 0$.

Now let us define the temporal expectation operation $\hat{E}^{\{\alpha\}}\{\cdot\}$ that is simply the sum of all sine waves component extractors in $L_{s_1^\varepsilon}(t, \tau)$:

$$\hat{E}^{\{\alpha\}}\{L_{s_1^\varepsilon}(t, \tau)\} \overset{\Delta}{=} R_{s_1^\varepsilon}(t, \tau) \overset{\Delta}{=} \sum_{\alpha \neq 0} \langle L_{s_1^\varepsilon}(u, \tau)e^{-j2\pi\alpha u}\rangle_u e^{j2\pi\alpha t} \tag{5}$$

$$= p_{s_1^\varepsilon}(t, \tau) = \sum_{\alpha \in I_1} R_{x^\varepsilon}^\alpha(\tau)e^{j2\pi\alpha t} \tag{6}$$

where $R_{s_1^\varepsilon}(t, \tau)$ is called the temporal moment function (TMF) of $s_1(t)$.

Note that the computation of the TMF must be accomplished from discrete samples of $s_1(t)$ with sampling period T_s. In addition, due to aliasing effects in both α and f parameters, T_s must be sufficiently small to avoid spectral aliasing, then $T_s \leq B_1/4$ where B_1 is bandwidth of $s_1(t)$. More details treatments about the continuous- and discrete-time time-series in the FOT probability framework context can be found in [4]-[8], and references therein.

3   Problem Formulation

Consider the following model of instantaneous mixtures:

$$y(t) = \mathbf{A}s(t) + b(t) \tag{7}$$

where $y(t) \in \mathbb{C}^M$ (or $\in \mathbb{R}^M$), $\mathbf{A} \in \mathbb{C}^{M \times N}(\mathbb{R}^{M \times N})$, $s(t) \in \mathbb{C}^N(\mathbb{R}^N)$ and $b(t) \in \mathbb{C}^M$ ($\mathbb{R}^M$) denote the observation vector, the unknown mixing matrix which characterizes the medium or the channel, the vector of unknown source signals and the additive stationary noise, respectively. The purpose of BSS problem is to identify the mixing matrix $\mathbf{A}$ and/or recover the source signals $s(t)$ from the sensor array observation $y(t)$. Thus, one should look for a separating matrix $\mathbf{B} \in \mathbb{C}^{N \times M}$ such that: $\mathbf{BA} = \mathbf{P}\Lambda$, where $\mathbf{P} \in \mathbb{P}^{N \times N}$, $\Lambda \in \mathbb{D}^{N \times N}$, $\mathbb{P}^{N \times N}$ is a set of permutation matrices and $\mathbb{D}^{N \times N}$ is a set of non-singular diagonal matrices.

The following assumptions are hold in through this paper: The mixing matrix $\mathbf{A}$ has full column rank ($M \geq N$). Each component of the source signal $s(t)$ is statistically mutually independent (over time), exhibits SOCS with at least one non-zero real number of the cycle frequency and may contain more than one cycle frequency. Moreover, the source signal may also share one or more common cycle frequencies. The additive stationary noise $b(t)$ may be Gaussian or non Gaussian, temporally white or colored, spatially uncorrelated or correlated from sensor to each other. The source signal $s(t)$ and noise $b(t)$ are statistically independent of each other.

4   Identifiability Condition and Proposed Algorithm

The concept of identifiability condition in BSS problem was first defined and introduced by L. Tong *et al.* in [1] with assuming that source signals are stationary. Moreover, another one was also presented by Abed-Meraim, K. *et al.* [11] in the case where source signals are assumed to be cyclostationary. It was shown in [11] that BSS can be achieved using the second order cyclic correlation matrices for a set of lags, if and only if there does not exist two distinct source signals $s_k(t)$, $s_l(t)$, $k \neq l, k, l = 1, \ldots, N$ whose cycle frequencies are the same ($\alpha_k = \alpha_l$) and whose cyclic autocorrelation vectors ρ_k and ρ_l ($\rho_k \triangleq [R_{s_k}^{\alpha_k}(0), \ldots, R_{s_k}^{\alpha_k}(\tau_p)]^T$) are linearly dependent.

For cyclostationary signal, the TMF, cumulant, etc are really the function of two independent variables, t and τ, and periodic or polyperiodic in t for each value of τ. Thus, the identifiability condition and hence, the algorithm in [11], that is only based on the cyclic correlation function depending on only τ for each cycle frequency, has certain limitations. Let us demonstrate this limitation by the following example. Given two distinct CS which have different cycle frequency sets but share a common cycle frequency. Their cyclic autocorrelation vectors at the same cycle frequency are linearly dependent. Therefore, according to the identifiability condition in [11], they can not be separated. However, in practice, such two CS can be totally separated. Thus, to overcome this limitation, we present here an identifiability condition in company with an algorithm, based on second order cyclostationary in order to limit a class of inseparable cyclostationary signal in BSS problem.

Theorem 1. (*A necessary and sufficient condition for BBS to be achievable*): *The source signals can be recovered*[1] *in BSS problem if and only if for all* $k \neq l, k, l = 1 \ldots N$, *there exist at least one value of* ε *and a pair* (t_a, τ_a), (t_b, τ_b) *(which are not identically at the same time), such that:*

i. $R_{s_k^\varepsilon}(t_a, \tau_a) R_{s_l^\varepsilon}(t_a, \tau_a) > 0$
ii. $R_{s_k^\varepsilon}(t_b, \tau_b)/R_{s_k^\varepsilon}(t_a, \tau_a) \neq R_{s_l^\varepsilon}(t_b, \tau_b)/R_{s_l^\varepsilon}(t_a, \tau_a)$

According to (5) and (6), it is easy to see that there always exist the value of (t_a, τ_a) such that the first condition in Theorem 1 is satisfied. Moreover, under the assumptions in Section 3, we obtain the following equation:

$$R_{y^\varepsilon}(t, \tau) \triangleq \hat{E}^{\{\alpha\}}\{y(t)y^{\varepsilon T}(t + \tau_1)\} = A\hat{E}^{\{\alpha\}}\{s(t)s^{\varepsilon T}(t + \tau_1)\}A^{\varepsilon T} \qquad (8)$$

$$= A R_{s^\varepsilon}(t, \tau) A^{\varepsilon T} = A\mathbf{diag}(R_{s_1^\varepsilon}(t, \tau), \ldots, R_{s_N^\varepsilon}(t, \tau))A^{\varepsilon T} \qquad (9)$$

where $R_{s^\varepsilon}(t, \tau) \triangleq \hat{E}^{\{\alpha\}}\{s(t)s^{\varepsilon T}(t + \tau_1)\}$, $s^{1T}(t) \triangleq s^H(t)$, $s^{-1T}(t) \triangleq s^T(t)$ and the superscript H denotes the complex conjugate transpose of a matrix. Note that the equations in (8) and (9) are called the identification equation of blind identification problem. Furthermore, the additive stationary noise component does not contribute to $R_{y^\varepsilon}(t, \tau)$, and the (k, l)th entry of the matrix $R_{y^\varepsilon}(t, \tau)$ is computed by:

$$R_{y^\varepsilon}(t, \tau)_{k,l} \triangleq \sum_{\alpha \in I} \langle y_k(u)y_l^\varepsilon(u + \tau_1)e^{-j2\pi\alpha u} \rangle_u e^{j2\pi\alpha t} \qquad (10)$$

$$= \sum_{i=1}^{N} a_{ki}a_{li}^\varepsilon R_{s_i^\varepsilon}(t, \tau) \qquad (11)$$

Proof of necessity: By contradiction, it suffices to show that if $R_{s_k^\varepsilon}(t, \tau) = \xi\beta R_{s_l^\varepsilon}(t, \tau)$ where $\xi = \pm 1$ and $\beta > 0$ for some k, l, then there exist a matrix B which can not be written as $\mathbf{B} = \mathbf{A}\mathbf{P}\Lambda$. Without loss of generality, let $k = 1, l = 2$, and with any mixing matrix $\mathbf{A}$ and source signal $s(t)$ satisfying the assumptions in Section 3 and the equation in (7), (8), (9), we now define another mixing matrix $\mathbf{B}$ and the source singal $\hat{s}(t)$ as follow:

$$B = A\mathbf{diag}\left(\frac{1}{\sqrt{\beta} + \xi m^2\sqrt{\beta}}\begin{bmatrix} m & 1 \\ \sqrt{\beta} & \xi m\sqrt{\beta} \end{bmatrix}, I_{N-2}\right) \qquad (12)$$

$$\hat{s}(t) = \mathbf{diag}\left(\begin{bmatrix} -\xi m\sqrt{\beta} & 1 \\ \sqrt{\beta} & m \end{bmatrix}, I_{N-2}\right)s(t) \qquad (13)$$

---

[1] One can be found a matrix B such that $\mathbf{BA} = \mathbf{P}\Lambda$ or $\mathbf{B} = \mathbf{A}\mathbf{P}\Lambda$.

In (12), if there exist the value of ξ, m such that $1+\xi m^2 = 0 \Rightarrow$ Q. E. D., else, according to (7), (12), (13), it is easy to verify that: $y(t) = \mathbf{A}s(t)+b(t) = \mathbf{B}\hat{s}(t)+b(t)$ and $R_{\hat{s}^\varepsilon}(t,\tau) = \mathbf{diag}\left(k(m^2+\xi)R_{s_1^\varepsilon}(t,\tau), (m^2+1)R_{s_2^\varepsilon}(t,\tau), \ldots, R_{s_N^\varepsilon}(t,\tau))\right)$. Then, $\mathbf{B}$ and $\hat{s}(t)$ satisfy the (8), (9) but, $\mathbf{B} \neq \mathbf{AP}\Lambda \Rightarrow$ Q. E. D.

Proof of sufficiency: One can be used the Lemma 4.1 in [15] to prove the sufficient part of this theorem.

Note that the proposed condition can not be satisfied if and only if there exist two distinct sources $s_k(t)$ and $s_l(t)$ that have the same cycle frequency sets and their cyclic autocorrelation vectors ρ_k and ρ_l are linearly dependent at each cycle frequency in the sets. Moreover, we assume that I can be known a priori or directly estimated from the data by using the existed algorithm in [17] or [18]. We now present an algorithm for blind identification of CS:

1. Choose some value t_i, $i = 1, \ldots, K$, estimate the set of matrices: $R_{y^\varepsilon}(t_i, 0)$ by using (8), (10).
2. Choose a set of real numbers $\{\beta_i\}$, $i = 1, \ldots, K$ [16] such that the matrix F, defined as follows:

$$F \overset{\Delta}{=} \sum_{i=1}^{K} \beta_i R_{y^\varepsilon}(t_i, 0) = \sum_{i=1}^{K} A\beta_i R_s(t_i, 0)A^H \tag{14}$$

   is a positive definite matrix.
3. Compute the Singular Value Decomposition (SVD) of F:

$$F = [U_{1:N}U_{N+1:M}]diag(d_1^2, ..., d_N^2, 0)[U_{1:N}U_{N+1:M}]^H \tag{15}$$

4. Data transformation:

$$W \overset{\Delta}{=} diag(1/d_1, ..., 1/d_N)U_{1:N}^H \tag{16}$$

$$z(t) \overset{\Delta}{=} Wy(t) = WAs(t) + Wb(t) \tag{17}$$

5. Choose some value of lag τ and t_i, compute the set of matrix $R = \{R_{z^\varepsilon}(t_i, \tau)\}$
6. An orthogonal matrix V is then obtained as diagonalizer of the set of matrix R. $R_k = VD_kV^H$
7. Channel estimation A: $\hat{A} = U_{1:N}diag(d_1, \ldots, d_N)V$.
8. Source estimation: $\hat{s}(t) = V^H z(t)$.

5   Simulations

In this section, we present three different simulations. For all simulations, two cyclostationary source signals, one is an AM signal which has only one cycle frequency $(2f_1)$ and the other is a stacked carrier AM signal which maybe have more than one cycle frequency [4] $(2f_2 + k/T_0, k = -n, \ldots, n)$, are created as follows:

$$s_1(t) = a_1(t)cos(2\pi f_1 t + \pi/6) \tag{18}$$

$$s_2(t) = \sum_{k=-n}^{n} a_2(t)cos(2\pi f_2 t + \pi k/T_0 t + \pi/8) \tag{19}$$

where $a_1(t), a_2(t)$ are the colored second-order wide-sense stationary signals which have been obtained by passing two independent white noise signals through the low-pass filter with transfer function $H_1(f), H_2(f)$, respectively. The mixing matrix is chosen randomly:

$$A = \begin{bmatrix} 0.9797 & 0.8757 \\ 0.7373 & 0.8714 \\ 0.1365 & 0.2523 \end{bmatrix} \tag{20}$$

The additive noise is chosen to be Gaussian temporally colored and spatially correlated ($b(t) = Qn(t)$ where $Q_{ij} = 0.6^{|i-j|}$ and $n(t)$ is Gaussian temporally colored). In order to evaluate the effectiveness of the proposed approach, one can measure the following index: "Mean rejection level" [2], defined as:

$$I_{perf} \triangleq \sum_{p \neq q} \frac{E\,|(BA)_{pq}|^2}{E\,|(BA)_{pp}|^2} \tag{21}$$

In addition, the snapshot size is 1000 samples. The mean rejection level is estimated by averaging 100 independent trials. The signal to noise ratio (SNR) is computed as $SNR = -10log_{10}\sigma^2$. We compare the performance of the proposed method with SOBI and JADE algorithms.

Example 1: Two CS have different cycle frequency set, $f_1 = 0.1/T_s, f_2 = 0.15/T_s, n = 0$.

Example 2: Two CS have the same cycle frequency set $f_1 = f_2 = 0.1/T_s$, $n = 0$ but the low-pass filter are different $H_1(f) \neq H_2(f)$.

Example 3: Two CS have different cycle frequency set but share a common cycle frequency $f_1 = f_2 = 0.1/T_s, n = 1$ and $T_0 = 20T_s$. The low-pass filters are the same.

Fig.1 shows a plots of "Mean rejection level" as a function of SNR for the various algorithms in the simulation example.

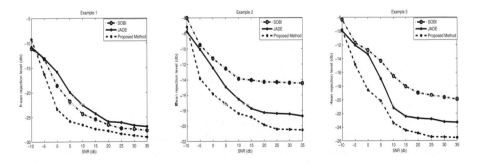

Fig. 1. Mean rejection level versus SNR

6   Conclusions

This paper briefly provides some relevant of the theory of the cyclostationary signal in the framework of the Fraction Of Time probability model. In addition, the new identifiability condition as well as an algorithm, based on the second order for blind separation of cyclostationary signals are also presented. Such a new method is shown to operate properly in the presence of stationary noise, which can be either Gaussian or non-Gaussian, spatially white or colored, correlated or uncorrelated from sensor to another. Simulation results have shown that the proposed algorithm performs better than SOBI and JADE algorithms.

References

1. Tong, L., Liu, R., Soon, V.C., Huang, Y.F.: Indeterminacy and identifiability of blind identification. IEEE Trans. Circuits Sys. 38(5), 499–509 (1991)
2. Belouchrani, A., Abed-Meraim, K., Cardoso, J.F., Moulines, E.: A blind source separation technique using second order statistics. IEEE Trans. Signal Process. 45(2), 434–444 (1997)
3. Cardoso, J.F.: Blind beam forming for non Gaussian signals. IEE proceedings 140(6), 362–370 (1993)
4. Gardner, W.A.: Statistical Spectral Analysis: A Nonprobabilistic Theory. Prentice Hall, Englewood Cliffs (1988)
5. Spooner, C.M.: Theory and application of higher order cyclostationarity, Dept. Elec. Comput. Eng., Univ. of CA (1992)
6. Gardner, W.A.: Cyclostationarity in Communications and Signal Processing. IEEE Press, Los Alamitos (1994)
7. Gardner, W.A., Napolitano, A., Paura, L.: Cyclostationarity: half a century of research. Signal Processing 86(4), 639–697 (2006)
8. Napolitano, A.: Cyclic higher-order statistics: input/output relations for discrete and continuous-time MIMO linear almost-periodically time-variant systems. Signal processing 42(2), 147–166 (1995)
9. Liang, Y.C., Leyman, A.R., Soong, B.H.: Blind source separation using second order cyclic statistics. In: Pro. SPWAC, Paris, France, pp. 57–60 (1997)
10. Ferreol, A., Chevalier, P.: On the behavior of current second and higher order blind sourceseparation methods for cyclostationary sources. IEEE Trans. Signal Process. 48(6), 1712–1725 (2000)
11. Abed-Meraim, K., Xiang, Y., Manton, J.H., Hua, Y.: Blind source separation using second-order cyclostationary statistics. IEEE Trans. Signal Process. 49(2), 694–701 (2001)
12. Jallon, P., Chevreuil, A.: Separation of instantaneous mixtures of cyclo-stationary sources. Signal Process. 87(11), 2718–2732 (2007)
13. Pham, D.T.: Blind separation of cyclostationary sources using joint block approximate diagonalization. In: Davies, M.E., James, C.J., Abdallah, S.A., Plumbley, M.D. (eds.) ICA 2007. LNCS, vol. 4666, pp. 244–251. Springer, Heidelberg (2007)
14. CheViet, N.A., El Badaoui, M., Belouchrani, A., Guillet, F.: Blind separation of cyclostationary sources using joint block approximate diagonalization. In: Proc. 5th IEEE SAM, Darmstadt, Germany, July 21-23, pp. 492–495 (2008)

15. Tong, L., Inouye, Y., Liu, R.W.: Waveform-preserving blind estimation of multiple idependent sources. IEEE Trans. Signal Process. 41(7), 2461–2470 (1993)
16. Tong, L., Inouye, Y., Liu, R.: A finite-step global convergence algorithm for the parameter estimation of multichannel MA processes. IEEE Trans. Signal Process. 40(10), 2547–2558 (1992)
17. Schell, S.V., Gardner, W.A.: Progress on signal selective direction finding. In: Proc. Fifth IEEE/ASSP Workshop Spectrum Estimation and Modelling, Rochester, NY, pp. 144–148 (1990)
18. Dandawaté, A.V., Giannakis, G.B.: Statistical tests for the presence of cyclostationarity. IEEE Trans. Signal Processing 42, 2355–2369 (1994)

A Novel Subspace Clustering Method for Dictionary Design

B. Vikrham Gowreesunker and Ahmed H. Tewfik

University of Minnesota
Dept. of Electrical and Computer Engineering
200 Union Street SE. Minneapolis, MN 55455

Abstract. In this work, we propose a novel subspace clustering method for learning dictionaries from data. In our method, we seek to identify the subspaces where our data lives and find an orthogonal set of vectors that spans each of those subspaces. We use an Orthogonal Subspace Pursuit (OSP) decomposition method to identify the subspaces where the observation data lives, followed by a clustering operation to identify all observations that lives in the same subspace. This work is motivated by the need for faster dictionary training methods in blind source separation (BSS) of audio sources. We show that our dictionary design method offer considerable computational savings when compared to the K-SVD[1] for similar performance. Furthermore, the training method also offers better generalizability when evaluated on data beyond the training set, and consequently is well suited for continuously changing audio scenes.

1 Introduction

In the problem of underdetermined blind source separation, it is common practice to assume sparsity in some representation in order to separate the sources. In the literature, one can find two approaches based on sparsity. The first is to use sparse "traditional" transforms such as the Short Time Fourier Transform (STFT) or Modified Discrete Cosine Transform (MDCT). The second is to use redundant representations. In recent years, sparse transforms have gain in popularity because they are computationally faster to compute and have demonstrated good performance. However, redundant representations can have sparser representation than STFT or MDCT and as such hold even more potential for source separation. In this work, we are interested in developing a training method to be used to learn redundant representations from data. These representations, also known as overcomplete dictionaries, have the potential to yield very sparse representations and could offer significant source separation performance improvements over traditional transforms, with the caveat that the dictionary has to be properly tuned to the underlying signals.

In our previous work on sparse decomposition and underdetermined BSS, we have found that the best dictionary training method available was the K-SVD [1] which we have used extensively in the past [2,3] for audio source separation. However, the computational complexity of the K-SVD makes it impractical for

T. Adali et al. (Eds.): ICA 2009, LNCS 5441, pp. 34–41, 2009.

audio separation for scenarios where the audio scene is continuously changing. In such cases a lower complexity online training method is required to keep the dictionary tuned to the signals. Given a matrix a collection of training signals, the goal of our algorithm is to identify subspaces in the data which can represent its most salient features. The collection of vectors spanning the subspaces would collectively represent our dictionary. The training takes place in two stages. The first stage involves using a sparse decomposition algorithm to find the representation of a vector in terms of the training set. The assumption here is that if multiple signals vectors live on the same subspace, then each one of these signals can be represented as a linear combination of the others. The second stage is a data clustering operation, where the vectors that lies on the learned subspace are identified and then removed from the training set.

In BSS, it is desirable to have a dictionary that can be adaptable to changing audio scene, i.e. it should be computationally tractable and be generalizability to signals beyond the training set. We show that that our subspace training method can run in minutes in cases where the K-SVD takes hours to train. Furthermore, we show that for training data, our method can deliver representations that are almost as sparse as the K-SVD and for data beyond the training set, our methods delivers a sparser representation. This generalizability property is very desirable for continuously changing audio. In section 2, we review the idea of sparse decomposition using overcomplete dictionaries and describe an Orthogonal Subspace Pursuit (OSP) method that we have found to give the sparsest representation of any iteratively greedy method for a generic dictionary. We then show how the OSP method can be combined with a data clustering method to produce a dictionary training method. In section 3, we describe how our method was evaluated in the context of sparse representation and source separation. In section 4, we discuss some work currently under investigation and the challenges that lies ahead.

2   Dictionary Learning Using Orthogonal Subspace Pursuit

Given a set of K observation vectors, $Y = \{y_i\}_{i=1}^{K}$, each of dimension $N \times 1$, the goal of a dictionary training algorithm is to find a set of potentially overcomplete basis vectors that are the common building blocks for these data. The idea of subspace clustering is based on the observation that if the data in our training set were derived from the same basis vectors, they will all lie on the plane spanned by those vectors. To identify the subspaces, we note that when several observation vectors lie on the same subspace as illustrated in fig 1, each vector can be expressed as a linear combination of the other vectors from the same subspace. Using a proper sparse decomposition algorithm, and the matrix $D = Y$ as a dictionary, we could find the representation of a vector from our training set as a linear combination of the columns of D, and consequently identifying an orthogonal set of vectors that span this subspace. To find all the subspaces within our data, we need to combine the subspace identification step with a clustering step where we find all the observation data that lies on the same subspace and remove them from the training set. This ensures a faster convergence.

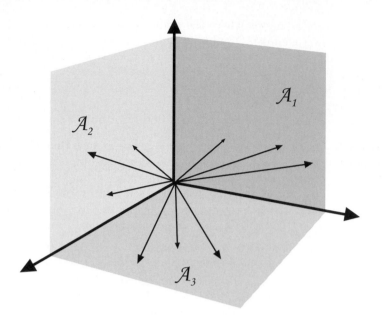

Fig. 1. Example of data living on subspaces, A_1, A_2, A_3

In section 2.1, we give a brief description of how the very popular K-SVD algorithm tackles the dictionary training task. In this paper, we have used an Orthogonal Subspace Pursuit (OSP) sparse decomposition algorithm, which we describe in section 2.2, to identify the subspaces. The OSP method, is based on the Orthogonal Least Square (OLS) procedure originally introduced in the context of regression and neural networks [4]. We found this method to give the sparsest representation of all the available sparse decomposition methods. In section 2.3, we describe the complete dictionary training method using subspace identification and clustering.

2.1   Prior Art: The K-SVD Algorithm

The K-SVD algorithm is a generalization of the K-Means method that is used to design optimal codebooks for vector quantization (VQ). The algorithm tries to solve the following problem:

$$min_{D,X}\{\|Y - DX\|_F^2\} \text{ s.t } \forall i \; \|x_i\|_0 \leq T_0, \tag{1}$$

where T_0 is a hard constraint on the maximum number of non-zero elements allowed, and D is the learned dictionary. This is solved iteratively in two stages, involving a sparse coding stage using a pursuit method, and a update stage where the dictionary atoms and the coefficients are sequentially updated using SVD operations.

2.2 Orthogonal Subspace Pursuit

The Orthogonal Subspace Pursuit (OSP) algorithm is an iteratively greedy pursuit method, similar to the Matching Pursuit (MP) and Orthogonal Matching Pursuit (OMP) [5], with a key difference in the directional update procedure. Let vector, y, be a vector of dimension $N \times 1$, and let D be an initial known overcomplete dictionary. A sparse decomposition algorithm seeks to approximate y in terms of a minimum number of columns of D. Let the dictionary be a collection of K atoms, d_k, of size $N \times 1$ each, where $K >> N$ such that $D = \{d_k\}_{k=1}^{K}$. Using a sparse decomposition algorithm, we can express the signal in terms of its approximation, and a residual such that $y = \hat{y}^m + r^m$ at m^{th} iteration. We denote Γ^m as the set of atoms that has been selected by the algorithm until iteration m, while Ω^m is the set of atoms that has not yet been selected. At all times, the union of Γ^m and Ω^m is the set of all dictionary indices. At each iteration, we have $\hat{y}^m = \sum_{i \in \Gamma^m} x_i d_i$, and r^m is the approximation error at the mth iteration, and is commonly known as the residual.

The OSP procedure involves two main steps, the first involves finding the most correlated atom from the dictionary, and the second involves a dictionary decorrelation step where the atoms that were not selected are decorrelated from previously chosen atoms. The algorithm is described below:

1. Initialization
   $m = 0, r^0 = y, \hat{y}^0 = 0, \Omega^0 = \{1, 2, 3, ..., K\}, \Gamma^0 = \oslash, W = \{d_i\}_{i \in \Omega^0}$
2. While stopping criteria are not met
   (a) $m = m + 1$
   (b) Pick atom with maximum correlation to residual
   $i^m = \arg_j \max |< w_j, r^{m-1} >|, j \in \Omega^{m-1}$
   (c) Remove atom index from set Ω and add to set Γ
   $\Omega^m = \Omega^{m-1} \otimes i^m; \Gamma^m = \Gamma^{m-1} \cup i^m$
   (d) Update residual and approximation
   $r^m = r^{m-1} - < w_{i^m}, r^{m-1} > w_{i^m}; \hat{y}^m = \hat{y}^{m-1} + < w_{i^m}, r^{m-1} > w_{i^m}$
   (e) Decorrelate remaining dictionary atoms from w_{i^m} such that for all $j \in \Omega^m$,
   $w_j = w_j - < w_j, w_{i^m} > w_{i^m}; w_j = w_j / norm(w_j)$
   (f) Check stopping criteria

The stopping criteria is an error threshold and a maximum number of iterations. Although this method is computationally more demanding than the MP or OMP, we found it to produce the sparsest representation. The algorithm will produce a set of orthogonal vectors that spans the subspace where y lives. In the next section, we describe how this can be used to learn dictionaries.

2.3 Dictionary Training Algorithm

For the dictionary training, we have the training set in matrix Y, and an initial dictionary, D, which we initialize to equal to Y. The algorithm has two stages, first we identify a subspace from the training data, and second we find all the training data that lives on this subspace and remove them from Y before looking for the next subspace. The algorithm can be describe as follows:

1. Initialize Algorithm
   $i = 0$, $D^0 = Y$,
2. Sparse coding
   (a) $i = i + 1$ Choose the ith vector, y_i, from the training set, and remove it from the dictionary
   (b) Find a representation of y_i in terms of D^i using the OSP method described above, where $D^i = D \otimes y_i$
   (c) Let S_i be the set of λ_i vectors that represents y_i
   (d) Find the SVD of S_i such that $U \Sigma V^T = S_i$
   (e) We define the subspace, A_i, as the first λ_i columns of U
3. Create data clusters
   (a) Find the projection of y_j onto the subspace of A_i, where $j \neq i$
   (b) Assign all y_j that lies in the subspace of A_i to the same cluster and remove them from training set, Y
4. The learned dictionary is A = the collection of subspaces $\{A_i\}_{i=1}^{M}$, where M is the number of subspaces

It should be noted that the only change done to the dictionary at each iteration is to ensure that the data, y_i is not part of the dictionary, otherwise the solution will be trivial. Our algorithm also includes some safeguards to ensure that the subspaces are legitimate. First, we require that the vectors lies on the subspace within a certain error threshold, to ensure robustness to noise. Second, if a data point cannot be represented by others in the training set, within a reasonable approximation error, it is not used.

3   Evaluation

Depending on the application, there are different ways of evaluating a dictionary. However, it is generally agreed that the signals should admit a sparse representation over the dictionary with the elements of the dictionary characterizing the most important features of the signals. Furthermore, in the context of blind source separation, it is of interest to have a dictionary that has some generalizability beyond the training data with some practical computation cost. In this section, we show that the proposed subspace approach offers better sparsity, generalizability and has significant computational advantage when compared to the K-SVD algorithm.

For the set of experiments below, we assume that we have three tracks of 10 seconds speech from male speakers, sampled at 16 kHz. We would like to design a dictionary that can efficiently capture the 3 sources. For evaluation, we split our data into 2 groups, a training set and a testing set. The training set comprises of the first T seconds of speech from each speaker and the testing set is the remaining (10 - T) seconds of speech. We train our dictionary using only the training set and evaluate it using both the training and testing sets. For each set, the data is formatted into frames of size, N, with a Hanning window and a

50% overlap. We compare two methods, the K-SVD and the subspace approach described above for N = 128, 256 and 512. The K-SVD algorithm can always be further optimized by changing the maximum number of coefficients or by using a good initialization of the dictionary. However, the results presented here was for dictionary initialization using data, and fair limit on the max number of coefficient from our experience with the data set. Although the actual numbers might improve with more optimization, the essence of the differences still holds true. Sparsity is evaluated by the number of non-zero coefficients in the coefficient vector and is presented here in the form of (mean +/- standard deviation). The computational complexity is described by the matlab runtime on 2.5 GHz PC and the approximation error is the 2-norm of the error, $\|y - \hat{y}\|_2$.

3.1   Sparsity Results

Once, we have our dictionary, the sparsity was evaluated using the Orthogonal Matching Pursuit (OMP) sparse decomposition algorithm, such that each data frame can be represented as $\hat{y}_i = Ax_i$, where x_i is the coefficient vector. The sparsity is evaluated from the mean and standard deviation of the number of non-zero elements in x_i for all elements in the training or test set. For N = 128 and 256, the first 5 seconds of each track was used for training and the remainder was used for testing. For N = 512, we had to use the first 8 seconds for training and only the last 2 seconds for testing, because the K-SVD requires that we have a lot more test data than the size of the dictionary. In table 1, we show the results when dictionary was evaluated for the training set. Clearly, the proposed subspace method has much lower computational complexity than the K-SVD and this difference is accentuated as the dimension increases to N = 512, which a commonly used frame length for 16 kHz audio. The mean number of non-zero coefficient is slightly better for K-SVD in the case of N = 256 and 512 but the standard deviation is much larger than for the Subspace method, which makes that the Subspace approach is more appealing. In table 2, we present the results for the testing set, and clearly, the subspace approach is much sparser than the K-SVD method, and hence has better generalizability beyond the training data. In section 3.2, we explore how the proposed dictionary training method was used in the context of BSS.

Table 1. Results for training set data. The convention μ +/- σ means that μ is the mean and σ is the standard deviation per frame.

Frame size	Method	No. of Non-zero Coef	Approx. Error	Training Time
N = 128	K-SVD	12.3 +/- 13.8	0.09 +/- 0.01	16 min
	Subspace	9.3 +/- 6.5	0.09 +/- 0.02	9 min
N = 256	K-SVD	16.2 +/- 24.0	0.09 +/- 0.02	32 min
	Subspace	18.9 +/- 12.3	0.09 +/- 0.02	4 min
N = 512	K-SVD	29.2 +/- 38.1	0.10 +/- 0.04	507 min
	Subspace	35.3 +/- 26.4	0.10 +/- 0.05	7 min

Table 2. Results for test set data. The convention $\mu +/- \sigma$ means that μ is the mean and σ is the standard deviation per frame.

Frame size	Method	No. of Non-zero coefficients	Approximation Error
N = 128	K-SVD	21.3 +/- 19.1	0.09 +/- 0.01
	Subspace	13.6 +/- 10.9	0.09 +/- 0.01
N = 256	K-SVD	49.3 +/- 41.1	0.10 +/- 0.02
	Subspace	41.1 +/- 25.7	0.10 +/- 0.01
N = 512	K-SVD	93.1 +/- 49.2	0.14 +/- 0.10
	Subspace	70.4 +/- 47.0	0.11 +/- 0.04

3.2 Blind Source Separation Results

We applied the proposed subspace learning method to train dictionaries for BSS of underdetermined instantaneous mixtures. We used the sources and mixtures from the development data made available by the stereo audio source separation campaign (Sisec 2008) concurrently organized with ICA 2009. The dictionaries were trained on the sources the same way as in section 3.1 and separation performance was evaluated on the test data, i.e. the last T seconds for the two frame sizes, N = 256 and N = 512. In figure 2, we compare the average Signal-to-Interference Ratio (SIR) improvement using our BSS algorithm [2] for two dictionaries, the K-SVD and the Subspace method, on test data.

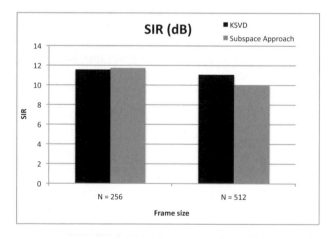

Fig. 2. Signal-to-Interference ratio for test data with frame size = 256, 512

We find the performance of both dictionaries to be practically the same when tested beyond the training data although we showed earlier that the Subspace method offered better generalizability. This is due to the SIR results being skewed by frames in the test data that couldn't be completely represented by the dictionary. This is a problem any training method faces when generalized to test data. One solution we have successfully used for BSS is to have the final dictionary be a union of the trained dictionary and a standard set of bases such as Discrete

Cosine Transform (DCT), which ensures that features that were not present in the training set can be represented. However, this discussion is more relevant to practical implementation of BSS than the training method itself. We foresee that a computationally tractable method like the subspace clustering method proposed here could be incorporated in a typical BSS algorithm by starting with an offline trained dictionary, and have that dictionary periodically updated every 20 to 30 seconds to track changes in the audio scene.

4   Conclusion

We proposed a novel dictionary training method that is based on the idea of identifying and extracting subspaces within the data. The merits of the algorithm is the relative low complexity when compared to algorithms like the K-SVD for comparable performance, and the generalizability to data beyond the training set. The target application for this dictionary training method is that of blind source separation of audio sources. In such applications, the ability to periodically update the dictionary is critical especially when the environment is changing, and all the sources are not known apriori. The algorithm has been successfully used in source separation and we are currently working on finding ways to better train a dictionary when only mixtures are available and developing a shift-invariant version of the method.

References

1. Aharon, M., Elad, M., Bruckstein, A.M.: The k-svd: An algorithm for designing of overcomplete dictionaries for sparse representation. IEEE Trans. on Signal Processing 54(11), 4311–4322 (2006)
2. Gowreesunker, B.V., Tewfik, A.H.: Blind source separation using monochannel overcomplete dictionaries. In: IEEE Int. Conf. Acoustics, Speech and Signal Processing (ICASSP), Las Vegas, USA, pp. 33–36 (March 2008)
3. Gowreesunker, B.V., Tewfik, A.H.: Two improved sparse decomposition methods for blind source separation. In: Davies, M.E., James, C.J., Abdallah, S.A., Plumbley, M.D. (eds.) ICA 2007. LNCS, vol. 4666, pp. 365–372. Springer, Heidelberg (2007)
4. Chen, S., Cowan, C.F., Grant, P.M.: Orthogonal least squares learning algorithm for radial basis function networks. IEEE transactions on neural networks 2(2), 302–309 (1991)
5. Mallat, S.: A Wavelet Tour of Signal Processing, 2nd edn. Academic Press, London (1999)

Comparison of Two Main Approaches
to Joint SVD

Gen Hori

Center for Intellectual Property Strategies, RIKEN, Saitama 351-0198, Japan
hori@brain.riken.jp

Abstract. Joint SVD is a problem of finding a pair of unitary matrices which simultaneously diagonalizes several (possibly non-square) matrices. This paper compares two main approaches to joint SVD problem, "approach via joint diagonalization" and "direct approach". The former is relatively easy to implement because we can make use of literature of joint diagonalization algorithms while the latter has advantages in numerical accuracy and flexibility to fit on-line applications. Numerical simulation for comparison using gradient-based algorithms verifies that the latter has advantage in numerical accuracy.

1   Introduction

Joint diagonalization has played an essential role in independent component analysis and blind signal separation with several target matrices including fourth order cumurant matrices[1][2] and time-delayed correlation matrices[3]. One of natural extensions of joint diagonalization is "joint SVD", a problem of simultaneous singular value decomposition of several (possibly non-square) real or complex matrices with a single pair of orthogonal or unitary matrices. Given a set of K real or complex $m \times n$ matrices

$$\{A_1, A_2, \ldots, A_K\},$$

joint SVD (or simultaneous SVD) is a problem of finding orthogonal or unitary matrices U and V which make

$$\{U^* A_1 V, U^* A_2 V, \ldots, U^* A_K V\}$$

as diagonal as possible simultaneously where $*$ denotes the Hermitian transpose. We say that a joint SVD problem is *left-exact* (*right-exact*) when all the given matrices share a common set of left (right) singular vectors. We say that the problem is *exact* when it is *left-exact* and *right-exact* and the left and right singular vectors have one-to-one correspondence over the given matrices. From a purely mathematical viewpoint, a joint SVD problem can be solved exactly only if the problem is *exact*, where the SVD of an arbitrary single matrix A_i gives the solution to the problem. From a practical viewpoint, however, the problem is rarely *exact* and we have to find optimal solution with respect to some diagonality criterion. One of the typical situations in practical applications is that additive

T. Adali et al. (Eds.): ICA 2009, LNCS 5441, pp. 42–49, 2009.

noise on observed matrices makes the problem *non-exact* although the problem is theoretically *exact*.

Joint SVD has found an application in image processing[4] and is expected to find more applications in signal processing and image processing. The purpose of this paper is to compare two main approaches to joint SVD problem, "approach via joint diagonalization" and "direct approach" described in the following sections, in aspects including numerical accuracy, computational complexity and flexibility to fit on-line applications. A gradient-based joint SVD algorithm of "direct approach" is derived and compared with another gradient-based joint SVD algorithm of "approach via joint diagonalization" by simulation.

The rest of the paper is organized as follows. Section 2 explains "approach via joint diagonalization". Section 3 explains "direct approach" and derives a gradient-based joint SVD algorithm of "direct approach". Section 4 carries out a numerical simulation to compare two main approaches from the viewpoint of numerical accuracy. Section 5 contains concluding remarks.

2   Approach via Joint Diagonalization

Consider two unitary joint diagonalization problems of sets of Hermitian matrices,

$$\{A_1 A_1^*, A_2 A_2^*, \ldots, A_K A_K^*\} \tag{1}$$

and

$$\{A_1^* A_1, A_2^* A_2, \ldots, A_K^* A_K\}, \tag{2}$$

and suppose that unitary matrices U and V are the solutions to the respective unitary joint diagonalization problems obtained using some previous joint diagonalization algorithm, then

$$\{U^* A_1 V, U^* A_2 V, \ldots, U^* A_K V\}$$

is expected to be the joint SVD of the given set of matrices,

$$\{A_1, A_2, \ldots, A_K\}.$$

The advantage of this "approach via joint diagonalization" is that it is relatively easy to implement because we can make use of rich literature of developed joint diagonalization algorithms. On the other hand, it is not expected to be numerically accurate because the matrix multiplications for making Hermitian matrices square the singular values and make condition numbers worse. Also the matrix multiplications increase computational complexity and it is difficult to fit on-line applications where given matrices are time-varying.

3   Direct Approach

Some recent works have been intended to develop new algorithms for joint SVD which do not depend on previous joint diagonalization algorithms. Pesquet-Popescu et al.[4] developed a joint SVD algorithm which is an extension of the Jacobi algorithm. We introduce a gradient-based joint SVD algorithm of "direct approach" in the following subsection. Comparing to "approach via joint diagonalization", "direct approach" is expected to be numerically more accurate, does not require computational cost to calculate matrix multiplication for making Hermitian matrices and is easier to fit on-line applications. The first subsection introduces matrix gradient flows for SVD and the second one extends them to matrix gradient flows for joint SVD.

3.1   Matrix Gradient Flows for SVD

Let $\mathfrak{U}(n)$ and $\mathfrak{u}(n)$ denote the sets of all the $n \times n$ unitary and skew Hermitian matrices respectively,

$$\mathfrak{U}(n) = \{\, U \in M(n, C) \mid U^*U = I_n \,\},$$
$$\mathfrak{u}(n) = \{\, X \in M(n, C) \mid X^* = -X \,\}.$$

Consider the set of all the $m \times n$ matrices which share the singular values with $A_0 \in M(m \times n, C)$,

$$\mathfrak{S} = \{\, A = U^*A_0V \mid U \in \mathfrak{U}(m), V \in \mathfrak{U}(n) \,\},$$

and a continuous dynamical system on the set,

$$A(t) = U(t)^*A_0V(t),$$

where $m \geq n$ (See [10] for detail of the structure of the set $\mathfrak{S}$). When $U(t)$ and $V(t)$ evolve as

$$\frac{d}{dt}U(t) = U(t)X(t), \quad U(0) = I_m, \tag{3}$$

$$\frac{d}{dt}V(t) = V(t)Y(t), \quad V(0) = I_n, \tag{4}$$

where $X(t) \in \mathfrak{u}(m)$ and $Y(t) \in \mathfrak{u}(n)$, we have

$$\dot{A} = \dot{U}^*A_0V + U^*A_0\dot{V} = (UX)^*A_0V + U^*A_0VY$$
$$= X^*(U^*A_0V) + (U^*A_0V)Y$$
$$= X^*A + AY = AY - XA,$$

therefore $A(t)$ evolves as

$$\frac{d}{dt}A(t) = A(t)Y(t) - X(t)A(t), \quad A(0) = A_0. \tag{5}$$

Suppose that ϕ is an arbitrary real-valued function defined on the set of $m \times n$ complex matrices,

$$\phi : M(m \times n, C) \to R.$$

The rest of this subsection is spent on the derivation of the gradient equation of the potential function $\varphi(U, V) = \phi(A) = \phi(U^*A_0V)$. In the derivation, we regard $M(m \times n, C)$ not as an mn-dimensional complex space C^{mn} but as a $2mn$-dimensional real space R^{2mn}. We denote the real and the imaginary parts of a complex number x by $\hat{x}$ and $\check{x}$ as well as $\operatorname{Re} x$ and $\operatorname{Im} x$ respectively. We use the relations such as $\operatorname{tr} AB = \operatorname{tr} BA$ and $\operatorname{Re} \operatorname{tr} X^* Y = \sum_{i,j} (\hat{x}_{ij} \hat{y}_{ij} + \check{x}_{ij} \check{y}_{ij})$.

We derive the gradient equation with respect to the inner product on $\mathfrak{u}(n)$ defined as

$$< X_1, X_2 >= \operatorname{Re} \operatorname{tr} X_1^* X_2.$$

We note that the inner product coincides with (the negative of) the Killing form in Lie group theory and the gradient method with respect to the inner product is equivalent to what is called "natural gradient" or "relative gradient" in ICA learning theory. It is easy to verify that the orthogonal projection to $\mathfrak{u}(n)$ with respect to the inner product is given by

$$\pi_{\mathfrak{u}(n)} A = \frac{1}{2}(A - A^*).$$

Using the chain rule, we have

$$\frac{d}{dt} \phi(A(t)) = \sum_{i,j} \left(\frac{\partial \phi(A)}{\partial \hat{a}_{ij}} \frac{d\hat{a}_{ij}}{dt} + \frac{\partial \phi(A)}{\partial \check{a}_{ij}} \frac{d\check{a}_{ij}}{dt} \right)$$

$$= \operatorname{Re} \operatorname{tr} \left(\frac{d\phi}{dA} \right)^* \frac{dA}{dt}$$

$$= \operatorname{Re} \operatorname{tr} \left(\frac{d\phi}{dA} \right)^* (AY - XA)$$

$$= \operatorname{Re} \operatorname{tr} \left(A^* \left(\frac{d\phi}{dA} \right) \right)^* Y - \operatorname{Re} \operatorname{tr} \left(\left(\frac{d\phi}{dA} \right) A^* \right)^* X$$

where $\left(\frac{d\phi}{dA} \right)$ denotes the matrix whose (i,j)-th entry is $\frac{\partial \phi(A)}{\partial \hat{a}_{ij}} + \frac{\partial \phi(A)}{\partial \check{a}_{ij}} i$. Let $\{X_k | k = 1, \ldots, m^2\}$ and $\{Y_l | l = 1, \ldots, n^2\}$ be the orthonormal basis of $\mathfrak{u}(m)$ and $\mathfrak{u}(n)$ on R respectively. The steepest ascent direction of $\phi(A)$ in terms of X and Y is calculated as

$$X = -\sum_{k=1}^{m^2} \left(\operatorname{Re} \operatorname{tr} \left(\left(\frac{d\phi}{dA} \right) A^* \right)^* X_k \right) X_k$$

$$= -\pi_{\mathfrak{u}(m)} \left(\left(\frac{d\phi}{dA} \right) A^* \right)$$

$$= \frac{1}{2} \left(A \left(\frac{d\phi}{dA} \right)^* - \left(\frac{d\phi}{dA} \right) A^* \right),$$

$$Y = \sum_{l=1}^{n^2} \left(\operatorname{Re} \operatorname{tr} \left(A^* \left(\frac{d\phi}{dA} \right) \right)^* Y_l \right) Y_l$$

$$= \pi_{\mathfrak{u}(n)} \left(A^* \left(\frac{d\phi}{dA} \right) \right)$$

$$= \frac{1}{2}\left(A^*\left(\frac{d\phi}{dA}\right) - \left(\frac{d\phi}{dA}\right)^* A \right).$$

Substituting these in (1), (2) and (3), we obtain the general gradient ascent equations in terms of U and V,

$$\frac{dU}{dt} = \frac{1}{2}U\left(A\left(\frac{d\phi}{dA}\right)^* - \left(\frac{d\phi}{dA}\right)A^* \right), \tag{6}$$

$$\frac{dV}{dt} = \frac{1}{2}V\left(A^*\left(\frac{d\phi}{dA}\right) - \left(\frac{d\phi}{dA}\right)^* A \right), \tag{7}$$

and the equation in terms of A,

$$\frac{dA}{dt} = \frac{1}{2}A\left(A^*\left(\frac{d\phi}{dA}\right) - \left(\frac{d\phi}{dA}\right)^* A \right) - \frac{1}{2}\left(A\left(\frac{d\phi}{dA}\right)^* - \left(\frac{d\phi}{dA}\right)A^* \right)A. \tag{8}$$

Example. Consider the gradient descent equation of

$$\psi(A) = \sum_{\substack{1 \le i \le m \\ 1 \le j \le n \\ i \ne j}} |a_{ij}|^2$$

which is expected to converge to the SVD of the initial matrix. This is equivalent to the gradient ascent equation of

$$\phi(A) = \sum_{1 \le j \le n} |a_{jj}|^2$$

because $\psi(A) + \phi(A)$ is invariant on $\mathfrak{S}$. We have

$$\frac{d\phi}{dA} = 2\mathrm{diag}(A)$$

where $\mathrm{diag}(A)$ denotes an $m \times n$ extended diagonal matrix whose diagonals are the same as A and off-diagonals are 0's, which yields with (6), (7) and (8),

$$\dot{U} = U((U^*A_kV)\mathrm{diag}(U^*A_kV)^* - \mathrm{diag}(U^*A_kV)(U^*A_kV)^*), \tag{9}$$

$$\dot{V} = V((U^*A_kV)^*\mathrm{diag}(U^*A_kV) - \mathrm{diag}(U^*A_kV)^*(U^*A_kV)), \tag{10}$$

$$\dot{A} = A(A^*\mathrm{diag}(A) - \mathrm{diag}(A)^*A) - (A\mathrm{diag}(A)^* - \mathrm{diag}(A)A^*)A. \tag{11}$$

This is equivalent to the SVD flow introduced by [11] where the initial matrix and the flow are on the set of real matrices. They proved that the flow converges to the SVD of the initial matrix as $t \to \infty$ if the initial matrix meets suitable conditions (See [11] for detail).

3.2 Matrix Gradient Flows for Joint SVD

This section proposes to use the gradient ascent equations of U and V of the potential function

$$\varphi(U,V) = \sum_{k=1}^{K} \phi_k(U^*A_kV) \tag{12}$$

to solve a joint SVD problem where each $\phi_k(A)$ is a diagonality criterion supposed to take its maximum when A is an extended diagonal matrix. Using (6) and (7), the gradient ascent equations of U and V of (12) are obtained as

$$\frac{dU}{dt} = \frac{1}{2}U\sum_{k=1}^{K}((U^*A_kV)(\frac{d\phi_k}{dA}(U^*A_kV))^* - (\frac{d\phi_k}{dA}(U^*A_kV))(U^*A_kV)^*), \quad (13)$$

$$\frac{dV}{dt} = \frac{1}{2}V\sum_{k=1}^{K}((U^*A_kV)^*(\frac{d\phi_k}{dA}(U^*A_kV)) - (\frac{d\phi_k}{dA}(U^*A_kV))^*(U^*A_kV)). \quad (14)$$

Example. To derive actual gradient equations for solving joint SVD problems, we define the diagonality criteria

$$\phi_k(A) = \sum_{1\leq j\leq n}|a_{jj}|^2, \qquad k = 1,\ldots,K,$$

which take their maxima when A is an extended diagonal matrix and substitute them in (13) and (14) to obtain

$$\dot{U} = U\sum_{k=1}^{K}((U^*A_kV)\text{diag}(U^*A_kV)^* - \text{diag}(U^*A_kV)(U^*A_kV)^*), \quad (15)$$

$$\dot{V} = V\sum_{k=1}^{K}((U^*A_kV)^*\text{diag}(U^*A_kV) - \text{diag}(U^*A_kV)^*(U^*A_kV)). \quad (16)$$

This can be regarded as a superposition of the SVD flows (9) and (10) and is expected to converge to the solution of a joint SVD problem.

4   Simulation

We compare two main approaches to joint SVD by simulation with gradient-based algorithms. Although the gradient-based algorithms have convergence too slow for practical use, comparison using them gives us information on the essential difference of two main approaches which is useful for further development of practical algorithms. We use (15) and (16) as "direct approach". As "approach via joint diagonalization", we apply gradient-based joint diagonalization algorithm introduced by [5],

$$\dot{U} = U\sum_{k=1}^{K}((U^*A_kV)\text{diag}(U^*A_kV) - \text{diag}(U^*A_kV)(U^*A_kV)), \quad (17)$$

to the two sets of Hermitian matrices (1) and (2). We apply two approaches with gradient-based algorithms to a toy *exact* joint SVD problem which consists of the following two 5×3 real matrices,

$$A_1 = U_0\,\text{diag}_{5\times 3}(1,2,3)\,V_0{}^*,$$
$$A_2 = U_0\,\text{diag}_{5\times 3}(3,2,1)\,V_0{}^*,$$

where U_0 and V_0 are 5×5 and 3×3 randomly generated orthogonal matrices which are close to identity matrices. The integration of the gradient equations is done by the Euler method with a stepsize 0.001. Fig. 1 shows the convergence of "direct approach"(solid lines) and "approach via joint diagonalization"(dot lines). The vertical axis shows the deviations of U and V from the true diagonalizers U_0 and V_0 respectively while the horizontal axis the iteration number of the Euler method. The deviation of V from the true diagonalizer V_0 is measured by $|\log(VV_0^*)|$ where log stands for the logarithm of a matrix and $|\cdot|$ is the Frobenius norm. Since U has the indeterminacy in its last columns, the deviation of U from U_0 is measured by $|\log(U(:,1:3)U_0(:,1:3)^*)|$ where $U(:,1:3)$ means the first three columns of U. The fast convergence of "approach via joint diagonalization" in early phase is mainly due to the squared singular values of the Hermitian matrices. From the late phase, we see that "direct approach" has advantage in numerical accuracy comparing to "approach via joint diagonalization".

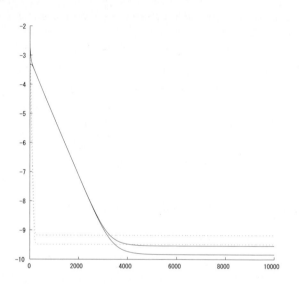

Fig. 1. Convergence of "direct approach"(solid lines) and "approach via joint diagonalization"(dot lines) – deviations from true diagonalizers versus iteration number

5   Concluding Remarks

Two main approaches to joint SVD, "approach via joint diagonalization" and "direct approach", were considered. The former is relatively easy to implement because we can depend largely on literature of existing joint diagonalization algorithms. On the other hand, the latter has advantage in computational complexity and flexibility to fit on-line application where target matrices are time-varying. Two main approaches were compared by simulation with gradient-based algorithms which gave some evidence that "direct approach" has advantage in numerical accuracy. Since the two approaches are formulated as optimization problems on the set of unitary matrices, combinations with the conjugate gradient method and the Newton method with orthogonality constraints such as introduced by [13][14] may bring us more practical algorithms. Total comparison of those methods including the Jacobi-based algorithms[1][2][4] is left to our future study.

References

1. Cardoso, J.-F., Souloumiac, A.: Blind beamforming for non Gaussian signals. IEEE Proc. -F 140, 362–370 (1993)
2. Cardoso, J.-F., Souloumiac, A.: Jacobi angles for simultaneous diagonalization. SIAM J. Mat. Anal. Appl. 17, 161–164 (1996)
3. Molgedey, L., Schuster, H.G.: Separation of a mixture of independent signals using time delayed correlations. Phys. Rev. Lett. 72, 3634–3637 (1994)
4. Pesquet-Popescu, B., Pesquet, J.-C., Petropulu, A.P.: Joint singular value decomposition - a new tool for separable representation of images. In: Proc. 2001 International Conference on Image Processing, vol. 2, pp. 569–572 (2001)
5. Hori, G.: A new approach to joint diagonalization. In: Proc. 2nd Int. Workshop on Independent Component Analysis and Blind Signal Separation (ICA 2000), Helsinki, Finland, pp. 151–155 (2000)
6. Hori, G.: A general framework for SVD flows and joint SVD flows. In: Proc. 2003 IEEE International Conference on Acoustics, Speech and Signal Processing, vol. 2, pp. 693–696 (2003)
7. Tanaka, T., Fiori, S.: Least squares approximate joint diagonalization on the orthogonal group. In: Proc. 2007 IEEE International Conference on Acoustics, Speech and Signal Processing, vol. 2, pp. 649–652 (2007)
8. Fiori, S.: Singular value decomposition learning on double Stiefel manifold. Intl. J. Neural Systems 13(2), 155–170 (2003)
9. Helmke, U., Moore, J.B.: Optimization and dynamical systems. Addison-Wesley, Reading (1994)
10. Helmke, U., Moore, J.B.: Singular value decomposition via gradient and self equivalent flows. Linear Algebra Appl. 69, 223–248 (1992)
11. Chu, M.T., Driessel, K.R.: The projected gradient method for least squares matrix approximations with spectral constraints. SIAM J. Numer. Anal. 27, 1050–1060 (1990)
12. Smith, S.T.: Dynamical systems that perform the singular value decomposition. System and Control Letters 16, 319–328 (1991)
13. Manton, J.H.: Optimisation algorithms exploiting unitary constraints. IEEE Trans. Signal Processing 50, 635–650 (2002)
14. Edelman, A., Arias, T.A., Smith, S.T.: The geometry of algorithms with orthogonality constraints. SIAM J. Matrix Anal. Appl. 20, 303–353 (1998)

Optimal Performance of Second-Order Multidimensional ICA

Dana Lahat[1,*], Jean-François Cardoso[2], and Hagit Messer[1]

[1] School of Electrical Engineering, Tel-Aviv University, 69978 Tel-Aviv, Israel
[2] LTCI, TELECOM ParisTech and CNRS, 46 rue Barrault, 75013 Paris, France

Abstract. Independent component analysis (ICA) and blind source separation (BSS) deal with extracting mutually-independent elements from their observed mixtures. In "classical" ICA, each component is one-dimensional in the sense that it is proportional to a column of the mixing matrix. However, this paper considers a more general setup, of *multidimensional* components. In terms of the underlying sources, this means that the source covariance matrix is block-diagonal rather than diagonal, so that sources belonging to the same block are correlated whereas sources belonging to different blocks are uncorrelated. These two points of view —correlated sources vs. multidimensional components— are considered in this paper. The latter offers the benefit of providing a unique decomposition. We present a novel, closed-form expression for the optimal performance of second-order ICA in the case of multidimensional elements. Our analysis is verified through numerical experiments.

Keywords: Independent component analysis, blind source separation, correlated sources, multidimensional components, performance analysis, joint block diagonalization.

1 Introduction

In their most basic setting, independent component analysis (ICA) and blind source separation (BSS) aim at extracting m mutually independent elements from m observed mixtures. The model is of T observations of an $m \times 1$ vector $\boldsymbol{x}(t)$, modeled as

$$\boldsymbol{x}(t) = A\boldsymbol{s}(t) \qquad 1 \leq t \leq T \tag{1}$$

where A is an $m \times m$ full-rank matrix and $\boldsymbol{s}(t)$ is a vector of independent *sources*. A natural extension of practical interest is to consider correlated sources. More specifically, this contribution[1] addresses the case where some entries of $\boldsymbol{s}(t)$ are correlated to some others. That is, the m sources can be partitioned into $n \leq m$

---

[*] This research was partially supported by the Chateaubriand Fellowship of the "service scientifique" of the French embassy in Israel.

[1] This contribution presents results of a paper under preparation, titled "Multidimensional ICA: Second-Order Methods and Applications".

T. Adali et al. (Eds.): ICA 2009, LNCS 5441, pp. 50–57, 2009.

groups with the sources of the same group being dependent while sources belonging to two different groups are independent. This is equivalent to considering multidimensional components, as detailed below.

In the literature, correlated sources have already been discussed (e.g. [1, 5, 6, 9, 12]). However, to the best of our knowledge, no complete performance analysis has been conducted. Here, we consider a second-order based method which extends the maximum likelihood (ML) treatment of [10, 11] to the case of correlated sources/multidimensional components and provides a complete asymptotic performance analysis for it. The closed-form expression for the error covariance matrix is the Cramér-Rao lower bound (CRLB) for the case of Gaussian sources.

2   Correlated Sources vs. Multidimensional Components

Two Points of View. In this work, we partition the columns of A into n blocks: $A = [A_1, \ldots, A_n]$. Denote $m(i)$ the number of columns in A_i so that $\sum_{i=1}^{n} m(i) = m$. We partition similarly the source vector as $s(t) = [s_1^\dagger(t), \ldots, s_n^\dagger(t)]^\dagger$, where $()^\dagger$ denotes transpose, and define the i-th component as $x_i(t) \triangleq A_i s_i(t)$. With these notations, the *multiplicative model* (1) becomes an *additive model*:

$$x(t) = \sum_{i=1}^{n} A_i s_i(t) = \sum_{i=1}^{n} x_i(t) . \tag{2}$$

By writing $x(t) = \sum_i A_i s_i(t)$ we express the model in terms of sources, while by writing $x(t) = \sum_i x_i(t)$ we express the model in terms of components.

We address the problem of blind separation of correlated sources defined as follows: find matrices A_i with $m(i)$ columns such that A is full rank and the corresponding source vectors s_i and s_j are uncorrelated for $i \neq j$. Note that if a given set of matrices A_i satisfies this requirement, then so does the set $\widetilde{A}_i = A_i Z_i$ where $\{Z_i\}$ is any set of $m(i) \times m(i)$ invertible matrices. Such a transform propagates into the sources as $\widetilde{s}_i = Z_i^{-1} s_i$. Hence, the scale indetermination of regular ICA is exacerbated into a problem where each s_i can be blindly recovered only up to some scaling matrix. However, the components themselves are unaffected by such a transform. Indeed, $\widetilde{x}_i = \widetilde{A}_i \widetilde{s}_i = (A_i Z_i)(Z_i^{-1} s_i) = A_i s_i = x_i$.

In the following, we keep on using both points of view. The *source model* is closer to the familiar ICA setting but has the drawback of resorting to matrices A_i which are not blindly identifiable. On the other hand, the components $x_i(t)$ are well defined and can be recovered blindly without scale ambiguity. Applications in which the component model is adequate are, for instance, spike detection in MEG data [8], fetal ECG detection [3] and separation of astrophysical emissions [4].

Projectors as Parameters. The fact that each A_i can be blindly identified up to a scaling factor means that only its column space has some significance. Therefore, the identifiable parameters are the (orthogonal) projectors onto Span(A_i), i.e. the subspace in which each component lives:

$$\Pi_i = A_i (A_i^\dagger A_i)^{-1} A_i^\dagger = A_i A_i^\sharp \tag{3}$$

with $A_i^\sharp = (A_i^\dagger A_i)^{-1} A_i^\dagger$ denoting the pseudo-inverse of A_i.

The component model $\boldsymbol{x}(t) = \sum_{i=1}^n \boldsymbol{x}_i(t)$ has no mixing matrix and therefore no separating matrix. This is unneeded anyway since we want to extract components, not sources. What we need is the (oblique) projector P_i onto $\mathrm{Span}(A_i)$ orthogonally to all other components. These are the operators satisfying $P_i A_j = \delta_{ij} A_i$, so that the i-th component is extracted by $P_i \boldsymbol{x}(t) = \boldsymbol{x}_i(t)$. An oblique projector can be expressed in terms of all orthogonal projectors as [3]

$$P_i = \Pi_i \left(\sum_{j=1}^n \Pi_j \right)^{-1}.$$

In summary, we have defined a linear model for $\boldsymbol{x}(t)$ as the sum of n independent components $\boldsymbol{x}_i(t)$ living in subspaces of dimensions $m(i)$. In order to recover the i-th component, one must estimate the (uniquely defined) oblique projector P_i.

3 Gaussian Likelihood for Multidimensional Components

Contrast Function. We generalize the method of Pham and Garat [11] and Pham and Cardoso [10], which is based on localized covariance matrices. The sample set $t = \{1, \dots, T\}$ is partitioned into Q domains $\mathcal{D}_q$, $q = 1, \dots, Q$ where domain q contains n_q samples (hence, $\sum_q n_q = T$). A simple Gaussian likelihood for the data is set up by assuming that model (2) holds, that $\boldsymbol{x}(t)$ is independent of $\boldsymbol{x}(t')$ if $t \neq t'$ and that, for any $t \in \mathcal{D}_q$, vector $\boldsymbol{s}_i(t)$ is normally distributed with zero mean and a covariance matrix denoted $P_{S,ii}^{(q)}$. Then, for any $t \in \mathcal{D}_q$, the covariance matrix of $\boldsymbol{s}(t)$ is

$$P_S^{(q)} \triangleq \begin{bmatrix} P_{S,11}^{(q)} & \mathbf{0} & \mathbf{0} \\ \mathbf{0} & \ddots & \mathbf{0} \\ \mathbf{0} & \mathbf{0} & P_{S,nn}^{(q)} \end{bmatrix} \triangleq \mathrm{blkdiag}\{P_{S,11}^{(q)}, \dots, P_{S,nn}^{(q)}\} \tag{4}$$

where $\mathrm{blkdiag}\{\}$ creates a block-diagonal matrix from the matrices in brackets and the covariance matrix of $\boldsymbol{x}(t)$, $t \in \mathcal{D}_q$ is $P_X^{(q)} = A P_S^{(q)} A^\dagger$.

Proceeding as in [10], the likelihood of this model depends on the parameter $\boldsymbol{\theta} = \left\{ A, \{P_S^{(q)}\}_{q=1}^Q \right\}$ and is found to be [10]

$$\log p(\boldsymbol{x}; \boldsymbol{\theta}) = -\sum_{q=1}^Q n_q D(\overline{P}_X^{(q)}, A P_S^{(q)} A^\dagger) + \mathrm{cst} \tag{5}$$

where we use the Kullback-based divergence between two $m \times m$ matrices

$$D(R_1, R_2) = \frac{1}{2} \left(\mathrm{tr}\left\{ R_1 R_2^{-1} \right\} - \log \det(R_1 R_2^{-1}) - m \right) \tag{6}$$

and the data enters only via a set of localized covariance matrices:

$$\overline{P}_X^{(q)} \triangleq \frac{1}{n_q} \sum_{t \in \mathcal{D}_q} \boldsymbol{x}(t) \boldsymbol{x}^\dagger(t) \ . \tag{7}$$

It is not hard to show that, for fixed A (hence for fixed $\boldsymbol{y}$, where $\boldsymbol{y}(t) = A^{-1}\boldsymbol{x}(t)$), (5) is minimized with respect to $P_S^{(q)}$ at point

$$\widehat{P}_S^{(q)} = \mathrm{blkdiag}\{\overline{P}_Y^{(q)}\} \ . \tag{8}$$

Therefore, $\max_{P_S^{(q)}} \log p(\boldsymbol{x}; \boldsymbol{\theta}) = -T\, C_{\mathrm{ML}}(A) + \mathrm{cst}$ where we define the contrast function

$$C_{\mathrm{ML}}(A) \triangleq \langle D(\overline{P}_Y^{(q)}, \mathrm{blkdiag}\{\overline{P}_Y^{(q)}\}) \rangle \tag{9}$$

with angle brackets meaning $\langle M^{(q)} \rangle \triangleq \frac{1}{T} \sum_{q=1}^Q n_q M^{(q)}$, that is, averaging across domains. Hence, minimizing $C_{\mathrm{ML}}(A)$ is equivalent to maximizing our likelihood. Clearly, $C_{\mathrm{ML}}(A)$ is also a criterion of joint block diagonalization (JBD) (see also [2,9]).

Estimating Equations. The relative variation of C_{ML} with respect to A can be formulated in full analogy to the uncorrelated sources case in [10]. One obtains the relative gradient with respect to A:

$$\nabla C_{\mathrm{ML}}(A) = \langle \mathrm{blkdiag}^{-1}\{\overline{P}_Y^{(q)}\} \, \overline{P}_Y^{(q)} \rangle - I \ . \tag{10}$$

Setting $\nabla C_{\mathrm{ML}}(A) = 0$ yields the *estimating equations* in terms of the estimated sources $\boldsymbol{y}(t)$:

$$I = \langle \mathrm{blkdiag}^{-1}\{\overline{P}_Y^{(q)}\} \, \overline{P}_Y^{(q)} \rangle \ . \tag{11}$$

Using the block-diagonal structure of $P_S^{(q)}$, eq. (11) also reads block-wise as

$$\langle \overline{P}_{Y_{ii}}^{(q)\,-1} \, \overline{P}_{Y_{ij}}^{(q)} \rangle = 0, \quad i \neq j \tag{12}$$

where i, j are understood as *block* indices. The main-diagonal blocks $i = j$ do not yield any constraints, reflecting the scale indetermination of the problem when expressed in terms of sources. While the form of (8)-(12) is as in [10] with *diag* replacing *blkdiag*, a formal proof of this extension is given in [7].

With some care, these estimating equations can be re-expressed (see [7] for details) in terms of components exclusively:

$$\langle \overline{P}_{X_i X_i}^{(q)\sharp} \overline{P}_{X_i X_j}^{(q)} \rangle = 0 \quad i \neq j \tag{13}$$

i.e. the components $\boldsymbol{x}_i$ estimated for the ML value of A are such that (13) holds.

4   Error Analysis

Quantifying the Separation Error. Given an estimate $\widehat{P}_i$ of the oblique projector on a component subspace, the estimated i-th component is

$$\widehat{\boldsymbol{x}}_i = \widehat{P}_i \boldsymbol{x} = (P_i + \delta P_i) \sum_{j=1}^{n} \boldsymbol{x}_j = \boldsymbol{x}_i + \sum_{j=1}^{n} \delta P_i \boldsymbol{x}_j \; . \tag{14}$$

Using $\sum_{i=1}^{n} P_i = I$, so that $\delta P_i = -\sum_{j \neq i} \delta P_j$, the error for component i is

$$\Delta \boldsymbol{x}_i(t) \overset{\triangle}{=} \widehat{\boldsymbol{x}}_i(t) - \boldsymbol{x}_i(t) = -\sum_{j \neq i}^{n} \mathcal{E}_{ji} \boldsymbol{x}_i(t) + \sum_{j \neq i}^{n} \mathcal{E}_{ij} \boldsymbol{x}_j(t) \tag{15}$$

where we have defined pairwise error terms:

$$\mathcal{E}_{ij} \overset{\triangle}{=} \delta P_i \, \Pi_j = (\widehat{P}_i - P_i) \, \Pi_j \; . \tag{16}$$

The first sum in the RHS of (15) is a *reconstruction error* for the i-th component[2]. The second term in the RHS of (15) is a sum over contaminations of the recovered i-th component by all other components.

For fixed $\mathcal{E}$ the relative mean square error (MSE) for component i, defined below as $\rho_i(\mathcal{E})$, is split (since components are mutually uncorrelated) into a sum of $\rho_{ij}(\mathcal{E})$. The contribution of each component to the MSE is

$$\rho_i(\mathcal{E}) \overset{\triangle}{=} \sigma_i^{-2} E \left\{ \frac{1}{T} \sum_{t=1}^{T} |\Delta \boldsymbol{x}_i(t)|^2 \right\} = \sum_{j=1}^{n} \rho_{ij}(\mathcal{E}) \tag{17}$$

where we denote $\sigma_i^2 \overset{\triangle}{=} \operatorname{tr} \left\{ \langle P_{X_i X_i}^{(q)} \rangle \right\}$ the total power of component i averaged over all domains. Some algebra [7] yields

$$\rho_{ii}(\mathcal{E}) = \sigma_i^{-2} \operatorname{tr} \left\{ \langle P_{X_i X_i}^{(q)} \rangle \sum_{j \neq i}^{n} \mathcal{E}_{ji}^{\dagger} \mathcal{E}_{ji} \right\} \; , \quad \rho_{ij}(\mathcal{E}) = \sigma_i^{-2} \operatorname{tr} \left\{ \langle P_{X_j X_j}^{(q)} \rangle \mathcal{E}_{ij}^{\dagger} \mathcal{E}_{ij} \right\} \; . \tag{18}$$

Define the covariance of the estimation error (15) as

$$\mathrm{MSE}_i \overset{\triangle}{=} \sum_{j=1}^{n} E \left\{ \rho_{ij}(\mathcal{E}) \right\} \; . \tag{19}$$

Taking expectation over $\mathcal{E}$, we are therefore led to evaluate [7]

$$E \left\{ \rho_{ii}(\mathcal{E}) \right\} = \sigma_i^{-2} \operatorname{tr} \left[(I \otimes \langle P_{X_i X_i}^{(q)} \rangle) \sum_{j \neq i}^{n} \operatorname{Cov}(\operatorname{vec}\{\mathcal{E}_{ji}^{\dagger}\}) \right]$$

$$E \left\{ \rho_{ij}(\mathcal{E}) \right\} = \sigma_i^{-2} \operatorname{tr} \left[(\langle P_{X_j X_j}^{(q)} \rangle \otimes I) \operatorname{Cov}(\operatorname{vec}\{\mathcal{E}_{ij}\}) \right] \tag{20}$$

where $\otimes$ denotes the Kronecker product.

---

[2] In classical ICA, it would correspond to an error in the scale of $\widehat{\boldsymbol{x}}_i$; in the more general case considered here, it is a more complicated form of error.

First-Order Error Analysis. The computation for the error analysis proceeds along the lines of first-order approximations [7]. We introduce

$$G_{ij}^{(q)} = \overline{P}_{X_i X_j}^{(q)} P_{X_j X_j}^{(q)\sharp} \ , \qquad H_{ij}^{(q)} = P_{X_j X_j}^{(q)\sharp} \otimes P_{X_i X_i}^{(q)} \tag{21}$$

which come out of Taylor expansions of the estimating equations (13). We find that [7]:

$$\begin{bmatrix} \mathrm{vec}\{\mathcal{E}_{ji}^{\dagger}\} \\ \mathrm{vec}\{\mathcal{E}_{ij}\} \end{bmatrix} = -\begin{bmatrix} \langle H_{ij}^{(q)} \rangle & I \\ I & \langle H_{ij}^{(q)\sharp} \rangle \end{bmatrix}^{-1} \begin{bmatrix} \mathrm{vec}\{\langle G_{ij}^{(q)} \rangle\} \\ \mathrm{vec}\{\langle G_{ji}^{(q)\dagger} \rangle\} \end{bmatrix} + o(1/\sqrt{T}) \ , \tag{22}$$

so that [7]

$$\mathrm{Cov}\left(\begin{bmatrix} \mathrm{vec}\{\langle G_{ij}^{(q)} \rangle\} \\ \mathrm{vec}\{\langle G_{ji}^{(q)\dagger} \rangle\} \end{bmatrix}\right) = \frac{1}{T} \begin{bmatrix} \langle H_{ij}^{(q)} \rangle & I \\ I & \langle H_{ij}^{(q)\sharp} \rangle \end{bmatrix} . \tag{23}$$

Hence,

$$\mathrm{Cov}\left(\begin{bmatrix} \mathrm{vec}\{\mathcal{E}_{ji}^{\dagger}\} \\ \mathrm{vec}\{\mathcal{E}_{ij}\} \end{bmatrix}\right) = \frac{1}{T} \begin{bmatrix} \langle H_{ij}^{(q)} \rangle & I \\ I & \langle H_{ij}^{(q)\sharp} \rangle \end{bmatrix}^{-1} + o(1/T) \ . \tag{24}$$

Since (15) is a linear expression in $\mathcal{E}$, substituting the relevant entries of (24) in (20) yields an explicit, closed-form expression for the normalized MSE of $\widehat{x}_i$. Since the error covariance was derived for the ML estimator of P_i, expression (24) is the CRLB on the estimation error of $\widehat{P}_i$ for the case of Gaussian components.

5   Numerical Results

We implemented an algorithm, based on the relative gradient, which solves eq. (12). This algorithm thus obtains the ML solution and, since the reduced log-likelihood is of the form (9), our algorithm is a joint block diagonalization (JBD) algorithm.

In the simulations, all model and analysis requirements hold and data are obtained from the Gaussian distribution. Therefore, we expect the CRLB to be achieved. We set $Q = 5$, $T = 8193$. In each scenario, different A and $P_S^{(q)}$ are chosen. Each scenario is simulated with 500 Monte-Carlo trials.

Fig. 1 compares the theoretical prediction of the error $\mathcal{E}$ given by eq. (22) with the measured value $\widehat{\mathcal{E}}$ (defined by eq. (16)) in the case of scenario #3 of Table 1. Due to lack of space, we only present the results for a specific entry, namely $\mathcal{E}_{12}(2,4)$ (other entries behave similarly). Fig. 1(a) plots $\widehat{\mathcal{E}}_{12}(2,4)$ vs. $\mathcal{E}_{12}(2,4)$. The spread of the values along the diagonal (the thin line represents $\mathcal{E} = \widehat{\mathcal{E}}$) shows good correspondence between the empirical values and the predicted ones. Fig. 1(b) compares the residual error $(\widehat{\mathcal{E}}_{12}(2,4) - \mathcal{E}_{12}(2,4))$ with $\widehat{\mathcal{E}}_{12}(2,4)$; the residual, which represents higher-order terms, is smaller than the first-order term, showing that the method does operate in the asymptotic regime here.

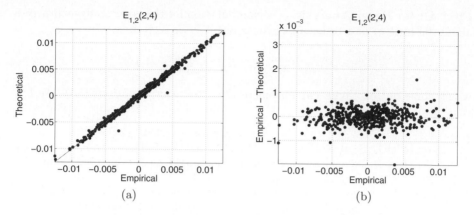

Fig. 1. Error terms: theoretical vs. empirical (a) $\mathcal{E}$ vs. $\widehat{\mathcal{E}}$ (b) $(\widehat{\mathcal{E}} - \mathcal{E})$ vs. $\widehat{\mathcal{E}}$

Table 1. Performance of Second-Order Multidimensional ICA

Scenario	Component dimensions	$\dfrac{\text{MSE(Empirical)}}{\text{MSE(Theoretical)}}$
1	5	1.054
	4	1.056
2	2	1.019
	3	1.012
	4	0.987
3	3	1.028
	2	1.065
	1	1.102
4	1	1.004
	1	1.008
	1	0.990

Table 1 lists the predicted MSE_i defined in (19) for several scenarios of component dimensions. The second column states the dimension of each component in the scenario. The third column gives the ratio of MSE_i, theoretical vs. simulated. Note that all values in the third column are close to 1. Therefore, the results in the "MSE(Empirical) / MSE(Theoretical)" column verify our analysis and show that the CRLB is achievable.

6   Summary

In this paper we present the concept of blind source separation of multidimensional components as a different perspective on the correlated sources model. We derive a source separation ML criterion for multidimensional components, based on second-order statistics. A closed-form expression for its performance is evaluated. By definition, this expression is the CRLB for the case of Gaussian components. Our derivation is verified in numerical simulations.

In [7], this performance is compared with the separation performance of classical joint diagonalization and our treatment of multidimensional components is shown to yield a significant gain in their separation.

References

1. Abed-Meraim, K., Belouchrani, A.: Algorithms for joint block diagonalization. In: Proceedings of EUSIPCO 2004, Vienna, Austria, pp. 209–212 (2004)
2. Bousbia-Salah, H., Belouchrani, A., Abed-Meraim, K.: Blind separation of convolutive mixtures using joint block diagonalization. In: Sixth International Symposium on Signal Processing and its Applications, vol. 1, pp. 13–16 (August 2001)
3. Cardoso, J.-F.: Multidimensional independent component analysis. In: Proc. IEEE International Conference on Acoustics, Speech and Signal Processing, 12-15 May, vol. 4, pp. 1941–1944 (1998)
4. Cardoso, J.-F., Martin, M., Delabrouille, J., Betoule, M., Patanchon, G.: Component separation with flexible models. Application to the separation of astrophysical emissions. In: IEEE J-STSP. Special Issue on: Signal Processing for Astronomical and Space Research Applications (October 2008), http://arxiv.org/abs/0803.1814
5. Févotte, C., Theis, F.J.: Pivot selection strategies in jacobi joint block-diagonalization. In: Davies, M.E., James, C.J., Abdallah, S.A., Plumbley, M.D. (eds.) ICA 2007. LNCS, vol. 4666, pp. 177–184. Springer, Heidelberg (2007)
6. Ghennioui, H., El Mostafa, F., Thirion-Moreau, N., Adib, A., Moreau, E.: A nonunitary joint block diagonalization algorithm for blind separation of convolutive mixtures of sources. Signal Processing Letters, IEEE 14(11), 860–863 (2007)
7. Lahat, D., Cardoso, J.-F., Messer, H.: Multidimensional ICA: Second-order methods and applications (in preparation)
8. Ossadtchi, A., Baillet, S., Mosher, J.C., Thyerlei, D., Sutherling, W., Leahy, R.M.: Automated interictal spike detection and source localization in magnetoencephalography using independent components analysis and spatio-temporal clustering. Clin. Neurophysiol. 115(3), 508–522 (2004)
9. Pham, D.T.: Blind separation of cyclostationary sources using joint block approximate diagonalization. In: Davies, M.E., James, C.J., Abdallah, S.A., Plumbley, M.D. (eds.) ICA 2007. LNCS, vol. 4666, pp. 244–251. Springer, Heidelberg (2007)
10. Pham, D.-T., Cardoso, J.-F.: Blind separation of instantaneous mixtures of non stationary sources. IEEE Trans. Sig. Proc. 49(9), 1837–1848 (2001)
11. Pham, D.-T., Garat, P.: Blind separation of mixtures of independent sources through a quasi- maximum likelihood approach. IEEE Trans. Sig. Proc. 45(7), 1712–1725 (1997)
12. Vollgraf, R., Obermayer, K.: Multi-dimensional ICA to separate correlated sources. In: Neural Information Processing Systems (NIPS 2001), vol. 14, pp. 993–1000. MIT Press, Cambridge (2001)

Using Signal Invariants
to Perform Nonlinear Blind Source Separation

David N. Levin

University of Chicago
d-levin@uchicago.edu

Abstract. Given a time series of multicomponent measurements $x(t)$, the usual objective of nonlinear blind source separation (BSS) is to find a "source" time series $s(t)$, comprised of statistically independent combinations of the measured components. In this paper, the source time series is required to have a density function in $(s, \dot{s})$-space that is equal to the product of density functions of individual components. This formulation of the BSS problem has a solution that is unique, up to permutations and component-wise transformations. Separability is shown to impose constraints on certain locally invariant (scalar) functions of x, which are derived from local higher-order correlations of the data's velocity $\dot{x}$. The data are separable if and only if they satisfy these constraints, and, if the constraints are satisfied, the sources can be explicitly constructed from the data. The method is illustrated by using it to separate two speech-like sounds recorded with a single microphone.

1 Introduction

Consider a time series of data $x(t)$, where x is a multiplet of N measurements (x_k for $k = 1, 2, \ldots, N$). The usual objectives of nonlinear BSS are: 1) determine if these data are instantaneous mixtures of N statistically independent source components $s(t)$

$$x(t) = f[s(t)], \tag{1}$$

where f is a possibly nonlinear, invertible N-component mixing function, and, if this is the case, 2) compute the mixing function. In other words, the problem is to find a coordinate transformation f^{-1} that transforms the observed data $x(t)$ from the measurement-defined coordinate system (x) on state space to a source coordinate system (s) in which the components of the transformed data are statistically independent. In the source coordinate system, let the state space probability density function (PDF) $\rho_S(s)$ be defined so that $\rho_S(s)ds$ is the fraction of total time that the source trajectory $s(t)$ is located within the volume element ds at location s. In the usual formulation of the BSS problem [1], the source components are required to be statistically independent in the sense that their state space PDF is the product of the density functions of the individual components

$$\rho_S(s) = \prod_{k=1}^{N} \rho_k(s_k), \tag{2}$$

T. Adali et al. (Eds.): ICA 2009, LNCS 5441, pp. 58–65, 2009.

However, it is well known that this criterion is so weak that this form of the BSS problem always has many solutions (see [2] and references therein).

The issue of non-uniqueness can be circumvented by considering the data's trajectory in $(s, \dot{s})$-space instead of state space (i.e., s-space). First, define the PDF $\rho_S(s, \dot{s})$ in this space so that $\rho_S(s, \dot{s})dsd\dot{s}$ is the fraction of total time that the location and velocity of the source trajectory are within the volume element $dsd\dot{s}$ at location $(s, \dot{s})$. Earlier papers [3,4] described a formulation of the BSS problem in which this PDF was required to be the product of the density functions of the individual components

$$\rho_S(s, \dot{s}) = \prod_{k=1}^{N} \rho_k(s_k, \dot{s}_k). \tag{3}$$

This type of statistical independence is satisfied by almost all classical non-interacting physical systems. Furthermore, because separability in $(s, \dot{s})$-space is a stronger requirement than separability in state space, the corresponding BSS problem can be shown to have a unique solution [3,4].

It was previously demonstrated [3,4] that the $(s, \dot{s})$-space PDF of a time series induces a Riemannian geometry on the data's state space, with the metric equal to the local second-order correlation matrix of the data's velocity. Nonlinear BSS can be performed by computing this metric in the x coordinate system (i.e., by computing the second-order correlation of $\dot{x}(t)$ at point x), as well as its first and second derivatives with respect to x. However, if the dimensionality of state space is high, it must be covered by a great deal of data in order to calculate these derivatives accurately. The current paper shows how to perform nonlinear BSS by computing higher-order local correlations of the data's velocity, instead of computing derivatives of its second-order correlation. This approach is advantageous because it requires less data for an accurate computation.

In addition to using a stronger criterion for statistical independence, there are technical differences between the proposed method and conventional ones. First of all, the technique in this paper exploits statistical constraints on the data that are *locally* defined in state space, in contrast to the usual criteria for statistical independence that are *global* conditions on the data time series or its time derivatives [5]. Furthermore, unlike many other methods [6], the mixing function is constructed in a "deterministic", non-parametric manner, without using probabilistic learning methods and without parameterizing it with a neural network architecture or other means.

The next section describes the theoretical framework of the method. Section 3 illustrates the method by using it to separate two simultaneous speech-like sounds that are recorded with a single microphone. The implications of this work are discussed in the last section.

2   Method

The local correlation of the data's velocity is

$$C_{kl...}(x) = <(\dot{x}_k - \bar{\dot{x}}_k)(\dot{x}_l - \bar{\dot{x}}_l)...>_x, \tag{4}$$

where $\bar{\dot{x}} = <\dot{x}>_x$ is the local time average of $\dot{x}$, where the bracket denotes the time average over the trajectory's segments in a small neighborhood of x, where $1 \leq k, l \leq N$, and where "..." denotes possible additional indices on the left side and corresponding factors of $\dot{x} - \bar{\dot{x}}$ on the right side. The definition of the PDF implies that this velocity correlation is one of its moments

$$C_{kl...}(x) = \frac{\int \rho(x, \dot{x})(\dot{x}_k - \bar{\dot{x}}_k)(\dot{x}_l - \bar{\dot{x}}_l)...d\dot{x}}{\int \rho(x, \dot{x})d\dot{x}}, \tag{5}$$

where $\rho(x, \dot{x})$ is the PDF in the x coordinate system. Incidentally, although Eq. (5) is useful in a formal sense, in practical applications, all required correlation functions can be computed directly from local time averages of the data (Eq. (4)), without explicitly computing the data's PDF. Also, note that the analogous velocity "correlation" with a single subscript vanishes identically.

Next, let $M(x)$ be a local $N \times N$ matrix, and use it to define transformed velocity correlations

$$I_{kl...}(x) = \sum_{1 \leq k', l', ... \leq N} M_{kk'}(x)M_{ll'}(x)...C_{k'l'...}(x), \tag{6}$$

where "..." denotes possible additional indices of I and C, as well as corresponding factors of $M(x)$. Because $C_{kl}(x)$ is positive definite at any point x, it is always possible to find an $M(x)$ such that

$$I_{kl}(x) = \delta_{kl} \tag{7}$$

$$\sum_{1 \leq m \leq N} I_{klmm}(x) = D_{kl}(x), \tag{8}$$

where $D(x)$ is a diagonal $N \times N$ matrix. As long as D is not degenerate, $M(x)$ is unique, up to arbitrary *local* permutations and reflections. In almost all realistic applications, the velocity correlations will be continuous functions of the state space coordinate x. Therefore, in any neighborhood of state space, there will always be a continuous solution for $M(x)$, and this solution is unique, up to arbitrary *global* reflections and permutations.

Now, imagine doing the same computation in some other coordinate system x'. An M-matrix that satisfies Eqs. (7, 8) in the x' coordinate system is given by

$$M'_{kl}(x') = \sum_{1 \leq m \leq N} M_{km}(x)\frac{\partial x_m}{\partial x'_l}(x'). \tag{9}$$

This can be understood in the following manner: because velocity correlations transform as contravariant tensors, the partial derivative factor transforms correlations from the x' coordinate system to the x coordinate system, and the factor $M(x)$ then transforms these correlations into the functions on the left sides of Eqs. (7, 8). All other solutions for $M'(x')$ differ from this one by global reflections and permutations. Similar reasoning shows that the functions $I'_{kl...}(x')$, derived by using Eq. (6) in the x' coordinate system, equal the functions $I_{kl...}(x)$, up to possible global permutations and reflections. In other words,

$$I'_{kl...}(x') = \sum_{1 \leq k', l',... \leq N} P_{kk'} P_{ll'} \ldots I_{k'l'...}(x), \tag{10}$$

where $P_{kk'}$ denotes an element of a product of permutation, reflection, and identity matrices. Essentially, the functions $I_{kl...}(x)$ transform as scalar functions on the state space, except for possible reflections and index permutations.

We now assume that the system is separable and derive some necessary conditions in the source coordinate system (s). Then, the above-described scalar functions are used to transfer these separability conditions to the measurement-defined coordinate system (x), where they can be tested with the data. In order to make the notation simple, it is assumed that $N - 2$. However, as described below, the methodology can be generalized in order to separate higher-dimensional data into possibly multidimensional source variables.

Separability implies that there is a transformation f^{-1} from the x coordinate system to a source coordinate system (s) in which Eq. (3) is true. Because of Eq. (5), the velocity correlation functions in the s coordinate system are products of correlations of independent sources

$$C_{S1...2...}(s_1, s_2) = C_{S1...}(s_1) C_{S2...}(s_2), \tag{11}$$

where $1 \ldots$ and $2 \ldots$ denote arbitrary numbers of indices equal to 1 and 2, respectively. It follows that, in the s coordinate system, Eqs. (7, 8) are satisfied by a block-diagonal matrix M, in which each "block" is the 1×1 M "matrix" satisfying Eqs. (7, 8) for one of the subsystems. Therefore, in the s coordinate system, the functions $I_{Skl...}(s)$, which are defined by Eq. (6) with all subscripts $kl \ldots$ equal to 1 (2), must equal the corresponding functions derived for subsystem 1 (2), and these latter functions depend on s_1 (s_2) alone. Although these constraints were derived in the s coordinate system, Eq. (10) implies that they are true in all coordinate systems, except for possible permutations and reflections. Therefore, in the measurement-defined coordinate system (x), the functions $I_{kl...}(x)$, which are defined by Eq. (6) with all subscripts equal to 1, must be functions of either $s_1(x)$ or $s_2(x)$. Likewise, the functions $I_{kl...}(x)$, which are defined by Eq. (6) with all subscripts equal to 2, must be functions of the other source variable ($s_2(x)$ or $s_1(x)$, respectively).

This coordinate-system-independent consequence of separability can be used to perform nonlinear BSS in the following manner:

1. Use Eq. (4) to compute velocity correlations, $C_{kl...}(x)$, from the data $x(t)$.
2. Use linear algebra to find a continuous matrix $M(x)$ that satisfies Eqs (7, 8) (or that satisfies similar algebraic constraints that determine $M(x)$ uniquely, except for permutations and reflections).
3. Use Eq. (6) to compute the functions $I_{kl...}(x)$.
4. Plot the values of the triplets

$$I_A(x) = \{I_{111}(x), I_{1111}(x), I_{11111}(x)\} \tag{12}$$

$$I_B(x) = \{I_{222}(x),\ I_{2222}(x),\ I_{22222}(x)\},\tag{13}$$

as x varies over the measurement-defined coordinate system.

5. If the plotted values of I_A and/or I_B *do not* lie in one-dimensional subspaces within the three-dimensional space of the plots, $I_A(x)$ and/or $I_B(x)$ cannot be functions of single source components ($s_1(x)$ or $s_2(x)$) as required by separability, and the data are not separable.

6. If the plotted values of both I_A and I_B *do* lie on one-dimensional manifolds, define one-dimensional coordinates (σ_A and σ_B, respectively) on those subspaces [7]. Then, compute the function $\sigma(x) = (\sigma_A(x), \sigma_B(x))$ that maps each coordinate x onto the value of σ, which parameterizes the point $(I_A(x), I_B(x))$. Notice that, because of the Takens' embedding theorem [8], x is invertibly related to the six components of $I_A(x)$ and $I_B(x)$, and, therefore, it is invertibly related to σ.

7. Transform the PDF (or correlations) of the measurements from the x coordinate system to the σ coordinate system. The data are separable if and only if the PDF factorizes (the correlations factorize) in the σ coordinate system.

The last statement can be understood in the following manner. As shown above, separability implies that $I_A(x)$ must be a function of a single source variable ($s_1(x)$ or $s_2(x)$), and the Takens theorem implies that this function is invertible. Because I_A is also an invertible function of σ_A, it follows that σ_A must be invertibly related to one of the source variables, and, in a similar manner, σ_B must be invertibly related to the other source variable. Thus, separability implies that σ_A and σ_B are themselves source variables. It follows that the data are separable if and only if the PDF factorizes in the σ coordinate system.

This procedure can be generalized to determine whether data with $N > 2$ are a mixture of two, possibly multidimensional source variables. Consider any partition of the x indices ($k = 1, 2, \ldots, N$) into two groups: d_A "A" indices and $d_B = N - d_A$ "B" indices. Let $I_A(x)$ ($I_B(x)$) be any set of more than $2d_A$ ($2d_B$) of the functions $I_{kl\ldots}(x)$ for which all subscripts belong to the A (B) group. Now, suppose that the data are separable. Then, there must be such a partition for which the values of $I_A(x)$ ($I_B(x)$) lie in a d_A-dimensional (d_B-dimensional) subspace, as x varies over the N-dimensional data space. Furthermore, if σ_A and σ_B are some coordinates on those subspaces, a set of source variables is given by $\sigma(x) = (\sigma_A(x), \sigma_B(x))$, the values of σ_A and σ_B that parameterize the point $(I_A(x),\ I_B(x))$. Therefore, to perform BSS: 1) systematically examine all possible index partitions and determine if the data-derived functions, $I_A(x)$ and $I_B(x)$, map x onto subspaces with the required dimensions; 2) if they do comprise such maps, construct the function $\sigma(x)$ and determine if the data's PDF factorizes in the σ coordinate system. The data are separable if and only if at least one such index partition leads to a $\sigma(x)$ that factorizes the PDF. If the data are separable, the same procedure can then be used to determine if each multicomponent independent variable (σ_A or σ_B) can be further separated into lower-dimensional independent variables.

3   Numerical Example: Separating Two Speech-Like Sounds Recorded with a Single Microphone

This section describes a numerical experiment in which two speech-like sounds were synthesized and then summed, as if they were simultaneously recorded with a single microphone. Each sound simulated an "utterance" of a vocal tract resembling a human vocal tract, except that it had fewer degrees of freedom (one degree of freedom instead of the 3-5 degrees of freedom of the human vocal tract). The methodology of Section 2 was blindly applied to a time series of two features extracted from the synthetic recording, in order to recover the time dependence of the state variable of each vocal tract (up to an unknown transformation on each voice's state space). BSS was performed with only 16 minutes of data, instead of the hours of data required to separate similar sounds using a differential geometric method [3,4].

The glottal waveforms of the two "voices" had different pitches (100 Hz and 160 Hz), and the "vocal tract" response of each voice was characterized by a damped sinusoid, whose amplitude, frequency, and damping were linear functions of that voice's state variable. For each voice, a 16 minute utterance was produced by using glottal impulses to drive the vocal tract's response, which was determined by the time-dependent state variable of that vocal tract. The state variable time series of each voice was synthesized by smoothly interpolating among successive states, which were randomly chosen at 100-120 msec intervals. The resulting utterances had energies differing by 2.4 dB, and they were summed and sampled at 16 kHz with 16-bit depth. Then, this "recorded" waveform was pre-emphasized and subjected to a short-term Fourier transform (using frames with 25 msec length and 5 msec spacing). The log energies of a bank of 12 mel-frequency filters between 0-8000 Hz were computed for each frame, and these were then averaged over pairs of consecutive frames. These log filterbank outputs were nonlinear functions of the two vocal tract state variables, which were chosen to be statistically independent of each other.

In order to blindly analyze these data, we first determined if any data components were redundant in the sense that they were simply functions of other components. Figure 1a shows the first three principal components of the log filterbank outputs during a typical short recording of the simultaneous utterances. Inspection showed that these data lay on an approximately two-dimensional surface within the ambient 12-D space, making it apparent that they were produced by a "hidden" dynamical system with two degrees of freedom. The redundant components were eliminated by using dimensional reduction (principal components analysis in small overlapping neighborhoods of the data) to establish a coordinate system x on this surface and to find $x(t)$, the trajectory of the recorded sound in that coordinate system. The next step was to determine if $x(t)$ was a nonlinear mixture of two source variables that were statistically independent of one another. Following steps 1-4 of the BSS procedure in Section 2, $x(t)$ of the entire recording was used to compute "invariants" $I_{kl...}(x)$ with up to five indices, and the related functions $I_A(x)$ and $I_B(x)$ were plotted, as illustrated in Fig. 1b. It was evident that the plotted values of both I_A and I_B did lie in

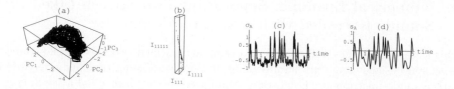

Fig. 1. (a) The first three principal components of log filterbank outputs of a typical short recording of two simultaneous speech-like sounds. (b) The distribution of the values of $I_A(x)$ (Eq. (12)), as x varied over the approximately two-dimensional manifold in (a). (c) The time dependence of one of the source variables, blindly computed from a typical five-second segment of the data's trajectory $x(t)$, by finding the coordinates of $I_A[x(t)]$ on the approximately one-dimensional manifold in (b). (d) The state variable time series originally used to generate one of the speech-like sounds during the five-second recording analyzed in (c).

approximately one-dimensional subspaces. Following step 6 of the BSS procedure, a dimensional reduction procedure [7] was used to define coordinates (σ_A and σ_B) on these one-dimensional manifolds, and a numerical representation of $\sigma(x) = (\sigma_A(x), \sigma_B(x))$ was derived. If the data were separable, σ must be a set of source variables, and $\sigma[x(t)]$ must describe the evolution of the underlying vocal tract states (up to invertible component-wise transformations). As illustrated in Figs. 1c-d, the time courses of the putative source variables $(\sigma_A[x(t)], \sigma_B[x(t)])$ did resemble the time courses of the state variables that were originally used to generate the voices' utterances (up to an invertible transformation on each state variable space). Thus, it was apparent that the information encoded in the time series of each vocal tract's state variable was blindly extracted from the simulated recording of the superposed utterances.

4   Discussion

In previous papers [3,4], the nonlinear BSS problem was formulated in (state, state velocity)-space, instead of state space as in conventional formulations. This approach is attractive because: 1) this type of statistical independence is satisfied by almost all classical non-interacting physical systems; 2) this form of the BSS problem has a unique solution in the following sense: either the data are inseparable, or they can be separated by a mixing function that is unique, up to permutations and transformations of independent source variables. This paper shows how to perform this type of nonlinear BSS by computing local higher-order correlations of the data's velocity $\dot{x}(t)$, instead of computing derivatives of its local second-order correlation as was previously proposed [3,4]. This is advantageous because it requires less data for an accurate computation, as demonstrated in a numerical example in which BSS was performed with minutes (instead of hours) of data.

The BSS procedure in Section 2 shows how to compute $(\sigma_A[x(t)], \sigma_B[x(t)])$, the trajectory of each independent subsystem in a specific coordinate system on that subsystem's state space. In many practical applications, a pattern recognition "engine" has been trained to recognize the meaning of trajectories of one subsystem (e.g., "A") in another coordinate system (e.g., s_A) on that subsystem's state space. In order to use this information, it is necessary to know the transformation to this coordinate system ($\sigma_A \rightarrow s_A$). For example, subsystem A may be the vocal tract of speaker A, and subsystem B may be a noise generator of some sort. In this example, we may have trained an automatic speech recognition (ASR) engine on the quiet speech of speaker A (or, equivalently, on the quiet speech of another speaker who mimics A in the sense that their state space trajectories are related by an invertible transformation when they speak the same utterances). In order to recognize the speaker's utterances in the presence of B, we must know the transformation from the vocal tract coordinates provided by BSS (σ_A) to the coordinates used to train the ASR engine (s_A). This mapping can be determined by using the training data to compute more than $2d_A$ invariants (like those in Eq. (6)) as functions of s_A. These must equal the invariants of one of the subsystems identified by the BSS procedure, up to a global permutation and/or reflection (Eq. (10)). This global transformation can be determined by permuting and reflecting the distribution of invariants produced by the training data, until it matches the distribution of invariants of one of the subsystems produced by the BSS procedure. Then, the mapping $\sigma_A \rightarrow s_A$ can be determined by finding paired values of σ_A and s_A that correspond to the same invariant values within these matching distributions. This type of analysis of human speech data is currently underway.

References

1. Jutten, C., Karhunen, J.: Advances in Nonlinear Blind Source Separation. In: Proceedings of the 4th International Symposium on Independent Component Analysis and Blind Signal Separation, Nara, Japan (April 2003)
2. Hyvärinen, A., Pajunen, P.: Nonlinear Independent Component Analysis: Existence and Uniqueness Results. Neural Networks 12, 429–439 (1999)
3. Levin, D.N.: Using state space differential geometry for nonlinear blind source separation. In: Davies, M.E., James, C.J., Abdallah, S.A., Plumbley, M.D. (eds.) ICA 2007. LNCS, vol. 4666, pp. 65–72. Springer, Heidelberg (2007)
4. Levin, D.N.: Using State Space Differential Geometry for Nonlinear Blind Source Separation. J. Applied Physics 103, art. no. 044906 (2008)
5. Lagrange, S., Jaulin, L., Vigneron, V., Jutten, C.: Analytical solution of the blind source separation problem using derivatives. In: Puntonet, C.G., Prieto, A.G. (eds.) ICA 2004. LNCS, vol. 3195, pp. 81–88. Springer, Heidelberg (2004)
6. Yang, H.H., Amari, S.-I., Cichocki, A.: Information-Theoretic Approach to Blind Separation of Sources in Non-linear Mixture. Signal Processing 64, 291–300 (1998)
7. Roweis, S.T., Saul, L.K.: Nonlinear Dimensionality Reduction by Locally Linear Embedding. Science 290, 2323–2326 (2000)
8. Sauer, T., Yorke, J.A., Casdagli, M.: Embedology. J. Statistical Physics 65, 579–616 (1991)

Bayesian Robust PCA for Incomplete Data

Jaakko Luttinen, Alexander Ilin, and Juha Karhunen

Helsinki University of Technology TKK
Department of Information and Computer Science
P.O. Box 5400, FI-02015 TKK, Espoo, Finland

Abstract. We present a probabilistic model for robust principal component analysis (PCA) in which the observation noise is modelled by Student-t distributions that are independent for different data dimensions. A heavy-tailed noise distribution is used to reduce the negative effect of outliers. Intractability of posterior evaluation is solved using variational Bayesian approximation methods. We show experimentally that the proposed model can be a useful tool for PCA preprocessing for incomplete noisy data. We also demonstrate that the assumed noise model can yield more accurate reconstructions of missing values: Corrupted dimensions of a "bad" sample may be reconstructed well from other dimensions of the same data vector. The model was motivated by a real-world weather dataset which was used for comparison of the proposed technique to relevant probabilistic PCA models.

1 Introduction

Principal component analysis (PCA) is a widely used method for data preprocessing (see, e.g., [1,2,3]). In independent component analysis (ICA) and source separation problems, PCA is used for reducing the dimensionality of the data to avoid overlearning, to suppress additive noise, and for prewhitening needed in several ICA algorithms [2,4]. PCA is based on the quadratic criteria of variance maximisation and minimisation of the mean-square representation error, and therefore it can be sensitive to outliers in the data. Robust PCA techniques have been introduced to cope this problem, see, for example, [4] and the references therein. The basic idea in robust PCA methods is to replace quadratic criteria leading to standard PCA by more slowly growing criteria.

PCA has a probabilistic interpretation as maximum likelihood estimation of a latent variable model called probabilistic PCA (PPCA) [5]. While PPCA is a rather simplistic model based on Gaussion assumptions, it can be used as a basis for building probabilistic extensions of classical PCA. Probabilistic models provide a principled way to cope with the overifitting problem, to do model comparison and to handle missing values. Probabilistic models for robust PCA have been introduced recently [6,7,8]. They treat possible outliers by using heavy-tailed distributions, such as Student-t or Laplacian, for describing the noise.

In this paper, we present a new robust PCA model based on the Student-t distribution and show how it can be identified for incomplete data, that is,

T. Adali et al. (Eds.): ICA 2009, LNCS 5441, pp. 66–73, 2009.
© Springer-Verlag Berlin Heidelberg 2009

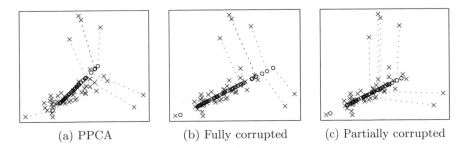

(a) PPCA   (b) Fully corrupted   (c) Partially corrupted

Fig. 1. Principal subspace estimation using (a) probabilistic PCA [5], (b) robust PCA assuming fully corrupted outliers [7] and (c) robust PCA assuming partially corrupted outliers. The crosses represent data points and the circles show their projections onto the found principal subspace.

datasets with missing values. We assume that the outliers can arise independently in each sensor (i.e. for each dimension of a data vector). This is different to the previously introduced techniques [6,7] which assume that all elements of an outlier data vector are corrupted. This work was inspired by our intention to apply a semi-blind source separation technique, called denoising source separation (DSS) to a weather dataset which is too much corrupted by outliers and missing values. We have earlier successfully applied DSS to exploratory analysis of global climate data [9].

Our modelling assumption can be more realistic for some datasets and therefore they can improve the quality of the principal subspace estimation and achieve better reconstructions of the missing values. The model can also be used to remove outliers by estimating the true values of their corrupted components from the uncorrupted ones. This is illustrated in Fig. 1 using an artificial two-dimensional data with a prominent principal direction and a few outliers. The subspace found by the simplest PCA model is affected by outliers, whereas robust techniques are able to find the right principal subspace. However, the reconstruction of the data is quite different depending on whether one assumes fully corrupted or partially corrupted outliers: Fully corrupted outliers can be reconstructed by projecting orthogonally onto the subspace, while improbable values of partially corrupted samples can be ignored and reconstructed based on the uncorrupted dimensions.

2   Model

Let us denote by $\{\boldsymbol{y}_n\}_{n=1}^N$ a set of M-dimensional observations $\boldsymbol{y}_n$. The data are assumed to be generated from hidden D-dimensional states $\{\boldsymbol{x}_n\}_{n=1}^N$ using the transformation:

$$\boldsymbol{y}_n = \boldsymbol{W}\boldsymbol{x}_n + \boldsymbol{\mu} + \boldsymbol{\epsilon}_n\,,$$

where $\boldsymbol{W}$ is a $M \times D$ loading matrix, $\boldsymbol{\mu}$ is a bias term and $\boldsymbol{\epsilon}_n$ is noise. Usually the dimensions fulfil $D < M < N$. The prior models for the latent variables are the same as in PPCA and we use conjugate prior for $\boldsymbol{\mu}$ and hierarchical prior for $\boldsymbol{W}$ as in [10] to diminish overfitting [11]:

$$p(\boldsymbol{X}) = \prod_{m=1}^{M} \prod_{n=1}^{N} \mathcal{N}(x_{mn}|0,1),$$

$$p(\boldsymbol{W}|\boldsymbol{\alpha}) = \prod_{m=1}^{M} \prod_{d=1}^{D} \mathcal{N}(w_{md}|0,\alpha_d^{-1}),$$

$$p(\boldsymbol{\mu}) = \prod_{m=1}^{M} \mathcal{N}(\mu_m|0,\beta^{-1}),$$

$$p(\boldsymbol{\alpha}) = \prod_{d=1}^{D} \mathcal{G}(\alpha_d|a_{\boldsymbol{\alpha}},b_{\boldsymbol{\alpha}}).$$

Hyperparameters $a_{\boldsymbol{\alpha}}$, $b_{\boldsymbol{\alpha}}$, and β are fixed to some proper values.

The noise term $\boldsymbol{\epsilon}_n$ is modelled using independent Student-t distributions for its elements. This is achieved by using a hierarchical model with extra variables u_{mn}:

$$p(\boldsymbol{Y},\boldsymbol{U}|\boldsymbol{W},\boldsymbol{X},\boldsymbol{\mu},\boldsymbol{\tau},\boldsymbol{\nu}) = \prod_{mn|\mathcal{O}_{mn}} \mathcal{N}\left(y_{mn}|\boldsymbol{w}_m^{\mathrm{T}}\boldsymbol{x}_n + \mu_m, \frac{1}{\tau_m u_{mn}}\right) \mathcal{G}(u_{mn}|\tfrac{\nu_m}{2}, \tfrac{\nu_m}{2}),$$

which yields a product of Student-t distributions $\mathcal{S}(y_{mn}|\boldsymbol{w}_m^{\mathrm{T}}\boldsymbol{x}_n + \mu_m, \frac{1}{\tau_m}, \nu_m)$ with degrees of freedom ν_m when $\boldsymbol{U}$ is marginalised out [12]. Here, $\mathcal{O}_{mn}$ denotes such indices that the corresponding y_{mn} is actually observed and $\boldsymbol{w}_m^{\mathrm{T}}$ is the m-th row of $\boldsymbol{W}$. Precision τ_m defines a scaling variable which is assigned a conjugate prior

$$p(\boldsymbol{\tau}) = \prod_{m=1}^{M} \mathcal{G}(\tau_m|a_{\boldsymbol{\tau}},b_{\boldsymbol{\tau}}),$$

with $a_{\boldsymbol{\tau}}$ and $b_{\boldsymbol{\tau}}$ set to proper values. Separate τ_m and ν_m are used for each dimension but with simple modifications the dimensions can have a common value. Especially for the precision $\boldsymbol{\tau}$, common modelling may prevent bad local minima. For the degrees of freedom $\boldsymbol{\nu}$ we set a uniform prior.

3   Posterior Approximation

Bayesian inference is done by evaluating the posterior distribution of the unknown variables given the observations. We use variational Bayesian approach to cope with the problem of intractability of the joint posterior distribution (see, e.g., [3, ch.10] for more details). The approximate distribution q is factorised with respect to the unknown variables as

$$\prod_{n=1}^{N} q(\boldsymbol{x}_n) \prod_{m=1}^{M} q(\boldsymbol{w}_m) \prod_{m=1}^{M} q(\mu_m) \prod_{m=1}^{M} q(\tau_m) \prod_{m=1}^{M} \prod_{n=1}^{N} q(u_{mn}) \prod_{d=1}^{D} q(\alpha_d)$$

and each factor $q(\boldsymbol{\theta}_i)$ is updated assuming the other factors are fixed. This is done by minimising the Kullback-Leibler divergence cost function. Using conjugate priors yields simple update rules presented in the appendix.

4   Experiments with Real-World Data

The proposed model was largely motivated by the analysis of real-world weather data from the Helsinki Testbed research project of mesoscale meteorology. The data consists of temperature measurements in Southern Finland over a period of almost two years with an interval of ten minutes, resulting in 89 000 time instances. Some parts of the data were discarded: Stations with no observations were removed and we used only the measurements taken in the lowest altitude in each location. The locations of the remaining 79 stations are shown in Fig. 2.

The quality of the dataset was partly poor. Approximately 35% of the data was missing and a large number of measurements were corrupted. Fig. 3 shows representative examples of measurements from four stations. The quality of the dataset can be summarised as follows: Half of the stations were relatively good, having no outstanding outliers and only short periods missing. More than 10 stations had a few outliers, similarly to the first signal from Fig. 3. Five stations had a large number of outliers, see the second signal in Fig. 3. The quality of the data from the rest of the stations was somewhat poor: The signals contained a small number of measurements and were corrupted by outliers, see the two signals at the bottom of Fig. 3.

Although the outliers may sometimes be easily distinguished from the data, removing them by hand requires a tedious procedure which turned out to be non-trivial in some cases. Therefore, we used the proposed robust PCA method as a preprocessing step which automatically solves the problems of outlier removal, dimensionality reduction and infilling missing values. To keep the preprocessing step simple, we did not take into account the temporal structure of the data.

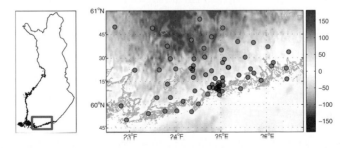

Fig. 2. The weather stations are shown as purple dots on the topographical map of the studied area. The colour represents the altitude above sea level in meters.

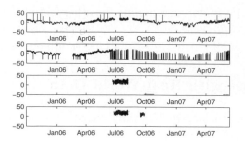

Fig. 3. Temperature data from four stations from the Helsinki Testbed dataset

In the presented experiment, we estimated the four-dimensional principal subspace of the data using the following models: probabilistic PCA [5], robust PPCA (RPCA-s) [7] and the robust model presented in this paper (RPCA-d). For RPCA-d, the degrees of freedom $\{\nu_m\}_{m=1}^{M}$ were modelled separately for each station whereas the precision $\tau_m = \tau$ was set to be common. Broad priors were obtained by setting $a_\alpha = b_\alpha = \beta = a_\tau = b_\tau = 10^{-3}$.

Fig. 4 presents the reconstruction of the missing data for the four signals from Fig. 3 using the compared techniques. The reconstructions obtained by PPCA and RPCA-s are clearly bad. Both models are over-fitted to outliers and to spontaneous correlations observed in scarce measurements from problematic stations. The methods reproduce accurately some outliers and generate new outliers in the place of missing values. In contrast, the results by RPCA-d are clearly much better: The outliers are removed and reasonable reconstructions of the missing values are obtained. Although the signals look rather similar in Fig. 4c (the analysed spatial area is small and the annual cycle is obviously the dominant pattern), the reconstructed signals look very plausible.

The loading matrix $\boldsymbol{W}$ obtained with the different techniques is also visualised in Fig. 4. Each column of $\boldsymbol{W}$ is a collection of weights showing the contribution of one principal component in reconstructing data in different spatial locations. The patterns shown in Fig. 4 are interpolations of the weights over the map of Southern Finland. The patterns produced by PPCA and RPCA-s clearly contain lots of artefacts: the components are over-fitted to the outliers registered in some weather stations. On the contrary, the components found by RPCA-d are much more meaningful (though they contain some artefacts due to problematic stations in the central area): The first component explains the dominant yearly and daily oscillations and the patterns associated with the rest of the principal components are very typical for PCA applied to spatially distributed data. Since the investigated area is rather small, the first principal component has similar loading for all weather stations. Note a clear coast line pattern in the second and the third components.

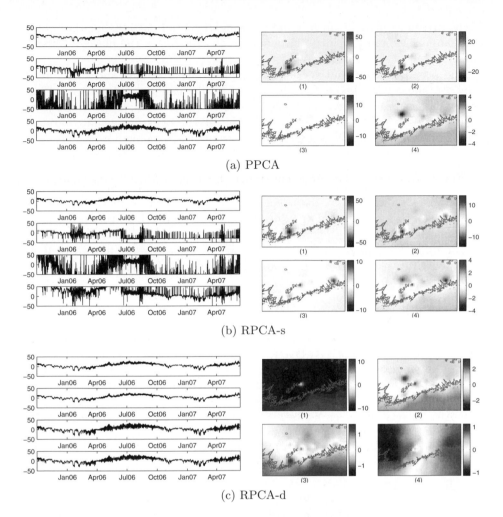

Fig. 4. Experimental results obtained for the Helsinki Testbed dataset with different models. Left: The reconstructions of the signals shown in Fig. 3. Right: The principal component loadings interpolated over the map of Southern Finland.

5   Conclusions

In this paper, we presented a probabilistic model for robust PCA which can be a useful tool for preprocessing incomplete data with outliers. The effect of outliers is diminished by using the Student-t distribution for modelling the ob servation noise. We showed that using a model with independent elements of the noise vector can be more appropriate for some real-world datasets. We tested the proposed method on a real-world weather dataset and compared our approach with the probabilistic PCA model [5] and robust PPCA assuming fully corrupted outlier vectors [7]. The experiment showed the superior performance of the presented model, which found meaningful spatial patterns for the principal components and provided reasonable reconstruction in the place of missing data.

The proposed algorithm is based on a probabilistic model and therefore it provides information about the uncertainty of the estimated parameters. The uncertainty information can be taken into account, for example, when the principal components are ordered according to the amount of explained data variance [11]. The model can easily be extended, for example, by taking into account the temporal structure of the data. This would result in better performance in the tasks of missing value reconstruction and outlier removal.

In our work, we use the proposed technique as a preprocessing step for further exploratory analysis of data. For example, one can investigate a principal subspace found for weather data in order to find meaningful weather patterns or to extract features which might be useful for statistical weather forecasts. This can be done, for example, by using rotation techniques closely related to ICA. We have earlier used this approach for analysis of global climate data [9].

References

1. Jolliffe, I.T.: Principal Component Analysis, 2nd edn. Springer, New York (2002)
2. Hyvärinen, A., Karhunen, J., Oja, E.: Independent Component Analysis. J. Wiley, Chichester (2001)
3. Bishop, C.M.: Pattern Recognition and Machine Learning (Information Science and Statistics). Springer, Secaucus (2006)
4. Cichocki, A., Amari, S.: Adaptive Blind Signal and Image Processing: Learning Algorithms and Applications. John Wiley & Sons, Inc., New York (2002)
5. Tipping, M.E., Bishop, C.M.: Probabilistic principal component analysis. Journal of the Royal Statistical Society Series B 61(3), 611–622 (1999)
6. Zhao, J., Jiang, Q.: Probabilistic PCA for t distributions. Neurocomputing 69, 2217–2226 (2006)
7. Archambeau, C., Delannay, N., Verleysen, M.: Robust probabilistic projections. In: Proceedings of the 23rd International Conference on Machine Learning (ICML 2006), pp. 33–40. ACM, New York (2007)
8. Gao, J.: Robust L1 principal component analysis and its Bayesian variational inference. Neural Computation 20(2), 555–572 (2008)
9. Ilin, A., Valpola, H., Oja, E.: Exploratory analysis of climate data using source separation methods. Neural Networks 19(2), 155–167 (2006)
10. Bishop, C.M.: Variational principal components. In: Proceedings of the 9th International Conference on Artificial Neural Networks (ICANN 1999), vol. 1, pp. 509–514 (1999)
11. Ilin, A., Raiko, T.: Practical approaches to principal component analysis in the presence of missing values. Technical Report TKK-ICS-R6, Helsinki University of Technology, Espoo, Finland (2008), http://www.cis.hut.fi/alexilin/
12. Liu, C., Rubin, D.B.: ML estimation of the t distribution using EM and its extensions, ECM and ECME. Statistica Sinica 5, 19–39 (1995)

Appendix: Update Rules

$q(\boldsymbol{x}_n) = \mathcal{N}(\boldsymbol{x}_n|\overline{\boldsymbol{x}}_n, \boldsymbol{\Sigma}_{\boldsymbol{x}_n})$, $q(\boldsymbol{w}_m) = \mathcal{N}(\boldsymbol{w}_m|\overline{\boldsymbol{w}}_m, \boldsymbol{\Sigma}_{\boldsymbol{w}_m})$ and $q(\boldsymbol{\mu}) = \mathcal{N}(\mu_m|\widetilde{\overline{\mu}}_m, \widetilde{\mu}_m)$ are Gaussian density functions updated as follows:

$$\boldsymbol{\Sigma}_{\boldsymbol{x}_n}^{-1} = \boldsymbol{I} + \sum_{m|\mathcal{O}_{mn}} \langle\tau_m\rangle\langle u_{mn}\rangle(\overline{\boldsymbol{w}}_m\overline{\boldsymbol{w}}_m^{\mathrm{T}} + \boldsymbol{\Sigma}_{\boldsymbol{w}_m})$$

$$\overline{\boldsymbol{x}}_n = \boldsymbol{\Sigma}_{\boldsymbol{x}_n} \sum_{m|\mathcal{O}_{mn}} \langle\tau_m\rangle\langle u_{mn}\rangle\overline{\boldsymbol{w}}_m(y_{mn} - \overline{\mu}_m)$$

$$\boldsymbol{\Sigma}_{\boldsymbol{w}_m}^{-1} = \mathrm{diag}\,\langle\boldsymbol{\alpha}\rangle + \langle\tau_m\rangle \sum_{n|\mathcal{O}_{mn}} \langle u_{mn}\rangle(\overline{\boldsymbol{x}}_n\overline{\boldsymbol{x}}_n^{\mathrm{T}} + \boldsymbol{\Sigma}_{\boldsymbol{x}_n})$$

$$\overline{\boldsymbol{w}}_m = \boldsymbol{\Sigma}_{\boldsymbol{w}_m}\langle\tau_m\rangle \sum_{n|\mathcal{O}_{mn}} \langle u_{mn}\rangle\overline{\boldsymbol{x}}_n(y_{mn} - \overline{\mu}_m)$$

$$\widetilde{\mu}_m^{-1} = \beta + \langle\tau_m\rangle \sum_{n|\mathcal{O}_{mn}} \langle u_{mn}\rangle$$

$$\overline{\mu}_m = \widetilde{\mu}_m\langle\tau_m\rangle \sum_{n|\mathcal{O}_{mn}} \langle u_{mn}\rangle \left(y_{mn} - \overline{\boldsymbol{w}}_m^{\mathrm{T}}\overline{\boldsymbol{x}}_n\right)$$

where $\langle\cdot\rangle$ denotes expectations over the approximate distribution.

Approximate $q(\tau_m) = \mathcal{G}(\tau_m|\breve{a}_{\tau_m}, \breve{b}_{\tau_m})$, $q(u_{mn}) = \mathcal{G}(u_{mn}|\breve{a}_{u_{mn}}, \breve{b}_{u_{mn}})$ and $q(\alpha_d) = \mathcal{G}(\alpha_d|\breve{a}_{\boldsymbol{\alpha}}, \breve{b}_{\alpha_d})$ are Gamma density functions updated as follows:

$$\breve{a}_{\tau_m} = a_{\boldsymbol{\tau}} + \tfrac{N_m}{2} \qquad \breve{b}_{\tau_m} = b_{\boldsymbol{\tau}} + \tfrac{1}{2} \sum_{n|\mathcal{O}_{mn}} \langle u_{mn}\rangle(e_{mn}^2 + \widetilde{\mu}_m + \widetilde{\xi}_{mn})$$

$$\breve{a}_{u_{mn}} = \tfrac{\nu_m}{2} + \tfrac{1}{2} \qquad \breve{b}_{u_{mn}} = \tfrac{\nu_m}{2} + \tfrac{1}{2}\langle\tau_m\rangle(e_{mn}^2 + \widetilde{\mu}_m + \widetilde{\xi}_{mn})$$

$$\breve{a}_{\boldsymbol{\alpha}} = a_{\boldsymbol{\alpha}} + \tfrac{M}{2} \qquad \breve{b}_{\alpha_d} = b_{\boldsymbol{\alpha}} + \tfrac{1}{2} \sum_{m=1}^{M} \langle w_{md}^2\rangle$$

where $\breve{a}_{u_{mn}}$ and $\breve{b}_{u_{mn}}$ are estimated only for observed y_{mn}, N_m denotes the number of observed values in the set $\{y_{mn}\}_{n=1}^N$, while e_{mn} and $\widetilde{\xi}_{mn}$ are shorthand notations for

$$e_{mn} = y_{mn} - \overline{\boldsymbol{w}}_m^{\mathrm{T}}\overline{\boldsymbol{x}}_n - \overline{\mu}_m$$

$$\widetilde{\xi}_{mn} = \overline{\boldsymbol{w}}_m^{\mathrm{T}}\boldsymbol{\Sigma}_{\boldsymbol{x}_n}\overline{\boldsymbol{w}}_m + \overline{\boldsymbol{x}}_n^{\mathrm{T}}\boldsymbol{\Sigma}_{\boldsymbol{w}_m}\overline{\boldsymbol{x}}_n + \mathrm{tr}(\boldsymbol{\Sigma}_{\boldsymbol{w}_m}\boldsymbol{\Sigma}_{\boldsymbol{x}_n})\,.$$

The degrees of freedom $\boldsymbol{\nu}$ are point-estimated in order to keep the posterior approximation analytically tractable. The maximum likelihood estimate is found by maximising the lower bound of the model loglikelihood. This yields

$$1 + \log\left(\tfrac{\nu_m}{2}\right) - \psi\left(\tfrac{\nu_m}{2}\right) + \tfrac{1}{N_m} \sum_{n|\mathcal{O}_{mn}} (\langle\log u_{mn}\rangle - \langle u_{mn}\rangle) = 0\,,$$

which can be solved using line search methods. One may try to start updating the hyperparameters $\boldsymbol{\alpha}$ and $\boldsymbol{\nu}$ after the iteration has already run for some time if the algorithm seems to converge to bad local optimum.

Independent Component Analysis (ICA) Using Pearsonian Density Function

Abhijit Mandal[1] and Arnab Chakraborty[2]

[1] Indian Statistical Institute, Kolkata 700108, India
abhi_r@isical.ac.in
[2] Statistical Consultant, Kolkata 700108, India
arnabc@stanfordalumni.org

Abstract. Independent component analysis (ICA) is an important topic of signal processing and neural network which transforms an observed multidimensional random vector into components that are mutually as independent as possible. In this paper, we have introduced a new method called *SwiPe-ICA* (Stepwise Pearsonian ICA) that combines the methodology of projection pursuit with Pearsonian density estimation. Pearsonian density function instead of the classical polynomial density expansions is employed to approximate the density along each one-dimensional projection using differential entropy. This approximation of entropy is more exact than the classical approximation based on the polynomial density expansions when the source signals are supergaussian. The validity of the new algorithm is verified by computer simulation.

Keywords: Independent component analysis, Pearsonian density function, blind signal separation.

1 Introduction

The ICA was started to solve the *blind source separation* [12] problem, *i.e.* given only the mixtures of a set of underlying signal sources, the aim is to retrieve the original signals, having unknown but independent distributions. Unlike Principal Component Analysis (PCA), which only decorrelates the data, ICA searches for that direction in the data-space, which are independent across statistical moments of all orders [1,8].

The problem of blind signal separation arises in many areas such as telecommunication, economic indicators, digital images and medical science. The references at the end list pointers to the vast literature.

In this paper we shall use Pearsonian density function for estimating density. This function has already been used for ICA [7]. However, our approach is completely different in the sense that our method strongly relies on stepwise entropy minimization.

T. Adali et al. (Eds.): ICA 2009, LNCS 5441, pp. 74–81, 2009.

2   ICA Model

Let $\mathbf{x} = (x_1, x_2, \cdots, x_m)^T$ be the observed random vector, and $\mathbf{s} = (s_1, s_2, \cdots, s_p)^T$ be the latent random vector with mutually independent components (IC). The ICA model assumes that

$$\mathbf{x} = \mathbf{As} , \tag{1}$$

where $\mathbf{A}$ is a constant $m \times p$ matrix which is unknown. We assume that $\mathbf{A}$ is of full column rank, *i.e.* the columns of $\mathbf{A}$ are linearly independent. We make a further assumption that the dimension of $\mathbf{x}$ and $\mathbf{s}$ are equal, *i.e.* $m = p$. So $\mathbf{A}$ is assumed to be non-singular. Non-Gaussianity of the ICs is necessary for identifiability of the model (1) [1].

Suppose $\mathbf{x}(t)$, $t = 1, 2, \cdots, n$, are the observed values of $\mathbf{x}$. Let us define $\mathbf{X}_{p \times n} = [\mathbf{x}(1), ..., \mathbf{x}(n)]$. Similarly, we have $\mathbf{s}(t)$'s and $\mathbf{S}_{p \times n}$. Then our model can be written as

$$\mathbf{X} = \mathbf{AS} . \tag{2}$$

Now the problem is to estimate the (weight) matrix $\mathbf{W}$ so that the components of

$$\hat{\mathbf{S}} = \mathbf{WX} \tag{3}$$

are statistically as independent among each other as possible.

3   Entropy and Projection Pursuit

Comon [12] has showed a general formulation for ICA using the entropy function. The differential entropy of a random vector $\mathbf{y} = (y_1, y_2, \cdots, y_n)^T$ is defined as

$$\mathbf{H}(\mathbf{y}) = -\int f(\mathbf{y}) \log f(\mathbf{y}) d\mathbf{y} , \tag{4}$$

where $f(\cdot)$ is the density of $\mathbf{y}$.

Negentropy, defined as $\mathbf{J}(\mathbf{y}) = \mathbf{H}(\mathbf{z}) - \mathbf{H}(\mathbf{y})$, is invariant under linear transformations. Here $\mathbf{z}$ is a Gaussian random vector with the same covariance matrix as $\mathbf{y}$. Negentropy is a measure of nongaussianity [4] of a random vector. Now we define the mutual information $\mathbf{I}$ among n uncorrelated random variables of a random vector $\mathbf{y}$ as

$$\mathbf{I}(y_1, y_2, \cdots, y_n) = \mathbf{J}(\mathbf{y}) - \sum_i \mathbf{J}(y_i) , \tag{5}$$

which is a information-theoretic measure of the independence among the random variables. Hence $\mathbf{W}$ in (3) can be determined so that the mutual information of the transformed components, $\hat{s}_i$, is minimized. It can be noted that the source signals are identifiable up to scale and permutation.

By (5), finding an invertible transformation $\mathbf{W}$ that minimizes the mutual information is roughly equivalent to finding direction in which the negentropy is maximized. This formulation of ICA also shows explicitly the connection between ICA and projection pursuit [13].

4   Approximating the Entropy Function

The estimation of entropy is very difficult in practice. Using the definition in (4) requires estimation of the density of $\mathbf{x}$, which is theoretically difficult and computationally demanding. Simpler approximations of entropy have been proposed both in the context of projection pursuit [11,13] and independent component analysis [12,14]. These approximations are usually based on approximating the density $f(x)$ using the polynomial expansions of Gram-Charlier or Edgeworth. The construction leads to the use of higher order moments, which provide a poor approximation of entropy. Finite sample estimations of higher order moments are highly sensitive to outliers, *i.e.* their values depend on only a few, possibly erroneous, observations with large values. This means that outliers may completely determine the estimates of moments. Again, even if the moments were estimated perfectly, they measure mainly the tails of the distribution, and are largely unaffected by the structure near the center of the distribution [5,6].

Here we shall estimate the density $f(\cdot)$ assuming that it satisfies the Pearsonian [9,10] condition

$$\frac{df}{dx} = \frac{(x - \alpha)f(x)}{b_0 + b_1 x + b_2 x^2} .$$ (6)

The general solution of the differential equation is

$$f(x) = C \exp\left[\int \frac{(x - \alpha)dx}{b_0 + b_1 x + b_2 x^2}\right] .$$ (7)

By choosing C suitably we can make $f(x)$ a probability density. The key idea behind estimating density using Pearsonian curves lies in estimating the parameters α, b_0, b_1 and b_2 from the data. In case of zero mean and the unit variance the moment estimate of the parameters become

$$\begin{aligned}
\widehat{b}_1 &= \widehat{\alpha} = -\mu_3(\mu_4 + 3)/D , \\
\widehat{b}_0 &= -(4\mu_4 - 3\mu_3{}^2)/D , \\
\widehat{b}_2 &= -(2\mu_4 - 3\mu_3{}^2 - 6)/D ,
\end{aligned}$$ (8)

where $D = 2(5\mu_4 - 6\mu_3{}^2 - 9)$ and μ_i is the ith central moment.

So the estimate of the density function becomes

$$\widehat{f}(x) = C \exp\left[\int \frac{(x - \widehat{\alpha})dx}{\widehat{b}_0 + \widehat{b}_1 x + \widehat{b}_2 x^2}\right] ,$$ (9)

where C is a suitable constant. In the next section we will see that we do not need to calculate C in our algorithm.

Let $\mathbf{y} = (y_1, y_2, \cdots, y_n)^T$ be a random sample of size n from a random variable $\mathbf{Y}$. Then we will estimate the entropy of $\mathbf{Y}$ as

$$H(\mathbf{Y}) = -\frac{1}{n}\sum_{i=1}^{n} \log \widehat{f}(y_i) .$$ (10)

Let us define

$$h(x) = \frac{x - \widehat{\alpha}}{\widehat{b}_0 + \widehat{b}_1 x + \widehat{b}_2 x^2} \ , \tag{11}$$

and

$$g(x) = \int h(x)dx = \log \hat{f}(x) - \log C \ . \tag{12}$$

So the entropy of $\mathbf{Y}$ in (10) can be written as

$$H(\mathbf{Y}) = -\frac{1}{n}\sum_{i=1}^{n} g(y_i) - \log C \ . \tag{13}$$

5 *SwiPe-ICA*

In this section we present a new algorithm called Stepwise Pearsonian ICA (*SwiPe-ICA*) which retains the stepwise nature of projection pursuit, and yet benefits from Pearsonian density estimation paradigm. The algorithm finds $\mathbf{W}$ row by row from the (centered and whitened) input mixture signals. To find each row we project the data along various directions orthogonal to the rows already found. The best direction is chosen to minimize the estimated entropy along that projection. While this skeletal structure is very similar to that of FastICA [2], *SwiPe-ICA* employs a completely different method for estimating the entropy.

To find the jth direction $\mathbf{a}_j$ we minimize $H(\mathbf{a}^{\mathbf{T}}\mathbf{X})$ or equivalently maximize $-nH(\mathbf{a}^{\mathbf{T}}\mathbf{X})$ with respect to $\mathbf{a}$ subject to the conditions

$$\mathbf{a}^T \mathbf{a} = 1 \text{ and } \mathbf{a}^T \mathbf{a}_k = 0 \ \ \forall k = 1, \cdots, j - 1, \tag{14}$$

where $\mathbf{a}_1, \cdots, \mathbf{a}_{j-1}$ are the directions already found. The Lagrangian multiplier method and Newton-Raphson iterations are used for this constrained maximization. At the jth step we will maximize

$$\phi_j(\mathbf{a}, \boldsymbol{\lambda}) = -nH(\mathbf{a}^T\mathbf{X}) + \lambda_j(\mathbf{a}^T\mathbf{a} - 1) + \sum_{k=1}^{j-1} \lambda_k \mathbf{a}^T \mathbf{a}_k \tag{15}$$

with respect to $\mathbf{a}$ and $\boldsymbol{\lambda} = (\lambda_1, \lambda_2, \cdots, \lambda_j)^T$, where $\boldsymbol{\lambda}$ is the vector of Lagrange multipliers. For simplicity we assume that $\partial\theta/\partial\mathbf{a}$ is negligible for the parameters $\theta = b_0, b_1, b_2$ and C (this assumption is supported by simulation results).

So at the jth step the iterations are applied to the equation $\mathbf{S}_j(\mathbf{a}, \lambda) = 0$, where

$$\mathbf{S}_j(\mathbf{a}, \lambda) = \begin{pmatrix} \partial\phi_j/\partial\mathbf{a} \\ \partial\phi_j/\partial\lambda_1 \\ \partial\phi_j/\partial\lambda_2 \\ \vdots \\ \partial\phi_j/\partial\lambda_{j-1} \\ \partial\phi_j/\partial\lambda_j \end{pmatrix} = \begin{pmatrix} \sum_{i=1}^{n} h(\mathbf{a}^T\mathbf{x}_i)\mathbf{x}_i + 2\lambda_j\mathbf{a} + \sum_{k=1}^{j-1} \lambda_k\mathbf{a}_k \\ \mathbf{a}_1^T\mathbf{a} \\ \mathbf{a}_2^T\mathbf{a} \\ \vdots \\ \mathbf{a}_{j-1}^T\mathbf{a} \\ \mathbf{a}^T\mathbf{a} - 1 \end{pmatrix} \ . \tag{16}$$

The $(r+1)^{th}$ iteration of the Newton-Raphson becomes

$$\mathbf{y}_j(r+1) = \mathbf{y}_j(r) - \boldsymbol{F}_j^{-1}(r)\mathbf{S}_j(r) , \qquad (17)$$

where $\mathbf{y_j}(r)$ and $\mathbf{S}_j(r)$ are the rth iteration of $(\mathbf{a}, \lambda_1, \lambda_2, \cdots, \lambda_j)^T$ and $\mathbf{S_j}(\mathbf{a}, \lambda)$ respectively, and $\boldsymbol{F}_j(r)$ are the rth iteration of

$$\boldsymbol{F}_j = \begin{pmatrix} \partial^2\phi_j/\partial\mathbf{a}^2 & \partial^2\phi_j/\partial\mathbf{a}\partial\lambda_1 & \cdots\cdots & \partial^2\phi_j/\partial\mathbf{a}\partial\lambda_j \\ \partial^2\phi_j/\partial\lambda_1\partial\mathbf{a} & \partial^2\phi_j/\partial\lambda_1^2 & \cdots\cdots & \partial^2\phi_j/\partial\lambda_1\partial\lambda_j \\ \vdots & \vdots & \vdots\ \vdots & \vdots \\ \partial^2\phi_j/\partial\lambda_j\partial\mathbf{a} & \partial^2\phi_j/\partial\lambda_j\partial\lambda_1 & \cdots\cdots & \partial^2\phi_j/\partial\lambda_j^2 \end{pmatrix} \qquad (18)$$

$$= \begin{pmatrix} \sum_{i=1}^n h'(\mathbf{a}^T\mathbf{x}_i)\mathbf{x}_i\mathbf{x}_i^T + 2\lambda_j I_p & \boldsymbol{\beta}^T \\ \boldsymbol{\beta} & O_{j\times j} \end{pmatrix} , \qquad (19)$$

where $\boldsymbol{\beta} = (\mathbf{a}_1, \mathbf{a}_2, \cdots, \cdots, \mathbf{a}_{j-1}, 2\mathbf{a})^T$, $O_{j\times j}$ is a zero matrix of order $j \times j$ and $h'(x) = \frac{d}{dx}h(x)$. Here we can simplify the computation of $\boldsymbol{F}_j^{-1}$ using the block structure of $\boldsymbol{F}_j$.

6   Results

6.1   Performance of the Pearsonian ICA in Audio Data

In this section we shall see how the Pearsonian ICA algorithm performs for an audio data set. The algorithm has been implemented in the Matlab programming language. The sounds that we have used are that of a car siren, a English news program in a television channel and a sound of piano. Each sound track is about 5 seconds long and contains 50000 sample points. Thus, $\mathbf{S}$ in (2) is of order 3×50000. The mixing matrix $\mathbf{A}$ is chosen randomly, where each of the 9 entries of $\mathbf{A}$ are independent and identically distributed (i.i.d.) uniform $(0,1)$.

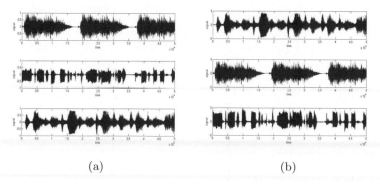

(a)   (b)

Fig. 1. The time series plots of the original (a) and the recovered signals (b)

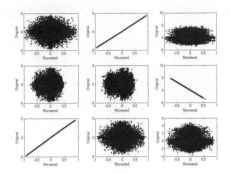

Fig. 2. The scatter plots of the original signals against the recovered signals

From figure (2) we can see that there is one and only one diagonal (approximately) plot in every row and column. The other plots show data clouds without much definite pattern. This implies that each recovered signal has high correlation with its corresponding source signal, while it is independent of the other source signals.

As the original sources are known one can use *signal-to-interface ratio* (SIR) [3] as a measure of performance of the ICA method, which is defined as

$$SIR(dB) = \frac{1}{n} \sum_{i=1}^{n} 10 \log_{10} \frac{max(\boldsymbol{P}_i)^2}{\boldsymbol{P}_i^T \boldsymbol{P}_i - max(\boldsymbol{P}_i)^2} \ , \tag{20}$$

where $\boldsymbol{P} = \mathbf{WA}$, $\boldsymbol{P}_i$ is the ith column of $\boldsymbol{P}$. The higher SIR is, the better the separation performance of the algorithm. Another commonly used index *cross-talking error* (CTE) [14], which is used to measure the accuracy of separating the independent components, is the distance between the overall transforming matrix $\boldsymbol{P}$ and an ideal permutation matrix (where every rows and columns contain exactly one element equal to unity and rest are zero). CTE is defined as

$$CTE = \sum_{i} \left(\sum_{j} \frac{|p_{ij}|}{max_{j'}|p_{ij'}|} - 1 \right) + \sum_{j} \left(\sum_{i} \frac{|p_{ij}|}{max_{i'}|p_{i'j}|} - 1 \right) \ , \tag{21}$$

where p_{ij} is the $(i, j)^{th}$ element of $\boldsymbol{P}$. The lower CTE is, the better the separation performance of the algorithm.

The values of these indexes in the above experiment are 49.6418 dB and 0.0274 respectively. The performance of *SwiPe-ICA* is not only much more better than the ICA methods based on the polynomial density expansions, it is also comparable with the FastICA method [2] particularly when the sources are supergaussian.

6.2 Comparison between Pearsonian ICA and FastICA

In the next experiment, we have used 3 artificially generated time series of 500 samples each. The analytical form along with their skewness and kurtosis are given in table 1, where the function $rem(x, y)$ denotes the remainder when x is divided by y, $rand(1, x)$ is an array of length x whose elements are random numbers from uniform $(0, 1)$, $I(\cdot)$ is an identity function and '$\otimes$' denotes the elementwise multiplication.

Table 1. The analytical form of signals along with their skewness and kurtosis

Source signal	Skewness	Kurtosis
Impulsive curve: $S_1(t) = ((rem(t, 23) - 11)/9)^5$	-0.011	2.353
Exponential decay: $S_2(t) = 5 \exp(-t/121)cos(37t)$	0.055	3.410
Spiky noise: $S_3(t) = (2I(rand(1, 500) < .5) - 1) \otimes \log(rand(1, 500))$	0.464	2.228

In this experiment we have used the FastICA algorithm taking different nonlinear contrast functions and compared their performances with our proposed algorithm by the indexes SIR and CTE. Each algorithm was repeated 500 times and the 9 entries of the mixing matrix were chosen randomly form uniform $(0, 1)$. The results are shown in the following table, where the median of CTE and SIR along with their first and third quantiles in brackets are mentioned.

Table 2. CTE and SIR of *SwiPe-ICA* and FastICA

Algorithm	CTE	SIR(dB)
SwiPe-ICA	0.3582 (0.2722, 0.4882)	27.9586 (24.7591, 30.5289)
FastICA with $g(u) = u^3$	0.7131 (0.5546, 1.0127)	21.1239 (18.4048, 23.6924)
FastICA with $g(u) = tanh(u)$	0.4226 (0.3325, 0.5601)	26.4373 (23.5400, 28.6691)
FastICA with $g(u) = u \exp(u^2/2)$	0.3813 (0.2894, 0.5155)	27.3453 (24.2308, 29.7747)

7 Some Insights behind the Efficiency of *SwiPe-ICA*

Although *SwiPe-ICA* is based on sample moments, Pearsonian density estimates are known to be "exceptionally accurate" for long tailed distributions [6]. So *SwiPe-ICA* has a significant advantage over other moment-based methods for supergaussian sources.

ICA does not require estimation of the source densities *per se*, rather it needs to estimate only the score function involving the derivatives of the densities. *SwiPe-ICA* avoids the instability of approximate differentiation by directly estimating the opposite of the score function and its derivative.

Pearsonian density estimation is somewhat complicated [6]. But since *SwiPe-ICA* works with the derivatives of the densities, much of the complexity is avoided.

8   Conclusion

In this paper we have proposed a new method for ICA involving density estimation by Pearsonian curves. The reason for using Pearsonian density function is that it gives a good estimate of the density function and the algorithm performs very well when the sources signals are supergaussian. This algorithm also gives good result when at most one of the sources is subgaussian, but simulations show that it is not so efficient in separating two subgaussian sources.

References

1. Hyvärinen, A.: Survey on independent component analysis. Neural Computing Surveys 2(94), 94–128 (1999)
2. Hyvärinen, A., Oja, E.: Independent component analysis: Algorithms and applications. Neural Networks 13(4), 411–430 (2000)
3. Ziehe, A., Muller, K.R.: TDSEP–an efficient algorithm for blind separation using time structure. In: Proc. Int. Conf. on Artificial Neural Networks (ICANN 1998), pp. 675–680 (1998)
4. Bell, A.J., Sejnowski, T.J.: An information-maximization approach to blind separation and blind deconvolution. Neural Computation 7, 1129–1159 (1995)
5. Pearson, E.S.: Some problems arising in approximating to probability distributions, using moments. Biometrika 50(1), 95–112 (1963)
6. Solomon, H., Stephens, M.A.: Approximations to density functions using pearson curves. Journal of the American Statistical Association 73(361), 153–160 (1978)
7. Karvanen, J., Eriksson, J., Koivunen, V.: Pearson System Based Method for Blind Separation. In: Proc. of 2nd International Workshop on Independent Component Analysis and Blind Signal Separation, pp. 585–590 (2000)
8. Cardoso, J.F., Comon, P.: Independent component analysis, a survey of some algebraic methods. In: 'Connecting the World' 1996 IEEE International Symposium on Circuits and Systems, ISCAS 1996, vol. 2 (1996)
9. Pearson, K.: Mathematical Contributions to the Theory of Evolution. XIX. Second Supplement to a Memoir on Skew Variation. Philosophical Transactions of the Royal Society of London. Series A, Containing Papers of a Mathematical or Physical Character 216, 429–457 (1916)
10. Kendall, M., Stuart, A.: The advanced theory of statistics, 4th edn. Distribution theory, vol. 1. Griffin, London (1977)
11. Jomes, M.C., Sibson, R.: What is projection pursuit? J. of the Royal Statistical Society, Ser. A 150, 1–36 (1987)
12. Comon, P.: Independent component analysis: A new concept? Signal Processing 36(3), 287–314 (1994)
13. Huber, P.J.: Projection pursuit. The Annals of Statistics 13(2), 435–475 (1985)
14. Amari, S., Cichocki, A., Yang, H.: A new learning algorithm for blind signal separation. Advances in neural information processing systems 8, 757–763 (1996)

An Analysis of Unsupervised Signal Processing Methods in the Context of Correlated Sources

Aline Neves[1,4], Cristina Wada[2,4], Ricardo Suyama[3,4], Romis Attux[2,4],
and João M.T. Romano[3,4]

[1] Engineering, Modeling and Applied Social Sciences Center, UFABC, Brazil
[2] Department of Computer Engineering and Industrial Automation
[3] Department of Microwave and Optics
[4] Laboratory of Signal Processing for Communications, UNICAMP, Brazil
aline.neves@ufabc.edu.br, {criswada,attux}@dca.fee.unicamp.br,
{suyama,romano}@dmo.fee.unicamp.br

Abstract. In light of the recently proposed generalized correlation function named correntropy, which exploits higher-order statistics and the time structure of signals, we have, in this work, two main objectives: 1) to give a new interpretation – founded on the relationships between the constant modulus (CM) and Shalvi-Weinstein criteria and between the latter and methods for ICA based on nongaussianity – to the performance of the constant modulus approach under dependent sources and 2) to analyze the correntropy in the context of blind deconvolution of i.i.d. and dependent sources, as well as to establish elements of a comparison between it and the CMA. The analyses and simulation results unveil some theoretical aspects hitherto unexplored.

Keywords: Blind deconvolution, correntropy, constant modulus criterion, Shalvi-Weinstein criteria, nongaussianity-based ICA.

1 Introduction

Interest in blind deconvolution (or equalization) techniques comes from the fact that their operation depends only on the channel output and some statistical characteristics of the transmitted signal. Considering classical techniques such as the constant modulus (CM) [1] or Shalvi-Weinstein (SW) [2] criteria, a good performance is guaranteed when the transmitted signal is composed of independent samples. However, when we consider, for example, the use of error-correcting codes, this condition is not satisfied, and the algorithms may present a non-satisfactory performance [3].

Recently, a new approach derived from the field of information-theoretic learning was proposed in a blind equalization context, taking into account the temporal structure of the transmitted signal. This measure, named correntropy, is conceived to deal with the presence of correlated sources, relaxing the restriction that was generally present when blind techniques were discussed [4].

In view of this new situation, it is interesting to note that much remains to be done in terms of analyzing classical blind equalization criteria, such as the CM approach,

T. Adali et al. (Eds.): ICA 2009, LNCS 5441, pp. 82–89, 2009.
© Springer-Verlag Berlin Heidelberg 2009

when a correlated signal is transmitted. Even with the development of blind source separation techniques, which enabled a new interpretation of blind equalization methods [5], the problem was not completely solved. In [6,7], the following conclusion is reached: the existence of dependence between source samples may change the positions of the CM error surface minima. Depending on the focused source, these minima can be associated with closed-eye situations even when exact channel inversion is possible. However, a complete explanation of such phenomenon was not provided.

In this paper, we propose a new interpretation of the performance of the CM criterion when applied to the recovery of correlated sources. This interpretation, which is founded on the equivalence between the CM and SW criteria and on the relationship between kurtosis and nongaussianity-based ICA, is tested in a number of representative situations. Afterwards, we present elements of a comparison between the CM algorithm (CMA) and a correntropy-based method for deconvolution of correlated and uncorrelated sources, emphasizing some pertinent aspects that were not discussed in [4]. These analyses provide theoretical elements to the blind equalization and blind source separation theories.

2 The Constant Modulus Criterion

The constant modulus (CM) criterion [1,7,8] is based on the idea of minimizing a dispersion of the absolute value of the equalizer output around a fixed value that depends on statistics of the transmitted signal, which gives rise to the following cost function:

$$J_{CM} = E\left[\left(|y(n)|^2 - R_2\right)^2\right], \text{ where } R_2 = \frac{E\left[|s(n)|^4\right]}{E\left[|s(n)|^2\right]}, \tag{1}$$

$y(n)$ is the equalizer output signal and $s(n)$ is the transmitted signal. The associated stochastic gradient search method, the constant modulus algorithm (CMA) [1,8], is given by:

$$\mathbf{w}_{n+1} = \mathbf{w}_n - \mu y_n \mathbf{x}_n \left(|y_n|^2 - R_2\right), \tag{2}$$

where $\mathbf{w}=[w_0\ w_1\ ...\ w_{L-1}]^T$ is the equalizer tap-weight vector, with length L, $\mathbf{x}_n=[x_n\ x_{n-1}\ ...\ x_{n-L+1}]^T$ is the equalizer input vector and μ is the step-size.

Analysis of the structure of (1) and of the dynamical behavior of (2) has been carried out considering, as a rule, that the transmitted signal is composed of independent and identically distributed (i.i.d.) samples. This assumption is valid in certain applications, but not in others, e.g. those including error-correcting codes or even those associated with certain nondigital signals.

Previous efforts to analyze the CM criterion for dependent data [7, 3] reached an important conclusion: dependent sources lead to modifications in the CM cost function, which, in some cases, give rise to minima that are inadequate from the standpoint of channel inversion [6]. Having this state of things in view, we shall, in section 4, propose an interpretation of the behavior of the constant modulus criterion under correlated sources that is based on two results: the equivalence between the CM and SW criteria and the relationship between the SW criterion and the idea of nongaussianity, which is a pillar of the independent component analysis (ICA).

3   Information-Theoretic Learning and Correntropy

Information-theoretic learning (ITL) [10] constitutes a relevant step towards an adaptive filtering framework effectively capable of making use of the potential brought by nonlinear structures and/or higher-order statistics. Among the techniques employed to perform ITL, those based on kernel methods have received a great deal of attention in the last years [11]. A consequence of these important efforts was the proposal of a generalized correlation function to which the name *correntropy* was associated [4]. This correlation function can be understood as a dependence measure that attempts to incorporate the benefits of information-theoretic entities.

It is possible to define correntropy as:

$$\hat{V}[m] = \frac{1}{N-m+1}\sum_{n=m}^{N} k\left(x_n - x_{n-m}\right), \tag{3}$$

where $k(.)$ denotes a kernel function, N is the size of the data window used to estimate correntropy and m is the lag being considered.

In [4], correntropy is employed to solve the blind equalization problem under the hypothesis that the information source is subject to some sort of coding. In such case, the source samples are temporally dependent, and the idea is exactly to incorporate the structure of this dependence in the equalization criterion. The used criterion is stated as follows:

$$J_c = \sum_{m=1}^{P}\left(V_s[m] - V_y[m]\right)^2, \tag{4}$$

where V_s is the correntropy of the source, V_y is the correntropy of the equalizer output and P is the number of lags. The associated adaptive algorithm is given by

$$\mathbf{w}_{n+1} = \mathbf{w}_n + \mu\sum_{m=1}^{P}\left(V_s[m] - V_y[m]\right)\frac{\partial V_y[m]}{\partial \mathbf{w}}. \tag{5}$$

To the best of our knowledge, a systematic analysis of the convergence of this technique under correlated and uncorrelated sources, including elements of comparison with more classical blind approaches, has not yet been performed. An objective of this work is to take a first step in this direction, as will be shown in section 5.

4   An Analysis of the Constant Modulus Criterion with Correlated Sources

As discussed in section 2, the problem of studying the CM criterion in the context of "non-i.i.d." sources is of great theoretical and practical relevance. We will now attempt to provide an explanation for the performance degradation verified in such context based on two pillars: the equivalence between the CM and SW criteria as established by Regalia [9] and the conceptual link between the SW approach and the idea of nongaussianity-based ICA.

In [9], Regalia proved that the CM and SW criteria have equivalent stationary points. Interestingly, an analysis of the SW theorem [2] from an ICA-oriented perspective reveals a relationship between the notion of describing zero-forcing (ZF) solutions in terms of an equality between the absolute values of the equalizer output and source kurtosis (under a power constraint) and the idea of seeking a condition that be as opposite as possible to the gaussianizing effect of the convolution [11]. When ZF solutions are not attainable, the family of SW criteria attempts to maximize the absolute value of the output kurtosis, in a clear parallel with the notion of seeking the recovery of a component that be as nongaussian as possible [5], i.e., that be "as close as possible", in some sense, to an underlying independent source.

When the transmitted signal is i.i.d., the entire line of reasoning makes sense, because this search for recovering an independent component will lead, at least approximately, to channel inversion. On the other hand, when the transmitted signal is no longer formed by i.i.d. samples, we are faced with an important question: will it be possible to maintain the link between the idea of inversion, which is the very essence of the equalization task, and the notion of seeking a nongaussian solution, which forms the core of the SW criteria?

Our approach to analyzing this problem follows exactly the outlined path: the use of the CM criterion is, implicitly, related to the idea of seeking the maximization of a measure of nongaussianity, which is related to the notion of estimating an independent component of a convolutional superposition. Therefore, when the CM is applied in the context of dependent sources, it should, in a certain sense, continue to approximate an independent component: therefore, it is possible that it attempts to equalize, when this is feasible, even the model that originates the sample dependence, which means that the obtained solutions will not be exclusively concerned with channel inversion, but will also "interpret" the source dependence as a mixing effect to be counterbalanced. In order to test the validity of this simple hypothesis, we shall now turn our attention to two representative scenarios.

4.1 First Scenario

In the first scenario, we assume that dependence between samples of the transmitted signal is generated by a linear finite impulse response (FIR) system $h(z)=1+0.4z^{-1}$ that acts as a stochastic process model or as a precoder. The input to this precoder is a sequence of i.i.d. +1/-1 samples. The channel is assumed to be $h(z)=1+0.6z^{-1}$.

In Fig. 1a, we present the contours of the CM cost function for two situations: when there is no precoder and when the precoder is applied. These contours reveal that the existence of dependence between samples indeed modifies the position of the minima, which gives support to the results reported in [7]. In accordance with the proposed interpretation, this modification comes from the fact that the CM solutions are attempting to equalize the combined precoder + channel impulse response. In order to test the validity of this idea, we present, in Fig. 1b, the contours of the CM cost function for two situations: dependent source + channel and i.i.d. source + precoder / channel cascade. The contours are identical, which reveals, in consonance with our line of reasoning, that both equalization problems are equivalent. It is important to remark that, in order to have a perfect superposition of both surfaces, it was necessary to apply a scaling factor to the precoder to make the constant R_2 equal in both cases. The constant

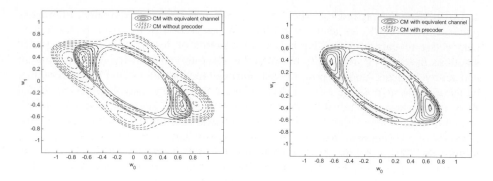

Fig. 1. Contours of CM cost function: (a) i.i.d and correlated sources, $h(z)=1+0.6z^{-1}$ (b) i.i.d source with $h(z)=1+1z^{-1}+0.24z^{-2}$ and correlated source with $h(z)=1+0.6z^{-1}$

influences only the magnitude of the equalizer parameter vector, there being no impact on the performance in terms of intersymbol interference removal.

4.2   Second Scenario

The second scenario is employed to illustrate an important point: although it is usual to speak of "correlated sources", when blind criteria are considered, the most relevant aspect is statistical dependence. The precoder, in this case, is an pass-filter with transfer function $h(z)=(1+2z^{-1})/(1+0.5z^{-1})$, which means that the generated source possesses uncorrelated - albeit not statistically independent – samples. Even with this white source, Fig. 2a reveals that there is a substantial modification in the contours of the CM cost function in comparison with the i.i.d. case. Furthermore, as shown in Fig. 2b, a comparison with the case in which an i.i.d. source is filtered with the combined precoder+channel response gives, once more, support to our interpretation: in view of the statistical dependence inherent to the source, the CM solutions will consider the all-pass precoder as additional system to be inverted.

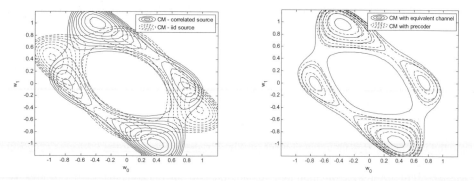

Fig. 2. Comparison of CM cost function for an i.i.d. source and a white but non-independent source, (a) same channel, $h(z)=1+0.6z^{-1}$ (b) i.i.d. source with equivalent channel (precoder+channel) and non-i.i.d. source with channel $h(z)=1+0.6z^{-1}$

5 CMA and Correntropy

As shown in section 3, correntropy is based on the expected value of a kernel function. Thus, while the CM criterion can be seen as a criterion that aims to recover the condition of independence between source samples, correntropy follows a "temporal version" of the pdf-matching (or, at least, higher-order moment-matching) philosophy that is similar, to a certain extent, to that associated with the Benveniste-Goursat-Ruget theorem [5]. The pdf can be estimated using Parzen's nonparametric approach, which leads to an efficient kernel method that, as shown in section 3, may exploit the time structure of the signal, in contrast with the standard CMA. In this section, we will compare and discuss the performance of both techniques when applied to the recovery of uncorrelated and dependent sources.

5.1 Uncorrelated Sources

The performance of the CMA when applied to the recovery of i.i.d. sources is a classical research topic [1,10]. Nevertheless, correntropy has not yet been analyzed in this context. Let us consider the transmission of a BPSK (*Binary Phase Shift Keying*) modulated signal. Following the definition given by (3), the theoretical correntropy function of this source is: $V_s[0]=k(0)$, and $V_s[m]=0.5k(0)+0.5k(2)$ for m≠0, where $k(.)$ is the Gaussian kernel.

Considering the simple FIR channel $h(z)=1+0.5z^{-1}$, Fig. 3a shows the performance obtained using CMA (2) and correntropy (5) for an average of 50 simulations with a signal to noise ratio (SNR) of 20dB. The CMA step-size parameter was 0.01, and correntropy was simulated considering P=5, N=100 and μ=0.3 (all choices were based on preliminary simulations). The equalizers had, in both cases, two taps initialized at **w** = $[1\ 0]^T$. It is interesting to observe that, in this case, CMA has a better performance. The algorithm converges faster, attains a lower value of ISI and oscillates much less than correntropy. In order to compare the points of minima of both criteria, Fig. 3b shows the contour curves of their error surfaces. Even though the surfaces present important differences, which is reflected, for instance, on their having completely different local minima, the global minima are close. Given these results, we conclude that, for this i.i.d. case, the CM approach was more efficient than the correntropy-based method in making an effective use of the higher order statistics of the involved signals.

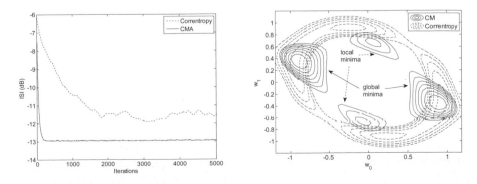

Fig. 3. Correntropy and CMA, $h(z)=1+0.5z^{-1}$, SNR=20dB (a) performance, (b) contour curves

Figure 4 considers the case of a white but not independent source. We use here the same example considered in section 4.2. In this case, correntropy is able to reduce some of the ISI but CMA is not, which reveals the importance of employing higher-order knowledge about the time structure of the transmitted signal. Correntropy was simulated with 5 taps, μ=0.08, P=5, N=100, while CMA had 7 taps and μ=0.0005.

5.2 Correlated Sources

Let us consider again a channel with transfer function $h(z)=1+0.5z^{-1}$. Fig. 5 shows the results obtained using an alternate mark inversion (AMI) source signal, instead of a BPSK modulated signal. The AMI signal is a correlated sequence drawn from the alphabet {-1,0,+1}, being its correntropy is given in [4]. Correntropy was simulated with P=5, N=100, μ=0.15, while CMA used a μ=0.005. Both filters had three coefficients.

In consonance with the second scenario shown in section 5.1, the correntropy-based method was the best solution. This can be explained once more by the fact that, in this case, only the correntropy is able to take into account the time structure of the signals: the CMA, as discussed in section 4, will treat the precoder as an additional distortion to be counterbalanced. The results reveal the importance of incorporating additional *a priori* information whenever one deals with "non i.i.d. sources". However, notice that this does not necessarily require the use of correntropy: it is possible to conceive a modified CMA to explore the statistical dependence.

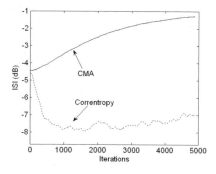

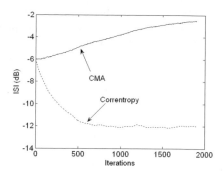

Fig. 4. Correntropy and CMA performance source codified by an all-pass filter

Fig. 5. Correntropy and CMA performance, AMI source, channel h=[1 0.5]

6 Conclusion

In this work, we provided a novel interpretation of the *modus operandi* of the constant modulus criterion under the hypothesis that the source has statistically dependent samples. The interpretation, which is founded on the equivalence between the CM and SW criteria and on nongaussianity-based ICA, allowed us to explain the modification in the position of the cost function minima (correspondent to a performance degradation) previously reported in the literature. We also presented an analysis of the correntropy for i.i.d. and statistically dependent sources. This analysis revealed that the CM approach employed the higher-order statistics of the involved signals in a more efficient way in the i.i.d. case. On the other hand, the *a priori* higher-order time structure information brought by the correntropy-based method is decisive when the sources are dependent.

There are many perspectives for future investigations, among which we highlight the realization of additional experiments to test the proposed interpretation of the CM criterion, an analysis of different kernels and parameters to improve the performance of the correntropy-based method under i.i.d. sources and the proposal of a modified CMA to include *a priori* information about the time structure of the relevant signals.

References

1. Godard, D.: Self recovering equalization and carrier tracking in two-dimensional data communication systems. IEEE Trans. on Communications 28(11), 1867–1875 (1980)
2. Shalvi, O., Weinstein, E.: New criteria for blind deconvolution of nonminimum phase systems (channels). IEEE Trans. on Information Theory 36(2), 312–321 (1990)
3. Axford, R., Milstein, L., Zeidler, J.: The effects of PN sequences on the misconvergence of the constant modulus algorithm (CMA). IEEE Trans. on Signal Processing 46(2), 519–523 (1998)
4. Santamaría, I., Pokharel, P., Principe, J.: Generalized correlation function: definition, properties, and application to blind equalization. IEEE Trans. on Signal Processing 54(6), 2187–2197 (2006)
5. Attux, R., Neves, A., Duarte, L., Suyama, R., Junqueira, C., Rangel, L., Dias, T., Romano, J.: On the relationship between blind equalization and blind source separation Part I: Foundations and Part II: Relationships. Journal of Communication and Information Systems (2009) (to appear)
6. Johnson, C., Schniter, P., Endres, T., Behm, J., Brown, D., Casas, R.: Blind equalization using the constant modulus criterion: a review. Proceedings of the IEEE 86(10), 1927–1950 (1998)
7. LeBlanc, J., Dogançay, K., Kennedy, R., Johnson, R.: Effects of input data correlation on the convergence of blind adaptive equalizers. IEEE International Conference on Acoustic, Speech and Signal Processing 3, 313–316 (1994)
8. Treichler, J., Agee, B.: New approach to multipath correction of constant modulus signals. IEEE Trans. on Acoustic, Speech and Signal Processing ASSP-31(2), 459–472 (1983)
9. Regalia, P.: On the equivalence between the Godard and Shalvi Weinstein schemes of blind equalization. Signal Processing 73(1-2), 185–190 (1999)
10. Principe, J., Xu, D., Fischer, J.: Information Theoretic Learning. In: Haykin, S. (ed.) Unsupervised Adaptive Filtering. Wiley, New York (2000)
11. Hyvarinen, A., Karhunen, J., Oja, E.: Independent Component analysis. Wiley Interscience, Hoboken (2001)

Probabilistic Formulation of Independent Vector Analysis Using Complex Gaussian Scale Mixtures

Jason A. Palmer[1,*], Ken Kreutz-Delgado[2], and Scott Makeig[1]

[1] Swartz Center for Computational Neuroscience
[2] Department of Electrical and Computer Engineering
University of California San Diego, La Jolla, CA 92093
{jason,scott}@sccn.ucsd.edu, kreutz@ece.ucsd.edu

Abstract. We propose a probabilistic model for the Independent Vector Analysis approach to blind deconvolution and derive an asymptotic Newton method to estimate the model by Maximum Likelihood.

1 Introduction

In this paper we propose a probabilistic model for the complex STFT coefficient vectors, which can be used in Independent Vector Analysis (IVA) [7] to perform blind deconvolution. We use adaptive source densities [11] that are dependent across frequencies [6,12] via complex Gaussian scale mixtures (CGSMs). These densities are shown to be the natural result of the sampling of Fourier coefficients of intermittent sources, in which a source is active only over a random fraction of each time window used to sample the DFT. We derive a Newton method for Maximum Likelihood estimation [2]. The IVA model was developed by Kim, Eltoft, Lee, and others [8,6,7].

Notation: We denote the imaginary number by $\jmath \triangleq \sqrt{-1}$.

2 Densities over Complex Vectors and STFT Coefficients

A probability density defined on a complex vector space is simply a joint density formulated over the real and imaginary parts of the vector. In particular, the probability density function is real valued. Integration over $\mathbf{C}^n$ is equivalent to integration over $\mathbf{R}^{2n}$. However, some real valued functions of complex vectors can be expressed in a simpler form in terms of the complex vectors $\mathbf{z} = \mathbf{x} + \jmath \mathbf{y}$ and $\mathbf{z}^* = \mathbf{x} - \jmath \mathbf{y}$, rather than the real and imaginary parts $\mathbf{x}$ and $\mathbf{y}$. And optimization of the real valued function may be more conveniently carried out in the complex space $\mathbf{C}^n$ using the Wirtinger calculus, instead of working in $\mathbf{R}^{2n}$ [9].

Fourier coefficients of a stationary time series are complex valued random variables. Stationarity implies that the covariance matrix of the real and imaginary parts has a particular form. Let $\mathbf{x}$ be an n dimensional complex valued random vector. Then if the real-valued random vectors $\mathbf{x}_R$, and $\mathbf{x}_I$ are jointly distributed as,

$$\begin{bmatrix} \mathbf{x}_R \\ \mathbf{x}_I \end{bmatrix} \sim \mathcal{N} \left(\begin{bmatrix} \mathbf{c}_R \\ \mathbf{c}_I \end{bmatrix}, \tfrac{1}{2} \begin{bmatrix} \boldsymbol{\Sigma}_R & -\boldsymbol{\Sigma}_I \\ \boldsymbol{\Sigma}_I & \boldsymbol{\Sigma}_R \end{bmatrix} \right)$$

* This research was partially supported by NSF grants ISS-0613595 and CCF-0830612.

T. Adali et al. (Eds.): ICA 2009, LNCS 5441, pp. 90–97, 2009.

for some complex vector $\mathbf{c}$ and Hermitian positive definite matrix $\boldsymbol{\Sigma}$, then we say $\mathbf{x}$ is *complex multivariate Normal* distributed, and write $\mathbf{x} \sim \mathcal{N}_C(\mathbf{c}, \boldsymbol{\Sigma})$.

The real domain probability density over $(\mathbf{x}_R, \mathbf{x}_I)$ can be written in terms of complex quantities $p(\mathbf{x}) \triangleq p(\mathbf{x}_R, \mathbf{x}_I)$ as,

$$p(\mathbf{x}) = \pi^{-n}(\det \boldsymbol{\Sigma})^{-1} \exp\left(-(\mathbf{x} - \mathbf{c})^H \boldsymbol{\Sigma}^{-1}(\mathbf{x} - \mathbf{c})\right)$$

In particular, the univariate complex Normal distribution is given by

$$\mathcal{N}_C(x; \mu, \sigma^2) = \pi^{-1}\sigma^{-2} \exp(-\sigma^{-2}|x - \mu|^2)$$

where x_R and x_I are independent $\mathcal{N}(\mu_R, \sigma^2/2)$ and $\mathcal{N}(\mu_I, \sigma^2/2)$ respectively.

Using the central limit theorem, it is shown that for stationary time series $y_i(t)$, the distribution of the Fourier coefficients $y_i(\omega_k)$, $k = 1, \ldots, N$, are independent complex multivariate Gaussian [4]. Specifically we have the following [4, Thm.4.4.1].

Theorem 1. *Let $\mathbf{x}(t)$ be a strictly stationary real vector time series with absolutely summable cumulant spectra of all orders. Suppose $2\omega_j$, $\omega_j \pm \omega_k \neq 0\,(mod\,2\pi)$ for $1 \leq j \leq k \leq N$. Let,*

$$\mathbf{d}^{(T)}(\omega) = \frac{1}{T}\sum_{\tau=0}^{T-1} \mathbf{x}(\tau)\exp(-\jmath\omega\tau)$$

Then the random vectors $\mathbf{d}^{(T)}(\omega_k)$, $k = 1, \ldots, N$, are asymptotically independent complex Normal random vectors, $\mathcal{N}_C(\mathbf{0}, \boldsymbol{\Sigma}(\omega_k))$. Also if $\omega = 0\,(mod\,2\pi)$ then $\mathbf{d}^{(T)}(\omega)$ is asymptotically $\mathcal{N}(\mathbf{c}, \boldsymbol{\Sigma}(0))$, independent of the previous estimates, and if $\omega = \pi\,(mod\,2\pi)$, then $\mathbf{d}^{(T)}(\omega)$ is asymptotically $\mathcal{N}(\mathbf{0}, \boldsymbol{\Sigma}(0))$, independent of the previous estimates.

The density of linear transformations can also be expressed simply in terms of complex quantities. If $\mathbf{b} \sim p_{\mathbf{b}}(\mathbf{b})$, then the density of $\mathbf{x} = \mathbf{A}\mathbf{b}$ is, with $\mathbf{W} \triangleq \mathbf{A}^{-1}$,

$$p_{\mathbf{x}}(\mathbf{x}) = \det\left(\mathbf{W}\mathbf{W}^H\right) p_{\mathbf{b}}(\mathbf{W}\mathbf{x})$$

3 Blind Deconvolution and Complex Gaussian Scale Mixtures

In the noiseless blind deconvolution problem [5], we have,

$$\mathbf{x}(t) = \sum_{\tau=0}^{\infty} \mathbf{A}(\tau)\mathbf{s}(t - \tau)$$

where the time series $s_i(t)$, $i = 1, \ldots, n$ are mutually independent. Using the discrete Fourier transform, this is expressed in the frequency domain as,

$$\mathbf{x}(\omega) = \mathbf{A}(\omega)\mathbf{s}(\omega)$$

where $\mathbf{x}(\omega) \triangleq \sum_\tau \mathbf{x}(\tau) \exp(-\jmath\omega\tau)$, etc. We consider the asymptotic distribution of the short time Fourier transform (STFT) coefficients, calculated by,

$$\mathbf{x}(\omega_k, t) \triangleq \mathbf{x}_{:kt} \triangleq \sum_{\tau=0}^{T-1} \mathbf{x}(t - \tau) \exp(-\jmath\omega_k\tau)$$

Since the sources $s_i(t)$ are independent, we have that $s_i(\omega_1)$ and $s_j(\omega_2)$ are independent for $i \neq j$, and any ω_1, ω_2. Thus the joint distribution over the sources is a product,

$$p\big(\mathbf{s}(\omega_1), \ldots, \mathbf{s}(\omega_N)\big) = \prod_{i=1}^{n} p\big(s_i(\omega_1), \ldots, s_i(\omega_N)\big)$$

By Theorem 1, for stationary sources, the joint distribution over the frequencies is the product of n univariate complex Gaussians,

$$p\big(s_i(\omega_1), \ldots, s_i(\omega_N)\big) = \prod_{k=1}^{N} \mathcal{N}_C\big(s_i(\omega_k); 0, \sigma_k^2\big)$$

where $\mathcal{N}_C(\mu, \sigma^2)$ is the univariate complex Normal distribution, which is a radially symmetric distribution over the real and imaginary parts of $s_i(\omega_k)$, with variance σ_k^2, which gives the spectral power at frequency ω_k. Theorem 1 holds for stationary time series, but most long term observations of sensor data are non-stationary.

A more realistic assumption is that the time series is stationary over contiguous blocks, switching at discrete (random) times among a set of stationary regimes. For each individual stationary regime, as the window size, and/or sampling rate tend to infinity, the Fourier coefficients will tend to complex Gaussians. In a switching time series, a window may contain two or more stationary regimes. If we let the sampling rate tend to infinity, the individual stationary segments in the window will still be asymptotically complex Normal, and the overall distribution of the coefficients in the window will be a convex sum of the two (or more) complex Normal random variables.

Since each window will contain a random size segment of a given stationary state, the distributions of the Fourier coefficients for each state will be complex Gaussian scale mixtures, with the Fourier coefficients at different frequencies being uncorrelated but dependent. The joint distribution is given by,

$$p\big(s_i(\omega_1), \ldots, s_i(\omega_N)\big) = \int_0^\infty \prod_{k=1}^{N} \mathcal{N}_C\big(s_i(\omega_k); 0, \xi\sigma_{i,k}^2\big) f(\xi)\, d\xi$$

The distribution of $\mathbf{s}_i$ is equivalent to the product of a non-negative mixing random variable with a complex Normal random vector, $\mathbf{s}_i = \xi^{1/2}\mathbf{z}_i$, where $\xi \sim p(\xi)$ and $\mathbf{z}_i \sim \mathcal{N}_C(0, \mathrm{diag}(\boldsymbol{\sigma}_i))$, and can be expressed in a form similar to that of real valued Gaussian scale mixtures, where the squared magnitude of the complex variable is substituted for the square of the real variable [12].

Knowing the distribution of the Fourier coefficients of the independent sources, we can calculate the distribution of the observation Fourier coefficients as,

$$p\big(\mathbf{x}(\omega_1), \ldots, \mathbf{x}(\omega_N)\big) = \left(\prod_{k=1}^{N} \det \mathbf{W}(\omega_k)\mathbf{W}(\omega_k)^H\right) \prod_{i=1}^{n} p_i\big(b_i(\omega_1), \ldots, b_i(\omega_N)\big)$$

The parameters of the unmixing system $\mathbf{W}(\omega)$, as well as the CGSM density models, for each source $i = 1, \ldots, n$ are adapted by Quasi Maximum Likelihood.

3.1   Generalized Inverse Gaussian and Generalized Hyperbolic

The Generalized Inverse Gaussian (GIG) density [3,6,12] is a convenient mixing density. It can be thought of as a combination and generalization of the gamma and inverse gamma distributions. The GIG density has the form,

$$\mathcal{N}^{\dagger}(\xi; \lambda, \tau, \upsilon) = \frac{\upsilon^{\lambda}}{2\, K_{\lambda}(\tau)}\, \xi^{\lambda-1} \exp\left(-\tfrac{1}{2}\tau\left((\upsilon\xi)^{-1} + \upsilon\xi\right)\right) \tag{1}$$

for $\xi > 0$, where K_{λ} is the Bessel K function, or modified Bessel function of the second kind. The moments of the Generalized Inverse Gaussian are easily found by direct integration, using the fact that (1) integrates to one,

$$E\{\xi^a\} = \upsilon^a \frac{K_{\lambda+a}(\tau)}{K_{\lambda}(\tau)} \tag{2}$$

Since υ is a scale parameter and we we will estimate the scale of the complex Normal directly, we will eliminate this redundancy by setting $\upsilon = 1$ in the mixing density.

The generalized hyperbolic distribution [3] is the complex Gaussian scale mixture arising from the GIG mixing density. We write the generalized hyperbolic density in the form,

$$\mathcal{GH}_C(\mathbf{b}_i; \lambda, \tau) = \frac{\tau^{n-\lambda}}{\pi^n K_{\lambda}(\tau)}\, \tau(\mathbf{b}_i)^{\lambda-n} K_{\lambda-n}\left(\tau(\mathbf{b}_i)\right) \tag{3}$$

where $\tau(\mathbf{b}_i) \triangleq \sqrt{\tau^2 + 2\tau \|\mathbf{b}_i\|^2}$.

The GIG density is conjugate to the Normal density, so the posterior density of ξ given $\mathbf{b}_i$ is also a GIG density. In particular the parameters of the posterior density of ξ are,

$$\lambda' = \lambda - n, \quad \tau' = \tau(\mathbf{b}_i), \quad \upsilon' = \tau/\tau(\mathbf{b}_i)$$

We can thus find the posterior expectation required for the EM algorithm using (2),

$$E\{\xi_i^{-1} \mid \mathbf{b}_i\} = \frac{\tau(\mathbf{b}_i)}{\tau}\, \frac{K_{\lambda-n-1}(\tau(\mathbf{b}_i))}{K_{\lambda-n}(\tau(\mathbf{b}_i))} \tag{4}$$

4   Probabilistic IVA and Maximum Likelihood

We shall assume for simplicity of presentation and implementation that $\omega_k \not\equiv 0 (\mathrm{mod}\,\pi)$ for $k = 1, \ldots, N$, so that all the STFT coefficients are all complex valued. Thus for each observation $\mathbf{x}_{:t}$ of STFT coefficients of the data, there is a set of random vectors of independent, zero mean sources $\mathbf{b}_{:kt}$, such that,

$$\mathbf{x}_{:kt} = \mathbf{A}_k \mathbf{b}_{:kt}, \quad k = 1, \ldots, N$$

The density of the observations $p(\mathbf{x}_{::t}) \triangleq p(\mathbf{x}_{:0t}, \mathbf{x}_{:1t} \ldots, \mathbf{x}_{:Nt})$ is given by,

$$p(\mathbf{x}_{::t}) = \left(\prod_{k=1}^{N} \det \mathbf{W}_k \mathbf{W}_k^H \right) \prod_{i=1}^{n} q_i \left(\mathbf{b}_{i:t} \right)$$

where $\mathbf{W}_k \triangleq \mathbf{A}_k^{-1}$, and $\mathbf{b}_{:kt} = \mathbf{W}_k \mathbf{x}_{:kt}$, $k = 1, \ldots, N$. The sources are modeled as mixtures of Gaussian Scale Mixtures, extending the real model in [11] to the complex dependent case. We define,

$$y_{ijkt} \triangleq \beta_{ijk}^{1/2} b_{ikt}$$

The source density models are defined as follows,

$$q_i \left(\mathbf{b}_{i:t} \right) = \sum_{j=1}^{m} \alpha_{ij} \left(\prod_{k=1}^{N} \beta_{ijk} \right) q_{ij} \left(\mathbf{y}_{ij:t} ; \tau_{ij} \right)$$

Each q_{ij} is a CGSM parameterized by ν_{ij}.

Thus the density of the observations $\mathbf{x}_{:::} \triangleq \{\mathbf{x}_{::t}\}$, $t = 1, \ldots, T$, is given by $p(\mathbf{x}_{:::}; \Theta) = \prod_{t=1}^{T} p(\mathbf{x}_{::t}; \Theta)$, and the parameters to be estimated are

$$\Theta = \left\{ \mathbf{W}_k, \alpha_{ij}, \beta_{ijk}, \tau_{ij} \right\}$$

for $i = 1, \ldots, n$, $j = 1, \ldots, m$, and $k = 1, \ldots, N$.

4.1   Maximum Likelihood

Given the STFT data $\mathbf{X} = \{\mathbf{x}_{:kt}\}$, $k = 1, \ldots, N$, $t = 1, \ldots, T$, we consider the ML estimate of $\mathbf{W}_k = \mathbf{A}_k^{-1}$, $k = 1, \ldots, N$. We employ the EM algorithm to allow the source density at each frequency to be a mixture of (multivariate intra-dependent) CGSMs. Thus we define the random index j_{it} ranging over $\{1, \ldots, m\}$ with probabilities α_{ij}, $j = 1, \ldots, m$, and we let $z_{ijt} = 1$ if $j_{it} = j$, and 0 otherwise. For the joint density of $\mathbf{X}$ and $\mathbf{Z}$, we have,

$$p(\mathbf{X}, \mathbf{Z}) = \prod_{t=1}^{T} \left(\prod_{k=1}^{N} \det \mathbf{W}_k \mathbf{W}_k^H \right) \prod_{i=1}^{n} \prod_{j=1}^{m} \alpha_{ij}^{z_{ijt}} \tilde{\beta}_{ij}^{z_{ijt}} q_{ij} \left(\mathbf{y}_{ij:t} \right)^{z_{ijt}}$$

where $q_{ij}(\mathbf{y}_{ij:t})$ is the jth mixture component of the dependent multivariate density model for the STFT coefficients of ith source, $\mathbf{b}_{i:t}$ is the vector of STFT coefficients for source i at time t, and $\tilde{\beta}_{ij} \triangleq \prod_{k=1}^{N} \beta_{ijk}$. We define,

$$f_{ij}(\|\mathbf{y}_{ij:t}\|) \triangleq g_{ij}(\|\mathbf{y}_{ij:t}\|^2) \triangleq - \log q_{ij}(\mathbf{y}_{ij:t})$$

For the log likelihood of the data then (scaled by $1/T$), which is to be maximized with respect to the parameters, we have,

$$L(\mathbf{W}) = \left(\sum_{k=1}^{N} \log \det \mathbf{W}_k \mathbf{W}_k^H \right) - \frac{1}{T} \sum_{t=1}^{T} \sum_{i=1}^{n} \sum_{j=1}^{m} \hat{z}_{ijt} \, g_{ij} \left(\sum_{\ell=1}^{N} \beta_{ij\ell} |b_{i\ell t}|^2 \right)$$

Note that $|b_{i\ell t}|^2 = (\mathbf{w}_{i\ell}\mathbf{x}_{:\ell t})(\mathbf{w}_{i\ell}\mathbf{x}_{:\ell t})^* = (\mathbf{w}_{i\ell}\mathbf{x}_{:\ell t})(\mathbf{w}_{i\ell}^*\mathbf{x}_{:\ell t}^*)$. The complex gradient, or Wirtinger conjugate derivative of L is thus,

$$\frac{\partial}{\partial \mathbf{W}_k^*} L(\mathbf{W}) \triangleq \mathbf{G}_k = \mathbf{W}_k^{-H} + \frac{1}{T}\sum_{t=1}^{T}\boldsymbol{\varphi}_{:kt}\mathbf{x}_{:kt}^H \tag{5}$$

for $k = 1,\ldots,N$, where,

$$\varphi_{ikt} \triangleq -b_{ikt}\gamma_{ikt}, \quad \gamma_{ikt} \triangleq \sum_{j=1}^{m} \hat{z}_{ijt}\,\beta_{ijk}\,g'_{ij}\big(\textstyle\sum_{\ell=1}^{N}\beta_{ij\ell}|b_{i\ell t}|^2\big) \tag{6}$$

If we multiply (5) by $\mathbf{W}_k^H\mathbf{W}_k$ on the right, we get,

$$\Delta\mathbf{W}_k \propto \left(\mathbf{I} + \frac{1}{T}\sum_{t=1}^{T}\boldsymbol{\varphi}_{:kt}\mathbf{b}_{:kt}^H\right)\mathbf{W}_k \tag{7}$$

This linear transformation of the complex gradient is a positive definite, and thus a valid descent direction. The direction (7) is known as the "natural gradient" [1].

4.2 Hessian and Newton Method

If we expand L in a second order Taylor series, $L(\mathbf{W} + d\mathbf{W})$, and solve for the maximizing $d\mathbf{W}$, we arrive at the equation,

$$\mathcal{H}_{11}(d\mathbf{W}_k^*) + \mathcal{H}_{12}(d\mathbf{W}_k) + \sum_{\ell\neq k}\mathcal{H}'_{11}(d\mathbf{W}_\ell^*) + \mathcal{H}'_{12}(d\mathbf{W}_\ell) = -\mathbf{G}_k \tag{8}$$

for $k = 1,\ldots,N$, where $\mathcal{H}_{11}$ denotes two consecutive conjugate derivatives of $\mathbf{W}_k$, $\mathcal{H}_{12}$ denotes a conjugate derivative of $\mathbf{W}_k$ followed by a non-conjugate derivative of $\mathbf{W}_k$, etc. The $\mathcal{H}'$ terms are derivatives with respect to $\mathbf{W}_k$ followed by derivatives with respect to $\mathbf{W}_\ell$ where $\ell \neq k$.

Now, we have,

$$\frac{\partial\varphi_{ikt}}{\partial[\mathbf{W}_k]_{rs}} = -\delta_{ir}\left(\gamma_{ikt} + \psi_{ikt}\right)x_{skt}$$

where δ_{ir} is the Kronecker delta symbol, and,

$$\psi_{ikt} \triangleq |b_{ikt}|^2 \sum_{j=1}^{m} \hat{z}_{ijt}\,\beta_{ijk}^2\,g''_{ij}\big(\textstyle\sum_{\ell=1}^{N}\beta_{ij\ell}|b_{i\ell t}|^2\big)$$

Similarly, we have,

$$\frac{\partial\varphi_{ikt}}{\partial[\mathbf{W}_k]_{rs}^*} = -\delta_{ir}\,\psi_{ikt}\left(b_{ikt}/|b_{ikt}|\right)^2 x_{skt}^*$$

Now let $\mathbf{B}_k = d\mathbf{W}_k$. Then we have,

$$\mathcal{H}_{11}(\mathbf{B}_k) + \mathcal{H}_{12}(\mathbf{B}_k^*) = -\mathbf{W}_k^{-H}\mathbf{B}_k^H\mathbf{W}_k^{-H}$$
$$- \left\langle \mathrm{diag}(\boldsymbol{\zeta}_{:kt})\mathbf{B}_k^*\mathbf{x}_{:kt}^*\mathbf{x}_{:kt}^H\right\rangle_T - \left\langle \mathrm{diag}(\boldsymbol{\gamma}_{:kt} + \boldsymbol{\psi}_{:kt})\mathbf{B}_k\mathbf{x}_{:kt}\mathbf{x}_{:kt}^H\right\rangle_T$$

where $\langle \, \cdot \, \rangle_T$ denotes the empirical average $\frac{1}{T} \sum \cdot$, and,

$$\zeta_{ikt} \triangleq \psi_{ikt} \left(b_{ikt}/|b_{ikt}| \right)^2$$

The terms involving $\mathcal{H}'_{11}$, etc., tend to zero since the STFT coefficients at different frequencies are orthogonal (though nevertheless dependent) by definition.

Thus (8) reduces to,

$$\mathbf{G}_k = \mathbf{W}_k^{-H} \mathbf{B}_k^H \mathbf{W}_k^{-H} + \Big\langle \mathrm{diag}(\boldsymbol{\zeta}_{:kt}) \mathbf{B}_k^* \mathbf{x}_{:kt}^* \mathbf{x}_{:kt}^H \Big\rangle_T + \Big\langle \mathrm{diag}(\boldsymbol{\gamma}_{:kt} + \boldsymbol{\psi}_{:kt}) \mathbf{B}_k \mathbf{x}_{:kt} \mathbf{x}_{:kt}^H \Big\rangle_T \tag{9}$$

We would like to solve this equation for $\mathbf{B}_k = d\mathbf{W}_k$.

If we define $\tilde{\mathbf{B}}_k \triangleq \mathbf{B}_k \mathbf{W}_k^{-1}$ and $\tilde{\mathbf{G}}_k \triangleq \mathbf{G}_k \mathbf{W}_k^H$, then (9) can be written,

$$\tilde{\mathbf{G}}_k = \tilde{\mathbf{B}}_k^H + \Big\langle \mathrm{diag}(\boldsymbol{\zeta}_{:kt}) \tilde{\mathbf{B}}_k^* \mathbf{b}_{:kt}^* \mathbf{b}_{:kt}^H \Big\rangle_T + \Big\langle \mathrm{diag}(\boldsymbol{\gamma}_{:kt} + \boldsymbol{\psi}_{:kt}) \tilde{\mathbf{B}}_k \mathbf{b}_{:kt} \mathbf{b}_{:kt}^H \Big\rangle_T \tag{10}$$

We find asymptotically for the diagonal elements,

$$\big[\tilde{\mathbf{G}}_k \big]_{ii} = [\tilde{\mathbf{B}}_k]_{ii}^* + E\{\psi_{ikt}|b_{ikt}|^2\}[\tilde{\mathbf{B}}_k]_{ii}^* + E\{(\gamma_{ikt} + \psi_{ikt})|b_{ikt}|^2\}[\tilde{\mathbf{B}}_k]_{ii}$$

The cross terms drop out since the expected value of $\psi_{ikt} b_{skt} b_{ikt}^*$ is zero for $i \neq s$ by the independence and zero mean assumption on the sources. Thus we have,

$$\big[\tilde{\mathbf{G}}_k \big]_{ii} = \big([\tilde{\mathbf{G}}_k]_{ii} - 1 + \eta_{ik} \big) [\tilde{\mathbf{B}}_k]_{ii} + (1 + \eta_{ik})[\tilde{\mathbf{B}}_k]_{ii}^*$$

where we define $\eta_{ik} \triangleq E\{\psi_{ikt}|b_{ikt}|^2\}$. Since $[\tilde{\mathbf{G}}_k]_{ii}$ is real, this equation implies that $[\tilde{\mathbf{B}}_k]_{ii}$ must be real, and,

$$[\tilde{\mathbf{B}}_k]_{ii} = \frac{[\tilde{\mathbf{G}}_k]_{ii}}{[\tilde{\mathbf{G}}_k]_{ii} + 2\eta_{ik}} \tag{11}$$

The equation for the off-diagonal elements in (10) is,

$$[\tilde{\mathbf{G}}_k]_{ij} = [\tilde{\mathbf{B}}_k]_{ji}^* + E\{\zeta_{ikt}\} E\{b_{jkt}^2\}^* [\tilde{\mathbf{B}}_k]_{ij}^* + E\{\gamma_{ikt} + \psi_{ikt}\} E\{|b_{jkt}|^2\}[\tilde{\mathbf{B}}_k]_{ij}$$

For circular complex sources such as the Fourier coefficients of stationary signals, we have $E\{b_{ikt}^2\} = 0$. We can thus solve the following system of equations for $[\tilde{\mathbf{B}}_k]_{ij}$,

$$\big[\tilde{\mathbf{G}}_k \big]_{ij} = [\tilde{\mathbf{B}}_k^*]_{ji} + \kappa_{ik}\sigma_{jk}^2 [\tilde{\mathbf{B}}_k]_{ij}$$

$$\big[\tilde{\mathbf{G}}_k^* \big]_{ji} = [\tilde{\mathbf{B}}_k]_{ij} + \kappa_{jk}\sigma_{ik}^2 [\tilde{\mathbf{B}}_k^*]_{ji}$$

where we define $\kappa_{ik} \triangleq E\{\gamma_{ikt} + \psi_{ikt}\}$ and $\sigma_{ik}^2 \triangleq E\{|b_{ikt}|^2\}$. Thus,

$$\big[\tilde{\mathbf{B}}_k \big]_{ij} = \frac{\kappa_{jk}\sigma_{ik}^2 [\tilde{\mathbf{G}}_k]_{ij} - [\tilde{\mathbf{G}}_k^*]_{ji}}{\kappa_{ik}\kappa_{jk}\sigma_{ik}^2\sigma_{jk}^2 - 1} \tag{12}$$

where $\tilde{\mathbf{G}}_k = \mathbf{I} + \frac{1}{T}\sum_{t=1}^{T} \boldsymbol{\varphi}_{:kt} \mathbf{b}_{:kt}^H$. With $\tilde{\mathbf{B}}_k$ defined by (11) and (12), we then put,

$$\Delta\mathbf{W}_k = \tilde{\mathbf{B}}_k \mathbf{W}_k \tag{13}$$

For the EM updates of α_{ij} and β_{ijk}, we have,

$$\alpha_{ij} = \frac{1}{T}\sum_{t=1}^{T} \hat{z}_{ijt}, \quad \beta_{ijk}^{-1} = \frac{1}{T}\sum_{t=1}^{T} \hat{z}_{ijt} \, E\{\xi_{ijt}^{-1}|\mathbf{b}_{i:t}\} |b_{ikt}|^2 \tag{14}$$

where the posterior expectation is given by equation (4).

5   Conclusion

We have formulated a probabilistic framework for IVA and developed an asymptotic Newton method to estimate the parameters by Maximum Likelihood with adaptive complex Gaussian scale mixtures. We have extended the IVA model by deriving the model probabilistically using complex Gaussian scale mixtures. We also allow sources to have arbitrary spectra by adapting the spectral power parameters β_{ijk}, and to assume multiple spectral regimes using a mixture of CGSMs in the source model. Here we have concentrated on Generalized Inverse Gaussian mixing density, but the framework can accommodate other CGSM families as well, e.g. Generalized Gaussian or Logistic densities [12]. The Newton method can be extended to handle non-circular sources as well, leading from equation (9) to a block 4×4 Hessian structure [10]. The probabilistic model also allows straightforward extension to an IVA mixture model.

References

1. Amari, S.-I.: Natural gradient works efficiently in learning. Neural Computation 10(2), 251–276 (1998)
2. Amari, S.-I., Chen, T.-P., Cichocki, A.: Stability analysis of learning algorithms for blind source separation. Neural Networks 10(8), 1345–1351 (1997)
3. Barndorff-Nielsen, O., Kent, J., Sørensen, M.: Normal variance-mean mixtures and z distributions. International Statistical Review 50, 145–159 (1982)
4. Brillinger, D.R.: Time Series: Data Analysis and Theory. SIAM, Philadelphia (2001); reproduction of original publication by Holden Day, Inc., San Francisco (1981)
5. Cichocki, A., Amari, S.: Adaptive Blind Signal and Image Processing. John Wiley & Sons, Ltd., West Sussex (2002)
6. Eltoft, T., Kim, T.-S., Lee, T.-W.: Multivariate Scale Mixture of Gaussians Modeling. In: Rosca, J.P., et al. (eds.) ICA 2006. LNCS, vol. 3889, pp. 799–806. Springer, Heidelberg (2006)
7. Kim, T., Attias, H., Lee, S.-Y., Lee, T.-W.: Blind source separation exploiting higher-order frequency dependencies. IEEE Transactions on Speech and Audio Processing 15(1) (2007)
8. Kim, T.-S., Eltoft, T., Lee, T.-W.: Independent vector analysis: An extension of ICA to multivariate components. In: Rosca, J.P., et al. (eds.) ICA 2006. LNCS, vol. 3889, pp. 165–172. Springer, Heidelberg (2006)
9. Kreutz-Delgado, K.: The complex gradient operator and the cr-calculus. Lecture Supplement to ECE275A, Ver. ECE275CG-F05v1.3c, University of California, San Diego, USA (2005), http://dsp.ucsd.edu/~kreutz/PEI05.html
10. Li, H., Adali, T.: Stability analysis of complex Maximum Likelihood ICA using Wirtinger calculus. In: Proceedings of the 33rd IEEE International Conference on Acoustics and Signal Processing (ICASSP 2008), Las Vegas, NV, pp. 1801–1804 (2008)
11. Palmer, J.A., Kreutz-Delgado, K., Makeig, S.: Super-gaussian mixture source model for ICA. In: Rosca, J.P., Erdogmus, D., Príncipe, J.C., Haykin, S. (eds.) ICA 2006. LNCS, vol. 3889, pp. 854–861. Springer, Heidelberg (2006)
12. Palmer, J.A., Kreutz-Delgado, K., Rao, B.D., Makeig, S.: Modeling and estimation of dependent subspaces with non-radially symmetric and skewed densities. In: Davies, M.E., James, C.J., Abdallah, S.A., Plumbley, M.D. (eds.) ICA 2007. LNCS, vol. 4666, pp. 97–104. Springer, Heidelberg (2007)

Modeling the Short Time Fourier Transform Ratio and Application to Underdetermined Audio Source Separation

Dinh-Tuan Pham[1], Zaher El-Chami[2], Alexandre Guérin[2],
and Christine Servière[3]

[1] Laboratory Jean Kuntzmann, CNRS - INPG - UJF Grenoble, France
Dinh-Tuan.Pham@imag.fr
[2] Orange Labs, Lannion, France
{zaher.elchami,alexandre.guerin}@orange.ft-group.com
[3] GIPSA-lab, CNRS - INPG Grenoble, France
christine.serviere@inpg.fr

Abstract. This paper presents the theoretical background for the Model Based Underdetermined Source Separation presented in [5]. We show that for a given frequency band, in contrast to customary assumption, the observed Short-Time Fourier Transform (STFT) ratio coming from one source is not constant in time, but is a random variable whose distribution we have obtained. Using this distribution and the Time-Frequency (TF) "disjoint" assumption of sources, we are able to obtain promising results in separating four audio sources from two microphones in a real reverberant room.

1 Introduction

Blind Source Separation (BSS) is a well known technique to recover multiple original sources from their mixtures. Applications of BSS can be found in different fields like acoustic, biomedical and even economic. In the acoustic domain, BSS aims at separating speech and/or audio sources that have been mixed and then captured by multiple microphones in usually reverberant environment (the problem is referred as the cocktail party). The difficulty of the problem lies in the nature of the mixing system which is not instantaneous but convolutive (due to numerous reflections) with long impulse responses. In real reverberant situation, such responses can have thousands of taps even with small sampling rate (2400 taps for 60-dB reverberation time[1] of 300ms with 8KHz sampling rate). The problem gets harder in the underdetermined case (number of sources greater than number of microphones): in this case, no unique algebraic solution exists unless additional information on the mixing matrix or on the sources is given.

Different approaches are proposed in the literature to separate underdetermined audio mixtures. Most of these methods operate in the frequency domain

---

[1] This is the time for which the sound energy has decayed by 60dB.

T. Adali et al. (Eds.): ICA 2009, LNCS 5441, pp. 98–105, 2009.
© Springer-Verlag Berlin Heidelberg 2009

(FD). The time-domain mixed signals $x_j(t)$ observed at the microphones are converted into frequency-domain time-series signals by a short-time Fourier transform (STFT):

$$X_j(t,\omega) = \sum_{k=0}^{L-1} w(k)x_j(t+k)e^{-i\omega k} = \sum_{p=1}^{N} X_{j,p}(t,\omega), \tag{1}$$

where $w(\cdot)$ is an analysis window (e.g. Hamming) and $X_{j,p}(t,\omega)$ is the contributed STFT of the pth source to the jth sensor, that is of $x_{j,p}(t) = (a_{j,p} \star s_p)(t)$, $a_{j,p}(\cdot)$ denoting the filter impulse response describing the effect of the pth source on the jth microphone and $\star$ denoting the convolution. Working in the FD has two major reasons. Firstly, it permits transforming convolutive mixtures into instantaneous ones in each frequency band. Specifically $X_{j,p}(t,\omega)$ may be approximated by $A_{j,p}(\omega)S_p(t,\omega)$ where $A_{j,p}$ is the frequency response of the filter $a_{j,p}$ and $S_p(t,\omega)$ is the STFT of $s_p(t)$. Secondly, the sparseness of speech signals becomes prominent in the Time-Frequency (TF) domain. This feature plays a crucial role in the underdetermined case: It entails that in the TF domain, the quasi-support[2] of the source would be nearly disjoint, hence in each TF bin there is a high probability that at most one source is dominant. Thus the sources can be separated by binary masks. Still it remains the problem of how to allocate each TF bin to the corresponding source. Applying the "disjoint" assumption, at each bin (t,ω), the last sum in (1) contains one single non negligible term, hence:

$$X_j(t,\omega) \approx X_{j,q}(t,\omega) \approx A_{jq}(\omega)S_q(t,\omega) \tag{2}$$

where q denoted the dominant source index at (t,ω). The ratio of the observed STFT at this point would then be:

$$R(t,\omega) = \frac{X_1(t,\omega)}{X_2(t,\omega)} \approx \frac{X_{1,q}(t,\omega)}{X_{2,q}(t,\omega)} \approx \frac{A_{1q}(\omega)}{A_{2q}(\omega)} \tag{3}$$

Being frequency and source dependent (but time independent), this ratio can be used in each frequency band to identify the sources in each TF bin. Most existing methods follows this approach. Note that, the modulus and argument of $R(t,\omega)$ are no other than the traditional Interchannel Level Difference (ILD) and Interchannel Phase Difference (IPD).

As we can see, underdetermined audio separation is based on two approximation: the sparseness or more exactly the "disjoint" assumption, which results in the first approximation in (2) and (3), and the second approximation in (2) or more exactly in (3). The validity of the sparseness assumption can be found in [2,?,4]. In this paper we will show that the second approximation in (2) and (3) unfortunately are not valid in realistic situations where the room impulse response often exceeds the STFT window length L. More precisely, we will show that for a fixed frequency band ω and even with only one active source $s_p(t,\omega)$, the observed STFT ratio $X_{1,p}(t,\omega)/X_{2,p}(t,\omega)$ can not be considered as constant

---

[2] By quasi-support we mean the set of TF bins for which the signal is not negligible.

in time, but is a random variable with an infinite variance. The logarithm of this ratio however has a finite variance and its mean and variance would tend to $\log[A_{1,p}(\omega)/A_{2,p}(\omega)]$ and 0 as $L \to \infty$ (as expected). But for L not large with respect to the impulse response length, the variance can be appreciable. We will derive formulas for the mean and variance of this "log ratio" and also its probability density function. Having this "log ratio" model and after estimating its parameters by an Expectation Maximization (EM) algorithm, soft masks can be derived to separate the sources. Note that Balan and Rosca [1] has also considered a probabilistic model for the observed STFT ratio, but they still rely on the approximation (3) while relaxing somewhat the "disjoint assumption".

2   The STFT Ratio Distribution

We consider the case where only one source s_p is active at the two microphones:

$$x_{j,p}(t) = (a_{j,p} \star s_p)(t) = \sum_{k=0}^{M-1} a_{jp}(k)s_p(t-k), \quad j = 1, 2.$$

The time-domain observed signals $x_{j,p}(t)$ are converted into frequency-domain time-series $X_{j,p}(t, \omega)$ signals using the STFT:

$$X_{j,p}(t, \omega) = \sum_{k=0}^{L-1} w(k)x_{j,p}(t+k)e^{-ik\omega} = \sum_{k=1-L}^{0} w(-k)e^{ik\omega}x_{j,p}(t-k)$$

As we can see from the above second right hand side, $X_{j,p}(t, \omega)$ can be considered as the output of the filter $w_\omega : w_\omega(k) = w(-k)e^{ik\omega}$ applied to the signal $x_{j,p}(t)$. At its turn, $x_{j,p}(t)$ is the filtered version of $s_p(t)$ by the filter $a_{j,p}$. Thus, for a given ω, $X_{j,p}(t, \omega)$ is the filtered version of $s_p(t)$ by the filter $a_{j,p} \star w_\omega$, which has Fourier transform $W(\omega - \cdot)A_{j,p}(\cdot)$, W denoting the Fourier transform of w.

2.1   Variance and Correlation

Note that the filter $a_{jp} \star w_\omega$ has support in $[1 - M, L - 1]$. Therefore if we assume that the signal s_p can be considered stationary in the time frame $[t - L + 1, t + M - 1]$ with power spectral density denoted by $D_p(t, \cdot)^3$, then $X_{1,p}(t, \omega)$ and $X_{2,p}(t, \omega)$ have variances

$$\sigma_{j,p}^2(t, \omega) = E|X_{j,p}(t, \omega)|^2 = \int_{-\pi}^{\pi} |W(\omega - \lambda)|^2 |A_{jp}(\lambda)|^2 D_p(t, \lambda) \frac{d\lambda}{2\pi}, \quad j = 1, 2$$

and (complex) covariance

$$\begin{aligned} \mathrm{cov}[X_{1,p}(t, \omega), X_{2,p}(t, \omega)] &= E|X_{1,p}(t, \omega)X_{2,p}^*(t, \omega)] \\ &= \int_{-\pi}^{\pi} |W(\omega - \lambda)|^2 A_{1p}(\lambda)A_{2p}^*(\lambda)D_p(l, \lambda) \frac{d\lambda}{2\pi}. \end{aligned}$$

---

[3] This assumption might not be quite realistic for large $L + M - 2$, however the calculations might still be valid by interpreting $D_p(t, \cdot)$ as the *average spectral density*.

Note that for large L, $|W|^2$ will have the shape of a Dirac: it is negligible outside in interval centered at 0 of length of the order $1/L$. Thus if we assume that the spectral density does not vary much in a interval of frequencies of length about $1/L$, we may pull the term $D_p(t, \lambda)$ outside the above integrals, which yields the approximations to th complex correlation between $X_{1,p}(t, \omega)$ and $X_{2,p}(t, \omega)$

$$\rho_p(t, \omega) \approx \frac{\int_{-\pi}^{\pi} |W(\omega - \lambda)|^2 A_{1p}(\lambda) A_{2p}^*(\lambda) d\lambda}{[\int_{-\pi}^{\pi} |W(\omega - \lambda)|^2 |A_{1p}(\lambda)|^2 d\lambda \int_{-\pi}^{\pi} |W(\omega - \lambda)|^2 |A_{2p}(\lambda)|^2 d\lambda]^{1/2}}$$

and to their variance ratio

$$\frac{\sigma_{1,p}^2(t, \omega)}{\sigma_{2,p}^2(t, \omega)} \approx \frac{\int_{-\pi}^{\pi} |W(\omega - \lambda)|^2 |A_{1p}(\lambda)|^2 d\lambda}{\int_{-\pi}^{\pi} |W(\omega - \lambda)|^2 |A_{2p}(\lambda)|^2 d\lambda}.$$

The above approximations are reasonable and will be made. It follows that the variance ratio and the correlation don't vary in time and depend only on the chosen analysis window and the filter impulse responses. Note that if we make the *unrealistic* assumption that M is much smaller than L, then A_{1p} and $A_{2,p}$ also don't vary much in a interval of length about $1/L$ and thus $|\rho_p(t, \omega)| \approx 1$. This means that $X_{1p}(t, \omega)/X_{1p}(t, \omega)$ is constant in time and we get the second approximation in (2) and (3) mentioned in the introduction.

2.2   Ratio Distribution

To simplify the notations, in this section we will omit the ω variable since it remains fixed throughout. By the approximations in previous section, the variance ratio and correlation do not depend on time, hence the time variable will also be dropped in their symbols. We assume that L is large enough so that, by the Central Limit Theorem, the pair $X_{1,p}(t)$ and $X_{2,p}(t)$ can be considered as Gaussian (complex circular) with variance $\sigma_{1,p}^2(t)$ and $\sigma_{2,p}^2(t)$ and correlation ρ_p. To compute the distribution of $X_{1,p}(t)/X_{2,p}(t)$, a trick is used to take the correlated item $X_{2,p}(t)$ out of $X_{1,p}(t)$. It can be checked that the difference $X_{1,p}(t)/\sigma_{1,p}(t) - \rho_p X_{2,p}(t)/\sigma_{2,p}(t)$ is uncorrelated with, hence independent of, $X_{2,p}(t)$ and has variance $1 - |\rho_p|^2$. Thus:

$$X_{1,p}(t)/\sigma_{1,p}(t) = \rho_p X_{2,p}(t)/\sigma_{2,p} + (1 - |\rho_p|^2)^{1/2} R_{1|2,p}(t)$$

where $R_{1|2,p}(t)$ is a complex standard normal variable independent of $X_{2,p}(t)$. Therefore

$$R(t) = \frac{X_{1,p}(t)}{X_{2,p}(t)} = \frac{\sigma_{1,p}}{\sigma_{2,p}} \left(\rho_p + \sqrt{1 - |\rho_p|^2} \right) \frac{R_{1|2,p}(t)}{X_{2,p}(t)/\sigma_{2,p}} \tag{4}$$

Both $R_{1|2,p}(t)$ and $X_{2,p}(t)/\sigma_{2,p}(t)$ are complex standard normal variables, hence are distributed as $\sqrt{V} e^{iU}$ where V is an exponential variable with unit mean and U is uniform in $[0, 2\pi]$. Thus the right hand side of (4) has the same distribution as:

$$\frac{\sigma_{1,p}}{\sigma_{2,p}} \left(\rho_p + \sqrt{1 - |\rho_p|^2} \right) \sqrt{R} e^{iU}$$

where R is the ratio of two independent exponential variables of same mean and U is uniform in $[0, 2\pi]$ (since the difference modulo 2π of 2 independent uniform variables in $[0, 2\pi]$ is again uniform in $[0, 2\pi]$). Explicit calculation shows that R admits the density function $1/(1 + r)^2$, hence it has an infinite mean but its square root has a finite mean. Therefore $R(t)$ has infinite variance. To avoid this problem, we propose to consider its logarithm, which is distributed as

$$\log(\sigma_{1,p}/\sigma_{2,p}) + i \arg \rho_p + \log[|\rho_p| + (1 - |\rho_p|^2)^{1/2}\sqrt{R}e^{iU}] \qquad (5)$$

($\arg z$ denoting the argument of the complex number z: $z = |z|e^{i \arg z}$), since $(U - \arg \rho_p) \bmod 2\pi$ has the same distribution as U.

The first two terms in (5) are non random and correspond to the mean of $\log[X_{1,p}(t)/X_{2,p}(t)]$ since the remainder term $\log[|\rho_p| + (1 - |\rho_p|^2)^{1/2}\sqrt{R}e^{iU}]$ has zero mean as it can be seen later. The joint density of the real and imaginary part of this complex variable can be obtained from the known joint density of the independent random variables R, U and the transformation $r, u \mapsto \log[|\rho_p| + (1 - |\rho_p|^2)^{1/2}\sqrt{r}e^{iu}] = x + iy$. An explicit (and tedious) calculation yields this joint density:

$$p_{|\rho_p|}(x, y) = \frac{1}{4\pi} \frac{1 - |\rho_p|^2}{(\cosh x - |\rho_p| \cos x)^2}. \qquad (6)$$

This density is an even function of both of its arguments, hence the corresponding complex variable has zero mean and uncorrelated real and imaginary parts.

Finally, the time set of $\log |R(t)|$ and $\arg R(t)$ have the joint density:

$$p_{|\rho_p|}(x - \log |r_p|, y - \arg r_p) \qquad (7)$$

where $r_p = (\sigma_{1,p}/\sigma_{2,p})e^{i \arg \rho_p}$. Note that r_p and $|\rho_p|$ may be viewed respectively as standing for the source s_p position in space and for the reverberation degree of the acoustic path between s_p and the set of microphones, so that $|\rho_p|$ tends to one when the mixture tends to be anechoic.

3   Application to Sparse Source Separation

In [5], we have used the density model given in (6) – (7) to separate audio sources in the underdetermined case. The method and results are briefly described here.

Under the "disjoint" assumption, at most one source is dominant in each TF bin, but we don't know which one it is. Thus given the one source log ratio density function in (6) –(7), we adopt the following model for the distribution of the real and imaginary parts of $\log R(t, \omega)$ for a given ω:

$$p(x, y | \boldsymbol{\rho}, \boldsymbol{r}, \boldsymbol{\mu}) = \sum_{p=1}^{N} \mu_p p_{|\rho_p|}(x - \log |r_p|, y - \arg r_p)$$

where $\boldsymbol{\rho} = (|\rho_1|, \ldots, |\rho_N|)$, $\boldsymbol{r} = (r_1, \ldots, r_N)$ and $\boldsymbol{\mu} = (\mu_1, \ldots, \mu_N)$ are the frequency dependent model parameters (the frequency variable ω is again omitted for simplicity), μ_p denoting the a priori probability that the pth source being dominant in the considered frequency band. As no model is known for the parameters r_p

and $|\rho_p|$ as function of frequency or that such a model is too complex, we choose to estimate them independently at each frequency band, for example, by the maximum likelihood method based on the data set $\{\log|R(t)|, \arg R(t)\}$ and the model (6) – (7). Once estimated, the *a posteriori probability* that p^{th} source is dominant at the TF point (t, ω) is given by:

$$\pi_p(t) = \frac{\mu_p p_{|\rho_p|}[\log|R(t)/r_p|, \arg R(t)/r_p]}{\sum_{i=1}^{N} \mu_i p_{|\rho_i|}[\log|R(t)/r_i|, \arg R(t)/r_i]}.$$

These *a posteriori* probabilities can be used to construct a binary or a soft mask to extract the pth source.

To maximize the likelihood, the EM (expectation-maximization) algorithm can be used. The hidden variable is taken as the indicator variable that indicates which source is dominant at each TF point. Of course, working in FD has the permutation ambiguities problem. These ambiguities should be aligned properly so that the separated frequency components that originate from the same source are grouped together, using for example the methods in [3].

4   Experiments and Results

We adopt the experimental setup (room dimensions and source/microphone positions) as described in Fig. 1. Further, development data can be downloaded from the Signal Separation Evaluation Campaign site at [6].

First, in order to highlight the *bad* approximation of $X_{1,p}(t, \omega)/X_{2,p}(t, \omega)$ by $A_{1p}(\omega)/A_{2p}(\omega)$ we compute numerically the correlation $\rho_p(\omega)$ between the twos STFT output $X_{1,p}(t, \omega)$ and $X_{2,p}(t, \omega)$ and their standard deviation ratio $\sigma_{1,p}(\omega)/\sigma_{2,p}(\omega)$. We choose $p = 1$, hence the computation is based on the mixing filters which relate the first source to the set of microphones (D=5cm, source angle = $-45°$). Theoretically, the "log ratio" $\log[X_{1,1}(t, \omega)/X_{2,1}(t, \omega)]$ has means $\log[\sigma_{1,1}(\omega)/\sigma_{2,1}(\omega)] + i \arg \rho_1(\omega)$ and its variability is controlled by $\rho_1(\omega)$ but is better measured by $\sqrt{1 - |\rho_1(\omega)|^2}$, since the later (by numerical calculation) varies more or less linearly with the standard deviations of its real

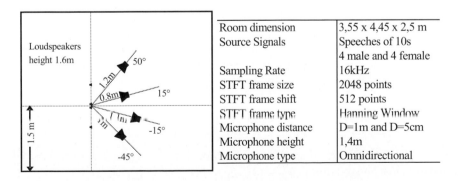

Fig. 1. Recording arrangement used for development data

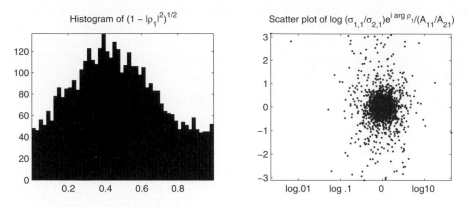

Fig. 2. Histogram of $\sqrt{1-|\rho_1|^2}$, left panel, and scatter plot of $\log[(\sigma_{1,1}/\sigma_{2,1})e^{i\arg\rho_1}/(A_{11}/A_{21})]$, right panel

and imaginary parts. Therefore we plot in Figure 2 the histogram of $\sqrt{1-|\rho_1|^2}$ (left panel) and the scatter plot of $\log[(\sigma_{1,1}/\sigma_{2,1})e^{i\arg\rho_1}/(A_{11}/A_{21})]$ (right panel, real and imaginary parts as abscissa and ordinate). One can see that for many frequency bins the variability of the "log ratio" is significant. In fact half of them have a $\sqrt{1-|\rho_1|^2}$ exceeding 0.4554. This corresponds to a standard deviation of $\log|X_{1,1}(t,\omega)/X_{2,1}(t,\omega)|$ of 0.53 and a difference of ± 0.53 in the logarithm corresponds to a 70% increase or 41% decrease in value. The means of the "log ratio" can also be quite different from $\log[A_{11}(\omega)/A_{21}(\omega)]$ for some frequencies.

Next, we examine the performance of the algorithm in section 3. We measure it by the improvement of the Signal to Interference Ratio (SIR) and Signal to distortion Ratio (SDR) as in [4]. For both criteria, larger number refers to better results. SIR improvement is calculated by the difference between the mean of the Input and Output SIR:

$$\text{InputSIR}_p = \frac{1}{2}\sum_{j=1}^{2} 10\log_{10}\frac{\sum_t |x_{j,p}(t)|^2}{\sum_t |\sum_{q\neq p} x_{j,q}(t)|^2}$$

$$\text{OutputSIR}_p = \frac{1}{2}\sum_{j=1}^{2} 10\log_{10}\frac{\sum_t |y_{j,pp}(t)|^2}{\sum_t |\sum_{q\neq p} y_{j,qp}(t)|^2}$$

where $y_{j,qp}$ is the inverse STFT of the observation $X_{j,q}$ masked by M_p[4]. The SDR for output p is defined by the mean of the power ratio between $x_{j,p}$ and the distortion in $y_{j,pp}$

$$\text{SDR}_i = \frac{1}{2}\sum_{j=1}^{2} 10\log_{10}\frac{\sum_t |\alpha_p x_{j,p}(t)|^2}{\sum_t |y_{j,pp}(t) - \alpha_p x_{j,p}(t)|^2}$$

where α_p is an adjusting factor for amplitude differences and is optimally obtained by least-mean-square: $\alpha_p = (\sum_t y_{j,pp}(t) x_{j,p}(t))/\sum_t |x_{j,p}(t)|^2$.

---

[4] $\text{M}_p(t,\omega) = \pi_p(t,\omega)$ is the separation soft mask that extracts the source p.

Table 1. Results for live recording with two microphone spacing (5cm, 1m) and two different types of speaker; Overall performance of the separation is presented in the last column

Mixture	4 females live recording								4 males live recording								Over all
	$D = 5\text{cm}$				$D = 1\text{m}$				$D = 5\text{cm}$				$D = 1\text{m}$				
Source	s_1	s_2	s_3	s_4	s_1	s_2	s_3	s_4	s_1	s_2	s_3	s_4	s_1	s_2	s_3	s_4	
SIR	7.8	8.3	8	7.1	9.3	7	9.9	9.8	7.8	6.6	7.3	7.2	8.2	6.8	9	8.4	8.1
SDR	6.7	1.5	6.4	8.8	6.8	7.1	6.7	5.9	4.5	3	7.8	7.2	5.2	6.4	6.7	6.7	6.1

Table 1 displays the performance results. As we can see the new algorithm gives promising results in a very chalenging situation (average SIR improvement of 8.1dB). These results compare favorably with other previous methods (see [5]). Our method can make use of soft mask which gives less distortion in the separated sources, in another term, less artefacts as it can be seen in the obtained SDR average (6.1dB).

5   Conclusion

In this paper, we show the non validity of approximating the one source observed STFT ratio by the filter frequency response ratio. Instead, the first ratio is a random variable, the distribution of which we have obtained in close form. A separation procedure based on this distribution model has been briefly described. Experiments are conducted in a real reverberant room where results show good performance for the new proposed algorithm.

References

1. Balan, R., Rosca, J.: Statistical Properties of STFT Ratios for two Channel Systems and Applications to Blind Source Separation. In: Proc. ICA 2000, Helsinki, Findland, pp. 429–434 (June 2000)
2. Yilmaz, O., Rickard, S.: Blind separation of speech mixtures via time-frequency masking. IEEE Transactions on Signal Processing 52(7), 1830–1847 (2004)
3. Sawada, H., Araki, S., Makino, S.: Measuring Dependence of Bin-wise Separated Signals for Permutation Alignment in Frequency-domain BSS. In: ISCAS, New Orleans, USA (May 2007)
4. Araki, S., Sawada, H., Makino, S.: K-means Based Underdetermined Blind Speech Separation. In: Makino, S., Lee, T.-W., Sawada, H. (eds.) Blind Speech Separation, pp. 243–270. Springer, New-York (2007)
5. El-Chami, Z., Pham, D.-T., Servière, Ç., Guérin, A.: A New-Model Based Underdetermined Source Separation. In: IWAENC, Seattle, USA (September 2008)
6. Signal Separation Evaluation Campaign (2008),
   http://sisec.wiki.irisa.fr/tiki-index.php

Minimum Determinant Constraint for Non-negative Matrix Factorization

Reinhard Schachtner[1,2], Gerhard Pöppel[2], Ana Maria Tomé[3],
and Elmar W. Lang[1]

[1] CIMLG / Biophysics, University of Regensburg, 93040 Regensburg, Germany
[2] Infineon Technologies AG, 93049 Regensburg, Germany
[3] DETI / IEETA, Universidade de Aveiro, 3810-Aveiro, Portugal
elmar.lang@biologie.uni-regensburg.de

Abstract. We propose a determinant criterion to constrain the solutions of non-negative matrix factorization problems and achieve unique and optimal solutions in a general setting, provided an exact solution exists. We demonstrate with illustrative examples how optimal solutions are obtained using our new algorithm *detNMF* and discuss the difference to NMF algorithms imposing sparsity constraints.

1 Introduction

Non-negative Matrix Factorization (NMF) has seen numerous applications since its invention by Paatero [1]. A couple of different cost functions as well as various optimization techniques have been presented in recent years [2], [6], [12]. However, uniqueness of NMF-solutions is still an open issue despite some recent attempts to deal with the subject [10], [11]. There are two popular routes to enforce uniqueness of the solutions. While in [4], arguments from sparse coding are invoked, the development of positive matrix factorization as surveyed in [14] rather proposes application-driven solutions requiring background knowledge. Both approaches are limited to special applications where specific information about the data is available or specific assumptions concerning the composition of data are necessary.

This letter suggests a more principled approach to uniqueness. Starting from a geometrical point of view similar to [16], we illustrate the problem of uniqueness and explain how, given no further restrictions, the determinant can be used to identify an optimal solution arising naturally among all possible exact non-negative decompositions of a given data matrix .

Recently, a similar concept has been proposed to solve spectral unmixing problems assuming a linear mixing model [18]. However, the additional summation constraint on the mixing coefficients and an approximation involving a PCA imposes unnecessary limitations and complications which distinguishes the approach in [18] from ours.

T. Adali et al. (Eds.): ICA 2009, LNCS 5441, pp. 106–113, 2009.
© Springer-Verlag Berlin Heidelberg 2009

2   Geometrical Approach

2.1   Problem Illustration

In the following, we consider an exact decomposition where each observed *signature* $\mathbf{X}_{n*}$ $(n = 1, \ldots, N)$ can be represented as a non-negative, linear combination of K non-negative *modes* $\mathbf{H}_{k*}$:

$$\mathbf{X}_{n*} = \sum_{k=1}^{K} W_{nk}\mathbf{H}_{k*} \tag{1}$$

We further assume the $\mathbf{H}_{k*}$ to be normalized, so that $\sum_j (H_{kj})^2 = 1$ for $k = 1, \ldots, K$. The goal of any NMF-algorithm then is the identification of such modes uniquely, given the data $\mathbf{X}$, the number of modes K and the non-negativity constraints. Note that uniqueness is limited by trivial scaling and permutation indeterminacies.

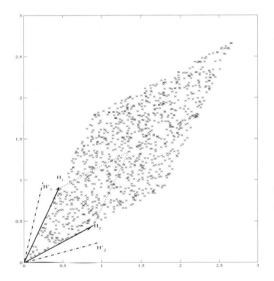

Fig. 1. 2D illustration of *detNMF*: Data were generated via $\mathbf{WH}$, using $\mathbf{H} = \sqrt{\frac{4}{5}} \begin{pmatrix} \frac{1}{2} & 1 \\ 1 & \frac{1}{2} \end{pmatrix}$ and arbitrary mixing coefficients $w_{nm} \in \{0, 2\}$. Two possible NMF-bases $(\mathbf{H}_{1*}, \mathbf{H}_{2*})$, $(\mathbf{H}'_{1*}, \mathbf{H}'_{2*})$ are indicated as well as the minimal determinant parallelepiped with normalized h_{km} and w_{nk}.

Geometrically, non-negativity manifests itself as follows: (see figure 1 for a two-dimensional example where $N = K = 2$, see also [14], [16])

- $\mathbf{X} \geq 0$:
  All data vectors $\{\mathbf{X}_{i*}\}_{i=1}^{N}$ constitute a cloud in the non-negative quadrant $(\mathbb{R}_0^+)^M$ of $\mathbb{R}^M$.

- **H $\geq$ 0:**
  Each basis vector $\mathbf{H}_{k*}$ points into $(\mathbb{R}_0^+)^M$; Normalized vectors $\mathbf{H}_{k*}$ span a K-dimensional subset of the non-negative quadrant $(\mathbb{R}_0^+)^M$ of $\mathbb{R}^M$.
- **W $\geq$ 0:**
  Positive coefficients W_{nk} for all k imply that a data vector $\mathbf{X}_{n*}$ is inside this K-dimensional subset, whereas $W_{nk} = 0$ for at least one k indicates that point X_{n*} lies on a K-1 dimensional peripheral surface. Here, we consider the surfaces as part of the interior of the subset.

Any set of K non-negative vectors $\{\mathbf{H}_{k*}\}_{k=1}^K$ which enclose the data $\mathbf{X}$ provides a valid solution to the NMF-problem $\mathbf{X} = \mathbf{WH}$. A natural choice of a unique solution, given the data $\mathbf{X}$, is a set of nonnegative modes $\{\mathbf{H}_{k*}\}_{k=1}^K$, which span a parallepiped with minimal volume containing $\mathbf{X}$. In other words, the non-negative K-tuple of N-dimensional vectors enclosing the data in the tightest possible way is our candidate for a unique solution. (See the pair $(\mathbf{H}_{1*}, \mathbf{H}_{2*})$ in figure 1.)

3   The Algorithm detNMF

3.1   Determinant Criterion

Let $P(\mathbf{H})$ be the parallelepiped spanned by the vectors $\mathbf{H}_{1*}, \ldots \mathbf{H}_{K*}$. If $\mathbf{H}$ is a square matrix, the volume of $P(\mathbf{H})$ is given by $vol(P) = |\det(\mathbf{H})|$, otherwise the volume of the parallelepiped can be obtained by $vol(P) = \sqrt{det(\mathbf{HH}^T)}$. Thus we seek non-negative modes $\mathbf{H}_{k*}$ such that $\det(\mathbf{HH}^T)$ is minimal. If we normalize each mode $\mathbf{H}_{k*}$ by $\left(\sqrt{\sum_j (H_{kj})^2}\right)^{-1}$, a minimal $P(\mathbf{H})$ is equivalent to minimal angles between the edges of $P(\mathbf{H})$. Below we provide the algorithm *detNMF* which directly implements this determinant criterion.

Assuming a noise-free model, there is at least one exact solution for the correct K. Unless the convex hull of the data is rotationally symmetric, there is an optimal solution among them selected by the determinant criterion (unique up to permutation). If there is rotational symmetry, there are several equivalent optimal solutions, which cannot be distinguished by the determinant criterion.

If the model $\mathbf{X} \approx \mathbf{WH}$ holds only approximately, a suitable tradeoff between the reconstruction accuracy and a minimal determinant must be found. In that case, the existence of unique solutions is of course arguable, and will not be discussed here (see [9]).

3.2   Update Rules

Consider the mean squared reconstruction error regularized by the minimal determinant criterion ($\alpha > 0$)

$$E_{det}(\mathbf{X}, \mathbf{WH}) = (NM)^{-1} \sum_{n=1}^N \sum_{m=1}^M (X_{nm} - [\mathbf{WH}]_{nm})^2 + \alpha \det(\mathbf{HH}^T), \quad (2)$$

The term $\det\left(\mathbf{H}\mathbf{H}^T\right)$ is differentiable, and its partial derivatives are given by

$$\frac{\partial \det(\mathbf{H}\mathbf{H}^T)}{\partial H_{km}} = 2 \det\left(\mathbf{H}\mathbf{H}^T\right)[(\mathbf{H}\mathbf{H}^T)^{-1}\mathbf{H}]_{km}. \tag{3}$$

Gradient descent then leads to the following multiplicative update rules (4)-(7) of the algorithm *detNMF* :

update:

$$H_{km} \leftarrow H_{km} \left(\frac{[\mathbf{W}^T\mathbf{X}]_{km}}{[\mathbf{W}^T\mathbf{W}\mathbf{H}]_{km}} - \alpha \cdot \det\left(\mathbf{H}\mathbf{H}^T\right)\frac{[(\mathbf{H}\mathbf{H}^T)^{-1}\mathbf{H}]_{km}}{[\mathbf{W}^T\mathbf{W}\mathbf{H}]_{km}} \right) \tag{4}$$

$$W_{nk} \leftarrow W_{nk} \cdot \frac{[\mathbf{X}\mathbf{H}^T]_{nk}}{[\mathbf{W}\mathbf{H}\mathbf{H}^T]_{nk}} \tag{5}$$

normalize:

$$H_{km} \leftarrow \frac{H_{km}}{\sqrt{\sum_m (H_{km})^2}} \tag{6}$$

$$W_{nk} \leftarrow W_{nk} \cdot \sqrt{\sum_m (H_{km})^2} \tag{7}$$

These update rules represent a modification of the well-known unconstrained NMF algorithm [2] to which they reduce in the limit $\alpha \to 0$. Note that the latter algorithm has been shown not to increase the reconstruction error [3]. General methods of how to regularize multiplicative updates with additional constraints are discussed in [6], [12] or [15].

As described in Section , any solution to an exactly solvable NMF-problem requires the modes $\mathbf{H}_{k*}$ to lie on the periphery of the data cloud. The NMF algorithm, initialized randomly, moves the modes $\mathbf{H}_{k*}$ outside of the given data cloud. For any $\alpha > 0$, the determinant constraint instead forces the basis vectors $\mathbf{H}_{k*}$ to move towards each other. These conflicting requirements can be controlled by choosing a small $\alpha > 0$ which should be increased only when all modes lie outside the data cloud. In simulations we observed that with α kept small enough so that the reconstruction error does not increase during an iteration step, very satisfactory results are obtained.

3.3 Geometrical Interpretation and the Multi-layer Technique

The determinant criterion can also be related to a practice called multi-layer technique [7], [8]. Such a sequential decomposition of nonnegative matrices as a cascade of unmixing systems usually yields a better performance than one single decomposition. The determinant criterion intuitively explains this effect by noting that

$$\det\left(\mathbf{H}\mathbf{H}^T\right) = \det\left(\mathbf{H}^{(L)}\ldots\mathbf{H}^{(1)}\mathbf{H}^{(1)T}\ldots\mathbf{H}^{(L)T}\right)$$
$$= \det\left(\mathbf{H}^{(L)}\mathbf{H}^{(L)T}\right)\ldots\det\left(\mathbf{H}^{(1)}\mathbf{H}^{(1)T}\right). \tag{8}$$

Normalizing the rows of $\mathbf{H}^{(l)}$ such that $(\sum_j (H_{kj}^{(l)})^2) = 1$ for all $k = 1, \ldots, K$ implies that $\det\left(\mathbf{H}^{(l)}\mathbf{H}^{(l)T}\right) \leq 1$ for every term $l = 1 \ldots L$ in the last line of (8). The larger L, the higher is the probability to obtain a solution with a small determinant in at least one iteration, and accordingly a small overall determinant results.

4   Illustrative Example

4.1   Unconstrained NMF versus detNMF

We tested the performance of the *detNMF* algorithm on artificial data. For this purpose, we used the following simulation setup:

- Generate $\mathbf{H}$ (see figure 2, left)

$$\mathbf{H} = \begin{pmatrix} \mathbf{H}_{1*} \\ \mathbf{H}_{2*} \\ \mathbf{H}_{3*} \end{pmatrix} = \begin{pmatrix} bowl \\ stair \\ block \end{pmatrix} \tag{9}$$

- Initialize $\mathbf{W}$ by non-negative random numbers

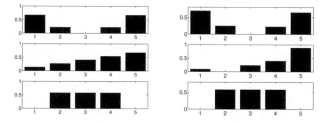

Fig. 2. *Left*: Original features *bowl* $\mathbf{H}_{1*}$(top), *stair* $\mathbf{H}_{2*}$(center), and *block* $\mathbf{H}_{3*}$(bottom), which are perfectly recovered by the detNMF algorithm. Each vector $\mathbf{H}_{k*}$ is a 5-dim analogon to the 2-dim vectors in figure 1. *Right*: Example of a valid solution, but with wrong features $\mathbf{H}'_{1*}, \mathbf{H}'_{2*}, \mathbf{H}'_{3*}$, obtained via unconstrained NMF.

The algorithm receives $\mathbf{X} = \mathbf{WH}$ and $K = 3$ as input. The matrices $\mathbf{H}$ and $\mathbf{W}$ are initialized with random positive numbers and then iteratively updated as described above until the normalized reconstruction error falls below a certain threshold (e.g. $(NM)^{-1}E(\alpha = 0) < 10^{-10}$). In the simulations, the algorithm *detNMF* always extracted the correct modes despite starting with random initializations as well as different original coefficient matrices $\mathbf{W}$ and varying numbers of individuals (e.g. $N = 100, 1000, 10000$).

During iteratively reconstructing the data matrix $\mathbf{X}$ with sufficient precision, the determinant criterion pushes the feature vectors towards the optimal solution with the smallest possible determinant. In contrast, the unconstrained version ($\alpha = 0$) converges to several different solutions, depending on the initialization. In figure 2 (right) we give an example of an exact nonnegative factorization of the data in $(\mathbf{W'}, \mathbf{H'})$ which does not reproduce the original modes $\mathbf{H}$ correctly. Note that $\mathbf{H'}_{2*} + 0.2\mathbf{H'}_{3*} \approx \mathbf{H}_{2*}$. Furthermore, note that $\det(\mathbf{HH}^T) = 0.18$ and $\det(\mathbf{H'H'}^T) = 0.31$. The basis $\mathbf{H'}$ is sufficient to explain all data by a non-negative superposition, but does not contain the correct solution which generated the data and which is characterized by a minimal determinant.

4.2 Determinant Criterion versus Sparseness Constraints

We now discuss an example in which the algorithm *detNMF* obtains the correct solution of a specified NMF problem, whereas a sparse approach [4], [13] fails to do so. For comparison, we chose the *nnsc*-algorithm, as described in [4] (*nnsc*: non-negative sparse coding), which minimizes the following objective function

$$E_{nnsc} = (NM)^{-1} \sum_{n=1}^{N} \sum_{m=1}^{M} (X_{nm} - [\mathbf{WH}]_{nm})^2 + \lambda \sum_{nk} W_{nk}. \qquad (10)$$

The term $\lambda \sum_{nk} W_{nk}, \lambda \geq 0$ penalizes large weights W_{nk}, hence a maximally sparse $\mathbf{W}$ corresponds to a solution where the columns $\mathbf{W}_{*k}$ contain as many zeros as possible.

Here we discuss a counter-example which demonstrates potential drawbacks of sparsity constraints. Using the same modes $\mathbf{H}_{k*}$ as in (eq. 9), we construct the coefficients in $\mathbf{W}$ as follows:

90% of the points are generated via $s \cdot (t \cdot \mathbf{H}_{1*} + (1-t) \cdot \mathbf{H}_{3*})$, where the parameter t is randomly drawn from a Gaussian distribution with $(\mu, \sigma) = (0.5, 0.03)$, and s is equally distributed between 0 and 1. The feature vectors $\mathbf{H}_{k*}$ constitute the edges of a tetrahedron (see figure 3, solid lines). By construction, the data has exactly three principal components related to nonzero eigenvalues. Most data points lie on a surface between two edges, hence two modes contribute roughly equally. The remaining 10% of the data are equally distributed on all surfaces spanned by the mode vectors.

After random initialization of $\mathbf{W}$ and $\mathbf{H}$ we applied both algorithms, *detNMF* and *nnsc*, until the reconstruction error in the objective function ($(NM)^{-1}E_{det}(\alpha = 0)$ in eq. 2 and $(NM)^{-1}E_{nnsc}(\lambda = 0)$ in eq. 10, respectively) was smaller than 10^{-10}. The *detNMF* algorithm recovered the correct modes. The *nnsc* algorithm, instead, produced a solution with a smaller value of $\sum_{n,k} W_{nk}$, but also a larger determinant. The final values of the constraints are given in table 1, while a $3D$-visualization of the data and the resulting modes are shown in figure 3. Note that while one mode $\mathbf{H'}_{i*}$ is oriented such as to satisfy the sparseness constraint, a second mode moves away from the given data points to satisfy the overall non-negativity constraint. Thus, if the data is not sparse, the decomposition achieved by a sparse NMF-algorithm does not necessarily yield the correct underlying causes. The minimal determinant constraint provides a more general approach, because it does not depend on the distribution of coefficients directly and obtains the true unique solutions even if the data are not sparse. In case of very sparsely encoded data where 90% of the data lie close to the edges H_{k*}, both algorithms achieve similar results as is seen also in Table 1.

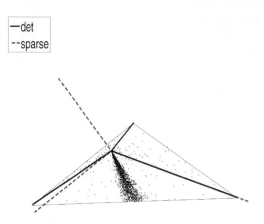

Fig. 3. 3D-visualization of the example (see text for details). Data vectors and modes are projected onto the three principal components of the data. In this space, the original modes H_{1*}, H_{2*}, H_{3*} constitute the edges of a tetrahedron with unit length, which are exactly recovered by the *detNMF* algorithm (solid lines). Obviously, the *nnsc*-NMF algorithm fails to position all basis vectors correctly. Note that the normalized modes deduced from the *nnsc* algorithm are drawn on an elongated scale (dashed lines) to render them visible. All modes intersect in a vertex at the origin. *Left*: top view, the origin vertex is in the center *Right*: side view, the left corner is the origin.

Table 1. Comparison of the values of the constraints at the optimal solutions gained by both algorithms, *nnsc* and *detNMF* for a non-sparse and a very sparse data set

	non-sparse		sparse	
	$\det(\mathbf{HH}^T)$	$\sum_{n,k} W_{nk}$	$\det(\mathbf{HH}^T)$	$\sum_{n.k} W_{nk}$
detnmf	0.1806	495.6798	0.1803	583.3226
nnsc	0.3096	487.7167	0.1804	583.2772

References

1. Paatero, P., Tapper, U.: Positive matrix factorization: a non-negative factor model with optimal utilization of error estimates of data values. Environmetrics 5, 111–126 (1994)
2. Lee, D., Seung, H.: Learning the parts of objects by non-negative matrix factorization. Nature 401, 788–791 (1999)
3. Lee, D., Seung, H.: Algorithms for Non-negative Matrix Factorization. In: NIPS, pp. 556–562 (2000)
4. Hoyer, P.: Non-Negative Sparse Coding. In: Proc. IEEE Workshop on Neural Networks for Signal Processing, Martigny, Switzerland, pp. 557–565 (2002)
5. Hoyer, P.: Non-negative Matrix Factorization with Sparseness Constraints. Journal of Machine Learning Research 5 (2004)
6. Cichocki, A., Zdunek, R., Amari, S.-i.: Csiszár's divergences for non-negative matrix factorization: Family of new algorithms. In: Rosca, J.P., Erdogmus, D., Príncipe, J.C., Haykin, S. (eds.) ICA 2006. LNCS, vol. 3889, pp. 32–39. Springer, Heidelberg (2006)

7. Stadlthanner, K., Theis, F.J., Puntonet, C.G., Górriz, J.-M., Tomé, A.M., Lang, E.W.: Hybridizing sparse component analysis with genetic algorithms for blind source separation. In: Oliveira, J.L., Maojo, V., Martín-Sánchez, F., Pereira, A.S. (eds.) ISBMDA 2005. LNCS (LNBI), vol. 3745, pp. 137–148. Springer, Heidelberg (2005)

8. Cichocki, A., Zdunek, R., Amari, S.: Nonnegative Matrix and Tensor Factorization. IEEE Signal Processing Magazine, 142–145 (January 2008)

9. Laurberg, H., Christensen, M.G., Plumbley, M.D., Hansen, L.K., Jensen, S.H.: Theorems on Positive Data: On the Uniqueness of NMF. Computational Intelligence and Neuroscience 2008, article ID 764206, doi:10.1155/2008/764206

10. Donoho, D., Stodden, V.: When does non-negative matrix factorization give a correct decomposition into parts? In: Proc. NIPS 2003. MIT Press, Cambridge (2004)

11. Theis, F.J., Stadlthanner, K., Tanaka, T.: First results on uniqueness of sparse non-negative matrix factorization. In: Proc. EUSIPCO, Eurasip (2005)

12. Berry, M.W., Browne, M., Langville, A.N., Pauca, V.P., Plemmons, R.J.: Algorithms and applications for approximate nonnegative matrix factorization. Computational Statistics & Data Analysis 52(1), 155–173 (2007)

13. Li, S., et al.: Learning Spatially Localized, Parts-Based Representation. In: Proc. IEEE Conf. Computer Vision and Pattern Recognition (2001)

14. Hopke, P.K.: A guide to positive matrix factorization. Clarkson University (2000), http://www.epa.gov/ttnamti1/files/ambient/pm25/workshop/laymen.pdf

15. Dhillon, I., Sra, S.: Generalized Nonnegative Matrix Approximations with Bregman Divergences. In: Proc. Neural Information Processing Systems, USA (2005)

16. Sajda, P., et al.: Recovery of constituent spectra in 3D chemical shift imaging using non-negative matrix fatorization. In: 4th International Symposium on Independent Component Analysis and Blind Signal Separation, pp. 71–76 (2003)

17. Juvela, M., Lehtinen, K., Paatero, P.: The use of Positive Matrix Factorization in the analysis of molecular line spectra. Mon. Not. R. Astron. Soc. 280, 616–626 (1996)

18. Miao, L., Qi, H.: Endmember extraction from highly mixed data using minimum volume constrained nonnegative matrix factorization. IEEE Trans. Geosci. Remote Sens. 45(3), 765–777 (2006)

Separation of Convolutive Mixtures with Hybrid Sources

Christine Servière

GIPSA-Lab, ENSIEG BP 46
38402 Saint-Martin d'Hères, France
christine.serviere@inpg.fr

Abstract. We propose in this paper a unique method to separate sources that may have different statistical properties, in the case of FIR convolutive mixtures. No constraint is necessary on the source statistics (i.i.d variables, Gaussian sources or temporally correlated sources..), nor on the number of each type of sources. On the contrary of previous works, no assumption of overdetermined mixtures is used. It relies on joint block-diagonalization of correlation matrices of some appropriate variables called differential complex signals, which are introduced in the paper.

1 Introduction

Blind source separation methods (BSS) consist in restoring a set of N unknown source signals from $M \geq N$ observed mixtures. Sources are only supposed to be mutually statistically independent. Many methods have been proposed for instantaneous mixtures but recovering the sources from their linear FIR convolutive mixtures remains a challenging problem. Several solutions have been addressed in the time-domain. In the first approach, authors assume each source to be non Gaussian, independent and identically distributed variables. Non-Gaussianity and independence are then used as separation criteria and therefore methods such as deflation approach, convolutive fast ICA or contrast maximization have been developed [1-4]. In the second approach, temporally correlated sources are considered. The FIR convolutive mixing model can be reformulated into an instantaneous one as suggested in [5-7] by introducing some appropriate variables. Then the problem comes down to separation of an instantaneous mixture of some new sources but some of then are now dependent. To solve this challenging problem, authors have generalized standard BSS methods for the instantaneous case by using a unitary or nonunitary joint block diagonalization scheme [6-8]. However this new instantaneous mixing model is full rank under the hypothesis of overdetermined model $M > N$. More precisely, a condition links the source number N, the sensor number M and the taps number of the mixing filter L.

We propose in this paper a unique method to separate sources that may have different statistical properties: white sources, i.i.d variables (eventually Gaussian) or temporally correlated sources. No constraint is necessary on the type of sources, the number of each type of sources or on M and N. So, we suppose M=N in all the paper. First the aim of BSS is presented in section 2. Then in section 3, we introduce the differential complex signal, denoted $\Delta s(t, k, P)$, as the difference between $s(t)$ and its

T. Adali et al. (Eds.): ICA 2009, LNCS 5441, pp. 114–121, 2009.

delayed version $s(t+P)exp(-j2\pi kP/L)$. The differential complex signal is linked to the Sliding Fourier Transform calculated at frequency bin k. Its temporal redundancy makes $\Delta s(t, k, P)$ correlated versus parameter k, whatever the original statistical property of $s(t)$. Differential methods have been already used in [9,10] but in a different way for undetermined mixtures and non-stationary signals. The goal was to hide stationary signals and keep non-stationary ones. In section 4, we show how the FIR convolutive mixing model can be reformulated into an instantaneous one by introducing the differential complex signals as new sources whose some of them are now dependent. Then the problem comes down to separation of an instantaneous mixture, invertible without supposing overdetermined mixtures. The problem is solved in section 5 by the joint block-diagonalization of a set of correlation matrices (NxN) instead of (ML')x(ML') where L' is an integer verifying: $ML' \geq N(L+L')$ in previous works. The proposed cost function is then easier to handle with, even for large dimension of the mixing filters L.

2 Aim of the BSS and Assumptions

Let us consider the standard convolutive mixing model with N inputs and N outputs. Each sensor $x_j(t)$ $(j=1, .., N)$ receives a linear convolution (noted *) of each source $s_i(t)$ $(i=1,...,N)$ at discrete time t:

$$x_j(t) = \sum_{i=1}^{N} h_{ij} * s_i(t) = \sum_{i=1}^{N} \sum_{k=0}^{K-1} h_{ij}(k) * s_i(t-k) \tag{1}$$

where h_{ij} represents the impulse response from source i to sensor j with K taps. The sources are only supposed to be uncorrelated.

As the inverse of mixing filters are not necessarily causal, the aim of BSS is to recover non-causal filters with impulse responses g_{ij} between sensor i and output j, such that the output vector $y(t)$ estimates the sources, up to a linear filter :

$$y_j(t) = \sum_{i=1}^{N} \sum_{k=-L/2}^{L/2} g_{ij}(k)x_i(t-k) \tag{2}$$

3 Differential Signals Constructed from Sliding Fourier Transforms

The Sliding Fourier Transform (SFT) of a discrete-time signal $s(t)$ is obtained by applying the L-point discrete Fourier Transform to moving windows of size P $(P<L)$, $[s(t), s(t+1),..., s(t+P-1)]$, i.e:

$$s(\omega_k,t) = \sum_{m=0}^{P-1} s(t+m)\exp(-j\omega_k m), k = 0,...,L-1, t = 0,1,... \tag{3}$$

where $\omega_k=2\pi k/L$ denotes the frequency bin. We assume that the number of frequencies, L, is greater than the window size P. This implies that the windows must be padded with L-P zeros before processing them. (3) introduces redundancy in the frequency-domain signal when considering overlapping consecutive windows at t and $t+1$. Indeed, $s(\omega_k, t)$ and $s(\omega_k, t+1)$ only differ in two samples $s(t)$ and $s(t+P)$. This redundancy can be exploited to express a relation between the time-domain signal $s(t)$ and the time-frequency signal (SFT) $s(\omega_k, t)$:

$$s(t) - s(t+P)e^{-j2\pi kP/L} = s(\omega_k, t) - s(\omega_k, t+1)e^{-j2\pi kP/L}, k = 0, ..., L-1 \quad (4)$$

This relation is used in section 4 and is the key point to reformulate the convolutive mixture (1) into an instantaneous one, without supposing overdetermined mixture and constraint between M, N and L. We suppose M=N in all the paper.

Denote $\Delta s(t, k, P)$ the differential complex signal between $s(t)$ and the delayed version $s(t+P)exp(-j2\pi kP/L)$, of parameters k and P. From (4), we have :

$$\Delta s(t, k, P) = s(t) - s(t+P)e^{-j2\pi kP/L}$$
$$= s(\omega_k, t) - s(\omega_k, t+1)e^{-j2\pi kP/L}, k = 0, ..., L-1 \quad (5)$$

The main property of the differential complex signal $\Delta s(t, k, P)$, which will be useful in the rest of the paper, is that $\Delta s(t, k, P)$ is correlated versus parameter k, whatever the statistical property of the signal $s(t)$: $E\{\Delta s(t, k_1, P)\Delta s(t, k_2, P)^*\}$ is not zero.

$$E\{\Delta s(t, k_1, P)\Delta s(t, k_2, P)^*\} = \Gamma_s(0)(1 + e^{-j2\pi(k_1 - k_2)P/L}) - \Gamma_s(P)(e^{-j2\pi k_1 P/L} + e^{j2\pi k_2 P/L})$$
$$E\{\Delta s(t, k, P)\Delta s(t, k, P)^*\} = 2(\Gamma_s(0) - \Gamma_s(P)\cos(2\pi kP/L)) \quad (6)$$

with : $\Gamma_s(\tau) = E\{s(t)s(t-\tau)^*\}$.

4   Reformulation of the Model

The BSS convolutive mixing model (1) can be reformulated, using the differential signals of sources $\Delta s_j(t, k, P)$ and observations $\Delta x_i(t, k, P)$. From the properties of the Fourier transforms, $x_i(\omega_k, t)$ are instantaneous mixtures of the frequency-domain sources $s_j(\omega_k, t,)$:

$$x_i(\omega_k, t) = \sum_{j=1}^{N} H_{ij}(\omega_k)s_j(\omega_k, t) \quad (7)$$

if P is chosen such that K<P<L, and $H_{ij}(\omega_k)$ is the L-point DFT of filters h_{ij}:

$$H_{ij}(\omega_k) = \sum_{m=0}^{P-1} h_{ij}(m)\exp(-j\omega_k m), k = 0, ..., L-1, \quad \omega_k = 2\pi k/L \quad (8)$$

Applying equation (5) to the differential observations signals and using (7), the model described in (1) is rewritten as :

$$\Delta x_i(t, k, P) = \sum_{j=1}^{N} H_{ij}(\omega_k)\Delta s_j(t, k, P) \quad (9)$$

Let us denote by $\Delta x(t, P)$ the vector of the differential observations signals and by $\Delta s(t, P)$ the vector of the differential sources signals, where :

$$\Delta x(t,P) = \left[\Delta x_1(t,k_0,P),...,\Delta x_1(t,k_{L-1},P),...,\Delta x_N(t,k_0,P),...,\Delta x_N(t,k_{L-1},P),\right]^T \quad (10)$$

$$\Delta s(t,P) = \left[\Delta s_1(t,k_0,P),...,\Delta s_1(t,k_{L-1},P),...,\Delta s_N(t,k_0,P),...,\Delta s_N(t,k_{L-1},P),\right]^T \quad (11)$$

The model described by (1) and (9) is rewritten in matrix form and reformulated into the instantaneous model:

$$\Delta x(t,P) = H\Delta s(t,P) \quad (12)$$

$$H = \begin{bmatrix} H_{11} & . & H_{1N} \\ . & . & . \\ H_{N1} & . & H_{NN} \end{bmatrix} \quad H_{ij} = \begin{bmatrix} H_{ij}(\omega_0) & 0 & O \\ 0 & . & 0 \\ 0 & 0 & H_{ij}(\omega_{L-1}) \end{bmatrix}$$

where H_{ij} is a (LxL) diagonal matrix.

The convolutive model states now : $\Delta x(t, P) = H\,\Delta s(t, P)$. Each component of $\Delta s(t, P)$, $\Delta s_i(t, k, P)$, is correlated versus parameter k (6), so that the correlation matrix of $[\Delta s_i(t, k_0, P),..., \Delta s_i(t, k_{L-1}, P)]$ is full for each source $i=1,...,N$, whatever the original temporal structure of the source and its statistical properties. We see from (6) that even the correlation matrix of a white signal is full (and also different for different values of P). The correlation matrix of $\Delta s(t, P)$ is then a block diagonal matrix.

5 Separating Algorithm

As proposed in [6-8], one possible way to tackle the problem is joint block-diagonalization of a set of n estimated correlation matrices of the observations $E\{\Delta x(t, P_i)\Delta x(t, P_i)^H\}$ for $i=1,...,n$. The assumption of non-stationary sources is usually exploited to form a set of different correlation matrices. In this paper, we see from (12) that the set may be also composed of correlation matrices calculated with different values of P.

Let us consider the set of matrices composed by the n correlation matrices $E\{\Delta x(t, P_i)\Delta x(t, P_i)^H\}, i=1,...,n$, where $(.)^H$ denotes the transpose conjugate operator.

$$E\{\Delta x(t,P)\Delta x(t,P)^H\} = HE\{\Delta s(t,P)\Delta s(t,P)^H\}H^H \quad (13)$$

$$\text{where : } E\{\Delta s(t,P)\Delta s(t,P)^H\} = \begin{bmatrix} D_{P,11} & 0 & 0 \\ 0 & D_{P,ii} & 0 \\ 0 & 0 & D_{P,NN} \end{bmatrix}$$

H is a (NL)x(NL) full rank matrix and G stands for its inverse. As proposed in [8] G is estimated from the non-unitary joint block diagonalization of the set of n matrices :

$$GE\{\Delta x(t, P_i)\Delta x(t, P_i)^H\}G^H, \text{ for } i=1,...,n.$$

Let us consider that : $M_P = E\{\Delta x(t, P)\Delta x(t, P)^H\}$. We propose to minimize the following cost function :

$$C(G) = \sum_{P=PI}^{Pn} \left\| Offdiag\{GM_pG^H\} \right\|_F^2 \tag{14}$$

where $Offdiag\{.\}$ denotes the zero-block-diagonal matrix and $\|.\|_F$ the Frobenius norm. This cost function has already been envisaged in [8] in a general way and leads to :

$$C(G) = \sum_{P=PI}^{Pn} \sum_{i,j=1(i\neq j)}^{N} \sum_{m,n=1}^{L} \left| g_i^m M_P \left(g_j^n \right)^H \right|^2 \tag{15}$$

where g_i^m stands for the m th row vector of matrix G_i if $G =[G_1 \ ... \ G_N]^T$ and G_i $i=1,..,N$, are (LxN) block matrices. Yet the interest of using the previous instantaneous model is that the new 'mixing matrix' H and its inverse G own a particular structure that can be introduced in the function (15) to simplify it. As H is constituted of square block-matrices H_{ij} $(i,j=1,N)$ which are all diagonal matrices, its inverse G has also the same particular structure:

$$G = \begin{bmatrix} G_{11} & . & G_{1N} \\ . & . & . \\ G_{N1} & . & G_{NN} \end{bmatrix} \quad\quad G_{ij} = \begin{bmatrix} G_{ij}(\omega_0) & 0 & 0 \\ 0 & . & 0 \\ 0 & 0 & G_{ij}(\omega_{L-1}) \end{bmatrix} \tag{16}$$

G_{ij} are (LxL) diagonal matrices and terms in G verify :

$$\begin{bmatrix} G_{11}(\omega_i) & . & G_{1N}(\omega_i) \\ . & . & . \\ G_{N1}(\omega_i) & . & G_{NN}(\omega_i) \end{bmatrix} \cdot \begin{bmatrix} H_{11}(\omega_i) & . & H_{1N}(\omega_i) \\ . & . & . \\ H_{N1}(\omega_i) & . & B_{NN}(\omega_i) \end{bmatrix} = Id, \omega_i = \omega_0,...,\omega_{L-1} \tag{17}$$

As many terms are null in G_{ij} and g_i^m the cost function may be simplified into :

$$C(G) = \sum_{P=PI}^{Pn} \sum_{i,j=1(i\neq j)}^{N} \sum_{m,n=1}^{L} \left| \tilde{g}_i^m \tilde{M}_P(m,n) \left(\tilde{g}_j^n \right)^H \right|^2 \tag{18}$$

where $\tilde{g}_i^m$ is the restriction of vector g_i^m to the N non-zero terms (instead of LN terms in g_i^m). From (16), we have:

$$\tilde{g}_i^m = [G_{i1}(\omega_{m-1}) \quad G_{i2}(\omega_{m-1}) \quad . \quad .G_{iN}(\omega_{m-1})] \tag{19}$$

$\tilde{M}_P(m,n)$ is a NxN matrix composed of elements of matrix M_P (NL)x(NL). If $M_P = [M_1 \ ... \ M_N]^T$ where M_i $i=1,..,N$, are block matrices of dimension LxNL, $\tilde{M}_P(m,n)$

is extracted from M_P by first cancelling (L-1) rows in each matrix M_i. These rows depend on m and correspond to null terms in vector g_i^m in the matrix product $g_i^m M_P$. Then if $M_P = [\ M'_1\ ...\ M'_N\ ]$ where $M'_i\ i=1,..,N$, are block matrices of dimension NLxL, (L-1) columns are cancelled in each matrix M'_i. The place of these columns depends on n and corresponds to null terms in vector $\left(g_j^n\right)^H$ in the matrix product $M_P\left(g_j^n\right)^H$. After these two operations, the matrix is contracted into a NxN matrix $\tilde{M}_p(m,n)$ where the null terms are eliminated. So, vectors $\tilde{g}_i^m, \tilde{g}_j^n$, and the (NxN) matrix $\tilde{M}_p(m,n)$ verify the scalar equality:

$$g_i^m M_P \left(g_j^n\right)^H = \tilde{g}_i^m \tilde{M}_P(m,n)\left(\tilde{g}_j^n\right)^H$$

The cost function can be rewritten on the following quadratic form : (20)

$$C(G) = \sum_{m=1}^{L}\sum_{i=1}^{N} \tilde{g}_i^m B(i,g)\left(\tilde{g}_i^m\right)^H$$

$$B(i,g) = \sum_{P=P1}^{Pn}\sum_{j=1(i\neq j)}^{N}\sum_{n=1}^{L} \tilde{M}_P(m,n)\left(\tilde{g}_j^n\right)^H \tilde{g}_j^n \left(\tilde{M}_P(m,n)\right)^H$$

As proposed in [8], the minimization of (20) is achieved by taking $\tilde{g}_i^m$ as the L unit eigenvectors associated with the L smallest eigenvalues of the NxN matrices $B(i,g)$. The minimization is achieved with an iterative procedure as $B(i,g)$ also depends on vectors $\tilde{g}_i^m$ for a given i.

The main interest of this approach versus previous ones [6-8] is that the matrices to diagonalized are NxN instead of (LN)x(LN), where N is the source number. The sources are not supposed to be temporally correlated or non-stationary. So the proposed method can be applied to hybrid sources and its effectiveness is shown in next section. The difference of the proposed method with a frequency-approach is that there is no permutation ambiguity as the criterion comes from the joint bock diagonalization of correlation matrices and not spectral ones.

6 Performance and Simulation

We present simulations to illustrate the effectiveness of the proposed algorithm. It is measured with the performance index I which shows the closeness of the global matrix GH with a true bock diagonal matrix. The better result is obtained when I is close to 0.

$$I = (1/(N(N-1))(\sum_{i=1}^{N}(\sum_{j=1}^{N}\frac{\|(GH)_{ij}\|^2}{\max_l \|(GH)_{il}\|^2}-1) + \sum_{j=1}^{N}(\sum_{i=1}^{N}\frac{\|(GH)_{ij}\|^2}{\max_l \|(GH)_{lj}\|^2}-1)$$

The index I is displayed in figure 1 versus the number of matrices M_P for the proposed method and compared with [8]. The case of 4 sensors is considered. The sources are mixed through FIR filters of K=4 coefficients in the model (1). The coefficients are random entries chosen from a Gaussian distribution with zero mean and unit variance. 40000 samples are used in the simulation and the index is then averaged on 50 realizations. Spatially and temporally uncorrelated noises are also added with a signal to noise ratio of 10dB. [8] needs temporally correlated signals and overdetermined mixtures. For L=4, only two sources are permitted. So, the performance index of [8] is calculated in the case of 4 mixtures of two speech signals (solid green line). The performance index of the proposed method is computed, considering 4 mixtures of 4 sources : two sources are Gaussian iid signals and two sources are the previous speech signals (solid blue line). The proposed method shows better results for the convergence speed and the residual error than method [8] (due to smaller matrices). The presence of iid sources does not prevent the separation to be achieved.

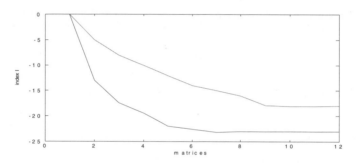

Fig. 1. Index performance versus number of matrices

To show that the method does not assume hypothesis on the type of statistics, we compute the performance index for 5 matrices with 4 different mixtures : 4 Gaussian white sources, 4 speech signals, 2 speech signals and 2 Gaussian iid. variables, 2 sinusoids and 2 uniform iid variables. The mixing matrix is computed as previously. The values of I shown in table 1 reveal that the index is quasi independent of the type of source statistics. [8] is not able to achieve the separation in these cases. Other simulations results prove that the performance index only depends on the signal to noise ratio and the source number.

Table 1. Performance index for hybrid sources

	4 Gaussian white sources	4 speech signals	2 speech signals + 2 Gaussian iid. variables	2 sinusoids and 2 uniform iid variables
I	-23.4dB	- 21.5dB	-23.1dB	-22.5dB

7   Conclusion

We focus on the separation of FIR convolutive mixtures of hybrid sources and propose a unique method to separate hybrid sources. The sources are allowed to have different statistics (i.i.d or temporally correlated variables, eventually Gaussian) and the number of each type of source is not necessary. It relies on joint block-diagonalization of correlation matrices of some appropriate variables called differential complex signals, defined in the paper. The differential complex signal is linked to the Sliding Fourier Transform. Its temporal redundancy makes it correlated versus an appropriate parameter whatever the original statistical property of the sources.

References

1. Comon, P.: Contrasts for multichannel blind deconvolution. IEEE Signal Processing Letters 3(7), 209–211 (1996)
2. Simon, C., Loubaton, P., Jutten, C.: Separation of a class of convolutive mixtures: a contrast function approach. Signal Processing 81, 883–887 (2001)
3. Thomas, J., Deville, Y., Hossseini, S.: Time –Domain Fast Fixed-Point Algorithms for Convolutive ICA. IEEE Signal Processing Letters 13(4), 228–231 (2006)
4. Tugnait, J.K.: Identification and deconvolution of multichannel linear non-Gaussian processes using HOS and inverse filter criteria. IEEE T. SP 45(3), 658–672
5. Ghorokhov, A., Loubaton, P.: Subspace based techniques for second order blind separation of convolutive mixtures with temporally correlated sources. IEEE Trans. Circuit Syst. 44(9), 813–820 (1997)
6. Bousbiah, H., Belouchrani, A., Abed-Meraim, K.: Jacobi-like algorithm for blind signal separation of convolutive mixtures. Electronic Letters 37, 1049–1050 (2001)
7. Févotte, C., Doncarli, C.: A unified presentation of blind source separation methods for convolutive mixtures using block-diagonalization. In: ICA 2003, Nara, Japan (2003)
8. Ghennioui, H., Fadaili, E., Thirion, N., Adib, A., Moreau, E.: A Nonunitary Joint Block Diagonalization Algorithm for Blind Separation of Convolutive Mixtures of Sources. IEEE Signal Processing Letters 14(11) (November 2007)
9. Deville, Y., Benali, M., Abrad, F.: Differential source separation for underdetermined instantaneous or convolutive mixtures. Signal Processing 84(10)
10. Choi, Cichocki, A., Deville, Y.: Differential decorrelation for nonstationary source separation. In: ICA 2001, Helsinki (2001)

Blind Extraction of Chaotic Sources from White Gaussian Noise Based on a Measure of Determinism

Diogo C. Soriano[1,3], Ricardo Suyama[2,3], and Romis Attux[1,3]

[1] Dept. of Computer Engineering and Industrial Automation
School of Electrical and Computer Engineering (FEEC) – University of Campinas
(Unicamp) – Caixa Postal 6101, CEP 13083-970 - Campinas – SP – Brazil
[2] Dept. of Microwave and Optics
School of Electrical and Computer Engineering (FEEC) – University of Campinas
(Unicamp) – Caixa Postal 6101, CEP 13083-970 - Campinas – SP – Brazil
[3] Lab. of Signal Processing for Communications (DSPCom)
School of Electrical and Computer Engineering (FEEC) – University of Campinas
(Unicamp) – Caixa Postal 6101, CEP 13083-970 - Campinas – SP – Brazil
{soriano,attux}@dca.fee.unicamp.br, suyama@dmo.fee.unicamp.br

Abstract. This work presents a new method to perform blind extraction of chaotic signals mixed with stochastic sources. The technique makes use of the features underlying the generation of chaotic sources to recover a signal that is "as deterministic as possible". The method is applied to invertible and underdetermined mixture models and illustrates the potential of incorporating such *a priori* information about the nature of the sources in the process of blind extraction.

Keywords: Blind source extraction, blind source separation, chaotic signals, recurrence maps.

1 Introduction

Nonlinear systems exhibit a very rich dynamical scenario, which includes possibilities such as convergence to fixed points, existence of limit-cycles, quasi-periodicity and chaos. This last kind of oscillation, in particular, is characterized by aperiodicity and sensitivity to initial conditions, aspects that can be easily mistaken for random behavior [1].

In fact, distinguishing chaotic from random signals is a far from a trivial task, which is reflected in the relatively small number of methods developed to solve it [2, 3]. Typically, these methods are less robust when the analyzed signals are mixtures of chaotic and random processes: in these cases, it is certainly of great use to rely on a preprocessing stage (e.g., a filtering process in frequency domain) in order to enhance specific features of each class of signals, although this can cause the loss of relevant information [4, 5, 6], since both signals have a broadband spectrum.

From a theoretical standpoint, the above described scenario is intimately related to the blind source extraction (BSE) problem, as its essence lies in the idea of recovering a specific set of signals of interest – usually the deterministic signals – from mixed versions of them. In order to achieve a successful extraction, there must be a criterion capable of establishing an effective distinction between signals belonging to different classes. A natural possibility would be to employ a classical blind approach such as

T. Adali et al. (Eds.): ICA 2009, LNCS 5441, pp. 122–129, 2009.
© Springer-Verlag Berlin Heidelberg 2009

the well-established independent component analysis (ICA), although it would not be capable of exploring the peculiar features of the problem, particularly, the fact that some signals are generated by a deterministic dynamical system. As a matter of fact, this is an instance in which *a priori* information about the sources is available, which is always something that widens the applicability of blind signal processing (research areas such as sparse component analysis attest this fact).

In this work, a new method for solving the BSE problem when chaotic and stochastic processes are mixed is presented. The technique explores the dynamic features underlying the generation of the chaotic sources to recover a signal that is "as deterministic as possible". The solution employs a recurrence map - a well-established tool for analyzing dynamical systems [3, 7] – to build a cost function capable of quantifying the degree of determinism of a given signal. This cost function is used to adapt linear separating systems under different signal and mixture models (both invertible and underdetermined), and a comparison with a classical ICA methodology is established.

2   Chaotic Signals and the Generation of a Recurrence Map

For our purposes, a chaotic signal should be simply understood as one generated by a chaotic system, which means that its properties are defined by the dynamical system that generates it and the behavior of its trajectories in the phase space [1]. Therefore, the first analytical step we must take is to reconstruct the associated phase-space from the observed signal. In view of Takens' embedding theorem [1, 3, 4, 6], reconstruction can be performed by defining a state vector $\mathbf{x}(k)$ such that

$$\mathbf{x}(k) = \begin{bmatrix} x(k) & x(k-\tau) & \cdots & x(k-(d_e-1)\tau) \end{bmatrix} \tag{1}$$

where d_e represents the embedding dimension – defined as the effective number of degrees of freedom of the dynamical system – and τ represents the delay between samples. Even though this trajectory may not be exactly the same as that generated by the system, it will be topologically equivalent thereto [1].

Since the approach we wish to present is based on a recurrence map, let us consider how this analytical tool - first proposed by Eckmann *et. al.* [3] – can be built and how it provides means to determine whether one deals with random or chaotic signals. Using the reconstructed state $\mathbf{x}(k)$, the recurrence map will be represented by an NxN matrix, where the element (i,j) will be zero in a grayscale representation (a black dot) whenever x(i) is sufficiently close to x(j), i.e., whenever |x(i)-x(j)|< ε.

The applicability of recurrence maps becomes clear when we compare maps obtained from signals of different natures. In Fig. 1, we present maps generated from a periodic signal (Fig. 1A), a chaotic signal (Fig. 1B), a random signal (Fig. 1C) and a mixture of chaotic and random signals (Fig. 1D). Notice how the patterns differ in their structure and their regularity: in fact, these aspects provide relevant information about the dynamical behavior. For example, long diagonal lines represent periodic states, while horizontal and vertical lines represent unchanged states.

In this work, the crucial point is that chaotic time series generate shorter diagonals than those associated with periodic signals, but longer than those associated with a random process [7], as can be observed in Fig. 1. This suggests that the "length" of diagonals in a recurrence map can be used to form a contrast function to separate a deterministic from a random signal. In the following, we shall develop this idea.

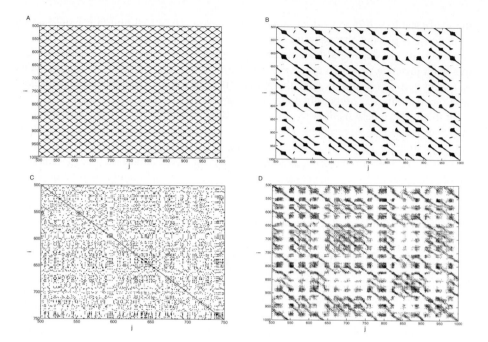

Fig. 1. Panels A, B, C, D show in detail, respectively, the recurrence maps ($N = 1000$ samples, $d_e = 3$, $\tau = 3$, $\varepsilon = 0.5$) from a periodic oscillation ($sin(10t)$), a chaotic Lorenz time series, a gaussian random source (zero mean and unitary variance) and a mixture of random and chaotic sources

3 Extracting Deterministic Sources

Let us consider that two sources - one being a chaotic signal $s_c(n)$ and the other being a stochastic signal $s_s(n)$ – are linearly mixed, giving rise to $\mathbf{x}(n) = \mathbf{A}\mathbf{s}(n)$, where $\mathbf{x}(n) = [x_1(n)\ x_2(n)]^T$ is the mixture vector, $\mathbf{A}$ is the 2x2 mixing matrix (which is assumed to have full rank) and $\mathbf{s}(n) = [s_1(n)\ s_2(n)]^T$ is the source vector. In this problem, the aim of the BSE task we wish to solve is to extract a source from the mixtures, which can be achieved by multiplying the mixture vector by a vector $\mathbf{w}$, so that the output vector yield $\mathbf{y}(n) = \mathbf{w}^T\mathbf{x}(n) = \mathbf{s}_c(n)$. Let us also assume, without loss of generality, that the mixing matrix is orthogonal (this can be achieved via a whitening procedure) and that $\mathbf{w}$ can be parameterized in terms of a single variable θ, i.e., $\mathbf{w} = [\cos\theta\ \sin\theta]^T$.

Classical ICA approaches look for solutions in θ that ensure maximal nongaussianity (which can be evaluated with the aid of the kurtosis of the output components) or, alternatively, by maximization of independence between the elements of the output vector (e.g. by minimizing a mutual information measure). It is also important to remark that ICA allows the recovery of the original sources up to scale and permutation ambiguities [8]. However, when it is known that one of the sources is, for example, a deterministic chaotic signal, it is viable to obtain the separating system within a very different framework, a framework based on *a priori* information, i.e., the deterministic nature of the signal. In this case, and assuming that the idea is to extract $s_c(n)$, it is possible to look for a separating vector *that maximizes the deterministic character of the estimated source*. An immediate consequence is that the permutation ambiguity should not exist, at least in its original form, since the measure will establish a difference between deterministic and stochastic sources.

The deterministic character of the outputs can be measured, in consonance with what was presented in section 2, by employing a recurrence map. A straightforward possibility is to analyze the occurrence of diagonal lines (specially "long" ones) in the map. In very simple terms, we will use the fact that "long diagonals" tend to characterize deterministic behavior to tailor a cost function capable of discriminating between signals of different natures. The rationale is that "long diagonals", in a certain sense, are indicative of temporal and spatial correlation caused by the deterministic generative law underlying $s_c(n)$, whereas the same does not hold, by definition, for a random signal. The reader interested in a more formal presentation of recurrence maps and of their relationship with the idea of correlation is referred to [2, 7].

In the exposed methodology, we aim to find the value of θ that provides the separating vector that maximizes the following cost function (based on the recurrence map of the output signal $y(n)$):

$$fit = \sum_{k=a}^{b} N_k \tag{2}$$

where N_k is the percentage of points in a diagonal of length k, which implicitly depends on the mixing matrix, on the reconstruction parameters and on the threshold distance ε.

4 Results

In order to analyze the validity of the proposed methodology, we shall turn our attention to distinct representative simulation scenarios. In all cases, we will consider that $s_c(n)$ is the first state variable of the emblematic Lorenz system [1] (pre-processed to have zero mean and unitary variance), and that the stochastic sources are always white gaussian processes (zero mean and 10 dB below the chaotic source in power). These choices were made after a sequence of initial tests, although they are by no means mandatory: tests with other chaotic systems (e.g. the Rossler dynamical system and the Logistic map) led to similar results.

The first set of simulations exactly corresponds to the example used in our previous discussion, being a typical case in which perfect inversion of the mixing matrix is possible. There are two sources, $s_c(n)$ and $s_s(n)$, and the method will be tested in its potential of recovering both the chaotic signal and the stochastic signal. In the second set of simulations, we will consider a case that is a "twofold taboo" in comparison with the standard signal model: there will be three sources and only two mixtures and, moreover, two sources will be gaussian processes with zero mean and unitary variance. In order to guide our analysis in this underdetermined context, we will consider as a reference the performance of the separating solutions obtained via the (supervised) Wiener paradigm [9]. The idea is to have a clearer idea of the influence of criterion and structure in the quality of the obtained estimates.

4.1 First Scenario

In the first scenario, we have two signals - $s_c(n)$ and $s_s(n)$ – and a full rank mixing matrix $\mathbf{A} = [\sin\theta\ -\cos\theta;\ \cos\theta\ \sin\theta]$, with $\theta = \text{pi}/6$. In order to recover the chaotic source, the extracting vector $\mathbf{w}$ should be chosen such that $y_1(n) = \mathbf{w}^T\mathbf{x}(n)$ be as deterministic as possible, using (2) to quantify the degree of determinism of the estimated signal.

The cost function (2) was evaluated taking into account diagonals of length 20 to 40 ($a = 20$ and $b = 40$). In Fig. 2, we present the values of the proposed cost function and also of two commonly used contrasts in classical ICA methods: the kurtosis of $y_1(n)$, and the mutual information between $y_1(n)$ and $y_2(n) = [-\sin\theta\ \cos\theta]\mathbf{x}(n)$ (evaluated with the estimator developed in [10]).

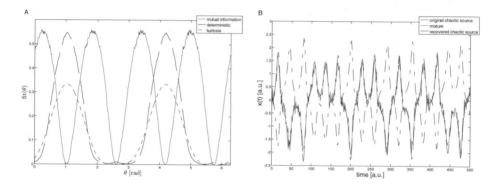

Fig. 2. Panel A – deterministic measure (2) of $y_1(n)$, kurtosis of $y_1(n)$ and mutual information between $y_1(n)$ and $y_2(n)$ for different θ values. Panel B - time series of the $s_c(n)$, $x_1(n)$ and the recovered $s_c(n)$ with inverted phase obtained from deterministic approach, (a.u. – arbitrary units).

The first important conclusion is that the proposed method has global optima at the solutions that lead to perfect inversion (up to a sign ambiguity), a feature shared by methods based on kurtosis and mutual information. These results reveal that the proposal fulfilled the essential "soundness requirements" of a separation method and had a performance equivalent to that obtained via classical ICA methods. Fig. 2B contains the time series of $s_c(n)$, $x_1(n)$ and the estimate of $s_c(n)$ (with inverted phase) provided by deterministic approach.

The return map for the recovered signal is shown in Fig. 3A, and is very similar to that shown in Fig. 1B. In Fig. 3B, we show the percentage of dots lying on diagonals of lengths between 20 and 40: notice how the mixed signal has a completely different behavior in this respect, whereas $s_c(n)$ and $y_1(n)$ have similar curves.

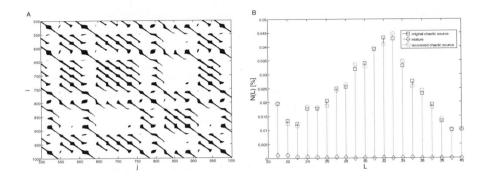

Fig. 3. Panel A - show in detail the recurrence map obtained from recovered chaotic source, (*N* = 1000, τ = 3, d_e = 3, ε = 0.5). Panel B – percentage of points (N(L)) in diagonals of length L for the signals $s_c(n)$, $x_1(n)$ and the estimate of $s_c(n)$.

4.2 Second Scenario

The analysis of the proposed method in the previous scenario essentially revealed the viability of its adoption, but the context of underdetermined separation (more sources than sensors) offers other important perspectives of application. In our analysis, we will assume that a chaotic source is mixed with two other stochastic (gaussian distributed) signals. Notice that the use of a mutual information measure is, in this case, made difficult due to the structural impossibility of reaching a condition of perfect inversion. A measure based on kurtosis can be directly applied in this case, but it is conceptually difficult to rely on its establishing any distinction in favor of the recovery of the deterministic signal.

Fig. 4A shows the three cost functions for this scenario (**A** = [0.2 0.7 0.4; 0.7 0.2 0.3]). Notice that the mutual information of the output is not a reliable criterion for

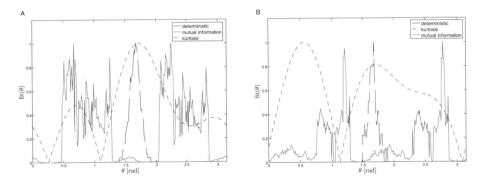

Fig. 4. Panel A – typical result of normalized deterministic measure of $y_1(n)$, kurtosis of $y_1(n)$ and mutual information between $y_1(n)$ and $y_2(n)$ for different θ values when the underdetermined case is considered. Panel B – the same simulation is performed, but the situation when kurtosis fail to recover the chaotic source is illustrated.

determining a separating matrix (due to the structural aspect discussed above), while nongaussianity, evaluated using kurtosis, has clear nulls related to the recovery of the stochastic sources (which are Gaussian) but does not have a maximum at the correct angle in all cases (as shown in Fig. 4B).

Table 1 shows the mean and standard deviation of the residual mean-square errors of the recovered chaotic source in 50 runs performed in different scenarios (for three approaches: Wiener, the proposed method and the kurtosis-based method). Notice how the values related to the deterministic approach are very similar to those associated with the Wiener paradigm. In fact, in 30% of the performed simulations, the kurtosis-based method led to an MSE 50% higher than that obtained via the Wiener approach, which never happened for the deterministic method. Even when the mixture matrix ($\mathbf{A} = [0.8\ 0.7\ 0.4;\ 0.7\ 0.3\ 0.5]$) gives rise to a relatively large minimum Wiener error, the performance of the proposed method is better than that of the kurtosis criterion.

Table 1. Mean square error (MSE) ± standard deviation (using 50 trials) for the recovered chaotic source estimated via the Wiener approach (MSEw), via the proposed deterministic measure (MSEd) and via an ICA method based on kurtosis (MSEk). The first and the second data rows consider, respectively, mixture matrices that provide a low and high minimum Wiener error.

MSEw	MSEd	MSEk
$0.063 \pm 2.83 \cdot 10^{-3}$	$0.069 \pm 6.17 \cdot 10^{-3}$	0.157 ± 0.321
$0.403 \pm 1.73 \cdot 10^{-2}$	$1.118 \pm 4.24 \cdot 10^{-2}$	1.268 ± 0.800

5 Discussions and Conclusions

This work presented a method to perform BSE when there are mixtures of chaotic and white gaussian processes. The results showed that the method was capable of producing solutions with a performance compatible with that of classical ICA under an invertible mixture model. Additionally, for more complicate (e.g. underdetermined) models, the *a priori* information incorporated by the approach was responsible for a significant degree of robustness, allowing an adequate estimate of the separating matrix even when only a small number of samples is available. The deterministic approach is not only restricted to gaussian processes, but can be employed for mixtures of uniform stochastic sources as well (data not shown). On the other hand, the proposal requires a higher amount of computational resources in comparison with standard separation/extraction solutions.

It is important to stress that is quite possible to achieve better criterions to solve this extraction problem based on other recurrence plot characteristics [7] or taking into account statistical properties of the signals [11]. Moreover, the blind extraction / separation of chaotic and stochastic sources is also important for being related to the practical problem of denoising chaotic time series, which is extremely relevant to the study of several dynamical processes [1,4,6]. This initial work was not devoted to solving this specific problem, but a more detailed investigation of the proposed method is certainly an interesting perspective for its application in denoising. In addition, it seems possible to adapt these denoising methods in order to improve the solution of extraction problem exposed here.

Finally, after this discussion, perhaps it can be stated that the main contribution of this work is not only a technique, but a distinct way of employing *a priori* information about certain signals that play a key role in several scenarios of theoretical and practical significance in dynamical systems. It is our belief that the implications and the potential of this "deterministic component analysis" deserve careful analysis.

Acknowledgments. The authors would like to thank FAPESP and CAPES for the financial support and Everton Z. Nadalin for his comments and suggestions.

References

1. Abarbanel, D.I.: Analysis of Observed Chaotic Data. Springer, New York (1996)
2. Grassberger, P., Procaccia, I.: Measuring the strangeness of strange attractors. Physica D 9, 189–208 (1983)
3. Eckmann, J.-P., Kamphorst, S.O., Ruelle, D.: Recurrence Plots of Dynamical Systems. Europhysics Letters 4, 973–977 (1987)
4. Landa, P.S., Rosenblum, M.G.: Time Series Analysis for System Identification and Diagnostics. Physica D 48, 232–254 (1991)
5. Badii, R., Broggi, G., Derighetti, B., Ravani, M., Rubio, M.A.: Dimension Increase in Filtered Chaotic Signals. Physical Review Letters 11, 979–982 (1988)
6. Kostelich, E.J., Schreiber, T.: Noise reduction in chaotic time-series data: a survey of common methods. Physical Review E 48(3), 1752–1763 (1993)
7. Marwan, N., Romano, M.C., Thiel, M., Kurths, J.: Recurrence plots for the analysis of complex systems. Physics Reports 438, 237–329 (2007)
8. Hyvärinen, A., Karhunen, J., Oja, E.: Independent Component Analysis. John Wiley & Sons, New-York (2001)
9. Haykin, S.: Adaptive Filter Theory. Prentice Hall, New Jersey (2002)
10. Darbellay, G.A., Vajda, I.: Estimation of the Information by Adaptive Partitioning of the Observation Space. IEEE Transactions on Information Theory 45, 1315–1321 (1999)
11. Rohde, G.K., Nichols, J.M., Dissinger, B.M., Bucholtz, F.: Stochastic analysis of recurrence plots with applications to the detection of deterministic signals. Physica D 237, 619–629 (2008)

Estimating Squared-Loss Mutual Information for Independent Component Analysis

Taiji Suzuki[1] and Masashi Sugiyama[2]

[1] Department of Mathematical Informatics, The University of Tokyo
7-3-1 Hongo, Bunkyo-ku, Tokyo 113-8656, Japan
s-taiji@stat.t.u-tokyo.ac.jp
[2] Department of Computer Science, Tokyo Institute of Technology
2-12-1 O-okayama, Meguro-ku, Tokyo 152-8552, Japan
sugi@cs.titech.ac.jp

Abstract. Accurately evaluating statistical independence among random variables is a key component of Independent Component Analysis (ICA). In this paper, we employ a squared-loss variant of mutual information as an independence measure and give its estimation method. Our basic idea is to estimate the *ratio* of probability densities directly without going through density estimation, by which a hard task of density estimation can be avoided. In this density-ratio approach, a natural cross-validation procedure is available for model selection. Thanks to this, all tuning parameters such as the kernel width or the regularization parameter can be objectively optimized. This is a highly useful property in unsupervised learning problems such as ICA. Based on this novel independence measure, we develop a new ICA algorithm named *Least-squares Independent Component Analysis* (LICA).

Keywords: Independent component analysis, Mutual information, Squared loss, Density ratio estimation, Cross-validation.

1 Introduction

The purpose of Independent Component Analysis (ICA) [1] is to obtain a transformation matrix that separates mixed signals into statistically-independent sources signals. A direct approach to ICA is to find a transformation matrix such that independence among separated signals is maximized under some independence measure such as *mutual information* (MI).

Various approaches to computing the independence measures from samples have been studied so far. A naive approach is to estimate probability densities based on parametric or non-parametric density estimation. However, finding an appropriate parametric model is not easy without strong prior knowledge and non-parametric estimation is not accurate in high-dimensional problems. Thus this naive approach is not so useful in practice. Another approach is to approximate the *negative entropy* based on the Gram-Charlier expansion [6,7,3] or the Edgeworth expansion [5]. An advantage of this approach is that a hard

T. Adali et al. (Eds.): ICA 2009, LNCS 5441, pp. 130–137, 2009.

Table 1. Summary of existing and proposed ICA methods

	Model selection	Distribution
Fast ICA (FICA) [2]	**Not Necessary**	Not Free
Natural-gradient ICA (NICA) [3]	**Not Necessary**	Not Free
Kernel ICA (KICA) [4]	Not Available	**Free**
Edgeworth-expansion ICA (EICA) [5]	**Not Necessary**	Nearly normal
Least-squares ICA (LICA) [proposed]	**Available**	**Free**

task of density estimation is not directly involved. However, these expansion techniques are based on the assumption that the target density is close to normal and violation of this assumption can cause large approximation error.

The above approaches are based on the probability densities of signals. Another line of research that does not explicitly involve probability densities employs *non-linear correlation*—signals are statistically independent if and only if all non-linear correlations among the signals vanish. Following this line, computationally efficient algorithms have been developed based on the fourth-order statistics [8,2]. However, these methods ignore higher-order correlation and thus could be inaccurate depending on the target distribution. To cope with this problem, the *kernel trick* has been applied in ICA [4], which allows us to evaluate all non-linear correlations efficiently. However, its practical performance depends on the choice of kernels (more specifically, the Gaussian kernel width) and there seems no theoretically-justified method to determine the kernel width. This is a critical problem in unsupervised learning problem such as ICA.

In this paper, we use a squared-loss variant of MI as an independence measure and give a novel method for estimating it. Our key idea is to estimate the ratio of probability densities contained in squared-loss MI (SMI) directly without going through density estimation. This allows us to avoid a hard task of density estimation. Another practically important advantage of this density-ratio approach is that a natural cross-validation (CV) procedure is available for model selection. Thus all tuning parameters such as the kernel width or the regularization parameter can be objectively optimized through CV.

From an algorithmic point of view, the density-ratio approach *analytically* provides a non-parametric estimate of SMI; furthermore its derivative can also be computed analytically and these useful properties are utilized in deriving a new ICA algorithm—the proposed method is named *Least-squares Independent Component Analysis* (LICA). Characteristics of existing and proposed ICA methods are summarized in Table 1, highlighting the advantage of the proposed LICA approach.

2   SMI Estimation for ICA

In this section, we formulate the ICA problem and introduce our independence measure, SMI. Then we give an estimation method of SMI and based on it we derive an ICA algorithm.

2.1   Problem Formulation

Suppose there is a d-dimensional random signal

$$\boldsymbol{x} = (x^{(1)}, \ldots, x^{(d)})^\top$$

drawn from a distribution with density $p(\boldsymbol{x})$, where $\{x^{(m)}\}_{m=1}^d$ are statistically independent of each other. Thus, $p(\boldsymbol{x})$ can be factorized as

$$p(\boldsymbol{x}) = \prod_{m=1}^d p_m(x^{(m)}).$$

We cannot directly observe the *source* signal $\boldsymbol{x}$, but a linearly mixed signal $\boldsymbol{y}$:

$$\boldsymbol{y} = \boldsymbol{A}\boldsymbol{x},$$

where $\boldsymbol{A}$ is a $d \times d$ invertible matrix called the *mixing matrix*. The goal of ICA is, given mixed signal samples $\{\boldsymbol{y}_i\}_{i=1}^n$, to obtain a *demixing matrix* $\boldsymbol{W}$ that recovers the original source signal $\boldsymbol{x}$—we denote the demixed signal by $\boldsymbol{z}$:

$$\boldsymbol{z} = \boldsymbol{W}\boldsymbol{y}.$$

The ideal solution is given by $\boldsymbol{W} = \boldsymbol{A}^{-1}$, but we can only recover it up to permutation and scaling of components of $\boldsymbol{x}$ due to non-identifiability of the ICA setup [1].

A direct approach to ICA is to determine $\boldsymbol{W}$ so that components of $\boldsymbol{z}$ are as independent as possible. Here, we adopt SMI as the independence measure:

$$I_s(Z^{(1)}, \ldots, Z^{(d)}) := \frac{1}{2} \int \left(\frac{q(\boldsymbol{z})}{r(\boldsymbol{z})} - 1 \right)^2 r(\boldsymbol{z}) d\boldsymbol{z}, \tag{1}$$

where $q(\boldsymbol{z})$ denotes the joint density of $\boldsymbol{z}$ and $r(\boldsymbol{z})$ denotes the product of marginal densities $\{q_m(z^{(m)})\}_{m=1}^d$:

$$r(\boldsymbol{z}) = \prod_{m=1}^d q_m(z^{(m)}).$$

Since I_s vanishes if and only if $q(\boldsymbol{z}) = r(\boldsymbol{z})$, the degree of independence among $\{z^{(m)}\}_{m=1}^d$ may be measured by SMI. Note that Eq.(1) corresponds to the f-*divergence* from $q(\boldsymbol{x})$ to $r(\boldsymbol{z})$ with the squared loss, while ordinary MI corresponds to the f-divergence with the log loss. Thus SMI could be regarded as a natural generalization of ordinary MI.

Based on the independence detection property of SMI, we try to find the demixing matrix $\boldsymbol{W}$ that minimizes SMI estimated from the demixed samples:

$$\{\boldsymbol{z}_i \mid \boldsymbol{z}_i = (z_i^{(1)}, \ldots, z_i^{(d)})^\top := \boldsymbol{W}\boldsymbol{y}_i\}_{i=1}^n.$$

Our key constraint when estimating SMI is that we want to avoid density estimation. Below, we show how this could be accomplished.

2.2   SMI Inference via Density Ratio Estimation

Using *convex duality* [9], we can express SMI as

$$I_s(Z^{(1)}, \ldots, Z^{(d)}) = \sup_g \left[\int \left(g(z)q(z) - \frac{1}{2}g(z)^2 r(z) \right) dz - \frac{1}{2} \right], \qquad (2)$$

where $\sup_g$ is taken over all measurable functions. Thus computing I_s is reduced to solving the following optimization problem:

$$\inf_g \left[\int \left(\frac{1}{2}g(z)^2 r(z) - g(z)q(z) \right) dz \right]. \qquad (3)$$

We can confirm that the optimal solution g^* of the problem (3) is given as

$$g^*(z) = \frac{q(z)}{r(z)}. \qquad (4)$$

Thus, solving the problem (3) amounts to inferring the density ratio (4).

However, directly solving the problem (3) is not possible due to the following two reasons. The first reason is that finding the minimizer over all measurable functions is not tractable in practice since the search space is too vast. To overcome this problem, we restrict the search space to some linear subspace $\mathcal{G}$:

$$\mathcal{G} = \{ \boldsymbol{\alpha}^\top \boldsymbol{\varphi}(z) \mid \boldsymbol{\alpha} = (\alpha_1, \ldots, \alpha_b)^\top \in \mathbb{R}^b \}, \qquad (5)$$

where $\boldsymbol{\alpha}$ is a parameter to be learned from samples, $^\top$ denotes the transpose of a matrix or a vector, and $\boldsymbol{\varphi}(z)$ is basis function such that

$$\boldsymbol{\varphi}(z) = (\varphi_1(z), \ldots, \varphi_b(z))^\top \geq \mathbf{0}_b \quad \text{for all } \boldsymbol{x}.$$

$\mathbf{0}_b$ denotes the b-dimensional vector with all zeros. Note that $\boldsymbol{\varphi}(z)$ could be dependent on the samples $\{z_i\}_{i=1}^n$, i.e., *kernel* models are also allowed. We explain how the basis functions $\boldsymbol{\varphi}(z)$ are chosen in Section 2.3.

The second reason why directly solving the problem (3) is not possible is that the true probability densities $q(z)$ and $r(z)$ contained in the density ratio (4) are unavailable. To cope with this problem, we approximate them by their empirical distributions—then the optimization problem is reduced to

$$\widehat{\boldsymbol{\alpha}} := \underset{\boldsymbol{\alpha} \in \mathbb{R}^b}{\operatorname{argmin}} \left[\frac{1}{2}\boldsymbol{\alpha}^\top \widehat{\boldsymbol{H}}\boldsymbol{\alpha} - \widehat{\boldsymbol{h}}^\top \boldsymbol{\alpha} + \frac{1}{2}\lambda \boldsymbol{\alpha}^\top \boldsymbol{\alpha} \right], \qquad (6)$$

where we included $\lambda \boldsymbol{\alpha}^\top \boldsymbol{\alpha}$ ($\lambda > 0$) for regularization purposes and

$$\widehat{\boldsymbol{H}} := \frac{1}{n^d} \sum_{i_1, \ldots, i_d = 1}^n \boldsymbol{\varphi}(z_{i_1}^{(1)}, \ldots, z_{i_d}^{(d)}) \boldsymbol{\varphi}(z_{i_1}^{(1)}, \ldots, z_{i_d}^{(d)})^\top, \quad \widehat{\boldsymbol{h}} := \frac{1}{n} \sum_{i=1}^n \boldsymbol{\varphi}(z_i^{(1)}, \ldots, z_i^{(d)}).$$

Differentiating the objective function (6) with respect to $\boldsymbol{\alpha}$ and equating it to zero, we can obtain an analytic-form solution as

$$\widehat{\boldsymbol{\alpha}} = (\widehat{\boldsymbol{H}} + \lambda \boldsymbol{I}_b)^{-1} \widehat{\boldsymbol{h}},$$

where I_b is the b-dimensional identity matrix. Thus, the solution can be computed very efficiently just by solving a system of linear equations. Using the solution $\widehat{\alpha}$, we can approximate SMI as

$$\widehat{I}_s = -\frac{1}{2} - \frac{1}{2}\widehat{\alpha}^\top \widehat{H}\widehat{\alpha} + \widehat{h}^\top \widehat{\alpha}. \tag{7}$$

Ordinary MI can also be estimated similarly using the density ratio [10]. However, the use of SMI is more advantageous due to the analytic-form solution.

2.3   Design of Basis Functions and Model Selection

As basis functions, we propose using a Gaussian kernel model:

$$\varphi_\ell(z) = \exp\left(-\frac{\|z - v_\ell\|^2}{2\sigma^2}\right) = \prod_{m=1}^{d} \exp\left(-\frac{(z^{(m)} - v_\ell^{(m)})^2}{2\sigma^2}\right), \tag{8}$$

where $\{v_\ell \mid v_\ell = (v_\ell^{(1)}, \ldots, v_\ell^{(d)})^\top\}_{\ell=1}^{b}$ are Gaussian centers randomly chosen from $\{z_i\}_{i=1}^{n}$—more precisely, we set $v_\ell = z_{c(\ell)}$, where $\{c(\ell)\}_{\ell=1}^{b}$ are randomly chosen from $\{1, \ldots, n\}$ without replacement. An advantage of the Gaussian kernel lies in the factorizability in Eq.(8), contributing to reducing the computation of the matrix $\widehat{H}$ significantly:

$$\widehat{H}_{\ell,\ell'} = \frac{1}{n^d} \prod_{m=1}^{d} \left[\sum_{i=1}^{n} \exp\left(-\frac{(z_i^{(m)} - v_\ell^{(m)})^2 + (z_i^{(m)} - v_{\ell'}^{(m)})^2}{2\sigma^2}\right)\right].$$

In the experiments, we fix the number of basis functions at

$$b = \min(100, n),$$

and choose the Gaussian width σ and the regularization parameter λ by CV with grid search as follows. First, the samples $\{z_i\}_{i=1}^{n}$ are divided into K disjoint subsets $\{\mathcal{Z}_k\}_{k=1}^{K}$ of (approximately) the same size (we use $K = 5$ in the experiments). Then an estimator $\widehat{\alpha}_{\mathcal{Z}_k}$ is obtained using $\{\mathcal{Z}_j\}_{j\neq k}$ (i.e., without $\mathcal{Z}_k$) and the approximation error for the hold-out samples $\mathcal{Z}_k$ is computed:

$$J_{\mathcal{Z}_k}^{(K\text{-CV})} = \frac{1}{2}\widehat{\alpha}_{\mathcal{Z}_k}^\top \widehat{H}\widehat{\alpha}_{\mathcal{Z}_k} - \widehat{h}^\top \widehat{\alpha}_{\mathcal{Z}_k},$$

where $|\mathcal{Z}_k|$ denotes the number of sample pairs in the set $\mathcal{Z}_k$. This procedure is repeated for $k = 1, \ldots, K$ and its average $J^{(K\text{-CV})}$ is outputted:

$$J^{(K\text{-CV})} = \frac{1}{K}\sum_{k=1}^{K} J_{\mathcal{Z}_k}^{(K\text{-CV})}.$$

For model selection, we compute $J^{(K\text{-CV})}$ for all model candidates (the Gaussian width σ and the regularization parameter λ in the current setting) and choose the model that minimizes $J^{(K\text{-CV})}$. We can show that $J^{(K\text{-CV})}$ is an almost unbiased estimate of the objective function in Eq.(3), where the 'almost'-ness comes from the fact that the number of samples is reduced in the CV procedure due to data splitting.

2.4   The LICA Algorithm

Finally, we show how the above SMI estimation idea could be employed in the context of ICA.

Here, we use a simple gradient technique for obtaining a minimizer of the estimated SMI. The update rule of the demixing matrix $\boldsymbol{W}$ is given by

$$\boldsymbol{W} \longleftarrow \boldsymbol{W} - \varepsilon \frac{\partial \widehat{I}_s}{\partial \boldsymbol{W}}, \tag{9}$$

where $\varepsilon \; (> 0)$ is the step size. We can show that the gradient is given by

$$\frac{\partial \widehat{I}_s}{\partial W_{\ell,\ell'}} = \frac{\partial \widehat{\boldsymbol{h}}^\top}{\partial W_{\ell,\ell'}} (-\widehat{\boldsymbol{\beta}} + 2\widehat{\boldsymbol{\alpha}}) + \widehat{\boldsymbol{\alpha}}^\top \frac{\partial \widehat{\boldsymbol{H}}}{\partial W_{\ell,\ell'}} (\widehat{\boldsymbol{\beta}} - \frac{3}{2}\widehat{\boldsymbol{\alpha}}), \tag{10}$$

where

$$\frac{\partial \widehat{h}_k}{\partial W_{\ell,\ell'}} = \frac{1}{n\sigma^2} \sum_{i=1}^{n} (z_i^{(\ell)} - v_k^{(\ell)})(u_k^{(\ell')} - y_i^{(\ell')}) \exp\left(-\frac{\|\boldsymbol{z}_i - \boldsymbol{v}_k\|^2}{2\sigma^2}\right),$$

$$\frac{\partial \widehat{H}_{k,k'}}{\partial W_{\ell,\ell'}} = \frac{1}{n^{d-1}} \prod_{m=1, m\neq\ell}^{d} \left[\sum_{i=1}^{n} \exp\left(-\frac{(z_i^{(m)} - v_k^{(m)})^2 + (z_i^{(m)} - v_{k'}^{(m)})^2}{2\sigma^2}\right)\right]$$

$$\times \left[\frac{1}{n\sigma^2} \sum_{i=1}^{n} \left((z_i^{(\ell)} - v_k^{(\ell)})(u_k^{(\ell')} - y_i^{(\ell')}) + (z_i^{(\ell)} - v_{k'}^{(\ell)})(u_{k'}^{(\ell')} - y_i^{(\ell')})\right)\right.$$

$$\left. \times \exp\left(-\frac{(z_i^{(\ell)} - v_k^{(\ell)})^2 + (z_i^{(\ell)} - v_{k'}^{(\ell)})^2}{2\sigma^2}\right)\right],$$

$$\boldsymbol{u}_\ell = \boldsymbol{y}_{c(\ell)}, \quad \boldsymbol{y}_i = (y_i^{(1)}, \ldots, y_i^{(d)})^\top, \quad \text{and} \quad \widehat{\boldsymbol{\beta}} = (\widehat{\boldsymbol{H}} + \lambda \boldsymbol{I}_b)^{-1} \widehat{\boldsymbol{H}} \widehat{\boldsymbol{\alpha}}.$$

In ICA, scaling of components of $\boldsymbol{z}$ can be arbitrary. This implies that the above gradient updating rule can lead to a solution with bad scaling, which is not preferable from a numerical point of view. To avoid numerical instability, we normalize $\boldsymbol{W}$ at each gradient iteration as

$$W_{\ell,\ell'} \longleftarrow \frac{W_{\ell,\ell'}}{\sqrt{\sum_{m=1}^{d} W_{\ell,m}^2}}. \tag{11}$$

The proposed ICA algorithm, which we call *Least-squares Independent Component Analysis* (LICA), is summarised below.

---

1. Initialize demixing matrix $\boldsymbol{W}$ and normalize it by Eq.(11).
2. Optimize Gaussian width σ and regularization parameter λ by CV.
3. Compute gradient $\frac{\partial \widehat{I}_s}{\partial \boldsymbol{W}}$ by Eq.(10).
4. Choose step-size ε such that $\widehat{I}_s$ (see Eq.(7)) is minimized (*line-search*).
5. Update $\boldsymbol{W}$ by Eq.(9).
6. Normalize $\boldsymbol{W}$ by Eq.(11).
7. Repeat 2.–6. until $\boldsymbol{W}$ converges.

---

3   Numerical Examples

In this section, we illustrate how our algorithm behaves using the following three 2-dimensional datasets:

(a) **Sub-Sub-Gaussians:** $p(\boldsymbol{x}) = U(x^{(1)}; -0.5, 0.5)U(x^{(2)}; -0.5, 0.5)$,
(b) **Super-Super-Gaussians:** $p(\boldsymbol{x}) = L(x^{(1)}; 0, 1)L(x^{(2)}; 0, 1)$,
(c) **Sub-Super-Gaussians:** $p(\boldsymbol{x}) = U(x^{(1)}; -0.5, 0.5)L(x^{(2)}; 0, 1)$,

where $U(x; a, b)$ $(a, b \in \mathbb{R}, a < b)$ denotes the uniform density on $[a, b]$ and $L(x; \mu, v)$ $(\mu \in \mathbb{R}, v > 0)$ denotes the Laplacian density with mean μ and variance v. Let the number of samples be $n = 300$ and we observe mixed samples $\{\boldsymbol{y}_i\}_{i=1}^n$ through the following mixing matrix:

$$A = \begin{pmatrix} \cos(\pi/4) & \sin(\pi/4) \\ -\sin(\pi/4) & \cos(\pi/4) \end{pmatrix}.$$

The observed samples are plotted in Figure 1.

Figure 2 depicts the value of estimated SMI (7) over iterations and Figure 3 depicts the elements of the demixing matrix $\boldsymbol{W}$ over iterations. The true independent directions as well as the estimated independent directions are plotted in Figure 1. The results show that estimated SMI decreases rapidly and good solutions are obtained for all the datasets.

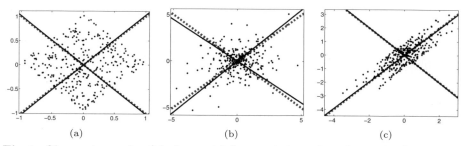

Fig. 1. Observed samples (black asterisks), true independent directions (red dotted lines) and estimated independent directions (blue solid lines)

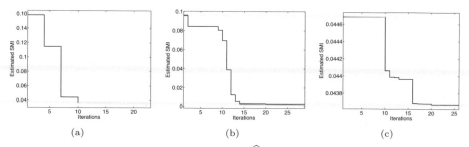

Fig. 2. Estimated SMI $\widehat{I}_s$ over iterations

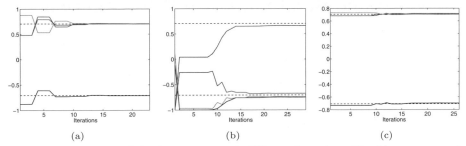

(a)   (b)   (c)

Fig. 3. The elements of the demixing matrix W over iterations. The blue, green, red, cyan lines correspond to $W_{1,1}$, $W_{1,2}$, $W_{2,1}$, and $W_{2,2}$, respectively. The black dotted lines denote the true values.

4   Conclusions

In this paper, we have proposed a new estimation method of a squared-loss variant of mutual information, and based on this, we developed an ICA algorithm. The proposed ICA method, named least-squares ICA (LICA), has several preferable properties, e.g., it is distribution-free and model selection by cross-validation is available. Our future work includes development of efficient optimization algorithm beyond gradient techniques.

References

1. Hyvärinen, A., Karhunen, J., Oja, E.: Independent Component Analysis. Wiley, New York (2001)
2. Hyvärinen, A.: Fast and robust fixed-point algorithms for independent component analysis. IEEE Transactions on Neural Networks 10(3), 626–634 (1999)
3. Amari, S., Cichocki, A., Yang, H.H.: A new learning algorithm for blind signal separation. In: Advances in Neural Information Processing Systems, pp. 757–763. MIT Press, Cambridge (1996)
4. Bach, F.R., Jordan, M.I.: Kernel independent component analysis. Journal of Machine Learning Research 3, 1–48 (2002)
5. Van Hulle, M.M.: Sequential fixed-point ICA based on mutual information minimization. Neural Computation 20(5), 1344–1365 (2008)
6. Cardoso, J.F., Souloumiac, A.: Blind beamforming for non-Gaussian signals. Radar and Signal Processing, IEE Proceedings-F 140(6), 362–370 (1993)
7. Comon, P.: Independent component analysis, a new concept? Signal Processing 36(3), 287–314 (1994)
8. Jutten, C., Herault, J.: Blind separation of sources, part I: An adaptive algorithm based on neuromimetic architecture. Signal Processing 24(1), 1–10 (1991)
9. Boyd, S., Vandenberghe, L.: Convex Optimization. Cambridge University Press, Cambridge (2004)
10. Suzuki, T., Sugiyama, M., Sese, J., Kanamori, T.: Approximating mutual information by maximum likelihood density ratio estimation. In: New Challenges for Feature Selection in Data Mining and Knowledge Discovery. JMLR Workshop and Conference Proceedings, vol. 4, pp. 5–20 (2008)

Complete Blind Subspace Deconvolution

Zoltán Szabó

Department of Information Systems, Eötvös Loránd University,
Pázmány P. sétány 1/C, Budapest H-1117, Hungary
szzoli@cs.elte.hu
http://nipg.inf.elte.hu

Abstract. In this paper we address the blind subspace deconvolution (BSSD) problem; an extension of both the blind source deconvolution (BSD) and the independent subspace analysis (ISA) tasks. While previous works have been focused on the undercomplete case, here we extend the theory to complete systems. Particularly, we derive a separation technique for the complete BSSD problem: we solve the problem by reducing the estimation task to ISA via linear prediction. Numerical examples illustrate the efficiency of the proposed method.

Keywords: Complete blind subspace deconvolution, separation principle, linear prediction, independent subspace analysis.

1 Introduction

Recently, research on independent component analysis (ICA) [1,2] and its extensions has gained much attention. One can think of ICA as a cocktail-party problem, where there are D *one-dimensional* sound sources and D microphones and the task is to estimate the original sources from the observed mixed signals. Nonetheless, applications in which only certain groups of sources are independent may be highly relevant in practice. In this case, the independent sources can be multidimensional. For instance, consider the generalization of the cocktail-party problem, where *independent groups* of people are talking separately about independent topics or more than one group of musicians are playing at the party. This problem is referred to as independent subspace analysis (ISA) [3].[1] Another extension of the original ICA task is the blind source deconvolution (BSD) problem. This problem emerges, for example, when a cocktail-party takes place in an *echoic* room. Several theoretical questions of ICA, ISA and BSD have already been addressed (see, e.g., [4], [5] and [6] for recent reviews, respectively), and numerous application areas show the potential of these fields including (i) remote sensing applications: passive radar/sonar processing, (ii) image-deblurring, image restoration, (iii) speech enhancement using microphone arrays, acoustics, (iv) multi-antenna wireless communications, sensor networks, (v) financial, gene and biomedical signal—EEG, ECG, MEG, fMRI—analysis, (vi) face view recognition, (vii) optics, (viii) seismic exploration.

---

[1] ISA is also called multidimensional independent component analysis and group ICA in the literature.

T. Adali et al. (Eds.): ICA 2009, LNCS 5441, pp. 138–145, 2009.

The simultaneous assumption of the two extensions, that is, ISA combined with BSD, has recently emerged in the literature. For example, at the cocktail-party, groups of people or groups of musicians may form *independent source groups* and *echoes* may be present. The task is called blind subspace deconvolution (BSSD). Probably one of the most exciting and fundamental hypotheses of the ICA research has been formed by [3]: the solution of the ISA problem can be *separated* to ICA and then clustering the ICA elements into statistically dependent subspaces. This ISA *separation principle* has been rigorously proven for some distribution types in [5], and forms the basis of the state-of-the-art ISA algorithms. Similar separation based techniques can be derived for the solution of the *undercomplete* BSSD task (uBSSD), where in terms of the cocktail-party problem there are more microphones than acoustic sources. It has been shown that the uBSSD problem can be reduced to ISA by means of temporal concatenation [5]. However, the associated ISA problem can easily become 'high dimensional'. The dimensionality problem can be circumvented by applying a linear predictive approximation (LPA) based reduction [7]. Here, we show that it is possible to extend the LPA idea to the *complete* BSSD task.[2] In the undercomplete case, the LPA based solution was based on the observation that the polynomial matrix describing the temporal convolution had, under rather general conditions[3], a polynomial matrix left inverse. In the complete case such an inverse doesn't exist in general. However, provided that the convolution can be represented by an infinite order autoregressive process, one can construct an efficient estimation method for the hidden components via an asymptotically consistent LPA procedure. This thought is used here to extend the technique of [7] to the complete case.

The paper is structured as follows: Section 2 formulates the problem domain. Section 3 shows how to transform the complete BSSD task into an ISA task via LPA. Section 4 contains numerical illustrations. Conclusions are drawn in Section 5.

2   The BSSD Model

Here, we define the BSSD task [5]. Assume that we have M hidden, independent, multidimensional *components* (random variables). Suppose also that only their

$$\mathbf{x}(t) = \sum_{l=0}^{L} \mathbf{H}_l \mathbf{s}(t - l) \tag{1}$$

convolutive mixture is available for observation, where $\mathbf{x}(t) \in \mathbb{R}^{D_x}$ and $\mathbf{s}(t)$ is the concatenation of the components $\mathbf{s}^m(t) \in \mathbb{R}^{d_m}$, that is $\mathbf{s}(t) = [\mathbf{s}^1(t); \ldots ; \mathbf{s}^M(t)] \in \mathbb{R}^{D_s}$ $(D_s = \sum_{m=1}^{M} d_m)$. Denoting the time-shift operation by z, one may write Eq. (1) compactly as

$$\mathbf{x} = \mathbf{H}[z]\mathbf{s}, \tag{2}$$

---

[2] The overcomplete BSSD task is challenging and as of yet no general solution is known.

[3] If the coefficients of the undercomplete polynomial matrix are drawn from a non-degenerate continuous distribution, such an inverse exists with probability one.

where the mixing process is described by the polynomial matrix $\mathbf{H}[z] := \sum_{l=0}^{L} \mathbf{H}_l z^l \in \mathbb{R}[z]^{D_x \times D_s}$. We assume that the components $\mathbf{s}^m$ are

1. independent: $I(\mathbf{s}^1, \ldots, \mathbf{s}^M) = 0$, where I denotes the mutual information,
2. i.i.d. (independent identically distributed) in t, and
3. there is at most one Gaussian component among $\mathbf{s}^m$s.

The goal of the BSSD problem is to estimate the original source $\mathbf{s}(t)$ by using observations $\mathbf{x}(t)$ only. While $D_x > D_s$ is the *undercomplete* case , $D_x = D_s$ is the *complete* one. The case $L = 0$ corresponds to the ISA task, and if $d_m = 1$ ($\forall m$) also holds, then the ICA task is recovered. In the BSD task $d_m = 1$ ($\forall m$) and L is a non-negative integer.

3   Method

Contrary to previous works [5,7] focusing on the undercomplete BSSD problem, in the present paper we address the complete task ($D = D_x = D_s$). We assume that the polynomial matrix $\mathbf{H}[z]$ is *invertible*, that is $\det(\mathbf{H}[z]) \neq 0$, for all $z \in \mathbb{C}, |z| \leq 1$. Let $E(\cdot)$ and $cov(\cdot)$ denote the expectation value, and the covariance of a random variable, respectively. Because the mean can be subtracted from the process and the transformation $\mathbf{x} = (\mathbf{H}[z]\mathbf{B}^{-1})(\mathbf{B}\mathbf{s})$ leads to the same observation, one may presume, without any loss of generality, that $\mathbf{s}$ is white:

$$E(\mathbf{s}) = \mathbf{0}, \quad cov(\mathbf{s}) = \mathbf{I}, \tag{3}$$

where $\mathbf{I}$ is the identity matrix. The invertibility of $\mathbf{H}[z]$ implies that the observation process $\mathbf{x}$ can be represented as an infinite order autoregressive (AR) process [8]:

$$\mathbf{x}(t) = \sum_{j=1}^{\infty} \mathbf{F}_j \mathbf{x}(t-j) + \mathbf{F}_0 \mathbf{s}(t). \tag{4}$$

By applying a finite order LPA approximation (fitting an AR process to $\mathbf{x}$), the innovation process $\mathbf{F}_0\mathbf{s}(t)$ can be estimated. The innovation can be seen as the observation of an ISA problem because components of $\mathbf{s}$ are independent: ISA techniques can be used to identify components $\mathbf{s}^m$. Choosing the order of the fitted AR process to $\mathbf{x}$ as $p = o(T^{\frac{1}{3}}) \xrightarrow{T \to \infty} \infty$, where T denotes the number of samples, guarantees that the AR approximation for the MA model is asymptotically consistent [9].

4   Illustrations

Here, we illustrate the efficiency of the proposed complete BSSD estimation technique. Test cases are introduced in Section 4.1. To evaluate the solutions we use the performance measure given in Section 4.2. Numerical results are presented in Section 4.3.

4.1   Databases

We define three databases to study our identification algorithm. The *smiley* test has 2-dimensional source components representing the 6 basic facial expressions ($d_m = 2$, $M = 6$). Sources $\mathbf{s}^m$ were generated by sampling 2-dimensional coordinates proportional to the corresponding pixel intensities (see Fig. 1(a)). In the *3D-geom* test $\mathbf{s}^m$s were random variables uniformly distributed on 3-dimensional geometric forms ($d_m = 3$). We chose 4 different components ($M = 4$) and, as a result, the dimension of the hidden source $\mathbf{s}$ is $D = 12$ (see Fig. 1(b)). Our *Beatles* test [5] is a non-i.i.d. example. Here, hidden sources are stereo Beatles songs.[4] 8 kHz sampled portions of two songs (A Hard Day's Night, Can't Buy Me Love) made the hidden $\mathbf{s}^m$s. Thus, the dimension of the components d_m was 2, the number of the components M was 2, and the dimension of the hidden source D was 4.

(a)   (b)

Fig. 1. Illustration of the *smiley* (a) and the *3D-geom* databases (b)

4.2   Performance Measure, the Amari-Index

Recovery of components $\mathbf{s}^m$ are subject to the ambiguities of the ISA task. Namely, components of equal dimension can be recovered up to permutation and invertible transformation within the subspaces [10]. For this reason, in the ideal case, the linear transformation $\mathbf{G}$ that optimally approximates the relation $\mathbf{s} \mapsto \hat{\mathbf{s}}$, where $\hat{\mathbf{s}}$ denotes the estimated hidden source, resides also within the subspaces and so it is a *block-permutation matrix*. This block-permutation structure can be measured by the ISA adapted version [11] of the Amari-error [12] normalized to the interval $[0,1]$ [13]. Namely, let us suppose that the source components are d-dimensional[5], and let us decompose matrix $\mathbf{G} \in \mathbb{R}^{D \times D}$ into blocks of size $d \times d$: $\mathbf{G} = [\mathbf{G}_{ij}]_{i,j=1,\dots,M}$. Let g_{ij} denote the sum of the absolute values of matrix $\mathbf{G}_{ij} \in \mathbb{R}^{d \times d}$. Now, the following term

$$r(\mathbf{G}) := \frac{1}{2M(M-1)} \left[\sum_{i=1}^{M} \left(\frac{\sum_{j=1}^{M} g_{ij}}{\max_j g_{ij}} - 1 \right) + \sum_{j=1}^{M} \left(\frac{\sum_{i=1}^{M} g_{ij}}{\max_i g_{ij}} - 1 \right) \right] \quad (5)$$

denotes the *Amari-index* that takes values in $[0,1]$: for an ideal block-permutation matrix $\mathbf{G}$ it takes 0; for the worst case it takes 1.

---

[4] See http://rock.mididb.com/beatles/

[5] The $d = d_m$ ($\forall m$) constraint was used only at the performance measurements (i.e., for the Amari-index).

4.3   Simulations

Results on databases *smiley*, *3D-geom* and *Beatles* are provided here. The Amari-index was used to measure the performance of the proposed complete BSSD method. For each individual parameter, 20 random runs were averaged. Our parameters are: T, the sample number of observations $\mathbf{x}(t)$, L, the parameter of the length of the convolution (the length of the convolution is $L + 1$), and λ, parameter of the invertible $\mathbf{H}[z]$. It is expected that if the roots of $\mathbf{H}[z]$ are close to the unit circle then our estimation will deteriorate, because the invertibility of $\mathbf{H}[z]$ comes to question. We investigated this by generating the polynomial matrix $\mathbf{H}[z]$ as follows:

$$\mathbf{H}[z] = \left[\prod_{l=0}^{L} (\mathbf{I} - \lambda \mathbf{O}_i z) \right] \mathbf{H}_0 \quad (|\lambda| < 1, \lambda \in \mathbb{R}), \tag{6}$$

where matrices $\mathbf{H}_0$ and $\mathbf{O}_i \in \mathbb{R}^{D \times D}$ were random orthogonal and the $\lambda \to 1$ limit was studied. 'Random run' means random choice of quantities $\mathbf{H}[z]$ and $\mathbf{s}$. The AR fit to observation $\mathbf{x}$ was performed by the method detailed in [14]. To study how the $o(T^{1/3})$ AR order (see Section 3) is exploited, the order of the estimated AR process was limited from above by $p_{max}(T) = 2\lfloor T^{\frac{1}{3} - \frac{1}{1000}} \rfloor$, and we used the Schwarz's Bayesian Criterion to determine the optimal p_{opt} order from the interval $[1, p_{max}(T)]$. The ISA subtask on the estimated innovation was carried out by the joint f-decorrelation method [15].

First we studied the Amari-index as a function of the sample size. For the *smiley* and *3D-geom* databases the sample number T varied between $2,000$ and $20,000$. The length of convolution varied as $L = 1, 2, 5, 10$. The λ parameter of $\mathbf{H}[z]$ was chosen as $0.4, 0.6, 0.7, 0.8, 0.85, 0.9$. Results are shown in Fig. 2(a)-(b). The estimation errors indicate that for $L = 10$ and about $\lambda = 0.85$ the estimation is still efficient, see Fig. 3 for an illustration of the estimated source components. The Amari-indices follow the power law $r(T) \propto T^{-c}$ ($c > 0$). The power law decline is manifested by straight line on log-log scale. The slopes of these straight lines are very close to each other. Numerical values for the estimation errors are given in Table 1. The estimated optimal AR orders are provided in Fig. 2(c). The figure demonstrates that as $\lambda \to 1$ the maximal possible order $p_{max}(T)$ is more and more exploited.

On the *Beatles* database the λ parameter was increased to 0.9, and the sample number T varied between $2,000$ and $100,000$. Results are presented in Fig. 2(d). According to the figure, for $L = 1, 2, 5$ the error of estimation drops for sample number $T = 10,000 - 20,000$, and for $L = 10$ the 'power law' decline of the Amari-index, which was apparent on the *smiley* and the *3D-geom* databases, also appears. Numerical values for the estimation errors are given in Table 1. On the *Beatles* test, the maximal possible AR order $p_{max}(T)$ was fully exploited on the examined parameter domain.

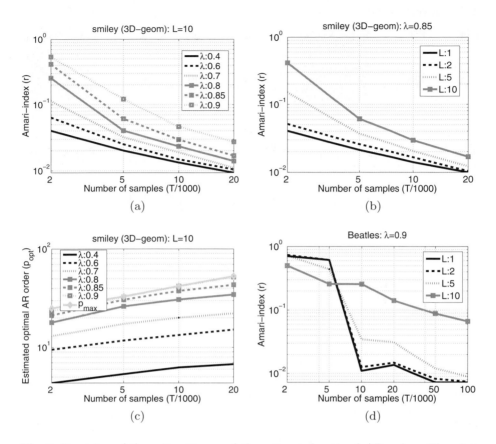

Fig. 2. Precision of the estimations and the estimated optimal AR orders. The plots are on log-log scale. (a), (b): on the *smiley* (*3D-geom*) database the Amari-index as a function of the sample number for different $\lambda \to 1$ parameter values of $\mathbf{H}[z]$ and convolution lengths, respectively. In (a): $L = 10$, in (b): $\lambda = 0.85$. (c): on the *smiley* (*3D-geom*) database the estimated AR order as a function of the sample number with $L = 10$ for different λ values. (d): the same as (b), but for the *Beatles* dataset with $\lambda = 0.9$. For graphical illustration, see Fig. 3. For numerical values, see Table 1.

Table 1. Amari-index in percentages on the *smiley, 3D-geom* ($\lambda = 0.85, T = 20,000$) and the Beatles dataset ($\lambda = 0.9, T = 100,000$) for different convolution lengths: mean$\pm$ standard deviation. For other sample numbers, see Fig. 2.

	$L = 1$	$L = 2$	$L = 5$	$L = 10$
smiley	0.99% ($\pm 0.11\%$)	1.04% ($\pm 0.09\%$)	1.22% ($\pm 0.15\%$)	1.69% ($\pm 0.26\%$)
3D-geom	0.42% ($\pm 0.06\%$)	0.54% ($\pm 0.05\%$)	0.88% ($\pm 0.14\%$)	1.15% ($\pm 0.24\%$)
Beatles	0.72% ($\pm 0.12\%$)	0.75% ($\pm 0.11\%$)	0.90% ($\pm 0.23\%$)	6.64% ($\pm 7.49\%$)

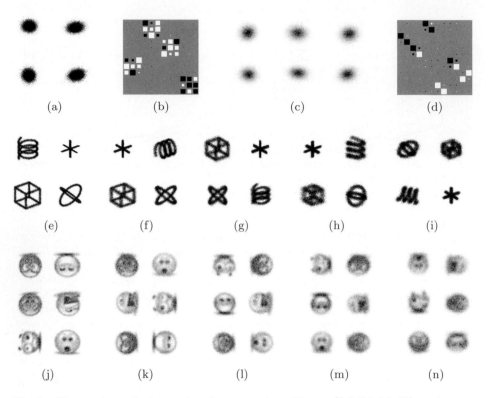

Fig. 3. Illustration of the estimations on the *3D-geom*[(a),(b),(e)-(i)] and *smiley*[(c),(d),(j)-(n)] datasets. Number of samples: $T = 20,000$. Length of the convolution: $L = 10$. In the first row: $\lambda = 0.4$. (a), (c): observed convolved signal $\mathbf{x}(t)$. (b), (d): Hinton-diagram of $\mathbf{G}$, ideally a block-permutation matrix with 2×2 and 3×3 blocks, respectively. (e)-(i), (j)-(n): estimated components $\hat{\mathbf{s}}^m$, recovered up to the ISA ambiguities from left to right for $\lambda = 0.4, 0.6, 0.7, 0.8, 0.85$. All the plotted estimations have average Amari-indices, see Fig. 2(a).

5   Conclusions

In this paper we focused on the complete case of the blind subspace deconvolution (BSSD) problem, a common extension of the independent subspace analysis (ISA) and the blind source deconvolution (BSD) tasks. We presented a separation technique for the solution of the complete BSSD task: the estimation task has been reduced to ISA via linear predictive approximation (LPA). We also demonstrated the efficiency of the algorithm on different datasets. Our simulations revealed that the error of the estimation of the hidden sources decreases in a power law fashion as the sample size increases. Interestingly, our algorithm recovered the sources when the assumptions of the BSSD problem were violated; that is in the case of the *Beatles* songs with non-i.i.d. sources. This result points to the ISA separation principle; one expects that it may be valid for a larger domain. Similar conjecture exists for joint block diagonalization [16] about the *global* minima.

Acknowledgments. This work has been supported by the National Office for Research and Technology.

References

1. Jutten, C., Hérault, J.: Blind separation of sources: An adaptive algorithm based on neuromimetic architecture. Signal Processing 24, 1–10 (1991)
2. Comon, P.: Independent component analysis, a new concept? Signal Processing 36, 287–314 (1994)
3. Cardoso, J.: Multidimensional independent component analysis. In: International Conference on Acoustics, Speech, and Signal Processing (ICASSP 1998), vol. 4, pp. 1941–1944 (1998)
4. Choi, S., Cichocki, A., Park, H., Lee, S.: Blind source separation and independent component analysis. Neural Information Processing - Letters and Reviews 6, 1–57 (2005)
5. Szabó, Z., Póczos, B., Lőrincz, A.: Undercomplete blind subspace deconvolution. Journal of Machine Learning Research 8, 1063–1095 (2007)
6. Pedersen, M.S., Larsen, J., Kjems, U., Parra, L.C.: A survey of convolutive blind source separation methods. In: Springer Handbook of Speech Processing. Springer, Heidelberg (2007)
7. Szabó, Z., Póczos, B., Lőrincz, A.: Undercomplete blind subspace deconvolution via linear prediction. In: Kok, J.N., Koronacki, J., Lopez de Mantaras, R., Matwin, S., Mladenič, D., Skowron, A. (eds.) ECML 2007. LNCS (LNAI), vol. 4701, pp. 740–747. Springer, Heidelberg (2007)
8. Fuller, W.A.: Introduction to Statistical Time Series. Wiley Interscience, Hoboken (1995)
9. Galbraith, J., Ullah, A., Zinde-Walsh, V.: Estimation of the vector moving average model by vector autoregression. Econometric Reviews 21(2), 205–219 (2002)
10. Theis, F.J.: Uniqueness of complex and multidimensional independent component analysis. Signal Processing 84(5), 951–956 (2004)
11. Theis, F.J.: Blind signal separation into groups of dependent signals using joint block diagonalization. In: Proceedings of International Society for Computer Aided Surgery (ISCAS 2005), Kobe, Japan, pp. 5878–5881 (2005)
12. Amari, S., Cichocki, A., Yang, H.H.: A new learning algorithm for blind signal separation. Advances in Neural Information Processing Systems 8, 757–763 (1996)
13. Szabó, Z., Póczos, B., Lőrincz, A.: Cross-entropy optimization for independent process analysis. In: Rosca, J.P., Erdogmus, D., Príncipe, J.C., Haykin, S. (eds.) ICA 2006. LNCS, vol. 3889, pp. 909–916. Springer, Heidelberg (2006)
14. Neumaier, A., Schneider, T.: Estimation of parameters and eigenmodes of multivariate autoregressive models. ACM Transactions on Mathematical Software 27(1), 27–57 (2001)
15. Szabó, Z., Lőrincz, A.: Real and complex independent subspace analysis by generalized variance. In: ICA Research Network International Workshop (ICARN 2006), pp. 85–88 (2006), http://arxiv.org/abs/math.ST/0610438
16. Abed-Meraim, K., Belouchrani, A.: Algorithms for joint block diagonalization. In: European Signal Processing Conference (EUSIPCO 2004), pp. 209–212 (2004)

Fast Parallel Estimation of High Dimensional Information Theoretical Quantities with Low Dimensional Random Projection Ensembles

Zoltán Szabó and András Lőrincz

Department of Information Systems, Eötvös Loránd University,
Pázmány P. sétány 1/C, Budapest H-1117, Hungary
szzoli@cs.elte.hu, andras.lorincz@elte.hu
http://nipg.inf.elte.hu

Abstract. The estimation of relevant information theoretical quantities, such as entropy, mutual information, and various divergences is computationally expensive in high dimensions. However, for this task, one may apply pairwise Euclidean distances of sample points, which suits random projection (RP) based low dimensional embeddings. The Johnson-Lindenstrauss (JL) lemma gives theoretical bound on the dimension of the low dimensional embedding. We adapt the RP technique for the estimation of information theoretical quantities. Intriguingly, we find that embeddings into extremely small dimensions, far below the bounds of the JL lemma, provide satisfactory estimates for the original task. We illustrate this in the Independent Subspace Analysis (ISA) task; we combine RP dimension reduction with a simple ensemble method. We gain considerable speed-up with the potential of real-time parallel estimation of high dimensional information theoretical quantities.

Keywords: Independent subspace analysis, random projection, pairwise distances, information theoretical estimations.

1 Introduction

The take-off of information theory goes back to the forties [1]. Tremendous applications have been developed ever since. The computation/estimation of information theoretical quantities (entropy, mutual information, divergence) is still slow. However, consistent estimation of these quantities is possible by nearest neighbor (NN) methods (see, e.g., [2]) that use the pairwise distances of sample points. Although search for nearest neighbors can also be expensive in high dimensions [3], low dimensional approximate isometric embedding of points of high dimensional Euclidean space can be addressed by the Johnson-Lindenstrauss Lemma [4] and the related random projection (RP) methods [5,6]. The RP approach proved to be successful, e.g., in classification, clustering, search for *approximate NN* (ANN), dimension estimation of manifolds, estimation of mixture of Gaussian models, compressions, data stream computation (see, e.g., [7]). We note that the RP approach is also related to compressed sensing [8].

T. Adali et al. (Eds.): ICA 2009, LNCS 5441, pp. 146–153, 2009.

In this paper we show a novel application of the RP technique: we estimate information theoretical quantities using the ANN-preserving properties of the RP technique. We illustrate the method on the estimation of Shannon's multidimensional differential entropy for the Independent Subspace Analysis (ISA) task [9]. The ISA problem extends Independent Component Analysis (ICA) [10] by allowing multidimensional independent components: at a cocktail party, ICA (ISA) is searching for (groups of) people talking independently. Another application area is image registration, where (i) information-theoretical similarity criterion can be advantageous, and (ii) high dimensional features should be handled [11,2] (work in progress). We note that RPs have been applied for ICA, but the underlying considerations differ from ours: [12] picks out random samples using Bernoulli variables and decreases the computational load on ICA, [13] uses RPs for preprocessing before principal component analysis.

The paper is structured as follows: Section 2 formulates the problem domain. In Section 3 the random projection technique is adapted to the estimation of information theoretical quantities and we use it for the estimation of multidimensional differential entropy. Section 4 contains the numerical illustrations. Conclusions are drawn in Section 5.

2   The ISA Model

Let us define the ISA task. Let us assume the observations $\mathbf{x}(t) \in \mathbb{R}^D, t = 1, 2, \ldots$ are linear mixtures of multidimensional independent sources, *components* $\mathbf{s}^m(t)$:

$$\mathbf{x}(t) = \mathbf{A}\mathbf{s}(t), \tag{1}$$

where $\mathbf{s}(t)$ concatenates components $\mathbf{s}^m(t) \in \mathbb{R}^{d_m}$; $\mathbf{s}(t) = [\mathbf{s}^1(t); \ldots; \mathbf{s}^M(t)] \in \mathbb{R}^D$ ($D = \sum_{m=1}^M d_m$). Our assumptions are the following:

1. components are (i) independent: $I(\mathbf{s}^1, \ldots, \mathbf{s}^M) = 0$, where I denotes the mutual information, (ii) i.i.d. (independent identically distributed) in t, and (iii) there is at most one Gaussian component among $\mathbf{s}^m$s.
2. The unknown $\mathbf{A} \in \mathbb{R}^{D \times D}$ *mixing matrix* is invertible.

In the ISA problem one estimates hidden source components ($\mathbf{s}^m$) from observations $\mathbf{x}(t)$ alone. (ICA problem: $\forall d_m = 1$). The ISA problem has ambiguities [14,15]: components of equal dimension can be determined up to permutation and up to invertible transformation within the subspaces. Thus, for ISA demixing matrix $\mathbf{W}_{\mathrm{ISA}}$ we have that $\mathbf{W}_{\mathrm{ISA}}\mathbf{A} \in \mathbb{R}^{D \times D}$ is a block-permutation (or block-scaling [16]) matrix. The block-permutation property and the quality of the ISA estimation can be measured by the ISA adapted and normalized Amari-error [17], the *Amari-index* (r) [18], which is 0 for perfect estimation and can not exceed 1.

3   Method

We present our RP based approach through the ISA problem. The ISA task can be viewed as the minimization of the mutual information between the estimated components, or equivalently as the minimization of the sum of Shannon's multi-dimensional differential entropies of the estimated components on the orthogonal group [19]:

$$J(\mathbf{W}) := \sum_{m=1}^{M} H(\mathbf{y}^m) \to \min_{\mathbf{W} \in \mathcal{O}^D}, \qquad (2)$$

where $\mathbf{y} = \mathbf{W}\mathbf{x}$, $\mathbf{y} = [\mathbf{y}^1; \ldots; \mathbf{y}^M]$, $\mathbf{y}^m \in \mathbb{R}^{d_m}$ and d_ms are given. It has been conjectured that the solution of the ISA task can be reduced to ICA followed by grouping of the non-independent ICA elements into ISA subspaces [9]. The conjecture has been rigorously proven by the ISA Separation Theorem for some distribution types [20]. It means that the demixing matrix assumes the form $\mathbf{W}_{\text{ISA}} = \mathbf{P}\mathbf{W}_{\text{ICA}}$ ($\hat{\mathbf{y}}_{\text{ISA}} = [\hat{\mathbf{y}}_{\text{ISA}}^1; \ldots; \hat{\mathbf{y}}_{\text{ISA}}^M] = \mathbf{P}\hat{\mathbf{y}}_{\text{ICA}}$, $\hat{\mathbf{y}}_{\text{ISA}}^m \in \mathbb{R}^{d_m}$), where the permutation matrix $\mathbf{P} \in \mathbb{R}^{D \times D}$ is to be determined. Estimation of cost function J involves multidimensional entropy estimation, which is computationally expensive in high dimensions, but can be executed by NN methods consistently [21,22]. It has been shown in [11] (in the field of image registration with high dimensional features) that the computational load can be decreased somewhat by (i) dividing the samples into groups and then (ii) computing the averages of the group estimates. We will combine this *parallelizable ensemble approach* with the ANN-preserving properties of RPs and get drastic savings. We suggest the following entropy estimation method[1], for each estimated ISA component $\mathbf{v} := \hat{\mathbf{y}}_{\text{ISA}}^m$: (i) divide the T samples $\{\mathbf{v}(1), \ldots, \mathbf{v}(T)\}$ into N groups indexed by sets $I_1, \ldots, I_N$ so that each group contains K samples, (ii) for all fixed groups take the random projection of $\mathbf{v}$ as $\mathbf{v}_{n,\text{RP}}(t) := \mathbf{R}_n \mathbf{v}(t)$ ($t \in I_n; n = 1, \ldots, N; \mathbf{R}_n \in \mathbb{R}^{d'_m \times d_m}$), (iii) average the estimated entropies of the RP-ed groups to get the estimation $\hat{H}(\mathbf{v}) = \frac{1}{N} \sum_{n=1}^{N} \hat{H}(\mathbf{v}_{n,\text{RP}})$. Our particular choice for $\mathbf{R}_n$ is given in Section 4.2. For the optimization of the estimated cost function $\hat{J}(\mathbf{P})$ one can apply (i) greedy optimization (exchange of 2 coordinates if it decreases $\hat{J}$), or (ii) global methods of higher computational burden, e.g., the cross-entropy (CE) method [23] adapted to permutation searches, because the estimation of $\hat{J}$ is quick.

4 Illustrations

Here, we illustrate the efficiency of the proposed RP based entropy estimation. Section 4.1 is about test cases. Numerical results are presented in Section 4.2.

4.1 Databases

We define three databases [20] to study our RP based ISA identification algorithm. In the *d-spherical* test hidden sources $\mathbf{s}^m$ were spherical random variables. Since spherical variables assume the form $\mathbf{v} = \rho \mathbf{u}$, where $\mathbf{u}$ is uniformly distributed on the d-dimensional unit sphere, and ρ is a non-negative scalar random variable independent of $\mathbf{u}$, they can be given by means of ρ (see Fig. 1(a)).

---

[1] The idea can be used for a number of information theoretical quantities, provided that they can be estimated by means of pairwise Euclidean distances of the samples.

In the *d-geom* dataset $\mathbf{s}^m$s were random variables uniformly distributed on d-dimensional geometric forms (see Fig. 1(b)). In the *all-k-independent* database, the *d*-dimensional hidden components $\mathbf{v} := \mathbf{s}^m$ were created as follows: coordinates v_i $(i = 1, \ldots, k)$ were independent uniform random variables on the set $\{0, \ldots, k\text{-}1\}$, whereas v_{k+1} was set to $mod(v_1 + \ldots + v_k, k)$. In this construction, every k-element subset of $\{v_1, \ldots, v_{k+1}\}$ is made of independent variables.

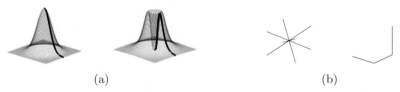

(a)   (b)

Fig. 1. Illustration of the (a): *d spherical* ($d = 2$), and (b). *d-geom* ($d = 3$) databases. ρ of the stochastic representation on the left (right): exponential with parameter $\mu = 1$ (lognormal with parameters $\mu = 0$, $\sigma = 1$).

4.2   Simulations

Results on databases *d-spherical*, *d-geom*, and *all-k-independent* are provided here. These experimental studies focused on the following issues:

1. What dimensional reduction can be achieved in the entropy estimation of the ISA problem by means of random projections?
2. What speed-up can be gained with the RP dimension reduction?
3. What are the advantages of our RP based approach in global optimization?

In our experiments the number of components was minimal ($M = 2$). We used the Amari-index to measure and compare the performance of the different methods. For each individual parameter, 50 random runs were averaged. Our parameters included T, the sample number of observations $\mathbf{x}(t)$ and d, the dimension of the components ($d = d_1 = d_2{}^2$). We also studied different estimations of the ISA cost function: we used the RADICAL procedure[3] [24] and the NN method [19] for entropy estimation and the Kernel Canonical Correlation Analysis (KCCA) [25] for mutual information estimation. The reduced dimension d' in RP and the optimization method (greedy, global (CE), NCut [26]) of the ISA cost were also varied in different tests. Random run means random choice of quantities $\mathbf{A}$ and $\mathbf{s}$. The ICA step was performed by the well-known fastICA method. The size of the randomly projected groups was set to $|I_n| = 2,000$, except for the case $d = 50$, when it was $5,000$. RP was realized by the *database-friendly projection* technique, i.e., the $r_{n,ij}$ coordinates of $\mathbf{R}_n$ were drawn independently from distribution $P(r_{n,ij} = \pm 1) = 1/2$, but more general constructions could also be used [5,6].

---

[2] This constraint was used only for the evaluation of the performance (Amari-index) of the algorithm.

[3] We chose RADICAL, because it is consistent, asymptotically efficient, converges rapidly, and it is computationally efficient.

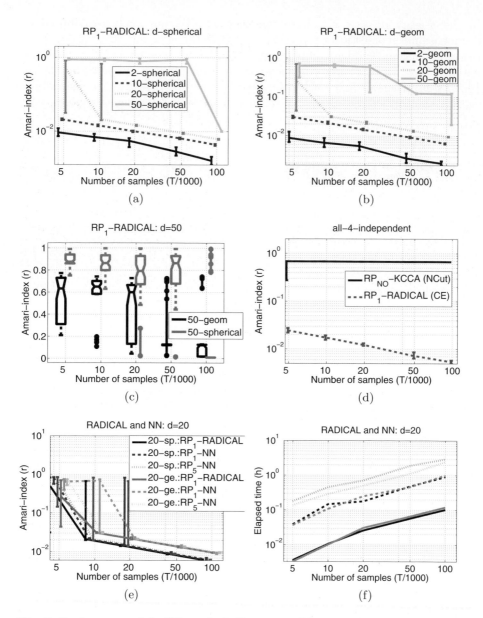

Fig. 2. Performance of the RP method. Notations: 'RP$_{d'}$ - method of cost estimation (method of optimization if not greedy)'. (a), (b): accuracy of the estimation versus the number of samples for the *d-spherical* and the *d-geom* databases on log-log scale. (c): notched boxed plots for $d = 50$, (d): Performance comparison on the *all-4-independent* database between the RP method using global optimization and the NCut based grouping of coordinates using the pairwise mutual information graph (on log-log scale). (e)-(f): Accuracy and computation time comparisons with the NN based method for the *20-spherical* and the *20-geom* databases (on log-log scale).

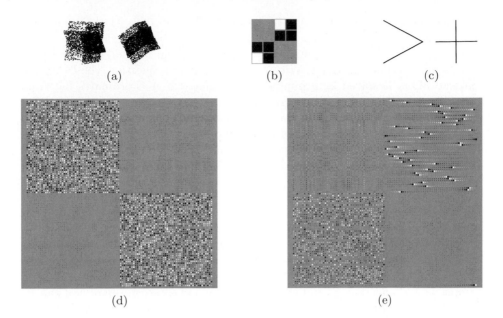

(a)   (b)   (c)

(d)   (e)

Fig. 3. Estimated components and Hinton-diagrams. Number of samples: $T = 100,000$. Databases *2-geom*: (a)-(c), *50-spherical*: (d), *50-geom*: (e). (a): observed signals $\mathbf{x}(t)$, (b): Hinton-diagram: the product of the mixing matrix of the ISA task and the estimated demixing matrix is approximately a block-permutation matrix with 2×2 blocks, (c): estimated components $\hat{\mathbf{s}}^m$, recovered up to the ISA ambiguities, (d)-(e): Hinton-diagrams of the *50-spherical* and the *50-geom* tests, respectively. Hinton-diagrams have average Amari-indices: for (b) 0.2%, for (d) 1%, for (e) 12%.

In the first study we were interested in the limits of the RP dimension reduction. We increased dimension d of the subspaces for the *d-spherical* and the *d-geom* databases ($d = 2, 10, 20, 50$) and studied the extreme case, the RP dimension d' was set to 1. Results are summarized in Fig. 2(a)-(b) with quartiles (Q_1, Q_2, Q_3). We found that the estimation error decreases with sample number according to a power law $[r(T) \propto T^{-c} \ (c > 0)]$ and the estimation works up to about $d = 50$. For the $d = 50$ case we present notched boxed plots (Fig. 2(c)). We show the quartiles and depict the outliers, i.e., those that fall outside of interval $[Q_1 - 1.5(Q_3 - Q_1), Q_3 + 1.5(Q_3 - Q_1)]$ by circles. According to the figure, the error of estimation drops for sample number $T = 100,000$ for both types of datasets: for databases *50-geom* and *50-spherical*, respectively, we have 5 and 9 outliers from 50 random runs and thus with probability 90% and 82%, the estimation is accurate. As for question two, we compared the efficiency (Q_1, Q_2, Q_3) of our method for $d = 20$ with the NN methods by RP-ing into $d' = 1$ and $d' = 5$ dimensions. Results are shown in Fig. 2(e)-(f).[4] The figure demonstrates that for database *20-geom* performances are similar, but for database *20-spherical*

---

[4] We note that for $d = 20$ and without dimension reduction the NN methods are very slow for the ISA tasks.

our method has smaller standard deviation for $T = 20,000$. At the same time our method offers 8 to 30 times speed-up at $T = 100,000$ *for serial implementations*. Figure 3 presents the components estimated by our method for dimensions $d = 2$ and $d = 50$, respectively. With regard to our third question, the ISA problem can often be solved by grouping the estimated ICA coordinates based on their mutual information. However, this method, as illustrated by (Q_1, Q_2, Q_3) in Fig. 2(d), does not work for our *all-4-independent* database. Inserting the RP based technique into global optimization procedure, we get accurate estimation for this case, too. CE optimization was used here. Results are presented in Fig. 2(d).

5   Conclusions

In this paper we have shown that random projections (RP) can be used for the estimation of information theoretical quantities. The underlying thought of our approach is that RP approximately preserves the Euclidean distances between sample points and that a number of information theoretical quantities can be estimated from the pairwise distances of sample points. The proposed technique has been demonstrated on the estimation of Shannon's multidimensional differential entropy for the solution of the Independent Subspace Analysis task. The promise of this work is a considerable speed-up that results from the *RP technique* and the *parallel nature of the ensemble method* that we applied. Promising applications emerge – among many others – in the field of image registration.

Acknowledgments. This work has been supported by the National Office for Research and Technology and by the EC NEST 'Perceptual Consciousness: Explication and Testing' grant under contract 043261. Opinions and errors in this manuscript are the author's responsibility, they do not necessarily reflect the opinions of the EC or other project members.

References

1. Shannon, C.: The mathematical theory of communication. Bell System Technical Journal 27, 623–656 (1948)
2. Neemuchwala, H., Hero, A., Zabuawala, S., Carson, P.: Image registration methods in high-dimensional space. Int. J. Imaging Syst. and Technol. 16(5), 130–145 (2007)
3. Arya, S., Mount, D.M., Netanyahu, N.S., Silverman, R., Wu, A.Y.: An optimal algorithm for approximate nearest neighbor searching in fixed dimensions. J. of the ACM 45(6), 891 (1998)
4. Johnson, W., Lindenstrauss, J.: Extensions of Lipschitz maps into a Hilbert space. Contemporary Mathematics 26, 189–206 (1984)
5. Arriga, R.I., Vempala, S.: An algorithmic theory of learning: Robust concepts and random projections. Machine Learning 63, 161–182 (2006)
6. Matoušek, J.: On variants of the Johnson-Lindenstrauss lemma. Random Structures and Algorithms 33(2), 142–156 (2008)

7. Vempala, S.S.: The Random Projection Method. DIMACS Series in Discrete Math., vol. 65 (2005), http://dimacs.rutgers.edu/Volumes/Vol65.html
8. Baraniuk, R., Davenport, M., DeVore, R., Wakin, M.: A simple proof of the restricted isometry property for random matrices (to appear, 2008)
9. Cardoso, J.: Multidimensional independent component analysis. In: ICASSP 1998, vol. 4, pp. 1941–1944 (1998)
10. Jutten, C., Hérault, J.: Blind separation of sources: An adaptive algorithm based on neuromimetic architecture. Signal Processing 24, 1–10 (1991)
11. Kybic, J.: High-dimensional mutual information estimation for image registration. In: ICIP 2004, pp. 1779–1782. IEEE Computer Society, Los Alamitos (2004)
12. Gaito, S., Greppi, A., Grossi, G.: Random projections for dimensionality reduction in ICA. Int. J. of Appl. Math. and Comp. Sci. 3(4), 154–158 (2006)
13. Bingham, E.: Advances in Independent Component Analysis with Applications to Data Mining. PhD thesis, Helsinki University of Technology (2003), http://www.cis.hut.fi/ella/thesis/
14. Theis, F.J.: Uniqueness of complex and multidimensional independent component analysis. Signal Processing 84(5), 951–956 (2004)
15. Theis, F.J.: Multidimensional independent component analysis using characteristic functions. In: EUSIPCO 2005 (2005)
16. Theis, F.J.: Blind signal separation into groups of dependent signals using joint block diagonalization. In: ISCAS 2005, Kobe, Japan, pp. 5878–5881 (2005)
17. Amari, S., Cichocki, A., Yang, H.H.: A new learning algorithm for blind signal separation. In: NIPS 1996, vol. 8, pp. 757–763 (1996)
18. Szabó, Z., Póczos, B., Lőrincz, A.: Cross-entropy optimization for independent process analysis. In: Rosca, J.P., Erdogmus, D., Príncipe, J.C., Haykin, S. (eds.) ICA 2006. LNCS, vol. 3889, pp. 909–916. Springer, Heidelberg (2006)
19. Póczos, B., Lőrincz, A.: Independent subspace analysis using k-nearest neighborhood distances. In: Duch, W., Kacprzyk, J., Oja, E., Zadrożny, S. (eds.) ICANN 2005. LNCS, vol. 3697, pp. 163–168. Springer, Heidelberg (2005)
20. Szabó, Z., Póczos, B., Lőrincz, A.: Undercomplete blind subspace deconvolution. J. of Machine Learning Res. 8, 1063–1095 (2007)
21. Kozachenko, L.F., Leonenko, N.N.: On statistical estimation of entropy of random vector. Problems Infor. Transmiss. 23(2), 95–101 (1987)
22. Hero, A., Ma, B., Michel, O., Gorman, J.: Applications of entropic spanning graphs. Signal Processing 19(5), 85–95 (2002)
23. Rubinstein, R.Y., Kroese, D.P.: The Cross-Entropy Method. Springer, Heidelberg (2004)
24. Learned-Miller, E.G., Fisher III, J.W.: ICA using spacings estimates of entropy. J. of Machine Learning Res. 4, 1271–1295 (2003)
25. Bach, F.R., Jordan, M.I.: Beyond independent components: Trees and clusters. J. of Machine Learning Res. 4, 1205–1233 (2003)
26. Póczos, B., Szabó, Z., Kiszlinger, M., Lőrincz, A.: Independent process analysis without a priori dimensional information. In: Davies, M.E., James, C.J., Abdallah, S.A., Plumbley, M.D. (eds.) ICA 2007. LNCS, vol. 4666, pp. 252–259. Springer, Heidelberg (2007)

Theoretical Analysis and Comparison of Several Criteria on Linear Model Dimension Reduction

Shikui Tu and Lei Xu

Department of Computer Science and Engineering,
The Chinese University of Hong Kong, Hong Kong, P.R. China
{sktu,lxu}@cse.cuhk.edu.hk

Abstract. Detecting the dimension of the latent subspace of a linear model, such as Factor Analysis, is a well-known model selection problem. The common approach is a two-phase implementation with the help of an information criterion. Aiming at a theoretical analysis and comparison of different criteria, we formulate a tool to obtain an order of their approximate underestimation-tendencies, i.e., AIC, BIC/MDL, CAIC, BYY-FA(a), from weak to strong under mild conditions, by studying a key statistic and a crucial but unknown indicator set. We also find that DNLL favors cases with slightly dispersed signal and noise eigenvalues. Simulations agree with the theoretical results, and also indicate the advantage of BYY-FA(b) in the cases of small sample size and large noise.

1 Introduction

Linear model is one of the most common modeling approaches to multivariate data in many scientific fields. Factor Analysis (FA)[1] is a such widely-used linear model that assumes the observations come from a linear mixture of some latent Gaussian factors with additive Gaussian noise. It is usually used for dimension reduction via detecting the hidden structures. Also, as recently revisited in [2], PCA is equivalent to a special case of FA [1] under the Maximum Likelihood (ML) principle. FA is extended to Independent Component Analysis (ICA)[3] by requiring higher order independence, no noise and square mixing matrix.

One of the fundamental tasks in FA modeling is determining the dimension of the latent subspace, i.e., the number of hidden factors. It is a model selection problem in machine learning. Also, it is addressed as the problem of detecting the number of signals through a noisy channel in signal processing [4,5,6,7,8]. One conventional approach is hypothesis tests based on the likelihood ratio statistic [9] and a subjective threshold. Another approach is the two-phase implementation that requires no subjective threshold with the help of an information criterion such as Akaike's Information Criterion (AIC)[10], Bozdogan's Consistent Akaike's Information Criterion (CAIC)[11], Schwarz's Bayesian Information Criterion (BIC)[12] (which coincides with Rissanen's Minimum Description Length (MDL)[13]), and Bayesian Ying-Yang (BYY) harmony learning criterion[14].

Following an early work [4] in signal processing literature, a framework was proposed in [5] for studying criteria such as AIC and MDL, with asymptotic

T. Adali et al. (Eds.): ICA 2009, LNCS 5441, pp. 154–162, 2009.
© Springer-Verlag Berlin Heidelberg 2009

bounds provided for overestimation and underestimation probabilities, which was further studied in [6,7]. Recently, the behaviors of AIC and MDL in a situation with high dimensional signals but relatively few samples were investigated in [8]. In this track [4,5,6,7,8], FA is considered in its special case of PCA, and the studies are focused on asymptotic properties, such as consistency and asymptotic normality, and the results were shown to be robust for non-Gaussian sources empirically[7]. However, in practical the sample size is finite or even small, and it is intractable to get an exact selection accuracies of different criteria. An easier way is to study their relative selection tendencies for a preliminary comparison.

This paper formulates a tool further developed from[5] for a theoretical comparison of typical criteria in terms of ordered approximate underestimation tendencies. It suffices to study a key statistic and an indicator set which is inherently associated with each criterion and depends on the distribution of samples. The order from weak to strong is shown to be AIC,BIC,CAIC and BYY-FA(a) under mild conditions, while DNLL is found to favor the cases with slightly dispersed signal and noise eigenvalues. Though analytically hard, BYY-FA(b) is shown to be empirically superior for those small-sample-size and large-noise cases.

The rest of the paper is organized as follows. In Section 2, we briefly review FA and several criteria. In Section 3, we formulate a tool for comparisons of different criteria via studying a key statistic and a crucial indicator set, and then conduct simulations in Section 4. The conclusion is made in Section 5.

2   Factor Analysis and Serval Model Selection Criteria

Factor Analysis. Assume $\mathbf{x}$ is an observed n-dimensional random variable, and it is distributed according to the following descriptions:

$$\mathbf{x} = \mathbf{A}\mathbf{y} + \boldsymbol{\mu} + \mathbf{e}, \; p(\mathbf{x}|\mathbf{y}) = G(\mathbf{x}|\mathbf{A}\mathbf{y} + \boldsymbol{\mu}, \boldsymbol{\Sigma}_e), p(\mathbf{y}) = G(\mathbf{y}|\mathbf{0}, \boldsymbol{\Sigma}_y),$$

$$\begin{cases} \boldsymbol{\Theta}_m = \{\mathbf{A}, \boldsymbol{\Sigma}_e\} \text{ if } \boldsymbol{\Sigma}_y = \mathbf{I}_m \text{ (the } m \times m \text{ identity matrix),} \quad \textit{for } \mathbf{FA(a)}; \\ \boldsymbol{\Theta}_m = \{\mathbf{A}, \boldsymbol{\Lambda}_m, \boldsymbol{\Sigma}_e\} \text{ if } \boldsymbol{\Sigma}_y = \boldsymbol{\Lambda}_m \text{ (diagonal) and } \mathbf{A}^T\mathbf{A} = \mathbf{I}_m, \textit{ for } \mathbf{FA(b)}; \end{cases} \quad (1)$$

$$p(\mathbf{x}) = \int p(\mathbf{x}|\mathbf{y})p(\mathbf{y})d\mathbf{y} = G(\mathbf{x}|\boldsymbol{\mu}, \boldsymbol{\Sigma}_x), \; \boldsymbol{\Sigma}_x = \mathbf{A}\mathbf{A}^T + \boldsymbol{\Sigma}_e$$

where $\mathbf{y}$ is an $m \times 1$ hidden factor, $\boldsymbol{\Theta}_m$ is the unknown parameter set including an $n \times m$ factor loading matrix $\mathbf{A}$ and a diagonal noise covariance matrix $\boldsymbol{\Sigma}_e$, and $G(\bullet|\boldsymbol{\mu}, \boldsymbol{\Sigma})$ denotes a Gaussian distribution with the mean vector $\boldsymbol{\mu}$ and the covariance matrix $\boldsymbol{\Sigma}$. The two formulations, i.e., FA(a) and FA(b), are equivalent under the Maximum Likelihood principle for parameter learning, but they are different under the BYY harmony learning [14] for selecting m which will be introduced in Section 3.3&4.1. In the sequel, we assume $\boldsymbol{\mu} = \mathbf{0}, \boldsymbol{\Sigma}_e = \sigma_e^2\mathbf{I}_n$.

Several Criteria and Two-phase Implementation. The task of FA modeling consists of parameter learning and selecting m, based on a sample set $\mathcal{X}_N = \{\mathbf{x}_t\}_{t=1}^N$, and it is tackled by the following two-phase implementation:

- **Phase I:**   Compute $\hat{\boldsymbol{\Theta}}_m = \hat{\boldsymbol{\Theta}}(\mathcal{X}_N, m)$ for each $m \in [m_{low}, m_{up}]$ with m_{low} and m_{up} given. Normally, $\hat{\boldsymbol{\Theta}}_m$ is the Maximum Likelihood (ML) estimator $\hat{\boldsymbol{\Theta}}_m^{ML} = \arg\max_{\boldsymbol{\Theta}_m} \ln p(\mathcal{X}_N|\boldsymbol{\Theta}_m) = \arg\min_{\boldsymbol{\Theta}_m} \mathcal{E}_L(\mathcal{X}_N|\boldsymbol{\Theta}_m)$, where $\mathcal{E}_L(\mathcal{X}_N|\boldsymbol{\Theta}_m) = -\frac{2}{N}\ln p(\mathcal{X}_N|\boldsymbol{\Theta}_m)$ is denoted as **NLL**(negative log-likelihood).

– **Phase II:** Estimate $\hat{m} = \arg\min_m \mathcal{E}_{Cri}(\mathcal{X}_N, \hat{\Theta}_m)$, where $\mathcal{E}_{Cri}$ is formulated according to a criterion (Cri), e.g.,

$$\mathcal{E}_{Cri}(\mathcal{X}_N, \hat{\Theta}_m) = \mathcal{E}_L(\mathcal{X}_N, \hat{\Theta}_m) + \rho_{cri} d_m,$$

$$d_m = nm + 1, \quad \rho_{cri} = \begin{cases} \rho_L = 0; & \text{for } \textbf{NLL} \\ \rho_{AIC} = \frac{2}{N}; & \text{for } \textbf{AIC} \\ \rho_{BIC} = \frac{\ln N}{N}; & \text{for } \textbf{BIC} \\ \rho_{CIC} = \frac{\ln N + 1}{N}; & \text{for } \textbf{CAIC} \end{cases} \quad (2)$$

3   Theoretical Analysis and Comparisons

3.1   A Tool for Comparisons

Based on Sec.2&[5], this subsection further formulate a tool aiming at analysis and comparisons of different criteria for FA modeling, and provides a summarized guidance for the detailed analysis in the subsequent subsections.

– **select m via discrete optimization.** Consider $S(\Theta_m, m)$ to be a family of statistical models $p(\mathbf{x}|\Theta_m)$ for FA(a) given in eq.(1) with $\Theta_m = \{A_{n \times m}, \sigma_e^2\}$. Given a criterion(Cri) with $\mathcal{E}_{Cri} = \mathcal{E}_{Cri}(\mathcal{X}_N, \hat{\Theta}(\mathcal{X}_N, m)) = \mathcal{E}_{Cri}(\mathcal{X}_N, m)$, an estimator of m^*(the underlying true dimension) is given by $\hat{m}(\mathcal{X}_N) = \arg\min_m \mathcal{E}_{Cri}$. To locate the minima w.r.t. discrete m, no derivative can be used. However, it is reasonable to study instead the backward difference function $\nabla_m \mathcal{E}_{Cri} = \mathcal{E}_{Cri}(\mathcal{X}_N, m) - \mathcal{E}_{Cri}(\mathcal{X}_N, m-1)$, as shown in Fig.1(a)(b).
– **from $\nabla_m \mathcal{E}_{Cri}$ to local preference.** It is intractable to study $\nabla_m \mathcal{E}_{Cri}$ as a function of $\mathcal{X}_N, m$. Fortunately, $\nabla_m \mathcal{E}_{Cri}$ from several criteria for FA can be formulated as $\nabla_m \mathcal{E}_{Cri}(\gamma_m, m)$(Fig.1(b)(c)), a function of m and a statistic γ_m given in eq.(8), which will be shown in Sec.3.2&3.3. The medium γ_m extracts and transmits sufficient information from samples to selecting m, and also determines the *local preference* over each $\{m-1, m\}$ as in Fig.1(d). Γ_m^* and its element γ_m^* are separately termed **indicator set** and **indicator** at m. Note that γ_m is closely related to the signal-to-noise ratio.
– **approximate underestimation tendency.** *Underestimation* refers to an event "$\hat{m} < m^*$". Considering the *Local preference* defined in Fig.1(d) over $\{m^* - 1, m^*\}$, if $\gamma_{m^*} \in \Gamma_{m^*}^+$, then $m^* - 1$ is preferred to m^*, which indicates that "$\hat{m} < m^*$" is *likely* to happen (though not guaranteed). Therefore, it is reasonable to approximate the underestimation tendency by the probability $Pr\{\gamma_{m^*} \in \Gamma_{m^*}^+\}$. Its exact evaluation is intractable for a finite or small N, but the relative tendencies of different criteria can be determined as follows.
– **A TOOL for comparisons.** Fixing $m = m^*$, assume $\nabla_m \mathcal{E}_{Cri_1}(\gamma_m)$ and $\nabla_m \mathcal{E}_{Cri_2}(\gamma_m)$, sketched in Fig.1(c), are strictly monotone decreasing in domain Γ_D with their indicators satisfying $\gamma_m^*(\text{Cri}_1) < \gamma_m^*(\text{Cri}_2)$. Actually, these assumptions hold for several criteria as in Sec.3.2. Then, $\Gamma_m^+(\text{Cri}_i) = (-\infty, \gamma_m^*(\text{Cri}_i)) \bigcap \Gamma_D$, $i = 1, 2$, and $Pr\{\gamma_m \in \Gamma_m^+(\text{Cri}_2)\} - Pr\{\gamma_m \in \Gamma_m^+(\text{Cri}_1)\} = Pr\{\gamma_m^*(\text{Cri}_1) < \gamma_m < \gamma_m^*(\text{Cri}_2)\} \geq 0$. So, "*approximately* the underestimation tendency of Cri_2 is stronger than that of Cri_1" or $Cri_1 \prec_u Cri_2$. Similar analysis on overestimation can be performed at $m = m^* + 1$.

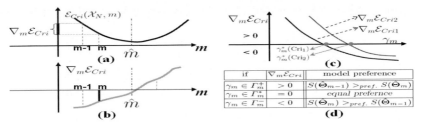

Fig. 1. For a given $\mathcal{X}_N$, graphs of $\mathcal{E}_{Cri}$ and $\nabla_m \mathcal{E}_{Cri}$ w.r.t. m are sketched in (a)&(b), while for two criteria, Cri_1 and Cri_2, the graphs of $\nabla_m \mathcal{E}_{Cri_1}, \nabla_m \mathcal{E}_{Cri_2}$ w.r.t. γ_m given m are sketched in (c), as well as its corresponding local preference defined in (d)

3.2   AIC, BIC, CAIC

Assume the eigenvalues of the sample covariance matrix, i.e., $S_N = \frac{1}{N}\sum_{t=1}^{N} \mathbf{x}_t \mathbf{x}_t^T$, are $\{s_i : 1 \le i \le n\}$ with $s_1 \ge \ldots \ge s_n$. The Maximum Likelihood (ML) estimate $\hat{\mathbf{\Theta}}_m^{ML}$ for FA(a) in eq.(1) is given to be ([1,4,2]):

$$\begin{cases} \hat{\mathbf{A}}_{n\times m}^{ML} = \mathbf{U}_{n\times m}(\mathbf{D}_m - \hat{\sigma}_e^2)^{\frac{1}{2}}\mathbf{R}^T, \quad \mathbf{D}_m = diag[s_1,\ldots,s_m], \\ \hat{\sigma}_e^{2,ML} = \frac{1}{n-m}\sum_{i=m+1}^{n} s_i, \end{cases} \tag{3}$$

where the i-th column of $\mathbf{U}_{n\times m}$ is the eigenvector of S_N corresponding to s_i, and $\mathbf{R}$ is an arbitrary rotation matrix. According to eq.(2) and eq.(3), the NLL and the difference functions of some criteria, are further formulated as:

$$\mathcal{E}_L(\mathcal{X}_N, \hat{\mathbf{\Theta}}_m^{ML}) = k \ln \sum_{i=m+1}^{n} s_i - k \ln k - \sum_{i=m+1}^{n} \ln s_i, \quad k = n - m, \tag{4}$$

$$\nabla_m \mathcal{E}_L(\gamma_m, m) \doteq \nabla_m \mathcal{E}_L(\mathcal{X}_N, \hat{\mathbf{\Theta}}_m^{ML}) = -(k+1)\ln\left(1 + \frac{\gamma_m - 1}{k+1}\right) + \ln \gamma_m \tag{5}$$

$$\frac{\partial \nabla_m \mathcal{E}_L(\gamma_m, m)}{\partial \gamma_m} = -\frac{k(\gamma_m - 1)}{(k + \gamma_m)\gamma_m} \le 0, \ \forall \gamma_m \in [1, +\infty). \tag{6}$$

$$\nabla_m \mathcal{E}_{Cri}(\gamma_m, m) \doteq \nabla_m \mathcal{E}_{Cri}(\mathcal{X}_N, \hat{\mathbf{\Theta}}_m^{ML}) = \nabla_m \mathcal{E}_L(\mathcal{X}_N, \hat{\mathbf{\Theta}}_m^{ML}) + n\rho_{cri} \tag{7}$$

where ρ_{cri} is given in eq.(2), and γ_m is explicitly formulated by

$$\gamma_m = \gamma_{m,m}, \quad \gamma_{i,m} = s_i/\mathcal{A}_{m+1}^n \ge 1, i = 1,\ldots,m; \quad \mathcal{A}_m^n = \frac{1}{n-m+1}\sum_{i=m}^{n} s_i, \tag{8}$$

Due to the space limit, all theoretical results are given without proofs.

Lemma 1. *(1). Given $\rho > 0$, the root γ^* of $\nabla_m \mathcal{E}_L(\gamma) = -n\rho$ is unique for $\gamma > 1$ and bounded in $(\gamma_{low}, \gamma_{up})$, where $\gamma_{low} = (k+1)C_0 - k$, and $\gamma_{up} = \gamma_{low} + \sqrt{2(k+1)C_0(C_0 - 1)}$, and $C_0 = \exp\{\frac{n\rho}{k}\}$, $k = n - m$. (2). For $\rho_1 > \rho_2 > 0$, we have $\gamma^*(\rho_1) > \gamma^*(\rho_2) > 1$.*

Remarks: Similar bounds were provided in[5,6] by two kinds of Taylor approximations w.r.t. two formulated variables separately, while Lemma 1(1) was derived by a second-order Taylor approximation (as in[6]) w.r.t to γ (as in[5]).

Theorem 1. *Since the indicator* $\gamma_m^*(Cri)$ *is a root of* $\nabla_m \mathcal{E}_{Cri}(\gamma_m) = 0$, *then*

1. $1 = \gamma_m^*(\text{NLL}) < \gamma_m^*(\text{AIC}) < \gamma_m^*(\text{BIC}) < \gamma_m^*(\text{CAIC})$, *if* $N \geq 8 > e^2$.
2. $\Gamma_m^+ = [1, \gamma_m^*(Cri))$, $\Gamma_m^- = (\gamma_m^*(Cri), +\infty)$, *and indicator set* $\Gamma_m^* = \{\gamma_m^*(Cri)\}$.
3. *Applying* $C_0 \approx 1 + \frac{n\rho}{k}$ *to* γ_{up} *in Lemma 1, we get a further approximation:*

$$\gamma_m^*(Cri) \approx 1 + (n + n/k) \cdot \frac{c}{N} + \frac{n}{k}\sqrt{2(k+1)\left(\frac{k}{n} + \frac{c}{N}\right)\frac{c}{N}} + O(\frac{c}{N}), \quad (9)$$

where $c = 2, \ln N, \ln N + 1$ *for* **AIC, BIC, CAIC** *separately, and* $k = n - m$.

Remarks: This theorem indicates: (1). NLL tends to select large m in probability one unless $\gamma_m = 1, \forall m > m^*$ or $s_i = \sigma^2$ $(\forall i \geq m^*)$ which requires $N \to +\infty$; (2). Fixing $m = m^*$, "AIC$\prec_u$BIC$\prec_u$CAIC" holds according to Sec.3.1.

3.3 DNLL and BYY-FA(a)

The likelihood-ratio test is a conventional approach to model selection in statistics [9]. The logarithm of the likelihood-ratio or the difference of the Negative-Log-likelihood (NLL) is denoted as **DNLL**, and the corresponding objective function is $\mathcal{E}_{\text{DNLL}}(\mathcal{X}_N, \hat{\Theta}_m^{ML}) = \nabla_m \mathcal{E}_L$, where $\nabla_m \mathcal{E}_L$ is given in eq.(5). Then,

$$\begin{aligned}\nabla_m(\mathcal{E}_{\text{DNLL}}(\mathcal{X}_N, \hat{\Theta}_m^{ML})) = \nabla_m^2 \mathcal{E}_L = -2(k+1)\ln\left(1 + \frac{\gamma_{m,m-1}}{k+1}\right) \\ + (k+2)\ln\left(1 + \frac{\gamma_{m-1,m} + \gamma_{m,m-2}}{k+2}\right) - \ln\frac{\gamma_{m-1,m}}{\gamma_{m,m}}\end{aligned} \quad (10)$$

where $\gamma_{m-1,m}, \gamma_{m,m}$ are formulated in eq.(8), and $\gamma_{m-1,m} \geq \gamma_{m,m} \geq 1$. According to the Formulation 1, $\gamma(\mathcal{X}_N, m)$ is generalized to a two-variable vector $(\gamma_{m-1,m}, \gamma_{m,m})$, and the indicator set Γ_m^* becomes a 2-dimensional boundary.

Theorem 2. *Define* $s_p, \ldots, s_q$ *to be "slightly dispersed", if* $|s_i - \mathcal{A}_p^q| < \delta$ *holds for any* $i \in [p, q]$ *and a very small* $\delta(> 0)$. *The criterion DNLL captures the variations of NLL, and especially at the unknown true dimension* m^* *we have*

1. *When* $m = m^*$: *If* $s_{m-1} \approx s_m \gg \mathcal{A}_{m+1}^n$, *then* $\gamma_{m-1,m} \approx \gamma_{m,m} \gg 1$, *which implies* $\nabla_m \mathcal{E}_{DNLL} < 0$, *i.e.,* m^* *is preferred to* $m^* - 1$.
2. *When* $m - 1 = m^*$: *If* $s_{m-1} \gg s_m \approx \mathcal{A}_{m+1}^n$, *then* $\gamma_{m-1,m} \gg \gamma_{m,m} \approx 1$, *which implies* $\nabla_m \mathcal{E}_{DNLL} > 0$, *i.e.,* m^* *is preferred to* $m^* + 1$.
3. *If* $s_1, \ldots, s_{m^*}$ *are slightly dispersed,* $s_{m^*+1}, \ldots, s_n$ *are also slightly dispersed, and* $s_{m^*} \gg s_{m^*+1}$, *then* m^* *is the global minimum of* $\mathcal{E}_{DNLL}$.

Remarks: Instead of strict mathematical formulations, the conditions in Theorem 2 are stated in an intuitive way. It implies DNLL favors slightly dispersed signal and noise eigenvalues, as well as a large signal-to-noise ratio (SNR). However, the conditions will be probably violated when N and SNR is small.

Another approach to tackling model selection problems is the Bayesian Ying-Yang (BYY) harmony learning theory[14]. We defer its detailed introduction to the next section. With γ_m again formulated in eq.(8), a BYY criterion (denoted as **BYY-FA(a)**) for FA(a) in eq.(1) as well as its difference function is

$$\mathcal{E}_H^a(\mathcal{X}_N, \hat{\Theta}_m^{ML}) = m\ln(2\pi e) + n\ln\left(\frac{1}{n-m}\sum_{i=m+1}^n s_i\right),$$

$$\nabla_m \mathcal{E}_H^a(\gamma_m, m) = \ln(2\pi e) - n\left\{\ln\left(1 + \frac{\gamma_m}{n-m}\right) + \ln\left(1 + \frac{1}{n-m}\right)\right\}, \tag{11}$$

Lemma 2. *According to eq.(11) and the tool defined in Sec.3.1, we have*

1. *Since the indicator $\gamma_m^*(H_a)$ is the root of $\nabla_m \mathcal{E}_H^a(\gamma_m) = 0$, then $\gamma_m^*(H_a) = (n - m + 1)\left[(2\pi e)^{\frac{1}{n}} - 1\right] + 1 > 1$, e.g., $\gamma_m^*(H_a) \approx 3.595$ when $n = 9, m = 3$.*
2. *$\Gamma_m^+ = [1, \gamma_m^*(H_a)), \Gamma_m^- = (\gamma_m^*(H_a), +\infty)$, the indicator set $\Gamma_m^* = \{\gamma_m^*(H_a)\}$.*

Theorem 3. *There exists an equivalent ρ_{H_a} for BYY-FA(a), and then we indirectly compare the indicator $\gamma_m^*(H_a)$ of BYY-FA(a) with $\gamma_m^*(Cri)$ of another criterion (Cri) by approximately comparing ρ_{H_a} with ρ_{cri} as follows:*

1. *ρ_{H_a} is bounded in $\left(\rho_{H_a}^{(low)}, \rho_{H_a}^{(up)}\right)$, where $\rho_{H_2}^{(up)} = \frac{n-m}{n}\ln c_n$ and $\rho_{H_a}^{(low)} = c_n + \frac{2c_n-1}{k-1} - \frac{\sqrt{2(k+1)c_n(c_n-1)+1}}{k-1}$, $c_n = \sqrt[n]{(2\pi e)}$, $k = n - m$.*
2. *Given $n, m, \exists N_{cri} > 1$ such that $\rho_{H_a} < \rho_{cri}$ iff $1 < N < N_{cri}$. Also, N_{cri} is lower bounded by N_{up}, which is the largest N that satisfies $\rho_{H_a}^{(up)} < \rho_{cri}$. E.g., $N_{up} = 14, 23, 31$ for AIC, BIC, and CAIC respectively, when $n = 9, m = 5$.*

Remarks: Consider $m = m^*$. (1). Since $\gamma_{m^*}^*(H_a)$ is irrelevant to N and γ_{m^*} is the ML estimator for the true unknown SNR $\gamma_o = \lambda_{m^*}/\sigma^2$, then BYY-FA(a) tends to underestimate m regardless of N as long as $\gamma_{m^*}^*(H_a) > \gamma_o$. (2). We compare BYY-FA(a) with other criteria (Cri), such as AIC, BIC and CAIC, directly by calculating each indicator $\gamma_{m^*}^*(Cri)$ as in Lemma 1&2 or indirectly in form of ρ_{cri} as in Theorem 3. (3). There exists a small N_{cri}, such that if $N < N_{cri}$ then BYY-FA(a)$\prec_u$Cri, otherwise Cri$\prec_u$BYY-FA(a), according to Sec.3.1.

4 Empirical Study and BYY-FA(b)

4.1 BYY-FA(b)

The criteria analyzed above are relatively easy for a theoretical analysis, while BYY Harmony Learning Theory on another formulation of FA, i.e., FA(b) in eq.(1), is difficult. However, via an empirical comparison, we still provide insights of its model selection performances.

Firstly proposed in 1995 and systematically developed in the past decade, Bayesian Ying-Yang (BYY) harmony learning theory is a general statistical learning framework that can handle both parameter learning and model selection under a best harmony principle. The BYY harmony learning leads us not only a set of new model selection criteria for typical structures, but also a class of automatic model selection algorithms. For more details, please refer to a recent systematic review[14].

FA(a) and FA(b) in eq.(1) are equivalent under ML principle but different under the BYY harmony learning theory [14]. The former leads to BYY-FA(a) in Sec.3.3, while the latter leads to BYY-FA(b) as follows, with a similar two-phase procedure (see eq.(7) in[14]) implemented,

$$\mathcal{E}_H^b = m \ln(2\pi e) + \ln|\mathbf{\Lambda}| + n \ln \sigma_e^2 + h^2 Tr\left[(\mathbf{A}\mathbf{\Lambda}\mathbf{A}^T + \sigma_e^2 \mathbf{I}_n)^{-1}\right]. \qquad (12)$$

4.2 Simulations

We design $3 \times 3 = 9$ cases of experimental environments by considering three levels of sample size N and noise σ_e^2 respectively, with $n = 9$ and $m^* = 3$ fixed. Three levels are $100, 50, 25$ for N or $0.1\lambda_{m^*}, 0.3\lambda_{m^*}, 0.5\lambda_{m^*}$ for σ_e^2 (equivalently $\gamma_o = \lambda_{m^*}/\sigma_e^2 = 10, 3.33, 2$), where λ_{m^*} is the m^*-th largest Gaussian signal's variance. We randomly generate samples according to each setting of FA in eq.(1) for each of 100 independent repeated runs, in which two-phase procedure is implemented by setting $[m_{low}, m_{up}] = [1,6]$ and randomly initializing $\mathbf{\Theta}_m$. The selection percentage rates are reported in Table 1. The indicators $\gamma_{m^*}^*(Cri)$ are approximately calculated by eq.(9), and $\gamma_{m^*}^*(H_a)$ by Lemma 2.

The simulations suggest the following observations. (1). The performances of all criteria are comparable when N, γ_o are large, but they decline at different speeds as N, γ_o reduce. (2). For a large $N(= 100)$, BIC and CAIC is consistent but AIC risks an overestimation. Let Cri be AIC, BIC or CAIC, and then $\gamma_{m^*}^*(Cri)$ grows as N reduces. When $\gamma_{m^*}^*(Cri)$ exceeds γ_o, Cri tends

Table 1. We report the percentage rates of model selection of 9 combinations in three categories, i.e., *underestimation*(**U**),*successful selection*(**S**) and *overestimation*(**O**). The indicator $\gamma_m^*(Cri)$ is calculated at $m = m^* = 3$. Note that $\gamma_m^*(Cri)$ by eq.(9) approximates γ_m^{num} well, where γ_m^{num} is the numerical solution of $\nabla_m \mathcal{E}_{Cri}(\gamma) = 0$.

(a). Sample size $N = 100$, $\gamma_o = \lambda_{m^*}/\sigma_e^2$ (3 levels)

noise level: Cri \ rates	$\gamma_o = 10$ U	S	O	$\gamma_o = 3.33$ U	S	O	$\gamma_o = 2$ U	S	O	$\gamma_m^*(Cri)$ approximated by eq.(9).	γ_m^{num} is the numerical sol.
AIC	0	99	1	0	**96**	4	0	**97**	3	$\gamma_3^*(AIC) \approx 1.87$	$\gamma_3^{num}(AIC) = 1.83$
BIC	0	100	0	1	99	4	9	91	0	$\gamma_3^*(BIC) \approx 2.50$	$\gamma_3^{num}(BIC) = 2.43$
CAIC	0	100	0	1	99	4	22	78	0	$\gamma_3^*(CAIC) \approx 2.72$	$\gamma_3^{num}(CAIC) = 2.65$
DNLL	2	98	0	39	61	0	63	27	0	not available	not available
BYY-FA(a)	0	100	0	30	70	0	98	2	0	$\gamma_3^*(H_a) = 3.59$	$\gamma_3^*(H_a) = 3.59$
BYY-FA(b)	0	100	0	1	99	0	1	95	4	not available	not available

(b). Sample size $N = 50$, $\gamma_o = \lambda_{m^*}/\sigma_e^2$ (3 levels)

(same as (a))	U	S	O	U	S	O	U	S	O	$\gamma_m^*(Cri)$ by eq.(9)	$\gamma_3^{num}(Cri)$
AIC	0	98	2	0	**99**	1	18	79	3	$\gamma_3^*(AIC) \approx 2.36$	$\gamma_3^{num}(AIC) = 2.30$
BIC	0	100	0	6	94	2	76	24	0	$\gamma_3^*(BIC) \approx 3.18$	$\gamma_3^{num}(BIC) = 3.10$
CAIC	0	100	0	20	80	0	91	9	0	$\gamma_3^*(CAIC) \approx 3.57$	$\gamma_3^{num}(CAIC) = 3.50$
DNLL	5	95	0	49	51	0	86	14	0	not available	not available
BYY-FA(a)	0	100	0	35	65	2	96	4	0	$\gamma_3^*(H_a) = 3.59$	$\gamma_3^*(H_a) = 3.59$
BYY-FA(b)	0	100	0	3	97	0	5	**84**	11	not available	not available

(b). Sample size $N = 25$, $\gamma_o = \lambda_{m^*}/\sigma_e^2$ (3 levels)

(same as (a))	U	S	O	U	S	O	U	S	O	$\gamma_m^*(Cri)$ by eq.(9)	$\gamma_3^{num}(Cri)$
AIC	1	92	7	13	85	2	58	40	2	$\gamma_3^*(AIC) \approx 3.21$	$\gamma_3^{num}(AIC) = 3.13$
BIC	1	99	0	49	51	0	94	6	0	$\gamma_3^*(BIC) \approx 4.15$	$\gamma_3^{num}(BIC) = 4.10$
CAIC	1	99	0	84	16	0	100	0	0	$\gamma_3^*(CAIC) \approx 4.88$	$\gamma_3^{num}(CAIC) = 4.91$
DNLL	11	89	0	62	38	0	89	11	0	not available	not available
BYY-FA(a)	1	92	7	35	63	2	85	12	3	$\gamma_3^*(H_a) = 3.59$	$\gamma_3^*(H_a) = 3.59$
BYY-FA(b)	0	99	1	6	**89**	5	21	**66**	13	not available	not available

to underestimates m, where AIC remains more robust. These agree with Theorem 1. (3). DNLL fails as N, γ_o reduce, which agrees with Theorem 2. (4). BYY-FA(a) tends to underestimate m when $\gamma_o < \gamma_{m^*}^*(H_b)$, which is worse than BIC and CAIC for $N = 100$ but better for $N = 25$. This coincides with Theorem 3. (5). BYY-FA(b) becomes evidently superior when $N \leq 50$ and $\gamma_o \leq 3.33$. For example, it improves by $4.7\%, 6.3\%, 65\%$ relative to AIC when $(N, \gamma_o) = (25, 3.33), (50, 2), (25, 2)$ respectively.

5   Conclusion

We have provided a preliminary theoretical comparison of several criteria based on the problem of selecting the hidden dimension of FA in its special case of PCA. It suffices to study a statistic and a crucial but unknown indicator set for each criterion. Due to the difficulty in exact evaluation of selection accuracy for a finite or small sample size N, the model selection behavior is preliminarily characterized by an order of the approximate underestimation tendencies, i.e., $\text{AIC} \prec_u \text{BIC} \prec_u \text{CAIC} \prec_u \text{BYY-FA(a)}$. DNLL requires a proper dispersion of signal and noise eigenvalues. The simulations agree with the theoretical results and also indicates that BYY-FA(b) becomes superior as N reduces and noise increases.

Acknowledgments. The work described in this paper was fully supported by a grant from the Research Grant Council of the Hong Kong SAR (Project No: CUHK4177/07E).

References

1. Anderson, T., Rubin, H.: Statistical inference in factor analysis. In: Proceedings of the third Berkeley symposium on mathematical statistics and probability, vol. 5, pp. 111–150 (1956)
2. Tipping, M.E., Bishop, C.M.: Mixtures of probabilistic principal component analyzers. Neural Computation 11(2), 443–482 (1999)
3. Hyvärinen, A.: Survey on Independent Component Analysis. Neural Computing Surveys 2, 94–128 (1999)
4. Wax, M., Kailath, T.: Detection of signals by information theoretic criteria. IEEE Trans. Acoustics, Speech and Signal Processing ASSP-33(2), 387 (1985)
5. Xu, W., Kaveh, M.: Analysis of the performance and sensitivity of eigendecomposition-based detectors. IEEE Transactions on Signal Processing 43(6), 1413–1426 (1995)
6. Liavas, A., Regalia, P.: On the behavior of information theoretic criteria for model order selection. IEEE Transactions on Signal Processing 49(8), 1689–1695 (2001)
7. Fishler, E., Poor, H.: Estimation of the number of sources in unbalanced arrays via information theoretic criteria. IEEE Transactions on Signal Processing 53(9), 3543–3553 (2005)
8. Nadakuditi, R., Edelman, A.: Sample eigenvalue based detection of high-dimensional signals in white noise using relatively few samples. IEEE Transactions on Signal Processing 56(7), 2625–2638 (2008)
9. Cox, D., Hinkley, D.: Theoretical Statistics. Chapman and Hall, London (1974)

10. Akaike, H.: A new look at the statistical model identification. IEEE Transactions on Automatic Control 19(6), 716–723 (1974)
11. Bozdogan, H.: Model selection and Akaike's Information Criterion (AIC): The general theory and its analytical extensions. Psychometrika 52(3), 345–370 (1987)
12. Schwarz, G.: Estimating the Dimension of a Model. The Annual of Statistics 6(2), 461–464 (1978)
13. Rissanen, J.: Modelling by the shortest data description. Automatica 14, 465–471 (1978)
14. Xu, L.: Bayesian ying yang system, best harmony learning, and gaussian manifold based family. In: Zurada, J.M., Yen, G.G., Wang, J. (eds.) Computational Intelligence: Research Frontiers. LNCS, vol. 5050, pp. 48–78. Springer, Heidelberg (2008)

Factor Analysis as Data Matrix Decomposition: A New Approach for Quasi-Sphering in Noisy ICA

Steffen Unkel and Nickolay T. Trendafilov

Department of Mathematics and Statistics,
Faculty of Mathematics, Computing & Technology,
The Open University, Walton Hall, Milton Keynes, MK7 6AA, United Kingdom
{S.Unkel,N.Trendafilov}@open.ac.uk

Abstract. In this paper, a new approach for quasi-sphering in noisy ICA by means of exploratory factor analysis (EFA) is introduced. The EFA model is considered as a novel form of data matrix decomposition. By factoring the data matrix, estimates for all EFA model parameters are obtained simultaneously. After the preprocessing, an existing ICA algorithm can be used to rotate the sphered factor scores towards independence. An application to climate data is presented to illustrate the proposed approach.

Keywords: Noisy ICA, Exploratory factor analysis, Less observations than variables, Factor rotation, Procrustes problems, Climate anomalies.

1 Introduction

As a preprocessing step in ICA, a mean-centered data matrix is often whitened or sphered to transform its covariance matrix to be the identity matrix. This operation was named quasi-sphering in the context of noisy ICA by [8]. To fit the noisy ICA model, exploratory factor analysis (EFA) is frequently used to perform quasi-sphering of the data [9,11]. Traditionally, EFA seeks estimates for a factor loadings matrix and a matrix of unique variances which give the best fit to the sample covariance/correlation matrix [12]. Sphered factor scores are obtained as a function of these estimates and the data.

In a number of modern applications, the number of variables p exceeds the number of available observations n. Thus, the sample covariance/correlation matrix is singular and methods such as e.g. maximum-likelihood factor analysis cannot be applied. In addition, numerical algorithms factorizing a $p \times p$ covariance/correlation matrix may become computationally slow if p is huge.

A new and efficient approach for fitting the EFA model is presented in this paper. The EFA model is considered as a specific form of data matrix decomposition. Estimates for all EFA model parameters are obtained simultaneously. Based on this initial solution, the JADE criterion [2] is optimized to estimate the mixing matrix and the independent sources. The proposed approach is applied to study winter climate anomalies in the Northern Hemisphere.

T. Adali et al. (Eds.): ICA 2009, LNCS 5441, pp. 163–170, 2009.

2 Statistical Models of Noisy ICA and EFA

Consider the noisy ICA model [8]:

$$\mathbf{x} = \mathbf{As} + \mathbf{u} , \tag{1}$$

where $\mathbf{x} \in \mathbb{R}^{p \times 1}$ is a random vector of manifest variables, $\mathbf{s} \in \mathbb{R}^{k \times 1}$ is a random vector of $k \leq p$ latent sources, $\mathbf{A} \in \mathbb{R}^{p \times k}$ is a mixing matrix with $\mathrm{rank}(\mathbf{A}) = k$, and $\mathbf{u} \in \mathbb{R}^{p \times 1}$ is a random vector of error terms. Assume that $\mathbf{s}$ consists of mutually independent sources of which at most one is Gaussian. Furthermore, let $\mathrm{E}(\mathbf{x}) = \mathrm{E}(\mathbf{s}) = \mathbf{0}$ and $\mathrm{E}(\mathbf{ss}') = \mathbf{I}_k$. Finally, suppose that $\mathrm{E}(\mathbf{su}') = \mathbf{0}_{k \times p}$ and $\mathbf{u} \sim \mathcal{N}_p(\mathbf{0}, \boldsymbol{\Psi}^2)$, where $\boldsymbol{\Psi}^2$ is a positive definite diagonal matrix.

The noisy ICA problem of obtaining approximately independent realizations of the sources can be transformed into a specific EFA task by showing the close connection between noisy ICA and EFA. Consider the EFA model [12]:

$$\mathbf{x} = \boldsymbol{\Lambda}\mathbf{f} + \boldsymbol{\Psi}\mathbf{u} , \tag{2}$$

where $\mathbf{f} \in \mathbb{R}^{k \times 1}$ is a vector of k common factors, $\boldsymbol{\Lambda} \in \mathbb{R}^{p \times k}$ is the loading matrix with $\mathrm{rank}(\boldsymbol{\Lambda}) = k$, $\mathbf{u} \in \mathbb{R}^{p \times 1}$ is a vector of unique factors, and $\boldsymbol{\Psi}$ is a $p \times p$ diagonal matrix of unique factor-pattern coefficients. In EFA it is assumed that $k \ll p$, where the choice of k is subject to some limitations [12], which will not be discussed here. Assume that $\mathrm{E}(\mathbf{x}) = \mathrm{E}(\mathbf{f}) = \mathrm{E}(\mathbf{u}) = \mathbf{0}$. Furthermore, let $\mathrm{E}(\mathbf{ff}') = \mathbf{I}_k$, $\mathrm{E}(\mathbf{uu}') = \mathbf{I}_p$ and $\mathrm{E}(\mathbf{fu}') = \mathbf{0}_{k \times p}$. Hence, all factors are uncorrelated with one another. Under these assumptions, the model in (2) represents an EFA model with uncorrelated or orthogonal (random) common factors.

For the random EFA model, it is often convenient to assume that not only $\mathbf{u}$ but also $\mathbf{f}$ and hence $\mathbf{x}$ are multinormally distributed. This assumption is usually made for purposes of statistical inference [12]. As the elements of $\mathbf{f}$ are uncorrelated, the assumption of normality means they are statistically independent random variables. Unlike EFA, noisy ICA assumes that the k sources in $\mathbf{s}$ are both mutually independent and non-normal, or that at least all but one are non-normal. Apart from this key difference, the EFA model in (2) is virtually identical to the noisy ICA model in (1), where the common factors $\mathbf{f}$ correspond to the sources $\mathbf{s}$ and the loadings $\boldsymbol{\Lambda}$ to the mixing matrix $\mathbf{A}$.

In the EFA model, the factor loadings are not unique. If $\mathbf{T}$ is a non-singular $k \times k$ matrix, (2) may be rewritten as

$$\mathbf{x} = \boldsymbol{\Lambda}\mathbf{T}\mathbf{T}^{-1}\mathbf{f} + \boldsymbol{\Psi}\mathbf{u} ,$$

which is a model with loading matrix $\boldsymbol{\Lambda}\mathbf{T}$ and common factors $\mathbf{T}^{-1}\mathbf{f}$. The assumptions about the random variables that make up the original model are not violated by this transformation. This means that there is an infinite number of factor loadings satisfying the original assumptions of the model.

As a remedy, parameter estimation is usually followed by some kind of rotation such that a certain simplicity criterion (e.g. Varimax) is optimized [12]. In noisy ICA, the assumption of both non-normal and independent common factors removes the rotational redundancy of the EFA model. The mixing matrix can be identified up to permutation and sign ambiguities and there is no need for 'simple structure' rotation. This is achieved by optimizing criteria that involve measures of departure from normality and/or independence [3,8].

One can start from an initial EFA solution, and rather than rotating the factor loadings towards simplicity, rotate the common factors orthogonally towards independence to achieve the ICA goal. From this point of view, ICA is just another method of factor rotation in EFA [7,8]. There are a number of problems to find factor scores following the classical EFA [12]. In the next section, a new approach to EFA is presented which facilitates the factor scores estimation.

3   Fitting the EFA Model by Data Matrix Decomposition

Given a multivariate sample of n independent observations on $\mathbf{x} = (x_1, \ldots, x_p)'$ $(n > p)$, the k-factor model in (2) can be written as

$$\mathbf{X} = \mathbf{F}\boldsymbol{\Lambda}' + \mathbf{U}\boldsymbol{\Psi} , \tag{3}$$

where $\mathbf{X} = (\mathbf{x}_1, \ldots, \mathbf{x}_p) \in \mathbb{R}^{n \times p}$ corresponds to the observed data matrix in which $\mathbf{x}_j = (x_{1j}, \ldots, x_{nj})'$ $(j = 1, \ldots, p)$, and $\mathbf{F} = (\mathbf{f}_1, \ldots, \mathbf{f}_k) \in \mathbb{R}^{n \times k}$ and $\mathbf{U} = (\mathbf{u}_1, \ldots, \mathbf{u}_p) \in \mathbb{R}^{n \times p}$ denote the unknown matrices of factor scores for the k common factors and p unique factors on n observations, respectively.

To facilitate notations, assume that the columns of $\mathbf{X}$, $\mathbf{F}$, and $\mathbf{U}$ are mean-centered and scaled to have unit length. Furthermore, suppose that rank$(\boldsymbol{\Lambda}) = k$, $\mathbf{F}'\mathbf{F} = \mathbf{I}_k$, $\mathbf{U}'\mathbf{U} = \mathbf{I}_p$, $\mathbf{U}'\mathbf{F} = \mathbf{0}_{p \times k}$, and that $\boldsymbol{\Psi}$ is a diagonal matrix.

The EFA model (3) and the assumptions imply the following model correlation structure $\mathbf{R}$ for the observed variables:

$$\mathbf{R} = \boldsymbol{\Lambda}\boldsymbol{\Lambda}' + \boldsymbol{\Psi}^2 . \tag{4}$$

In EFA, the most common approach is to find the pair $\{\hat{\boldsymbol{\Lambda}}, \hat{\boldsymbol{\Psi}}\}$ which gives the best fit, for certain k, to the sample correlation matrix $\mathbf{C} = \mathbf{X}'\mathbf{X}$ with respect to some discrepancy measure. The process of finding this pair is called factor extraction. Various factor extraction methods have been proposed [12].

After estimates $\hat{\boldsymbol{\Lambda}}$ and $\hat{\boldsymbol{\Psi}}$ have been found, common factor scores can be computed as a function of $\mathbf{X}$, $\hat{\boldsymbol{\Lambda}}$, and $\hat{\boldsymbol{\Psi}}$ in a number of ways [12]. For the EFA model (3), [1] proposed the following set of factor scores:

$$\hat{\mathbf{F}} = \mathbf{X}\hat{\boldsymbol{\Psi}}^{-2}\hat{\boldsymbol{\Lambda}} \left(\hat{\boldsymbol{\Lambda}}'\hat{\boldsymbol{\Psi}}^{-2}\mathbf{C}\hat{\boldsymbol{\Psi}}^{-2}\hat{\boldsymbol{\Lambda}} \right)^{-\frac{1}{2}} , \tag{5}$$

which satisfies the correlation-preserving constraint $\mathbf{F}'\mathbf{F} = \hat{\mathbf{F}}'\hat{\mathbf{F}} = \mathbf{I}_k$. Note that (5) is undefined if $\boldsymbol{\Psi}$ is singular, a situation not uncommon in practice.

In formulating EFA models, the standard approach is to embed the data in a replication framework by assuming the observations are realizations of random variables. For $n > p$, [4] formulated the EFA model directly in terms of the data instead. Let $||\mathbf{Z}||_F = \sqrt{\mathrm{trace}(\mathbf{Z}'\mathbf{Z})}$ denote the Frobenius norm of $\mathbf{Z}$ and consider minimizing the following least squares goodness-of-fit criterion [4]:

$$||\mathbf{X} - \mathbf{F}\mathbf{\Lambda}' - \mathbf{U}\mathbf{\Psi}||_F^2 \qquad (6)$$

subject to $\mathrm{rank}(\mathbf{\Lambda}) = k$, $\mathbf{F}'\mathbf{F} = \mathbf{I}_k$, $\mathbf{U}'\mathbf{U} = \mathbf{I}_p$, $\mathbf{U}'\mathbf{F} = \mathbf{0}_{p \times k}$, and $\mathbf{\Psi}$ being a $p \times p$ diagonal matrix. Thus, the observations $\mathbf{X}$ are in the space spanned by the common and unique factor scores. The loss function (6) is bounded below [5]. To optimize the objective function (6), [4] proposed an algorithm of an alternating least squares (ALS) type.

If the number of variables exceeds the number of observations ($p > n$), the sample correlation matrix is singular. Then, the standard maximum-likelihood factor analysis or generalized least squares factor analysis cannot be applied. Maximum-likelihood fit to such rank-deficient correlation matrices by the EFA correlation structure $\mathbf{\Lambda}\mathbf{\Lambda}' + \mathbf{\Psi}^2$ was considered in [13]. They look to approximate a singular correlation matrix by a positive definite one having the specific form $\mathbf{\Lambda}\mathbf{\Lambda}' + \mathbf{\Psi}^2$ and assuming $\mathbf{\Psi}^2$ positive definite.

Alternatively, one can use the least squares approach, which does not need $\mathbf{C}$ to be invertible. It can be formulated as the following optimization problem [10]:

$$\min_{\mathbf{\Lambda},\mathbf{\Psi}} ||\mathbf{C} - \mathbf{R}||_F^2 \ . \qquad (7)$$

However, there is a conceptual difficulty in adopting the approach to EFA introduced by [13], or solving the least squares problem (7). The EFA correlation structure $\mathbf{\Lambda}\mathbf{\Lambda}' + \mathbf{\Psi}^2$ is a consequence of the accepted EFA model (2) and the assumptions made for its parameters. When the number of variables exceeds the number of observations, the classical constraint $\mathbf{U}'\mathbf{U} = \mathbf{I}_p$ cannot be fulfilled as $\mathbf{U}'\mathbf{U}$ has at most rank n ($< p$). Furthermore, algorithms factorizing a $p \times p$ matrix $\mathbf{C}$ may become computationally slow if p is huge. In the case of high-dimensional data and $n \ll p$, taking a $n \times p$ data matrix as an input for EFA is a reasonable alternative.

With $\mathbf{U}'\mathbf{U} \neq \mathbf{I}_p$, the EFA correlation structure can be written as

$$\mathbf{R} = \mathbf{\Lambda}\mathbf{\Lambda}' + \mathbf{\Psi}\mathbf{U}'\mathbf{U}\mathbf{\Psi} \ . \qquad (8)$$

In order to preserve the standard EFA correlation structure (4), the more general constraint $\mathbf{U}'\mathbf{U}\mathbf{\Psi} = \mathbf{\Psi}$ is introduced. The immediate consequence from this new constraint is that the existence of unique factors with zero variances is acceptable in the EFA model when $n < p$. With classical EFA ($n > p$), there is a long-standing debate about the acceptance of zero entries in $\mathbf{\Psi}^2$. It seems that an universal EFA model covering both cases $n > p$ and $p > n$ should allow $\mathbf{\Psi}^2$ to be positive *semi*-definite. This modified EFA problem will fit the singular correlation $p \times p$ matrix of rank at most n by the sum $\mathbf{\Lambda}\mathbf{\Lambda}' + \mathbf{\Psi}_n^2$ of two positive semi-definite $p \times p$ matrices with ranks k and n (at most), respectively.

By defining the block matrices $\mathbf{B} = [\mathbf{F} : \mathbf{U}]$ and $\mathbf{A} = [\mathbf{\Lambda} : \mathbf{\Psi}]$ of dimensions $n \times (k + p)$ and $p \times (k + p)$, respectively, (6) can be rewritten as

$$||\mathbf{X} - \mathbf{B}\mathbf{A}'||_F^2 = ||\mathbf{X}||_F^2 + \mathrm{trace}(\mathbf{B}'\mathbf{B}\mathbf{A}'\mathbf{A}) - 2\mathrm{trace}(\mathbf{B}'\mathbf{X}\mathbf{A}) \ , \qquad (9)$$

which is optimized subject to the new constraint $\mathbf{BB'} = \mathbf{I}_n$. The middle term in (9), trace($\mathbf{B'BA'A}$), can be written as

$$\text{trace}(\mathbf{B'BA'A}) = \text{trace}\left\{\begin{bmatrix} \mathbf{F'} \\ \mathbf{U'} \end{bmatrix} [\mathbf{F} : \mathbf{U}] \begin{bmatrix} \mathbf{\Lambda'} \\ \mathbf{\Psi} \end{bmatrix} [\mathbf{\Lambda} : \mathbf{\Psi}]\right\} =$$

$$\text{trace}\left\{\begin{bmatrix} \mathbf{I}_k & \mathbf{0}_{k \times p} \\ \mathbf{0}_{p \times k} & \mathbf{U'U} \end{bmatrix} \begin{bmatrix} \mathbf{\Lambda'\Lambda} & \mathbf{\Lambda'\Psi} \\ \mathbf{\Psi\Lambda} & \mathbf{\Psi}^2 \end{bmatrix}\right\} = \text{trace}\left\{\begin{bmatrix} \mathbf{\Lambda'\Lambda} & \mathbf{\Lambda'\Psi} \\ \mathbf{U'U\Psi\Lambda} & \mathbf{U'U\Psi}^2 \end{bmatrix}\right\} =$$

$$\text{trace}(\mathbf{\Lambda'\Lambda}) + \text{trace}(\mathbf{\Psi}_n^2) \ ,$$

showing that trace($\mathbf{B'BA'A}$) does not depend on $\mathbf{F}$ and $\mathbf{U}$. Hence, as with the standard Procrustes problem, minimizing (9) over $\mathbf{B}$ is equivalent to the maximization of trace($\mathbf{B'XA}$). For this problem a closed-form solution via the singular value decomposition (SVD) of $\mathbf{XA}$ exists [6].

After solving the Procrustes problem for $\mathbf{B} = [\mathbf{F} : \mathbf{U}]$, one can update the values of $\mathbf{\Lambda}$ and $\mathbf{\Psi}$ by $\mathbf{\Lambda} = \mathbf{X'F}$ and $\mathbf{\Psi} = \text{diag}(\mathbf{U'X})$ using the identities

$$\mathbf{F'X} = \mathbf{F'F\Lambda'} + \mathbf{F'U\Psi} \Longrightarrow \mathbf{F'X} = \mathbf{\Lambda'} \ , \tag{10}$$

$$\mathbf{U'X} = \mathbf{U'F\Lambda'} + \mathbf{U'U\Psi} \Longrightarrow \mathbf{U'X} = \mathbf{\Psi} \text{ (and thus diagonal)} \ , \tag{11}$$

which follow from the EFA model (3) and the new constraint $\mathbf{U'U\Psi} = \mathbf{\Psi}$ imposed. The matrix of factor loadings $\mathbf{\Lambda}$ is required to have full column rank k. Constructing it as $\mathbf{\Lambda} = \mathbf{X'F}$ gives rank($\mathbf{\Lambda}$) = min{rank($\mathbf{X}$), rank($\mathbf{F}$)}. Assuming that rank($\mathbf{X}$) $\geq k$, the ALS algorithm preserves the full column rank property of $\mathbf{\Lambda}$. The whole ALS process is continued until the loss function (9) cannot be reduced further.

4   Rotation towards Independence

Finding estimates for uncorrelated factor scores $\hat{\mathbf{F}}$ and loadings $\hat{\mathbf{\Lambda}}$ is a standard EFA problem. To solve the corresponding noisy ICA problem one needs to go one step further. The common factor scores should be independent. For this reason, they are rotated towards independence, that is,

$$\hat{\mathbf{S}} = \hat{\mathbf{F}}\mathbf{T} \ , \tag{12}$$

for some orthogonal matrix $\mathbf{T}$, where $\hat{\mathbf{S}}$ denotes the estimate for the matrix of independent scores of the n observations on k sources. To find the matrix $\mathbf{T}$ that leads to approximately independent scores, various ICA algorithms can be applied [8], as e.g. JADE, FastICA, et cetera. Once the optimal rotation matrix $\hat{\mathbf{T}}$ and hence $\hat{\mathbf{S}}$ has been found, the noisy ICA mixing matrix is obtained by $\hat{\mathbf{A}} = \hat{\mathbf{\Lambda}}\mathbf{T}$.

5   Application

5.1   Data

In this section, the proposed approach is applied to climate data from the National Center for Environmental Prediction/National Center for Atmospheric

Research (NCEP/NCAR) reanalysis project. The data set consists of winter monthly sea-level pressures (SLP) over the Northern Hemisphere north of 20°N. The winter season is defined by the months December, January, and February. The data set spans the period December 1948 to February 2006 ($n = 174$ observations) and is available on a regular grid with a 2.5°latitude × 2.5°longitude resolution ($p = 29 \times 144 = 4176$ variables).

Prior to the analysis the data was preprocessed as follows. First, the mean annual cycle was calculated by averaging the monthly data over the years. Anomalies were then computed as departures from the mean annual cycle. To account for the converging longitudes poleward, an area weighting was finally performed by multiplying each grid point by the square root of the cosine of the corresponding latitude. These weighted sea-level pressure anomalies are considered in the following.

5.2   Results

By applying the EFA procedure described in section 3, five factors were extracted which account for 60.2% of the total variance in the data. This choice is dictated by the need for a balance between explained variance and spatial scales. Extracting more factors increases the explained variance but includes more small scales. Five factors are found to provide a fine balance.

For $k = 5$ and twenty random starts, the procedure requires on average 90 iterations, taking about 20 minutes in total using MATLAB 7.1 on a PC operating with an Intel Pentium 4 CPU having 2.4 GHz clock frequency and 1 GB of RAM. For comparison, factorizing a 4176×4176 covariance matrix and finding a numerical solution for the least squares optimization problem (7) based on an iterative Newton-Raphson procedure takes about 2.5 hours.

The JADE procedure [2] is used to obtain the matrix of approximately independent factor scores $\hat{\mathbf{S}} = \hat{\mathbf{F}}\mathbf{T}$ and an estimate of the ICA mixing matrix $\hat{\mathbf{A}} = \hat{\mathbf{\Lambda}}\mathbf{T}$.

By means of the mixing or loading matrix, noisy ICA provides a method of describing spatial patterns of winter sea-level pressures. For each factor, there is a loading for each manifest variable, and because variables are gridpoints it is possible to plot each loading on a map at its corresponding gridpoint, and then draw contours through geographical locations having the same coefficient values. Compared to a loading matrix with 4176 loadings for each factor, this spatial map representation greatly aids interpretation, as is illustrated in Figure 1.

These plots give the maps of loadings, arbitrarily renormalized to give 'round numbers' on the contours. The plots (i) and (ii) in Figure 1 represent the first and second column of the 4176×5 ICA mixing matrix, respectively. Winter months having large positive scores for the factors will tend to have high loadings and hence high SLP values, where loadings on the map are positive, and low SLP values at gridpoints where the coefficients are negative. Since the data is mean-centered the loadings can be interpreted as covariances between the observed variables and the common factors.

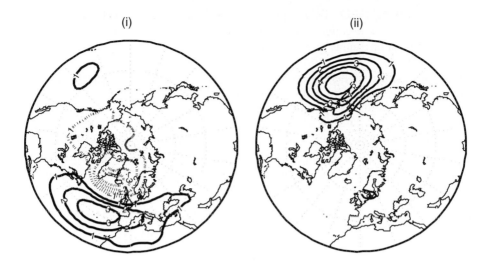

Fig. 1. Spatial map representation of the first (i) and second (ii) column of the ICA mixing matrix for $k = 5$ factors applied to winter SLP data. Positive contours are represented by solid lines and negative contours by dotted lines.

Spatial map (i) shows the North Atlantic Oscillation (NAO). The NAO is a climatic phenomenon in the North Atlantic Ocean of fluctuations in the difference of sea-level pressure between the Icelandic low and the Azores high. The second ICA pattern (ii) yields the North Pacific Oscillation (NPO) or Pacific pattern, a monopolar structure sitting over the North Pacific. Applying ICA to winter mean sea-level pressure anomalies shows that the NAO and the Pacific pattern correspond to two approximately independent factors.

6   Conclusions

This paper proposes a new method for sphering in ICA in the presence of noise based on EFA. The EFA is considered as a novel form of data matrix decomposition. To preserve the EFA model correlation structure for cases where the number of observations is less than the number of variables, a new model constraint is introduced which allows the matrix of unique-factor variances to be positive semi-definite. By formulating the EFA model directly in terms of the data and optimizing a least squares loss function, estimates for all EFA model parameters are obtained simultaneously. For high-dimensional data, this approach is computationally more efficient than factorizing a matrix of second-order cross products. Based on the initial EFA solution, the common factor scores are rotated towards independence to implement the noisy ICA model.

The proposed approach is applied to the monthly winter sea-level pressure field over the Northern Hemisphere for the period 1948-2006. It is shown that noisy ICA is able to identify physically interpretable spatial patterns, namely the NAO and the NPO. This paper gives evidence that noisy ICA can be of value in climate research for separating underlying anomaly signals that may be generated by independent physical processes.

Acknowledgments. The authors would like to thank the anonymous reviewers for their helpful comments and suggestions.

References

1. Anderson, R.D., Rubin, H.: Statistical inference in factor analysis. In: Neyman, J. (ed.) Proceedings of the 3rd Berkeley Symposium on Mathematical Statistics and Probability, vol. V, pp. 111–150. University of California Press, Berkeley (1956)
2. Cardoso, J.-F., Souloumiac, A.: Blind beamforming for non-Gaussian signals. IEE Proceedings-F 140, 362–370 (1993)
3. Davies, M.: Identifiability issues in noisy ICA. IEEE Signal Processing Letters 11, 470–473 (2004)
4. De Leeuw, J.: Least squares optimal scaling of partially observed linear systems. In: Van Montfort, K., Oud, J., Satorra, A. (eds.) Recent Developments on Structural Equation Models: Theory and Applications, pp. 121–134. Kluwer Academic Publishers, Dordrecht (2004)
5. Golub, G.H., Van Loan, C.F.: Matrix Computations, 3rd edn. The John Hopkins University Press, Baltimore (1996)
6. Gower, J.C., Dijksterhuis, G.B.: Procrustes Problems. Oxford University Press, Oxford (2004)
7. Hastie, T., Tibshirani, R., Friedman, J.H.: The Elements of Statistical Learning: Data Mining, Inference, and Prediction, 3rd edn. Springer, New York (2001)
8. Hyvärinen, A., Karhunen, J., Oja, E.: Independent Component Analysis. John Wiley & Sons, New York (2001)
9. Ikeda, S.: ICA on noisy data: a factor analysis approach. In: Girolami, M. (ed.) Advances in Independent Component Analysis, pp. 201–215. Springer, Berlin (2000)
10. Jöreskog, K.G.: Factor analysis by least-squares and maximum likelihood methods. In: Enslein, K., Ralston, A., Wilf, H.S. (eds.) Mathematical methods for digital computers, vol. 3, pp. 125–153. John Wiley & Sons, New York (1977)
11. Kawanabe, M., Murata, N.: Independent component analysis in the presence of Gaussian noise based on estimating functions. In: Proceedings of the 2nd International Workshop on Independent Component Analysis and Blind Signal Separation (ICA 2000), Helsinki, Finland, pp. 39–44 (2000)
12. Mulaik, S.A.: The Foundations of Factor Analysis. McGraw-Hill, New York (1972)
13. Robertson, D., Symons, J.: Maximum likelihood factor analysis with rank-deficient sample covariance matrices. Journal of Multivariate Analysis 98, 813–828 (2007)

Signal Subspace Separation Based on the Divergence Measure of a Set of Wavelets Coefficients

Vincent Vigneron[1] and Aurélien Hazan[2]

[1] IBISC CNRS FRE 3190, Université d'Evry, France
vincent.vigneron@ibisc.univ-evry.fr
[2] CES CNRS-UMR 8174 Equipe SAMOS-MATISSE, Paris, France
ahazan@univ-paris1.fr

Abstract. The various scales of a signal maintain relations of dependence the ones with the others. Those can vary in time and reveal speed changes in the studied phenomenon. In the goal to establish these changes, one shall compute first the wavelet transform of a signal, on various scales. Then one shall study the statistical dependences between these transforms thanks to an estimator of mutual information (MI) called *divergence*. The time-scale representation of the sources representation shall be compared with the representation of the mixtures according to delay in time and in frequency.

Keywords: Blind sources separation, divergence, Gaussian signal, wavelet transform.

1 Introduction

A growing interest is evident in investigating the dependence relationships between complex data such as curves, spectra, time series or more generally signals. In these cases, each observation consists of values of dependent variables which are usually function of time.

This paper examines the dynamic interactions, and more specifically, the impacts of changes in a set of wavelet coefficients by using wavelet analysis. As a semi-parametric method, wavelets analysis might be superior to detect the chaotic patterns in the non-coherent markets.

The paper begins with an information analysis of statistical dependencies between wavelet coefficients providing from signals. These intra and interscale dependencies are measured using mutual information (MI). MI depends strongly on the choice on the wavelet filters. Such dependencies have been studied intensively in [9]. Wavelet analysis takes contact with self-similar fractals and iterative analysis, through other techniques of functional approximation, such as radial basis functions, etc. although we do not want to enter in the arsenal of modern tools.In the study of time series it is crucial to understand what is dependent

T. Adali et al. (Eds.): ICA 2009, LNCS 5441, pp. 171–178, 2009.

and what independent of the temporal and space scales. The wavelet transform (WT) nearly decorrelates many time series and can be viewed as a Karhunen-Loève transform. Nevertheless, significant dependencies still exist between the wavelet coefficients.

Most algorithms focus on a certain type of dependencies, which it attempts to capture using a relatively simple and tractable model, such as the Karhunen-Loève transform (KLT), the discrete Fourier transform (DFT), and the discrete wavelet transform (DWT). Among them KLT is the most effective algorithm with minimal reconstruction error. The times series dataset is transformed into an orthogonal feature space in which each variable is orthogonal to the others. The time series dataset can be approximated by a low-rank approximation matrix by discarding the variables with lower energy [1].

DWT and DFT are powerful signal processing techniques and both of them have fast computational algorithms. DFT maps the time series data from the time domain to the frequency domain, and the fast Fourier transform algorithm (FFT) can compute the DFT coefficients in $\mathcal{O}(mn \log n)$ time. Unlike DFT which takes the original time series from the time domain and transforms it into the frequency domain, DWT transforms the time series from time domain into time-frequency $t - f$ domain.

The fact that the wavelet transform (WT) has the property of time-frequency localisation of the time series means that most of the times series energy can be represented by only a few wavelet coefficients. Chan and Fu used the Haar wavelet for time series representation and showed (classification) performance improvement over DFT [4]. Popivanov and Miller proposed an algorithm using the Daubechies wavelet for time series classification [13]. Lin *et al.* proposed an iterative clustering algorithm exploring the multi-scale property of wavelets [8]. Numerous other techniques for time series data reduction have been proposed such as regression tree [2], piecewise linear segmentation [7], etc. These algorithms work well for time series with few dimensions because the high correlation among time series data makes it possible to remove huge amount of redundant information. But for clustering algorithms with unlabeled data, determining the dimensionality of the feature dimensionality becomes more difficult. To our personnal knowledge, the feature dimension needs to be decided by the user.

The aim of this paper is to propose a time-series feature extraction algorithm using orthogonal wavelet capable to test for the presence of a dependence structure. The problem of determining the feature dimensionality is circumvented by choosing the appropriate scale of the WT. An ideal feature extraction technique has the ability to efficiently reduce the data while preserving the properties of the original data. However, infomation is lost in dimension reduction. The rest of the paper is organized as follows. Section 2 is a reminder on multiresolution analysis. Section 3 gives the basis for supporting our feature extraction algorithm. Section 4 contains a comprehensive experimental evaluation of the proposed algorithm. We conclude the paper by summarizing the main contribution in section 5.

2   A Refresher on Wavelet Representations

WT preserved the broad trend of the input sequence in an approximation part, while the localized changes are kept in some detail parts. More details about WT can be found in [5].

For short, a wavelet is a smooth and quickly vanishing oscillating function with good localisation properties in both frequency and time, this is more suitable for approximating time series data that contain regional structures [11,6]. To efficiently calculate the WT for signal processing, Mallat introduced the multiresolution analysis (MRA) and designed a family of fast algorithms based on [11]. With MRA, a signal can be viewed as being composed of a smooth background and fluctuations (also called *details*) on top of it. The distinction between the smooth part and the details is determined by the resolution, that is the scale below which the details of a signal cannot be discerned. At a given resolution, a signal is approximated by ignoring all fluctuations below that scale. We can progressively increase the resolution: finer details are then added to the coarser description, providing a better approximation of the signal

$$X = A_J + \sum_{j=1}^{J} D_j, \tag{1}$$

where A_j and D_j are respectively the approximation and the detail at level j of the signal X. In other words, any time series can be written as the sum of orthogonal signals. Each signals lies in a common space denoted by V_0 and are of lenght n [12]. But A_j and D_j belong to spaces V_j and W_j respectively. This sequence of nested approximation spaces (V_j) involved in the multiresolution framework is such that $V_J \subset V_{J-1} \subset \ldots \subset V_0$. The space W_j is an orthogonal complement of V_{j-1} in V_j, *i.e.* $V_j = V_{j-1} \oplus W_{j-1}$. Then by defining

$$\underbrace{V_0 \oplus W_0}_{V_1} \oplus W_1 \oplus \ldots \oplus W_{j-1} = V_j \tag{2}$$

any signal belonging to V_j (resp. W_j) can be viewed as an approximation (resp. detail) signals like A_j (resp. D_j). From a signal processing point of view, the approximation coefficients within lower scales correspond to the lower freequency part of the signal. Hence the first few coefficients A_j constitute a noise-reduced signal. Thus keeping these coeffciients will not loose much information from the original time series X. Hence, normally, the first coefficients are chosen as the features: they retain the entire information of X at a particular level of granularity. The task of choosing the first few wavelet coefficients is circumvented by choosing a particular scale. The candidate selection of feature dimensions is reduced from $\{1, 2, \ldots, n\}$ to $\{2^0, 2^1, \ldots, 2^{J-1}\}$

For our tests, we used the Haar wavelet which has the fastest transform algorithm and is the most popularly used orthogonal wavelet. Fig. 1 plots two wavelets: the Haar on the left and the Daubechies (db2) series on the right. Conceptually, these mother wavelet functions are analogous to the impulse response of a band-pass filter.

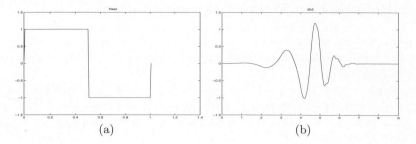

Fig. 1. Shapes of 2 wavelets commonly used in wavelet analysis. The sharp corners enable the transform to match up with local details that cannot be observed when using Fourier transform that matches only sinusoidal shapes. (a) The Haar wavelet (b) the Daubechies wavelet.

So far, for many time series, the construction is not very good. In this case, no effort was made to do a good job of the decomposition, but merely to perform one MRA and to make it comprehensive.

3   Methods

One frequent assumption in the wavelet domain is the absence of correlations between coefficients. This assumption should be questionned. In the literature devoted to the statistical physics, the dependence between scales is given a precise meaning, and models of the random processes are debated.

Consider two signals for which the same wavelet decomposition is performed. Feature extraction consists to select the set of *wavelet coefficients* which will maximise the *separation* between the subspaces occupied by the signals. These features can be interpreted as being representative of the dissimilarities between the signals if we generate a set of image patterns that exhibit an increase in separation as measured between signals. Such a problem is considered from the point of view of utilizing a matrix transformation to generate images which maximize the interset distance between signals while keeping the intraset distance of the subspaces constant. The separation of pattern classes can be measured in terms of quantities other than euclidean distances. A more abstract concept of distance is the *divergence* (see more details in [3,14]).

Consider 2 subspaces Ω_1 and Ω_2, governed by the probability densities $p_1(\mathbf{x}) - p(\mathbf{x}|\Omega_1)$ and $p_2(\mathbf{x}) = p(\mathbf{x}|\Omega_2)$ respectively. The divergence between the two subspaces is given by

$$J_{12} = \int_{\mathbf{x}} (p_1(\mathbf{x}) - p_2(\mathbf{x})) \log \frac{p_1(\mathbf{x})}{p_2(\mathbf{x})} d\mathbf{x}. \tag{3}$$

The divergence can be used as a criterion for generating an optimum set of features. We are seeking a transformation matrix A which yields image patterns of lower dimensionality, i.e. $\mathbf{y} = A\mathbf{x}$, where $\mathbf{y}$ is a m-vector, $\mathbf{x}$ is a n-vector, and

A is a $m \times n$ matrix whose rows are the linearly independent vectors $a_i, i = 1, \ldots m < n$. The divergence between the image patterns is given by

$$J'_{12} = \int_{\mathbf{y}} (p_1(\mathbf{y}) - p_2(\mathbf{y})) \log \frac{p_1(\mathbf{y})}{p_2(\mathbf{y})} d\mathbf{y}. \tag{4}$$

Assume that the wavelet coefficients in the classes Ω_1 and Ω_2 are Gaussian distributed $N(\mathbf{m}_1, C_1)$ and $N(\mathbf{m}_2, C_2)$ respectively. The means vectors of the image are given by $\mathbf{m}'_1 = A\mathbf{m}_1, \mathbf{m}'_2 = A\mathbf{m}_2$, and the covariance matrices are $C'_1 = AC_1A^T$, $C'_2 = AC_2A^T$.

Under these conditions, the divergence of the image coefficients is given by

$$J'_{12} = \frac{1}{2}\text{trace}[C'^{-1}_2 C'_1 + C'^{-1}_1 C'_2] - m + \frac{1}{2}\text{trace}[(C'^{-1}_1 + C'^{-1}_2)\mathbf{m}'\mathbf{m}'^{T}] \tag{5}$$

where $\mathbf{m} = A(\mathbf{m}_1 - \mathbf{m}_2)$. The trace of a matrix can be rewritten in terms of the eigenvalues [10]:

$$J'_{12} = \frac{1}{2}\sum_{i=1}^{m}(\lambda_i + \frac{1}{\lambda_i}) - m + \frac{1}{2}\lambda_{m+1} + \frac{1}{2}\lambda_{m+2}, \tag{6}$$

where λ_i are the eigenvalues of $C'^{-1}_2 C'_1$, $\lambda_{m+1}, \lambda_{m+2}$ are the eigenvalues of $C'^{-1}_1 \mathbf{m}'\mathbf{m}'^{T}$ and $C'^{-1}_2 \mathbf{m}'\mathbf{m}'^{T}$ respectively. The differential of (6) is

$$dJ'_{12} = \frac{1}{2}\sum_{i=1}^{m}(1 - \lambda_i^{-2})d\lambda_i + \frac{1}{2}d\lambda_{m+1} + \frac{1}{2}d\lambda_{m+2}. \tag{7}$$

From the eigenvalues equations $(AC_2A^T)^{-1}(AC_1A^T)\mathbf{e}_i = \lambda_i\mathbf{e}_i$, where $\mathbf{e}_i$ is the eigenvector of $C'^{-1}_2 C'_1$ associated with λ_i, the differential gives:

$$dA(C_1A^T - \lambda_iC_2A^T)\mathbf{e}_i + (AC_1 - \lambda_iAC_2)(dA^T)\mathbf{e}_i$$
$$= -(AC_1A^T - \lambda_iAC_2A^T)d\mathbf{e}_i + (d\lambda_i)AC_2A^T\mathbf{e}_i. \tag{8}$$

C'_1 and C'_2 are symmetrical, then the eigenvectors will be mutually orthogonal with respect to C'_2. Suppose we can find a complete set of eigenvectors such $d\mathbf{e}_i$ can be rewritten $d\mathbf{e}_i = \sum_{j=1}^{m} c_{ji}\mathbf{e}_j$, where the $\mathbf{e}_j$ are normalized with respect to C'_2, that is $\mathbf{e}_jC'_2\mathbf{e}_j = 1$. Substituting in (8) yields:

$$dA(C_1A^T - \lambda_iC_2A^T)\mathbf{e}_i + (AC_1 - \lambda_iAC_2)(dA^T)\mathbf{e}_i$$
$$= \sum_{i}^{m} c_{ji}(\lambda_j - \lambda_i)AC_2A^T\mathbf{e}_j + AC_2A^T\mathbf{e}_i(d\lambda_i). \tag{9}$$

Equation (9) results in

$$d\lambda_i = \mathbf{e}_i(dA)(C_1A^T - \lambda_iC_2A^T)\mathbf{e}_i. \tag{10}$$

The differentials $d\lambda_{m+1}$ and $d\lambda_{m+2}$ can be determined in a similar manner. Similarly we obtain:

$$d\lambda_{m+1} = 2\mathbf{e}_{m+1}(dA)(\mathbf{m}\mathbf{m}^{T}A^T - \lambda_{m+1}C_1A^T)\mathbf{e}_{m+1} \tag{11}$$

and

$$d\lambda_{m+2} = 2\mathbf{e}_{m+2}(dA)(\mathbf{m}\mathbf{m}^{T}A^T - \lambda_{m+2}C_2A^T)\mathbf{e}_{m+2}. \tag{12}$$

Inserting (11) and (12) into (7) yields in terms of the trace:

$$dJ'_{12} = \text{trace}[(dA)\Lambda],\tag{13}$$

where

$$\Lambda = \sum_{i=1}^{m}(1-\lambda_i^{-}2)(C_1A^T - \lambda_i C_2 A^T)\mathbf{e}_i\mathbf{e}_i^T +$$
$$+ (\mathbf{mm}^T A^T - \lambda_{m+1}C_1A^T)\mathbf{e}_{m+1}\mathbf{e}_{m+1}^T + (\mathbf{mm}^T A^T - \lambda_{m+2}C_2A^T)\mathbf{e}_{m+2}\mathbf{e}_{m+2}^T.$$

Since dA is arbitrary, a necessary conditions for J'_{12} to be an extremum is that Λ be equal to zero. In the particular case $C_1 = C_2 = C$ and $\mathbf{m}_1 \neq \mathbf{m}_2$, Λ reduces to $\Lambda = 2(\mathbf{mm}^T A^T - \lambda_{m+1}CA^T)\mathbf{e}_{m+1}\mathbf{e}_{m+2}^T$ and the necessary condition for J_{12} to be an extremum is that $\mathbf{mm}^T A^T \mathbf{e}_{m+1}\mathbf{e}_{m+2}^T = \lambda_{m+1}CA^T\mathbf{e}_{m+1}\mathbf{e}_{m+2}^T$. In the general case, the solution $\Lambda = 0$ can be found by using a steepest ascent approach in which we increment A by $\delta\Lambda$ where δ is some suitable convergence factor so that J_{12} increases:

$$A^{(t+1)} = A^{(t)} + \delta\Lambda^{(t)},\tag{14}$$

where t is an iteration index. Applying the transformation $\mathbf{y} = A\mathbf{x}$ to the patterns results in a satisfying m-dimensional image patterns. These patterns exhibit the expected separating properties.

4   Computer Simulations

In this simulation, we consider $N = 3$ source signals of length $T = 2048$ and $M = 3$ mixtures. The sources are defined by $s_1(t) = \sin(250\pi t^2)$, $s_2(t) = \sin(150\pi t^3)$, $s_3(t) = \sin(350\pi t)$, $t \in [0,1]$. The mixing matrix is unknown.

We consider the complex Morlet mother wavelet defined as : $\psi(t) = \frac{1}{\sqrt{\pi f_b}}e^{2i\pi f_c t}e^{-\frac{t^2}{f_b}}$. The mixtures are represented in the time-scale plane in Figure 2.a. Figure 2.b shows sources after performing the projection $A\mathbf{x}$ in the same plane.

We now summarize the results, obtained on a subset of the full 1024×1024 MI matrix, for clarity reasons. Fig.3(a) represents the p-values matrix (p_{ij}) for the statistical test of independence between two coefficients c_{j_1,k_1} and c_{j_2,k_2}. Low values in dark stand for values such that it is unlikely that the coefficients indexed are independent. After thresholding by the $a = 5\%$ significance level, the couples that failed to pass the test are shown in light color by Fig.3(b).

Remark that the main diagonal is rejected, which is consistent with the fact that a coefficient is hardly independent with itself. Now, other unexpected structures appear along the $y = ax$ line, for different slopes though. Their presence can be explained thanks to the wavelet coefficient dependence tree, displayed in a decimated way by Fig.3(d). A connection between two nodes means that two coefficients failed to pass the independence test, and we can note that nodes pertaining to a given scale s are connected with their closest neighbour. Such nodes correspond precisely to the $y = ax$ lines with $a \neq 1$ in the rejection matrix.

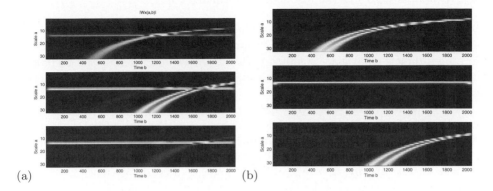

Fig. 2. (a) Time-scale transform of mixtures using complex Morlet wavelet. (b) Time-scale transform of estimated sources using complex Morlet wavelet.

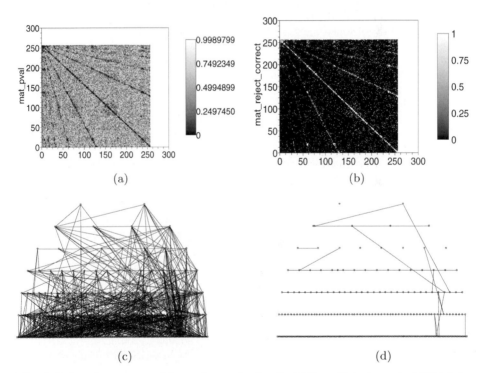

Fig. 3. Inferred structure of dependence, for the first 256 coefficients, out of 1024 (a) p-value matrix (p_{ij}), for the test against the null hypothesis of independence between two DWT coefficients indexed by i, j (b) rejection matrix, $r_{ij} = 1$ if the two coefficients cannot be considered as independent (c) Wavelet coefficients dependence graph (d) Decimated dependence graph

5   Conclusion

We emphasize in this paper that the separation of sources having different time-scale representations can be realized. Proposed solution is based on the projection in a high-dimensional space spanned with the wavelets chosen for the decomposition. We have illustrated the effectiveness of the proposed method thanks to computer simulations.

A deep study of noisy mixture can be interesting for future work since the wavelet transform can be used both for noise cancellation and source separation. A more challenging case will be also when the mixing system is a convolutive one, i.e., when the sources are mixed through a linear filtering operation.

References

1. Antoulas, A.: Encyclopedia for the Life Sciences, contribution 6.43.13.4 Frequency domain representation and singular value decomposition. UNESCO (2002)
2. Armstrong, J.S., Andress, J.G.: Exploratory analysis of marketing data: Trees vs. regression. Journal of Marketing Research 7, 487–492 (1970)
3. Basseville, M.: Distance measure for signal processing and pattern recognition. Signal Process 18(4), 349–369 (1989)
4. Chan, K.P., Fu, A.W.: Efficient time series matching by wavelets. In: Proceedings of the 15th International Conference on Data Engineering, pp. 126–133 (1999)
5. Daubechies, I.: Ten lectures on wavelets, society for industrial and applied mathematics (1992)
6. Daubechies, I., Mallat, S., Willsky, A.S.: Introduction to the spatial issue on wavelet transforms and multiresolution signal analysis. IEEE Trans. Inform. Theory, 529–532 (1992)
7. Lemire, D.: A better alternative to piecewise linear time series segmentation. In: SIAM Data Mining 2007 (2007)
8. Lin, J., Vlachos, M., Keogh, E., Gunopulos, D.: Iterative incremental clustering of time series (2004)
9. Liu, J., Moulin, P.: Information-theoretic analysis of interscale and intrascale dependencies between image wavelet coefficients. IEEE Transactions on Image Processing 10(11), 1647–1658 (2001)
10. Lütkepohl, H.: Handbook of Matrices, 1st edn. Wiley, Chichester (1997)
11. Mallat, S.: A theory for multiresolution signal decomposition: the wavelet representation. IEEE Trans. Pattern Anal. and Mach. Intelligence 11(7), 674–693 (1989)
12. Misiti, M., Misiti, Y., Oppenheim, G., Poggi, J.-M.: Clustering signals using wavelets. In: Sandoval, F., Prieto, A.G., Cabestany, J., Graña, M. (eds.) IWANN 2007. LNCS, vol. 4507, pp. 514–521. Springer, Heidelberg (2007)
13. Popivanov, I., Miller, R.J.: Similarity search over time-series data using wavelets. In: ICDE, pp. 212–221 (2002)
14. Rényi, A.: On measures on entropy and information. 4th Berkeley Symp. Math. Stat. and Prob. 1, 547–561 (1961)

ICA by Maximizing Non-stability

Baijie Wang[1], Ercan E. Kuruoglu[2], and Junying Zhang[1]

[1] Xidian University, P.O BOX 175, No. 2 South TaiBai Road, Xi'an, 710071, China
[2] ISTI-CNR via G. Moruzzi 1, 56124, Pisa, Italy
wangbaijie@gmail.com, ercan.kuruoglu@isti.cnr.it,
jyzhang@mail.xidian.edu.cn

Abstract. We propose a new approach for ICA by maximizing the non-stability contrast function in this paper. This new version of ICA is motivated by the Generalized Central Limit Theorem (GCLT), an important extension of classical CLT. We demonstrate that the classical ICA based on maximization of non-Gaussianity is a special case of the new approach of ICA we introduce here which is based on maximization of non-Stability with certain constraints. To be able to quantify non-stability, we introduce a new measure of stability namely Alpha-stable negentropy. A numerical gradient ascent algorithm for the maximization of the alpha-stable negentropy with the objective of source separation is also introduced in this paper. Experiments show that ICA by maximum of non-stability performs very successfully in impulsive source separation problems.

Keywords: ICA, non-stability, alpha-stable negentropy, source separation, impulsive signals.

1 Introduction

Independent Component Analysis (ICA) is a statistical method for separating a mixture signal into subcomponents with the assumption that the source signals are mutually independent and non-Gaussian except possibly one. Non-Gaussianity principle based on the Central Limit Theorem (CLT) is one important approach for separation. As a classic result in probability theory, Central Limit Theorem (CLT) tells us that mixed signals are closer to Gaussian than the original sources. Researchers in classical ICA use the maximum non-Gaussianity criterion to obtain independent sources [7], [8], [9].

Generally, negentropy or its approximations such as kurtosis based on the second- and fourth-order cumulants [8] are used as the contrast function, based on which algorithms, like FastICA [6], were developed. However, most impulsive signals do not possess finite second or higher-order moments, and classical ICA based on maximization of non-Gaussianity is not suitable to solve impulsive source separation problems. Limited works exist in the literature on source separation of impulsive signals. In [10], the Minimum Dispersion Criterion approach employed an l_p-norm contrast function is for the separation of alpha-stable distributed sources. Our work differs from such work in that we challenge the non-Gaussianity based ICA assumption and our method is designed not only for alpha-stable signals but is valid for any impulsive source mixture.

T. Adali et al. (Eds.): ICA 2009, LNCS 5441, pp. 179–186, 2009.
© Springer-Verlag Berlin Heidelberg 2009

Generalized Central Limit Theorem (GCLT) seen as an extension of CLT without the constraint of limit variance states that the sum of independent and identically distributed random variables with or without finite variance converges to an alpha stable distribution, for a large number of variables [2],[3].

Alpha stable distribution is a class of probability distributions that allows skewness and heavy tails. It has four parameters $\alpha, \beta, \lambda, \delta$ and is best described by its characteristic function showed below:

$$\varphi(z) = \exp\left\{ j\delta z - \gamma |z|^{\alpha} \left[1 + j\beta sign(z)\omega(z,\alpha)\right]\right\} \text{ where, } \omega(z,\alpha) = \begin{cases} \tan\dfrac{\alpha\pi}{2}, & if \ \alpha \neq 1 \\ \dfrac{2}{\pi}\log|z| & if \ \alpha = 1 \end{cases}$$

$$\text{and } 0 < \alpha \leq 2, -1 \leq \beta \leq 1, \gamma > 0, -\infty \leq \delta \leq \infty .$$

Note that the Gaussian distribution is a special case of alpha stable distribution when α equals 2, and this is important to understand the relationship between this new version of ICA and classical ICA discussed below.

According to GCLT, the sum of even two independent random variables (with no condition on the variance) is more likely to be alpha stable distributed than each of the two variables. In parallel to the motivation of the non-Gaussianity maximization based ICA, we can assert that to include the case of impulsive signals, we should define the component separation problem with the maximization of the non-stability. Hence, motivated by the generalized CLT, we propose a new version of ICA, where the non-Gaussianity principle is replaced by a new general non-stability principle. In this paper, we propose the use of alpha-stable negentropy as a new measure of non-stability. Through maximizing the non-stability contrast function, it finds the independent components.

2   Maximum Non-stability ICA Method

2.1   Definition

The model in maximum non-stability ICA is the same with that in classical ICA, which is:

$$x = As \tag{1}$$

What's different here is that source vectors $s_1, ..., s_n$ in s may have large or even infinite variance. Here, we should restrict $s_1, ..., s_n$ not to be alpha stable distribution because of the stability property, which is the property that the sum of two independent stable random variables with the same characteristic exponent is again stable and has the same characteristic exponent [3]. Similar to the classical ICA case in which Gaussian variables are inseparable, alpha stable source signals with the same characteristic exponent are inseparable.

Although alpha stable sources with different characteristic exponents may also be separable, but for simplicity we assume that $s_1, ..., s_n$ not to be alpha stable here. Moreover, A is thought to be square in this paper also for the sake of simplicity.

Our objective is to estimate the mixing matrix A, and then we can compute its inverse, say W, which separates the mixture signals into independent components by:

$$y = Wx \qquad (2)$$

2.2 Non-stability Principle

In this section, we are going to demonstrate that the non-stability principle is a meaningful way for separation and how it works in maximum non-stability ICA.

In the beginning, let us rewrite Eq. (2) in another way:

$$y_i = w_i x \qquad (3)$$

where w_i is the *ith* row vector in W. Now, if w_i was the *ith* row of the inverse of A in Eq. (1), y_i is the one of the independent components. Similar to classical ICA [4], we make a change of variables, defining $z_i = w_i A$ where z is a row vector. Then, equation Eq. (3) can be changed into:

$$y_i = w_i x = w_i As = z_i s \qquad (4)$$

From Eq. (4), we see y_i is a linear combination of s_i where s_i is the *ith* row vector in s. According to the GCLT, $z_i s$ is more stable than each s_i unless $z_i s$ is equal to one of the independent components in s. Similar to the classical ICA, we can find a w_i which maximizes the non-stability of $w_i x$ and it gives us one independent component. In the next section, we define a new measure of non-stability as a contrast function.

2.3 Measure of Non-stability

2.3.1 Negentropy and Alpha-Stable Negentropy

Various measures exist to quantify the Gaussianity of a distribution, among which negentropy is one of the most frequently used which provides us an information theoretic measure of non-Gaussianity. Negentropy is based on the differential entropy, which is closely tied with the idea of (Shannon) entropy.

Definition: Let X be a random variable with the probability density function $f(x)$ whose support vector is I, then the differential entropy $h(x)$ is defined as:

$$h(x) = -\int_I f(x) \log f(x) dx \qquad (5)$$

As in entropy for discrete distributions, differential entropy is a measure of uncertainty, for continuous distributions.

The classical (Gaussian) negentropy is defined as:

$$J(y) = h(y_{gauss}) - h(y) \qquad (6)$$

where y_{gauss} is a Gaussian random variable of the same covariance matrix as **y** [2]. Seen as an extension of the classical negentropy, we define alpha-stable negentropy J

in Eq. (7) as a measure of the distance of a distribution from alpha-stable distribution that is a measure of non-stability.

$$J(y) = \left| h(y_{\alpha - stable}) - h(y) \right| \tag{7}$$

where $y_{\alpha-stable}$ is an alpha-stable random variable with the parameters $\alpha, \beta, \gamma, \delta$ estimated from **y**. Here, **y** is not necessarily an alpha stable distribution, and more strictly in this initial study **y** is constrained not to be alpha stable distribution. But, as an impulsive signal it should have the characteristics of a heavy tailed distribution, Therefore, we estimate $\alpha, \beta, \gamma, \delta$ from it. From the definition, alpha-stable negentropy will be non-zero except the case that y is an alpha stable distribution. Alpha-stable negentropy can calculate the distance from the observed data to alpha stable data in entropy, so it can be used as a measure of the non-stability. By maximizing (7) we arrive at a signal which is as far as possible from the alpha-stable distribution with the parameters computed from the data.

It is necessary to explain the relationship between alpha-stable negentropy and the classical negentropy here. As mentioned above, Gaussian distribution is a special case of the alpha stable distribution when α equals 2. So, if we restrict α in $y_{\alpha-stable}$ in Eq. (7), alpha-stable negentropy will reduce to negentropy. Moreover, the relationship between maximum non-stability ICA and classical ICA can be understood by the similar way. If we add the finite variance constraint to GCLT, the limit distribution tends to be Gaussian distribution instead of alpha-stable distribution, and GCLT reduces to CLT. Then maximum non-stability ICA based on GCLT will also reduce to classical ICA.

2.3.2   Calculation of Alpha-Stable Negentropy

From the definition in Eq. (7), two differential entropies of $y_{\alpha-stable}$ and **y** should be calculated separately. In practice, prior knowledge of **y** is usually rare so we always do not know the exact distribution of **y**, so generally the histogram is used in computing the differential entropy of **y**.

As in the differential entropy of $y_{\alpha-stable}$, the calculation can be divided into two parts: the estimation of parameter: $\alpha, \beta, \gamma, \delta$ from y, and the numerical calculation of the differential entropy. There are some effective approaches which can well estimate the parameters from **y**, such maximum likelihood, quantiles, empirical characteristic function method and method of moments using fractional or lograritmic moments [1]. Then, the probability density can be calculated evaluating Fourier transform [5] from the characteristic function with parameters $\alpha, \beta, \gamma, \delta$. Finally, we get the differential entropy of $y_{\alpha-stable}$ through numerical integration and the alpha-stable negentropy is calculated from the difference of these two quantities.

2.4   New ICA Using Numerical Gradient Ascent Algorithm

In the previous section, we introduced a new measure of non-stability, alpha-stable negentropy, and how to calculate it. Now we need an algorithm to maximize this contrast function Eq. (7). In this first paper on the topic we provide a simple numerical gradient ascent algorithm for the maximization of the alpha-stable negentropy. Moreover, some issues to simplify this problem in this beginning work should be explained before further discussion. First, the data we are dealing with is of zero mean and no noise at all. Second, whitening is not included in this work because of the large or even infinite variance of the signals.

Gradient ascent algorithm finds the local maximum of the contrast function Eq. (7), and it takes steps proportional to the gradient:

$$w_i^{N+1} = w_i^N + \lambda \Delta w = w_i^N + \lambda \nabla J(w_i^N), \ N \geq 1 \ \text{and} \ \lambda > 0 \tag{8}$$

where ∇w and $\nabla J(w_i^N)$ are the gradient of contrast function and it equals

$$[\frac{d(J(w_i^N x))}{dw_{i1}}, \frac{d(J(w_i^N x))}{dw_{i2}}, \dots, \frac{d(J(w_i^N x))}{dw_{in}}], N \geq 1 \tag{9}$$

The reason we use a numerical method is that the gradient of contrast function Eq. (7) can only be calculated via the numerical differentiation by Eq. (9) at present. Full steps of the numerical gradient ascent algorithm are shown as follows:

1. Initialize a unit vector w_i^1 in Eq. (4) randomly.
2 Compute the Δw using Eq. (9)
3. Let $w_i^{N+1} = w_i^N + \lambda \Delta w$, $N \geq 1$ and $\lambda > 0$
4. If not converged, go to step 2.

Here, convergence means w_i^{N+1} and w_i^N point to the same direction. λ is chosen to be a small constant number.

3 Experiment

In this section, we examine maximum non-stability ICA through an impulsive source separation problem. Two source signals with 10000 sample data each are generated by $\alpha\_stable(\alpha_1,0,\gamma_1,0) + \alpha\_stable(\alpha_2,0,\gamma_2,0)$ separately. In source one, $\alpha_1 = 1, \alpha_2 = 0.7, \gamma_1 = 100, \gamma_2 = 10$, and in source two $\alpha_1 = 0.5, \alpha_2 = 0.6, \gamma_1 = 200, \gamma_2 = 100$. A 2 by 2 mixing matrix is chosen randomly. Note that the ICA by maximizing non-Gaussianity is not suitable to this problem because the source signals do not possess the second-or higher-order moments which can be used as contrast functions in classical ICA.

Fig. 1 shows the source signals and the mixture signals. Applying the numerical gradient ascent algorithm to this problem, the performance of maximum non-stability ICA can be shown in Fig. 2, Fig. 3 and Fig. 4.

In Fig. 2, lines representing the directions of separation vectors are plotted for this two-dimension problem. Red lines are right directions of the separation vectors while green lines the estimated directions. We can observe that they match well, which means that maximum non-stability ICA behaves effectively in this impulsive source separation problem. A visual observation of how this algorithm works can be seen in Fig. 3 in which the alpha-stable negentropy versus the angle between the direction of

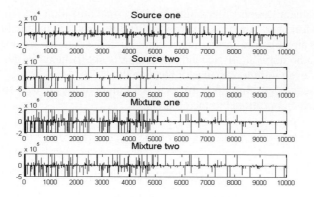

Fig. 1. Along with the vertical direction, two original source signals and two mixture signals are plotted

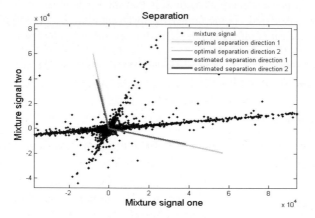

Fig. 2. Scatter dispersion of the mixture signals as well as optimal (deep color *lines*) and estimated (*light color lines*) separation directions of w_i are plotted

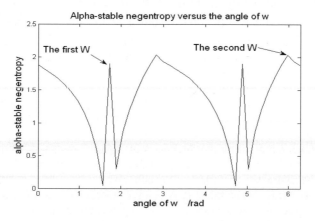

Fig. 3. Alpha-stable negentroepy versus the angel of w_i

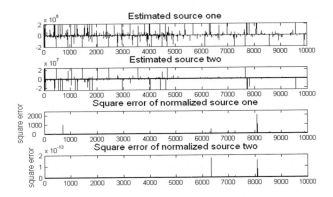

Fig. 4. Estimated source signals from maximum non-stability ICA and the square error between normalized source signals and estimated ones

w_i and the horizontal axis. Four maxima can be clearly seen from this curve. The first maximum refers to θ_1 which is the angle of the first estimated direction w_1, and the third maximum give us $\theta_1 + \pi$ or $-w_1$. Both w_1 and $-w_1$ are regarded to be estimated direction because they give the same independent component with the difference of the sign. This is an ambiguity of both classical ICA and this maximum non-stability ICA. Same explanation suits the second and the fourth maximum. Finally, the estimated source signals and square error between original and estimated sources with the normalized variance are plotted in Fig. 4. And the mean square errors for normalized s_1 and s_2 are 0.4806 and 3.75e-017 respectively.

New ICA performs successfully in solving this impulsive source separation problem, but some drawbacks still exist. Firstly, numerical calculation of the alpha-stable negentropy leads to a large computational cost. Secondly, numerical calculation introduces errors in the result, so large integration range and small integration step are necessary to protect the accuracy.

4   Conclusion

Motivated by the Generalized Central Limit Theorem (GCLT), in this paper, we illustrate that classical ICA is one special case of maximum non-stability ICA and the non-Gaussianity principle should be replaced by the non-stability principle. Meanwhile, a new method of non-stability called alpha-stable negentropy is also proposed as the contrast function in maximum non-stability ICA. Experiments using a numerical gradient ascent algorithm based on this new contrast function show that maximum non-stability ICA performs successfully in impulsive source separation problems. However, large computation cost and numerical error accumulation call for better methods for the maximization of the alpha-stable negentropy.

References

1. Kuruoglu, E.E.: Density Parameter Estimation of Skewed alpha-Stable Distribution. IEEE transactions on signal processing 49(10), 2192–2201 (2001)
2. Nikias, C.L., Shao, M.: Signal Processing with Alpha-Stable Distributions and Applications. Wiley, New York (1995)
3. Samorodnitsky, G., Taqqu, M.: Stable non-Gaussian random processes. Chapman and Hall, New York (1994)
4. Hyvärinen, A., Oja, E.: Independent Component Analysis: Algorithms and Applications. Neural Networks 13, 411–430 (2004)
5. Nolan, J.P.: Numerical Calculation of Stable Densities and Distribution Functions. Stochastic Models 13(4), 759–774 (1997)
6. Hyvärinen, A.: Fast and Robust Fixed-Point Algorithms for Independent Component Analysis. IEEE Transactions on Neural Networks 10(3), 626–634 (1999)
7. Hyvärinen, A., Karhunen, J., Oja, E.: Independent Component Analysis. John Wiley & Sons, Chichester (2001)
8. Comon, P.: Independent Component Analysis, A New Concept? Signal Processing 36, 287–314 (1994)
9. Hyvärinen, A.: Survey on Independent Component Analysis. Neural Computing Surveys 2, 94–128 (1999)
10. Sahmoudi, M., Abed-Meraim, K., Benidir, M.: Blind Separation of Impulsive Alpha-Stable Sources Using Minimum Dispersion Criterion. IEEE Signal Processing Letters 12(4), 281–284 (2005)

On Optimal Selection of Correlation Matrices for Matrix-Pencil-Based Separation

Arie Yeredor

School of Electrical Engineering, Tel-Aviv University, Israel
arie@eng.tau.ac.il

Abstract. The Matrix-Pencil approach to blind source separation estimates the mixing matrix from the Generalized Eigenvalue Decomposition (GEVD), or Exact Joint Diagonalization, of two "target-matrices". In a Second-Order-Statistics framework, these target-matrices are two different correlation matrices (e.g., at different lags, taken over different time-intervals, etc.), attempting to capture the diversity of the sources (e.g., diverse spectra, different nonstationarity profiles, etc.). A central question in this context is how to best choose these target-matrices, given a statistical model for the sources. To answer this question, we consider a general paradigm for the target-matrices, viewed as two "generalized correlation" matrices, whose structure is governed by two selected "Association-Matrices". We then derive an explicit expression (assuming Gaussian sources) for the resulting Interference-to-Source Ratio (ISR) in terms of the Association-Matrices. Subsequently, we show how to minimize the ISR with respect to these matrices, leading to optimized selection of the matrix-pair for GEVD-based separation.

1 Introduction and Problem Formulation

Perhaps one of the most conceptually appealing and computationally simple approaches to Blind Source Separation (BSS) is to base the estimation of the unknown mixing matrix on a pair of "target-matrices", estimated from the observed mixtures. Such a framework is sometimes called a Matrix-Pencil approach [10] or a Generalized Eigenvalue Decomposition (GEVD) approach [6], essentially leading to Exact Joint Diagonalization (EJD) of the two target-matrices. This approach basically relies on the understanding that a whitening (sphering) operation, which consists of diagonalizing the mixtures' covariance matrix, does about "half" of the separation - namely, it enables to estimate the mixing matrix up to some unitary matrix factor. Then, a single "second matrix" is usually sufficient for providing the "second half" (the unitary factor) of the estimate.

To be more specific, we consider the standard linear, instantaneous, square and noiseless BSS model $\boldsymbol{X} = \boldsymbol{AS}$, where $\boldsymbol{S}$ is a $K \times N$ matrix composed of K statistically independent, zero-mean $N \times 1$ source signals $\boldsymbol{S} = [\boldsymbol{s}_1 \ \boldsymbol{s}_2 \ \cdots \ \boldsymbol{s}_K]^T$; $\boldsymbol{A}$ is the unknown $K \times K$ mixing matrix; and $\boldsymbol{X}$ is the $K \times N$ matrix of observed mixtures. The two $K \times K$ target-matrices (or Matrix-Pencil), denoted $\widehat{\boldsymbol{R}}_1$ and $\widehat{\boldsymbol{R}}_2$, are usually empirical estimates of $\boldsymbol{R}_1 = \boldsymbol{A}\boldsymbol{D}_1\boldsymbol{A}^T$ and $\boldsymbol{R}_2 = \boldsymbol{A}\boldsymbol{D}_2\boldsymbol{A}^T$, where

T. Adali et al. (Eds.): ICA 2009, LNCS 5441, pp. 187–194, 2009.

$\boldsymbol{D}_1$ and $\boldsymbol{D}_2$ are diagonal by virtue of the sources' statistical independence. As mentioned above, $\boldsymbol{R}_1$ is usually [6] the observations' covariance. $\boldsymbol{R}_2$ is selected according to the sources' statistical model: For example, when the temporal structure of each source is composed of independent, identically-distributed (iid) samples, $\boldsymbol{R}_2$ can be taken as a linear combination of certain cumulants matrices of the observations (e.g., [6,4]) or as an off-origin Hessian of their joint log-characteristic function [11].

In many other cases of interest, both $\boldsymbol{R}_1$ and $\boldsymbol{R}_2$ can be based on Second-Order Statistics (SOS). To define a general framework, let $\boldsymbol{P}_1$ and $\boldsymbol{P}_2$ denote some arbitrary $N \times N$ matrices, which we term "Association-Matrices", and let the estimated "generalized correlation matrices" $\widehat{\boldsymbol{R}}_1$ and $\widehat{\boldsymbol{R}}_2$ be given by

$$\widehat{\boldsymbol{R}}_1 = \boldsymbol{X}\boldsymbol{P}_1\boldsymbol{X}^T \qquad\qquad \widehat{\boldsymbol{R}}_2 = \boldsymbol{X}\boldsymbol{P}_2\boldsymbol{X}^T. \tag{1}$$

By proper choice of the Association Matrices $\boldsymbol{P}_1$ and $\boldsymbol{P}_2$, several particular classical choices of $\widehat{\boldsymbol{R}}_1$ and $\widehat{\boldsymbol{R}}_2$ may be identified. For example, when $\boldsymbol{P}_1 = \frac{1}{N}\boldsymbol{I}$, $\widehat{\boldsymbol{R}}_1$ obviously becomes the observations' empirical correlation matrix. Then, if $\boldsymbol{P}_2$ is taken as an all-zeros matrix with a sequence of M non-zero values of $\frac{1}{M}$ somewhere along its main diagonal, $\widehat{\boldsymbol{R}}_2$ becomes the empirical correlation estimated over the respective time-segment, as used for non-stationary sources, e.g., in [6,7,8]. If $\boldsymbol{P}_2$ is taken as an all-zeros matrix with $\frac{1}{N-|\ell|}$ along its ℓ-th diagonal[1], $\widehat{\boldsymbol{R}}_2$ becomes the unbiased estimate of the observations' lagged correlation at lag ℓ, as used for stationary sources, e.g., in [7,2]. Likewise, if $\boldsymbol{P}_2$ is a general Toeplitz matrix, $\widehat{\boldsymbol{R}}_2$ can be regarded as a linear combination of estimated correlation matrices at different lags, or, alternatively, as the correlation matrix between linearly filtered versions of the observations, as used, e.g., in [10]. Spectral matrices, time-frequency matrices [3], or cyclic correlation matrices [1] can also be obtained by setting $\boldsymbol{P}_2$ to the appropriate transformation matrices.

However, in many cases of interest the source signals are not truly stationary, block-stationary, cyclostationary, etc. Even when they are, it is not clear that the respective Association Matrices mentioned above are optimal for the Matrix-Pencil approach in these cases. Thus, two important interesting questions arise in this general framework:

1. How does the resulting separation performance depend on $\boldsymbol{P}_1$ and $\boldsymbol{P}_2$?
2. Can $\boldsymbol{P}_1$ and $\boldsymbol{P}_2$ be chosen so as to optimize the performance, given a particular statistical model for the sources?

Our goal in this paper is to answer these two questions.

To this end, we assume a rather general statistical model for the sources: Our only restrictive assumption is that the sources are all (jointly) Gaussian (and, of course, independent). For convenience, we also assume zero-mean. Yet, we make no restrictions on the temporal covariance structures. We denote the covariance of the k-th source as a general $N \times N$ matrix $\boldsymbol{C}_k \triangleq E[\boldsymbol{s}_k\boldsymbol{s}_k^T]$, $k = 1, 2, ..., K$.

---

[1] The ℓ-th diagonal of $\boldsymbol{P}_2$ is the diagonal extending from $P_2[\ell+1, 1]$ to $P_2[N, N-\ell]$ for $\ell \geq 0$, or from $P_2[1, 1-\ell]$ to $P_2[N+\ell, N]$ for $\ell \leq 0$.

For example, for stationary sources the respective $\boldsymbol{C}_k$ would have a Toeplitz structure, whereas for block-stationary sources they would have a block-Toeplitz structure, etc. But we do not restrict the structure of $\boldsymbol{C}_k$ in any way, hence we do not confine the results to any particular temporal model of the sources.

2   A Small-Errors Perturbation Analysis for the EJD

Let $\boldsymbol{R}_q \overset{\triangle}{=} E[\boldsymbol{X}\boldsymbol{P}_q\boldsymbol{X}^T] = \boldsymbol{A}E[\boldsymbol{S}\boldsymbol{P}_q\boldsymbol{S}^T]\boldsymbol{A}^T \overset{\triangle}{=} \boldsymbol{A}\boldsymbol{D}_q\boldsymbol{A}^T$ ($q = 1, 2$) denote the true generalized-correlation matrices, estimated by $\widehat{\boldsymbol{R}}_1$ and $\widehat{\boldsymbol{R}}_2$ (resp.). Under some commonly met conditions (see, e.g., [12]), the closed-form EJD of $\widehat{\boldsymbol{R}}_1$ and $\widehat{\boldsymbol{R}}_2$ can be readily obtained from the eigen-decomposition of $\widehat{\boldsymbol{Q}} \overset{\triangle}{=} \widehat{\boldsymbol{R}}_1\widehat{\boldsymbol{R}}_2^{-1}$: The eigenvectors matrix $\widehat{\boldsymbol{A}}$ which satisfies $\widehat{\boldsymbol{Q}}\widehat{\boldsymbol{A}} = \widehat{\boldsymbol{A}}\widehat{\boldsymbol{D}}$ (where $\widehat{\boldsymbol{D}}$ is some diagonal matrix), also satisfies $\widehat{\boldsymbol{R}}_1 = \widehat{\boldsymbol{A}}\widehat{\boldsymbol{D}}_1\widehat{\boldsymbol{A}}^T$ and $\widehat{\boldsymbol{R}}_2 = \widehat{\boldsymbol{A}}\widehat{\boldsymbol{D}}_2\widehat{\boldsymbol{A}}^T$, where $\widehat{\boldsymbol{D}}_1$ and $\widehat{\boldsymbol{D}}_2$ are diagonal matrices, such that $\widehat{\boldsymbol{D}} = \widehat{\boldsymbol{D}}_1\widehat{\boldsymbol{D}}_2^{-1}$.

When the estimates $\widehat{\boldsymbol{R}}_1$ and $\widehat{\boldsymbol{R}}_2$ are exact and the model is identifiable, the resulting $\widehat{\boldsymbol{A}}$ coincides with $\boldsymbol{A}$ (up to the inevitable scale and permutation ambiguities). Naturally, however, departure of $\widehat{\boldsymbol{R}}_1$, $\widehat{\boldsymbol{R}}_2$ from their true values inflicts errors on $\widehat{\boldsymbol{A}}$. A common measure of the estimation error (useful in the BSS context) is to consider the resulting overall mixing-unmixing matrix $\boldsymbol{T} \overset{\triangle}{=} \widehat{\boldsymbol{A}}^{-1}\boldsymbol{A}$. Assuming, just for simplicity of notations, that the scaling and permutation ambiguities have been resolved, $\boldsymbol{T}$ would ideally be the identity matrix. However, due to estimation errors in $\widehat{\boldsymbol{A}}$, its off-diagonal elements $T[k,\ell]$ ($k \neq \ell$) would not vanish, and would reflect a residual mixing *per realization*. Their second moments, $E[T^2[k,\ell]]$, are usually called the ℓ-to-k ISR (denoted $ISR_{k,\ell}$), assuming that all sources have equal energy[2].

Our goal in this section is to quantify the effect of estimation errors in $\widehat{\boldsymbol{R}}_1$ and $\widehat{\boldsymbol{R}}_2$ on all $T[k,\ell]$. We begin by observing a very appealing invariance property of $\boldsymbol{T}$ in this context: Given a specific realization $\boldsymbol{S}$ of the sources, the same value of $\boldsymbol{T}$ would be obtained (in our framework) with *any* (nonsingular) $\boldsymbol{A}$.

To observe this, assume first that $\boldsymbol{A} = \boldsymbol{I}$, and let us denote the estimated $\widehat{\boldsymbol{R}}_1$ and $\widehat{\boldsymbol{R}}_2$ in this case as $\widehat{\boldsymbol{R}}_{q(I)} = \boldsymbol{X}\boldsymbol{P}_q\boldsymbol{X}^T = \boldsymbol{S}\boldsymbol{P}_q\boldsymbol{S}^T$ ($q = 1, 2$). Likewise, denote $\widehat{\boldsymbol{Q}}_{(I)} \overset{\triangle}{=} \widehat{\boldsymbol{R}}_{1(I)}\widehat{\boldsymbol{R}}_{2(I)}^{-1}$, with eigenvectors and eigenvalues matrices $\widehat{\boldsymbol{A}}_{(I)}$ and $\widehat{\boldsymbol{D}}_{(I)}$ (resp.): $\widehat{\boldsymbol{Q}}_{(I)}\widehat{\boldsymbol{A}}_{(I)} = \widehat{\boldsymbol{A}}_{(I)}\widehat{\boldsymbol{D}}_{(I)}$. Evidently, the matrix $\boldsymbol{T}$ in this case is given by $\boldsymbol{T}_{(I)} \overset{\triangle}{=} \widehat{\boldsymbol{A}}_{(I)}^{-1}\boldsymbol{A} = \widehat{\boldsymbol{A}}_{(I)}^{-1}\boldsymbol{I} = \widehat{\boldsymbol{A}}_{(I)}^{-1}$. Now consider a general mixing matrix $\boldsymbol{A}$. We then have $\widehat{\boldsymbol{R}}_q = \boldsymbol{X}\boldsymbol{P}_q\boldsymbol{X}^T = \boldsymbol{A}\boldsymbol{S}\boldsymbol{P}_q\boldsymbol{S}^T\boldsymbol{A}^T = \boldsymbol{A}\widehat{\boldsymbol{R}}_{q(I)}\boldsymbol{A}^T$ ($q = 1, 2$), and, consequently, $\widehat{\boldsymbol{Q}} = \boldsymbol{A}\widehat{\boldsymbol{Q}}_{(I)}\boldsymbol{A}^T$. It is readily observed that the matrix $\boldsymbol{A}\widehat{\boldsymbol{A}}_{(I)}$ is the eigenvectors matrix of $\widehat{\boldsymbol{Q}}$ (with eigenvalues matrix $\widehat{\boldsymbol{D}}_{(I)}$), since

$$\widehat{\boldsymbol{Q}}\boldsymbol{A}\widehat{\boldsymbol{A}}_{(I)} = \boldsymbol{A}\widehat{\boldsymbol{Q}}_{(I)}\boldsymbol{A}^{-1}\boldsymbol{A}\widehat{\boldsymbol{A}}_{(I)} = \boldsymbol{A}\widehat{\boldsymbol{Q}}_{(I)}\widehat{\boldsymbol{A}}_{(I)} = \boldsymbol{A}\widehat{\boldsymbol{A}}_{(I)}\widehat{\boldsymbol{D}}_{(I)}, \tag{2}$$

---

[2] If the sources have different energies, each $ISR_{k,\ell}$ should be normalized by the ratio between the respective energies, so as to reflect the mean residual energy ratios.

so $\widehat{A} = A\widehat{A}_{(I)}$, and therefore $T = \widehat{A}^{-1}A = \widehat{A}_{(I)}^{-1} = T_{(I)}$, which establishes the invariance of T in A. This appealing property allows us to analyze the matrix T under the simple non-mixing condition $A = I$, knowing that the same result would hold with any other invertible mixing matrix A.

Recall the definition of the diagonal matrices $D_1 \overset{\triangle}{=} E[SP_1S^T]$ and $D_2 \overset{\triangle}{=} E[SP_2S^T]$, where (at least) the latter is invertible, and denote $D \overset{\triangle}{=} D_1D_2^{-1}$. Assume a non-mixing condition ($A = I$), thus denoting

$$\widehat{R}_1 = D_1 + \mathcal{E}_1 \qquad\qquad \widehat{R}_2 = D_2 + \mathcal{E}_2 = D_2(I + D_2^{-1}\mathcal{E}_2), \qquad (3)$$

where $\mathcal{E}_1 \ll D_1$ and $\mathcal{E}_2 \ll D_2$ are the respective estimation errors, assumed "small". Applying a small-errors analysis, which neglects second (and higher) order terms in $\mathcal{E}_1, \mathcal{E}_2$, we get

$$\begin{aligned}\widehat{Q} = \widehat{R}_1\widehat{R}_2^{-1} &= (D_1 + \mathcal{E}_1)(I + D_2^{-1}\mathcal{E}_2)^{-1}D_2^{-1} \\ &\approx (D_1 + \mathcal{E}_1)(I - D_2^{-1}\mathcal{E}_2)D_2^{-1} \approx D + \mathcal{E}_1D_2^{-1} - D\mathcal{E}_2D_2^{-1}. \quad (4)\end{aligned}$$

In the error-free case the eigenvectors matrix of $\widehat{Q}$ would be the true mixing-matrix $\widehat{A} = A = I$, and the eigenvalues matrix would be D. Let us denote by $\mathcal{E}$ and Δ the resulting respective errors in these matrices (namely, $\widehat{A} = I + \mathcal{E}$, $\widehat{D} = D + \Delta$), such that $\widehat{Q}(I + \mathcal{E}) = (I + \mathcal{E})(D + \Delta)$. Substituting (4) we have (again, using the small-errors assumption)

$$\widehat{Q}(I + \mathcal{E}) \approx (D + \mathcal{E}_1D_2^{-1} - D\mathcal{E}_2D_2^{-1})(I + \mathcal{E}) \approx D + \mathcal{E}_1D_2^{-1} - D\mathcal{E}_2D_2^{-1} + D\mathcal{E}$$

and

$$(I + \mathcal{E})(D + \Delta) \approx D + \mathcal{E}D + \Delta. \qquad (5)$$

Equating these terms we get

$$D\mathcal{E} - \mathcal{E}D = D\mathcal{E}_2D_2^{-1} - \mathcal{E}_1D_2^{-1} + \Delta. \qquad (6)$$

Eventually we would have $T = \widehat{A}^{-1} \approx I - \mathcal{E}$, and since we are only interested in the off-diagonal terms of T, we may ignore the unknown Δ in (6). Denoting by $d_q[k]$ the $[k,k]$-th element of D_q ($q = 1, 2$), we have, from (6)

$$\mathcal{E}[k, \ell]\left(\frac{d_1[k]}{d_2[k]} - \frac{d_1[\ell]}{d_2[\ell]}\right) = \frac{d_1[k]}{d_2[k]d_2[\ell]}\mathcal{E}_2[k, \ell] - \frac{1}{d_2[\ell]}\mathcal{E}_1[k, \ell] \quad 1 \leq k \neq \ell \leq K. \quad (7)$$

Applying some straightforward algebraic manipulations, we end up with the following expression for $T[k, \ell] = -\mathcal{E}[k, \ell]$:

$$T[k, \ell] = \frac{d_1[k]\mathcal{E}_2[k, \ell] - d_2[k]\mathcal{E}_1[k, \ell]}{d_1[\ell]d_2[k] - d_1[k]d_2[\ell]} \quad 1 \leq k \neq \ell \leq K, \qquad (8)$$

which establishes the explicit dependence of T (under the small-errors assumption) on the generalized-correlations' estimation errors $\mathcal{E}_1$ and $\mathcal{E}_2$ under the non-mixing condition $A = I$.

3 Explicit Expressions for the ISR

We now turn to calculate the second moments of the off-diagonal elements of T (namely, the ISRs), based on (8) and on the second-order statistics of $\mathcal{E}_1$ and $\mathcal{E}_2$. All expressions will be given in terms of the sources' covariance matrices $\{C_k\}$ and the selected "Association Matrices" P_1 and P_2. We begin with expressions for the coefficients $d_q[k]$ $(q = 1, 2)$:

$$d_q[k] = E[s_k^T P_q s_k] = Tr\{P_q C_k\} \qquad k = 1, 2, \ldots, K. \tag{9}$$

Next, consider the zero-mean error-matrices $\mathcal{E}_1 = \widehat{R}_1 - D_1$, $\mathcal{E}_2 = \widehat{R}_2 - D_2$ (defined under the non-mixing condition $X = S$). For computing the ISR from (8) we need to know the variances and covariances of $\mathcal{E}_1[k, \ell]$ and $\mathcal{E}_2[k, \ell]$. From the non-mixing condition and the diagonality of D_1, D_2, we note that

$$\mathcal{E}_1[k, \ell] = s_k^T P_1 s_\ell \text{ and } \mathcal{E}_2[k, \ell] = s_k^T P_2 s_\ell. \tag{10}$$

Assume now that the source signals s_k and s_ℓ $(k \neq \ell)$ are Gaussian (and independent), and consider the two zero-mean random variables defined as $s_k^T P_p s_\ell$ and $s_k^T P_q s_\ell$ for some $p, q \in \{1, 2\}$. Their covariance is given by

$$E[s_k^T P_p s_\ell s_k^T P_q s_\ell] = \sum_{i,j,m,n=1}^{N} E\left[s_k[i]P_p[i,j]s_\ell[j]s_k[m]P_q[m,n]s_\ell[n]\right]$$

$$= \sum_{ijmn} P_p[i,j]P_q[m,n]E\left[s_k[i]s_k[m]s_\ell[j]s_\ell[n]\right] = \sum_{ijmn} P_p[i,j]P_q[m,n]C_k[i,m]C_\ell[j,n]$$

$$= \sum_{ijmn} P_p[i,j]C_\ell[j,n]P_q^T[n,m]C_k[m,i] = Tr\{P_p C_\ell P_q^T C_k\}, \tag{11}$$

where we have used the zero-mean, (joint) Gaussianity and statistical independence of the sources in the transition on the second line, and where $P_q^T[n,m]$ denotes the $[n, m]$-th element of P_q^T (which is the $[m, n]$-th element of P_q).

Applying this result for calculating the variances of $\mathcal{E}_1[k, \ell]$, $\mathcal{E}_2[k, \ell]$ and their covariance (by substituting $(p, q) = (1, 1), (2, 2), (1, 2)$, resp.), we obtain

$$E\left[\mathcal{E}_1^2[k, \ell]\right] = Tr\{P_1 C_\ell P_1^T C_k\} \quad , \quad E\left[\mathcal{E}_2^2[k, \ell]\right] = Tr\{P_2 C_\ell P_2^T C_k\} \tag{12}$$

$$E[\mathcal{E}_1[k, \ell]\mathcal{E}_2[k, \ell]] = Tr\{P_1 C_\ell P_2^T C_k\}. \tag{13}$$

Finally, by substituting (12), (13) and (9) into (8), we end up with

$$ISR_{k,\ell} = E[T^2[k, \ell]] = \Big(Tr^2\{P_1 C_k\} Tr\{P_2 C_\ell P_2^T C_k\} -$$

$$-2\, Tr\{P_1 C_k\} Tr\{P_2 C_k\} Tr\{P_1 C_\ell P_2^T C_k\} + Tr^2\{P_2 C_k\} Tr\{P_1 C_\ell P_1^T C_k\}\Big) /$$

$$/ \left(Tr\{P_1 C_\ell\} Tr\{P_2 C_k\} - Tr\{P_1 C_k\} Tr\{P_2 C_\ell\}\right)^2, \quad 1 \leq k \neq \ell \leq K. \tag{14}$$

Although derived under a non-mixing $(A = I)$ assumption, due to the invariance property this expression holds with *any* (nonsingular) mixing matrix A.

4   Optimizing the Association-Matrices

Having obtained an explicit expression for the ISRs, an interesting question that comes to mind is - which Association-Matrices would be optimal, given the sources' covariance matrices. Note that the ISR expression (14) is a ratio of fourth-order multinomials in the elements of P_1 and P_2, and as such is rather involved for direct minimization. However, assuming one of the matrices (say P_2) to be fixed, both the numerator and denominator of (14) are quadratic in the elements of the other (P_1). The ratio of two quadratic expressions can admit a closed-form minimizing solution, but if we treat all elements of the matrix P_1 as N^2 free unknowns, the resulting solution would be prohibitively computationally demanding (requiring the eigendecomposition of an $N^2 \times N^2$ matrix).

We therefore prefer to restrict the structure of P_1, so as to depend on a small number of parameters. One such plausible restriction is to constrain P_1 to be a symmetric band-Toeplitz matrix with L free diagonal values, and optimize the ISR with respect to these values. Note that this interpretation makes the estimated $\widehat{R}_1$ a linear combination of "standard" (symmetrized) lagged correlation matrices, where the coefficients of the linear combination are the parameters along the respective diagonals of P_1.

To this end, suppose that P_1 is parameterized by an $L \times 1$ vector $\boldsymbol{\theta}$, such that $P_1[k, \ell] = \theta[|k - \ell| + 1]$ for $0 \le |k - \ell| \le L - 1$ (and equals zero elsewhere). In this case we may note the following (for any $N \times N$ matrices Q, G, H):

$$Tr\{P_1 Q\} = \boldsymbol{\delta}^T \boldsymbol{\theta}, \text{ with } \delta[\ell] \overset{\triangle}{=} Tr\{J_\ell Q\} \quad \ell = 1, \dots, K \qquad (15)$$

$$Tr\{P_1 G P_1^T H\} = \boldsymbol{\theta}^T F \boldsymbol{\theta}, \text{ with } F[k, \ell] \overset{\triangle}{=} Tr\{J_\ell G J_k H\}, \quad k, \ell = 1, \dots, K$$

where J_ℓ is a symmetric Toeplitz matrix, which is all-zeros except for 1-s along its $\pm(\ell - 1)$-th diagonals. Expressions in this form are required in (14) with $Q = C_k$, $Q = C_\ell$, $Q = C_\ell P_2^T C_k$, $G = C_\ell$ and $H = C_k$. Consequently, we can express each $ISR_{k,\ell}$ of (14) in terms of $\boldsymbol{\theta}$ as

$$ISR_{k,\ell} = \frac{\boldsymbol{\theta}^T U(P_2, C_k, C_\ell) \boldsymbol{\theta}}{\boldsymbol{\theta}^T V(P_2, C_k, C_\ell) \boldsymbol{\theta}}, \qquad (16)$$

where the $L \times L$ matrices $U(P_2, C_k, C_\ell)$ and $V(P_2, C_k, C_\ell)$ are obtained from substituting (15) in (14) (explicit expressions are omitted due to lack of space).

We may now seek the optimal $\boldsymbol{\theta}$, so as to minimize any chosen element of the ISR matrix. By some kind of "poetic justice", minimization of (16) with respect to $\boldsymbol{\theta}$ requires yet another GEVD - this time the target matrices are U and V, and the minimizing $\boldsymbol{\theta}$ can be easily shown[3] to be the eigenvector of VU^{-1} associated with the smallest eigenvalue.

Once the minimizing $\boldsymbol{\theta}$ is found, P_1 can be constructed, and may then switch roles with P_2, so as to serve as the fixed matrix, enabling further minimization of $ISR_{k,\ell}$ with respect to P_2. The procedure may be iterated, giving rise to an

---

[3] Using Lagrange multipliers for constrained minimization of $\boldsymbol{\theta}^T U \boldsymbol{\theta}$ s.t. $\boldsymbol{\theta}^T V \boldsymbol{\theta} = 1$.

alternating-directions type optimization. Note that normally, this is an off-line procedure, which can be done in preparation for the on-line separation, whenever the covariance matrices of the sources are known in advance.

Two important comments are in order:

- The above procedure optimizes $\boldsymbol{P}_1$ and $\boldsymbol{P}_2$ for a *specific* element $ISR_{k,\ell}$ of the $K \times K$ ISR matrix. In other words, $\boldsymbol{P}_1$ and $\boldsymbol{P}_2$ which have been optimized for some (k,ℓ) might, in general, not be optimal for other (k,ℓ) pairs. Still, they can often do much better (for all ISRs) than just using the "standard" Association-Matrices mentioned in the Introduction. Note that given the closed-form ISR expressions (14), (16), it is also possible to minimize any measure of the overall ISR (e.g., $\sum_{k \neq \ell} ISR_{k,\ell}$) with respect to both $\boldsymbol{P}_1$ and $\boldsymbol{P}_2$, but this minimization problem would probably no longer admits an appealing closed-form solution.
- Optimal SOS-based separation usually requires the use of optimally-weighted Approximate Joint Diagonalization (AJD) of *several* "generalized correlation" matrices (or an equivalent solution), and cannot be attained with a simple matrix-pair EJD-based solution. For example, when the sources are stationary Autoregressive (AR) processes of some maximal order p, it has been explicitly shown (e.g., [9]) that Weighted AJD of $p + 1$ correlation matrices is required for attaining the optimal ISR. Moreover, an induced Cramér-Rao Lower Bound (iCRLB) on the ISR matrix for this case was derived [5], and the WASOBI algorithm [9] (based on $p + 1$ lagged correlation matrices) was shown to asymptotically attain this bound for all (k,ℓ) pairs *simultaneously*. Naturally, this is beyond the capability of a Matrix-Pencil based algorithm, whose attainable ISR would never be smaller (usually be higher) than the bound for each (k,ℓ) pair, as we demonstrate immediately.

5   Simulation Results

To enable comparison to the ISR bound in [5], we simulated AR(3) sources, $\boldsymbol{s}_1$, $\boldsymbol{s}_2$ and $\boldsymbol{s}_3$ with poles at $(-0.7, 0.2 \pm 0.9j)$, $(-0.8, 0.6 \pm 0.7j)$ and $(-0.9, 0.7 \pm 0.7j)$ (resp.), each of length $N = 500$, driven by white Gaussian noise. The 3×3 mixing matrix $\boldsymbol{A}$ was set randomly with standard Gaussian elements drawn independently. Each experiment was repeated 1000 times, where each trial consisted of estimating the generalized-correlation matrices, applying GEVD to obtain $\widehat{\boldsymbol{A}}$, and then resolving the scaling and permutation ambiguities to obtain $\boldsymbol{T} = \widehat{\boldsymbol{A}}^{-1} \boldsymbol{A}$ nearest to $\boldsymbol{I}$. The off-diagonal terms of $\boldsymbol{T}$ were squared and averaged over the 1000 trials to obtain the empirical $ISR_{k,\ell}$ for all $1 \leq k \neq \ell \leq 3$.

We conducted three experiments, differing only in the selection of the generalized correlation matrix-pair. In the first experiment we used the "standard" empirical correlation matrices at lags zero and one, denoted $\widehat{\boldsymbol{R}}_r[0]$ and $\widehat{\boldsymbol{R}}_x[1]$. In the second experiment we used target matrices obtained using the Association Matrices $\boldsymbol{P}_1$ and $\boldsymbol{P}_2$ obtained by optimizing $ISR_{1,2}$. In the third experiment we repeated the same with $\boldsymbol{P}_1$, $\boldsymbol{P}_2$ optimized for $ISR_{3,2}$. $\boldsymbol{P}_1$ and $\boldsymbol{P}_2$ were taken as symmetric band-Toeplitz matrices with seven nonzero diagonals (namely, $L = 4$), enabling linear combinations of all lagged correlations from lag zero to three.

The results are summarized in Table 1, showing good agreement (in each experiment) between the empirical ISRs (in parentheses) and their theoretically predicted values (using (14)). We observe that the theoretical (and empirical) results are never lower than the iCRLB, yet the optimized $ISR_{1,2}$ and $ISR_{3,2}$ (in the respective experiments) are significantly closer to the bound than their unoptimized values (in the first experiment, which used "standard" correlations).

Table 1. Theoretical and empirical results in terms of $(ISR_{k,\ell})^{-1}$ in [dB]. For each experiment, the empirical results are shown in parentheses next to the predicted theoretical results. The optimized ISRs ($ISR_{1,2}$ in the second and $ISR_{3,2}$ in the third experiment) are seen to be close to the induced CRLB, shown on the second column.

(k,ℓ)	iCRLB [5]	Using $\hat{\boldsymbol{R}}_x[0], \hat{\boldsymbol{R}}_x[1]$	Optimized for $ISR_{1,2}$	Optimized for $ISR_{3,2}$
$(1,2)$	33.3	27.7 (27.1)	33.2 (33.0)	31.1 (30.8)
$(1,3)$	42.4	40.4 (39.1)	41.9 (41.3)	41.5 (40.5)
$(2,1)$	32.3	27.0 (26.9)	30.6 (31.0)	32.0 (31.7)
$(2,3)$	27.3	24.6 (23.1)	26.1 (26.1)	26.1 (25.3)
$(3,1)$	34.2	26.7 (26.7)	31.7 (31.6)	33.9 (33.8)
$(3,2)$	25.1	16.2 (15.3)	15.5 (14.7)	24.3 (23.7)

References

1. Abed-Meraim, K., Xiang, Y., Manton, J., Hua, Y.: Blind source separation using second-order cyclostationary statistics. IEEE Trans. Sig. Proc. 49, 694–701 (2001)
2. Belouchrani, A., Abed-Meraim, K., Cardoso, J.F., Moulines, E.: A blind source separation technique using second-order statistics. IEEE Trans. Sig. Proc. 45, 434–444 (1997)
3. Belouchrani, A., Amin, M.: Blind source separation based on time-frequency signal representations. IEEE Trans. Sig. Proc. 46, 2888–2897 (1998)
4. Cardoso, J.F., Souloumiac, A.: Blind beamforming for non Gaussian signals. IEE - Proceedings -F 140, 362–370 (1993)
5. Doron, E., Yeredor, A., Tichavsky, P.: A Cramér-Rao-induced bound for blind separation of stationary parametric Gaussian sources. IEEE Sig. Proc. Letters 14, 417–420 (2007)
6. Parra, L., Sajda, P.: Blind source separation via generalized eigenvalue decomposition. Journal of Machine Learning Research 4, 1261–1269 (2003)
7. Parra, L., Spence, C.: Convolutive blind source separation of non-stationary sources. IEEE Trans. Speech and Audio Proc., 320–327 (2000)
8. Pham, D.T., Cardoso, J.F.: Blind separation of instantaneous mixtures of nonstationary sources. IEEE Trans. Sig. Proc. 49, 1837–1848 (2001)
9. Tichavsky, P., Doron, E., Yeredor, A., Nielsen, J.: A computationally affordable implementation of an asymptotically optimal BSS algorithm for AR sources. In: Proc. EUSIPCO 2006 (2006)
10. Tomé, A.: The generalized eigendecomposition approach to the blind source separation problem. Digital Signal Processing 16, 288–302 (2006)
11. Yeredor, A.: Blind source separation via the second characteristic function. Signal Processing 80, 897–902 (2000)
12. Yeredor, A.: On using exact joint diagonalization for non-iterative approximate joint diagonalization. IEEE Sig. Proc. Letters 12, 645–648 (2005)

ICA with Sparse Connections: Revisited

Kun Zhang[1], Heng Peng[2], Laiwan Chan[3], and Aapo Hyvärinen[1,4]

[1] Dept of Computer Science & HIIT, University of Helsinki, Finland
[2] Dept of Mathematics, Hong Kong Baptist University, Hong Kong
[3] Dept of Computer Science and Engineering, Chinese University of Hong Kong
[4] Dept of Mathematics and Statistics, University of Helsinki, Finland

Abstract. When applying independent component analysis (ICA), sometimes we expect the connections between the observed mixtures and the recovered independent components (or the original sources) to be sparse, to make the interpretation easier or to reduce the random effect in the results. In this paper we propose two methods to tackle this problem. One is based on adaptive Lasso, which exploits the L_1 penalty with data-adaptive weights. We show the relationship between this method and the classic information criteria such as BIC and AIC. The other is based on optimal brain surgeon, and we show how its stopping criterion is related to the information criteria. This method produces the solution path of the transformation matrix, with different number of zero entries. These methods involve low computational loads. Moreover, in each method, the parameter controlling the sparsity level of the transformation matrix has clear interpretations. By setting such parameters to certain values, the results of the proposed methods are consistent with those produced by classic information criteria.

1 Introduction

Independent component analysis (ICA) aims at recovering latent independent sources from their observable linear mixtures [4]. Denote by $\mathbf{x} = (x_1, ..., x_n)^T$ the vector of observable signals. $\mathbf{x}$ is assumed to be generated by $\mathbf{x} = \mathbf{As}$, where $\mathbf{s} = (s_1, ..., s_n)^T$ has mutually independent components. For simplicity we assume the number of observed signals is equal to that of the independent sources. Under certain conditions on the mixing matrix $\mathbf{A}$ and the distributions of s_i, ICA applies a linear transformation on $\mathbf{x}$, i.e., $\mathbf{y} = \mathbf{Wx}$, and tunes the de-mixing matrix $\mathbf{W}$ to make the components of $\mathbf{y} = (y_1, ..., y_n)^T$ mutually as independent as possible; finally y_i provide an estimate of the original sources s_i.

We sometimes prefer the transformation matrix (the de-mixing matrix $\mathbf{W}$ or mixing matrix $\mathbf{A}$) to be sparse, under the condition that y_i are independent, for reliable parameter estimation, or for an easier interpretation purpose [5,11] For example, when performing LiNGAM (short for linear, non-Gaussian, acyclic models) causality analysis based on ICA [8], we prefer $\mathbf{W}$ to be sparse, since the LiNGAM analysis requires that $\mathbf{W}$ can be permuted to lower triangularity.

Generally speaking, sparsity of the transformation matrix can be easily achieved. One can simply resort to the hard thresholding (which sets small coefficients to

T. Adali et al. (Eds.): ICA 2009, LNCS 5441, pp. 195–202, 2009.

zero), sparse priors [5], the SCAD penalty [11] (which corresponds to an improper prior). Wald test can also be used to set insignificant connections to zero [8]. The problem with these methods is how to determine the free parameter in these methods which controls the level of sparsity. Moreover, for most of them, it is unclear if the subset of the non-zero entries of the transforation matrix could be found consistently (e.g., the estimated subset converges to the correct one in probability when the data follow the model and the sample size grows infinite). On the other hand, one may exploit traditional information criteria, such as BIC [7] and AIC [1], to find the subset of non-zero coefficients in the transformation matrix. The properties of model selection based on information criteria have been well studied. For example, BIC can select the true model consistently, while the model selected by AIC has good prediction performance. Unfortunately, this model selection approach requires exhaustive search over all possible models, which usually involves two stages (training all models followed by comparison of the criteria) and is computationally intensive. Generally speaking, in ICA, the transformation matrix has many entries, and the space of candidate models is too large. Consequently this approach is not practical.

We propose two methods to do ICA with sparse connections which combine the strengths of the two model selection approaches mentioned above. The first one is based on adaptive Lasso [14], which exploits modified L_1 penalties and was recently proposed for variable selection in linear regression. We relate adaptive Lasso with the traditional information criteria, and show how to select the penalization parameter in adaptive Lasso to make its model selection results consistent with those based on information criteria. As L_1 penalties are not differentiable at zero, optimization involving such penalties based on gradients is generally troublesome. We further propose a very simple, yet effective scheme to solve this problem. The second method is based on optimal brain surgeon (OBS) [3] for network pruning. We also show the relationship between this approach and model selection based on traditional information criteria.

2   ICA Based on Maximum Likelihood

Since we will develop ICA with sparse connections by maximizing the penalized likelihood, in this section we briefly review the derivation of ICA algorithms from a maximum likelihood point of view [6]. Denote by f_i the density functions of s_i. The log likelihood of the observed data $\mathbf{x}$ is

$$L_T = \sum_{t=1}^{T} \sum_{i=1}^{n} \log f_i(y_{i,t}) + T \log |\det \mathbf{W}|, \qquad (1)$$

where T denotes the sample size. Note that if f_i are not given (say, if it is estimated from data), the scale of y_i and $\mathbf{W}$ estimated by maximizing the above likelihood is arbitrary, due to the scaling indeterminacy in ICA solutions. This can be avoided by constraining the variances of y_i or by keeping certain entries of $\mathbf{W}$ (or $\mathbf{A}$) constant. One scheme is to maximize the likelihood using gradient (or natural gradient) based methods: $\frac{1}{T} \cdot \frac{\partial L_T}{\partial \mathbf{W}} = -E\{\boldsymbol{\psi}(\mathbf{y})\mathbf{x}^T\} + [\mathbf{W}^T]^{-1}$, or, $\frac{1}{T} \cdot$

$\frac{\partial L_T}{\partial \mathbf{A}} = [\mathbf{A}^T]^{-1} \cdot [E\{\boldsymbol{\psi}(\mathbf{y})\mathbf{y}^T\} - \mathbf{I}]$, where $\boldsymbol{\psi}(\mathbf{y}) = (\psi_1(y_1), \cdots, \psi_n(y_n))^T$ with $\psi_1(y_1) = -\frac{f_i'(y_i)}{f(y_i)}$, and in each iteration the variance of each y_i is normalized.

Alternatively, one may incorporate the constraint $E\{y_i^2\} = 1$ using the regularization technique. The objective function to be maximized then becomes

$$J_T = \sum_{t=1}^{T}\sum_{i=1}^{n}\{\log f_i(y_{i,t})\} + T\log|\det \mathbf{W}| - \beta \sum_{j=1}^{n}(E\{y_i^2\} - 1)^2, \qquad (2)$$

where β is a regularization parameter. In our experiments we used $\beta = 1$. The gradients of the above function w.r.t. $\mathbf{W}$ and $\mathbf{A}$ can be easily derived.

3 ICA with Sparse Connections Based on Adaptive Lasso

We first propose to achieve the sparsity of the transformation matrix by penalized maximum likelihood. The penalty term we adopt is based on adaptive Lasso [14]. We will show that the result of this penalization method is consistent with that based on traditional information criteria, by setting the penalization parameter to certain given values.

3.1 Idea of Adaptive Lasso

Here we assume that the model under consideration satisfies some regularity conditions including identification conditions for the parameters $\boldsymbol{\theta}$, the consistency of the estimate $\hat{\boldsymbol{\theta}}$ when the sample size T tends to infinity, and the asymptotical normality of $\hat{\boldsymbol{\theta}}$. The penalized likelihood can be written as

$$pL(\boldsymbol{\theta}) = L(\boldsymbol{\theta}) - \lambda p_\lambda(\boldsymbol{\theta}), \qquad (3)$$

where $L(\boldsymbol{\theta})$ is the log likelihood, $\boldsymbol{\theta}$ contains the parameters (which are not redundant), and $p_\lambda(\boldsymbol{\theta}) = \sum_i p_\lambda(\theta_i)$ is the penalty.

The L_1 penalty is well known for producing sparse and continuous estimates [10]. However, it also causes bias in the estimate of significant parameters, and more importantly, it could select the true model consistently only when the data satisfy certain conditions [13]. Adaptive Lasso [14] was proposed to overcome the disadvantage of the L_1 penalty. In adaptive Lasso, $p_\lambda(\boldsymbol{\theta}) = \sum_i \hat{c}_i|\theta_i|$, with $\hat{c} = 1/|\hat{\boldsymbol{\theta}}|^\gamma$, where $\gamma > 0$ and $\hat{\boldsymbol{\theta}}$ is a consistent estimator to $\boldsymbol{\theta}$. In this way, the strength for penalizing different parameters may be different, depending on the magnitude of their estimate. It was shown that under some regularity conditions and the condition $\lambda_T/\sqrt{T} \to 0$ and $\lambda_T T^{(\gamma-1)/2} \to \infty$ (the subscript T in λ_T is used to indicate the dependence of λ on T), the adaptive Lasso estimate is consistent in model selection.

3.2 Relating Adaptive Lasso to Information Criteria

Let us focus on the case $\gamma = 1$ of adaptive Lasso, meaning that

$$p_\lambda(\theta_i) = \hat{c}_i|\theta_i| = |\theta_i|/|\hat{\theta}_i|, \qquad (4)$$

where $\hat{\boldsymbol{\theta}}$ can be any consistent estimator, e.g., the maximum likelihood estimator. After the convergence of the adaptive Lasso procedure, insignificant parameters become zero, and $p_\lambda(\theta_i) = 0$ for such parameters. On the other hand, the "oracle property" [2] holds for adaptive Lasso with suitable λ, meaning that the pointwise asymptotic distribution of the estimators is the same as if the true underlying model were given in advance. Significant parameters are then changed very little by the penalty, when the sample size is not small. Consequently, at convergence, $p_\lambda(\theta_i) = |\hat{\theta}_{i,ALasso}|/|\hat{\theta}_i| \approx 1$ for non-zero parameters, where $\hat{\theta}_{i,ALasso}$ denotes the adaptive Lasso estimator. In other words, the penalty $p_\lambda(\theta_i)$ *indicates whether the parameter θ_i is active or not.* Suppose the parameters considered are not redundant. $\sum_i p_\lambda(\theta_i)$ is then an approximator of the number of free parameters, denoted by D, in the resulting model. Recall that the traditional information criteria for model selection can be written as

$$\mathrm{IC}_D = -L(\hat{\boldsymbol{\theta}}_{D,ML}) + \lambda_{IC} D \qquad (5)$$

The BIC [7] and AIC [1] criteria are obtained by setting the value of λ_{IC} to

$$\lambda_{BIC} = 1/2 \cdot \log T, \quad \text{and} \quad \lambda_{AIC} = 1, \qquad (6)$$

respectively. Relating Eq. 5 to the penalized likelihood Eq. 3, one can see that by setting λ in adaptive Lasso considered here to λ_{IC} in Eq. 5 (which may be λ_{BIC}, λ_{AIC}, etc.), the model selection result of adaptive Lasso would be consistent with that obtained by minimizing the information criterion corresponding to λ_{IC}.

We give the following remarks for model selection based on adaptive Lasso. First, when the initialized model is very large (i.e., it involves very many parameters), $\hat{\boldsymbol{\theta}}$ may be too rough due to finite sample effects, and it is useful to update $\hat{\boldsymbol{\theta}}$ using a consistent estimator when a smaller model is derived. Second, in practice, especially when the sample size is not large, adaptive Lasso still causes bias in the estimate of significant parameters: usually the adaptive Lasso estimator still gives $p_\lambda(\theta_i) = |\hat{\theta}_{i,ALasso}|/|\hat{\theta}_i| < 1$ for significant parameters. Therefore, at convergence, the penalty $p_\lambda(\boldsymbol{\theta}) = \sum_i p_\lambda(\theta_i)$ is expected to be a little smaller than the number of parameters that are set to zero. To achieve that, we should give a heavier weight for the penalization term. That is, λ_T should be a little larger than the recommended values given above. (Or equivalently, $\hat{c}_i$ should be a little larger than $1/|\hat{\theta}_i|$.) In our experiments, we set $\lambda = 1.5\lambda_{BIC} = \frac{1.5}{2}\log T$ to achieve the BIC-like model selection.

3.3   ICA with Sparse Connections Based on Adaptive Lasso

It is obvious that without specifying the variance of y_i or specifying certain entries of $\mathbf{W}$ (or $\mathbf{A}$), applying adaptive Lasso will make the involved parameters smaller and smaller. To avoid that, we can either normalize the variance of y_i in each iteration or enforce $E\{y_i^2\} = 1$ by using Eq. 2 as the objective function. Here we adopt the former scheme. Consequently, the objective function for ICA with a sparse de-mixing matrix is the penalized likelihood:

$$pL_T = \sum_{t=1}^{T}\sum_{i=1}^{n} \log f_i(y_{i,t}) + T\log|\det \mathbf{W}| - \lambda \sum_{i,j=1}^{n} |w_{ij}|/|\hat{w}_{ij}|, \qquad (7)$$

where $\hat{w}_{ij}$ are entries of $\hat{\mathbf{W}}$, which is an estimate of $\mathbf{W}$ obtained by conventional ICA. Similarly, replacing $|w_{ij}|/|\hat{w}_{ij}|$ in Eq. 7 with $|a_{ij}|/|\hat{a}_{ij}|$ will produce ICA with a sparse mixing matrix. Note that unlike other methods, here λ is easily determined, based on the relationship between adaptive Lasso and information criteria discussed in Subsection 3.2

Now we aim to maximize the above penalized likelihood. Since the L_1 function is not differentiable at 0, gradient-based methods could not be directly applied for optimization involving L_1 penalties. Most existing methods for such optimization are not easy to implement or could not set insignificant parameters to 0 exactly. We propose a very simple but effective way for this problem.

3.4   A Simple Approach for Optimization Involving L_1 Penalties

The difficulties in optimization involving L_1 penalties are caused by the "sudden change" of the L_1 function. We can then consider such penalties as ravines that are parallel to some axes. The so-called adaptive step size technique [9], which was originally proposed for accelerating the optimization procedure in neural networks learning, can then be exploited for optimization involving such penalties. Note that for a ravine in the objective function parallel to an axis, use of an appropriate individual step size is equivalent to re-scaling the ravine. Moreover, if two successive updates of a given parameter are performed in the same/opposite directions, the step size should be increased/decreased. Consequently, the parameters that should be shrunk to 0 by L_1 penalties will gradually stop oscillation and converge to 0, due to the diminishing step size.

Suppose we aim to maximize the objective function J (Eq. 7, in this case). With an adaptive step size, the change of the parameter θ_i in the kth iteration is given by $\triangle\theta_i^{(k)} = \eta_i^{(k)}(\frac{\partial J}{\partial\theta_i})^{(k)}$, where the step size for parameter θ_i depends on the successive signs of the gradient: $\eta_i^{(k)} = \eta_i^{(k-1)}u$, if $(\frac{\partial J}{\partial\theta_i})^{(k)} \cdot (\frac{\partial J}{\partial\theta_i})^{(k-1)} > 0$, and $\eta_i^{(k)} = \eta_i^{(k-1)}d$, if $(\frac{\partial J}{\partial\theta_i})^{(k)} \cdot (\frac{\partial J}{\partial\theta_i})^{(k-1)} < 0$, with $u > 1$ and $d < 1$. We used $u = 1.1$ and $d = 0.5$ in experiments, and found that they work quite well.

4   ICA with Sparse Connections Based on Optimal Brain Surgeon

Sometimes we may want to obtain the solution path of ICA with sparse connections, which gives all possible solutions with different sparsity levels we are interested in. This can be achieved by using optimal brain surgeon (OBS) [3] for network pruning. We further show the relationship between the stopping criterion of OBS and traditional information criteria, and show how to make OBS produce similar results as information criteria do.

4.1   Optimal Brain Surgeon

Suppose we aim to maximize the objective function J. Assuming that the change of J around its (local) optimum is nearly quadratic in the perturbation of its parameters, i.e., $\delta J = -\frac{1}{2}\delta\boldsymbol{\theta}^T \mathbf{H}\delta\boldsymbol{\theta}$, where $\delta\boldsymbol{\theta}$ denotes the perturbation of the parameters and $-\mathbf{H}$ is the Hessian matrix (containing all second order derivatives). We are looking for a set of parameters whose deletion causes the least change in the value of J.

Mathematically, the least change in J caused by eliminating θ_q can be written as $\min_{\delta\boldsymbol{\theta}}\{\frac{1}{2}\delta\boldsymbol{\theta}^T\mathbf{H}\delta\boldsymbol{\theta}\}$, subject to $\mathbf{e}_q^T\delta\boldsymbol{\theta} + \theta_q = 0$, where $\mathbf{e}_q$ is the unit vector with only the qth element being 1. Using the Lagrangian multiplier, one can find that the optimal weight change and the resulting change in J are

$$\delta\boldsymbol{\theta} = -\frac{\theta_q}{[\mathbf{H}^{-1}]_{qq}}\mathbf{H}^{-1}\cdot\mathbf{w}_q, \quad \text{and} \quad S_q = \frac{1}{2}\frac{\theta_q^2}{[\mathbf{H}^{-1}]_{qq}}, \qquad (8)$$

where S_q is called the saliency of the θ_q. OBS finds the q that gives the smallest saliency and prunes it. If J is not very close to quadratic, one needs to adjust the remaining parameters to maximize J, after pruning a parameter and recalculating other parameters according to Eq. 8. We repeat the above pruning procedure util the smallest saliency of remaining parameters is larger than T_h, a threshold whose determination is discussed below. One advantage of OBS is that it does not cause any bias in the estimate of the remaining parameters.

4.2   Relating Stopping Criterion of OBS to Information Criteria

Suppose the objective function J is the log-likelihood of the data. We make the following assumptions. *1.* The information criterion Eq. 5 for model selection has no local minimum. *2.* For the OBS procedure, J is well approximated by a quadratic form, and no parameter pruned earlier becomes significant in a smaller model. Under assumption 1, the model selected by minimizing the information criterion has D^* free parameters if $IC_{D^*} > IC_{D^*-1}$ and $IC_{D^*} > IC_{D^*+1}$. According to Eq. 5, this gives $L(\hat{\boldsymbol{\theta}}_{D^*+1,ML}) - L(\hat{\boldsymbol{\theta}}_{D^*,ML}) < \lambda_{IC}$ while $L(\hat{\boldsymbol{\theta}}_{D^*,ML}) - L(\hat{\boldsymbol{\theta}}_{D^*-1,ML}) > \lambda_{IC}$. Assumption 2 implies that $L(\hat{\boldsymbol{\theta}}_{D+1,ML}) - L(\hat{\boldsymbol{\theta}}_{D,ML})$ is actually the smallest saliency S_q when we eliminate a parameter among all the $D+1$ parameters. One can then see that *by setting the threshold T_h for stopping the OBS procedure to λ_{IC} in the information criterion, OBS gives the same model selection result as the corresponding information criterion does.*

4.3   ICA with Entries Pruned by OBS

Entries of the ICA transformation matrix can be pruned by OBS, with Eq. 2 as the objective function. Note that Eq. 2 has incorporated the constraint on the scale of y_i. Due to space limitation, the calculation of the Hessian matrix, as well as how to avoid the heavy computational load in calculating its inverse, is not given here. Note that the quadratic approximation of the objective function may not be very accurate. Consequently, after pruning a parameter and updating others according to Eq. 8, one needs to update remaining parameters to reach the (local) optimum, by making use of the gradient of Eq. 2; alternatively, the Newton-Raphson method may be adopted, as the Hessian matrix has been calculated in the OBS stage.

5   Experiments

We first illustrate the performance of the proposed methods by simulation studies. These methods can carry out both ICA with a sparse de-mixing matrix and ICA with a sparse mixing matrix. Here only the former is demonstrated. We randomly generated a 5×5 lower-triangular matrix $\mathbf{W}$. The magnitude of its non-zero entries is uniformly distributed between 0.1 and 1. The mixing matrix was constructed as $\mathbf{A} = \mathbf{W}^{-1}$. The five sources s_i were obtained by passing independent Gaussian i.i.d. signals through power nonlinearities with the exponent between 1.5 and 2. The variances of the sources are randomly chosen between 0.2 and 1. The observations were generated by $\mathbf{x} = \mathbf{A s}$. We examined two cases in which the sample size is 200 and 500, respectively. ICA with a sparse de-mixing matrix based on adaptive Lasso and that based on OBS, proposed above, were used to separate such mixtures. To make their results similar to that given by BIC, for the former method, we set the penalization parameter $\lambda = 1.5\lambda_{BIC} = \frac{1.5}{2} \log T$, and for the latter one, we set the threshold $T_h = \lambda_{BIC} = \frac{\log T}{2}$.

We repeated the simulation for 40 trials. The percentages of lost connections (non-zero connections that were wrongly set to 0) and spurious connections (zero entries that were not set to 0) are summarized in Table 1. One can see that there are very few entries of $\mathbf{W}$ wrongly identified. As the sample size increases, the error rate diminishes. This coincides with the fact the BIC is consistent in model selection. Fig. 1(a) gives some of w_{ij} in the training process of ICA with sparse $\mathbf{W}$ based on adaptive Lasso in a typical run, while (b) plots the solution path of the OBS-based method for small parameters in a typical run (it shows the

Table 1. Percentages of lost non-zero entries and spurious connections (40 trials)

Sample size T	200		500	
Method	ALasso-based	OBS-based	ALasso-based	OBS-based
Lost connections (%)	2.83%	5%	1%	1.17%
Spurious connections (%)	3%	2%	0.33%	0.25%

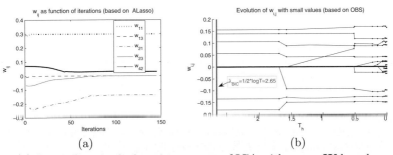

(a)   (b)

Fig. 1. (a) Some of w_{ij} in the learning process of ICA with sparse $\mathbf{W}$ based on adaptive Lasso (T=200). (b) A typical solution path of ICA with sparse $\mathbf{W}$ based on OBS (T=200). For clarity, only small weigts are shown.

solution of w_{ij} for each possible T_h between 0 and λ_{BIC}). Clearly the pruning result does not depend solely on the magnitudes of the parameters.

We also applied the proposed methods for ICA with sparse $\mathbf{W}$ to separate the 14-dimensional financial returns used in [12]. To obtain BIC-like model selection results, we used the same settings as in the simulations above. The resulting $\mathbf{W}$ could not be permuted to lower triangularity, meaning that LiNGAM [8] does not hold for this data set. This is consistent with the claim in [12].

6 Conclusion and Discussions

We have proposed two methods to perform ICA with a sparse transformation matrix (the mixing matrix $\mathbf{A}$ or de-mixing matrix $\mathbf{W}$). The methods are based on the adaptive Lasso penalty and the optimal brain surgeon technique, respectively. We have shown how to relate the proposed methods to model selection based on traditional information criteria (e.g., BIC and AIC). The proposed methods involve comparatively light computational load, and most importantly, one can easily determine the parameters that control the level of sparsity to make the model selection results consistent with those based on information criteria.

References

1. Akaike, H.: Information theory and an extension of the maximum likelihood principle. In: Proc. 2nd Int. Symp. on Information Theory, pp. 267–281 (1973)
2. Fan, J., Li, R.: Variable selection via nonconcave penalized likelihood and its oracle properties. J. Amer. Statist. Assoc. 96, 1348–1360 (2001)
3. Hassibi, B., Stork, D.G.: Second order derivatives for network pruning: Optimal brain surgeon. In: NIPS 5, pp. 164–171. Morgan Kaufmann, San Francisco (1993)
4. Hyvärinen, A., Karhunen, J., Oja, E.: Independent Component Analysis. John Wiley & Sons, Inc., Chichester (2001)
5. Hyvärinen, A., Karthikesh, R.: Imposing sparsity on the mixing matrix in independent component analysis. Neurocomputing 49, 151–162 (2002)
6. Pham, D.T., Garat, P.: Blind separation of mixture of independent sources through a quasi-maximum likelihood approach. IEEE Trans. on Signal Processing 45(7), 1712–1725 (1997)
7. Schwarz, G.: Estimating the dimension of a model. The Annals of Statistics 6, 461–464 (1978)
8. Shimizu, S., Hoyer, P.O., Hyvärinen, A., Kerminen, A.J.: A linear non-Gaussian acyclic model for causal discovery. JMLR 7, 2003–2030 (2006)
9. Silva, F.M., Almeida, L.B.: Acceleration techniques for the backpropagation algorithm. In: Neural Networks, pp. 110–119. Springer, Heidelberg (1990)
10. Tibshirani, R.: Regression shrinkage and selection via the lasso. Journal of the Royal Statistical Society 58(1), 267–288 (1996)
11. Zhang, K., Chan, L.-W.: ICA with sparse connections. In: Corchado, E., Yin, H., Botti, V., Fyfe, C. (eds.) IDEAL 2006. LNCS, vol. 4224, pp. 530–537. Springer, Heidelberg (2006)
12. Zhang, K., Chan, L.: Minimal nonlinear distortion principle for nonlinear independent component analysis. JMLR 9, 2455–2487 (2008)
13. Zhao, P., Yu, B.: On model selection consistency of lasso. JMLR 7, 2541–2563 (2006)
14. Zou, H.: The adaptive lasso and its oracle properties. Journal of the American Statistical Association 101(476), 1417–1429 (2006)

Source Separation of Phase-Locked Subspaces

Miguel Almeida* and Ricardo Vigário

Adaptive Informatics Research Centre
Department of Information and Computer Science
Helsinki University of Technology, Finland
{miguel.almeida,ricardo.vigario}@hut.fi
http://www.hut.fi/en

Abstract. ICA can be interpreted as the minimization of a disorder parameter, such as the sum of the mutual informations between the estimated sources. Following this interpretation, we present a disorder parameter minimization algorithm for the separation of sources organized in subspaces when the phase synchrony within a subspace is high and the phase synchrony across subspaces is low. We demonstrate that a previously reported algorithm for this type of situation has poor performance and present a modified version, called Independent Phase Analysis (IPA), which drastically improves the quality of results. We study the performance of IPA for different numbers of sources and discuss further improvements that are necessary for its application to real data.

Keywords: Phase-locking, synchrony, blind source separation (BSS), independent component analysis (ICA), subspaces, multiple runs, cost smoothness.

1   Introduction

The interest of the scientific community in synchrony phenomena has risen in recent years. Synchrony has been observed in a multitude of oscillating physical processes, including electrical circuits, laser beams and human neurons [1]. This behaviour is usually not due to a strong interaction forcing in-phase oscillations, but rather a consequence of a weak interaction that slowly drifts the relative phase values of the oscillators toward one another.

Our particular motivation for studying synchrony phenomena comes from the human brain. It has been discovered that, during a motor task, several brain regions oscillate coherently with one another [2,3]. There are multiple indications that several pathologies, including Alzheimer, Parkinson and autism, are associated with a disruption in the synchronization profile of the brain (see [4] for a review)

When trying to detect and quantify synchrony in real applications, it is important to have access to the time evolution of the individual oscillators. Otherwise, synchrony measures will not be accurate (midrange values of synchrony become

---

* Corresponding author.

T. Adali et al. (Eds.): ICA 2009, LNCS 5441, pp. 203–210, 2009.

more likely than high or low values). In many real applications such as EEG and MEG, the signals from individual oscillators, henceforth denominated sources, are not directly measurable. Instead, the analyst only has access to a superposition of the sources. For example, in brain electrophysiological signals (EEG and MEG), the signals measured in one sensor contain components coming from several brain regions [5].

Blind source separation (BSS) addresses these issues. The typical instantaneous linear mixing assumption is valid in this situation, because most of the energy is in frequencies below 1 KHz, and the quasi-static approximation of Maxwell's equations holds [6]. However, for example, independence of the sources is not valid, because phase-locked sources are not independent. We now address how to correctly separate the sources in this context to avoid the erroneous detection of spurious synchronization. The separation algorithm we propose uses solely the phase information of the signals, since signals may exhibit synchrony even when their amplitudes are uncorrelated [8].

It should be emphasized that the algorithm presented here assumes nothing specific of brain signals, and should be applicable to any situation where phase-locked sources are mixed approximately linearly and where noise levels are low.

2   Background, Motivation and Algorithm

Given two oscillators with phases $\phi_j(t)$ and $\phi_k(t)$ obtained through the Hilbert transform [9] for $t = 1, \ldots, T$, the Phase Locking Factor (PLF) is defined as

$$\varrho_{jk} = \left| \frac{1}{T} \sum_{t=1}^{T} e^{i[\phi_j(t) - \phi_k(t)]} \right| = \left| \left\langle e^{i(\phi_j - \phi_k)} \right\rangle \right|, \tag{1}$$

where $\langle \cdot \rangle$ denotes a time average operation. It is easy to see that $0 \leq \varrho_{jk} \leq 1$. $\varrho_{jk} = 1$ corresponds to two oscillators that are perfectly synchronized, i.e., that have a constant phase lag. Note that this lag may be non-zero, which allows for a non-instantaneous interaction. $\varrho_{jk} = 0$ is attained if the two oscillators are not synchronized as long as the observation period T is sufficiently long.

It is important to understand the effect of a linear mixture on the phase-locking between signals. Such effect can be intuitively described as "tending toward partial synchrony": If some sources have very low synchrony (PLF ≈ 0), their mixtures have a higher PLF, since each source is present in all mixed signals. If some sources have very high synchrony (PLF ≈ 1), their mixtures have a lower PLF, because each mixed signal has components from sources that were not phase-locked. These statements are illustrated in Fig. 1. Note that significant partial synchrony is present in all pairs of mixture signals.

We now discuss how linearly mixed signals $\mathbf{y}$ can be separated. Denote the extracted sources by $\mathbf{z} = \mathbf{W}^{\mathsf{T}} \mathbf{y}$. One classic formulation of ICA [7] is the minimization of the mutual information I of the extracted sources,

$$I(\mathbf{z}) = \sum_j H[z_j] - H[\mathbf{y}] - \log |\det \mathbf{W}|, \tag{2}$$

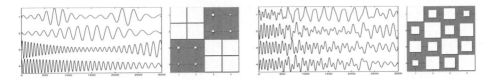

Fig. 1. Simulated data set used for comparing IPA with the logarithm and cosine cost functions: original sources *(far left)* and PLFs between them *(middle left)*; mixed signals *(middle right)* and PLFs between them *(far right)*. The area of each white square is proportional to the corresponding pairwise PLF. Sources 1 and 2 have a mutual constant phase difference, as do sources 3 and 4.

where $\mathbf{W}$ is a matrix to be determined such that $\mathbf{W}^\mathsf{T}\mathbf{A}$ (where $\mathbf{A}$ is the mixing matrix) is close to a permutation of a diagonal matrix. The entropy of the observations $H[\mathbf{y}]$ can be dropped since it does not depend on $\mathbf{W}$. The sum of the estimated source entropies $\sum_j H[z_j]$ can be considered a disorder or complexity measure, and $-\log|\det\mathbf{W}|$ can be viewed as a penalty term preventing degenerate solutions with $\mathbf{W}$ close to singular. This motivated the original formulation of Independent Phase Analysis (IPA) [10]. The disorder parameter P_{log} was given by

$$P_{log} = -\sum_{j,k} \varrho_{jk} \log \varrho_{jk}, \tag{3}$$

where ϱ_{jk} is the PLF between estimated sources j and k. P_{log} is non-negative and attains its minimum value of 0 only if all the ϱ_{jk} are either zero or one.[1] In other words, P_{log} is low when the estimated sources are either non-synchronized or fully synchronized. This suggests that minimization of P_{log} can separate sources that have PLFs close to one or zero. Thus the cost function to be minimized was

$$C_{log} = -\sum_{j,k} \varrho_{jk} \log \varrho_{jk} - \log|\det\mathbf{W}|, \tag{4}$$

where the penalty term $-\log|\det\mathbf{W}|$ prevents degenerate solutions which trivially have $\varrho_{jk} = 1$. Each column of $\mathbf{W}$ is constrained to have unit norm, to prevent trivial decreases of the penalty term. This cost function was used in [10].

We present results showing that this formulation of IPA gives poor results. Using the data set in Fig. 1, a typical solution given by this formulation of IPA is presented in the left half of Fig. 2. The key to understanding these poor results is the presence of a PLF value of zero (between estimated sources 2 and 4), when the real sources do not have any zero values of PLF. Since the disorder parameter P_{log} falls sharply close to $\varrho_{jk} = 0$, this situation is a sharp local minimum of C_{log}. Intuitively, once the algorithm finds a PLF of zero, it "stays there", preventing the optimization of the remaining PLF values.

Given the above considerations we propose that the disorder parameter P should still have minima for $\varrho_{jk} = 0, 1$ but these minima should not be sharp.

---

[1] Note that P_{log} cannot be interpreted as an entropy because $\sum_{j,k} \varrho_{jk} \neq 1$ in general.

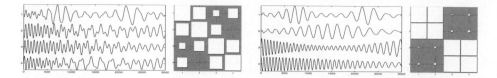

Fig. 2. Results of the IPA algorithm. Estimated sources for the logarithm cost function *(far left)* and PLFs between them *(middle left)*. Estimated sources for the cosine cost function *(middle right)* and PLFs between them *(far right)*. For the results of the cosine cost function, the permutation, scaling and sign of the extracted sources were manually manually. Results with $\lambda = 0.2$. The Amari Performance Index was 0.87 for the logarithm and 0.10 for the cosine cost functions, corresponding to the highest bins of Fig. 3.

Also, we propose that the disorder parameter be normalized regarding the total number of pairs of oscillators N^2 to make it scale better for large N. We thus propose a new disorder parameter, and a corresponding cost function, given by

$$P_{cos} = \frac{1}{N^2} \sum_{j,k} (1 - cos(2\pi \varrho_{jk})) \tag{5}$$

$$C_{cos} = (1 - \lambda)\frac{1}{N^2} \sum_{j,k} (1 - cos(2\pi \varrho_{jk})) - \lambda \log |\det \mathbf{W}|, \tag{6}$$

where λ controls the relative weight of the penalty term versus the disorder parameter. P_{cos} still has minima for $\varrho_{jk} = 0, 1$ but the derivative at those points is zero, allowing for a smoother optimization of the PLF values.

The gradient of C_{cos} relative to an entry w_{ij} of the weight matrix is given by

$$\frac{\partial C_{cos}}{\partial w_{ij}} = 4\pi \sum_{k=1}^{N} [\sin(2\pi \varrho_{jk})] \left\langle \sin(\Psi_{jk} - \Delta\phi_{jk})\frac{Y_i}{Z_j} \sin(\psi_i - \phi_j) \right\rangle - \left[\mathbf{W}^{-\mathsf{T}}\right]_{ij} \tag{7}$$

where ϱ_{jk} is the PLF between estimated sources j and k, $Y_i = |\tilde{y}_i|$ where $\tilde{y}_i$ is the analytic signal of the i-th measurement (obtained from the Hilbert transform [9]), $Z_j = |\tilde{z}_j|$ where $\tilde{z}_j$ is the analytic signal of the j-th estimated source, $\psi_i = \text{angle}(\tilde{y}_i)$ is the phase of the i-th measurement, $\phi_j = \text{angle}(\tilde{z}_j)$ is the phase of the j-th estimated source, $\Delta\phi_{jk} = \phi_j - \phi_k$ is the phase difference of estimated sources j and k, $\Psi_{jk} = \text{angle}\left(\langle e^{i\Delta\phi_{jk}}\rangle\right)$ is the average phase difference between estimated sources j and k, and $\left[\mathbf{W}^{-\mathsf{T}}\right]_{ij}$ is the (i, j) element of the inverse of $\mathbf{W}^\mathsf{T}$. Each column of $\mathbf{W}$ must be constrained to have unit norm to prevent trivial decreases of the penalty term.

3   Results

We present results that demonstrate that this smoother cost function drastically improves the quality of the separation. Gradient descent with adaptive step

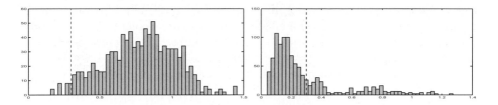

Fig. 3. Histogram of the Amari Performance Index (API) for the logarithm cost function *(left)* and for the cosine cost function *(right)*, and threshold for good separations (Amari Performance Index = 0.3, *dashed vertical line*). The logarithm cost function achieves only 2% of good separations, versus 71% for the cosine cost function. Results for $\lambda = 0.2$.

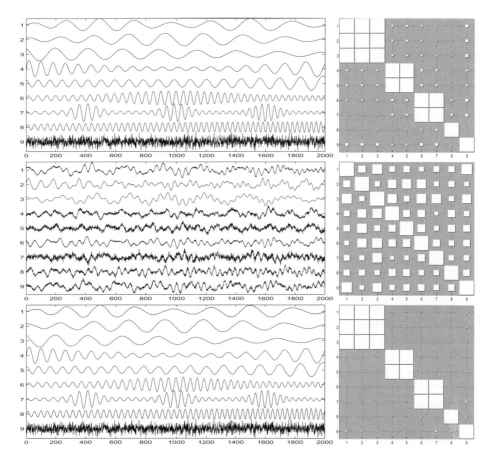

Fig. 4. Results of one run of IPA: original sources *(top left)* and PLFs between them *(top right)*; mixed signals *(middle left)* and PLFs between them *(middle right)*; extracted sources *(bottom left)* and PLFs between them *(bottom right)*. Results obtained for $\lambda = 0.2$, after manually compensating for permutation, scaling and sign of the extracted sources. The Amari Performance Index was 0.06.

sizes was used to perform the optimization, using MATLAB. This takes about 10 minutes on a 1.6 GHz machine. The data used are depicted in Fig. 1. We artificially simulate a noiseless mixture of four sources in two clusters of size 2 each. The mixing matrix was chosen randomly but constrained to be orthogonal (this is not strictly necessary; see Sec. 4), producing the measurements in the bottom of Fig. 1. We then run the IPA algorithm 1000 times for each of the two cost functions with random initializations of $\mathbf{W}$.

Typical results are shown, in Fig. 2, for the logarithm and the cosine cost function. To assess the quality of the source extraction, we use the Amari performance index (API) [11], which measures how close $\mathbf{W}^T\mathbf{A}$ is to a permutation of a diagonal matrix. The API is non-negative and the lower its value, the better the separation quality. We show histograms of the API for the logarithm and cosine cost functions in Fig. 3. The histograms show that the use of C_{cos} yields a drastic improvement in the quality of the results, relative to C_{log}. If we consider that an API below 0.3 indicates a good separation, in 71% of the tests using C_{cos} yielded good results, compared to just 2% for C_{log}.

A second experiment involved a larger number of sources with a more complex cluster structure. The data used and the corresponding results are shown in Fig. 4. It can be seen that the source separation was very good (API = 0.06). Naturally, since the number of parameters to optimize is N^2, the optimization procedure took considerably longer in this case, but on a 1.6 GHz machine it still took less than an hour. The majority of the runs yielded good separations, although the percentage of good separations is slightly less than for the small data set.

4   Discussion

These results show that IPA with the cosine cost function can successfully extract mixed sources based on their phase synchronization properties. The results of the multiple runs show also that most of the runs yield good separations, and the few times that IPA yields bad results can be circumvented by running it a few times and keeping the most consistent solutions.

It could be argued that since the used sources have distinct frequencies, a simple frequency filtering algorithm could separate the signals. Such a procedure would not be able to disambiguate the signals within each cluster. Also, in real applications, IPA should prove useful for signals which are not synchronized but have overlapping frequency spectra. It could also be argued that other source separation techniques are able to separate this kind of signals. We investigated this for FastICA [7] and TDSEP [12] on the small data set (results are not shown due to lack of space). FastICA fails to separate the sources because they are not independent. TDSEP sometimes gives results as good as IPA and sometimes not, depending on the specific sources used. This is only an empiric finding so far and we are actively studying which sources cause TDSEP to fail.

The results presented in Fig. 4 suggest that IPA will not stop when the correct PLF values are found, but will actually overtrain and yield more extreme values, although the quality of the separation is unharmed. In fact, preliminary results (not shown) suggest that IPA's performance is quite good if the sources' PLFs are above approximately 0.9 or below approximately 0.1. The decrease of performance of IPA for sources with PLF values far from 1 or 0 suggests that further improvements are required before IPA is applicable to noisy mixtures in real situations, where a) noise corrupts our estimate of the PLF values and b) even the true, noiseless underlying sources are probably not fully synchronized or fully desynchronized.

IPA has only one parameter, λ, which controls the relative weight of the disorder parameter and the penalty term. We found the algorithm to be very robust to changes in this parameter, with values between $\lambda = 0.1$ and $\lambda = 0.7$ yielding similar results. For example, using $\lambda = 0.5$ yields 65% of good separations instead of 71%. Note that if $\lambda = 0$, and if the sources have PLFs of 1 or 0, the trivial solutions with $\mathbf{W}$ singular have the same cost value as the correct solution. Therefore, even small values of λ are enough to differentiate these two situations.

Interestingly, the optimization procedure has two different time scales: usually, the first tens of iterations of gradient descent are enough to separate the subspaces from one another (up to around 20 iterations for $N = 4$, and around 100 for $N = 9$). Even for $N = 9$ this usually takes no more than a few minutes. After this initial phase, several hundred iterations of gradient descent are necessary to separate the sources within each subspace, and convergence is slower. This suggests that using more advanced optimization algorithms might prove very useful in speeding up IPA and refining the extracted sources.

Currently, IPA works well for mixing matrices not far from orthogonal, in particular for matrices with a low value of the eigenvalue ratio $\frac{|\text{eig}_{max}|}{|\text{eig}_{min}|}$, but its performance decays for matrices with higher ratios. We believe this limitation can be avoided by making the penalty term depend on the extracted sources alone instead of the demixing matrix $\mathbf{W}$.

5   Conclusion

We have presented an algorithm to separate phase-locked sources from linear mixtures. We have shown that using a cosine cost function in IPA yields drastically better separation results than those of the previous version, making IPA a valid choice for separation of sources that are either synchronized or de-synchronized, in noise-free situations. Further improvements are necessary to improve convergence speed and to deal with more complex real-world applications.

Acknowledgments. MA is funded by scholarship SFRH/BD/28834/2006 of the Portuguese Foundation for Science and Technology. This study was partially funded by the Academy of Finland through its Centres of Excellence Program 2006-2011.

References

1. Pikovsky, A., Rosenblum, M., Kurths, J.: Synchronization: A Universal Concept in Nonlinear Sciences. Cambridge University Press, Cambridge (2001)
2. Palva, J.M., Palva, S., Kaila, K.: Phase Synchrony Among Neuronal Oscillations in the Human Cortex. Journal of Neuroscience 25, 3962–3972 (2005)
3. Schoffelen, J.M., Oostenveld, R., Fries, P.: Imaging the Human Motor System's Beta-Band Synchronization During Isometric Contraction. NeuroImage 41, 437–447 (2008)
4. Uhlhaas, P.J., Singer, W.: Neural Synchrony in Brain Disorders: Relevance for Cognitive Dysfunctions and Pathophysiology. Neuron 52, 155–168 (2006)
5. Nunez, P.L., Srinivasan, R., Westdorp, A.F., Wijesinghe, R.S., Tucker, D.M., Silberstein, R.B., Cadusch, P.J.: EEG Coherency I: Statistics, Reference Electrode, Volume Conduction, Laplacians, Cortical Imaging, and Interpretation at Multiple Scales. Electroencephalography and clinical Neurophysiology 103, 499–515 (1997)
6. Vigário, R., Särelä, J., Jousmäki, V., Hämäläinen, M., Oja, E.: Independent Component Approach to the Analysis of EEG and MEG Recordings. IEEE Transactions On Biomedical Engineering 47, 589–593 (2000)
7. Hyvärinen, A., Karhunen, J., Oja, E.: Independent Component Analysis. John Wiley & Sons, Chichester (2001)
8. Rosenblum, M.G., Pikovsky, A.S., Kurths, J.: Phase Synchronization of Chaotic Oscillators. Physical Review Letters 76, 1804–1807 (1996)
9. Oppenheim, A.V., Schafer, R.W., Buck, J.R.: Discrete-Time Signal Processing. Prentice-Hall International Editions, Englewood Cliffs (1999)
10. Schleimer, J.H., Vigário, R.: Order in Complex Systems of Nonlinear Oscillators: Phase Locked Subspaces. In: Proceedings of the European Symposium on Neural Networks (2007)
11. Amari, S., Cichocki, A., Yang, H.H.: A New Learning Algorithm for Blind Signal Separation. Advances in Neural Information Processing Systems 8, 757–763 (1996)
12. Ziehe, A., Müller, K.-R.: TDSEP - an Efficient Algorithm for Blind Separation Using Time Structure. In: Proceedings of the International Conference on Artificial Neural Networks (1998)

Online Subband Blind Source Separation for Convolutive Mixtures Using a Uniform Filter Bank with Critical Sampling

Paulo B. Batalheiro[1], Mariane R. Petraglia[2], and Diego B. Haddad[3]

[1] State University of Rio de Janeiro, CTC/FEN/DETEL
20559-900, Rio de Janeiro, Brazil
bulkool@pads.ufrj.br
[2] Federal University of Rio de Janeiro, PEE/COPPE
CP 68504, 21945-970, Rio de Janeiro, Brazil
mariane@pads.ufrj.br
[3] Federal Center for Technological Education, CEFET-RJ
CP 26041-271, Nova Iguau, RJ, Brazil
diego@pads.ufrj.br

Abstract. Adaptive subband structures have been proposed with the objective of increasing the convergence speed and/or reducing the computational complexity of adaptation algorithms for applications which require a large number of adaptive coefficients. In this paper we propose an online blind source separation method for convolutive mixtures which employs a real-coefficient uniform subband structure with critical sampling and extra filters that cancel aliasing between adjacent channels. Since the separation filters in the subbands work at reduced sampling rates, the proposed method presents smaller computational complexity and larger steady-state signal to noise interference ratio when compared to the corresponding fullband algorithm.

1 Introduction

Blind source separation (BSS) techniques have been extensively investigated in the last decade, allowing the extraction of the signal of a desired source $s_q(n)$ from mixed signals of more than one source $x_p(n)$ without any other knowledge of the original sources, such as their positions or spectral contents, nor of the mixing process. Examples of applications of BSS are speech enhancement/recognition (*cocktail party* problem) and digital communication, among others. The mixtures can be classified as linear or non-linear and instantaneous or convolutive. This paper considers convolutive mixtures of speech signals, which takes into account the reverberation in echoic ambients. In such cases, finite impulse response (FIR) separation filters of large orders are usually required, making the separation task very complex. In order to solve such problem, several time-domain and frequency-domain methods based on independent component analysis (ICA) have been proposed in the literature.

T. Adali et al. (Eds.): ICA 2009, LNCS 5441, pp. 211–218, 2009.
© Springer-Verlag Berlin Heidelberg 2009

Some of these solutions employ FIR separation filters, whose coefficients are estimated with an ICA-based algorithm directly in the time-domain. In real applications, the separation filters have thousands of coefficients and, therefore, the BSS algorithms present large computational complexity, slow convergence and undesired whitening effect in the sources estimations [1]. In order to ease such difficulties, frequency-domain BSS methods were proposed, where the convolutions become products, and the convolutive mixtures can be treated as instantaneous mixtures in each frequency bin [2]. The disadvantages of such methods are the scaling and permutation problems among the bins, besides the need of using long windows of data for implementing high-order filters. Due to the non-stationarity of the speech signals and mixing systems, the estimates of the needed statistics for each bin might not be correct for long window data. Such disadvantages can degrade severely the performance of the frequency-domain algorithms. There are also techniques which combine the time and frequency domain solutions to improve the BSS performance and reduce its computational complexity [3]. In this scenery, subband methods have been proposed mainly due to their characteristics of breaking the high-order separation filters into independent smaller-order filters and of allowing the reduction of the sampling rate. Such methods usually employ oversampled uniform filter banks with complex-coefficients [4].

In this paper we propose an online subband BSS method which employs real-coefficients critically-sampled uniform filter banks and reduced-order separation FIR filters. The separation filters coefficients at the different subbands are adjusted independently by a time-domain adaptation algorithm [1], which employs second-order statistics. Extra filters are used in the proposed subband algorithm in order to cancel aliasing among adjacent bands [5].

2   BSS for Convolutive Mixtures

Fig. 1 illustrates a blind source separation system, where the number of sources is equal to the number of microphones. Considering that the unknown mixture system can be modeled by a set of FIR filters of length U (convolutive linear mixtures), the signals captured by the microphones $x_p(n)$ can be written as

$$x_p(n) = \sum_{q=1}^{P} \sum_{k=0}^{U-1} g_{pq}(k)s_q(n-k) \tag{1}$$

where y_{pq} is the filter that models the echo path from the qth source to the pth sensor, and P is the number of sources and sensors (determined BSS).

In the BSS problem, the coefficients of the separation filters w_{qp} (of length S) are estimated through an adaptive algorithm, so that their output signals become mutually independent, with the qth output given by

$$y_q(n) = \sum_{p=1}^{P} \sum_{k=0}^{S-1} w_{qp}(k)x_p(n-k). \tag{2}$$

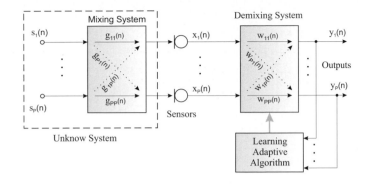

Fig. 1. Linear MIMO configuration for fullband BSS

3   Subband BSS Algorithm for Convolutive Mixtures

Subband BSS methods have been recently proposed with the objective of reducing the computational complexity and improving the adaptation convergence rate, while keeping the aliasing effects negligible and maintaining enough number of samples for estimating the statistics of the subband signals [4]. Usually, such results are achieved using oversampled filter banks. In this paper, we propose the use of a critically sampled structure that uses extra filters in the decomposition of the observed signals for canceling the aliasing among adjacent subbands [5]. The idea is to exploit the characteristics of better convergence rate and reduced computational complexity of the corresponding adaptive subband algorithm [5].

Figure 2 shows the kth channel of the linear TITO (two sources and two sensors) configuration for the M-channel subband BSS. In this structure each observed signal $(x_q(n))$ is decomposed by direct-path filters $(h_{k,k}(n))$ and by extra filters $(h_{k,k-1}(n)$ and $h_{k,k+1}(n))$. The resulting signals $(x_q^{k,k}(n)$, $x_q^{k,k-1}(n)$ and $x_q^{k,k+1}(n))$ are down-sampled by the critical decimation factor (M) and passed by separation filters $(w_{qp}^k(n)$, $w_{qp}^{k-1}(n)$ and $w_{qp}^{k+1}(n))$. The corresponding output signals are up-sampled and recombined by the synthesis filters $(f_k(n))$ to restore the fullband output signals (estimated sources, $y_q(n)$). Considering perfect reconstruction (PR) filter banks, this structure is able of exactly modeling any FIR unmixing system [5].

Assuming that $h_p(n)$ is the impulse response of a prototype filter of length N_P that allows perfect reconstruction in a cosine modulated multirate system of M bands [6], the analysis and synthesis filters are given, respectively, by [7]:

$$h_k(n) = 2h_p(n)\cos\left[\frac{\pi}{M}(k+0.5)(n - \frac{N_P - 1}{2}) + \theta_k\right], \tag{3}$$

$$f_k(n) = 2h_p(n)\cos\left[\frac{\pi}{M}(k+0.5)(n - \frac{N_P - 1}{2}) - \theta_k\right], \tag{4}$$

where $\theta_k = (-1)^k \frac{\pi}{4}$, for $0 \le k \le M - 1$ and $0 \le n \le N_P - 1$. The filters $h_{k,i}(n)$ of Fig. 2, which decompose the observed signals $x_q(n)$, have impulse responses

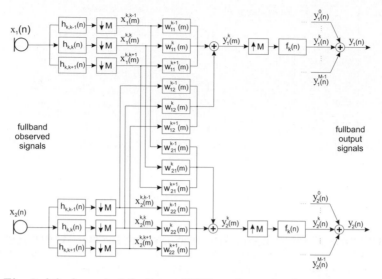

Fig. 2. kth channel of the linear TITO configuration for subband BSS

$h_{k,i}(n) = h_k(n) * h_i(n)$. The number of coefficients of each separation subfilter at the kth subband should be at least $K = \left\lceil \frac{S+N_P}{M} \right\rceil$ [5].

The observed subband signals of Fig. 2 can be expressed as

$$x_p^{k,i}(m) = \mathbf{x}_p^T(m)\mathbf{h}_{k,i}, \tag{5}$$

where $\mathbf{x}_p(m) = [x_p(mM), x_p(mM - 1), \ldots, x_p(mM - N_H + 1)]^T$ is the vector that contains the latest $N_H = 2N_p - 1$ samples of the pth sensor signal and $\mathbf{h}_{k,i} = [h_{k,i}(0), h_{k,i}(1), \ldots, h_{k,i}(N_H - 1)]^T$ is the vector that contains the N_H coefficients of the analysis filter $h_{k,i}(n)$.

For colored and non-stationary signals, such as speech signals, the BSS problem can be solved by diagonalizing the output correlation matrix considering multiple blocks in different time instants (TDD - *Time-Delayed Decorrelation*). In this section we extend the online wideband solution proposed in [1] to the subband domain, employing multirate processing. The method derived in [1] is based on second-order statistics and explores two characteristics of the source signals simultaneously: nonwhiteness and nonstationarity. From Fig. (2), considering that there is no overlap among the frequency responses of non-adjacent filters $h_k(n)$, the qth output signals at the kth subband are given by

$$y_q^k(m) = \sum_{p=1}^{P} \sum_{i=k-1}^{k+1} [\mathbf{w}_{qp}^i]^T \mathbf{x}_p^{k,i}(m), \tag{6}$$

where $\mathbf{x}_p^{k,i}(m)$ is the vector that contains the latest K samples of the pth sensor subband signal $x_p^{k,i}(m)$, and $\mathbf{w}_{qp}^i = [w_{qp}^i(0), w_{qp}^i(1), \ldots, w_{qp}^i(K-1)]^T$ is the vector that contains the K coefficients of the subband separation subfilters $w_{qp}^i(m)$. The vector $\mathbf{x}_p^{k,i}(m)$ can be expressed as

$$\mathbf{x}_p^{k,i}(m) = \mathbf{X}_p(m)\mathbf{h}_{k,i} \tag{7}$$

with the $K \times N_H$ matrix $\mathbf{X}_p(m)$ given by

$$\mathbf{X}_p(m) = \begin{bmatrix} x_p(mM) & x_p(mM-1) & \cdots & x_p(mM-N_H+1) \\ x_p((m-1)M) & x_p((m-1)M-1) & \cdots & x_p((m-1)M-N_H+1) \\ \vdots & \vdots & \ddots & \vdots \\ x_p((m-K+1)M) & x_p((m-K+1)M-1) & \cdots & x_p((m-K+1)M-N_H+1) \end{bmatrix}. \quad (8)$$

In the generic block time-domain subband BSS algorithm, defining N as the block size $(N \geq D)$ and D as the number of blocks which are used in the correlation estimates $(1 \leq D \leq K)$, the kth output vectors of block index ℓ can be written as

$$[\mathbf{y}_q^k(\ell)]^T = [y_q^k(\ell K), y_q^k(\ell K+1), \ldots, y_q^k(\ell K+N-1)] = \sum_{p=1}^{P} \sum_{i=k-1}^{k+1} [\mathbf{w}_{qp}^i]^T \, \hat{\mathbf{X}}_p^{k,i}(\ell) \quad (9)$$

with the $K \times N$ matrix $\hat{\mathbf{X}}_p^{k,i}(\ell) = [\mathbf{X}_p(\ell K), \mathbf{X}_p(\ell K+1), \cdots, \mathbf{X}_p(\ell K+N-1)] \, \mathbf{H}_{k,i}$, where the $N_H N \times N$ matrix $\mathbf{H}_{k,i}$ has the first column formed by the coefficients of $h_{k,i}(n)$ followed by $(N-1)N_H$ zeros, and the following columns are circularly shifted (by N_H positions) versions of the previous columns. The $D \times N$ matrices $\mathbf{Y}_q^k(\ell)$, formed by D subsequent output vectors, can be expressed as

$$\mathbf{Y}_q^k(\ell) = \sum_{p=1}^{P} \sum_{i=k-1}^{k+1} \mathbf{W}_{qp}^i(\ell) \, \mathbf{X}_p^{k,i}(\ell) \quad (10)$$

with $\mathbf{X}_p^{k,i}(\ell) = \left[\hat{\mathbf{X}}_p^{k,i}(\ell), \; \hat{\mathbf{X}}_p^{k,i}(\ell-1) \right]^T$ and $\mathbf{W}_{pq}^i(\ell)$ a $D \times 2K$ Sylvester-type matrix given by

$$\mathbf{W}_{qp}^i(\ell) = \begin{bmatrix} w_{pq}^i(0) \; w_{pq}^i(1) & \cdots & w_{pq}^i(K-1) & 0 & \cdots & 0 & 0 \\ 0 & w_{pq}^i(0) \; w_{pq}^i(1) & \cdots & w_{pq}^i(K-1) \; 0 & \cdots & 0 \\ \vdots & \ddots & \ddots & \ddots & \ddots & \ddots & 0 & 0 \\ 0 & \cdots & 0 & w_{pq}^i(0) & w_{pq}^i(1) & \cdots w_{pq}^i(K-1) \; 0 \end{bmatrix}. \quad (11)$$

Combining the P outputs at each subband, Eq. (10) can be expressed concisely as

$$\mathbf{Y}^k(\ell) = \left[\mathbf{Y}_1^k(\ell), \; \cdots, \; \mathbf{Y}_P^k(\ell) \right]^T = \sum_{i=k-1}^{k+1} \mathbf{W}^i(\ell) \mathbf{X}^{k,i}(\ell), \quad (12)$$

where

$$\mathbf{X}^{k,i}(\ell) = \left[\mathbf{X}_1^{k,i}(\ell), \; \ldots, \; \mathbf{X}_P^{k,i}(\ell) \right]^T, \; \mathbf{W}^i(\ell) = \begin{bmatrix} \mathbf{W}_{11}^i(\ell) & \cdots & \mathbf{W}_{1P}^i(\ell) \\ \vdots & \ddots & \vdots \\ \mathbf{W}_{P1}^i(\ell) & \cdots & \mathbf{W}_{PP}^i(\ell) \end{bmatrix}. \quad (13)$$

In matrix formulation, the online subband BSS cost function is given by

$$\Im^k(\ell) = \log(\det(\mathrm{bdiag}(\mathbf{Y}^k(\ell)[\mathbf{Y}^k(\ell)]^T))) - \log(\det(\mathbf{Y}^k(\ell)[\mathbf{Y}^k(\ell)]^T)), \quad (14)$$

where $\mathrm{bdiag}(\mathbf{A})$ is the operator that zeroes all the submatrices that are not located in the main diagonal of $\mathbf{A}$.

Applying the natural gradient method to the cost function of Eq. (14), we get

$$\nabla_{\mathbf{W}}^{GN} \Im(\ell) = 2\mathbf{W}^k(\ell) \left\{ \mathbf{R}_{yy}^k(\ell) - \text{bdiag}\left(\mathbf{R}_{yy}^k(\ell)\right) \right\} \left\{ \text{bdiag}\left(\mathbf{R}_{yy}^k(\ell)\right) \right\}^{-1} \quad (15)$$

where $\mathbf{R}_{yy}^k(\ell) = \mathbf{Y}^k(\ell)[\mathbf{Y}^k(\ell)]^T$.

The online algorithm for adjusting the coefficients of the separation subfilters, considering a TITO system, is given by (omitting the index ℓ for easing the notation)

$$\mathbf{W}^k(\ell) = \mathbf{W}^k(\ell-1) - 2\mu \begin{bmatrix} [\mathbf{R}_{y_1y_1}^k]^{-1}[\mathbf{R}_{y_2y_1}^k]^T\mathbf{W}_{21}^k & [\mathbf{R}_{y_1y_1}^k]^{-1}[\mathbf{R}_{y_2y_1}^k]^T\mathbf{W}_{22}^k \\ [\mathbf{R}_{y_2y_2}^k]^{-1}[\mathbf{R}_{y_1y_2}^k]^T\mathbf{W}_{11}^k & [\mathbf{R}_{y_2y_2}^k]^{-1}[\mathbf{R}_{y_1y_2}^k]^T\mathbf{W}_{12}^k \end{bmatrix} \quad (16)$$

where $\mathbf{R}_{y_qy_p}^k(\ell)$ of dimension $D \times D$ is a submatrix of $\mathbf{R}_{yy}^k(\ell)$ of Eq. (15) and μ is the step-size of the adaptation algorithm.

Due to redundancies in $\mathbf{W}_{qp}^k(\ell)$ (see Eq. (11)) and for convergence reasons [1], only the first K elements of the first-line of this matrix are updated at each iteration. In order to reduce the computational complexity of the algorithm, the normalization factor $[\mathbf{R}_{y_qy_q}^k(\ell)]^{-1}$ in Eq. (16) can be simplified considering a single output block [8], that is, $\mathbf{R}_{y_qy_q}^k(\ell) \approx [\mathbf{y}_q^k(\ell)]^T\mathbf{y}_q^k(\ell)\mathbf{I}$ with $\mathbf{y}_q^k(\ell)$ given in Eq. (9).

For high-order separation filters and straigthforward implementation, the overall number of multiplications per block (NMPB) required by the proposed subband BSS algorithm, considering only the dominant terms, is given by

$$\text{NMPB}_{SB} \approx \frac{P^2(12MK^3 - 8K^3)}{M}. \quad (17)$$

The corresponding expression for the fullband algorithm is

$$\text{NMPB}_{FB} \approx 4P^2S^3. \quad (18)$$

4   Experimental Results

In all experiments, two speech signals (of 60 s up to 75 s of duration) sampled at $F_s = 16 \ kHz$ were used (a female and a male voice, both in English language). Such signals were convolved with artificial impulse responses, obtained considering a room of dimensions 3.55 m × 4.55 m × 2.5 m (with reverberation time around 250 ms) [9]. Such impulse responses were truncated, considering only their first U samples. The distance between the two microphones was 5 cm and the sources were positioned at 1 m of distance from the center point of the microphones, at directions of -50^0 and 45^0.

In this experiment we compare the performances of the fullband algorithm presented in [1] and of the proposed subband online algorithm (Sec. 3), considering mixture filters of different lengths: $U = 256$, 512 and 1024. The lengths S of the fullband separation filters were fixed at the same value of the mixing filters.

The uniform subband structure was implemented using cosine modulated filter banks with $M = 2$, 4, 8 and 16 bands, $N_p = 16M$, all yielding perfect reconstruction. Table 1 presents the length K of the separation subfilters $w_{pq}^i(m)$ and the adaptation step-sizes used in the fullband ($M = 1$) and subband simulations for

Table 1. Parameters for the fullband and subband algorithms

	Length K of separation subfilters			Step-size
M	$U = 256$	$U = 512$	$U = 1024$	μ
1	256	512	1024	2.5×10^{-4}
2	144	272	528	5×10^{-4}
4	80	144	272	1×10^{-3}
8	48	80	144	2×10^{-3}
16	32	48	80	3×10^{-3}

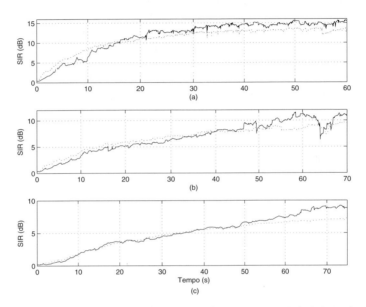

Fig. 3. SIR evolution for fullband (dotted line) and subband (solid line) algorithms with different mixing filters lengths: (a) U=256, (b) U=512, and (c) U=1024

different mixture filters. These step-size values were chosen such that all algorithms presented similar convergence rates. The number of time-lags used in the block correlation calculations was $D = K$ and the length of the output signal blocks was $N = 2K$.

Figure 3 shows the signal to interference ratio (SIR) evolution and Table 2 contains the steady-state SIR (corresponding to their average values over the latest 10 iterations) for fullband and subband algorithms, and present the number of multiplications per block (NMPB) according to Eqs. (17) and (18). From this table and Fig. 3, it can be observed that as the order of the mixture system increases (corresponding to more reverberation), the advantages of the subband structure over the fullband structure become more evident, resulting in a significantly larger final SIR and smaller processing time when compared to the fullband algorithm.

During our experiments we monitored the correlations among the estimates of the sources in the several subbands in order to avoid permutation problems.

Table 2. Number of multiplications per block and Steady-state SIR (in dB)

$S=U$	$M = 1$ NMPB	SIR	$M = 2$ NMPB	SIR	$M = 4$ NMPB	SIR	$M = 8$ NMPB	SIR	$M = 16$ NMPB	SIR
256	2.7×10^8	13.0	9.6×10^7	13.7	2.1×10^7	13.9	4.9×10^6	13.3	1.5×10^6	16.0
512	2.1×10^9	9.7	6.4×10^8	9.0	1.2×10^8	10.0	2.2×10^7	10.0	5.1×10^6	11.9
1024	1.7×10^{10}	7.2	4.7×10^9	6.6	8.1×10^8	7.4	1.3×10^8	7.5	2.3×10^7	8.8

5   Conclusion

In this paper we proposed a new subband blind source separation algorithm that employs uniform filter banks to decompose the signals of the sensors. The separation filters, applied to the subband observed signals, work at the critical sampling rate. The adaptation is performed by a natural-gradient type algorithm in each subband. The use of extra filters in the subband algorithm avoided aliasing effects, resulting in a subband BSS algorithm with steady-state performance similar to the one of fullband algorithm. Computer simulations with speech signals were presented, showing the advantages of the subband structure with respect to processing time and final signal to noise interference ratio over the fullband algorithm.

References

[1] Buchner, H., Aichner, R., Kellermann, W.: A Generalization of Blind Source separation Algorithms for Convolutive Mixtures Based on Second-Order Statistics. IEEE Trans. on Speech and Audio Process 13, 120–134 (2005)
[2] Saruwatari, H., Kawamura, T., Nishikawa, T., Lee, A., Shikano, K.: Blind Source separation Based on a Fast-Convergence Algorithm Combining ICA and Beamforming. IEEE Trans. Audio, Speech, and Language Process 14(2), 666–678 (2006)
[3] Nishikawa, T., Saruwatari, H., Shikano, K.: Comparison of Time-Domain ICA, Frequency-Domain ICA and Multistage ICA for Blind Source Separation. In: Proc. European Signal Processing Conf., pp. 15–18 (2006)
[4] Araki, S., Makino, S., Aichner, R., Nishikawa, T., Saruwatari, H.: Subband-Based Blind Separation for Convolutive Mixtures of speech. IEICE Trans. Fundamentals E88-A, 3593–3603 (2005)
[5] Petraglia, M., Batalheiro, P.: Filter Bank Design for a Subband Adaptive Filtering Structure With Critical Sampling. IEEE Trans. on Circuits And Systems I-Fundamental Theory And Applications 51(6), 1194–1202 (2004)
[6] Nguyen, T.: Digital filter banks design - quadratic constrained formulation. IEEE Transactions on Signal Processing 43, 2103–2108 (1995)
[7] Vaidyanathan, P.: Multirate Systems and Filter Banks. Prentice-Hall, Englewood Cliffs (1993)
[8] Aichner, R., Buchner, H., Kellermann, W.: Exploiting Narrowband Efficiency for Broadband Convolutive Blind Source Separation. EURASIP Journal on Applied Signal 2007, 1–9 (2006)
[9] http://sassec.gforge.inria.fr/

Cebysev Coefficients – Based Algorithm for Estimating the ICA Model

Doru Constantin

University of Pitesti
Department of Computer Science
Address Str. Tg. din Vale, No.1, Romania
cdomanid@yahoo.com

Abstract. The aim of the present paper is to propose a new algorithm
for the estimation of the ICA model, an algorithm based on Cebysev
method. The first sections briefly present the standard FastICA algorithm based on the Newton method and a new version of the FastICA
algorithm. The proposed algorithm to estimate the independent components use a superior order method as Cebysev coefficients technique.
The final section presents the results of a comparative analysis experimentally derived conclusions concerning the performance of the proposed
method. The tests were performed for signals source separation purposes.

Keywords: Independent Component Analysis, Blind Source Separation,
Numerical Methods.

1   Introduction

An important problem arising in signal processing, mathematical statistical and
neural networks is represented by the need of getting adequate representations of
multidimensional data. The problem can be stated in terms of finding a function
f such that the n dimensional transform defined by $s = f(x)$ possesses some
desired properties, where x is a m dimensional random vector. Being given its
computational simplicity, frequently the linear approach is attractive, that is the
transform is

$$s = Wx \tag{1}$$

where W is a matrix to be optimally determined from the point of view of a
pre-established criterion.

There are a long series of methods and principles already proposed in the literature for solving the problem of fitting an adequate linear transform for multidimensional data [1,4], as for instance, Principal Component Analysis (PCA), factor
analysis, projection methods and Independent Component Analysis (ICA).

The aim of Independent Component Analysis is to determine a transform
such that the components $s_i, i = \overline{1..n}$ becomes statistically independent, or at
least almost statistically independent. In order to find a suitable linear transform
to assure that (1) $s_i, i = \overline{1..n}$ become 'nearly' statistically independent several

T. Adali et al. (Eds.): ICA 2009, LNCS 5441, pp. 219–226, 2009.

methods have been developed so far. Some of them, as for instance Principal Component Analysis and factor analysis are second order approaches, that is they use exclusively the information contained by the covariance matrix of the random vector x, some of them, as for instance the projection methods and blind deconvolution are higher order methods that use an additional information to reduce the redundancies in the data. Independent Component Analysis has became one of the most promising approaches in this respect and, consists in the estimation of the generative model of the form $x = As$, where the $s = (s_1 s_2, \ldots s_n)^T$ are supposed to be independent, and A is the mixing matrix $m \times n-$ dimensional of the model. The data model estimation in the framework of independent component analysis is stated in terms of a variational problem formulated on a given objective function. The aim of the research reported in this paper is to introduce a new version of the FastICA algorithm; an algorithm that is based on Cebysev coefficients and to analyze the performances of the algorithm in signal applications.

2   Fixed-Point ICA Based on Cebysev Coefficients

2.1   Cebysev Method for the ICA Model

We consider the following equation:

$$f(w) = 0 \tag{2}$$

where f is a function, $f : I \to R, I = [w_a, w_b]$, I is an real interval and suppose that $f \in C([w_a, w_b])$, where $f \in C^n([w_a, w_b])$ means that the function f have continuous derivatives of order n.

Let $f \in C^{p+1}([w_a, w_b])$ and suppose that $w^* \in [w_a, w_b]$ is a solution for equation (2). We also suppose that $f'(w) \neq 0$ on the interval $[w_a, w_b]$. Then there is an inverse function of f denoted by φ and defined on $[f(w_a), f(w_b)]$. The function φ is a differentiable function by order $p + 1$.

Using the Taylor formula we have:

$$\varphi(z) = \varphi(y) + \sum_{i=1}^{p} \frac{\varphi^{(i)}(y)}{i!}(z - y)^i + \frac{\varphi^{(p+1)}(\eta)}{(p+1)!}(z - y)^{p+1}, \ \eta \in (y, z) \tag{3}$$

Taken $z = 0$ and $y = f(w)$ we observe that $\varphi(0) = w^*$, because $f(w) = y \Leftrightarrow w = \varphi(y)$ and $\eta \subset (f(w), f(w^*))$. Thus, there is one $\xi \in (w, w^*)$ with $f(\xi) = \eta$. Denoting by $\theta_i(w) = \varphi^i[f(w)]$ the Cebysev coefficients, we obtain:

$$w^* = w + \sum_{i=1}^{p}(-1)^i \frac{\theta_i(w)}{i!}(f(w))^i + (-1)^{p+1}\frac{\varphi^{(p+1)}[f(\xi)]}{(p+1)!}(f(w))^{p+1} \tag{4}$$

We denote by:

$$T(w) = w + \sum_{i=1}^{p}(-1)^i \frac{\theta_i(w)}{i!}(f(w))^i \tag{5}$$

and take T as being the iteration function. Thus, we obtain the Cebysev method for solving the equation (2):

$$w_{k+1} = T(w_k) \tag{6}$$

As compared to the Newton method, the convergence rate of the iterative method (6) is at least of order two. The Cebysev method has a convergence rate equal with $p+1$ [9]. Thus we get an iterative scheme yielding to an improved convergence rate in estimating the independent components.

Remark 1. The Cebysev coefficients may be calculated by successive derivates of the identity $\varphi[f(w)] \equiv w$. We obtain:

$$\varphi'[f(w)]f'(w) = 1$$
$$\varphi''[f'(w)^2] + \varphi'[f(w)]f''(w) = 0 \tag{7}$$

...

and it results the following expression that allows us to recursively extract the Cebysev coefficients values:

$$\theta_1(w)f'(w) = 1$$
$$\theta_2(w)[f'(w)^2] + \theta_1(w)f''(w) = 0 \tag{8}$$

...

Remark 2. The first two values of the Cebysev coefficients sequence are:

$$\theta_1(w) = f'(w)^{-1}$$
$$\theta_2(w) = -f''(w)f'(w)^{-3} \tag{9}$$

Remark 3 (Particular Cases). The recursively formulas may be written for particular values of the p index:

$$[case\ p=1]: w_{k+1} = w_k - \frac{f(w_k)}{f'(w_k)}$$
$$[case\ p=2]: w_{k+1} = w_k - \frac{f(w_k)}{f'(w_k)} - \frac{f''(w_k)f(w_k)^2}{f'(w_k)^3} \tag{10}$$

that is just the Newton method (p=1) and the Cebysev method with convergence rate equal three (p=2).

Remark 4. For two w_{n+1} and w_n consecutive approximations of the sequence from the (4) relation, taking into consideration the context of the ICA model in which the w (weighting vector) which leads us to the independent components are normalized, we have:

$$\|w_{n+1} - w_n\| = \sqrt{2(1- <w_{n+1}, w_n>)} \tag{11}$$

From the established (11) relation it results that at the convergence of the (4) sequence we obtain: $\|w_{n+1} - w_n\| \to 0$, meaning that from (11) we have $(1- <w_{n+1}, w_n>) \to 0 \Rightarrow <w_{n+1}, w_n> \to 1$.

Remark 5. In the case of the ICA model the f function is given by $f(w) = F^*(w)$, $F^*(w) = E\{zg(w^Tz)\} - \beta w$, where z is the observations vector, g is a function properly chosen, and β is a real constant.

2.2 Estimation Algorithm of the ICA Model by Cebysev Coefficients

In this article we want to achieve the establishment of the algorithm which estimates the independent components through successive approximations using as objective function the negentropy, which is largely used in the field.
We obtain:

$$F^*(w) = E\{zg(w^Tz)\} - \beta w = 0 \tag{12}$$

where β is a real constant, $\beta = E\{w_0^T zg(w_0^T z)\}$, where w_0 is the critical value of w.

For the (12) equation, using the Cebysev coefficients, the approximations sequence can be written by:

$$w_{k+1} = T(w_k)$$

where

$$T(w) = w + \sum_{i=1}^{p}(-1)^i \frac{\theta_i(w)}{i!}[F^*(w)]^i, \ \theta_i(w) = \varphi^i[F^*(w)]$$

and w is normalized at each iterative step, $w \leftarrow w/\|w\|$.

According to the observation 4, the convergence condition of the approximations sequence is given by $\langle w_{n+1}, w_n \rangle \to 1$, where w_{n+1} and w_n are two normalized consecutive values of the sequence. A detailed version of the FastICA algorithm based on Cebysev coefficients (CCM - Cebysev coefficients method) is described as follows.

CCM Algorithm - The FastICA algorithm based on Cebysev coefficients for estimating several independent components

Step 1 : Center the data to mean.
Step 2 : Apply the whitening transform to data (z).
Step 3 : Select the number of independent components n and set counter $r \leftarrow 1$.
Step 4 : Select the initial guess of unit norm for w_r.
Step 5 : Apply the updating rules:

$$w_r \leftarrow w_r + \sum_{i=1}^{p}(-1)^i \frac{\theta_i(w_r)}{i!}[F^*(w_r)]^i \tag{13}$$

where $\theta_i(w_r) = \varphi^i[F^*(w_r)]$, $F^*(w_r)$ is defined in (12) and g is defined as in [1].
Step 6 : Do the orthogonalization transform:

$$w_r \leftarrow w_r - \sum_{j=1}^{r-1}(w_r^T w_j)w_j \tag{14}$$

Step 7 : Let $w_r \leftarrow w_r / \|w_r\|$.

Step 8 : If w_r has not converged ($\|w_r{}^{k+1} - w_r{}^k\| > \varepsilon$, where ε is a small real constant), go back to step 5.

Step 9 : Set $r \leftarrow r + 1$. If $r \leq n$ then go to step 4.

3   Experimental Analysis

The assessment of the performances of the proposed algorithm for the determining of the independent components is achieved in problems of signals recognition.

We define an absolute mean sum error (AbsMSE) for comparing the accuracy of matching between original signals and restored signals. Then AbsMSE is defined as follows:

$$AbsMSE = \sum_{i=1}^{N} |s_i - s\_estimated_i| / N \qquad (15)$$

where s_i and $s\_estimated_i$ represent the i-th pixel values for original and restored signals, respectively, and N is the total number of pixels.

All the presented tests comprise the recognition performances of the independent components using as an objective function the negentropy for which they used one at a time in the approximation the three functions adopted in the field [1]. In a comparative study the proposed method based on Cebysev coefficients has recognition performances of the original signals which are better, more accurate than the implemented methods, such as FastICA based on the secant method [3] or the gradient type method.

3.1   Experimentally Derived Conclusions on the Performance of the Algorithm in the Case of the Mixtures of the Signals

Test I. We consider as observation data two signals which are mixed and recorded based on two independent components.

In this first test, the original sources are signals generated using the Matlab functions, and the results obtained after applying the algorithm based on Cebysev coefficients show a recognition performance similar to the standard FastICA

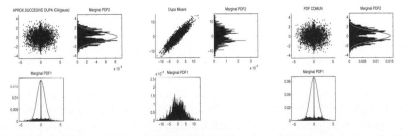

Fig. 1. Source Signals Discovered by the Algorithm (Left: 3 images), The Mixed Signals (Middle: 3 images) and Original Signals (Right: 3 images)

Table 1. AbsMSE of versions of the FastICA Algorithm for experimental test I

FastICA	Test I	Test I	Test I
Basic Method	tanh (g_1)	exp (g_2)	kurt (g_3)
Newton	0.010692	0.010703	0.010688
Secant	0.010726	0.010672	0.012785
Gradient	0.011879	0.011130	0.011668
Cebysev Coeff.	0.010666	0.010665	0.010668

method based on the Newton and to the method FastICA method based on the secant method. The source signals discovered by the algorithm, the mixed signals and the source signals generated by Matlab subjected to the analysis procedure in independent components are represented in figure 1. In the respective figure we can notice the marginal densities corresponding to the two signals as well as the joint density which is common to the mixtures for the source signals discovered by the algorithm, for the mixed signals and for the source signals, respectively.

The results of the test regarding the appliance of the algorithm based on successive approximations are given in table 1.

Test II. This test resembles the anterior test with the difference that it uses, as original signals, the uniform distribution signals.

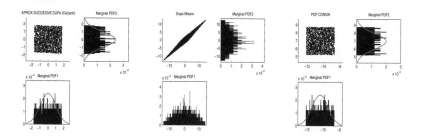

Fig. 2. Source Signals (uniform) Discovered by the Algorithm (Left: 3 images), The Mixed Signals (uniform) (Middle: 3 images) and Original Signals (uniform) (Right: 3 images)

The figure 2 comprise the original source signals, the mixed signals and the source signals discovered by the algorithm for the uniform signals case. In the respective figure, again, we can notice the marginal densities corresponding to the two signals as well as the joint density which is common to the mixtures for the source signals discovered by the algorithm, for the mixed signals and for the source signals, respectively. The results obtained after the comparative study regarding the proposed method and other methods used in the estimation of the ICA model, are similar to the ones from the first test conform with table 2.

Table 2. AbsMSE of versions of the FastICA Algorithm for experimental test II

FastICA	Test II	Test II	Test II
Basic Method	tanh (g_1)	exp (g_2)	kurt (g_3)
Newton	0.023866	0.023166	0.024166
Secant	0.023145	0.022754	0.023145
Gradient	0.023074	0.022991	0.023993
Cebysev Coeff.	0.022985	0.022981	0.022984

3.2 Experimentally Derived Conclusions on the Performance of the Algorithm in the Case of the Mixtures of the Face Images

Test III. The achieved test refers to the capacity of the proposed algorithm of recognizing independent image faces from images of the mixed faces which can represent joint and superimposed faces as well as deteriorated images subjected to restoration.

Fig. 3. Original Faces Discovered by the Algorithm (Left: 2 images), The Mixed Faces for Face Recognition (Middle: 2 images) and Original Faces for Face Recognition (Right: 2 images)

Table 3. AbsMSE of versions of the FastICA Algorithm for experimental test III

FastICA	TestIII	TestIII	TestIII
Basic Method	tanh (g_1)	exp (g_2)	kurt (g_3)
Newton	0.004482	0.004482	0.006113
Secant	0.005818	0.005945	0.005479
Gradient	0.005303	0.005486	0.005589
Cebysev Coeff.	0.005266	0.005376	0.005095

In this test we considered again the bidimensional case with two mixed images over which we apply the deterioration algorithm of the original faces. The original image faces, the mixed image faces and the image faces discovered by the proposed algorithm are in figure 3. Just as the anterior examples, the obtained results offer good recognition performances of the original faces, showing also in this case a qualitative superiority in the recognition (the results are presented in table 3) compared with other used methods and anterior specified.

4   Summary and Conclusions

In this article we developed an algorithm for estimating the independent components based on an iterative scheme that use Cebysev coefficients yielding to a superior convergence rate. We derived a suitable algorithm and supplied a comparative analysis of its recognition capacities against the previously developed algorithm. In order to derive conclusive remarks the tests were performed on different signal samples.

References

1. Hyvarinen, A., Karhunen, J., Oja, E.: Independent Component Analysis. John Wiley & Sons, Chichester (2001)
2. Constantin, D., State, L.: A Comparative Analysis on a Class of Numerical Methods for Estimating the ICA Model. International Journal of Computers, Communications & Control III, 15–17 (2008); ISSN 1841-9836, E-ISSN 1841-9844, Vol. III (2008), Suppl. issue: Proceedings of ICCCC 2008, Baile Felix, Oradea, Romania, 218–222, May 15-17 (2008)
3. Constantin, D., State, L.: Successive Approximations- Based Algorithm for Independent Component Analysis. In: Proceedings of the Third International Conference on Digital Telecommunications (ICDT 2008). IEEE Computer Society Press, Bucharest (2008)
4. Constantin, D., State, L.: A version of the fastICA algorithm based on the secant method combined with simple iterations method. In: Elmoataz, A., Lezoray, O., Nouboud, F., Mammass, D. (eds.) ICISP 2008 2008. LNCS, vol. 5099, pp. 312–320. Springer, Heidelberg (2008)
5. Hyvarinen, A., Oja, E.: Independent component analysis: algorithms and applications. Neural Networks 13, 411–433 (2000)
6. Hyvarinen, A.: Survey on Independent Component Analysis. Neural Computing Surveys 2, 94–128 (1999)
7. Constantin, D., State, L.: A Comparative Analysis on a Class of Numerical Methods for Estimating the ICA Model. In: Proceedings of the The IEEE 2nd International Conference on Computers, Communications & Control (ICCCC 2008), Baile Felix, Oradea, Romania, May 15-17 (2008)
8. Maruster, S.: Numerical methods for solving the nonlinear equations. Tehnica Ed., Bucharest (1981)

Histogram Based Blind Identification and Source Separation from Linear Instantaneous Mixtures

Konstantinos I. Diamantaras[1] and Theophilos Papadimitriou[2]

[1] Department of Informatics,
TEI of Thessaloniki,
Sindos 57400, Greece
kdiamant@it.teithe.gr
[2] Department of Int. Economic Relat. and Devel.,
Democritus University of Thrace
Komotini 69100, Greece
papadimi@ierd.duth.gr

Abstract. The paper presents a new geometric method for the blind identification of linear instantaneous MIMO systems driven by multi-level inputs. The number of outputs may be greater than, equal to, or even less than the number of sources. The sources are then extracted using the identified system parameters. Our approach is based on the fact that the distribution of the distances between the cluster centers of the observed data cloud reveals the mixing vectors in a simple way. In the noiseless case the method is deterministic, non-iterative and fast: it suffices to calculate the histogram of these distances. In the noisy case, the core algorithm must be combined with efficient clustering methods in order to yield satisfactory results for various SNR levels.

1   Introduction

Blind Source Separation refers to the recovering of n unknown signals using only the mixtures observed at m sensors. The term "blind" refers to the fact that the underlying mixing operator is unknown as well. According to the mixing process the models can be divided into memoryless linear mixture BSS (also known as instantaneous BSS) and convolutive mixture BSS (also referred to as Multi-Input Multi-Output (MIMO) Blind Deconvolution/Equalization). In the former the mixing operator is a constant matrix and there is no time shift of the source signals. Instantaneous BSS can't tolerate source multipath dispersion caused by reflections from obstacles between the source and the observation. Such multipath phenomena often appear in many applications in mobile communications, acoustics etc and they are modeled by MIMO convoluted systems.

The BSS methods can be divided into two major groups according to statistical moments: a) Higher-order methods (HOS), and b) Second-order methods (SOS). The methods of the first group are based on the optimization of higher order statistical functions [1,2]. Typically these methods try to separate sources based on their statistical independence. HOS methods can treat systems

T. Adali et al. (Eds.): ICA 2009, LNCS 5441, pp. 227–234, 2009.
© Springer-Verlag Berlin Heidelberg 2009

driven by independent and identically distributed random variables (i.i.d.). The methods of the latter group reduce the strong hypothesis of source statistical independence to source orthogonality. These methods exploit the diversity of the sources, either in the time or in the frequency domain. In [3] Tong showed that BSS can be achieved through the eigendecomposition of a spatial covariance matrix. However full separation of the sources requires eigendecomposition with distinct eigenvalues. Belouchrani in [4] suggested the joint diagonalization of a set of time-lagged covariance matrices so that there is less chance that a matrix contains close-value eigenvalues.

A third approach for BSS is based on the geometric properties of the output signal constellation. In [5] Diamantaras e.a. investigated the problem of blind separation and identification of an instantaneous Multi-Input Single-Output (MISO) system driven by binary antipodal signals. The system filter is recursively deflated, yielding in the final step: the filter and the source signals. The same idea was further developped by Li e.a. in [6]. The blind separation of convolutive MISO systems driven by binary sources was explored in [7]. The method is also based on the sequential deflation of the mixing filter, through the exploration of the observation signal succession rules. So long as the observation signal is rich enough, these methods yield accurate results.

In this paper we present a geometric method for blind identification and separation of multilevel signals, such as PAM coded sources. The proposed method is based on the distances between the centers of the clusters formed in the data constellation. The paper is organized as follows: in Section 2 we formally describe the problem and the basic assumptions; in Section 3 we present the new method blind identification and separation method starting with the noiseless case and extending our results to the noisy case. In Section 4 we present simulations results on two different types of data (a) under-determined noiseless 4-levels PAM source signals and (b) over-determined noisy 4-levels PAM source signals. Finally we conclude in Section 5.

2   Problem Formulation

Consider the m-dimensional instantaneous mixture of n real source signals $s_1(k)$, ..., $s_n(k)$,

$$\mathbf{x}(k) = \mathbf{a}_1 s_1(k) + \cdots + \mathbf{a}_n s_n(k) \tag{1}$$

where $\mathbf{x} = [x_1, \cdots, x_m]^T$. There is no restriction on the length m of the observation vector: it may be greater than, equal to, or less than the number of sources n. In any case, the real mixing vectors $\mathbf{a}_i \in \mathbb{R}^m$, are unknown and we assume that the sources take L values from a finite set

$$V = \{v_0, \cdots, v_{L-1}\}$$

where consecutive values in V have equal distances D, so $v_p = v_0 + pD$, $p = 0, \cdots, L-1$. Such source signals appear for example, in digital communications: Pulse Amplitude Modulated (PAM) signals with L levels take values from the set $V = \{-(L-1)/2, -(L-3)/2, \cdots, (L-3)/2, (L-1)/2\}$, with $D = 1$. Also square QAM signals with constellation size L^2 can be split into two real PAM signals with L levels (real and imaginary part of QAM). In this case, the mixing model (1) with n complex QAM inputs and m complex observations corresponds to a real model with $2n$ inputs and $2m$ observations:

$$\text{Re}\{\mathbf{x}\}(k) = \sum_{i=1}^{n} \text{Re}\{\mathbf{a}_i\}\text{Re}\{s_i\}(k) - \text{Im}\{\mathbf{a}_i\}\text{Im}\{s_i\}(k)$$

$$\text{Im}\{\mathbf{x}\}(k) = \sum_{i=1}^{n} \text{Re}\{\mathbf{a}_i\}\text{Im}\{s_i\}(k) + \text{Im}\{\mathbf{a}_i\}\text{Re}\{s_i\}(k)$$

Another example area is digital imaging where the luminance values are typically in the set $V = \{0, 1, 2, \cdots, 255\}$. Without loss of generality, in the sequel we shall assume that $D = 1$, so

$$v_p = v_0 + p, \quad p = 0, \cdots, L - 1. \tag{2}$$

Since the input values come from a finite alphabet, the values of $\mathbf{x}$ also belong to a finite set

$$S = \{\mathbf{c}_1, \cdots, \mathbf{c}_N\}. \tag{3}$$

Each member of S will be called a *center* and comes from a linear combination of the form

$$\mathbf{c}_p = \mathbf{a}_1 v_1(p) + \cdots + \mathbf{a}_n v_n(p) \tag{4}$$

for some $v_1(p), ..., v_n(p) \in V$. For our subsequent analysis it is essential that we make the following assumption

Assumption 1. *There are no repeated centers:*

$$\mathbf{c}_p = \mathbf{c}_q \Leftrightarrow v_1(p) = v_1(q), \cdots, v_n(p) = v_n(q) \tag{5}$$

Therefore, there will be $N = L^n$ distinct centers.

If the mixing vectors are in *general position* then the above assumption is true.

3   Counting Frequencies of Center Differences

Our proposed method uses the distribution of the centers and in particular, the frequencies of the center differences, in order to recover the mixing vectors and then reconstruct the sources. Subtracting any two centers $\mathbf{c}_p$, $\mathbf{c}_q$, we obtain

$$\mathbf{d}_{p,q} = \mathbf{c}_p - \mathbf{c}_q = \mathbf{a}_1 \beta_1^{(p,q)} + ... + \mathbf{a}_n \beta_n^{(p,q)} \tag{6}$$

where the values $\beta_i^{(p,q)} = v_i(p) - v_i(q)$, $i = 1, \cdots, n$, belong to the following set:

$$\Delta = \{v_i - v_j \mid \text{any } v_i, v_j \in V\}$$
$$= \{-L + 1, -L + 2, \cdots, 0, \cdots, L - 2, L - 1\}$$

Straightforward combinatorial analysis shows that the v-difference $\beta = 0$ can be generated in L different ways,

$$0 = v_0 - v_0 = v_1 - v_1 = \ldots = v_{L-1} - v_{L-1},$$

In general, the v-difference $\beta = \pm\delta$ can be generated in $L - \delta$ different ways,

$$\pm\delta = \pm(v_\delta - v_0) = \pm(v_{\delta+1} - v_1) = \ldots = \pm(v_{L-1} - v_{L-1-\delta}).$$

Assumption 2. *The input sequence* $\mathbf{s}(k) = [s_1(k), \cdots, s_n(k)]^T$, $k = 1, \cdots, K$, *is rich enough so that it contains all possible n-tuples of input values in* V^n.

Assumption 3. *The mixing vectors* $\mathbf{a}_i$ *are in general position so that every distinct sum* $\mathbf{d} = \sum_{i=1}^{n} \mathbf{a}_i \beta_i$ *is generated by a unique n-tuple* $(\beta_1, \cdots, \beta_n) \in \Delta^n$

$$\sum_{i=1}^{n} \mathbf{a}_i \beta_i = \sum_{i=1}^{n} \mathbf{a}_i \beta_i' \Leftrightarrow \beta_1 = \beta_1', \cdots, \beta_n = \beta_n',$$

$$\beta_i, \beta_i' \in \Delta . \tag{7}$$

Let us now take all the differences $\mathbf{d} = \mathbf{c}_p - \mathbf{c}_q$ between the members of S. These differences belong to the set

$$\Gamma = \left\{ \mathbf{d} = \sum_{i=1}^{n} \mathbf{a}_i \beta_i \mid \beta_i \in \Delta \right\}$$

Since, by assumption 3, each difference $\mathbf{d}$ is generated by a unique set of values β_1, ..., β_n, the frequency by which $\mathbf{d}$ occurs is the product $f_1 \times f_2 \times \cdots \times f_n$ where f_i is the number of different ways the v-difference β_i occurs. So the most common difference is $\mathbf{d} = 0$ generated by $\beta_1 = \beta_2 = \cdots = \beta_n = 0$ occuring $L \times L \times \cdots \times L = L^n$ times. The next most common differences are

$$\mathbf{d} = \mathbf{a}_p \cdot (\pm 1) + \sum_{i \neq p} \mathbf{a}_i \cdot 0 = \pm\mathbf{a}_p, \quad p = 1, \cdots, n$$

all of which occur $L^{(n-1)} \times (L-1)$ times, since $\beta = 0$ occurs in L different ways while $\beta = \pm 1$ occurs in $L - 1$ ways.

Next in frequency come the differences $\mathbf{d} = \pm\mathbf{a}_p \pm \mathbf{a}_q$, any p, q, each of which occurs $L^{(n-2)} \times (L-1)^2$ times, and then come the differences $\mathbf{d} = \pm 2\mathbf{a}_p$, any p, each of which occurs in $L^{(n-1)} \times (L-2)$ times, etc.

According to the above results it is clear that *all* the mixing vectors appear in the most frequent center differences. In particular, excluding the most common difference $\mathbf{d} = 0$, the next $2n$ most frequent differences are $\pm\mathbf{a}_p$ occuring $(L-1) \times L^{(n-1)}$ times. The permutation and sign ambiguity appears in all blind separation problems and can not be treated unless further information is given about the source signals.

Once we obtain the estimates (up to a permutation and sign)

$$\hat{\mathbf{a}}_p = \mathbf{a}_p \text{ or } -\mathbf{a}_p$$

the sources $s_i(k)$ are estimated by finding the source values v_i that minimize the reconstruction error:

$$\hat{\mathbf{s}}(k) = \arg \min_{[v_1, \cdots, v_n]^T} \left[\mathbf{x}(k) - \sum_{i=1}^{n} \hat{\mathbf{a}}_i v_i \right]^2$$

Therefore, we come to the following algorithm for separating the two sources based on the frequency of the center differences:

Algorithm 1. *Blind Separation for Noiseless Data*

1. *Using the observed signal $\mathbf{x}(k)$ form the set of centers $S = \{\mathbf{c}_1, \cdots, \mathbf{c}_N\}$*
2. *Compute the differences $\mathbf{d}_{p,q} = \mathbf{c}_p - \mathbf{c}_q$ for each pair of centers $\mathbf{c}_p \neq \mathbf{c}_q \in S$*
3. *Compute the histogram of the differences $\mathbf{d}_{p,q}$ (count how many times each difference appears)*
4. *Set $\pm\hat{\mathbf{a}}_p$, $p = 1, \cdots, n$, equal to the $2n$ most frequent differences.*

3.1 Adding Noise

In many practical cases the observed data are corrupted by noise described by the following data model

$$\mathbf{x}(k) = \sum_{i=1}^{n} \mathbf{a}_i s_i(k) + \mathbf{e}(k) \tag{8}$$

where the extra term $\mathbf{e}(k)$ corresponds to gaussian additive noise with zero mean and variance $\sigma^2 \mathbf{I}$. Even if the number of observation centers is known a priori, the localization of their correct position is a non-trivial task. The noisy observations of every class form a gaussian "bell" under which the correct observation center is hidden. The greater part of the observations are accumulated around the correct center, while the observations that are heavily altered by the noise tend to disperce from it. This assumption is the basic idea of our noise removal scheme: the closest observation pairs should be near a true observation center. This method can succesfully treat cases where the noise variance is smaller than the minimum distance between the observation centers. In the opposite case, the center localization may be erroneous. So, first we collect the closest observation pairs, and then we cluster them in N distinct classes. The mean observation of every class is the estimation of an observation center $\bar{\mathbf{c}}_i$ $i = 1, ..., N$. The estimated obervations are used to compute the differences $\bar{\mathbf{d}}_{p,q} = \bar{\mathbf{c}}_p - \bar{\mathbf{c}}_q$ for each pair of centers $\bar{\mathbf{c}}_p \neq \bar{\mathbf{c}}_q$. As there is an error in the estimation of the centers, there is an error propagated in the differences as well. Consequently, it is impossible to directly use the frequency method. Instead, we perform clustering in the differences, in order to calculate the differences histogram. The method can be summarized in following algorithm

Algorithm 2. *Blind Separation for Noisy Data*

1. *Using the observed signal $\mathbf{x}(k)$ form the approximated set of the centers $S = \{\bar{\mathbf{c}}_1, \cdots, \bar{\mathbf{c}}_N\}$*
2. *Compute the approximated differences $\bar{\mathbf{d}}_{p,q} = \bar{\mathbf{c}}_p - \bar{\mathbf{c}}_q$ for each pair of centers $\bar{\mathbf{c}}_p \neq \bar{\mathbf{c}}_q \in S$*

3. *Cluster the differences.*
4. *Compute the histogram of the differences* $\bar{\mathbf{d}}_{p,q}$ *(count how many times each difference appears)*
5. *Set* $\pm\hat{\mathbf{a}}_p$, $p = 1, \cdots, n$, *equal to the 2n most frequent differences.*

4   Results

4.1   Noiseless Simulation

We tested the proposed method on a system with $n = 4$ source 4-PAM signals, and one observation signal in an error-free environment. The mixing parameters were $[a_1, a_2, a_3, a_4] = [0.1739, 0.4259, 0.5305, -0.7120]$. The dataset consisted of $N = 2000$.

Using the histogram (Fig. 1) we collected eight($= 2n$) differences with the second higher frequency (the higher frequency corresponds to the zero difference). These correspond to the mixing parameters. The blind identification and the source signals reconstruction are perfect in the noiseless scenario.

4.2   Noisy Simulation

The noiseless scenario is very unrealistic since in real world applications there are various sources of noise. In order to simulate these situations, we conducted

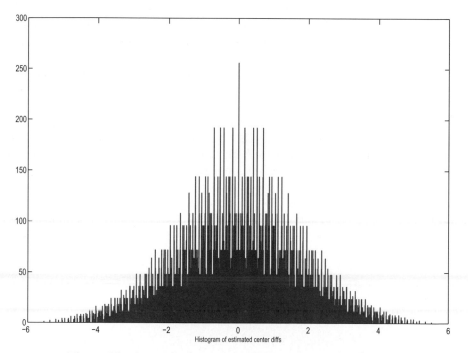

Fig. 1. The frequency histogram of the observation differences

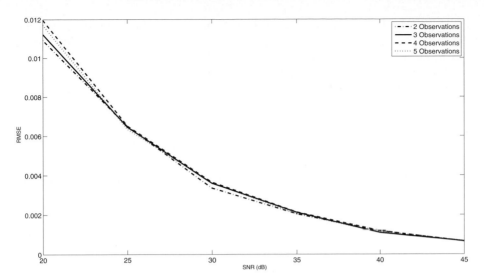

Fig. 2. The identification accuracy of the proposed method for various SNR levels

Table 1. The mean Symbol Error Rate (%) of the second scenario for various noise levels

	SNR					
Observation Signals	20dB	25dB	30dB	35dB	40dB	45dB
2	0.146	0.018	0	0.066	0	0
3	0	0	0	0	0	0
4	0.001	0	0	0	0	0
5	0	0	0	0	0	0

a second set of experiments, where gaussian additive noise was added to the mixtures. We created 4 different systems with two 4-PAM sources of 1000 samples, and two, three, four and five observation signals. The simulations were performed in a Monte Carlo framework with 100 tries for each set. Additive noise was injected to the observed signals with SNR varying from 20 to 45 dB. The mixing parameters were fixed for every system that we tested. The 2x2 system was generated by $\mathbf{a}_1 = [-0.611, 0.956]$ and $\mathbf{a}_2 = [-0.791, 0.294]$, the 2x3 system was generated by $\mathbf{a}_1 = [0.966, 0.920, 0.355]$ and $\mathbf{a}_2 = [0.259, 0.392, -0.935]$, the 2x4 system was generated by $\mathbf{a}_1 = [0.935, 0.827, -0.885, -0.776]$ and $\mathbf{a}_2 = [0.353, 0.562, -0.465, 0.631]$, and the 2x5 system was generated by the filters $\mathbf{a}_1 = [-0.516, 0.671, -0.946, -0.483, 0.831]$ and $\mathbf{a}_2 = [0.857, -0.742, -0.324, -0.876, -0.557]$.

The identification accuracy was counted using RMSE (Root Mean Square Error) defined as

$$RMSE = \sqrt{\frac{1}{m}\sum_{i=1}^{m}(a_i - \hat{a}_i)^2} \tag{9}$$

and the separation accuracy was counted using SER (Symbol Error Rate).

The RMSE of the filter estimation (Fig. 2) shows that the accuracy estimation is independent of the number of observation sources, since the performance is almost identical in every case we tested. One the other hand, the estimation error decreases as the SNR level decreases.

5   Conclusions

In this paper we presented a new blind identification method for MIMO instantaneous systems driven by multi-level sources. The method can be directly used for the blind separation of the mixtures. The method is based on the geometric characteristics of the observed data constellation and in particular on the distribution of the differences between centers. The core of the method is non-iterative. The suppression of the noise effect in the mixtures is based on clustering techniques. The method was tested in various noiseless and noisy simulation environment with success.

References

1. Comon, P.: Independent component analysis, a new concept? Signal Processing 36, 287–314 (1994)
2. Hyvärinen, A., Karhunen, J., Oja, E.: Independent Component Analysis. Wiley, New York (2001)
3. Tong, L., Liu, R., Soon, V., Huang, Y.: Indeterminacy and identifiability of blind identification. IEEE Trans. Circuits Syst. I 38(5), 499–509 (1991)
4. Belouchrani, A., Abed-Meraim, K., Cardoso, J.F., Moulines, E.: A Blind Source Separation Technique Using Second-Order Statistics. IEEE Trans. Signal Processing 45, 434–444 (1997)
5. Diamantaras, K.I., Chassioti, E.: Blind Separation of N Binary Sources from one Observation: A Deterministic Approach. In: Proceedings of the Second International Workshop on Independent Component Analysis and Blind Source Separation, Helsinki, Finland, pp. 93–98 (June 2000)
6. Li, Y., Cichocki, A., Zhang, L.: Blind Separation and Extraction of Binary Sources. IEICE Transactions on Fundamentals E86-A(3), 580–589 (2003)
7. Diamantaras, K., Papadimitriou, T.: Blind deconvolution of multi-input single-output systems with binary sources. IEEE Transactions on Signal Processing 54(10), 3720–3731 (2006)

Blind Source Separation of Post-Nonlinear Mixtures Using Evolutionary Computation and Gaussianization

Tiago M. Dias[1,3], Romis Attux[2,3], João M.T. Romano[1,3], and Ricardo Suyama[1,3]

[1] Department of Microwave and Optics
[2] Department of Computer Engineering and Industrial Automation
[3] DSPCOM – Laboratory of Signal Processing for Communications
School of Electrical and Computer Engineering, University of Campinas (UNICAMP)
C.P. 6101, CEP 13083-970, Campinas – SP, Brazil
tdias@decom.fee.unicamp.br, {rsuyama,romano}@dmo.fee.unicamp.br,
attux@dca.fee.unicamp.br

Abstract. In this work, we propose a new method for source separation of post-nonlinear mixtures that combines evolutionary-based global search, gaussianization and a local search step based on FastICA algorithm. The rationale of the proposal is to attempt to obtain efficient and precise solutions using with parsimony the available computational resources, and, as shown by the simulation results, this aim was satisfactorily fulfilled.

Keywords: Nonlinear blind source separation, post-nonlinear models, gaussianization, evolutionary algorithms, artificial immune systems.

1 Introduction

The problem of blind source separation (BSS) is related to the recovery of source signals only by using information contained in mixed versions of them. Until the end of the last decade, the great majority of the proposed BSS techniques [1] were designed to solve the standard BSS problem, that is to say, the case in which the mixture model is linear and instantaneous. However, in some applications [2], such as digital satellite communications, as a consequence of the nonlinear character of the sensors, the use of the linear BSS framework may lead to unsatisfactory results, which motivates the adoption of nonlinear mixing models.

An inherent difficulty associated with the separation of nonlinear mixtures comes from the fact that, in contrast with the linear case, it is not always possible to ensure that the model be separable solely by means of an independent component analysis (ICA) approach. Thus, one can conclude that the nonlinear BSS problem cannot be solved in a general fashion: it is necessary to make hypotheses about the source and/or the mixture model [1].

A strategy that is viable and relevant from a practical standpoint is to restrict the mixture model to a class of separable nonlinear systems [1] [3]. In accordance with this idea, the so-called Post-Nonlinear (PNL) model [3] emerges as an important option, especially because of its separability and its practical applicability [3]. The

T. Adali et al. (Eds.): ICA 2009, LNCS 5441, pp. 235–242, 2009.
© Springer-Verlag Berlin Heidelberg 2009

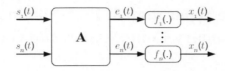

Fig. 1. The PNL mixing system

PNL mixture model is formed by a two-stage mixing system composed of a linear mixture followed by memoryless nonlinear functions, as shown in Fig. 1.

In [3], Taleb and Jutten proposed a paradigm for inverting the action of a PNL mixture system that was based on minimization of the mutual information between the source estimates. Despite its theoretical soundness, this approach suffers from two major practical drawbacks. The first one comes from the fact that the evaluation of the mutual information demands estimation of the marginal entropies, which may be a rather complex task. The second one is related to the presence of local minima in the mutual information-based cost function, which makes separating system adaptation via gradient-based algorithms very complicated. Therefore, in this strategy, the success of any paradigm is closely related to the efficacy of the entropy estimator and to the potential of avoiding local minima.

Having Taleb and Jutten's ideas in mind and taking the two-stage PNL mixing system as a study model, we propose a novel BSS strategy that employs gaussianization techniques [6] and a hybrid search algorithm, combining an evolutionary optimization tool with the well-known FastICA algorithm [1]. In this twofold approach, the FastICA algorithm is devoted to finding the solution to the linear part of the problem. On the other hand, evolutionary algorithms are used to optimize the parameters of the nonlinear part of the separating system, significantly reducing the chance of local convergence.

The work is structured as follows. In Section 2, the fundamentals of the problem of separating PNL mixtures are discussed. In Section 3, we present our proposed algorithm. Simulations results are shown and discussed in Section 4. Finally, in Section 5, we expose our conclusions.

2   Problem Statement

Let $s = [s_1, s_2, ..., s_N]^T$ denote N mutually independent sources and $x = [x_1, x_2, ..., x_N]^T$ be a vector of N mixtures of the source signals i.e., $x = \varphi(s)$. BSS techniques are used to recover the source signals based only on the observed samples of the mixtures and on a minimal amount of statistical information about the signals.

When the mapping $\varphi(.)$ is linear and instantaneous, this becomes the classical linear BSS problem, and the following model holds: $x = As$, where A is the NxN mixing matrix. In this case, separation algorithms based on ICA (Independent Component Analysis) aim to find a linear separating matrix W such that the vector $y = Wx$ has statistically independent elements. Ideally, the separating matrix W will be, up to scaling factors and permutations, the inverse of the mixing matrix A [1].

However, if $\varphi(.)$ is a nonlinear function, one should consider using a NBSS (Nonlinear Blind Separation Signal) approach. In this case, the independence hypothesis, which is the basis of ICA, may no longer be enough to obtain the original sources. Thus, it becomes essential to deal with a more restricted class of tractable mixtures, among which it is possible to highlight the two-stage PNL mixing system [3].

The PNL model divides the mixture into a linear and a nonlinear part, which means that one can consider the existence of two distinct but interrelated problems. Mathematically, the function $\varphi(.)$ is given by a composition of a nonlinear function $f(.)$ and a linear matrix A, and the model becomes: $x = f(As)$, where $f = [f_1(.), f_2(.), ..., f_N(.)]^T$ denotes the nonlinearities applied to each output of the linear mixing stage.

Source separation of PNL mixtures can be achieved by considering a full invertible separating system as depicted in Fig. 2, the output of which is $y = Wg(x)$ [5], being $g(.)=[g_1(.), g_2(.), ..., g_N(.)]^T$ a set of nonlinear functions that must be precisely adjusted in order to invert the action of the nonlinearity $f(.)$.

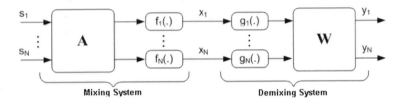

Fig. 2. The PNL mixing and separating systems

3 Nonlinear Source Separation Using a Gaussianization Strategy and an Artificial Immune System

An analysis of Fig. 2 reveals that BSS of PNL mixtures can be achieved by seeking a linear separating matrix W and a nonlinear function $g(.)$ able to counterbalance the action of the mixing system, given by a linear matrix A and nonlinearities $f(.)$. Therefore, we propose a solution that divides the problem into two parts: adaptation of the separating matrix W via the efficient local search method known as FastICA and choice of the parameters of $g(.)$ using an artificial immune system (AIS), with a significant potential of finding good optima employing a criterion based on the idea of Gaussianization. This approach combines the relative simplicity of a well-established adaptive methodology with the remarkable capacity of multimodal optimization that is characteristic of the adopted AIS.

Gaussianization, first introduced by Solé-Casals and Chen [5][6], is an approach whose *modus operandi* can be understood in light of the central limit theorem. This theorem states, in simple terms, that a linear mixture of nongaussian signals tends towards a gaussian signal. This tendency is more pronounced if the number of signals present in the mixture is increased. Applying these notions in the context of the problem of BSS of PNL mixtures, one can notice that the linear mixing stage has a *Gaussianizing effect*. Moreover, as depicted in Fig. 3, the "significantly Gaussian" vector $e=[e_1, e_2, ..., e_N]^T$ goes through a nonlinear stage, the effect of which tends to be of causing the signals to be *less Gaussian* or *more nongaussian*. In summary, the separating system input vector x will tend, once more, to be nongaussian.

This state of things opens a compelling perspective: what if the design of $g(.)$ is guided by the notion of recovering a condition of maximum Gaussianity? In such case, the nonlinearities will have, to a certain extent, an effect contrary to that of $f(.)$. As a consequence, the residual mixture will be linear or almost linear, which means that the use of a conventional matrix W should conclude the separating task. These are, in a nutshell, the concepts that give support to the use in BSS of the idea of Gaussianization.

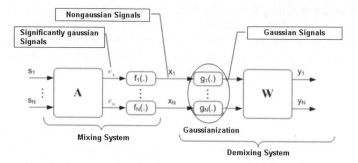

Fig. 3. Gaussianization Strategy

Mathematically, the gaussianization of a random variable x can be achieved by defining a transformation $g(.)$ that produces a Gaussian random variable $z=g(x)$. If z has unit variance and zero mean, its cumulative distribution is given by:

$$G(z_i) = \int_{-\infty}^{z_i} (1/\sqrt{2\pi}) \cdot e^{-\alpha^2/2} d\alpha.$$ (1)

The gaussian transformation can then be written as: $g_i = G^{-1}(P_i(x_i))$. This equation comes from the fact that the application of the cumulative distribution $P_i(.)$ of the mixtures to the mixture signals engenders a random variable with uniform distribution and that the application of the inverse of the Gaussian cumulative distribution $G^{-1}(.)$ to the uniform random variable originates a Gaussian random variable. This combination allows to the nongaussian vector x to be transformed into a Gaussian vector z.

In order to quantify the degree of (non)gaussianity of a signal, we shall use a measure of negentropy. By definition, the negentropy of a Gaussian signal is null [7], which means that it should be possible to carry out Gaussianization by finding a transformation $g(.)$ that produces an output negentropy as close as possible to zero. This leads to the following adaptation criterion, where H represents entropy:

$$min\ N_G(g(x)) = min\ \{H(g(x)) - H(x)\}.$$ (2)

being $z=g(x)$ a Gaussian random variable with the same correlation matrix as x.

As stated in equation (2), negentropy calculation requires entropy estimation which, as a rule, leads to pdf estimation. Alternatively, one can make use of a method based on polynomial moments [1]:

$$N_G(g(x)) = \alpha\ \{E[Q(g(x))] - E[Q(x)]\}^2.$$ (3)

where α is a constant and $Q(.)$ is a nonlinear even function.

The proposed criterion has a strong nonlinear character, which indicates that the optimization problem may have a pronounced multimodal character. This is one of the main concerns in the context of PNL models, as stated in previous works [11, 12, 13]. This fact justifies our adoption of an evolutionary search technique that establishes a fine balance between local and global search mechanisms and that, moreover, has been successfully applied in the context of BSS [8]: the artificial immune system known as opt-aiNet [9].

3.1 The opt-aiNet

The *Optimization Version of an Artificial Immune Network* (opt-aiNet), first proposed in [9], is a search algorithm founded on the combination of three concepts derived from modern theories about the immune system: clonal selection, affinity maturation and the concept of immune network [10].

In simple terms, the immune system can be understood as being composed of cells and molecules that carry receptors for antigens (disease-causing agents). When these cells recognize a given antigen, they are stimulated to proliferate. During the replication process, mutations occur in an inverse proportion to the degree of pathogen recognition (this process is intimately related to the clonal selection principle). Useful receptors tend to be kept, which means that the defense system will be able to respond with more efficiency to an invasion by the same or by a similar pathogen. Furthermore, the opt-aiNet also incorporates mechanisms of diversity and population size control that are inspired by the immune network theory, which supports the idea that the immune system possesses "eigenbehaviors" even in the absence of antigens [10].

Some important points should be taken into account when implementing the opt-aiNet algorithm: 1) The fitness function, which is the one being optimized, is, in fact, a measure of affinity between antibody and antigen; 2) Each solution corresponds to the information contained in a given receptor (network cell); 3) The affinity between cells is measured by a simple Euclidean distance. A pseudo-code for the opt-aiNet algorithm is shown in Table 1.

Table 1. Opt-aiNet Pseudo-code

1. Randomly initialize a population of cells (N_o)
2. **While** stopping criterion is not met **do**
    a) Determine the fitness of each network cell and normalize the vector of fitnesses;
    b) Generate a number N_c of clones for each network cell;
    c) Mutate each clone proportionally to its parent cell's fitness, keeping the parent cell.
    d) Determine the fitness of all cells.
    e) For each clone, select the cell with highest fitness and calculate the average fitness of the selected population.
    f) **If** the average error of the population is not significantly different from the previous iteration, then continue. **Else**, return to step 2.a)
    g) Determine the affinity of all cells in the network. Suppress all but the highest fitness of those cells whose affinities are less than the suppression threshold σ_s and determine the number of network cells, named memory cells, after suppression.
    h) Introduce a percentage β % of randomly generated cells and return to step 2.
3. EndWhile

Important definitions concerning the use of the opt-aiNet algorithms are as follows:

Coding: the nonlinear function $g(.)$ is modeled as a set of 5th-order polynomials with odd powers: $g_i(x_i) = c_{i1}x_i^5 + c_{i2}x_i^3 + c_{i3}x_i$. Since a crucial requirement of the PNL model is that each function be monotonic, the coefficients of each polynomial are restricted to be positive. Network cells are defined as vectors of $g(.)$ coefficients: $c_{ij}=[c_{11}, c_{21}, c_{31}, c_{21}, c_{22}, c_{23}, ...]$; where i represents the mixture and j represents the coefficient.

Fitness function: This is the algorithm's objective function given by: $1/N_G(g(x))$.

The opt-aiNet parameters are: N_0: number of individuals in the population; N_c: number of clones; σ_s: suppression threshold and β: mutation rate.

4 Simulations

In order to study the proposed methodology, this section provides simulation results based on two scenarios. In consonance with the structure of the PNL separating system, described by the input-output relationship $y = Wg(x)$, the adaptation process takes place, at each iteration, in separate stages, one in which the separating linear matrix W is adapted using the FastICA, and another in which the nonlinear functions $g(.)$ are updated with the aid of evolutionary algorithms. The standard method was tested with 10 trials run with the following parameters: $N_0=5$, $N_c=5$, $\beta=50$, $\sigma_s=3$.

4.1 Simulation Scenarios

The first PNL-based scenario is composed of two uniformly-distributed sources (between $[-1, 1]$) mixed with the following linear and nonlinear stages:

$$A = \begin{bmatrix} 1 & 0.6 \\ 0.5 & 1 \end{bmatrix} \quad ; \quad \begin{aligned} f_1(e_1) &= \tanh(2e_1) \\ f_2(e_2) &= 2\sqrt[5]{e_2} \end{aligned} \tag{4}$$

The average speed, the average time spent per iteration (all the simulations were performed in the same machine, an Athlon64 3000+ with 1GB RAM) and the average time to converge are depicted in Table 2. In addition to that, mean-square errors (MSE) between sources and their estimates are shown in Table 3.

Table 2. Metrics related to convergence and results for the first scenario (Performance Results)

Metrics	
Convergence speed (iterations)	1750
Time spent per iteration (ms)	222
Average time to converge (min)	6.5

Table 3. Average MSE results for the first scenario (Solution Quality)

MSE (x 10^{-3})	
y_1	23.6758
y_2	39.3453

Tables 2 and 3 show that the combination between the FastICA algorithm as a local search tool and the opt-aiNet as an evolutionary global search strategy leads to satisfactory results for an execution time that is reasonable in view of, for instance, that associated with the mutual information-based approach presented in [8]. As far as the estimation error is concerned, this new method did provide solutions with a small MSE.

The second scenario is simulated using the following linear and nonlinear characteristics:

$$
A = \begin{bmatrix} 1 & 0.6 & 0.5 \\ 0.5 & 1 & 0.4 \\ 0.4 & 0.6 & 1 \end{bmatrix} \quad ; \quad \begin{aligned} f_1(e_1) &= 2\sqrt[3]{e_1} \\ f_2(e_2) &= 2\sqrt[3]{e_2} \\ f_3(e_3) &= 2\sqrt[3]{e_3} \end{aligned} \quad (5)
$$

Once again, Table 4 and Table 5 reveal that a satisfactory performance in terms of MSE was reached, with an execution time that is very promising in comparison with that of [8]. In comparison with the previous scenario, it is possible to verify that there was reduction in the convergence speed and an increase in time spent per iteration, which leads to a higher total execution time. These results are expected due to the increase in the order of the problem to be solved.

Table 4. Metrics related to results for the second scenario (Performance Results)

Metrics	
Convergence speed (iterations)	1350
Time spent per iteration (ms)	430
Average time to converge (min)	9.7

Table 5. Average MSE results for the second scenario (Solution Quality)

MSE (x 10^{-3})	
y_1	25.2899
y_2	41.9756
y_3	89.0077

Since both scenarios have mixtures composed of a relatively small number of source signals, the applicability of the central limit theorem is, to a certain extent, restricted. This means that better results could be obtained for scenarios with more sources: this is, in our opinion, the main path to improving the results obtained in this work.

In view of the initial character of this work, our main objective was not to compare the proposed methodology to other approaches, but to analyze its applicability. In this sense, it is possible to state that the above presented results are quite encouraging, although preliminary efforts indicate that the Gaussianization-based solution still requires a certain degree of improvement if one aims to reach the same quality of results of, for instance, the evolutionary mutual information-based approach presented in [8].

5 Conclusions

When dealing with the problem of BSS of post-nonlinear mixtures, researchers usually face two problems: 1) the difficulty of estimating the nonlinearities; and 2) the multimodal character of the underlying optimization task. This article proposes a new method that is capable of meeting these requirements using the FastICA as a local search tool and an evolutionary gaussianization method to allow the adaptation of the nonlinear separating stage. The presented initial results show that the novel solution reaches an interesting tradeoff between performance, speed of convergence and computational effort, which leads us to the conclusion that the idea of allying an evolutionary algorithm to a FastICA-based refinement stage may be decisive in the process of implementing practical and efficient nonlinear separating systems.

References

1. Hyvärinen, A., Oja, E.: Independent Component Analysis. John Wiley and Sons, Chichester (2001)
2. Arons, B.: A review of the cocktail party effect. Journal of the American Voice Input/Output Society 12, 35–50 (2006)
3. Taleb, A., Jutten, C.: Source Separation in Post-Nonlinear Mixtures. IEEE Transactions on Signal Processing 47(10), 2807–2820 (1999)
4. Picinbono, B., Barret, M.: Nouvelle Présentation de la Méthode du Maximum d'Entropie. Traitement du Signal 7(2), 153–158 (1990)
5. Solé-Casals, J., Babaie-Zadeh, M., Jutten, C., Pham, D.T.: Improving Algorithm Speed in PNL Mixture Separation and Wiener System Inversion. In: Proceedings of the Fourth International Workshop on Independent Component Analysis and Blind Signal Separation, ICA, Nara, Japan (2003)
6. Chen, S.S., Gopinath, R.A.: Gaussianization. Technical Report, IBM T. J. Watson Research Center (2000)
7. Lee, T.W., Girolami, M., Bell, A.J., Sejnowski, T.J.: A Unifying Information-Theoretic Framework for Independent Component Analysis. Computers and Mathematics with Applications 39(11), 1–21 (2000)
8. Duarte, L.T., Suyama, R., de Faissol Attux, R.R., Von Zuben, F.J., Romano, J.M.T.: Blind source separation of post-nonlinear mixtures using evolutionary computation and order statistics. In: Rosca, J.P., Erdogmus, D., Príncipe, J.C., Haykin, S. (eds.) ICA 2006. LNCS, vol. 3889, pp. 66–73. Springer, Heidelberg (2006)
9. de Castro, L.N., Timmis, J.: An Artificial Immune Network for Multimodal Function Optimization. In: Proceedings of the IEEE Congress on Evolutionary Computation (2002)
10. de Castro, L.N., Timmis, J.: Artificial Immune Systems: a New Computational Intelligence Approach. Springer, Heidelberg (2002)
11. Ziehe, A., Kawanabe, M., And Harmeling, S.: Separation of post-nonlinear mixtures using ACE and Temporal Decorrelation. In: Proc. Int. Workshop on Independent Component Analysis and Blind Signal Separation – ICA, pp. 433–438 (2001)
12. Zhang, K., Chan, L.-W.: Extended Gaussianization Method for Blind Separation of Post-Nonlinear Mixtures. Neural Computation 17(2), 425–452 (2005)
13. Rojas, F., Puntonet, C.G., Rodriguez-Alvarez, M., Rojas, I., Martin-Clemente, R.: Blind source separation in post-nonlinear mixtures using competitive learning, simulated annealing, and a genetic algorithm. IEEE transactions on systems, man and cybernetics. Part C, Applications and reviews 34(4), 407–416 (2004)

Adapted Deflation Approach for Referenced Contrast Optimization in Blind MIMO Convolutive Source Separation

Rémi Dubroca, Christophe De Luigi, and Eric Moreau

University of Sud Toulon Var, ISITV
LSEET UMR-CNRS 6017
av. G.Pompidou, BP56,
F-83162 La Valette du Var Cedex, France
dubroca@univ-tln.fr, deluigi@univ-tln.fr, moreau@univ-tln.fr

Abstract. The paper deals with the problem of the blind source separation after a MIMO convolutive mixture. The source separation is done with successive extractions of one source signal that is a deflation approach. The extraction stage is performed using a MISO equalizer and based on a third order tensor decomposition. Computer simulations illustrate the good behavior and the usefulness of our algorithm in comparison with algorithms based on fourth order tensor.

Keywords: Contrast Functions, Blind Source Separation, Higher Order Statistics, Tensor Decomposition.

1 Introduction

We consider the blind source separation in a MIMO context. In this case non observable source signals are mixed through an unknown multidimensional convolutive channel. The goal of separation consists of recovering all the source signals. In our study, we proceed in a sequential way, that is a deflation approach based on successive extractions of one source [1].

One extraction stage consists of optimizing criteria based on high order statistics, see e.g. [2,3,4], involving high order dependence of the criteria on the parameters. Through the use of so-called reference signals, new solutions were proposed that consider high order statistics with a quadratic dependence of searched parameters [5,6].

In this paper, first we introduce tensor algebra and best rank-1 tensor approximation which are the basis of our algorithm. Then we expose the link with source separation and define a new criterion based on high order statistics but showing a cubic dependence on parameters thanks to reference signals. The optimization of the proposed criterion is based on best rank-1 approximation of third order tensors. We next perform the source separation in a new algorithm

T. Adali et al. (Eds.): ICA 2009, LNCS 5441, pp. 243–250, 2009.

based on a deflation approach and adapted to the reference signals. Finally, computer simulations are used to show the usefulness and the good behavior of our approach.

2   Tensor Algebra

In the following, tensors are written as calligraphic letters. For the following definitions, we refer to [7].

The scalar product $\langle \boldsymbol{A}, \boldsymbol{B} \rangle$ of two tensors $\boldsymbol{A}, \boldsymbol{B} \in \mathbb{R}^{I_1 \times I_2 \times I_3}$ is defined as

$$\langle \boldsymbol{A}, \boldsymbol{B} \rangle \triangleq \sum_{i_1} \sum_{i_2} \sum_{i_3} \mathcal{A}_{i_1 i_2 i_3} \mathcal{B}_{i_1 i_2 i_3}.$$

The Frobenius norm of $\boldsymbol{A}$ is then $\|\boldsymbol{A}\| \triangleq \sqrt{\langle \boldsymbol{A}, \boldsymbol{A} \rangle}$.

The 1-mode product of $\boldsymbol{A}$ by a matrix $\mathbf{U} \in \mathbb{R}^{J_1 \times I_1}$, denoted by $\boldsymbol{A} \times_1 \mathbf{U}$, is an $(J_1 \times I_2 \times I_3)$-tensor of which the entries are given by

$$(\boldsymbol{A} \times_1 \mathbf{U})_{j_1 i_2 i_3} \triangleq \sum_{i_1} \mathcal{A}_{i_1 i_2 i_3} \mathbf{U}_{j_1 i_1}.$$

We define similarly the 2-mode and the 3-mode products and the following notation, if $I_1 = I_2 = I_3 = J_1$:

$$\boldsymbol{A} \times^3 \mathbf{U} \triangleq \boldsymbol{A} \times_1 \mathbf{U} \times_2 \mathbf{U} \times_3 \mathbf{U}.$$

In the matrix unfoldings $\mathbf{A}_{(1)} \in \mathbb{R}^{I_1 \times (I_2 I_3)}$, $\mathbf{A}_{(2)} \in \mathbb{R}^{I_2 \times (I_3 I_1)}$ and $\mathbf{A}_{(3)} \in \mathbb{R}^{I_3 \times (I_1 I_2)}$ of $\boldsymbol{A}$, the component $\mathcal{A}_{i_1 i_2 i_3}$ is respectively at the $(i_1, (i_2 - 1)I_3 + i_3)$, $(i_2, (i_3 - 1)I_1 + i_1)$, $(i_3, (i_1 - 1)I_2 + i_2)$ coordinates. The High Order Singular Value Decomposition (HOSVD) of $\boldsymbol{A}$ [7] is given by:

$$\boldsymbol{A} = \boldsymbol{S} \times_1 \mathbf{U}^{(1)} \times_2 \mathbf{U}^{(2)} \times_3 \mathbf{U}^{(3)}, \tag{1}$$

where $\boldsymbol{S} \in \mathbb{R}^{I_1 \times I_2 \times I_3}$ and $\mathbf{U}^{(n)} \in \mathbb{R}^{I_n \times I_n}, n \in [1, 2, 3]$ are unitary matrix.

A rank-1 third order tensor $\boldsymbol{B} \in \mathbb{R}^{I_1 \times I_2 \times I_3}$ is equal to the outer product of 3 vectors $\mathbf{u}^{(1)}, \mathbf{u}^{(2)}, \mathbf{u}^{(3)}$ of size I_1, I_2, I_3 respectively. $\mathcal{B}_{i_1 i_2 i_3} = \mathbf{u}^{(1)}_{i_1} \mathbf{u}^{(2)}_{i_2} \mathbf{u}^{(3)}_{i_3}$ for all values of the indices, which is written as $\boldsymbol{B} = \mathbf{u}^{(1)} \circ \mathbf{u}^{(2)} \circ \mathbf{u}^{(3)}$.

The best rank-1 approximation of a third order tensor can be described as follows. Given $\boldsymbol{A} \in \mathbb{R}^{I_1 \times I_2 \times I_3}$, find a scalar λ and unit-norm vectors $\mathbf{u}^{(1)}, \mathbf{u}^{(2)}, \mathbf{u}^{(3)}$ such that the rank-1 tensor $\tilde{\boldsymbol{A}} \in \mathbb{R}^{I_1 \times I_2 \times I_3}$, $\tilde{\boldsymbol{A}} \triangleq \lambda \mathbf{u}^{(1)} \circ \mathbf{u}^{(2)} \circ \mathbf{u}^{(3)}$ minimizes the least-squares cost function

$$f(\tilde{\boldsymbol{A}}) = \|\boldsymbol{A} - \tilde{\boldsymbol{A}}\|^2.$$

This can be solve [8] using a higher-order extension of the power method (HOPM) for matrices.

HOPM
Input: $\boldsymbol{\mathcal{A}} \in \mathbb{R}^{I_1 \times I_2 \times I_3}$
Output: $\tilde{\boldsymbol{\mathcal{A}}} \in \mathbb{R}^{I_1 \times I_2 \times I_3}$

1. Initial values: $\mathbf{u}_0^{(n)}$ is the first column of $\mathbf{U}^{(n)}$ in the HOSVD decomposition, $n \in [1, 2, 3]$.
2. Until convergence:
   - $\tilde{\mathbf{u}}_{k+1}^{(1)} = \boldsymbol{\mathcal{A}} \times_2 \mathbf{u}_k^{(2)^T} \times_3 \mathbf{u}_k^{(3)^T}$
     $\lambda_{k+1}^{(1)} = \|\tilde{\mathbf{u}}_{k+1}^{(1)}\|$ and $\mathbf{u}_{k+1}^{(1)} = \tilde{\mathbf{u}}_{k+1}^{(1)}/\lambda_{k+1}^{(1)}$
   - $\tilde{\mathbf{u}}_{k+1}^{(2)} = \boldsymbol{\mathcal{A}} \times_1 \mathbf{u}_{k+1}^{(1)^T} \times_3 \mathbf{u}_k^{(3)^T}$
     $\lambda_{k+1}^{(2)} = \|\tilde{\mathbf{u}}_{k+1}^{(2)}\|$ and $\mathbf{u}_{k+1}^{(2)} = \tilde{\mathbf{u}}_{k+1}^{(2)}/\lambda_{k+1}^{(2)}$
   - $\tilde{\mathbf{u}}_{k+1}^{(3)} = \boldsymbol{\mathcal{A}} \times_1 \mathbf{u}_{k+1}^{(1)^T} \times_2 \mathbf{u}_{k+1}^{(2)^T}$
     $\lambda_{k+1}^{(3)} = \|\tilde{\mathbf{u}}_{k+1}^{(3)}\|$ and $\mathbf{u}_{k+1}^{(3)} = \tilde{\mathbf{u}}_{k+1}^{(3)}/\lambda_{k+1}^{(3)}$

   Solutions : $\mathbf{u}^{(1)}, \mathbf{u}^{(2)}, \mathbf{u}^{(3)}, \lambda$.
3. $\tilde{\boldsymbol{\mathcal{A}}} = \lambda \mathbf{u}^{(1)} \circ \mathbf{u}^{(2)} \circ \mathbf{u}^{(3)}$.

3   Source Separation

We consider the following noise free convolutive mixing model

$$\mathbf{x}(n) = \sum_{k \in \mathbb{Z}} \mathbf{M}(k)\mathbf{s}(n - k) \triangleq \{\mathbf{M}(z)\}\mathbf{s}(n), \tag{2}$$

where $\mathbf{x}(n)$ is the $(N, 1)$ observation vector $(N \in \mathbb{N}, N \geqslant 2)$, $\mathbf{s}(n)$ is the $(K, 1)$ source vector $(K \in \mathbb{N}^*)$, and $\mathbf{M}(n)$ is the (N, K) matrix corresponding to the impulse response of the convolutive mixing system, whose transfer function is denoted by $\mathbf{M}(z) = \sum_{n \in \mathbb{Z}} \mathbf{M}(n)z^{-n}$. To achieve the blind extraction, we assume that the LTI mixing system is stable, left invertible and the polynomial matrix z-transform $\mathbf{M}(z)$ is irreducible, that involves $N > K$. In the MISO context, the aim is to estimate one row of the separating matrix $\mathbf{W}(z)$, that is a $(1, N)$ vector filter $\mathbf{w}(z)$, called an equalizer, such that the scalar signal

$$y(n) = \sum_{k \in \mathbb{Z}} \mathbf{w}(k)\mathbf{x}(n - k) \tag{3}$$

restores one of the components $s_i(n)$, $i \in \{1, \ldots, K\}$, of the source vector. In this context it is classical to define the corresponding $(1, K)$ global vector filter $\mathbf{g}(z)$ by its impulse response $\mathbf{g}(n) \triangleq \sum_{k \in \mathbb{Z}} \mathbf{w}(k)\mathbf{M}(n - k)$. Hence we have

$$y(n) = \sum_{k \in \mathbb{Z}} \mathbf{g}(n - k)\mathbf{s}(k) \triangleq \{\mathbf{g}(z)\}\mathbf{s}(n) . \tag{4}$$

In this paper, we assume that the source signals $s_i(n)$, $i \in \{1, \ldots, K\}$ are zero-mean, unit variance, i.i.d. random signals and statistically mutually independent, at least up to the order of the considered cumulants. As the source signals are unobservable, there exist some inherent undetermined factors in their estimation. The extraction is done when the global filter reads

$$\exists i_0 \in \{1, \ldots, K\}, \exists l \in \mathbb{Z} \quad g_i(n) \triangleq (\mathbf{g}(n))_i = \alpha \delta_{n-l} \delta_{i-i_0} \tag{5}$$

where $\alpha \in \mathbb{R}, \alpha \neq 0$. The above relation is called the "equalization condition" and expresses the fact that $y(n)$ is equal to one source signal, $s_{i_0}(n - l)$ up to a delay.

We consider the definition of contrasts within the classical context of i.i.d. source signals given in [5] and we propose an new contrast function which is cubic w.r.t. the searched parameters.

Proposition 1. *Using the following fourth-order cross-cumulant, where $z(n)$ is a given reference signal,*

$$\kappa_{3,4,z}\{y(n)\} \triangleq \mathrm{Cum}\{y(n), y(n), y(n), z(n)\} , \tag{6}$$

the function

$$\mathcal{C}_{3,4,z}\{y(n)\} \triangleq |\kappa_{3,4,z}\{y(n)\}| , \tag{7}$$

is a contrast for $y(n)$ in the case of i.i.d. source signals.

The proof is reported in a forthcoming paper. The reference signal has to depend linearly on source signals. In practice, we choose $z(n)$ as the first observation signal. In the same way we consider the extraction based on fourth order cumulant [2,3] without reference, with our notation

$$\mathcal{C}_{4,4}\{y(n)\} \triangleq |\mathrm{Cum}\{y(n), y(n), y(n), y(n)\}|.$$

4   Proposed Algorithm

4.1   Extraction Stage

We assume that the mixing filter admits a MIMO-FIR left inverse filter of length D, which can be considered causal because of the delay ambiguity. The row vectors which define the impulse response can be stacked in the following $(1, QD)$ row vector:

$$\underline{\mathbf{w}} \triangleq (\mathbf{w}(0) \ldots \mathbf{w}(D-1)). \tag{8}$$

We also define the $(QD, 1)$ column vector of observations

$$\underline{\mathbf{x}}(n) \triangleq (\mathbf{x}(n)^T \mathbf{x}(n-1)^T \ldots \mathbf{x}(n-D-1)^T)^T . \tag{9}$$

It is then easily seen that $y(n) = \underline{\mathbf{w}}\,\underline{\mathbf{x}}(n)$. Considering the covariance matrix $\mathbf{R}_{\underline{\mathbf{x}}} = \mathrm{E}\{\underline{\mathbf{x}}(n)\underline{\mathbf{x}}(n)^T\}$, we have $\mathrm{E}\{|y(n)|^2\} = \underline{\mathbf{w}}\mathbf{R}_{\underline{\mathbf{x}}}\underline{\mathbf{w}}^T$.

Now using the multilinearity property of cumulants, we have

$$\kappa_{3,4,z}\{y(n)\} = \sum_{i,j,k} \underline{w}_i \underline{w}_j \underline{w}_k \mathrm{Cum}\{\underline{x}_i(n), \underline{x}_j(n), \underline{x}_k(n), z(n)\}. \tag{10}$$

Thus this relation can be written as a third order tensor decomposition

$$\kappa_{3,4,z}\{y(n)\} = \mathcal{C}_3 \times^3 \underline{\mathbf{w}} \quad \text{where} \quad (\mathcal{C}_3)_{i,j,k} = \text{Cum}\{\underline{x}_i(n), \underline{x}_j(n), \underline{x}_k(n), z(n)\}. \tag{11}$$

The optimization of the contrast function in (7) under the unit power constraint is equivalent to the maximization of $|\mathcal{C}_3 \times^3 \underline{\mathbf{w}}|$ under the constraint $\underline{\mathbf{w}}\mathbf{R_x}\underline{\mathbf{w}}^T = 1$. To ensure that we have a unique solution, $\underline{\mathbf{w}}$ is projected onto the signal subspace. So we decompose $\mathbf{R_x} = \mathbf{UDU}^T$ through SVD, we define $\mathbf{P} = \mathbf{UD}^{1/2}$ and $\mathbf{Q} = \mathbf{D}^{-1/2}\mathbf{U}^T$ in order to project onto the signal subspace: $\tilde{\underline{\mathbf{w}}} = \underline{\mathbf{w}}\mathbf{P}$ and $\tilde{\mathcal{C}}_3 = \mathcal{C}_3 \times^3 \mathbf{Q}$. Thus the optimization is equivalent to the maximization of $|\tilde{\mathcal{C}}_3 \times^3 \tilde{\underline{\mathbf{w}}}|$ under the constraint $\tilde{\underline{\mathbf{w}}}\,\tilde{\underline{\mathbf{w}}}^T = 1$. We propose to realize the above maximization, noted $E_{3,HOPM}$, searching for the best rank-1 approximation of tensor $\tilde{\mathcal{C}}_3$, using the HOPM algorithm. Similarly, the algorithm realizing the maximization of $|\tilde{\mathcal{C}}_4 \times^4 \tilde{\underline{\mathbf{w}}}|$, and using HOPM is called $E_{4,HOPM}$.

4.2   Deflation Approach

After the extraction of one source, we consider now an iterative way, called deflation, to extract all the sources [1,5]. At each stage, we cancel the contributions of the extracted source to the observations using a least mean squares (LMS) method. Then, the resulting problem becomes the separation of the $K - 1$ remaining sources knowing N observations. By an iterative process, we extract all the sources that is to say, the separation of the sources. The deflation principle is explained for two sources in the figure 1.

Consider the matrix $\mathbf{T(M)}$ $QD \times K(L + D - 1)$

$$\mathbf{T(M)} \triangleq \begin{pmatrix} \mathbf{M}(0) & \cdots & \mathbf{M}(L-1) & 0 & \cdots & 0 \\ 0 & \mathbf{M}(0) & \cdots & \mathbf{M}(L-1) & \ddots & \vdots \\ \vdots & \ddots & \ddots & \ddots & \ddots & 0 \\ 0 & \cdots & 0 & \mathbf{M}(0) & \cdots & \mathbf{M}(L-1) \end{pmatrix},$$

and the column vector

$$\underline{\mathbf{s}}(n) \triangleq (\mathbf{s}(n)^T \mathbf{s}(n-1)^T \ldots \mathbf{s}(n-L-D+2))^T,$$

we get $\underline{\mathbf{x}}(n) = \mathbf{T(M)}\underline{\mathbf{s}}(n)$ and $\mathbf{R_x} = \mathbf{T(M)}\mathbf{R_{\underline{s}}}\mathbf{T(M)}^T$, where $\mathbf{R_{\underline{s}}} \triangleq E\{\underline{\mathbf{s}}(n)\underline{\mathbf{s}}(n)^T\}$. If $D \geqslant K(L-1)$, the rank of $\mathbf{R_x}$ is equal to the rank of $\mathbf{R_{\underline{s}}}$. When we cancel the contributions of one source to the observations, this source disappears from the mixture. $L + D - 1$ components of $\underline{\mathbf{s}}(n)$ vanish, and the rank of $\mathbf{R_x}$ fall by the same number. At the P-th extraction, $P - 1$ deflation steps have been done and the rank of $\mathbf{R_x}$ is reduced of $(P-1)(L+D-1)$.

We propose here a method to avoid the rank reduction in the mixing matrix. In our algorithm, the reference directs the maximization to the extraction of a source included in the reference. This property is used in a "fixed point" like method in [5] to make the extraction better through iteration. To extract a new source after a first extraction, we use an observation signal as reference, that does not depend on the last extracted source. We obtain this modified observation signal in cancelling the last extracted source contributions to the observations using a LMS method. It's the same principle as the deflation, but here we propose, not to modify the whole observations but the observations seen in the reference of our criterion. Thus, we still want to extract one of the K sources using N observations, and the rank of $\mathbf{R_x}$ is not modified. In referenced contrast, this method can be seen as a deflation of the reference.

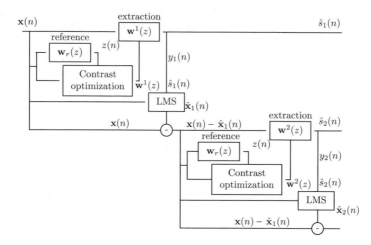

Fig. 1. Deflation principle for 2 sources

Deflation algorithm

Input: $\underline{\mathbf{x}}^0(n) = \underline{\mathbf{x}}(n)$, $r_0 = \text{rank}(\mathbf{R}_{\underline{\mathbf{x}}^0})$ For $i \in \{1, \dots, K\}$:

1. Decompose $\mathbf{R}_{\underline{\mathbf{x}}^{i-1}}$ in singular values, keep the r_{i-1} most important. Extract a source from $\mathbf{x}^{i-1}(n)$ thanks to the method from 4.1, giving $y_i(n)$
2. Deflation stage:
   – substract $y_i(n)$ from $\underline{\mathbf{x}}^{i-1}(n)$ by LMS method, giving $\underline{\mathbf{x}}^i(n)$
   – $r_i = r_{i-1} - (P-1)(L+D-1)$.
3. Output: $y_i(n)$

---

5 Simulation Results

We now propose computer simulations to illustrate the usefulness of the algorithm. We consider real-valued binary source signals (they take their values in {-1,1} with equal probabilities) and real-valued mixing systems. We compare the $E_{3,HOPM}$ and the unreferenced $E_{4,HOPM}$ using the classical deflation approach and the $E_{3,HOPM}$ using our proposed deflation of reference approach. In the $E_{3,HOPM}$, we realize two iterations of a "fixed point" like method. All the results presented below consist in Monte-Carlo simulations involving 100 realizations. At each run, the mixing system (according to a normal distribution) and the N_e sources samples have been drawn randomly. The quality of extraction is measured thanks to a performance index derived from [9] and defined by:

$$\text{ind}(\mathbf{g}) \triangleq \sum_{i \in \{1,\dots,N\}} \left(\frac{\sum\limits_{k \in \mathbb{Z}} |g_i(k)|^2}{\max\limits_{i \in \{1,\dots,N\}} \sum\limits_{k \in \mathbb{Z}} |g_i(k)|^2} \right) - 1. \tag{12}$$

We consider here a mixture of $K = 3$ source signals. The length of the mixing filter is $L = 3$, the number of observations is $N = 4$ and $D = 6$. In figure 2 we present the performances of the extraction for the $E_{3,HOPM}$ and $E_{4,HOPM}$ versus the number of sources samples N_e using the classical deflation method. In figure 3, we show the performances of the $E_{3,HOPM}$ using our deflation of reference method.

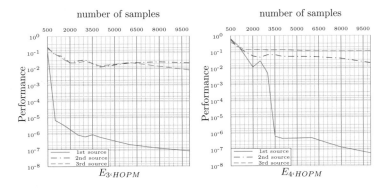

Fig. 2. Deflation, known rank: performance versus number of samples for $E_{3,HOPM}$ and $E_{4,HOPM}$

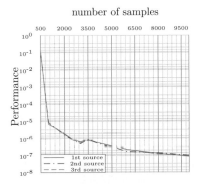

Fig. 3. Deflation of reference: performance versus number of samples for $E_{3,HOPM}$

As classically observed, the successive classical deflation stages lead to degradation of performances (see figure 2). Our proposed approach, directly using the reference signal for the deflation, shows always good performances for all extracted source signals (see figure 3).

6   Conclusion

We have proposed a new referenced contrast function optimized using a best rank-1 tensor decomposition and a new deflation approach for referenced contrast to solve the blind MIMO convolutive separation problem. Computer simulations illustrate interesting features and good performances in comparison with the classical approach of deflation and with the classical kurtosis based contrast.

References

1. Loubaton, P., Regalia, P.: Blind deconvolution of multivariate signals: A deflation approach. In: IEEE International Conference on Communications, ICC 1993 (1993)
2. Tugnait, J.K.: Identification and Deconvolution of Multichannel Linear Non-Gaussian Processes Using Higher Order Statistics and Inverse Filter Criteria. IEEE Transactions on Signal Processing 45(3), 658–672 (1997)
3. Simon, C., Loubaton, P., Jutten, C.: Separation of a class of convolutive mixtures: a contrast function approach. Signal Processing 81, 883–887 (2001)
4. Inouye, Y., Sato, T.: Iterative algorithms based on multistage criteria for multichannel blind deconvolution. IEEE Transactions on Signal Processing 47(6), 1759–1764 (1999)
5. Castella, M., Rhioui, S., Moreau, E., Pesquet, J.C.: Quadratic Higher-Order Criteria for Iterative Blind Separation of a MIMO Convolutive Mixture of Sources. IEEE Transactions on Signal Processing 55(1), 218–232 (2007)
6. Kawamoto, M., Kohno, K., Inouye, Y.: Eigenvector Algorithms Incorporated With Reference Systems for Solving Blind Deconvolution of MIMO-IIR Linear Systems. IEEE Signal Processing Letters 14(12), 996–999 (2007)
7. De Lathauwer, L., De Moor, B., Vandewalle, J.: A Multilinear Singular Value Decomposition. SIAM Journal on Matrix Analysis and Applications 21(4), 1253–1278 (2000)
8. De Lathauwer, L., De Moor, B., Vandewalle, J.: On the Best Rank-1 and Rank-$(R_1,R_2, ...,R_N)$ Approximation of Higher-Order Tensors. SIAM Journal on Matrix Analysis and Applications 21(4), 1324–1342 (2000)
9. Ghennioui, H., Fadaili, E.M., Thirion-Moreau, N., Adib, A., Moreau, E.: A Nonunitary Joint Block Diagonalization Algorithm for Blind Separation of Convolutive Mixtures of Sources. IEEE Signal Processing Letters 14(11), 860–863 (2007)

Blind Order Detection and Parameter Estimation of MISO-FIR Channels

Carlos Estêvão R. Fernandes*, Pierre Comon, and Gérard Favier

I3S Laboratory, University of Nice Sophia Antipolis, CNRS,
2000 route des Lucioles, 06903 Sophia Antipolis Cedex, France
{cfernand,pcomon,favier}@i3s.unice.fr

Abstract. We propose a blind algorithm to determine the order and estimate the coefficients of multiple-input single-output (MISO) communication channels. The proposed order detection method exploits the sensitiveness of a Chi-square test statistic to the non Gaussianity of a stochastic process. Order detection is coupled with channel parameter estimation in a nested-loop operation based on a deflation-type technique using the 4th-order output cumulants. Successively treating shorter and shorter channels, we can also determine the number of sources. Simulation results illustrate the performance of the proposed algorithm.

Keywords: Blind identification, channel parameter estimation, Chi-square test, high-order statistics (HOS), MISO channels, order detection.

1 Introduction

In wireless, satellite and radiocommunication systems, the transmission channel is typically characterized by multipath propagation, inducing frequency-selectiveness and intersymbol interference (ISI). To ensure correct information recovery and avoid performance limitations, it is necessary to reduce or suppress the ISI. Most channel parameter estimation algorithms require knowledge about the channel order (at least an upper bound) [11]. Channel order mismatches may yield bit error rate (BER) floors and signal-to-noise ratio (SNR) penalties [8].

A vast amount of papers can be found in the literature about the blind identification of overdetermined mixtures (those with more sensors than sources), including the instantaneous [4] and the convolutive cases [12,15,16]. More recently, underdetermined mixtures have also been treated [2,7], but the single-input case has received less attention. In this paper, we address the problem of determining the order and estimating the coefficients of finite impulse response (FIR) channels using only the output signals of a multiple-input single-output (MISO) communication system, which is an underdetermined convolutive mixture.

---

* C.E.R. Fernandes is with the Teleinformatics Engineering Dept. of the Federal University of Ceará, Fortaleza, Brazil, and is supported by the CAPES agency of the Brazilian government under the PNPD program.

T. Adali et al. (Eds.): ICA 2009, LNCS 5441, pp. 251–258, 2009.

The main contribution of this paper is the introduction of a blind procedure that detects the presence of signal sources in a MISO mixture, determines channel orders and estimates channel coefficients. The proposed blind channel order detection method is based on a Chi-square hypothesis test using the multivariate estimator of the 4th-order output cumulants. We use a scalar Chi-square test statistic, which is sensitive to the non Gaussianity of a stochastic process, enabling us to detect the order of a FIR channel model by measuring the energy of output cumulants. Based on a cumulant matrix rank-one approximation, a blind channel identification method is also introduced. Using a deflation approach, shorter and shorter channels are identified using the residual 4th-order cumulant information. This deflation-type technique was introduced in [9], in the context of 2×1 MISO channels, with a different order detection method.

The rest of this paper is organized as follows: in section 2, we present the MISO channel model and define output cumulant vectors. In section 3, we propose a channel order detection method based on 4th-order output cumulants. In section 4, we propose a new blind MISO channel identification algorithm and introduce a deflation-type detector that performs combined order detection and parameter estimation. In section 5, computer simulation results illustrate the performance of the proposed algorithm in terms of parameter estimation. In section 6, we draw conclusions and set out perspectives.

2   MISO Channel Model and 4th-Order Output Cumulants

Let us consider a multiuser communication system with an unknown number Q of co-channel users located far apart each other[1] and one single receive antenna. After sampling at the symbol rate, the equivalent baseband output signal $y(n)$ is given as:

$$x_q(n) = \sum_{\ell=0}^{L_q} h_q(\ell) s_q(n - \ell), \quad h_q(0) = 1,$$

$$y(n) = \sum_{q=1}^{Q} x_q(n) + v(n),$$

(1)

where $h_q(\ell)$ are the complex coefficients of the equivalent discrete MISO channel impulse response, including transmit and receive filters. The channel order is asociated with the channel memory and generally expressed in terms of the symbol period. Denoting by L_q the order of channel q, its memory is given by $L_q + 1$. The following assumptions hold:

A1 : The non-observable discrete input signals $s_q(n)$, $q \in [1, Q]$, are complex-valued, non-measurable, ergodic, stationary, mutually (spatially and temporally) independent and identically distributed with symmetric distribution. In addition, we assume that $\mathbb{E}\left\{s_q(n)\right\} = 0$, $\mathbb{E}\left\{|s_q(n)|^2\right\} = 1$ and $\gamma_{4,s_q} \triangleq \mathbb{E}\left\{|s_q(n)|^4\right\} - |\mathbb{E}\left\{s_q(n)^2\right\}|^2 - 2\left(\mathbb{E}\left\{|s_q(n)|^2\right\}\right)^2 \neq 0$.

---

[1] Transmitted signals share the same carrier frequency and use physically different channels.

A2 : The additive noise $v(n)$ is normally distributed, zero-mean with unknown autocorrelation function, and assumed to be independent from $s(n)$.

A3 : The FIR filter representing each channel is assumed to be causal with order L_q, $q \in [1, Q]$, i.e. $h_q(\ell) = 0$, $\forall \ell \notin [0, L_q]$, and $h_q(\ell) \neq 0$ for $\ell = L_q$ and $\ell = 0$.

A4 : All channel orders are bounded by a known value K, i.e. $K > L_q$ $\forall q \in [1, Q]$; without loss of generality, we consider that $K > L_1 > \ldots > L_Q$, so that $L_1 = \max\limits_{1 \leq q \leq Q} (L_q)$.

The 4th-order output cumulants, defined in [3], can be expressed as the sum of the marginal cumulants, i.e. $c_{4,y}(\tau_1, \tau_2, \tau_3) = \sum_{q=1}^{Q} C_q(\tau_1, \tau_2, \tau_3)$, where

$$C_q(\tau_1, \tau_2, \tau_3) = \gamma_{4,s_q} \sum_{\ell-0}^{L_q} h_q^*(\ell) h_q(\ell + \tau_1) h_q^*(\ell + \tau_2) h_q(\ell + \tau_3), \quad q \in [1, Q], \quad (2)$$

where $\gamma_{4,s_q} = c_{4,s_q}(0, 0, 0)$. From A3, we get:

$$C_q(\tau_1, \tau_2, \tau_3) = 0, \quad \forall \, |\tau_1|, |\tau_2|, |\tau_3| > L_q. \quad (3)$$

We can define a 4th-order output cumulant vector $\mathbf{c}_k = \sum_{q=1}^{Q} \mathbf{c}_{k,q}$ for each $k \in [1, K]$, where the pth-element of vector $\mathbf{c}_{k,q} \in \mathbb{C}^{P \times 1}$ is given by:

$$\left[\mathbf{c}_{k,q} \right]_p = C_q(i_p - 1, j_p - 1, k - 1), \quad q \in [1, Q], \; p \in [1, P], \quad (4)$$

with $(i_p, j_p) \in \mathcal{J}$, where the index set $\mathcal{J} = \{(i_1, j_1); \ldots; (i_P, j_P)\}$ is formed of strictly positive integer numbers $i_p, j_p \leq K$. Note from (3) that the marginal 4th-order output cumulants are zero whenever either i_p, j_p, or k are larger than the channel memory $L_q + 1$, and hence, $\mathbf{c}_{k,q} = \mathbf{0}_P$, $\forall k > L_q + 1$. Thus:

$$\mathbf{c}_k = \begin{cases} \mathbf{0}_P, & \text{if } k > L_1 + 1 \\ \mathbf{c}_{k,1}, & \text{if } L_2 + 1 < k \leq L_1 + 1 \\ \mathbf{c}_{k,2} + \mathbf{c}_{k,1}, & \text{if } L_3 + 1 < k \leq L_2 + 1 \\ \quad \vdots & \quad \vdots \\ \mathbf{c}_{k,Q} + \ldots + \mathbf{c}_{k,1}, & \text{if } k \leq L_Q + 1. \end{cases} \quad (5)$$

In most of the real-life situations, the true values of the output cumulants are not available and have to be estimated from the N output signal measurements $y(n)$, $n = 0, \ldots, N - 1$. Due to the ergodicity assumption, the 2nd- and 4th-order moments can be estimated by means of time averages. By combining moment estimates, it is straightforward to derive a cumulant estimator that is asymptotically unbiased and consistent [3,10].

Let us now define the following real-valued vector:

$$\mathbf{z}_k = \left[\text{Re}(\mathbf{c}_k)^\mathsf{T} \quad \text{Im}(\mathbf{c}_k)^\mathsf{T} \right]^\mathsf{T} \in \mathbb{R}^{2P \times 1}, \quad (6)$$

where the operators $\text{Re}(\cdot)$ and $\text{Im}(\cdot)$ return the real and imaginary parts of the vector argument, respectively. Consider the estimator $\hat{\mathbf{z}}_k$ with covariance matrix $\boldsymbol{\Sigma}_k \triangleq \mathbb{E} \left\{ (\mathbf{z}_k - \hat{\mathbf{z}}_k)(\mathbf{z}_k - \hat{\mathbf{z}}_k)^\mathsf{T} \right\} \in \mathbb{R}^{2P \times 2P}$, which can be readily deduced from the *circular* and the *noncircular* 2nd order moments $\mathbf{V}_k \triangleq$

$\mathbb{E}\left\{(\hat{\mathbf{c}}_k - \mathbf{c}_k)(\hat{\mathbf{c}}_k - \mathbf{c}_k)^{\mathsf{H}}\right\}$ and $\mathbf{W}_k \triangleq \mathbb{E}\left\{(\hat{\mathbf{c}}_k - \mathbf{c}_k)(\hat{\mathbf{c}}_k - \mathbf{c}_k)^{\mathsf{T}}\right\}$, which are the $P \times P$ positive-definite Hermitian and complex symmetric covariance matrices of the estimator $\hat{\mathbf{c}}_k$, respectively.

We can now define the scalar multivariate function $\xi_k \triangleq (\mathbf{z}_k - \hat{\mathbf{z}}_k)^{\mathsf{T}}$ $\boldsymbol{\Sigma}_k^{-1}(\mathbf{z}_k - \hat{\mathbf{z}}_k)$ [14]. Expressions for computing the covariance matrices are available for symmetrically distributed sources [1] and also in the general case [5]. These expressions, commonly used in the context of Gaussianity tests [6], are very useful for algorithmic purposes when only output measurements are available.

3   Channel Order Detection

The complex multivariate cumulant estimator $\hat{\mathbf{c}}_k$ is shown to be asymptotically Gaussian with mean equal to $\mathbf{c}_k$ and covariance matrix $\mathbf{V}_k$, i.e. $\hat{\mathbf{c}}_k \sim \mathcal{N}(\mathbf{c}_k, \mathbf{V}_k)$ [3]. As a consequence, we have $\hat{\mathbf{z}}_k \sim \mathcal{N}(\mathbf{z}_k, \boldsymbol{\Sigma}_k)$, as $N \to \infty$. Hence, $\hat{\mathbf{z}}_k$ can be viewed as an asymptotically Gaussian random vector, which can be standardized as[2] $\boldsymbol{\omega}_k = \boldsymbol{\Sigma}_k^{-1/2}(\mathbf{z}_k - \hat{\mathbf{z}}_k) \in \mathbb{R}^{d_k}$, so that $\boldsymbol{\omega}_k$ is asymptotically normal with zero mean and an identity covariance matrix.

Therefore, the scalar random variable ξ_k can be rewritten as $\xi_k = \boldsymbol{\omega}_k^{\mathsf{T}}\boldsymbol{\omega}_k$. Since $\boldsymbol{\omega}_k \sim \mathcal{N}(\mathbf{0}, \mathbf{I})$ as $N \to \infty$, we conclude that ξ_k asymptotically follows a Chi-square distribution with d_k degrees of freedom [13], i.e. $\xi_k \sim \mathcal{X}_{(d_k)}^2$, so that $\mu_{\xi_k} = \mathbb{E}\{\xi_k\} = d_k$ and $\sigma_{\xi_k}^2 = \mathbb{E}\left\{(\xi_k - \mu_{\xi_k})^2\right\} = 2d_k$.

3.1   Order Detection Algorithm

Note from (5) and (6) that $\mathbf{z}_k = \mathbf{0}$ for $k > L_1 + 1$. Replacing $\boldsymbol{\Sigma}_k$ by its estimate, $\boldsymbol{\omega}_k$ becomes $\hat{\boldsymbol{\omega}}_k = -\hat{\boldsymbol{\Sigma}}_k^{-1/2}\hat{\mathbf{z}}_k$, for $k > L_1 + 1$, and we can define:

$$\rho_k \triangleq \hat{\boldsymbol{\omega}}_k^{\mathsf{T}}\hat{\boldsymbol{\omega}}_k = \hat{\mathbf{z}}_k^{\mathsf{T}} \hat{\boldsymbol{\Sigma}}_k^{-1} \hat{\mathbf{z}}_k, \quad \text{for } k > L_1 + 1. \tag{7}$$

Since $\hat{\mathbf{z}}_k$ is a consistent estimator with asymptotically zero bias, ρ_k is asymptotically $\mathcal{X}_{(d_k)}^2$ for $k > L_1 + 1$. This property enables us to detect the presence of signal sources having nonzero 4th-order cumulants at $k \le L_1 + 1$. In this latter case, ρ_k has a *non-central* Chi-square distribution, $nC\mathcal{X}_{(d_k)}^2(\lambda_k)$, with d_k degrees of freedom and parameter $\lambda_k = \mathbf{z}_k^{\mathsf{T}}\boldsymbol{\Sigma}_k^{-1}\mathbf{z}_k$, so that $\mathbb{E}\{\rho_k\} = \lambda_k + d_k$.

Hypothesis test
In the sequel, a hypothesis test is defined, in order to determine the channel order. The null and alternative hypotheses, $H_0(k)$ and $H_1(k)$, respectively, are:

---

[2] We avoid zero elements in the vector $\hat{\mathbf{z}}_k$ by eliminating its m_k entries corresponding to the imaginary part of purely real-valued cumulants. In order to avoid the inversion of an ill-conditioned covariance matrix, we perform the eigenvalue decomposition of $\hat{\boldsymbol{\Sigma}}_k$ and then find its inverse removing the eigenvectors associated with the μ_k smallest eigenvalues, so that its condition number is smaller that a certain threshold ρ. So, the actual dimension of vector $\boldsymbol{\omega}_k$ is $d_k = 2P - m_k - \mu_k$.

$$H_0(k): \ k > L_1 + 1 \quad \Rightarrow \quad \rho_k \sim \mathcal{X}^2_{(d_k)} \quad \Rightarrow \quad \rho_k < \eta_k$$

$$H_1(k): \ k = L_1 + 1 \quad \Rightarrow \quad \rho_k \sim_{nC} \chi^2_{(d_k)}(\lambda_k) \quad \Rightarrow \quad \rho_k \geq \eta_k$$

Under $H_0(k)$, we have $\mathbb{E}\{\rho_k\} = d_k$ and we should expect that $\rho_k < \eta_k$, where η_k is a decision threshold associated with the number d_k of degrees of freedom of the test statistic ρ_k. Under $H_1(k)$, $\mathbb{E}\{\rho_k\} = \lambda_k + d_k$, and we should get $\rho_k \geq \eta_k$. Starting with $k = K$, the test is performed until we find the largest k for which the null hypothesis $H_0(k)$ is rejected, i.e. $\rho_k \geq \eta_k$, which implies $\hat{L}_1 = k - 1$. The non-rejection of $H_0(k)$ for a given k ($\rho_k < \eta_k$) induces a new test on ρ_{k-1}. Knowing the distribution of ρ_k, a decision threshold η_k can be established in order to ensure that, for each $k \in [L_q+2, K]$, the probability of rejection of $H_0(k)$ remains under an acceptable level α, i.e. $\mathrm{P}[\rho_k \geq \eta_k | H_0(k)] \leq \alpha$. This subject is not discussed here, but is addressed in a full-length version of this paper.

4   Blind MISO Channel Identification Algorithm

Equation (5) shows that, if $k = L_1 + 1$ then $\mathbf{c}_k$ can be written only in terms of the coefficients $h_1(\ell)$, since $\hat{\mathbf{c}}_{L_1+1,1} = \hat{\mathbf{c}}_{L_1+1}$. This allows us to estimate the channel coefficients of source $q = 1$. Using the estimated channel coefficients, the marginal cumulants of source $q = 1$ can be approximately calculated for all $k \in [1, L_1 + 1]$. Then, by subtracting $\hat{\mathbf{c}}_{k,1}$ from $\hat{\mathbf{c}}_k$, we get an identical situation with $Q-1$ sources. The algorithm is stopped when the residual cumulants contain no useful information. The following steps are repeated for each $q \in [1, Q]$:

1. Channel order detection: determine $L_q + 1$;
2. Blind channel identification: estimate channel coefficients $\hat{h}_q(\ell)$, $\ell \in [0, L_q]$;
3. Estimation of marginal cumulants: reconstruct $\hat{\mathbf{c}}_{k,q}$ for all $k \in [1, L_q + 1]$ using the estimated channel coefficients.

Before proceeding to user $q+1$, the marginal contribution of user q is subtracted from the estimated output cumulant vector $\hat{\mathbf{c}}_k$.

Parameter estimation based on a rank-1 approximation
Assuming known the channel order L_q and the marginal 4th-order cumulants $C_q(i_p - 1, j_p - 1, L_q)$, we now discuss the estimation of the channel coefficients $\hat{h}_q(\ell)$, with $(i_p, j_p) \in \mathcal{J}$, $q \in [1, Q]$. From (2), we know that $C_q(i_p-1, j_p-1, L_q) = \gamma_{4,s_q} h_q^*(0) h_q(i_p - 1) h_q^*(j_p - 1) h_q(L_q)$, $q \in [1, Q]$, and hence

$$\mathbf{c}_{L_q+1,q} = \gamma_{4,s_q} h_q^*(0) h_q(L_q) \mathbf{g}^{(q)}, \tag{8}$$

where $[\mathbf{g}^{(q)}]_p = h_q(i_p - 1) h_q^*(j_p - 1)$, $p \in [1, P]$. In order to recover the channel parameters, we need to impose some minimal conditions on the index set $\mathcal{J}$ of the cumulants utilized by the algorithm. Simple conditions ensuring correct parameter estimation are $i_p = p$ and $j_p = 1$, $\forall\, p \in [1, P]$, with $P = K$. Using such an index set, we get $[\mathbf{g}^{(q)}]_p = h_q(p - 1) h_q^*(0)$ and (8) becomes $\mathbf{c}_{L_q+1,q} =$

$\gamma_{4,s_q} h_q^{*^2}(0) h_q(L_q) \mathbf{h}^{(q)}$, with $\mathbf{h}^{(q)} = [h_q(0), \ldots, h_q(L_q), 0, \ldots, 0]^\mathsf{T} \in \mathbb{C}^{K \times 1}$. Thus we can construct the matrix $\mathbf{C}_q \in \mathbb{C}^{K \times K}$, as follows:

$$\mathbf{C}_q = \mathbf{c}_{L_q+1,q} \mathbf{c}_{L_q+1,q}^\mathsf{H} = \gamma_{4,s_q}^2 |h_q(0)|^4 |h_q(L_q)|^2 \mathbf{h}^{(q)} \mathbf{h}^{(q)\mathsf{H}}, \tag{9}$$

which is clearly a rank-1 matrix. A solution to (9) is obtained, up to a complex scaling factor, by computing the eigenvector associated with the largest eigenvalue of $\mathbf{C}_q$. By imposing the constraint $h_q(0) = 1$ we avoid the trivial solution and eliminate the scaling ambiguity.

Recomposition of marginal cumulants

Using (5) with $k = L_q + 1$, $q \in [1, Q]$, we can estimate the marginal cumulants of source q as follows:

$$\hat{\mathbf{c}}_{L_q+1,q} = \hat{\mathbf{c}}_{L_q+1} - \sum_{i=1}^{q-1} \bar{\mathbf{c}}_{L_q+1,i}. \tag{10}$$

where $\bar{\mathbf{c}}_{L_q+1,i}$ are the reconstructed cumulant vectors obtained from (2) using the previously estimated coefficient vectors $\hat{\mathbf{h}}_i$, $i \in [1, q-1]$. To achieve this step, the kurtosis of source q needs to be estimated, which can be done from (8), using the estimated marginal cumulants and the estimated channel parameters of source q.

5   Simulation Results

In order to evaluate the proposed detection/identification method, we make use of the two following criteria, which are defined for each signal source $q \in [1, Q]$:

i. The empirical probability of detection, defined as R_q/R, where R_q is the number of Monte Carlo simulations in which the estimated channel order matches its true value ($\hat{L}_q = L_q$) and R is the total number of simulations;
ii The normalized mean squared error (NMSE):

$$\text{NMSE}(q) = \frac{1}{R_q} \sum_{r=1}^{R_q} \frac{\left\| \widehat{\mathbf{h}}_q^{\langle r \rangle} - \mathbf{h}_q \right\|^2}{\|\mathbf{h}_q\|^2}, \quad q \in [1, Q] \tag{11}$$

where $\widehat{\mathbf{h}}_q^{\langle r \rangle}$ is the estimated channel vector of source q at the simulation r.

We consider a MISO channel with $Q = 2$ users, $L_1 = 2$, $L_2 = 1$, and one single receive antenna. It is assumed that channel coefficients do not vary within the duration of one time-slot (N symbols). The results below were obtained with the coefficient vectors $\mathbf{h}_1 = [1.0, \ 1.35 - 0.57\jmath, \ -0.72 + 1.49\jmath]^\mathsf{T}$ and $\mathbf{h}_2 = [1.0, \ -1.14 + 0.23\jmath]^\mathsf{T}$, which have been randomly generated from a continuous complex Gaussian distribution. The channel order upper bound is $K = 4$. Input signals are QPSK modulated and $R = 300$ time-slots have been used.

First, with a fixed SNR of 40dB, we used an output sample data length (N) varying from 10^3 to 10^4. In Fig. 1 (left), notice that detection performance for

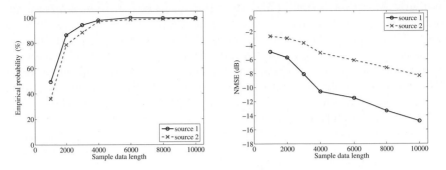

Fig. 1. Blind MISO channel order identification performance: empirical probability of detection (*left*) and NMSE (*right*) as a function of the sample data length, with SNR=40dB ($L_1 = 3$, $L_2 = 1$ and $K = 4$)

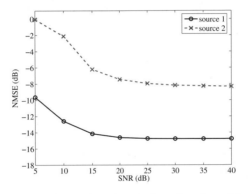

Fig. 2. SNR × NMSE: blind MISO channel identification with $N = 10000$ symbols ($L_1 = 3$, $L_2 = 1$ and $K = 4$)

both sources is satisfactory as soon as the sample data length is greater than $N = 3000$. As expected, and due to consistency of cumulant estimates, the channel identification error tends to zero as the sample size N tends to infinity, as shown in Fig. 1 (right).

In fig. 2, we show the results for the same experiment with $N = 10^4$ and SNR varying from 5 to 40dB. Note that the proposed method is quite robust with respect to additive Gaussian noise at moderate and high SNR levels. For low SNR levels the channel estimation performance is significantly degraded. In both fig. 1 and fig. 2 we note a severe performance loss between sources 1 and 2 due to the error propagation caused by the deflation technique.

6   Conclusion

In this paper, we have proposed a channel order detection algorithm based on HOS hypothesis testing, relying on some properties of the 4th-order output cumulants. Moreover, a new blind algorithm has been introduced, combining order detection and parameter estimation in the context of frequency-selective MISO-FIR channels. Based on a deflation technique, the proposed algorithm successively detects the signal sources, determines the order of their individual transmission channel and estimates the associated channel coefficients, testing for the presence of shorter and shorter channels. Computer simulation results have been shown to illustrate the performance of the proposed method.

The case of channels with the same order needs further investigation. Cumulants are also sensitive to the nonlinearity of a stochastic process, and the output of a linear MISO-FIR filter with i.i.d. inputs cannot be obtained by linear filtering of a single i.i.d. input. Exploiting this property, we can detect the number of sources when channels may have the same length.

References

1. Albera, L., Comon, P.: Asymptotic performance of contrast-based blind source separation algorithms. In: 2nd IEEE SAM Sig. Process. Workshop (2002)
2. Albera, L., Ferréol, A., Comon, P., Chevalier, P.: Blind Identif. of Overcomplete Mixtures of Sources (BIOME). Lin. Algebra Appl. 391(11), 1–30 (2004)
3. Brillinger, D.R.: Time Series: Data analysis and theory. Holden-Day (1981)
4. Cardoso, J.-F.: Source separation using higher order moments. In: Proc. of IEEE ICASSP, pp. 2109–2112 (1989)
5. Comon, P., Chevalier, P., Capdevielle, V.: Performance of contrast-based blind source separation. In: Proc. of SPAWC, pp. 345–348 (1997)
6. Comon, P., Deruaz, L.: Normality tests for coloured samples. In: Proc. IEEE-ATHOS Workshop on HOS, pp. 217–221 (1995)
7. De Lathauwer, L., De Moor, B., Vandewalle, J., Cardoso, J.-F.: Independent component analysis of largely underdetermined mixtures. In: 4th Int. Symp. on ICA, pp. 29–34 (2003)
8. Delmas, J.-P., Gazzah, H., Liavas, A.P., Regalia, P.A.: Statistical analysis of some second-order methods for blind channel identification/equalization with respect to channel undermodeling. IEEE Trans. on Sig. Process. 48(7), 1984–1998 (2000)
9. Fernandes, C.E.R., Comon, P., Favier, G.: Order detection and blind identification of 2 × 1 MISO channels. In: Proc. of IEEE ICASSP, vol. 3, pp. 753–756 (2007)
10. Lacoume, J.-L., Amblard, P.-O., Comon, P.: Statistiques d'ordre supérieur pour le traitement du signal. Masson (1997) (in French)
11. Loubaton, P., Moulines, E., Regalia, P.: Subspace method for blind identification and deconvolution. In: Signal processing advances in wireless and mobile communications, vol. II, pp. 63–112. Prentice-Hall, Englewood Cliffs (2000)
12. Moulines, E., Duhamel, P., Cardoso, J.-F., Mayrargue, S.: Subspace methods for blind ident. of multich. FIR filters. IEEE Trans. Sig. Process. 43(2), 516–525 (1995)
13. Muirhead, R.J.: Aspects of multivariate statistical theory, vol. XIX, p. 673. John Wiley & Sons, New York (1982)
14. Porat, B., Friedlander, B.: Performance analysis of parameter estimation algorithms based on high-order moments. J. Adapt. Control Sig. Process. 3, 191–229 (1989)
15. Tong, L., Perreau, S.: Multichannel blind identification: From subspace to maximum likelihood methods. Proceedings of the IEEE 86, 10 (1998)
16. Tong, L., Xu, G., Kailath, T.: Blind identication based on second-order statistics: A time domain approaches. IEEE Trans. on Inform. Theory 40(2), 340–349 (1994)

Hierarchical Extraction of Independent Subspaces of Unknown Dimensions

Peter Gruber[1], Harold W. Gutch[2], and Fabian J. Theis[2,3]

[1] Computational Intelligence Group, Institute for Biophysics,
University of Regensburg, 93040 Regensburg, Germany
[2] Max Planck Institute for Dynamics and Self-Organization, Göttingen, Germany
[3] CMB, Institute for Bioinformatics and Systems Biology,
Helmholtz Zentrum München, Germany
fabian.theis@helmholtz-muenchen.de
http://cmb.helmholtz-muenchen.de

Abstract. Independent Subspace Analysis (ISA) is an extension of Independent Component Analysis (ICA) that aims to linearly transform a random vector such as to render groups of its components mutually independent. A recently proposed fixed-point algorithm is able to locally perform ISA if the sizes of the subspaces are known, however global convergence is a serious problem as the proposed cost function has additional local minima. We introduce an extension to this algorithm, based on the idea that the algorithm converges to a solution, in which subspaces that are members of the global minimum occur with a higher frequency. We show that this overcomes the algorithm's limitations. Moreover, this idea allows a blind approach, where no a priori knowledge of subspace sizes is required.

Assuming an independent random vector $\mathbf{S}$ that is mixed by an unknown mixing matrix $\mathbf{A}$, Independent Component Analysis (ICA) denotes the task of recovering $\mathbf{S}$, given only the mixed signals, $\mathbf{X} = \mathbf{AS}$. It is known [1] that under mild assumptions, ICA has a solution that is unique up to the obvious indeterminacies of permutation and scaling. Since we are operating blindly i.e. we only see the mixed data set $\mathbf{X}$ and not $\mathbf{S}$, we cannot know if $\mathbf{S}$ actually follows the ICA assumption of statistical independence. ICA only guarantees reconstruction for data that follows the model, so it can be used to analyze only a subset of all random variables, and one cannot even tell in advance if a given data set falls into this subset or not.

Independent Subspace Analysis (ISA) extends the ICA model by allowing dependencies in the source data set $\mathbf{S}$ as introduced by Cardoso and others [2,3]. In the ISA model, $\mathbf{S}$ does not consist anymore of one dimensional independent random variables $S_1, \ldots, S_N$, but rather of random vectors $\mathbf{S}_1, \ldots, \mathbf{S}_N$ such that the random vectors $\mathbf{S}_i$ are mutually independent, but dependencies within the components $\mathbf{S}_i$ are allowed. A commonly used ad-hoc approach to solve the ISA problem is to simply apply an ICA algorithm — with the reasoning that it will find a representation that is as independent as possible — and then to perform

T. Adali et al. (Eds.): ICA 2009, LNCS 5441, pp. 259–266, 2009.

clustering on the resulting sources in order to group components that still show non-neglectable dependencies [4]. This however depends strongly on the specific ICA algorithm; for some (JADE, Infomax) there exists strong evidence that such an approach is feasible [4]. In general however it is unclear if the ICA algorithm can perform the first task of separating any data set as much as possible.

In this contribution we use clustering in the space of linear subspaces to address problems with the FastISA approach [5]. At first we apply it to the independent subspace extraction problem to enhance the stability of the algorithm and to access the reliability of the result. This is done using the results from multiple runs with different initializations and/or bootstrapping. As a second step we apply clustering techniques to the collected subspaces of multiple runs with different extraction parameters, especially the subspace dimensions. The resulting centroids of the clustering then exhibit the independent subspace structure of the data and hence lead to a full independent subspace analysis.

1   Independent Subspace Analysis (ISA)

Generalizing ICA to deal with the case of inherent dependencies within the sources, a first approach can be described as follows: Given an N-dimensional signal $\mathbf{X}$, find an invertible $N \times N$ matrix $\mathbf{A}$ such that $\mathbf{AX} = (\mathbf{S}_1^\top, \ldots, \mathbf{S}_k^\top)^\top$ with mutually independent $\mathbf{S}_j$. It is easy to see that this description is not sufficient, as for any $\mathbf{X}$, choosing $\mathbf{A} = \mathbf{I}_N$ and then $k = 1$ and $\mathbf{S}_1 = \mathbf{X}$ fulfills this condition: Here $\mathbf{X}$ can be seen as one large signal, independent of the (non-existing) rest. Clearly this is not a desired result, so additional constraints are required. The additional requirement of *irreducibility* of the extracted sources $\mathbf{S}_k$ imposes sufficient restrictions to overcome this drawback: A signal $\mathbf{Y}$ is called irreducible if there is no $\mathbf{A}$ such that $\mathbf{AY} = (\mathbf{S}_1^\top, \mathbf{S}_2^\top)^\top$ with $\mathbf{S}_1$ and $\mathbf{S}_2$ being independent. Under the additional assumption of irreducibility of the extracted sources, ISA indeed has a unique solution up to the obvious indeterminacies, choice of basis within the subspaces and order of the subspaces [6].

1.1   The FastISA Algorithm

The idea of the FastISA algorithm as introduced by Köster and Hyvärinen [5] can be described as searching for linear subspaces of $\mathbb{R}^n$ such that the norms of the projections of the mixed signals onto these subspaces are stochastically independent. Given an N-dimensional mixed signal $\mathbf{X}$ and known subspace dimensions $N_1, \ldots, N_l$ (where $\sum_j N_j = N$), FastISA searches for matrices $\mathbf{w}_j \in \mathbb{R}^{N \times N_j}$, $(j = 1, \ldots, l)$ such that the norms of the projections $u_j := \|\mathbf{w}_j^\top \mathbf{X}\|_2$ are independent. The algorithm estimates the demixing matrix $\mathbf{W} = (\mathbf{w}_1, \ldots, \mathbf{w}_l)^\top$ by blockwise updates of its possibly multidimensional rows $\mathbf{w}_i^\top$ using the update step

$$\mathbf{w}_j^{\text{new}} = E\{\mathbf{X}(\mathbf{w}_j^\top \mathbf{X})^\top g(\|\mathbf{w}_j^\top \mathbf{X}\|_2^2)\} - E\{g(\|\mathbf{w}_j^\top \mathbf{X}\|_2^2) + 2(\mathbf{w}_j^\top \mathbf{X})^2 g'(\|\mathbf{w}_j^\top \mathbf{X}\|_2^2)\}\mathbf{w}_j$$

where $g(.)$ and $g'(.)$ are the first and second derivatives of a nonlinear 'score' function $G(.)$. For our simulations we chose $G(x) = \sqrt{x + \gamma}$ where γ is a small, arbitrary constant to aid with stability, and it was chosen to be 0.1. Assuming whitened sources and recordings, which does not lose any generality in offline ISA algorithms, the demixing matrix is known to be orthonormal, so $\mathbf{W}$ is orthonormalized after each update step in the algorithm. The algorithm terminates if the change $\|\mathbf{W}^{\mathrm{new}}\mathbf{W}^{\top}\|$ falls below a threshold. Here $\|.\|$ denotes the anti-blockwise Frobenius norm, that is the sum of the squares of only the values not on the block diagonal. It is known [5] that this algorithm locally converges against a demixing matrix $\mathbf{W}$, so FastISA can be used exactly this way if and only if one already has a good estimate of an actual solution.

1.2   Using Grassmannian Clustering for Global Search and Independent Subspace Extraction

The FastISA cost function is prone to local minima, so it alone cannot be used to solve the ISA problem. To overcome these limitations, we propose a two-level approach. The first step consists of running FastISA multiple times with different initial conditions. We keep only those FastISA outputs (*candidate demixings*) where the algorithm terminated by reaching a fixed point. The algorithm will only sometimes converge to the global minimum of the cost function, but will converge to local minima otherwise. The second step therefore consists of choosing which of the reconstructions represent valid reconstructions. We perform clustering on the linear subspaces defined by the rows of the estimated demixing matrices. We here observe that among the set of candidate demixings those that are real demixings form a rather dense cluster, while the other suggestions are somewhat scattered.

In toy data experiments we found that for typical parameters (total dimension ≤ 10, number of samples ≤ 1000) the correct reconstruction is found in only very few cases ($\leq 5\%$), see figure 1(a). Hence traditional outlier detection algorithms are not very effective since the majority of results are in fact outliers, but we can exploit the fact that the correct cluster is far more densely packed than the rest of the subspaces. To achieve this we calculate the distances of each subspace to all others and select those subspaces having a maximal variance in those distances. This technique yields very good results, see figure 1(b).

We only get good clustering if our choice of a subspace dimension was valid, i.e. if it can be written as a sum of some of the N_j. If this is not the case, that is, if there is no k-dimensional data subset in our original signals, then essentially all candidate demixings are equally incorrect, and we find that we are unable to perform the clustering as mentioned above. This can be used to estimate dimensionality and can effectively be detected in the variance of the resulting cluster, see figure 2.

One approach for Independent Subspace Analysis is the common deflationary algorithm from projection pursuit, but now using Independent Subspace Extraction as basis. Additionally, the FastISA algorithm can easily be extended to not only decompose the data set into two subspaces of given size, but into an arbitrary number of subspaces, as long as their total dimension matches the source dimensionality, see figure 3(b) for the result of such runs.

Input: $\mathbf{X}$ data matrix of dimension d with n samples
Output: set S of subspaces

`collect`:
$i \leftarrow \lfloor d/2 \rfloor$
repeat
1   use FastISA to extract subspaces s_1, s_2 of dimension i and $d - i$
2   add s_1, s_2 to S
3   run `collect` on $P_{s_2}\mathbf{X}$ to get a set of subspaces T (P_{s_2} is the projection onto the subspace s_2)
4   add $P_{s_2}^{-1}s$ (canonical embedding) to S for all s in T
5    $i \leftarrow i - 1$
until $i = 0$;

Algorithm 1. Collect candidate subspaces

All these algorithms suffer from the very frequent local minima in the FastISA algorithm. Hence we propose a completely different independent subspace analysis method based on the FastISA algorithm. At first we recursively collect a big number of candidate subspaces using algorithm 1 and then use clustering (here k-means) on this set. The resulting cluster centroids then will be the sought after independent subspaces. For results of this algorithm see figure 4.

1.3   Metric and Clustering in ISA Subspaces

For clustering we need a notion of similarity between subspaces. The standard approach is to define a metric. The subspaces resulting from an ISA analysis are best viewed as elements of the total Grassmannian space $\bigcup \mathcal{G}_i(\mathbb{R}^n) = \mathcal{G}(\mathbb{R}^n)$ where $\mathcal{G}_i(S) = \{s \text{ subspace } S \mid \dim(s) = i\}$ are the subspaces of dimension i.

The usual metric [7] on the total Grassmannian space is the one induced by the Frobenius norm of the linear projections onto the subspaces:

$$d(\mathbf{x}, \mathbf{y}) := \|\mathbf{x}\mathbf{x}^\top - \mathbf{y}\mathbf{y}^\top\|_F = \sqrt{<\mathbf{x}\mathbf{x}^\top, \mathbf{x}\mathbf{x}^\top - \mathbf{y}\mathbf{y}^\top> - <\mathbf{y}\mathbf{y}^\top, \mathbf{x}\mathbf{x}^\top - \mathbf{y}\mathbf{y}^\top>}$$

where $\mathbf{x}$ and $\mathbf{y}$ are represented by orthogonal bases — note that orthogonal base transformation does not change the right hand side. $\|\mathbf{P}\|_F$ denotes the Frobenius norm and $<\mathbf{P}, \mathbf{Q}>$ the scalar product induced by the Frobenius norm, i.e. the standard scalar product. This metric has the property that subspaces of different dimension have at least distance 1 even if they are contained in each other. For ISA purposes this is not desired and we use a corrected 'metric'

$$d_c(\mathbf{x}, \mathbf{y}) := (<\mathbf{x}\mathbf{x}^\top, \mathbf{x}\mathbf{x}^\top - \mathbf{y}\mathbf{y}^\top><\mathbf{y}\mathbf{y}^\top, \mathbf{x}\mathbf{x}^\top - \mathbf{y}\mathbf{y}^\top>)^{\frac{1}{4}}$$

which is equal to 0 if and only if one of the parameters is a subspace of the other. It is not a metric anymore because it satisfies neither positive-definiteness nor the triangle-equation on $\mathcal{G}$. However it is easily seen that on every $\mathcal{G}_i$ both metrics coincide.

The k-means algorithm is a iterated 2 step minimization of the quadratic quantisation error by replacing all the elements of a cluster by its associated

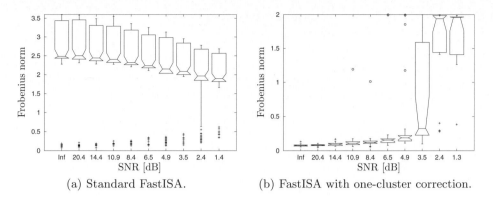

(a) Standard FastISA.   (b) FastISA with one-cluster correction.

Fig. 1. Comparison of standard FastISA and the one-cluster extension on a 5 dimensional data set with 2 independent signals of dimension 2 and 3. The comparison is done with varying levels of additional white noise between no noise and noise with SNR = 1.34. The performance is measured by the anti-blockwise Frobenius norm and the box plot shows the statistics over 80 runs.

cluster centroid. In the first step the clusters are selected by searching for the closest centroid and in the second step the centroids are updated based on the current cluster associations. For the calculation of centroids we have to solve an optimization problem of the form: $f(\mathbf{y}) = \sum_{\mathbf{x}} d(\mathbf{x}, \mathbf{y})^2$. This optimization problem is solved by the eigenvectors of $\sum_{\mathbf{x}} \mathbf{x}\mathbf{x}^\top$ [8,9]. For the corrected metric the situation is more complicated and we have to resort to a gradient descent approach. For $f_c(\mathbf{y}) = \sum_{\mathbf{x}} d_c(\mathbf{x}, \mathbf{y})^2$ we can calculate $\nabla f_c(\mathbf{y}) = \sum_{\mathbf{x}} d_c(\mathbf{x}, \mathbf{y})^{-2}(4 < \mathbf{x}\mathbf{x}^\top, \mathbf{x}\mathbf{x}^\top - \mathbf{y}\mathbf{y}^\top > -2\|\mathbf{x}\mathbf{x}^\top - \mathbf{y}\mathbf{y}^\top\|_F \mathbf{x}\mathbf{x}^\top)\mathbf{y}$. Since both metrics are related, we can use the eigenvector solution as a starting point for the gradient descent. Using simulations we found that the corrections for the centroid are quite small and

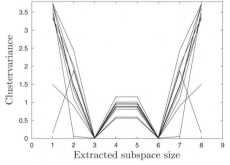
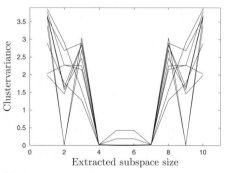

(a) Data set containing subspaces of dimension 3 and 6.

(b) Data set containing subspaces of dimension 2, 4 and 5.

Fig. 2. The subspace dimension detection capability of the one-cluster algorithm is illustrated on 9 and 11 dimensional data sets with 1000 samples. Each test is run 10 times. Note the minima at all combinations of input subspace dimensions 2,4,5,6 = 2+4,7 = 2+5,9 = 4+5 in the second example.

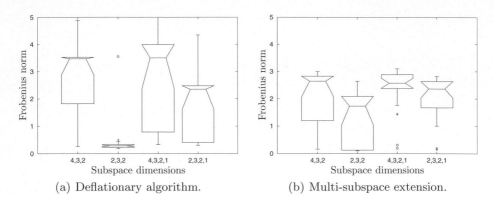

(a) Deflationary algorithm.   (b) Multi-subspace extension.

Fig. 3. The deflationary and the multi-subspace algorithm are applied to different data sets with 1000 samples each (the input subspace dimensions are indicated at the x-axis). The multi-subspace algorithm was run with the one-cluster outlier removal on 100 simple runs. The experiment is repeated 20 times and the box plot shows the statistics of the runs.

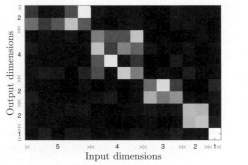
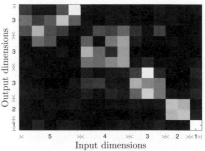

(a) The algorithm extracted subspaces of   (b) The algorithm extracted subspaces of
dimension 2,4,2,2 and 1.   dimension 3,3,3,2 and 1.

Fig. 4. Independent subspace analysis of a 15 dimensional data set with 1000 samples containing 1, 2, 3, 4 and 5 dimensional independent subspaces. The result from 2 different runs are shown. The anti-blockwise Frobenius norms are 0.51 and 1.00 respectively. As comparison: The anti-blockwise Frobenius norm of a JADE run using the mixing matrix to find the correct permutation is 0.67. The graphic shows the absolute value of **WA** where **W** is the estimated demixing matrix. The input dimensions are at the x-axis and the output dimensions are at the y-axis.

can often be neglected (especially in the starting steps of the k-means clustering algorithm).

Clustering in the full Grassmannian space also exhibits the problem of estimating the dimensions of the centroid subspaces. This is a typical dimension estimation problem and we use a Minimum Description Length (MDL) estimator [10,11] as in [12] to select an appropriate dimension. For this we consider the sum of the associated projectors $\mathbf{V}_j = \sum_{\mathbf{x}} \mathbf{x}\mathbf{x}^\top$ for $\mathbf{x} \in S_j$ of each subset in one cluster $S_j \subset \mathcal{G}$ and view this as a covariance matrix of a random vector where the number of samples matches the cluster size. The MDL value is calculated as:

$$p = \operatorname*{argmin}_{p=1,\ldots,m} - \ln \frac{\prod_{j=p+1}^{m} \delta_j^{\frac{1}{m-p}}}{\sum_{j=p+1}^{m} \frac{\delta_j}{m-p}} + \left(\frac{pm - \frac{p+p^2}{2} + 1}{(m-p)n} \right) \left(\frac{1}{2} + \ln \gamma - \frac{1}{2p} \sum_{j=1}^{p} \ln \sqrt{\frac{2\delta_j^2}{n}} \right)$$

where δ_j are the singular values of the covariance matrix and γ a parameter relating the variances to coding lengths (typical values are around 32 which corresponds with a bit depth of the input data of around 50).

2   Simulations

All the toy data in the simulations was constructed similarly as in the works of Köster [5]. At first we evaluate the stability of the FastISA algorithm on a 2-independent subspace problem, i.e. a data set containing 2 independent subspaces. We use generated data with different levels of additional noise and multiple runs to access the gain in stability we get from clustering in independent subspace extraction. Figure 1(a) shows the result of the simple FastISA algorithm and figure 1(b) the results using a one cluster/outlier detection enhanced algorithm described in section 1.2. One can see that the one-cluster algorithm greatly enhances the stability: For example the anti-blockwise Frobenius norm is below 0.1 even for SNR > 4 for the enhanced version whereas the original algorithm does not exceeds norm 2 on average. This makes the algorithm a viable alternative for real world examples.

In the next experiment we try to detect the dimensionality of the contained subspaces. Each test is run 10 times and the cluster variance of the one-cluster is analyzed, see figure 2 for the result. The minima of the cluster variances are clearly at the input subspace dimensions and at non-irreducible combinations thereof.

In a further experiment we compare the deflationary approach to the multi-subspace extension of the FastISA algorithm with one-cluster outlier detection. In figure 3 both algorithms are applied to data sets with varying subspace structure. It is clear that both suffer from the poor convergence properties of the FastISA algorithm in these cases, but the deflationary algorithm outperforms the multi-subspace version. The result is that only in one case (input subspace dimensions 2,3,2) the deflationary approach can extract the correct subspaces on average. In all other cases only a few runs can identify the input subspaces.

Finally, we illustrate the effectiveness of the collect and cluster algorithm on a 15 dimensional data set. The results of two runs are shown in figure 4. The input subspaces are clearly identified but due to the incorrect dimension estimation of the cluster centroids not all input dimensions are recovered.

3   Outlook

We evaluated different techniques to enhance the FastISA algorithm for ISA problems by proposing to use clustering techniques in Grassmannian space.

A crucial part of the clustering in total Grassmannian space is the correct selection of the dimension of the centroid. Especially with the corrected metric this stems from the inclusion property of the metric which often results in overestimating the subspace dimension. Enhancing this would result in a more reliable independent subspace analysis. In the future we want to investigate the feasibility of a divide and conquer approach as described above. We also want to employ other clustering techniques to the collected subspace set to reduce the number of parameters of the independent subspace analysis. Here we want to look especially into hierarchical clustering and softclustering to eliminate the need for prior knowledge of the number of subspaces. Softclustering would also mitigate the centroid dimensionality problem mentioned above.

All algorithms are available as MATLAB/OCTAVE code at `http://www-aglang.uni-regensburg.de`

In future we plan to evaluate and use the algorithm on real world data from functional MRI data containing blocks of dependent multidimensional sources.

References

1. Comon, P.: Independent component analysis - a new concept? Signal Processing 36, 287–314 (1994)
2. Cardoso, J.: Multidimensional independent component analysis. In: Proc. of ICASSP 1998, Seattle (1998)
3. Hyvärinen, A., Hoyer, P.: Emergence of phase and shift invariant features by decomposition of natural images into independent feature subspaces. Neural Computation 12(7), 1705–1720 (2000)
4. Theis, F.: Towards a general independent subspace analysis. In: Proc. NIPS 2006 (2007)
5. Hyvärinen, A., Köster, U.: Fastisa: A fast fixed-point algorithm for independent subspace analysis. In: ESANN, pp. 371–376 (2006)
6. Gutch, H., Theis, F.J.: Independent subspace analysis is unique, given irreducibility. In: Davies, M.E., James, C.J., Abdallah, S.A., Plumbley, M.D. (eds.) ICA 2007. LNCS, vol. 4666, pp. 49–56. Springer, Heidelberg (2007)
7. Edelman, A., Arias, T., Smith, S.: The geometry of algorithms with orthogonality constraints. SIAM Journal on Matrix Analysis and Applications 20(2), 303–353 (1999)
8. Gruber, P., Theis, F.: Grassmann clustering. In: Proc. EUSIPCO 2006, Florence, Italy (2006)
9. Sarlette, A., Sepulchre, R.: Consensus optimization on manifolds (to appear, 2008)
10. Vitányi, P., Li, M.: Minimum description length induction, bayesianism, and kolmogorov complexity. IEEE Transactions on Information Theory 46, 446–464 (2000)
11. Vetter, R., Vesin, J.M., Celka, P., Renevey, P., Krauss, J.: Automatic nonlinear noise reduction using local principal component analysis and MDL parameter selection. In: Proceedings of the IASTED International Conference on Signal Processing Pattern Recognition and Applications (SPPRA 2002), Crete, pp. 290–294 (2002)
12. Gruber, P., Stadlthanner, K., Böhm, M., Theis, F.J., Lang, E.W., Tomé, A.M., Teixeira, A.R., Puntonet, C.G., Saéz, J.M.G.: Denoising using local projective subspace methods. Neurocomputing (2006)

On IIR Filters and Convergence Acceleration for Convolutive Blind Source Separation

Diego B. Haddad[1], Mariane R. Petraglia[2], and Paulo B. Batalheiro[3]

[1] Federal Center for Technological Education, CEFET-RJ
CP 26041-271 , Nova Iguaçu, Brazil
diego@pads.ufrj
[2] Federal University of Rio de Janeiro, PEE/COPPE
68504, 21945-970, Rio de Janeiro, Brazil
mariane@pads.ufrj.br
[3] State University of Rio de Janeiro, CTC/FEN/DETEL
20559-900, Rio de Janeiro, Brazil
bulkool@pads.ufrj.br

Abstract. It is desirable that online configurations of convolutive source separation algorithms present fast convergence. In this paper, we propose two heuristic forms of increasing the convergence speed of a source separation algorithm based on second-order statistics. The first approach consists of using time-varying learning factors, while the second approach employs a recursive estimation of the short-time autocorrelation functions of the outputs. We also verify, through experiments, whether the cost function considered in the derivation of the algorithm yields, in general, good selection of IIR filters to perform the separation.

1 Introduction

A MIMO (multiple-input multiple-output) linear model accurately approximates several transfer functions encountered in practice. Among them are the acoustics transfer-function of a room and the existing multiple paths in telecommunications systems. Considering a MIMO system with P inputs, we can decompose each of its outputs in P components, each of them originating from one of the inputs. This fact allows us to describe each output as a mixture of its inputs. To estimate the inputs of the MIMO system (or mixture system) from its outputs only (without any knowledge about the spectra of the sources nor the mixing system) is referred in the literature as the problem of Blind Source Separation (BSS) or *cocktail party*. This expression refers to the human ability of maintaining a dialog in a party, despite the fact that other conversations are taking place.

An instantaneous mixture model does not contemplate the presence of different delays from the signals to the sensors, nor reverberation or multiple-paths. In this paper we focus on blind source separation techniques considering the more general case of convolutive mixtures, called Convolutive Blind Source Separation (CBSS). A robust technique for the CBSS problem was presented in [1], and is called Generalized Blind Source Separation (GBSS) algorithm.

T. Adali et al. (Eds.): ICA 2009, LNCS 5441, pp. 267–273, 2009.

Most online applications of the GBSS algorithm require fast convergence. However, the convergence rate obtained with such approach is not good enough in many cases. In this paper we propose two techniques with the objective of reducing this problem: i) use of time-varying and distinct learning factors for the different coefficients and ii) recursive estimation of the required statistics, reusing the data. Experiments with IIR separation filters are also included.

2   GBSS Method

The GBSS method employs a linear MIMO separation system composed of FIR filters which, when applied to the mixtures, present, as outputs, estimations of the sources. The coefficients of the separation filters are obtained through the minimization of a cost function. Considering that there are P sources to be separated from the same number P of sensor signals, the nth sample of the qth output can be expressed as

$$y_q(n) = \sum_{p=1}^{P} \sum_{k=0}^{L-1} w_{pq}(k)x_p(n-k), \tag{1}$$

where L is the length of the filters (w_{pq}) of the separation system. Define the $N \times 2L$ matrix

$$\mathbf{W}_{pq} = \begin{bmatrix} w_{pq(0)} & 0 & \cdots & 0 \\ w_{pq}(1) & w_{pq}(0) & \ddots & \cdots \\ \vdots & w_{pq}(1) & \ddots & 0 \\ w_{pq}(L-1) & \vdots & \ddots & w_{pq}(0) \\ 0 & w_{pq}(L-1) & \ddots & w_{pq}(1) \\ \vdots & \vdots & \ddots & \vdots \\ 0 & \cdots & 0 & w_{pq}(L-1) \\ 0 & \cdots & 0 & 0 \\ \vdots & \cdots & \vdots & \vdots \\ 0 & \cdots & 0 & 0 \end{bmatrix}, \tag{2}$$

with N being the data block size. Let $\mathbf{Y}_q(m)$ be the matrix that contains the samples of the mth block of the qth estimated output, given by

$$\mathbf{Y}_q(m) = \begin{bmatrix} y_q(mL) & \cdots & y_q(mL-D+1) \\ y_q(mL+1) & \ddots & y_q(mL-D+2) \\ \vdots & \ddots & \vdots \\ y_q(mL+N-1) & \cdots & y_q(mL-D+N) \end{bmatrix}, \tag{3}$$

where D is the number of blocks which are used in the correlation estimates, and the matrix $\mathbf{X}_p(m)$ of dimension $L \times 2L$ be formed by samples of the pth mixture, such that $\mathbf{Y}_q(m) = \sum_{p=1}^{P} \mathbf{X}_p(m)\mathbf{W}_{pq}$. Combining the input and output matrices of all channels, we get $\mathbf{X}(m) = [\mathbf{X}_1(m) \cdots \mathbf{X}_P(m)]$ and $\mathbf{Y}(m) = [\mathbf{Y}_1(m) \cdots \mathbf{Y}_P(m)]$. Defining the matrix $\mathbf{W}$ with the coefficients of all separation filters as

$$\mathbf{W} = \begin{bmatrix} \mathbf{W}_{11} & \cdots & \mathbf{W}_{1P} \\ \vdots & \ddots & \vdots \\ \mathbf{W}_{P1} & \cdots & \mathbf{W}_{PP} \end{bmatrix}, \tag{4}$$

we can express the output signals in the compact form:

$$\mathbf{Y}(m) = \mathbf{X}(m)\mathbf{W}. \tag{5}$$

The cost function of the online GBSS method is given by

$$\Im(m) = \log\left\{\det\left[\mathrm{bdiag}\left(\mathbf{Y}^T(m)\mathbf{Y}(m)\right)\right]\right\} - \log\left[\det\left(\mathbf{Y}^T(m)\mathbf{Y}(m)\right)\right], \tag{6}$$

where $(\cdot)^T$ denotes the transpose operation, and $\mathrm{bdiag}(\mathbf{A})$ is the operator that zeroes the sub-matrices out of the main diagonal of the matrix $\mathbf{A}$.

For simplicity, let us suppose $P = 2$. The natural gradient for the above cost function is given by (omitting the index m for easing the notation)

$$\nabla_{\mathbf{W}}^{GN} = 2\begin{bmatrix} \mathbf{W}_{12}\mathbf{R}_{\mathbf{y}_2\mathbf{y}_1}\mathbf{R}_{\mathbf{y}_1\mathbf{y}_1}^{-1} & \mathbf{W}_{11}\mathbf{R}_{\mathbf{y}_1\mathbf{y}_2}\mathbf{R}_{\mathbf{y}_2\mathbf{y}_2}^{-1} \\ \mathbf{W}_{22}\mathbf{R}_{\mathbf{y}_2\mathbf{y}_1}\mathbf{R}_{\mathbf{y}_1\mathbf{y}_1}^{-1} & \mathbf{W}_{21}\mathbf{R}_{\mathbf{y}_1\mathbf{y}_2}\mathbf{R}_{\mathbf{y}_2\mathbf{y}_2}^{-1} \end{bmatrix}, \tag{7}$$

where $\mathbf{R}_{\mathbf{y}_p\mathbf{y}_q} = \mathbf{Y}_p^T(m)\mathbf{Y}_q(m)$. The simplified GBSS algorithm (of the NLMS-type) is obtained substituting $\mathbf{R}_{\mathbf{y}_p\mathbf{y}_p}^{-1}$ by the scalar $1/\left[(\mathbf{y}_\mathbf{p})^T\mathbf{y}_\mathbf{p}\right]$, where $\mathbf{y}_\mathbf{p}$ is the first column of $\mathbf{Y}_q(m)$.

The update equation, employing a learning factor μ, is given by

$$\mathbf{W}(m) = \mathbf{W}(m-1) - \mu\nabla_{\mathbf{W}}^{GN}. \tag{8}$$

The matrix $\mathbf{W}$ is Toeplitz (see Eq. (2)), and hence presents redundancies. Therefore, only the first L elements of the first column of the the sub-matrices $\mathbf{W}_{pq}$ are updated.

3   Approach 1: Varying Learning Factors

The GBSS method employs a fixed learning factor μ, that is identical for all parameters $w_{pq}(k)$. Since hundreds, or even thousands, of coefficients are usually required, it is expected that the use of distinct learning factors for the different parameters and of some heuristics for varying such factors, along the adaptation process, could accelerate the convergence of the algorithm [2], [3].

Our first proposal (**VM**-I) was to verify the change in each parameter $w_{pq}(k)$ between two consecutive iterations. If such change at a given iteration has the same sign of the previous iteration, then the learning factor of the corresponding coefficient, denoted by $\mu_{pq}^i(k)$ for the ith iteration, was multiplied by a constant α slightly larger than 1. Otherwise, $\mu_{pq}^i(k)$ was divided by α.

In our simulations, we verified that the **VM**-I algorithm presented a convergence rate larger than the conventional algorithm. However, after convergence, we verified a degradation in the algorithm performance, since some of the learning factors became very large, causing instability.

In order to avoid such instability problem, we limited the values of the learning factors to a constant $C\mu_0$, where C is an arbitrary positive constant and μ_0 is the initial learning factor value. This modification generated the algorithm referred to as **VM**-II. In our simulations, the use of small values for C resulted in more regular convergence, even though the convergence speed was slower. This new algorithm also presented a degraded convergence (not as bad as the **VM**-I algorithm) when compared to the original algorithm. Such degradation is a result of the use of a larger learning factor, which tends to increase the final signal-to-interference ratio (SIR).

Based on the above observations, we propose a modification in the algorithm by introducing a variable limit for $\mu_{pq}^i(k)$. Such limit assumes larger values in the beginning of the adaptation process, to guarantee fast convergence, and decreases after some time, to avoid performance degradation due to use of a large learning factor. We denote the resulting algorithm as **VM**-III. In this algorithm, the variable limit to the learning factor is: $C(i) = 1 + (C_0 - 1)e^{-\lambda(i-1)}$, where i is the iteration number and λ is a new parameter that controls the rate with which each $C(i)$ tends to 1. We observe that $C(1) = C_0$ and that $\lim_{i \to \infty} C(i) = 1$, that is, after increasing the initial convergence, the algorithm tends to behave like the conventional GBSS algorithm.

4 Approach 2: Reusing the Data

In the online implementation of the above BSS algorithms, the matrices $\mathbf{R}_{\mathbf{y_p y_p}}$ are obtained at each iteration from the actual data block only, corresponding to noisy estimates of the autocorrelation matrices. A form of introducing information from other data blocks, without increasing significantly the computational complexity, is to update such estimates in a recursive manner, such as

$$\mathbf{R}_{\mathbf{y_p y_p}}^{(new)} = \gamma \mathbf{R}_{\mathbf{y_p y_p}}^{(old)} + (1 - \gamma)\mathbf{R}_{\mathbf{y_p y_p}}^{(current)}, \tag{9}$$

where γ, known as forgetting factor, is a positive constant smaller than (but close to) 1. We denote the algorithm that incorporates such recursion as GBSS$_{rec}$. Since in such algorithm the data from previous blocks is reused through the above recursion, the resulting convergence rate is improved with respect to that of the conventional GBSS method.

5 Selection of IIR Filters

FIR filters are usually employed in the separation techniques of CBSS, due to their well-behaved characteristics. In this paper we exam the possibility of employing the cost function of Eq. (6) to select IIR filters for the separation system.

The calculation of the gradient for IIR filters is a complex procedure, but it might be worth if the minima of such function correspond to maxima of the SIR.

We propose to evaluate such possibility by varying the coefficients of an IIR filter, obtaining the corresponding output signal and the SIR. Comparing the SIR and the cost function, we can verify if the minimization of the cost function leads us to the correct selection of IIR filters for the separation problem. The results of such evaluation will be shown in Subsection 6.3.

6   Simulation Results

In all experiments, speech signals (of 10 s up to 20 s of duration) sampled at $F_s = 8\ kHz$ were used.

6.1   Experiment 1

In the online implementations, the simplified GBSS algorithm presented better results and tended to be more robust than the non-simplified algorithm. Therefore, we employed the learning factor proposed in **VM**-III in the simplified algorithm. Using the first 512 samples of the transfer functions of a simulated room (obtained from [5] and resampled at 8 kHz) at two different source locations, we generated two mixtures from the signals of two sources. A comparison of the SIR evolutions of the conventional and proposed simplified GBSS algorithms is presented in Fig. 1. A significant improvement in the performance of the **VM**-III algorithm when compared to the simplified GBSS can be noticed. This behavior was observed with large variations in the parameters μ, C_0, α and λ.

6.2   Experiment 2

Choosing four different combinations of the sources positions (with the transfer functions of [5], resampled at 8 kHz and truncated to 512 coefficients) and using $\gamma = 0.8$ for the GBSS$_{rec}$, we obtained the SIR evolutions of the original and proposed algorithm, shown in Fig. 2. We verified in all four different experiments that the GBSS$_{rec}$ algorithm presented better performance than the simplified GBSS algorithm.

6.3   Experiment 3

In this experiment, we used a mixture system with FIR filters of length $U = 3$ and considered FIR and IIR separation filters. Separation FIR filters with $L = 2$ coefficients were optimized by the GBSS algorithm, resulting in a SIR equal to 16.27 dB. First-order IIR separation filters, with transfer functions $W_{pq}(z) = b_{pq}/(1 + a_{pq}z^{-1})$, were also designed. The optimal values for a_{pq} and for b_{pq} were searched in the intervals $[-0.99, 0.99]$ and $[-3, 3]$, respectively. In our simulations we verified that, in general, the IIR system obtained with the cost function of Eq. (6) is not optimum. The most frequent problems found in the filters search

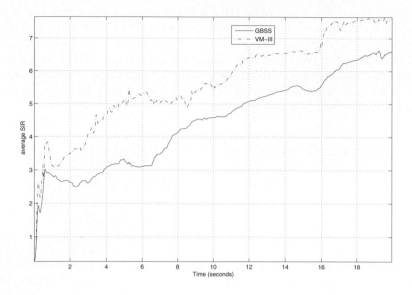

Fig. 1. Average SIR (in dB) of the simplified GBSS and of the proposed **VM**-III algorithm for $\alpha = 1.25$, $\mu = 5 \times 10^{-4}$, $C_0 = 10$, $L = 512$ and $\lambda = 0.005$

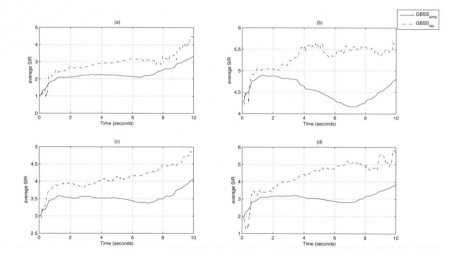

Fig. 2. Average SIR (in dB) for the simplified GBSS and proposed GBSS$_{rec}$ for $\gamma = 0.8$ and $\mu = 10^{-4}$ and four different mixtures

were: i) "local" minima of the cost function did not correspond to maxima of the SIR; ii) "local" maxima of the SIR corresponded to global minima of the cost function; iii) "local" minima of the cost function corresponded to "local" maxima of the SIR. Figure 3 illustrates the occurrence of problem (iii) during the optimization of the parameters of the filter W_{22}.

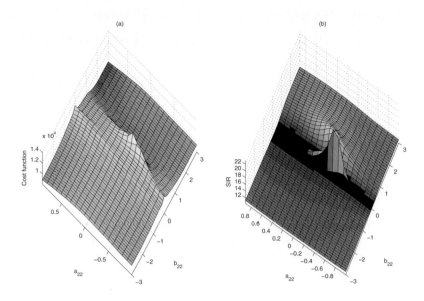

Fig. 3. (a) Cost function values for parameters a_{22} and b_{22}; (b) SIR (in dB) for parameters a_{22} and b_{22}

7 Conclusions

In this paper, we proposed two techniques to speed up the convergence of a blind source separation algorithm. The first technique employed variable learning factor heuristics to improve the initial convergence rate, resulting in the **VM**-III algorithm. The second technique employed recursive estimations for the short-time correlations, resulting in the GBSS$_{rec}$ method. The combination of both techniques seems to be promising for improving the performance of the convolutive blind source separation algorithm proposed in [1]. We also verified, through simulations, that the cost function employed in this algorithm is not adequate for the selection of IIR separation filters.

References

[1] Buchner, H., Aichner, R., Kellermann, W.: A Generalization of Blind Source separation Algorithms for Convolutive Mixtures Based on Second-Order Statistics. IEEE Trans. on Speech and Audio Process. 13, 120–134 (2005)
[2] Jacobs, R.A.: Increased rates of convergence through learning rate adaptation. Neural Networks 1, 295–307 (1988)
[3] Haykin, S.: Neural Networks, 2nd edn. Prentice Hall, Englewood Cliffs (1999)
[4] Haykin, S.: Haykin, Adaptive Filter Theory, 3rd edn. Prentice Hall, Englewood Cliffs (1996)
[5] http://sassec.gforge.inria.fr/

A Cascade System for Solving Permutation and Gain Problems in Frequency-Domain BSS*

Daniel Iglesia, Héctor J. Pérez-Iglesias, and Adriana Dapena

Departamento de Electrónica y Sistemas
Universidade da Coruña, Facultad de Informática
Campus de Elviña 5, 15071, A Coruña, Spain
{dani,hperez,adriana}@udc.es

Abstract. This paper presents a novel technique for separating convolutive mixtures of statistically independent non-Gaussian signals. The time-domain convolution is transformed into several instantaneous mixtures in the frequency-domain. The separation of these mixtures is performed in two steps. First, the instantaneous mixture at the frequency of reference is solved using JADE and the other mixtures are then separated using the Mean Square Error (MSE) criterion. We also present a novel method to select the frequency of reference.

1 Introduction

Blind source separation (BSS) consists of recovering a set of unknown signals (sources) from mixtures recorded by an array of sensors (observations). The term *blind* (or unsupervised) refers to the fact that the sources and the mixing system are completely unknown [2]. In many real-world applications, the sources are received at the antennas by multiple paths [8] and the observations, $x_i(n)$, $i = 1, 2, \cdots, M$, are expressed as convolutive mixtures of the sources $s_i(n)$, $i = 1, 2, \cdots, N$. We have

$$\mathbf{x}(n) = \sum_{l=-\infty}^{\infty} \mathbf{A}(l)\mathbf{s}(n - l) \qquad (1)$$

where $\mathbf{A}(l)$ is an unknown $M \times N$ matrix representing the mixing system whose elements, $a_{ij}(l)$, model the propagation conditions from the j-th source to the i th observation. One way of solving the convolutive problem consists in transforming the convolutive mixture in several instantaneous mixtures by using the Fourier transform given by

$$x_i[\omega_k, n] = \sum_{t=0}^{L-1} x_i(n(L-1) + t)e^{-j\omega_k t}, \quad n = 0, 1, \ldots \qquad (2)$$

* This work has been supported by Xunta de Galicia, Ministerio de Educación y Ciencia of Spain and FEDER funds of the European Union under grants number PGIDT06TIC10501PR and TEC2007-68020-C04-01.

T. Adali et al. (Eds.): ICA 2009, LNCS 5441, pp. 274–281, 2009.

where $\omega_k = 2\pi k/L, k = 0, 1, ..., L - 1$ denotes the frequency and L denotes the number of points in the discrete Fourier transform (DFT). Using the DFT properties, the frequency-domain observation vector at each frequency is an instantaneous mixture of the frequency-domain sources given by

$$\mathbf{x}[\omega_k, n] = \mathbf{A}[\omega_k]\mathbf{s}[\omega_k, n] \tag{3}$$

where $\mathbf{A}[\omega_k]$ represents the mixing coefficients in the frequency domain. As a result, we can recover the frequency-domain sources by using a MIMO system with output

$$\mathbf{y}[\omega_k, n] = \mathbf{W}^H[\omega_k]\mathbf{x}[\omega_k, n] \tag{4}$$

where $\mathbf{W}[\omega_k]$ is the $M \times N$ coefficients matrix which is typically obtained using BSS algorithms proposed in the case of instantaneous mixtures. Notice that if the separating system at each frequency is computed independently, the frequency-domain sources can be recovered in a different order (permutation indeterminacy) and with different gain (gain indeterminacy). When this occurs, the sources cannot be recovered using the inverse DFT (IDFT).

The classical scheme for solving permutation/gain indeterminacy consists of including additional stages in the separating system. For a review of this approaches the reader is referred to [8]. Capdevielle et al. [3] have proposed a method to avoid the permutation indeterminacy by recovering the continuity of the frequency spectra. Due to the independence of the sources, the cross-correlation between the frequency-domain outputs corresponding to different sources is zero and, therefore, the frequency-domain outputs corresponding to the same source can be determined by maximizing the cross-correlation between them. This idea has also been used in [7,10]. The disadvantage of this method is its high computational cost since the Fourier transform is calculated with a window shift of one point. In addition, it can be only used for temporally-colored sources. Computationally less expensive methods for solving the permutation indeterminacy have been proposed in [6,9] for temporally-white sources. The idea is also to cluster the outputs taking into account the statistical dependence between frequency-domain outputs corresponding to the same source but the DFT is applied to non-overlapped windows: Mejuto et al. [9] have proposed maximizing 4th-order cross-cumulants while Dapena et al. [6] utilize cross-correlation.

The main contribution of this work is to present in Section 2 a separating system that does not suffer from permutation/gain indeterminacy. We will show that if the frequency-domain sources corresponding to the same time-domain source are correlated, permutation/gain indeterminacy can be avoided by selecting the separating coefficients in order to minimize the mean squared error (MSE) between the frequency-domain outputs in adjacent frequencies. Using this strategy, all the frequency-domain sources are extracted with the same order and gain, and the sources can be recovered using the frequency-to-time transform. Section 3 focuses on the problem of selecting an adequate frequency of reference. Section 4 presents some simulation results and Section 5 contains the conclusions.

2   The FD-BSS Cascade System

In this section, we propose a novel Frequency-Domain Blind Source Separation (FD-BSS) system that does not suffer from permutation/gain indeterminacy. We assume the following conditions:

C1. The convolutive mixture can be transformed into several instantaneous mixtures using the DFT given in equation (2).

C2. The cross-correlation between the sources in different frequencies (ω_k and ω_r) satisfies

$$E[s_i[\omega_k, n]s_i^*[\omega_r, n]] \neq 0 \tag{5}$$

$$E[s_i[\omega_k, n]s_j^*[\omega_r, n]] = 0 \quad \text{if } i \neq j \tag{6}$$

where the cross-correlation is estimated by sample averaging over the R samples

$$E[s_i[\omega_k, n]s_j^*[\omega_r, n]] = \frac{1}{R} \sum_{n=0}^{R-1} s_i[\omega_k, n]s_j^*[\omega_r, n] \tag{7}$$

C3. The mixing matrix $\mathbf{A}[\omega]$ in all the frequencies is invertible.

The spectral correlation C2 is crucial to avoid the permutation/amplitude ambiguities. In section 4 we will show that signals typically used in digital communication satisfied this condition.

Figure 1 shows the proposed separating system. Each particular observation, $x_j(t), j = 1, ..., M$, is split into non-overlapped segments of L points and the L-DFT is computed over them. The next step consists in computing the separating

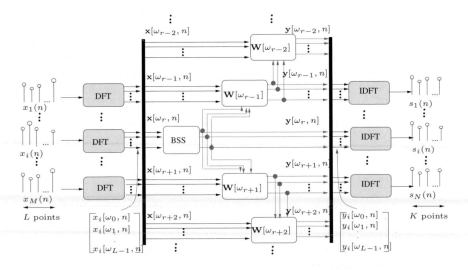

Fig. 1. Scheme of the FD-BSS Cascade System

coefficients at *one* specific frequency, denoted by ω_r, by using a given unsupervised algorithm whose output, $\mathbf{y}[\omega_r, n]$, will be used in the adjacent frequencies $r - 1$ and $r + 1$. The separating matrices in these frequencies are computed by using the MSE criterion defined by

$$\mathbf{W}[\omega_k] = \arg \min_{\mathbf{W}[\omega_k]} J(\mathbf{W}[\omega_k])$$

$$= \arg \min_{\mathbf{W}[\omega_k]} E[|\mathbf{y}[\omega_k, n] - \mathbf{\Delta}_k \, \mathbf{y}[\omega_r, n]|^2], \quad k \in \{r - 1, r + 1\} \quad (8)$$

where $\mathbf{\Delta}_k$ is an $N \times N$ diagonal matrix that must be determined by taking into account the statistics of the sources. Notice that condition C2 guarantees that the frequency-domain sources corresponding to the same time-domain sources are correlated and, therefore, the criterion (8) can be used to measure the dependence between $\mathbf{y}[\omega_k, n]$ and $\mathbf{y}[\omega_r, n]$.

The solution to the MSE criterion (8) is given by

$$\mathbf{W}_o[\omega_k] = \mathbf{\Delta}_k^* \, \mathbf{R_x}^{-1}[\omega_k, \omega_k] \mathbf{R_{xy}}[\omega_k, \omega_r] \quad (9)$$

$$\text{where } \mathbf{R_x}[\omega_k, \omega_k] = E[\mathbf{x}[\omega_k, n]\mathbf{x}^H[\omega_k, n]]$$

$$\mathbf{R_{xy}}[\omega_k, \omega_r] = E[\mathbf{x}[\omega_k, n]\mathbf{y}^H[\omega_r, n]] \quad (10)$$

If $\mathbf{R_x}[\omega_k, \omega_k]$ is not full rank, a pseudo-inverse is computed instead of the inverse in equation (9).

The outputs of these frequency $\mathbf{y}[\omega_k, n] = \mathbf{W}_o^H[\omega_k]\mathbf{x}[\omega_k, n]$ are now used in the adjacent frequencies. The last stage of our separating system consists of computing the IDFT of each output in all the frequencies, $y_i[\omega_k, n]$, $k = 0, ..., L - 1$, to obtain the estimation of the sources in the time domain.

2.1   The Matrix $\mathbf{\Delta}_k$

A crucial task in this approach is to determine the matrix $\mathbf{\Delta}_k$ in order to ensure the same permutation and gain in all the frequencies. Using $\mathbf{x}[\omega_k, n] = \mathbf{A}[\omega_k]\mathbf{s}[\omega_k, n]$ and $\mathbf{y}[\omega_r, n] = \mathbf{G}[\omega_r]\mathbf{s}[\omega_r, n]$, the matrix $\mathbf{R_{xy}}[\omega_k, \omega_r]$ can be written as

$$\mathbf{R_{xy}}[\omega_k, \omega_r] = \mathbf{A}[\omega_k]\mathbf{R_s}[\omega_k, \omega_r]\mathbf{G}^H[\omega_r] \quad (11)$$

where $\mathbf{R_s}[\omega_k, \omega_r] = E[\mathbf{s}[\omega_k, n]\mathbf{s}^H[\omega_r, n]]$. By a similar reasoning, the other term in equation (10) can be written as follows

$$\mathbf{R_x}[\omega_k, \omega_k] = \mathbf{A}[\omega_k]\mathbf{R_s}[\omega_k, \omega_k]\mathbf{A}^H[\omega_k] \quad (12)$$

where $\mathbf{R_s}[\omega_k, \omega_k] = E[\mathbf{s}[\omega_k]\mathbf{s}^H[\omega_k]]$. Substituting (11) and (12) in (9), the separating matrix takes the form

$$\mathbf{W}_o[\omega_k] = \mathbf{\Delta}_k^* \, (\mathbf{A}[\omega_k]\mathbf{R_s}[\omega_k, \omega_k]\mathbf{A}^H[\omega_k])^{-1}(\mathbf{A}[\omega_k]\mathbf{R_s}[\omega_k, \omega_k]\mathbf{G}^H[\omega_k])$$

$$= \mathbf{\Delta}_k^* \mathbf{A}^{-H}[\omega_k]\mathbf{R_s}^{-1}[\omega_k, \omega_k]\mathbf{R_s}[\omega_k, \omega_r]\mathbf{G}^H[\omega_r] \quad (13)$$

where $\mathbf{A}^{-H}[\omega_k] = (\mathbf{A}^H[\omega_k])^{-1}$. Using this equation we can express the gain matrix at the k-th frequency as follows

$$\begin{aligned}
\mathbf{G}[\omega_k] &= \mathbf{W}_o^H[\omega_k]\mathbf{A}[\omega_k] \\
&= \mathbf{\Delta}_k \ \mathbf{G}[\omega_r]\mathbf{R_s}^H[\omega_k,\omega_r]\mathbf{R_s}^{-H}[\omega_k,\omega_k] = \mathbf{\Delta}_k\mathbf{G}[\omega_r]\mathbf{P}[\omega_k,\omega_r]
\end{aligned} \quad (14)$$

where $\mathbf{R_s}^{-H}[\omega_k,\omega_k]=(\mathbf{R_s}[\omega_k,\omega_k]^H)^{-1}$ and $\mathbf{P}[\omega_k,\omega_r]=\mathbf{R_s}^H[\omega_k,\omega_r]\mathbf{R_s}^{-H}[\omega_k,\omega_k]$. In particular, when the condition C2 is verified, $\mathbf{R_s}[\omega_k,\omega_r]$ and $\mathbf{R_s}[\omega_k,\omega_k]$ are diagonal matrices with entries $E[s_i[\omega_k,n]s_i^*[\omega_r,n]]$ and $E[|s_i[\omega_k,n]|^2]$, respectively. This implies that $\mathbf{P}[\omega_k,\omega_r]$ is a diagonal matrix

$$\mathbf{P}[\omega_k,\omega_r] = \begin{bmatrix} \frac{E[s_1^*[\omega_k,n]s_1[\omega_r,n]]}{E[|s_1[\omega_k,n]|^2]} & 0 & \cdots & 0 \\ 0 & \frac{E[s_2^*[\omega_k,n]s_2[\omega_r,n]]}{E[|s_2[\omega_k,n]|^2]} & \cdots & 0 \\ \vdots & & \cdots & \vdots \\ 0 & 0 & \cdots & \frac{E[s_N^*[\omega_k,n]s_N[\omega_r,n]]}{E[|s_N[\omega_k,n]|^2]} \end{bmatrix} \quad (15)$$

In order to obtain $\mathbf{G}[\omega_k] = \mathbf{G}[\omega_r]$ in expression (14), the matrix $\mathbf{\Delta}_k$ must be computed by inverting the matrix $\mathbf{P}[\omega_k,\omega_r]$ given in equation (15).

3   The Reference Frequency

A new question arises when the convolutive problem is transformed into instantaneous mixtures: not all instantaneous separation algorithms are designed for separating mixtures of complex-valued sources and the convergence of some algorithms, like Complex FastICA [1], depend on the selection of some non-linearity functions and step-size parameters.

In order to separate the sources at the reference frequency, we have selected the blind identification algorithm by Joint Approximate Diagonalization of Eigenmatrices (JADE) proposed in [4] which works with complex-valued signals and whose convergence does not depend on non-linearities. The JADE algorithm can be described by the following steps[1]:

Step 1. Compute the whitening matrix $\mathbf{U}[\omega_r]$ from the sample covariance $\mathbf{R_x}[\omega_r,\omega_r]$ and obtain the whitened process $\mathbf{z}[\omega_r] = \mathbf{U}[\omega_r]\mathbf{x}[\omega_r]$.

Step 2. Compute the 4th-order cumulants matrices $\mathbf{C}^{k,l} = cum(\mathbf{z}[\omega_r], \mathbf{z}^H[\omega_r], z_k[\omega_r], z_l^*[\omega_r])$, $k,l = 1,...,M$ of $\mathbf{z}[\omega_r]$ with dimension $M \times M$.

Step 3. Compute the $M^2 \times M^2$ matrix $\mathbf{B} = [\hat{\mathbf{c}}^{1,1}, \hat{\mathbf{c}}^{1,2}, ..., \hat{\mathbf{c}}^{1,M}, ..., \hat{\mathbf{c}}^{M,M}]$ where $\hat{\mathbf{c}}^{i,j} = [(\mathbf{c}_1^{i,j})^T, (\mathbf{c}_2^{i,j})^T, ..., (\mathbf{c}_M^{i,j})^T]^T$ is an $M^2 \times 1$ vector formed by the columns of $\mathbf{C}^{i,j}$ ($\mathbf{c}_k^{i,j}$ denotes the k-th column of $\mathbf{C}^{i,j}$).

Step 4. Perform the eigendecomposition of $\mathbf{B}$ and select the N more significant eigenpairs $\{\lambda_i, \mathbf{m}_i\}, i = 1,...,N$. From these eigenpairs, compute the $M \times M$ matrices $\mathbf{M}_i = \lambda_i[\mathbf{m}_{i(1:M)} \ \mathbf{m}_{i(M+1:2M)}....]$, $i = 1,...,N$.

---

[1] A Matlab implementation of JADE is available on *www.tsi.enst.fr/˜cardoso/guidesepsou.html*

Step 5. Diagonalize jointly the set of matrices $\mathbf{M}_i$, $i = 1, ..., N$ to obtain the matrix $\mathbf{W}[\omega_r]$. This step is implemented by extending the single-matrix Jacobi technique to several matrices as described in [5].

Step 6. Estimate the mixing matrix $\mathbf{A}[\omega_r]$ as $\hat{\mathbf{A}}[\omega_r] = \mathbf{U}^H[\omega_r]\mathbf{W}[\omega_r]$ and recover the frequency-domain sources as $\mathbf{y}[\omega_r, n] = \hat{\mathbf{A}}^{-1}[\omega_r]\mathbf{x}[\omega_r, n]$.

JADE provides an adequate separation in most case but we have observed that the degree of separation depends on the eigenvalue spread of the matrix diagonalized in Step 5. We propose the following procedure to select the reference frequency:

– Set as reference frequency $r = 0$.

Step 1. Compute JADE for the frequency ω_r and evaluate the eigenvalue spread of the matrix diagonalized in Step 5. We define the eigenvalue spread as follows

$$e_i = \sum_{i=1} |\lambda_i - \lambda_{i+1}| \tag{16}$$

where λ_i, $i=1, 2..$ denote the eigenvalues of this matrix, with $\lambda_i > \lambda_{i+1}$. The case of N matrices to diagonalize, we take the averaged eigenvalue spread.

Step 2. If the eigenvalue spread is greater than a threshold β, use this frequency as reference. If this not the case, set $r = r + 1$ and go to Step 1.

If the eigenvalue spread is less than β for all the frequencies, we set r to the frequency with the highest eigenvalue spread. Note that this procedure increases the computational load because JADE is used in several frequencies. As shown in Section 4, however, this scheme significantly improves the performance of the FD-BSS Cascade system.

4 Simulation Results

In this section we will illustrate the behavior of the proposed separating system. We have used complex-valued temporally-white sources (4-QAM) divided in blocks of $K = 11$ symbols. The number of blocks has been 5,000. The mixing system is modeled as FIR filters of $P = 6$ taps with randomly generated coefficients. The L-DFT has been applied to non-overlapped windows of $L = K + P - 1 = 16$ points of the observations.

First, we will determine the matrix $\boldsymbol{\Delta}_k$ that must be used for temporally-white signals. Note that the sources have been transmitted in block of K points and, therefore, the sources at the k-th frequency have the form

$$s_i[\omega_k, n] = \sum_{t=0}^{K-1} s_i(n(K-1)+t)e^{-j\omega_k t} = \mathbf{f}_k^H \mathbf{s}_i(n) \tag{17}$$

where $\mathbf{f}_k = \mathbf{f} = [1, e^{j2\pi k/L}, \cdots, e^{j2\pi k(K-1)/L}]^T$ and $\mathbf{s}_i(n) = [s_i(n), s_i(n+1), ..., s_i(n+K-1)$. Using these expressions, it is straightforward to determine that the matrix $\boldsymbol{\Delta}_k$ must be selected as follows

$$\mathbf{\Delta}_k = \frac{E[|s_i[\omega_k, n]|^2]}{E[s_i^*[\omega_k, n]s_i[\omega_r, n]]}\mathbf{I} = \frac{K}{\mathbf{f}_k^H \mathbf{f}_r}\mathbf{I} = \frac{K}{\sum_{m=0}^{K-1} e^{-j2\pi(r-k)m/L}}\mathbf{I} \qquad (18)$$

Another interesting question is how to select the parameter β. We have performed 1,000 independent simulations varying the reference frequency from 0 to 15 and we have evaluated the eigenvalue spread (equation (16)) in Step 5 of JADE and the final error probability. The results are plotted in Figure 2. Note that a poor performance has been obtained in some cases where the eigenvalue spread is close to 0.01. For this reason, we have selected $\beta = 0.02$.

Now, we will compare the performance obtained using the following strategies:

- The mixtures in all the frequencies are separated by using JADE. In this case, the permutation/gain ambiguity is not eliminated.
- The mixtures in all the frequencies are separated by using JADE. The permutation indeterminacy is solved using the method proposed in [9] and the gain indeterminacy is solved using the method proposed in [6].
- The mixtures in all the frequencies are separated by using JADE. Both the permutation and the gain indeterminacies are solved using the approach proposed in [6].
- FD-BSS Cascade system with $r = 0$.
- FD-BSS Cascade system selecting the reference frequency using the procedure described in Section 3 with $\beta = 0.02$.

Table 1 shows the error probability and the time required to recover 5,000 symbols (measured using Matlab code running in a Centrino Duo Processor at 1.66 GHz). We have averaged the results of 1,000 independent simulations. The low error probability and computational load of the FD-BSS Cascade system is apparent. Notice that using $\beta = 0.02$ the time remains low and the probability error is considerably reduced.

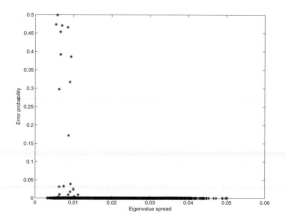

Fig. 2. Error probability versus eigenvalue spread

Table 1. Comparison between frequency-domain BSS approaches in terms of error probability and processing time

	JADE	JADE & Permutation [9] & Gain [6]	JADE & Permutation [6] & Gain [6]	FD-BSS Cascade ($r = 0$)	FD-BSS Cascade($\beta = 0.02$)
Error probability	0.67	1.40×10^{-3}	5.40×10^{-3}	2.30×10^{-3}	8.45×10^{-7}
Time (sec.)	1.9440	2.3216	2.1953	0.1950	0.3230

5 Conclusions

In this paper we have presented a simple strategy to separate convolutive mixtures of statistically independent sources. We have shown that in applications where there is statistical dependence between the frequency-domain sources in different frequencies, it is possible to solve the permutation/gain indeterminacy by using a simple MSE-based unsupervised strategy. The basic idea is to separate the instantaneous mixture at a specific frequency using a BSS algorithm and use these signals to extract the frequency-domain sources in the other frequencies. Finally, the sources are recovered by applying the IDFT to the frequency-domain outputs.

References

1. Bingham, E., Hyvärinen, A.: A Fast Fixed-point Algorithm for Independent Component Analysis of Complex-valued Signals. International Journal of Neural Systems 10(1), 1–8 (2000)
2. Cao, X., Wen-Liu, R.: General Approach to Blind Source Separation. IEEE Transactions on Signal Processing 44(4), 562–571 (1996)
3. Capdevielle, V., Serviére, C., Lacoume, J.L.: Blind Separation of Wide-band Sources in the Frequency Domain. In: Proc. ICASSP 1995, pp. 2080–2083 (1995)
4. Cardoso, J.F., Souloumiac, A.: Blind Beamforming for non-Gaussian Signals. IEE-Proceedings-F 140(6), 362–370 (1993)
5. Cardoso, J.F., Souloumiac, A.: Jacobi Angles for Simultaneous Diagonalization. SIAM Journal on Matrix Analysis and Applications 17(1) (January 1996)
6. Dapena, A., Castedo, L.: A Novel Frequency Domain Approach for Separating Convolutive Mixtures of Temporally White Signals. Digital Signal Processing 13(2), 304–316 (2003)
7. Gelle, G., Colas, M., Servière, C.: BSS for Fault Detection and Machine Monitoring Time or Frequency Domain Approach? In: Proc. ICA 2000, pp. 555–560 (June 2000)
8. Pedersen, M., Larsen, J., Kjems, U., Parra, L.: A Survey of Convolutive Blind Source Separation Methods. In: Springer Handbook on Speech Processing and Speech Communication. Springer, Heidelberg (2007)
9. Mejuto, C., Dapena, A., Castedo, L.: Frequency-domain Infomax for Blind Separation of Convolutive Mixtures. In: ICA 2000, pp. 315–320 (2000)
10. Servière, C.: Separation of Speech Signals under Reverberant Conditions. In: EUSIPCO 2004, pp. 1693–1696 (2004)

A Method for Filter Shaping in Convolutive Blind Source Separation

Radoslaw Mazur and Alfred Mertins

Institute for Signal Processing, University of Lübeck, 23538 Lübeck, Germany
{mazur,mertins}@isip.uni-luebeck.de

Abstract. An often used approach for separating convolutive mixtures is the transformation to the time-frequency domain where an instantaneous ICA algorithm can be applied for each frequency separately. This approach leads to the so called permutation and scaling ambiguity. While different methods for the permutation problem have been widely studied, the solution for the scaling problem is usually based on the minimal distortion principle. We propose an alternative approach that shapes the unmixing filters to have an exponential decay which mimics the form of room impulse responses. These new filters still add some reverberation to the restored signals, but the audible distortions are clearly reduced. Additionally the length of the unmixing filters is reduced, so these filters will suffer less from circular-convolution effects that are inherent to unmixing approaches based on bin-wise ICA followed by permutation and scaling correction. The results for the new algorithm will be shown on a real-world example.

1 Introduction

The blind source separation (BSS) problem has been widely studied for the instantaneous mixing case and several efficient algorithms exist [1,2,3]. However, in a real-world scenario in an echoic environment, the situation becomes more difficult, because the signals arrive several times with different time lags, and the mixing process becomes convolutive. Although some time-domain methods for solving the convolutive mixing problem exist [4,5], the usual approach is to transform the signals to the time-frequency domain, where the convolution becomes a multiplication [6] and each frequency bin can be separated using an instantaneous method. This simplification has a major disadvantage though. As every separated bin can be arbitrarily permuted and scaled, a correction is needed. When the permutation is not correctly solved the separation of the entire signals fails. A variety of different approaches has been proposed to solve this problem utilizing either the time structure of the signals [7,8,9] or the properties of the unmixing matrices [10,11,12]. When the scaling is not corrected, a filtered version of the signals is recovered. In [13,14] the authors proposed a postfilter method that aims to recover the signals as they have been recorded at the microphones, accepting the distortions of the mixing system while not adding new ones. This concept appears to be quite reasonable, but the desired goal is

T. Adali et al. (Eds.): ICA 2009, LNCS 5441, pp. 282–289, 2009.

often not exactly achieved in practice due to circular convolution artifacts that stem from the bin-wise independent design of the unmixing filters which does not obey the filter-length constraints known from fast-convolution algorithms. This problem has been addressed in [15], where the authors applied a smoothing to the filters in the time domain in order to reduce the circular-convolution effects.

In this paper, we propose a new method for solving the scaling ambiguity with the aim of shaping the unmixing filters to have an exponential decay. This mimics the behavior of room impulse responses and reduces the reverberation. In order to achieve this, we calculate the dependency between the scaling factors and the impulse responses of the unmixing filterbank and calculate the scaling factors that shape the desired form.

2 The Framework for Mixing and Blind Unmixing

The instantaneous mixing process of N sources into N observations can be modeled by an $N \times N$ matrix $\mathbf{A}$. With the source vector $\boldsymbol{s}(n) = [s_1(n), \ldots, s_N(n)]^T$ and negligible measurement noise, the observation signals are given by

$$\boldsymbol{x}(n) = [x_1(n), \ldots, x_N(n)]^T = \mathbf{A} \cdot \boldsymbol{s}(n). \tag{1}$$

The separation is again a multiplication with a matrix $\mathbf{B}$:

$$\boldsymbol{y}(n) = \mathbf{B} \cdot \boldsymbol{x}(n) \tag{2}$$

with $\boldsymbol{y}(n) = [y_1(n), \ldots, y_N(n)]^T$. The only source of information for the estimation of $\mathbf{B}$ is the observed process $\boldsymbol{x}(n)$. The separation is successful when $\mathbf{B}$ can be estimated so that $\mathbf{BA} = \mathbf{D\Pi}$ with $\mathbf{\Pi}$ being a permutation matrix and $\mathbf{D}$ being an arbitrary diagonal matrix. These two matrices stand for the two ambiguities of BSS. The signals may appear in any order and can be arbitrarily scaled. For the separation we use the well known gradient-based update rule according to [1].

When dealing with real-world acoustic scenarios it is necessary to consider the reverberation. The mixing system can be modeled by FIR filters of length L. Depending on the reverberation time and sampling rate, L can reach several thousand. The convolutive mixing model reads

$$\boldsymbol{x}(n) = \mathbf{H}(n) * \mathbf{s}(n) = \sum_{l=0}^{L-1} \mathbf{H}(l)\mathbf{s}(n-l) \tag{3}$$

where $\mathbf{H}(n)$ is a sequence of $N \times N$ matrices containing the impulse responses of the mixing channels. For the separation we use FIR filters of length $M \geq L - 1$ and obtain

$$\boldsymbol{y}(n) = \mathbf{W}(n) * \boldsymbol{x}(n) = \sum_{l=0}^{M-1} \mathbf{W}(l)\boldsymbol{x}(n-l) \tag{4}$$

with $\mathbf{W}(n)$ containing the unmixing coefficients.

Using the short-time Fourier transform (STFT), the signals can be transformed to the time-frequency domain, where the convolution approximately becomes a multiplication [6]:

$$\boldsymbol{Y}(\omega_k, \tau) = \boldsymbol{W}(\omega_k)\boldsymbol{X}(\omega_k, \tau), \quad k = 0, 1, \ldots, K - 1, \tag{5}$$

with K being the FFT length. The major benefit of this approach is the possibility to estimate the unmixing matrices for each frequency independently, however, at the price of possible permutation and scaling in each frequency bin:

$$\boldsymbol{Y}(\omega_k, \tau) = \boldsymbol{W}(\omega_k)\boldsymbol{X}(\omega_k, \tau) = \boldsymbol{D}(\omega_k)\boldsymbol{\Pi}(\omega_k)\boldsymbol{S}(\omega_k, \tau) \tag{6}$$

where $\boldsymbol{\Pi}(\omega)$ is a frequency-dependent permutation matrix and $\boldsymbol{D}(\omega)$ an arbitrary diagonal scaling matrix.

The correction of the permutation is essential, because the entire unmixing process fails if different permutations occur at different frequencies. A number of approaches has been proposed to solve this problem. [7,8,9,10,11,12].

When the scaling ambiguity is not solved, filtered versions of the sources are recovered. A widely used approach has been proposed in [13]. The authors aimed to recover the signals as they were recorded at the microphones accepting all filtering done by the mixing system. A similar technique has been proposed in [14] under the paradigm of the minimal distortion principle, which uses the unmixing matrix

$$\boldsymbol{W}'(\omega) = \mathrm{dg}(\boldsymbol{W}^{-1}(\omega)) \cdot \boldsymbol{W}(\omega) \tag{7}$$

with $\mathrm{dg}(\cdot)$ returning the argument with all off-diagonal elements set to zero. However, as mentioned in the introduction, the independent filter design for each frequency bin may result in severe circular convolution artifacts in the final unmixed time-domain signals. In this paper, we therefore propose a method to re-scale the frequency components in such a way that the resulting unmixing filters obey a desired decay behavior. This new approach will be described in the next section.

3   Filter Shaping

The proposed method is to introduce a set of scaling factors $c(\omega)$ for the unmixed frequency components that ensure that the unmixing filters obey a desired decay behavior. The motivation for this comes from the fact that the impulse responses achieved by the minimal distortion principle have a quite arbitrary form. In particular, they often show many large coefficients after the main peak, which results in a significant amount of added reverberation and can even lead to problems of circular-convolution artifacts. For addressing both above-mentioned problems we propose to shape the unmixing filters to have an exponential decay. This reduces the perceived echoes as well as the problems of circular convolution.

In Fig. 1 the overall BSS system is shown. It consists of $N \times N$ single channels as depicted in Fig. 2. In this representation the permutation has already been corrected. The dependency of time-domain filter coefficients of a filter vector

Fig. 1. Overview of frequency-domain BSS

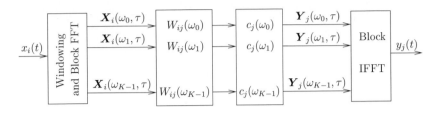

Fig. 2. Data flow from input i to output j

$\boldsymbol{w}_{ij}$ and scaling factors $\boldsymbol{c}_j = [c_j(\omega_0), c_j(\omega_1), \ldots, c_j(\omega_{K-1})]^T$ for output j can be calculated as follows:

$$
\begin{aligned}
\boldsymbol{w}_{ij} &= \sum_l \boldsymbol{E}_l \cdot \mathcal{F}^{-1} \cdot \boldsymbol{C}_j \cdot \boldsymbol{W}_{ij} \cdot \mathcal{F} \cdot \boldsymbol{D}_l \cdot \delta \\
&= \sum_l \boldsymbol{E}_l \cdot \mathcal{F}^{-1} \cdot \operatorname{diag}(\mathcal{F} \cdot \boldsymbol{D}_l \cdot \delta) \cdot \boldsymbol{W}_{ij} \cdot \boldsymbol{c}_j \\
&= \boldsymbol{V}_{ij} \cdot \boldsymbol{c}_j
\end{aligned}
\tag{8}
$$

where $\operatorname{diag}(\cdot)$ converts a vector to a diagonal matrix. The term δ is a unit vector containing a single one and zeros otherwise. $\boldsymbol{D}_l$ is a diagonal matrix containing the coefficients of the STFT analysis window shifted to the lth position according to the STFT window shift. $\mathcal{F}$ is the DFT matrix. $\boldsymbol{W}_{ij}$ is a diagonal matrix containing the frequency-domain unmixing coefficients. $\boldsymbol{c}_j$ is a vector of the sought scaling factors, and $\boldsymbol{C}_j$ is a diagonal matrix made up as $\boldsymbol{C}_j = \operatorname{diag}(\boldsymbol{c}_j)$. $\boldsymbol{E}_l$ is a shifting matrix corresponding to $\boldsymbol{D}_l$, defined in such a way that the overlapping STFT blocks are correctly merged. Note that for real-valued signals and filters, the above equation can be modified to exploit the conjugate symmetry in the frequency domain.

Using the formulation of [16,17] a desired impulse response $\boldsymbol{d}_{d_{ij}}$ can now be expressed as

$$
\boldsymbol{d}_{d_{ij}} = \operatorname{diag}(\boldsymbol{\gamma}_{d_{ij}}) \cdot \boldsymbol{V}_{ij} \cdot \boldsymbol{c}_j
\tag{9}
$$

with $\boldsymbol{\gamma}_{d_{ij}} = [\gamma_{d_{ij}}(0), \gamma_{d_{ij}}(1), \ldots, \gamma_{d_{ij}}(M-1)]$ a vector with the desired shape of the unmixing filter. Here we use a two-sided exponentially decaying window

$$
\gamma_{d_{ij}}(n) = \begin{cases} 10^{q_1(n_o - n)} & \text{for } 0 \leq n \leq n_0 \\ 10^{q_2(n - n_o)} & \text{for } n_0 \leq n \end{cases}
\tag{10}
$$

with n_0 being the position of the maximum of $|\boldsymbol{w}_{ij}|$. The factors q_1 and q_2 have been chosen heuristically as $q_1 = -0.1$ and $q_2 = -0.05$.

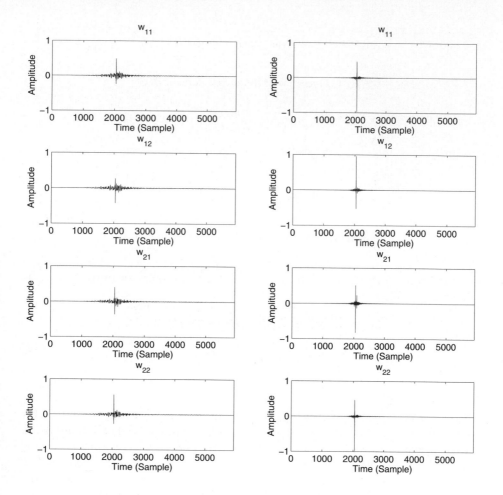

Fig. 3. Comparison of filter sets using the minimal distortion principle (left) and the new method (right)

An undesired part is formulated analogously as

$$d_{u_{ij}} = \mathrm{diag}(\boldsymbol{\gamma}_{u_{ij}}) \cdot \boldsymbol{V}_{ij} \cdot \boldsymbol{c}_j \tag{11}$$

with

$$\gamma_{u_{ij}}(n) = \begin{cases} 10^{q_3(n_o - n)} & \text{for } 0 \le n \le n_0 \\ 10^{q_4(n - n_o)} & \text{for } n_0 \le n \end{cases} \tag{12}$$

Here the factors have been chosen heuristically to be $q_3 = 0.001$ and $q_4 = 0.0005$.

As the filters corresponding to the same output channel have the same scaling factors, $c_j(\omega)$ has to be optimized simultaneously for these filters. Therefore $\boldsymbol{V}_{ij}$ corresponding to the same output j are stacked to $\bar{\boldsymbol{V}}_j$. The same applies for $\boldsymbol{\gamma}_{d_{ij}}$ and $\boldsymbol{d}_{d_{ij}}$ which are stacked into $\bar{\boldsymbol{\gamma}}_{d_j}(n)$ and $\bar{\boldsymbol{d}}_{d_j}$ respectively.

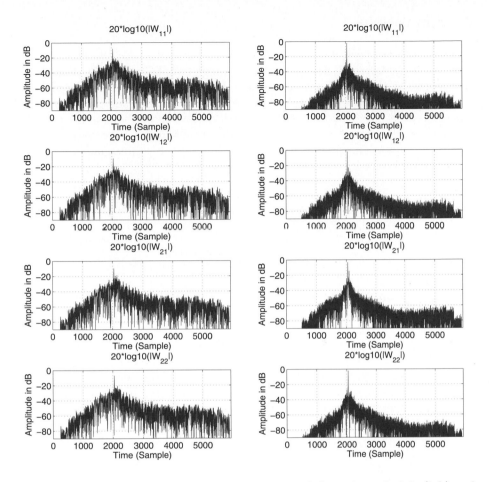

Fig. 4. Magnitudes of filters designed via the minimal distortion principle (left) and the new method (right)

Now the matrices $\bar{A}$ and $\bar{B}$ can be calculated as in [18]:

$$d_u^H d_u = d_{u_j}^H \cdot \bar{V}_j^H \cdot \mathrm{diag}(\gamma_{u_j}^H) \cdot \mathrm{diag}(\gamma_{u_j}) \cdot \bar{V}_j \cdot d_{u_j} = d_{u_j}^H \cdot \bar{A} \cdot d_{u_j} \quad (13)$$

$$d_d^H d_d = d_{d_j}^H \cdot \bar{V}_j^H \cdot \mathrm{diag}(\gamma_{d_j}^H) \cdot \mathrm{diag}(\gamma_{d_j}) \cdot \bar{V}_j \cdot d_{d_j} = d_{d_j}^H \cdot \bar{B} \cdot d_{d_j} \quad (14)$$

Finally, the optimal scaling factors c_{opt} are the solution of the generalized eigenvalue problem [18]

$$\bar{B} \cdot c_{opt} = \bar{A} \cdot c_{opt} \cdot \lambda_{max} \quad (15)$$

with λ_{max} being the largest eigenvalue and c_{opt} being the corresponding eigenvector.

4   Simulations

Simulations have been done on real-world data available at [19]. This data set consists of eight-seconds long speech recordings sampled at 8 kHz. The chosen parameters were a Hann window of length 2048, a window shift of 256, and an FFT-length of $K = 4096$. Every frequency bin has been separated using 200 iterations of the gradient-based rule from [1]. As the original sources are available for the considered data set, the permutation problem has been ideally solved, so that permutation ambiguities do not influence the results, and the scaling problem can be studied exclusively.

Table 1. Comparison of the signal-to-interference ratios in dB between the minimal distortion principle and the new algorithm

	Left	Right	Overall
MDP	16.07	16.72	16.41
New Alg.	24.88	28.81	26.75

In Figs. 3 and 4 the filters designed with the traditional method (7) and the proposed method are shown, respectively. The main difference is the clearly visible and significantly bigger main peak and the faster decay of the impulse responses designed with our method. As one can observe by comparing Fig. 4, the energy difference between the main peak and the tail of the impulse response could be increased by about 25 dB.

The new filters are also able to significantly enhance the separation performance as shown in Table 1.

5   Conclusions

In this paper, we have proposed the use of the scaling ambiguity of convolutive blind source separation for shaping the unmixing filters. We calculate a set of scaling factors that shape exponentially decaying impulse responses with less reverberation. On a real-world example, the energy decay could be improved by 25dB, which also translated into better signal-to-interference ratios.

References

1. Amari, S., Cichocki, A., Yang, H.H.: A new learning algorithm for blind signal separation. In: Touretzky, D.S., Mozer, M.C., Hasselmo, M.E. (eds.) Advances in Neural Information Processing Systems, vol. 8, pp. 757–763. MIT Press, Cambridge (1996)
2. Hyvärinen, A., Oja, E.: A fast fixed-point algorithm for independent component analysis. Neural Computation 9, 1483–1492 (1997)

3. Cardoso, J.F., Soulomiac, A.: Blind beamforming for non-Gaussian signals. Proc. Inst. Elec. Eng., pt. F. 140(6), 362–370 (1993)
4. Douglas, S.C., Sawada, H., Makino, S.: Natural gradient multichannel blind deconvolution and speech separation using causal FIR filters. IEEE Trans. Speech and Audio Processing 13(1), 92–104 (2005)
5. Aichner, R., Buchner, H., Araki, S., Makino, S.: On-line time-domain blind source separation of nonstationary convolved signals. In: Proc. 4th Int. Symp. on Independent Component Analysis and Blind Signal Separation (ICA 2003), Nara, Japan, pp. 987–992 (April 2003)
6. Smaragdis, P.: Blind separation of convolved mixtures in the frequency domain. Neurocomputing 22(1-3), 21–34 (1998)
7. Rahbar, K., Reilly, J.: A frequency domain method for blind source separation of convolutive audio mixtures. IEEE Trans. Speech and Audio Processing 13(5), 832–844 (2005)
8. Anemüller, J., Kollmeier, B.: Amplitude modulation decorrelation for convolutive blind source separation. In: Proceddings of the second international workshop on independent component analysis and blind signal separation, pp. 215–220 (2000)
9. Mazur, R., Mertins, A.: Solving the permutation problem in convolutive blind source separation. In: Davies, M.E., James, C.J., Abdallah, S.A., Plumbley, M.D. (eds.) ICA 2007. LNCS, vol. 4666, pp. 512–519. Springer, Heidelberg (2007)
10. Sawada, H., Mukai, R., Araki, S., Makino, S.: A robust and precise method for solving the permutation problem of frequency-domain blind source separation. IEEE Trans. Speech and Audio Processing 12(5), 530–538 (2004)
11. Wang, W., Chambers, J.A., Sanei, S.: A novel hybrid approach to the permutation problem of frequency domain blind source separation. In: Puntonet, C.G., Prieto, A.G. (eds.) ICA 2004. LNCS, vol. 3195, pp. 532–539. Springer, Heidelberg (2004)
12. Mukai, R., Sawada, H., Araki, S., Makino, S.: Blind source separation of 3-d located many speech signals. In: 2005 IEEE Workshop on Applications of Signal Processing to Audio and Acoustics, pp. 9–12 (October 2005)
13. Ikeda, S., Murata, N.: A method of blind separation based on temporal structure of signals. In: Proc. Int. Conf. on Neural Information Processing, pp. 737–742 (1998)
14. Matsuoka, K.: Minimal distortion principle for blind source separation. In: Proceedings of the 41st SICE Annual Conference, vol. 4, pp. 2138–2143, August 5-7 (2002)
15. Sawada, H., Mukai, R., de la Kethulle, S., Araki, S., Makino, S.: Spectral smoothing for frequency-domain blind source separation. In: International Workshop on Acoustic Echo and Noise Control (IWAENC 2003), pp. 311–314 (September 2003)
16. Arslan, G., Evans, B.L., Kiaei, S.: Equalization for discrete multitone transceivers to maximize bit rate. IEEE Trans. on Signal Processing 49(12), 3123–3135 (December)
17. Kallinger, M., Mertins, A.: Multi-channel room impulse response shaping - a study. In: Proc. IEEE Int. Conf. Acoust., Speech, and Signal Processing, Toulouse, France, vol. V, pp. 101–104 (May 2006)
18. Melsa, P., Younce, R., Rohrs, C.: Impulse response shortening for discrete multitone transceivers. IEEE Trans. on Communications 44(12), 1662 1672 (1996)
19. http://www.kecl.ntt.co.jp/icl/signal/sawada/demo/bss2to4/index.html

Cumulative State Coherence Transform for a Robust Two-Channel Multiple Source Localization

Francesco Nesta[1,2], Piergiorgio Svaizer[1], and Maurizio Omologo[1]

[1] Fondazione Bruno Kessler - Irst, Trento Italy
[2] UNITN, Trento Italy
{nesta,svaizer,omologo}@fbk.eu
http://shine.itc.it/people/nesta

Abstract. This work presents a novel robust method for a two-channel multiple Time Difference of Arrival (TDOA) estimation. The method is based on a recursive frequency-domain Independent Component Analysis (ICA) and on the novel State Coherence Transform (SCT). ICA is computed at different independent time-blocks and the obtained demixing matrices are used to generate observations of the propagation model of the intercepted sources. For the assumed time-frequency sparse dominance of the recorded sources, the observed propagation models are likely to represent all the active sources. The global coherence of the models is evaluated by a cumulated SCT, which provides a precise TDOA estimation for all the sources. Experimental results show that an accurate localization of 7 closely-spaced sources is possibile given only few seconds of data even in the case of low SNR. Experiments also show the advantage of the proposed strategy when compared with other popular two-microphone GCC-PHAT based methods.

Keywords: Blind source separation (BSS), TDOA estimation, independent component analysis (ICA), multiple speaker localization.

1 Introduction

Multiple speaker localization is a difficult problem which has interesting implications in the field of acoustic signal processing. A high reverberation time, the presence of strong enviromental noise, and spatial ambiguity make the localization task harder, especially when just two microphones are used. The problem of multiple TDOA estimation is also relevant in the Blind Source Separation (BSS) community since knowledge of the TDOAs is useful for solving the "permutation problem" [1]. Recent works show that the frequency-domain BSS is strictly connected with the propagation models of the sources [2] and that the localization of multiple sources based on the Independent Component Analysis outperforms other popular narrow-band techniques such as MUSIC [3].

Our earlier work [4] showed that a joint TDOA estimation can be performed for all the sources with a State Coherence Transform (SCT) applied to the demixing matrices obtained by the Independent Component Analysis. The SCT

T. Adali et al. (Eds.): ICA 2009, LNCS 5441, pp. 290–297, 2009.

is invariant to the permutations and insensitive to the phase-wrapping ambiguity of frequencies affected by spatial aliasing, and thus can be applied even to widely-spaced microphones. In this work we extend the SCT to a cumulative SCT (cSCT) which, according to a sparse-dominance assumption of the sources, enables the estimation of the TDOA of many sources by using only two microphones.

In the next section we recall the physical interpretation of the ICA when applied in frequency-domain and the formulation of the SCT. Following, the cumulative SCT is discussed. Finally, in section 4, experimental results show the effectiveness of the proposed method when compared to other GCC-PHAT [5] based popular methods.

2   Physical Interpretation of the Demixing Matrices of Frequency-Domain BSS

In frequency-domain the signals observed by the microphones can be modeled using a time-frequency representation computed by a short-time Fourier analysis. According to the convolutive model for the observed mixture, each time-frequency component can be considered as a linear combination of the time-frequency components of the original source signals. In matrix notation one can write:

$$\mathbf{y}(k,\tau) = \mathbf{H}(k)\mathbf{x}(k,\tau) \tag{1}$$

where $\mathbf{y}(k,\tau)$ are the observed mixtures, $\mathbf{x}(k,\tau)$ are the original signals, τ is the time instant at which each frequency is evaluated according to the time-frame shifting, k is the frequency bin index and $\mathbf{H}(k)$ is a mixing matrix. A complex-valued ICA is applied to the time-series of each frequency. Then the original components can be retrieved by computing a demixing matrix $\mathbf{W}(k)$ which is an estimate of the matrix $\mathbf{H}(k)^{-1}$ up to scaling and permutation ambiguities:

$$\overline{\mathbf{x}}(k,\tau) = \Lambda(k)\mathbf{P}(k)\mathbf{W}(k)\mathbf{y}(k,\tau) \tag{2}$$

where $\Lambda(k)$ and $\mathbf{P}(k)$ are a complex-valued scaling matrix and a permutation matrix, respectively.

In the ideal case, neglecting the reverberation, we can assume the sources to be in free-field conditions. Thus the signals observed at the microphones can be considered to be a sum of delayed and scaled version of the orignal source signals according to the relative position of the sources with respect to the microphones. For the case of two channels, in frequency-domain each mixing matrix can be modelled as:

$$\mathbf{H}(k) = \begin{pmatrix} |h_{11}(k)|e^{-j\varphi_{11}(k)} & |h_{12}(k)|e^{-j\varphi_{12}(k)} \\ |h_{21}(k)|e^{-j\varphi_{21}(k)} & |h_{22}(k)|e^{-j\varphi_{22}(k)} \end{pmatrix} \tag{3}$$

$$\varphi_{iq}(k) = 2\pi f_k T_{iq} \tag{4}$$

where T_{iq} is the propagation time from the q-th source to the i-th microphone and f_k is the frequency, in Hz, associated to the k-th frequency bin. Our earlier work [6] has shown that the elements of each row of $\mathbf{W}(k)$ can be directly used to obtain the observations of the ideal propagation models. These are expressed for the two sources by the ratios computed as follows:

$$r_1(k) = -\frac{w_{12}(k)}{w_{11}(k)}, \quad r_2(k) = -\frac{w_{22}(k)}{w_{21}(k)} \tag{5}$$

Such ratios are scaling invariant and their phase is expected to vary linearly with frequency, according to the TDOAs of the sources. Neglecting the wave attenuation, the ideal propagation model of each source can be represented as:

$$c(k, \tau) = e^{-j2\pi f_k \tau} \tag{6}$$

where τ is the TDOA of the source. Thus, for each frequency, an observation of the ideal propagation model can be obtained by normalizing the ratios $r_i(k)$ with respect to their magnitude:

$$\bar{r}_i(k) = \frac{r_i(k)}{||r_i(k)||} \tag{7}$$

In [7] we showed that a joint multiple TDOA estimation can be performed by using a State Coherence Transform (SCT) which was formulated as follows:

$$SCT(\tau) = \sum_k \sum_{i=1}^{N} \left[1 - g\left(\frac{||c(k, \tau) - \bar{r}_i(k)||}{2} \right) \right] \tag{8}$$

where N is the number of the observed states for each frequency and $g(\cdot)$ is a function of the euclidean distance. In this work we select a non linear function defined as:

$$g(x) = tanh(\alpha \cdot x) \tag{9}$$

where α is chosen according to the distance between the microphones, as described in [7].

3   Cumulative SCT for Multiple TDOA Estimation

Theoretically, the SCT is able to estimate a number of TDOAs equal to the number of microphones. However, we can assume that for each time-frequency block only two sources are dominant (sparse dominance). Then by computing ICA in different time-frequency blocks we would observe states which represent the propagation models of all the active sources. For every time-block and any frequency k each ratio $\bar{r}_i(k)$ is an observation of the propagation model of the $i-$th source among the N dominant ones. Then, for the case of two microphones the coherence of all the observed states can be globally evaluated on a multiplicity of time blocks by using a cumulative SCT (cSCT), which is formulated as follows:

$$cSCT(\tau) = \sum_b \sum_k \sum_{i=1}^{2} \left[1 - tanh\left(\alpha \frac{||c(k, \tau) - \bar{r}_i^b(k)||}{2} \right) \right] \tag{10}$$

where $\bar{r}_i^b(k)$ is the normalized state obtained for the $k - th$ frequency bin in the time block b. The cumulative SCT analysis requires to evaluate the ratios $\bar{r}_i^b(k)$ by applying ICA to short time-blocks. However the accuracy of the ICA estimator decreases with the length of the observed data and thus an appropriate choice for the block size is needed. In [8] we showed that a recursive approach across the frequency can be exploited to increase the ICA accuracy when short signals are observed (e.g. 300ms). For the physical interpretation of the frequency-domain BSS, the demixing matrices are expected to vary smoothly across the frequencies. Such a property can be used to recursively initalize each ICA with a smooth estimation of $\mathbf{W}(f)$. The initialization decreases the probability that ICA could converge to wrong local minima generated from the observed limited amount of data. A detailed description of the methods can be found at [6] and [8].

4   Experimental Results

The algorithm has been coded both in matlab and C++ and works in real-time on a normal laptop. To perform an on-line TDOA estimation rather than applying directly the formula (10), we evaluate the SCT for each single block and we recursively average the envelopes over the time as follows:

$$\overline{cSCT}_b(\tau) = \frac{1}{b}SCT_b(\tau) + \frac{(b-1)}{b}\overline{cSCT}_{b-1}(\tau) \qquad (11)$$

where b is the time-block for which the $SCT_b(\tau)$ is evaluated. We summarize in pseudo-code the main steps of the implemented algorithm:

> *apply the Short-Time Fourier Transform to the recorded signals*
> *subdivide each time-frequency series in b_max blocks*
> ***for** b=1 **to** b_max*
>   ***for** k=maximum_frequency **to** 1*
>   *compute the matrix W(k) by the recursive-ICA for the b-th block*
>   *compute the normalized ratios $\bar{r}_i(k)$ as in (5) and (7)*
>   ***end***
>   *compute the SCT as in (9)*
>   *compute the cSCT as in (11)*
>   *extract the TDOAs corresponding to the peaks of the cSCT envelope*
> ***end***

In this experiment the algorithm has been evaluated for the estimation of the TDOAs of 7 loudspeakers playing simultaneously sound files of about 10 seconds: 3 male utterances, 3 female utterances, 1 pop song. The mixture signals were produced by separately recording the spatial image of each source over the two microphones then summing the spatial images of all sources over each channel. The loudspeakers have been uniformly spaced with an average angular distance of about 13° and located at an average distance of about 1 meter from the center of the two microphones. Recordings were performed in a room with $T_{60} = 700ms$ with a sampling rate of $fs = 16kHz$ and the FFT analysis was performed with

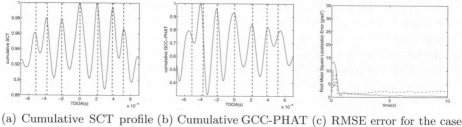

(a) Cumulative SCT profile (b) Cumulative GCC-PHAT (c) RMSE error for the case
(SNR=20dB)   profile (SNR=20dB)   of SNR=20dB

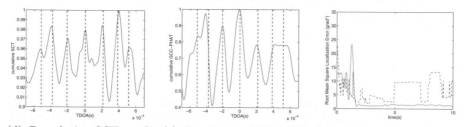

(d) Cumulative SCT profile (e) Cumulative GCC-PHAT (f) RMSE error for the case
(SNR=5dB)   profile (SNR=5dB)   of SNR=5dB

Fig. 1. Comparison between the cSCT and the cumulative GCC-PHAT. The red dotted lines are the true expected TDOAs. (c) and (f) show the RMSE localization error for the cumulative GCC-PHAT (blue dotted line) and the cSCT (solid red line).

an Hanning window of 2048 samples and a frame-shifting of 512 samples. The length of the time-block used for the ICA and the SCT analysis was $300ms$. The signals were recorded with two microphones spaced of $0.26m$; according to the sound speed and to the maximum admissible time-delay, the SCT was computed for 180 values of τ in the range from -13 to +13 samples.

In [3] it has been shown that ICA outperforms the MUSIC algorithm in the DOA estimation for the case of multiple sources. In such a work closely spaced microphones were used to avoid the frequency-phase ambiguity introduced by the spatial aliasing. On the contrary, the SCT does not suffer of such ambiguity since the TDOA is evaluated according to the complex-valued model of the acoustic propagation. Moreover, it has been shown that the SCT can be considered a multisource extension of the GCC-PHAT (see [4]). Thus, the method has been compared with two other GCC-PHAT based approach: 1) TDOAs selected by a cumulative GCC-PHAT 2) TDOAs selected by Time-frequency histogram (TFH) [9]. For both the methods the GCC-PHAT was computed over frames of 4096 points with step of 256 points. Moreover, in order to obtain a theoretical resolution of 180 possible time-delays, an interpolation was applied. The cumulative GCC-PHAT was obtained by a recursive averaging of the envelopes similarly to formula (11). Figure 1(subpictures (a)(b)(d)(e)) shows the final envelopes associated with the cumulative SCT and with the cumulative GCC-PHAT, when the signals are affected by an Additive White Gaussian Noise (AWGN) resulting

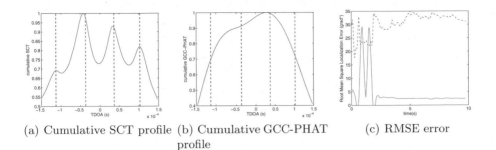

(a) Cumulative SCT profile   (b) Cumulative GCC-PHAT profile   (c) RMSE error

Fig. 2. Comparison between the cSCT and the cumulative GCC-PHAT for 4 sources recorded with microphones spaced of $0.05m$. In the envelopes the red dotted lines are the true expected TDOAs. In (c) is plotted the RMSE localization error for the cumulative GCC-PHAT (blue dotted line) and the cSCT (solid red line).

in a SNR of $20dB$ and $5dB$. For the case of lower SNR both the envelopes show clear peaks at values close to the corresponding expected TDOAs (red dotted lines). In fact since the sources are sparse in time-frequency domain the cumulative GCC-PHAT is able to intercept the coherence of the observed propagation models across all the time-frequency points. In this condition the cumulative GCC-PHAT can be assumed to be a good approximation of the cumulative SCT. However as the noise increases the cSCT clearly outperforms the GCC-PHAT. Infact, whilst the GCC-PHAT uses the normalized cross-spectrum as observation of the propagation models, the cSCT uses the demixing matrix obtained by the ICA stage and it is consequently less sensitive to the noise. In subpictures (c) and (f) of figure 1 the Root Mean Square localization Error is plotted. The corresponding directions of arrival were computed according to the geometrical information and using the TDOAs estimated at each time block. It is worth noting that for both the cases the cSCT converges to a small error and just in few seconds.

Another important advantage of the cSCT is the increased spatial resolution of the TDOA estimator when microphones are closely spaced. In figure 2 we show the resulting envelopes obtained with the cumulative GCC-PHAT and the cSCT, when 4 sources are recorded by two microphones spaced of $0.05m$. Since the phase difference between the propagation models of different sources is reduced, the resolution of the resulting cumulative GCC-PHAT is too low to enable the discrimination of peaks related to different sources. On the other hands, the non linear mapping of the cSCT reduces the intereference effect between the propagation models belonging to different sources and consequently the resolution of the envelope is increased.

A comparison between the proposed method with a GCC-PHAT based time-frequency histogram (TFH) [9] is also provided. For the TFH all the TDOAs chosen from the GCC-PHAT of each frame are pooled in a histogram where the maxima are expected to correspond to the most probable TDOAs. In order to have a direct comparison between the two methods, we similarly computed

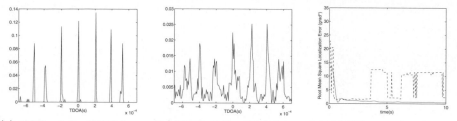

(a) TFH of the TDOAs se-
lected by the cSCT profile
(SNR=20dB)

(b) TFH of the TDOAs se-
lected by the GCC-PHAT of
profile (SNR=20dB)

(c) RMSE error for the case
SNR=20dB

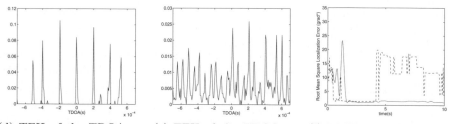

(d) TFH of the TDOAs se-
lected by the cSCT profile
(SNR=5dB)

(e) TFH of the TDOAs se-
lected by the GCC-PHAT of
profile (SNR=5dB)

(f) RMSE error for the case
SNR=5dB

Fig. 3. Comparison between the cSCT and the cumulative GCC-PHAT. In the envelopes the red dotted lines are the true expected TDOAs. In (c) and (f) is plotted the RMSE localization error for the cumulative GCC-PHAT (blue dotted line) and the cSCT (solid red line).

the histogram of the TDOAs obtained at each time-block, using the peaks of the cSCT. Figure 3 shows the normalized histograms of the selected TDOAs for both the methods. We observe that when the GCC-PHAT is used, the histogram is very noisy. Even for high SNR the noiseness of the histogram make the correct peak selection hard and the resulting averaged localization error is unstable during the time. In the case of the cSCT the stability of the TDOA estimates produce clearer and more stable results.

5   Conclusions

This work introduced a new method to perform multiple TDOA estimation by using only two microphones. The SCT and the recursive ICA are combined to obtain a robust TDOA estimator for many active sources. Experimental results show that the method is robust even in presence of strong noise and enables the discrimination of 7 active sources recorded under challenging conditions. Future investigation will concern the use of the proposed strategy as a starting point for a robust underdetermined blind source separation.

References

1. Saruwatari, H., Kurita, S., Takeda, K., Itakura, F., Nishikawa, T., Shikano, K.: Blind source separation combining independent component analysis and beamforming. EURASIP J. Appl. Signal Process. 2003(1), 1135–1146 (2003)
2. Sawada, H., Araki, S., Mukai, R., Makino, S.: Grouping separated frequency components by estimating propagation model parameters in frequency-domain blind source separation. IEEE Transactions on Audio, Speech, and Language Processing 15(5), 1592–1604 (2007)
3. Sawada, H., Mukai, R., Makino, S.: Direction of arrival estimation for multiple source signals using independent component analysis. In: Proceedings of ISSPA, vol. 2, pp. 411–414 (July 2003)
4. Nesta, F., Omologo, M., Svaizer, P.: A novel robust solution to the permutation problem based on a joint multiple TDOA estimation. In: Proceedings of IWAENC, Seattle, USA (September 2008)
5. Knapp, C.H., Carter, G.C.: The generalized correlation method for estimation of time delay. IEEE Transactions on Acoustics, Speech, and Signal Processing 24, 320–327 (1976)
6. Nesta, F., Omologo, M., Svaizer, P.: Separating short signals in highly reverberant environment by a recursive frequency-domain BSS. In: Proceedings of HSCMA, Trento, Italy (May 2008)
7. Nesta, F., Omologo, M., Svaizer, P.: Multiple TDOA estimation by using a state coherence transform for solving the permutation problem in frequency-domain BSS. In: Proceedings of MLSP, Cancun, Mexico (October 2008)
8. Nesta, F., Svaizer, P., Omologo, M.: A BSS method for short utterances by a recursive solution to the permutation problem. In: Proceedings of SAM, Darmstadt, Germany (July 2008)
9. Aarabi, P., Mavandadi, S.: Multi-source time delays of arrival estimation using conditional time-frequency histograms. Information Fusion 4(2), 111–122 (2003)

Least Square Joint Diagonalization of Matrices under an Intrinsic Scale Constraint

Dinh-Tuan Pham[1] and Marco Congedo[2]

[1] Laboratory Jean Kuntzmann, cnrs - Grenoble INP - UJF, Grenoble, France
Dinh-Tuan.Pham@imag.fr
http://www-ljk.imag.fr/membres/Dinh-Tuan.Pham
[2] GIPSA-lab, cnrs - UJF - Univ. Stendhal - Grenoble INP, Grenoble, France

Abstract. We present a new algorithm for approximate joint diagonalization of several symmetric matrices. While it is based on the classical least squares criterion, a novel intrinsic scale constraint leads to a simple and easily parallelizable algorithm, called LSDIC (Least squares Diagonalization under an Intrinsic Constraint. Numerical simulations show that the algorithm behaves well as compared to other approximate joint diagonalization algorithms.

1 Introduction

The diagonalization of a matrix and the joint diagonalization of two matrices are well-established concepts in linear algebra and their use in engineering is ubiquitous. Approximate joint diagonalization (AJD) refers to the problem of diagonalizing more then two matrices simultaneously. Considerable interest for AJD followed the discovery that it yields a solution for independent component analysis (e.g., JADE [2]) and second-order blind source separation (e.g., SOBI [1], see also [8]).

The classical linear instantaneous BSS problem assumes a mixing model of the form $\mathbf{x}(t) = \mathbf{A}s(t)$, where $\mathbf{x}$ is a K-vector holding the sensor measurements, $\mathbf{A}$ is the $K \times K$ mixing matrix and $\mathbf{s}$ is a K-vector holding the source processes. The task is to recover the sources out of a scaling and permutation indeterminacies from the observation $\mathbf{x}$, assuming no knowledge of $\mathbf{A}$ and of the sources distribution. A simple approach to the source separation problem is to consider a set of matrices $\{\mathbf{C}_1, \ldots, \mathbf{C}_N\}$ consisting of statistics of the observations (of second order in most cases) which are estimates of matrices of the form $\mathbf{A}\mathbf{D}_n\mathbf{A}^T$, where $\mathbf{D}_n$ is a diagonal matrix with k-th diagonal element depending only on the distribution of the k-th source. The AJD then seeks a matrix $\mathbf{B}^T$ such that all N congruence transformations $\mathbf{B}^T\mathbf{C}_n\mathbf{B}$, $n = 1, \ldots, N$, are as diagonal as possible. Therefore it provides an estimate of the inverse of $\mathbf{A}$ (up to a scaling and a permutation) and the BSS problem is solved by $\mathbf{s}(t) = \mathbf{B}^T\mathbf{x}(t)$.

Several iterative algorithms have been developed to solve the AJD problem. A simple way to proceed is to minimize the criterion

$$\sum_{n=1}^{N} \|\mathbf{Off}(\mathbf{B}^T\mathbf{C}_n\mathbf{B})\|^2 \tag{1}$$

T. Adali et al. (Eds.): ICA 2009, LNCS 5441, pp. 298–305, 2009.
© Springer-Verlag Berlin Heidelberg 2009

where **Off** denotes the operator which retains only the off diagonal of its matrix argument and $\|\mathbf{M}\| = [\mathrm{tr}(\mathbf{M}\mathbf{M}^T)]^{1/2}$ denotes the Frobenius norm of the matrix $\mathbf{M}$. However, without any restriction, one would end up with the trivial solution $\mathbf{B} = \mathbf{0}$. In [3], $\mathbf{B}$ is restricted to the orthogonal group and an efficient algorithm based on Givens rotations is presented. This restriction allows BSS only after whitening the sensor measurements, but whitening is known to affect adversely the quality of the separation forcing the data covariance to be exactly diagonal at the expenses of the input matrix set. In [9,5] it is required instead that $\mathbf{Diag}(\mathbf{B}^T\mathbf{C}_0\mathbf{B}) = \mathbf{I}$, with $\mathbf{C}_0$ being some positive definite matrix, **Diag** denoting the operator which cancels the off diagonal elements of its matrix argument. This amounts to a normalization of the columns of $\mathbf{B}$ (a scaling). The algorithms are slow because for each iteration one needs to repeat K times the search for the eigenvector associated with the smallest eigenvalue of a $K \times K$ matrix. To avoid degeneracy, as well as the trivial solution, in [6] a term proportional to $-\log|\det(\mathbf{B})|$ is added to the criterion, albeit yielding an even slower algorithm.

Another line of research has focused on multiplicative algorithms with superlinear convergence. Such algorithms update the $\mathbf{C}_n$ matrices at each iteration and typically are faster than those minimizing the off-criterion. In [7] the introduction of a different criterion leads to a very efficient multiplicative Givens-like algorithm. However, it allows only positive definite input matrices and such requirement may be cumbersome in some BSS applications. The notion of criterion is dropped altogether in [10], where a multiplicative algorithm based on heuristics is described instead.

In this paper we will elaborate further upon the off criterion (1), expanding results presented in [4], where it has been shown that its minimizer under certain constraint satisfies a nested system of K generalized Rayleigh quotients. We propose a pseudo-Newton fixed point algorithm to solve such system which requiring no eigenvalue-eigenvector decomposition and no $\mathbf{C}_n$ matrices updates. We discuss the local stability of the algorithm, i.e., its convergence near the solution. The algorithm, named LSDIC (Least Squares Diagonalization under an Intrinsic Constraint) has complexity per iteration similar to multiplicative algorithms. As in the case of other algorithms minimizing the off-criterion, the convergence is linear, thus overall execution time may be superior to multiplicative algorithms. However LSDIC is naturally parallelizable not only with respect to N, like multiplicative algorithms, but also with respect to K, allowing computational super-efficiency in massive parallel computing architectures. Furthermore, unlike multiplicative algorithms (with the exception of [7]), it accommodates complex mixing and/or sources.

2   The Criterion

We consider the joint approximation diagonalization of a set of N symmetric $K \times K$ matrices $\mathbf{C}_1, \dots, \mathbf{C}_N$. We consider the least square criterion (1) subjected to the constraint: $\sum_{n=1}^{N}(\mathbf{b}_k^T\mathbf{C}_k\mathbf{b}_k)^2 = 1$, $k = 1, \dots, K$ ($\mathbf{b}_k$ denoting the k-th column of $\mathbf{B}$). Unlike other constraints, this is a new intrinsic constraint that does not favor a particular matrix in the above set or make use of another matrix outside this set.

Since $\sum_{n=1}^{N}(\mathbf{b}_k^T\mathbf{C}_n\mathbf{b}_k)^2 = \sum_{n=1}^{N}\mathbf{b}_k^T\mathbf{C}_n\mathbf{b}_k\mathbf{b}_k^T\mathbf{C}_n\mathbf{b}_k$ the constraint can be rewritten as $\mathbf{b}_k^T\mathbf{M}(\mathbf{b}_k)\mathbf{b}_k = 1$, $k = 1,\ldots,K$, where

$$\mathbf{M}(\mathbf{b}_k) = \sum_{n=1}^{N}\mathbf{C}_n\mathbf{b}_k\mathbf{b}_k^T\mathbf{C}_n. \qquad (2)$$

Further, $\|\mathbf{Off}(\mathbf{B}^T\mathbf{C}_n\mathbf{B})\|^2 = \|\mathbf{B}^T\mathbf{C}_n\mathbf{B}\|^2 - \sum_{k=1}^{K}(\mathbf{b}_k^T\mathbf{C}_n\mathbf{b}_k)^2$ and $\|\mathbf{B}^T\mathbf{C}_n\mathbf{B}\|^2 = \mathrm{tr}(\mathbf{B}^T\mathbf{C}_n\mathbf{B}\mathbf{B}^T\mathbf{C}_n\mathbf{B})$, therefore, the criterion (1) under the above constraint reduces to $\mathrm{tr}[\mathbf{B}^T\mathbf{M}(\mathbf{B})\mathbf{B}] - K$ where $\mathbf{M}(\mathbf{B})$ is defined as in (2) but with $\mathbf{b}_k$ replaced by $\mathbf{B}$. Note that $\mathbf{M}(\mathbf{B})$ can also be computed as $\mathbf{M}(\mathbf{B}) = \sum_{k=1}^{K}\mathbf{M}(\mathbf{b}_k)$.

One can turn the minimization problem of $\mathrm{tr}[\mathbf{B}^T\mathbf{M}(\mathbf{B})\mathbf{B}]$ under the constraint $\mathbf{b}_k^T\mathbf{M}(\mathbf{b}_k)\mathbf{b}_k = 1$, $k = 1,\ldots,K$, into one without constraint, as follows. Note that for an arbitrary matrix $\mathbf{B}$, the matrix $\mathbf{B}\mathbf{D}^{-1/4}(\mathbf{B})$, where $\mathbf{D}(\mathbf{B})$ is the diagonal matrix with diagonal elements

$$d(\mathbf{b}_k) = \mathbf{b}_k^T\mathbf{M}(\mathbf{b}_k)\mathbf{b}_k, \quad k = 1,\ldots,K, \qquad (3)$$

will satisfy the constraint. Replacing $\mathbf{B}$ in the criterion $\mathrm{tr}[\mathbf{B}^T\mathbf{M}(\mathbf{B})\mathbf{B}]$ by $\mathbf{B}\mathbf{D}^{-1/4}(\mathbf{B})$ then yields the criterion

$$\mathcal{C}(\mathbf{B}) = \mathrm{tr}[\mathbf{D}^{-1/2}(\mathbf{B})\mathbf{B}^T\tilde{\mathbf{M}}(\mathbf{B})\mathbf{B}] = \sum_{n=1}^{N}\mathrm{tr}\{[\mathbf{D}^{-1/2}(\mathbf{B})\mathbf{B}^T\mathbf{C}_n\mathbf{B}]^2\}, \qquad (4)$$

where

$$\tilde{\mathbf{M}}(\mathbf{B}) = \sum_{n=1}^{N}\mathbf{C}_n\mathbf{B}\mathbf{D}^{1/2}(\mathbf{B})\mathbf{B}^T\mathbf{C}_n = \sum_{k=1}^{K}\mathbf{M}(\mathbf{b}_k)/d^{1/2}(\mathbf{b}_k).$$

Therefore one may simply minimize $C(\mathbf{B})$ given in (4) *without any constraint*.

3   Gradient of the Criterion

To compute the gradient of the criterion (4) at the point $\mathbf{B}$ we shall perform its Taylor expansion around $\mathbf{B}$ up to first order. Let $\boldsymbol{\Delta}$ be a small increment of $\mathbf{B}$ then for any symmetric matrix $\mathbf{E}$, one has

$$\mathrm{tr}\{[\mathbf{E}(\mathbf{B}+\boldsymbol{\Delta})^T\mathbf{C}_n(\mathbf{B}+\boldsymbol{\Delta})]^2\}$$
$$= \mathrm{tr}[(\mathbf{E}\mathbf{B}^T\mathbf{C}_n\mathbf{B} + \mathbf{E}\mathbf{B}^T\mathbf{C}_n\boldsymbol{\Delta} + \mathbf{E}\boldsymbol{\Delta}^T\mathbf{C}_n\mathbf{B} + \mathbf{E}\boldsymbol{\Delta}^T\mathbf{C}_n\boldsymbol{\Delta})^2]$$
$$= \mathrm{tr}[(\mathbf{E}\mathbf{B}^T\mathbf{C}_n\mathbf{B})^2] + 4\mathrm{tr}(\mathbf{E}\mathbf{B}^T\mathbf{C}_n\mathbf{B}\mathbf{E}\mathbf{B}^T\mathbf{C}_n\boldsymbol{\Delta}) + O(\|\boldsymbol{\Delta}\|^2), \qquad (5)$$

where $O(\|\boldsymbol{\Delta}\|^2)$ denotes a term of the same order as $\|\boldsymbol{\Delta}\|^2$ as $\boldsymbol{\Delta} \to 0$. By the same calculation with $\mathbf{E}$ being the identity matrix and $\mathbf{B}$ and $\boldsymbol{\Delta}$ replaced by their k-th rows $\mathbf{b}_k$ and $\boldsymbol{\delta}_k$, one gets, after summing up with respect to n:

$$d(\mathbf{b}_k + \boldsymbol{\delta}_k) = d(\mathbf{b}_k) + 4\mathbf{b}_k^T\mathbf{M}(\mathbf{b}_k)\boldsymbol{\delta}_k + O(\|\boldsymbol{\delta}_k\|^2).$$

Therefore

$$\mathbf{D}(\mathbf{B} + \boldsymbol{\Delta}) = \mathbf{D}(\mathbf{B}) + 4\mathbf{Diag}[\mathbf{P}^T(\mathbf{B})\boldsymbol{\Delta}] + O(\|\boldsymbol{\Delta}_k\|^2)$$

where

$$\mathbf{P}(\mathbf{B}) = [\,\mathbf{M}(\mathbf{b}_1)\mathbf{b}_1 \quad \cdots \quad \mathbf{M}(\mathbf{b}_K)\mathbf{b}_K\,].$$

Then from the first order Taylor expansion $(x + \epsilon)^{-1/2} = x^{-1/2} - \frac{1}{2}x^{-3/2}\epsilon + O(|\epsilon|^2)$, one gets

$$\mathbf{D}^{-1/2}(\mathbf{B} + \boldsymbol{\Delta}) = \mathbf{D}^{-1/2}(\mathbf{B}) - 2\mathbf{D}^{-3/2}(\mathbf{B})\mathbf{Diag}[\mathbf{P}^T(\mathbf{B})\boldsymbol{\Delta}] + O(\|\boldsymbol{\Delta}\|^2) \quad (6)$$

Applying now (5) with $\mathbf{E} = \mathbf{D}^{-1/2}(\mathbf{B} + \boldsymbol{\Delta}) = \mathbf{D}^{-1/2}(\mathbf{B}) + O(\|\mathbf{D}\|)$ yields

$$\mathrm{tr}\{[\mathbf{D}^{-1/2}(\mathbf{B} + \boldsymbol{\Delta})(\mathbf{B} + \boldsymbol{\Delta})^T\mathbf{C}_n(\mathbf{B} + \boldsymbol{\Delta})]^2\} =$$
$$\mathrm{tr}\{[\mathbf{D}^{-1/2}(\mathbf{B} + \boldsymbol{\Delta})\mathbf{B}^T\mathbf{C}_n\mathbf{B}]^2\} + 4\mathrm{tr}[\mathbf{D}^{-1/2}(\mathbf{B})\mathbf{B}^T\mathbf{C}_n\mathbf{B}\mathbf{D}^{-1/2}(\mathbf{B})\mathbf{B}^T\mathbf{C}_n\boldsymbol{\Delta}]$$
$$+ O(\|\boldsymbol{\Delta}\|^2).$$

But by (6)

$$\mathrm{tr}\{[\mathbf{D}^{-1/2}(\mathbf{B} + \boldsymbol{\Delta})\mathbf{B}^T\mathbf{C}_n\mathbf{B}]^2\} = \mathrm{tr}\{[\mathbf{D}^{-1/2}(\mathbf{B})\mathbf{B}^T\mathbf{C}_n\mathbf{B}]^2\}$$
$$- 4\mathrm{tr}\{\mathbf{B}^T\mathbf{C}_n\mathbf{B}\mathbf{D}^{-1/2}(\mathbf{B})\mathbf{B}^T\mathbf{C}_n\mathbf{B}\mathbf{D}^{-3/2}(\mathbf{B})\mathbf{Diag}[\mathbf{P}(\mathbf{B})\boldsymbol{\Delta}]\} + O(\|\boldsymbol{\Delta}\|^2).$$

Therefore combining the above results and summing up with respect to n, one gets:

$$\mathcal{C}(\mathbf{B} + \boldsymbol{\Delta}) = \mathcal{C}(\mathbf{B}) + 4\mathrm{tr}[\mathbf{D}^{-1/2}(\mathbf{B})\mathbf{B}^T\tilde{\mathbf{M}}(\mathbf{B})\boldsymbol{\Delta}]$$
$$- 4\mathrm{tr}\{\mathbf{B}^T\tilde{\mathbf{M}}(\mathbf{B})\mathbf{B}\mathbf{D}^{-3/2}(\mathbf{B})\mathbf{Diag}[\mathbf{P}^T(\mathbf{B})\boldsymbol{\Delta}]\} + O(\|\boldsymbol{\Delta}\|^2).$$

Finally, noting that $\mathrm{tr}[\mathbf{U}\mathbf{Diag}(\mathbf{V})] = \mathrm{tr}[\mathbf{Diag}(\mathbf{U})\mathbf{Diag}(\mathbf{V})] = \mathrm{tr}[\mathbf{Diag}(\mathbf{U})\mathbf{V}]$, one gets

$$\mathcal{C}(\mathbf{B} + \boldsymbol{\Delta}) = \mathcal{C}(\mathbf{B}) + 4\mathrm{tr}\{\mathbf{D}^{-1/2}(\mathbf{B})[\mathbf{B}^T\tilde{\mathbf{M}}(\mathbf{B}) - \boldsymbol{\Gamma}(\mathbf{B})\mathbf{P}^T(\mathbf{B})]\boldsymbol{\Delta}\} + O(\|\boldsymbol{\Delta}\|^2).$$

where

$$\boldsymbol{\Gamma}(\mathbf{B}) = \mathbf{D}^{-1}(\mathbf{B})\mathbf{Diag}[\mathbf{B}^T\tilde{\mathbf{M}}(\mathbf{B})\mathbf{B}].$$

The above relation shows that the gradient of the criterion $\mathcal{C}$ is

$$\mathcal{C}'(\mathbf{B}) = 4[\tilde{\mathbf{M}}(\mathbf{B})\mathbf{B} - \mathbf{P}(\mathbf{B})\boldsymbol{\Gamma}(\mathbf{B})]\mathbf{D}^{-1/2}(\mathbf{B})].$$

Setting the gradient to 0 yields the equations

$$\mathbf{M}(\mathbf{b}_k)\mathbf{b}_k = \frac{\mathbf{b}_k^T\mathbf{M}(\mathbf{b}_k)\mathbf{b}_k}{\mathbf{b}_k^T\tilde{\mathbf{M}}(\mathbf{B})\mathbf{b}_k}\tilde{\mathbf{M}}(\mathbf{B})\mathbf{b}_k, \qquad k = 1, \ldots, K$$

which means that $\mathbf{b}_k$ is a generalized eigenvector of $\mathbf{M}(\mathbf{b}_k)$ relative to $\tilde{\mathbf{M}}(\mathbf{B})$, with eigenvalue $[\mathbf{b}_k^T\mathbf{M}(\mathbf{b}_k)\mathbf{b}_k]/[\mathbf{b}_k^T\tilde{\mathbf{M}}(\mathbf{B})\mathbf{b}_k]$.

4   Pseudo Newton Algorithm

We have thought of implementing quasi Newton algorithm to minimize the criterion (4), which would require the calculation of an approximate Hessian. Although such calculation is theoretically possible, the result is quite involved so that the corresponding algorithm is very complex and costly computationally. Therefore we shall replace the true Hessian with some reasonable positive definite matrix, which we call pseudo Hessian. Let $\text{vec}(\mathbf{B})$ denotes the vector formed by stacking the columns of $\mathbf{B}$, then the pseudo Newton algorithm can be expressed as $\text{vec}(\mathbf{B}) \leftarrow \text{vec}(\mathbf{B}) - \mathcal{H}^{-1}(\mathbf{B})\text{vec}[\mathcal{C}'(\mathbf{B})]$ where $\mathcal{H}(\mathbf{B})$ is the pseudo Hessian. We take

$$\mathcal{H}(\mathbf{B}) = 4 \begin{bmatrix} \dfrac{\tilde{\mathbf{M}}(\mathbf{B})}{d^{1/2}(\mathbf{b}_1)} & 0 & \cdots & 0 \\ 0 & \ddots & \ddots & \vdots \\ \vdots & \ddots & \ddots & 0 \\ 0 & \cdots & 0 & \dfrac{\tilde{\mathbf{M}}(\mathbf{B})}{d^{1/2}(\mathbf{b}_K)} \end{bmatrix}.$$

This is a reasonable choice as it is positive definite and is of about the same order of magnitude as the Hessian (the expansion of $\mathcal{C}(\mathbf{B} + \boldsymbol{\Delta})$ up to second order in $\boldsymbol{\Delta}$ contains actually the positive term $4\text{tr}[\boldsymbol{\Delta}^T \mathbf{D}^{-1/2}(\mathbf{B})\tilde{\mathbf{M}}(\mathbf{B})\boldsymbol{\Delta}]$). But the main reason for making the above choice is that the corresponding algorithm reduces to a very simple fixed point iteration:

$$\mathbf{B} \leftarrow \tilde{\mathbf{M}}^{-1}(\mathbf{B})\mathbf{P}(\mathbf{B})\boldsymbol{\Gamma}(\mathbf{B})\mathbf{D}^{-1/2}(\mathbf{B}),$$

or explicitly

$$\mathbf{b}_k \leftarrow \frac{\mathbf{b}_k^T \tilde{\mathbf{M}}(\mathbf{B})\mathbf{b}_k}{d(\mathbf{b}_k)}\tilde{\mathbf{M}}^{-1}(\mathbf{B})\mathbf{M}(\mathbf{b}_k)\mathbf{b}_k, \quad k = 1, \dots, K.$$

Further, since the criterion (4) is scale invariant, we may drop the factor $\mathbf{b}_k^T \tilde{\mathbf{M}}(\mathbf{B})\mathbf{b}_k/d(\mathbf{b}_k)$ and renormalize $\mathbf{b}_k$ so that $d(\mathbf{b}_k) = 1$. This leads to the algorithm:

$$\mathbf{b}_k \leftarrow \mathbf{M}^{-1}(\mathbf{B})\mathbf{M}(\mathbf{b}_k)\mathbf{b}_k, \quad \mathbf{b}_k \leftarrow \mathbf{b}_k/[\mathbf{b}_k^T \mathbf{M}(\mathbf{b}_k)\mathbf{b}_k]^{1/4}, \quad k = 1, \dots, K,$$

assuming that $\mathbf{B}$ has been previously normalized so that $\tilde{\mathbf{M}}(\mathbf{B}) = \mathbf{M}(\mathbf{B})$.

It is worthwhile to note that the normalization of $\mathbf{b}_k$ already requires the calculation of $\mathbf{M}(\mathbf{b}_k)\mathbf{b}_k$, hence the computation of the unnormalized new $\mathbf{b}_k$ requires only a pre-multiplication with $\tilde{\mathbf{M}}^{-1}(\mathbf{B})$. Thus our algorithm can be implemented as follows.

1. Initialization: Start with some unnormalized $\mathbf{B}$. For $k = 1, \dots, K$, compute (in parallel): $\mathbf{M}_k = \sum_{n=1}^{N} \mathbf{C}_n \mathbf{b}_k \mathbf{b}_k^T \mathbf{C}_n$, $\mathbf{p}_k = \mathbf{M}_k \mathbf{b}_k$, $s_k = (\mathbf{b}_k^T \mathbf{p}_k)^{1/2}$ and normalize $\mathbf{b}_k \leftarrow \mathbf{b}_k/s_k^{1/2}$, $\mathbf{p}_k \leftarrow \mathbf{p}_k/s_k^{3/2}$.
2. Iteration: while not converge do

– Compute $\mathbf{M} = \sum_{k=1}^{K} \mathbf{M}_k/s_k$, then make the Cholesky decomposition $\mathbf{M} = \mathbf{R}^T \mathbf{R}$ ($\mathbf{R}$ upper triangular).

– For $k = 1, \ldots, K$, compute (in parallel): $\mathbf{b}_k = \mathbf{R}^{-1}(\mathbf{R}^{T-1}\mathbf{p}_k)$, $\mathbf{M}_k = \sum_{n=1}^{N} \mathbf{C}_n \mathbf{b}_k \mathbf{b}_k^T \mathbf{C}_n$, $\mathbf{p}_k = \mathbf{M}_k \mathbf{b}_k$, $s_k = (\mathbf{b}_k^T \mathbf{p}_k)^{1/2}$ and normalize $\mathbf{b}_k \leftarrow \mathbf{b}_k/s_k^{1/2}$, $\mathbf{p}_k \leftarrow \mathbf{p}_k/s_k^{3/2}$. (Note that multiplication by the inverse of a triangular matrix is done by solving a linear triangular system.)

The above fixed point algorithm is locally stable (i.e. convergent when started near a solution) if the matrix $\mathbf{I} - \mathcal{H}^{-1}(\mathbf{B})\mathcal{C}''(\mathbf{B})$, where $\tilde{\mathcal{C}}''(\mathbf{B})$ denotes the true Hessian, has all its eigenvalues of modulus strictly less than 1. The convergence speed is controlled by the maximum modulus of the eigenvalues. For the (quasi-) Newton algorithm, $\mathcal{H}$ is (nearly) equal to $\mathcal{C}''(\mathbf{B})$, hence the matrix $I - \mathcal{H}^{-1}(\mathbf{B})\mathcal{C}''(\mathbf{B})$ is (nearly) zero and the algorithm has (almost) quadratic convergence. For the pseudo Newton algorithm, if the chosen $\mathcal{H}(\mathbf{B})$ is not too far from $\mathcal{C}''(\mathbf{B})$, one may still expect convergence although the speed is only linear. Simulations indeed show that our pseudo Newton algorithm converges generally well. Note that (global) convergence is not guaranteed even in the quasi-Newton algorithm, as the starting point may be too far from the solution. But since $\mathcal{H}(\mathbf{B})$ is positive definite, one can always reduce the step size, i.e. by taking the new vec($\mathbf{B}$) as vec($\mathbf{B}$) $- \lambda \mathcal{H}^{-1}(\mathbf{B})\mathcal{C}'(\mathbf{B})$, with $\lambda \in (0,1]$, so that the criterion is decreased at each iteration. The algorithm would then converge at least to a local minimum point.

5   Simulation

We compare our LSDIC algorithm to the well-established FFDIAG algorithm of [10] and QDIAG of [9]. We plan to perform a more comprehensive comparison and to publish that in a longer article elsewhere.

Square diagonal matrices with each diagonal entry distributed as a χ-square random variable with one degree of freedom are generated. Each of these matrices, named $\mathbf{D}_n$, may represent the error-free covariance matrix of independent standard Gaussian processes. The noisy input matrices are obtained as $\mathbf{C}_n = \mathbf{A}\mathbf{D}_n\mathbf{A}^T + \mathbf{N}_n$ with symmetric noise matrix $\mathbf{N}_n$ having entries randomly distributed as a Gaussian with zero mean and standard deviation σ. Furthermore, the diagonal elements of $\mathbf{N}_n$ are taken unsigned so to obtain (in general) positive definitive input matrices $\mathbf{C}_n$. The parameter σ controls the overall signal to noise ratio of the input matrices. Two different values will be considered, of which one ($\sigma = 0.01$) represents a small amount of noise closely simulating the exact joint diagonalization (JD) case and the other ($\sigma = 0.05$) simulating the approximate JD case. Two kinds of mixing matrix $\mathbf{A}$ are considered. In the general case mixing matrix $\mathbf{A}$ is obtained as the pseudo-inverse of a matrix with unit norm row vectors which entries are randomly distributed as a standard Gaussian; in this case the mixing matrix may be badly conditioned and we can evaluate the robustness of the AJD algorithms with respect to the conditioning of the mixing matrix. We also consider the case in which $\mathbf{A}$ is a random orthogonal matrix; in this case we can evaluate their robustness with respect to noise.

For QDIAG a normalization matrix $\mathbf{C}_0$ such that $\mathbf{Diag}(\mathbf{B}^T\mathbf{C}_0\mathbf{B}) = \mathbf{I}$ needs to be specified; we use the sum of the input matrix for $\mathbf{C}_0$. As it is well known given true mixing $\mathbf{A}$, each AJD algorithm estimates demixing matrix $\mathbf{B}^T$, which should approximate the inverse of actual $\mathbf{A}$ out of row scaling and permutation. Then, the global matrix $\mathbf{G} = \mathbf{B}^T\mathbf{A}$ should equal a scaled permutation matrix. At each (simulation) repetition we compute the performance index as

$$\text{Performance Index} = \frac{2(K-1)\sum_{i=1}^{K}\sum_{j=1}^{K} G_{ij}^2}{\sum_{i=1}^{K} \max_{j=1,\ldots,K} G_{ij}^2 + \sum_{j=1}^{K} \max_{i=1,\ldots,K} G_{ij}^2}. \tag{7}$$

This index is positive and reaches its maximum 1 if and only if $\mathbf{G}$ has only one non-null elements in each row and column, i.e., if $\mathbf{B}^T$ has been estimated exactly out of the usual row scaling and permutation ambiguities. We computed the means and standard deviations obtained across 250 repetitions for 30 input matrices of dimension 15×15. For each simulation set we then computed all pair-wise bi-directional student-t tests for the null hypothesis $\mu_j = \mu_k, j \neq k$, where μ_j $(j = 1, 2, 3)$ denotes the performance means of the j-th AJD methods. We corrected the resulting p-values for the number of comparisons (3 method pairs) using Bonferroni's method. Results are presented in table 1.

Table 1. Mean and standard deviation (in parentheses) of the performance index (7) attained by QDIAG [9], FFDIAG [10] and our LSDIC algorithm across 250 repetitions of the simulation with $K = 15$ and $N = 30$. The higher the mean and the lower the standard deviation, the better the performance. Legend: $<$ ($>$) indicates that the mean performance of the AJD method is significantly worse (better) as compared to the other two methods as seen with a student-t test with 248 degrees of freedom and setting the type I (false positive) error rate to 0.05 after Bonferroni correction (bi-directional $p(t) < 0.017$).

	Orthogonal Mixing (good conditioning)		Non-Orthogonal Mixing (variable conditioning)	
	$\sigma = 0.01$	$\sigma = 0.05$	$\sigma = 0.01$	$\sigma = 0.05$
QDIAG	0.99976054	0.93320283$^<$	0.99976047	0.93436055
	(0.00015259)	(0.05647088)	(0.00015269)	(0.05463265)
FFDIAG	0.99975514	0.94904938	0.98244331$^<$	0.88311444$^<$
	(0.00015743)	(0.04617571)	(0.05888834)	(0.13230755)
LSDIC	0.99976258	0.95775684	0.99976252	0.95796237$^>$
	(0.00014988)	(0.03721794)	(0.00014992)	(0.03702060)

To see how badly conditioned are our (non orthogonal) mixing matrices, we have computed their condition number for matrix inversion with respect to the Frobenius norm, which for a square matrix $\mathbf{A}$ is given by $\|\mathbf{A}\|\|\mathbf{A}^{-1}\|$. This number is always not less than 1 and a high values indicate bad conditioning. In our 250 repetitions of the simulation, we found that the logarithms of the condition number of our mixing matrices have mean 4.90, standard deviation 1.00, minimum 3.44 and maximum 9.03.

In the case of orthogonal mixing the three algorithms display nearly identical results in the low-noise simulation ($\sigma = 0.01$), whereas in the noisy simulation ($\sigma = 0.05$) QDIAG performs significantly worse then both FFDIAG and LSDIC. In the case of non-orthogonal mixing matrices (general case) FFDIAG performs significantly worse than both QDIAG and LSDIC regardless the noise level, whereas LSDIC significantly outperformed both QDIAG and FFDIAG in the high-noise ($\sigma = 0.05$) case. The lower mean and larger standard deviation displayed by FFDIAG in the non-orthogonal case reflects occasional failing in estimating correctly the demixing matrix B due to the ill-conditioning of the mixing matrix. On the other hand QDIAG appears little robust with respect to noise regardless the conditioning of the mixing matrix. These simulations show the good behavior of LSDIC both in low and high noise conditions and its robustness with respect of the ill-conditioning of the mixing matrix.

6 Conclusion

We have proposed a new algorithm for joint approximated diagonalization of a set of matrices, based on the common least squares criterion but with a novel intrinsic scale constraint. The algorithm is simple, generally fast and is easy to parallelize.

References

1. Belouchrani, A., Abed-Meraim, K., Cardoso, J.-F., Moulines, R.: A blind source separation technique using second-order statistics. IEEE Trans. Signal Process 45(2), 434–444 (1997)
2. Cardoso, J.-F., Souloumiac, A.: Blind beamforming for non-gaussian signals. IEE Proc.-F (Radar and Signal Process) 140(6), 362–370 (1993)
3. Cardoso, J.-F., Souloumiac, A.: Jacobi angles for simultaneous diagonalization. SIAM Journal on Matrix Analysis and Applications 17(1), 161–164 (1996)
4. Congedo, M., Pham, D.-T.: Least-squares joint diagonalization of a matrix set by a congruence transformation. In: Singaporean-French IPAL (Image Processing and Application Laboratory) Symposium (February 2009)
5. Dégerine, S., Kane, E.: A comparative study of approximate joint diagonalization algorithms for blind source separation in presence of additive noise. IEEE Trans. on Signal Process 55(6), 3022–3031 (2007)
6. Li, X.-L., Zhang, X.D.: Nonorthogonal joint diagonalization free of degenerate solution. IEEE Transactions on Signal Processing 55(5), 1803–1814 (2007)
7. Pham, D.T.: Joint approximate diagonalization of positive definite matrices. SIAM Journal on Matrix Analysis and Applications 22(4), 1136–1152 (2001)
8. Pham, D.-T., Cardoso, J.-F.: Blind separation of instantaneous mixtures of non stationary sources. IEEE Trans. on Signal Process 42(9), 1837–1848 (2001)
9. Vollgraf, R., Obermayer, K.: Quadratic optimization for simultaneous matrix diagonalization. IEEE Trans. on Signal Process 54(9), 3270–3278 (2006)
10. Ziehe, A., Laskov, P., Nolte, G., Müller, K.-R.: A fast algorithm for joint diagonalization with non-orthogonal transformations and its application to blind source separation. Journal of Machine Learning Research 5, 801–818 (2004)

Iterative-Shift Cluster-Based Time-Frequency BSS for Fractional-Time-Delay Mixtures

Matthieu Puigt and Yannick Deville

Laboratoire d'Astrophysique de Toulouse-Tarbes, Université de Toulouse, CNRS
14 Av. Edouard Belin, 31400 Toulouse, France
mpuigt@ast.obs-mip.fr, ydeville@ast.obs-mip.fr

Abstract. We recently proposed several time-frequency blind source separation methods applicable to attenuated and delayed mixtures. They were limited by heuristics and only processed integer-sample time shifts. In this paper, we propose and test improvements of these approaches based on modified clustering techniques for scale coefficients and a new iterative method to estimate possibly non-integer and large delays.

1 Introduction

Blind source separation (BSS) consists in estimating a set of N unknown sources from P mixtures of these sources. Most of the approaches that have been developed to this end are based on Independent Component Analysis [1]. Since the last decade, some methods using Sparse Component Analysis, e.g. time-frequency (TF) analysis, have been proposed [1]. Some of these approaches like DUET and its extensions [2,3] need the sources to be (approximately) disjoint in the analysis domain (WDO assumption). DUET separates attenuated and delayed (AD) mixtures, a simple class of convolutive mixtures physically corresponding to anechoic propagation, but cannot estimate delays higher than one sample. On the contrary, we recently proposed several AD TF-BSS methods, i.e. AD-TIFROM and AD-TIFCORR [4,5], which only need tiny TF zones where a source occurs alone in order to estimate the mixing matrix. These methods deal with a much wider range of time shifts (typically 0 to 200 samples) than DUET but they only estimate integer-sample time shifts and need a user-defined threshold in order to distinguish columns of the mixing matrix, which is a major limitation since the adequate values of this threshold depend on the unknown mixing parameters. Lastly, Arberet *et al.* also proposed an AD-TF method, called DEMIX Anechoic [6], which can be wiewed as a *hybrid* combination of DUET and AD-TIFROM/CORR: like DUET, it assumes approximate WDO (and deals with underdetermined non-integer-sample AD mixtures) but adds a confidence measure of the "single-source quality", as proposed in [4,5]. However, the tests in [6] are only performed with less than 3-sample time shifts.

As a consequence of the above analysis, we here propose several improvements to our previous methods: the first one consists in solving the problem of the above user-defined threshold, thanks to a clustering approach. We also propose a new iterative method to estimate time shifts, applicable to their fractional values. We

T. Adali et al. (Eds.): ICA 2009, LNCS 5441, pp. 306–313, 2009.
© Springer-Verlag Berlin Heidelberg 2009

thus process the same class of mixtures as DUET and DEMIX Anechoic, while dealing with a much larger range of delays.

2   Problem Statement

In this paper, we assume that N unknown sources $s_j(n)$ are mixed through AD propagation channels and added, thus providing a set of N mixed signals $x_i(n)$:

$$x_i(n) = \sum_{j=1}^{N} a_{ij} s_j(n - n_{ij}) \qquad j \in \{1 \dots N\}, \tag{1}$$

where $a_{ij} > 0$ are constant scale coefficients and n_{ij} are first supposed to be integer time shifts for the sake of clarity. We then propose an extension to non-integer ones in Subsect. 5.2. By applying the approach used in [4], we handle the scale/filter and permutation indeterminacies inherent in the BSS problem: we rewrite (1) with respect to the contributions of the sources in the first observation, up to an arbitrary permutation $\sigma(.)$. These contributions read

$$s'_j(n) = a_{1,\sigma(j)} \, s_{\sigma(j)} \left(n - n_{1,\sigma(j)} \right) \tag{2}$$

and (1) may be rewritten as

$$x_i(n) = \sum_{j=1}^{N} a_{i,\sigma(j)} \, s_{\sigma(j)} \left(n - n_{i,\sigma(j)} \right) = \sum_{j=1}^{N} b_{ij} \, s'_j \left(n - \mu_{ij} \right) \tag{3}$$

with

$$b_{ij} = \frac{a_{i,\sigma(j)}}{a_{1,\sigma(j)}} \text{ and } \mu_{ij} = n_{i,\sigma(j)} - n_{1,\sigma(j)}. \tag{4}$$

Equation (3) reads in the Fourier domain

$$X_i(\omega) = \sum_{j=1}^{N} b_{ij} \, e^{-j\omega\mu_{ij}} \, S'_j(\omega) \qquad i \in \{1 \dots N\}. \tag{5}$$

We therefore aim at estimating the following mixing matrix

$$B(\omega) = \left[b_{ij} e^{-j\omega\mu_{ij}} \right] \qquad i, j \in \{1 \dots N\}. \tag{6}$$

3   Overall Structure of AD-TIFCORR

As stated in Sect. 1, we recently proposed two AD TIme-Frequency BSS approaches resp. based on Ratios Of Mixtures and CORRelation. Here, we introduce improvements, applicable to both methods, that we describe in the framework of AD-TIFCORR [5]. The considered TF transform of the signals is the Short Time Fourier Transform (STFT). Our approach uses the following definitions and assumptions.

Definition 1. *A "TF analysis zone" is a series of adjacent TF windows (n, ω).*
More precisely, they can have any shape in the TF plane but we focus on *constant-frequency* (CF) and *constant-time* (CT) analysis zones which can be resp. defined as series of M time-adjacent and M' frequency-adjacent TF windows (n, ω). We resp. denote them (T, ω) and (n, Ω).

Definition 2. *A source is said to be "isolated" in a TF analysis zone if only this source has a non-zero power (i.e. mean-square value) in this TF zone.*

Definition 3. *A source is said to be "accessible" in the TF domain if there exist at least one TF zone where it is isolated.*

Assumption 1. *i) each source is accessible in the TF domain and ii) there exist no TF zone where the TF transforms of all sources are equal to zero everywhere*[1].

For any couple of signals u_1 and u_2, we define the cross-correlation of the TF transforms of these signals over the considered CF analysis zone (T, ω) as

$$R_{u_1 u_2}(T, \omega) = \frac{1}{M} \sum_{p=1}^{M} U_1(n_p, \omega) U_2^*(n_p, \omega), \tag{7}$$

where the superscript $*$ denotes complex conjugate. The corresponding correlation coefficient reads

$$r_{u_1 u_2}(T, \omega) = \frac{R_{u_1 u_2}(T, \omega)}{\sqrt{R_{u_1 u_1}(T, \omega) R_{u_2 u_2}(T, \omega)}}. \tag{8}$$

For each analysis zone (T, ω), the vector consisting of TF values $S_j(n_p, \omega)$ of any source signal $s_j(n)$ is denoted $V_{s_j}(T, \omega)$ hereafter.

Assumption 2. *Over each analysis zone (T, ω), the non-zero vectors $V_{s_j}(T, \omega)$ are linearly independent (if there exist at least two such vectors in this zone)*[2].

The AD-TIFCORR method aims at estimating the mixing matrix defined in (6), i.e. the scale coefficients b_{im} and the associated time shifts μ_{im}. It is composed of three main stages, preceded by a pre-processing stage:

1. The pre-processing stage consists in deriving the STFTs of the mixed signals.
2. We detect single-source CF analysis zones, where we estimate the scale coefficients b_{im}.
3. We detect single-source CT analysis zones, where we estimate the time shifts $\mu_{im'}$ and couple then to b_{im}. The basic version of Steps 2 and 3 and their improved version proposed in this paper are resp. detailed in Sect. 4 and 5.
4. In the combination stage, we eventually derive the output signals. They may be obtained in the frequency domain by computing

$$\underline{Y}(\omega) = B^{-1}(\omega) \underline{X}(\omega), \tag{9}$$

where $\underline{Y}(\omega) = [Y_1(\omega) \cdots Y_N(\omega)]^T$ and $\underline{X}(\omega) = [X_1(\omega) \cdots X_N(\omega)]^T$.

---

[1] Assumption 1 ii) is only made for the sake of simplicity: it may be removed in practice, thanks to the noise contained by real recordings [5].

[2] In [5], we assumed the sources to be uncorrelated but we recently proved [7] that, thanks to Assumption 2, TIFCORR can also be applied to correlated sources.

4   Detection of Single-Source Zones and Identification of Mixing Parameters in "Basic" AD-TIFCORR

Equation (5), written in the frequency domain, remains almost exact when expressed in the TF domain (provided the temporal width of TF windows in the STFTs is significantly higher than the applied time shifts n_{ij}), i.e.

$$X_i(n, \omega) = \sum_{j=1}^{N} b_{ij}\, e^{-j\omega\mu_{ij}}\, S'_j(n, \omega) \qquad i \in \{1 \ldots N\}. \tag{10}$$

Applying the general proof of [7] directly shows that a necessary and sufficient condition for a source to be isolated in a TF analysis zone (T, ω) is

$$|r_{x_1 x_i}(T, \omega)| = 1 \qquad \forall i \in \{2 \ldots N\}. \tag{11}$$

In practice, we compute the mean $\overline{|r_{x_1 x_i}(T, \omega)|}$ over i of $|r_{x_1 x_i}(T, \omega)|$ and sort the TF zones (T, ω_l) according to decreasing values of $\overline{|r_{x_1 x_i}(T, \omega_l)|}$. If a source $S_k(n, \omega)$ is isolated in a TF zone (T, ω_l) then (10) and (7) yield

$$\left| \frac{R_{x_i x_1}(T, \omega_l)}{R_{x_1 x_1}(T, \omega_l)} \right| = \left| \frac{a_{ik}}{a_{1k}} e^{-j\omega(n_{ik}-n_{1k})} \right| = \left| b_{im} e^{-j\omega\mu_{im}} \right| = b_{im} \tag{12}$$

with b_{im} and μ_{im} defined by (4) and $k = \sigma(m)$. The basic procedure for estimating all parameters b_{im} applies as follows: we successively study the analysis zones in the above sorted list, we estimate a column of parameters b_{im} according to (12) and keep it if its distance with respect to each previously found column is above a user-defined threshold ϵ_1, which is a major limitation of the method, since its value has to be selected with respect to the values of *unknown* b_{im}. We stop the procedure when N columns have been found.

The estimation of time shifts is based on the phase of ratios $\alpha_i(n, \omega)$ of TF transforms of mixtures, defined as

$$\alpha_i(n, \omega) = \frac{X_i(n, \omega)}{X_1(n, \omega)} = \frac{\sum_{j=1}^{N} a_{ij}\, e^{-j\,\omega n_{ij}}\, S_j(n, \omega)}{\sum_{j=1}^{N} a_{1j}\, e^{-j\,\omega n_{1j}}\, S_j(n, \omega)}. \tag{13}$$

If a source $S_k(n, \omega)$ is isolated in a TF zone $(n_{p'}, \Omega)$, then the curve associated to the variations of the phase of $\alpha_i(n_{p'}, \omega_{l'})$ in this zone is a line and its slope is equal to (the opposite of) the associated time shift $\mu_{im'}$ (with $k = \sigma(m')$). We therefore consider each CT zone $(n_{p'}, \Omega)$ and compute the regression line and the corresponding error associated to the variations of the above phase. The CT zones with a low error are single-source and columns of $\mu_{im'}$ are set to (the opposite of) the integers which are the closest to the slopes of the corresponding lines. A procedure described in [5] then maps each column of time shifts $\mu_{im'}$ to a previously estimated column of scale coefficients b_{im}. A set of time shifts $\mu_{im'}$ is thus associated to each parameter b_{im}. We then derive the histograms of these parameters $\mu_{im'}$, independently for each index i and each index m. We eventually keep the peak value in each histogram as the estimate of μ_{im}.

5   Proposed Improvements

5.1   Clustering Techniques for Scale Coefficients b_{im}

We now present improvements of the AD-TIFCORR method, which can also be applied to the basic AD-TIFROM methods defined in [4]. The first one consists in modifying the detection and estimation stages for scale coefficients. Indeed, when many sources are mixed or time shifts are quite large, we faced cases when a "false" single-source TF zone is detected and the associated column of b_{im} is kept, thus yielding poor performance. We therefore proposed [4] a simple clustering method, called CBCE (Cardinality-Based Cluster Elimination) hereafter, that solves this problem but still needs the user-defined threshold ϵ_1 defined in Sect. 4 in order to separate clusters.

We here solve both above problems, i.e. we do not depend on the value of a *user-defined* threshold to distinguish columns of b_{im} and we cancel false single-source zones effects, by using the K-means clustering technique [8] or its modified version K-medians [9]. Our new approach thus operates as follows: we first determine all TF zones such that

$$\overline{|r_{x_1 x_i}(T,\omega)|} \geq 1 - \epsilon_2, \tag{14}$$

we keep at most the 1000 "best" zones (T,ω) satisfying (14) if they exist, and we then process their columns of coefficients $\widehat{b}_{im}$ with K-means or K-medians, hence denoting these methods *Selective K-means* (resp. *K-medians*) hereafter. Indeed, when many zones meet (14) (typically from several thousands to several tens of thousands), we noticed that the *standard* K-means/medians estimate the coefficients b_{im} with a poor accuracy, because the number of "coarse" estimates is much higher than the number of "accurate" ones[3]. This result is very important because a lot of WDO-based BSS methods, e.g. [3], use K-means on all the analysis domain and could be highly improved as well with our selective approach.

5.2   An Iterative Method for Estimating Time Shifts μ_{im}

We presented in Sect. 4 an approach for estimating time shifts. However, in some cases, this method does not yield estimated time shifts *exactly* equal to the theoretical ones. In order to improve it, we propose an iterative method that operates as follows:

1. in a first step, we estimate time shifts μ_{im} in constant-time TF zones, as explained in Sect. 4. These estimated values, associated to scale coefficients b_{im}, are denoted $\widehat{\mu}_{im}$ hereafter.

---

[3] As an alternative, one can reduce the value of ϵ_2 in (14). However, in this case, we noticed that when the time shifts were quite large, some sources could be unaccessible. Limiting the number of TF zones to 1000 is thus an acceptable trade-off: on the one hand, when there exist many single-source zones, the 1000 "best" zones yield a high accuracy in the estimation of coefficients b_{im}. On the contrary, when there are few of them, the threshold ϵ_2 guarantees a quite good "single-source quality".

2. For each of these estimates, we reduce the time lag of the associated source in observations, by applying a time shift of $\widehat{\mu}_{im}$ samples to $x_1(n)$, i.e. we create the modified observations defined as

$$\begin{cases} \widetilde{x_1}(n) = x_1(n - \widehat{\mu}_{im}) \\ \widetilde{x}_i(n) = x_i(n) \end{cases}.$$ (15)

If the considered estimated time shift $\widehat{\mu}_{im}$, associated to a source $s_k(n)$, is equal to the corresponding theoretical value, then, from the point of view of this source $s_k(n)$, the BSS problem (15) is instantaneous. On the contrary, if the estimated and theoretical values of μ_{im} are not equal, there exist a residual time shift $\tau_{im} \neq 0$ in the signals $\widetilde{x_1}(n)$ and $\widetilde{x}_i(n)$ so that

$$\frac{R_{\widetilde{x}_i \widetilde{x_1}}}{R_{\widetilde{x_1} \widetilde{x_1}}} = b_{im} e^{-j\omega\tau_{im}}.$$ (16)

3. We apply the procedure described in Sect. 4 to the time-shifted observations $\widetilde{x_1}(n)$ and $\widetilde{x}_i(n)$, in order to estimate τ_{im} and we derive $\widehat{\mu}_{im} + \widehat{\tau}_{im}$ which is a closer estimate of μ_{im} than $\widehat{\mu}_{im}$.
4. Steps 2 and 3 are performed I times ($I \geq 1$) or until

$$\forall i, m \in \{2, \ldots, N\}, \quad \widehat{\tau}_{im} = 0.$$ (17)

This iterative method not only provides a means to improve the estimation of time shifts but also to estimate non-integer ones. Indeed, let us suppose that $\mu_{im} \notin \mathbb{N}$. What we can compute with the basic or the above improved method is an estimate of the integer part of μ_{im}, denoted $\widehat{\lfloor \mu_{im} \rfloor}$ hereafter. For each estimate $\widehat{\lfloor \mu_{im} \rfloor}$, there exists a residual fractional time shift ζ_{im} such that

$$\mu_{im} = \widehat{\lfloor \mu_{im} \rfloor} + \zeta_{im}.$$ (18)

We then propose to run a slightly modified version of the above iterative procedure: in Step 2, we apply a delay of $\widehat{\lfloor \mu_{im} \rfloor}$ samples to (15) and we oversample by a factor F (e.g. $F = 10$) the observations $\widetilde{x_1}(n)$ and $\widetilde{x}_i(n)$. As a consequence, at the end of the iterative procedure, we ideally estimate ζ_{im} with an accuracy of about F^{-I}, where I is the above-defined number of iterations.

6 Experimental Results

We test the performance of both basic and improved methods described in Sect. 4 and 5, with 2 sets of $N = 2$ sources of English speech signals sampled at 20 kHz that we previously used in [4]. The performance achieved in each test is measured by the overall output signal-to-interference-ratio achieved by this system, denoted SIR^{out} below [4], and the *success rate*, i.e. the percentage of cases when estimated integer time shifts are *exactly* equal to theoretical ones [4]. We consider symmetrical mixing matrices defined as

$$A(\omega) = \begin{bmatrix} 1 & \lambda e^{-j\omega\eta} \\ \lambda e^{-j\omega\eta} & 1 \end{bmatrix}, \qquad \lambda = 0.5, 0.9 \text{ and } \eta = 0, 10, 20, 200.$$ (19)

Our method uses TF transforms of observations twice, when resp. considering CF and CT analysis zones. Therefore, we independently use two sets of parameters, resp. associated to the above two types of analysis zones, i.e.

- We here denote d (resp. d') the number of samples of observed signals $x_i(n)$ in each time window of STFTs used in CF (resp. CT) analysis zones.
- ρ (resp. ρ') is the temporal overlap between the time windows in the STFTs used in CF (resp. CT) analysis zones.

All these parameters take the following values: $d = 256$, $M = 10$, $\rho = 75\%$, $d' = 2^m$, $M' = 2^{m-7}, 2^{m-6}$ or 2^{m-5}, $\rho' = 50, 75$ or 90% with $m = 9 \dots 14$. The above-defined thresholds are fixed to $\epsilon_1 = 0.15$ and $\epsilon_2 = 1.5e - 2$ while the number I of iterations, when used, for estimating time shifts is set to 1.

Figure 1(a) provides the inverse of the mean of the Frobenius norm of the difference between the estimated and actual matrices of parameters b_{im}, over the sets of sources and the mixing parameter λ, with respect to the time shift η and the clustering technique (the basic approach [5] without clustering is also tested). This figure shows that Selective K-means and K-medians outperform all other techniques. In particular, when $\eta = 0$, they provide estimates 100 times more accurate than non-selective K-means and K-medians, i.e. which only test (14). Lastly, Selective K-medians yields slightly better estimates than Selective K-means for $\eta \neq 0$. As a consequence, only this approach is tested below.

We showed [4] that our AD TF-BSS methods are mainly sensitive to the number d' of samples in STFTs computations. Figure 1(b) therefore shows the mean performance obtained by the AD-TIFCORR method with both basic and iterative methods for estimating μ_{im}, with respect to d'. This figure shows that the proposed approach improves SIR^{out} and success rates (above 90%) for the lowest values of d' (below 8192).

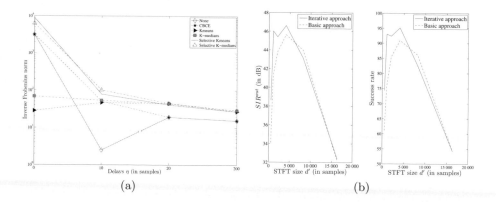

(a)   (b)

Fig. 1. (a) Inverse of the mean Frobenius norm of the difference between estimated and actual values of matrix $[b_{im}]$ vs applied time shift η. (b) *Left:* mean SIR^{out} vs STFT size d'. *Right:* mean success rate vs STFT size d'.

We now test our Selective K-medians-based iterative AD-TIFCORR method for non-integer-sample AD mixtures of the above sets of sources, with the following mixing parameters in (19): $\lambda = 0.9$ and $\eta = 30.5$. The parameters in CF zones are the same as above and, for the sake of brevity, we fixed the CT parameters to $d' = 2048$, $M' = 16$ and $\rho' = 75\%$, and the oversampling factor to $F = 10$. The proposed approach succeeded in estimating the mixing matrix with both sets of sources: SIR^{out} were then resp. equal to 48.7 and 37.9 dB, while "Basic" AD-TIFCORR [5] yielded SIR^{out} resp. equal to 12.3 and 10.5 dB.

7   Conclusion and Extensions

We introduced improvements for our previous AD TF-BSS methods based on single-source analysis zones: we used the proposed Selective K-means or K-medians clustering techniques for estimating scale coefficients and an iterative method for estimating possibly non-integer time shifts. We presented various aspects of the experimental performance of the approach, showing the relevance of the proposed improvements. In our future investigations, we will aim at performing a more detailed characterization of their performance, at studying the underdetermined case and at extending our method to general convolutive mixtures.

References

1. O'Grady, P.D., Pearlmutter, B.A., Rickard, S.: Survey of sparse and non-sparse methods in source separation. International Journal of Imaging Systems and Technology 15, 18–33 (2005)
2. Yilmaz, Ö., Rickard, S.: Blind separation of speech mixtures via time-frequency masking. IEEE Transactions on Signal Processing 52(7), 1830–1847 (2004)
3. Saab, R., Yilmaz, Ö., McKeown, M.J., Abugharbieh, R.: Underdetermined sparse blind source separation with delays. In: Proc. of SPARS 2005 (2005)
4. Puigt, M., Deville, Y.: Time-frequency ratio-based blind separation methods for attenuated and time-delayed sources. MSSP 19(6), 1348–1379 (2005)
5. Puigt, M., Deville, Y.: A time-frequency correlation-based blind source separation method for time-delayed mixtures. In: Proc. of ICASSP 2006, vol. 5, pp. 853–856 (2006)
6. Arberet, S., Gribonval, R., Bimbot, F.: A robust method to count and locate audio sources in a stereophonic linear anechoic mixture. In: Proc. of ICASSP 2007, vol. 3, pp. 745–748 (2007)
7. Deville, Y., Bissessur, D., Puigt, M., Hosseini, S., Carfantan, H.: A time-scale correlation-based blind separation method applicable to correlated sources. In: Proc. of ESANN 2006, pp. 337–344 (2006)
8. Xui, R., Wunsch II, D.: Survey of clustering algorithms. IEEE Transactions on Neural Networks 16(3), 645–678 (2005)
9. Bradley, P.S., Mangasarian, O.L., Street, W.N.: Clustering via concave minimization. In: Proc. of NIPS 1996, vol. 9, pp. 368–374 (1996)

Blind Non-stationnary Sources Separation by Sparsity in a Linear Instantaneous Mixture

Bertrand Rivet

GIPSA-lab, CNRS UMR-5216, Grenoble INP,
Domaine Universitaire, BP 46, 38402 Saint Martin d'Hères cedex, France
bertrand.rivet@gipsa-lab.inpg.fr

Abstract. In the case of a determined linear instantaneous mixture, a method to estimate non-stationnary sources with non activity periods is proposed. The method is based on the assumption that speech signals are inactive in some unknown temporal periods. Such silence periods allow to estimate the rows of the demixing matrix by a new algorithm called Direction Estimation of Separating Matrix (DESM). The periods of sources inactivity are estimated by a generalised eigen decomposition of covariance matrices of the mixtures, and the separating matrix is then estimated by a kernel principal component analysis. Experiments are provided with determined mixtures, and shown to be efficient.

1 Introduction

Blind source separation consists of estimating unknown signals (denoted sources) from mixtures of them without prior knowledge neither about the nature of mixing function nor about the sources. When involved sources have specific properties, the source separation can be based on them leading thus to semi-blind source separation methods. For instance when speech signals are present among the sources, non-stationarity [10,13], or sparse decomposition in a specific basis [17,1,2] have been exploited. In parallel, the bi-modal (audio-visual) nature of speech was used [14,16,11,12]. Audiovisual speech source separation is based on the strong links which exist between the sound produced by a speaker and visual speech signals, in particular speaker's lips movement: these methods exploit the complementarity and the redundancy of these modalities.

The proposed method is based on the "sparsity" of speech signals. Indeed, they are highly non-sationnary: there are a lot of lapses of time during which the signal power is negligible compared its averaged power, for instance time between words. The proposed method draws from [12] where the silent moments of a speaker (estimated by a purely visual voice activity detection [3]) are used to identify the function which allows to extract this specific speech signal. Even if this audiovisual approach is efficient (even in the convolutive case), it requires a specific device to record simultaneously audio and video signals by microphones and camera, respectively. In the present study, the instantaneous mixing case is addressed by a purely acoustic method which estimates jointly the voice non-activity periods and the separation matrix.

T. Adali et al. (Eds.): ICA 2009, LNCS 5441, pp. 314–321, 2009.

This paper is organised as follows. Section 2 presents the proposed approach to exploit the natural speech sparsity, while Section 3 describes DESM algorithm to estimate sources. Numerical experiments and results are given in Section 4 before conclusions and perspectives in Section 5.

2   Exploitation of Natural Speech "Sparsity"

In this section, the general framework of source separation with instantaneous mixture is recalled before introducing the proposed method to exploit the natural sparsity of speech.

Let $\mathbf{s}(t) \in \mathbb{R}^{N_s}$ denotes the N_s dimensional column vector of source signals whose j-th component is $s_j(t)$. With instantaneous mixtures, observations $x_i(t)$ are expressed as a linear combination of sources $s_j(t)$: $x_i(t) = \sum_j a_{i,j} s_j(t)$, or with matrix notation

$$\mathbf{x}(t) = A\,\mathbf{s}(t), \qquad (1)$$

where $A \in \mathbb{R}^{N_m \times N_s}$ denotes the mixing matrix whose (i,j)-th entry is $a_{i,j}$ and $\mathbf{x}(t) \in \mathbb{R}^{N_m}$ is the N_m dimensional column vector of the observations. In this study, the determined case is considered: the number of mixtures N_m is so equal to the number of sources N_s. The source estimation problem is then equivalent to estimate a separating matrix $B \in \mathbb{R}^{N_s \times N_s}$ such that

$$\mathbf{y}(t) = B\,\mathbf{x}(t) \qquad (2)$$

is a vector whose components are the estimate of sources $s_i(t)$.

The independant component analysis (ICA) [6,4], which exploits the mutual independance between the sources, was widely used to solve this problem. Recently sparsity was introduced in source separation [8]. For instance, methods proposed in [17,1,2] are based on the assumption that, in some basis, there exist some parts where at most one source is present at the time allowing thus to estimate the mixing matrix. Indeed, if at time index τ, only source $s_n(\tau)$ is active (i.e., $\forall i \neq n, s_i(\tau) = 0$) then $\mathbf{x}(\tau) = \mathbf{a}_n s_n(\tau)$. In other words, mixtures $\mathbf{x}(\tau)$ are proportional to n-th column $\mathbf{a}_n$ of mixing matrix A. It is thus possible to estimate all the columns of the mixing matrix, and to express the separating matrix B as the inverse of the estimated mixing matrix.

The proposed method is quite different since it is based on the assumption that there exist some time indexes where at least one source is inactive: i.e. for $t = \tau$, $\exists n \,/\, s_n(\tau) = 0$. Let suppose, in this section, that all the sources are stationary excepted one, let say $s_1(t)$ without any loose of generality. Let R_1 denotes the covariance matrix of observations $\mathbf{x}(t)$ computed for all time indexes t, and let R_2 denotes covariance matrix of observations computed during an inactivity period of source $s_1(t)$. The proposed method is based on the general eigenvalue decomposition of couple (R_2, R_1) [15]. It is easy to check that (R_2, R_1) admits only two disctint generalised eigenvalues: 1 degenerated $N_s - 1$ times (whose eigensubspace $\mathcal{E}$ is a hyperplan complementary to $\mathbf{a}_1$), and 0 whose generalised eigenvector $\mathbf{v}$ is orthogonal to $\mathcal{E}$. Thus the projection of observations

$\mathbf{x}(t)$ on generalised eigenvector $\mathbf{v}$ allows to extract source $s_1(t)$ by cancelling the contribution of other sources: $\forall i \neq 1$, $\mathbf{v}^T \mathbf{a}_i = 0$, since $\mathbf{a}_i$ for $i \neq 1$ are in hyperplan $\mathcal{E}$.

This method allows first to detect if source $s_1(t)$ vanishes by testing generalised eigenvalues and then to extract source $s_1(t)$ when it is active by projecting the observations on the generalised eigenvector associated with the generalised eigenvalue equal to zero.

3   DESM Algorithm

In the previous section, only one source was considered to be non-stationnary with non active periods. However, several sources can be non active (possibly in different periods): for instance if the mixtures contain several speech sources. Moreover, the inactivity periods are unknown. In this section, the proposed Direction Estimation of Separating Matrix (DESM) algorithm is presented: it extends the previously proposed method (Section 2) to extract from the mixture all the sources with non active periods.

To detect time periods where at least one source is non active, we proposed to compute the generalised decomposition of couple $\{(R_2(\tau), R_1)\}_\tau$ where R_1 is the covariance matrix of observations $\mathbf{x}(t)$ estimated with all time samples and $R_2(\tau)$ is the covariance of observations $\mathbf{x}(t)$ estimated on windowed samples around τ (typically, this window is about 100 milliseconds). The generalised eigen decompositions of $\{(R_2(\tau), R_1)\}_\tau$ provide

$$R_2(\tau)\,\Phi(\tau) = R_1\,\Phi(\tau)\,\Lambda(\tau), \tag{3}$$

where $\Lambda(\tau)$ is a diagonal matrix whose diagonal terms $\lambda_1(\tau) \leq \cdots \leq \lambda_{N_s}(\tau)$ are the generalised eigenvalues and $\Phi(\tau)$ is an orthonormal matrix whose columns $\phi_i(\tau)$ are the generalised eigenvectors. Thus at τ time, if N sources are inactive then N generalised eigenvalues are null whose associated generalised eigenvectors defined a subspace orthogonal to the subspace spanned by the $N_s - N$ active sources.

The proposed DESM algorithm can thus be decomposed in two steps:

1. the first one is to detect the periods where at least one source is inactive by testing the generalised eigenvalues $\{\lambda_1(\tau)\}_\tau$: if $\lambda_1(\tau) \leq \eta$, where η is a threshold chosen a priori, then the algorithm decides that at least one source was inactive during time window centred on τ. Let $\Theta = \{\tau \mid \lambda_1(\tau) \leq \eta\}$ be the set of time indexes where at least one source is inactive (the cardinal of Θ is N_τ). This provides a set of vectors $\{\phi_1(\tau)\}_{\tau \in \Theta}$ defined as the set of the first generalised eigenvector with $\tau \in \Theta$. These vectors are mainly aligned in the directions which allows to extract corresponding sources (Fig. 2). These direction are the rows of the separating matrix.
2. then the DESM algorithm estimates these directions thanks to a kernel principal component analysis (kernel PCA) [9,7], where the kernel is chosen as

$$k\big(\phi_1(t), \phi_1(t')\big) = k_{t,t'} \overset{\triangle}{=} \begin{cases} \frac{\phi_1^T(t)\phi_1(t') - \cos\theta_0}{1 - \cos\theta_0}, & \text{if } \phi_1^T(t)\phi_1(t') \geq \cos\theta_0 \\ 0, & \text{else} \end{cases} \tag{4}$$

with t and t' in Θ, and where θ_0 is an angle which is chosen *a priori*. Kernel PCA consists in performing an eigen decomposition of matrix $K \in \mathbb{R}^{N_\tau \times N_\tau}$ whose (i,j)-th entry is $k_{i,j}$:

$$K = \Psi \Delta \Psi^T, \tag{5}$$

where Δ is a diagonal matrix of eigenvalues of K, and Ψ is an orthonormal matrix whose columns are eigenvectors of K. Let $W = [\psi_1, \cdots, \psi_{N_s}]$ be the matrix composed by the concatenation of N_s eigenvectors ψ_i associated with the N_s largest eigenvalues.

The separation matrix is then obtained by

$$B = W^T K V, \tag{6}$$

where $V = [\phi_1(t \in \Theta)]$ is the matrix obtained by the concatenation of generalised eigenvector associated with the smallest generalised eigenvalue $\lambda_1(t)$ (3) with $t \in \Theta$. The sources are finally estimated thanks to

$$\hat{s}(t) = B\mathbf{x}(t), \tag{7}$$

for all time indexes t, including those when sources are active.

Finally, DESM algorithm which allows to extract non-stationnary sources with inactive periods, is summarised in Algorithm 1.

Algorithm 1. DESM algorithm

1: Compute covariance matrix R_1 from all time samples
2: **for** each τ **do**
3:     Compute covariance matrix $R_2(\tau)$ with time window centred on τ
4:     Compute generalised eigen decomposition (3) of couple $(R_2(\tau), R_1)$
        $\Rightarrow (\Phi(\tau), \Lambda(\tau))$
5: **end for**
6: Estimate $\Theta = \{\tau \mid \lambda_1(\tau) \leq \eta\}$
7: Compute matrix K defined by (4)
8: Perform eigen decomposition (5) of $K \Rightarrow (\Psi, \Delta)$
9: Compute $W = [\psi_1, \cdots, \psi_{N_s}]$ and $V = [\phi_1(t \in \Theta)]$
10: Compute $B = W^T K V$ (6)
11: Estimate sources by $\hat{s}(t) = B\mathbf{x}(t)$

Note that using generalised eigenvalues of couple $(R_1, R_2(\tau))$ in stage 4, instead of using simple eigenvalues of $R_2(\tau)$, overcomes the problem of relative power of sources. In particular when some of the sources are definitely less powerful than others, using eigenvalues can lead to consider that these sources are inactive.

4   Numerical Experiments

In this section, the principle of the proposed DESM algorithm is illustrated to extract speech signals from linear instantaneous mixtures of audio sources (*i.e.* speech and musical sources). The sources are from two databases: the first one is composed of 18 French sentences read by French speakers (males and females), the second one is composed of music signals. All signals were sampled at 16kHz. In the different tested configurations, the sources are randomly chosen and the entries of the mixing matrix are randomly chosen from a uniform random variable distributed from -1 to 1. For each configuration (*i.e.* for each number of sources) 100 different mixtures were tested.

In the first experiment, the extraction of two speech signals from three mixtures is illustrated (Fig. 3). One of the source is thus a musical signal without inactive periods, the second in this example. First of all, the estimation of non-activity periods thanks to generalised eigen decomposition is illustrated on Fig. 1. As one can see on the top plot, which represents the power of the three sources computed with a time-sliding window of 100ms, the two speech sources have (possibly overlapped) non-activity periods while the musical source has its short term power almost constant. It is quite interesting to note that the smallest generalised eigenvalue $\lambda_1(t)$ (bottom plot) allows to detect these inactivity periods, without labbeling which speech signals are inactive. Moreover, generalised eigenvectors $\phi_1(t)$ (Fig. 2) are mainly in two directions corresponding to the rows of the separating matrix that extract the two speech signals. Fig. 3 shows that the proposed DESM algorithm is efficient to extract speech signals ($\hat{s}_1(t)$ et $\hat{s}_2(t)$). Moreover, the third estimated source is still a mixture of the three sources since kernel PCA of matrix K_1 only presents two significant eigenvalues. More generaly, the number of significant eigenvalues of matrix K_1 could be used to estimate the number of speech sources to only extract these sources.

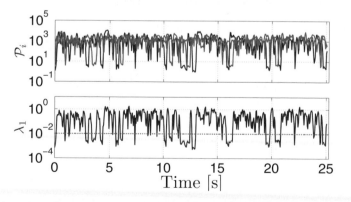

Fig. 1. Estimation of non-activity periods thanks to DESM algorithm. Top figure shows the power of the three sources on a time-sliding window of 100ms (blue, green and red curves for the 1st, 2nd and 3rd source, respectively). Bottom figure shows the smallest generalised eigenvalue (3) $\lambda_1(\tau)$ (blue curve) as well as chosen threshold η (red dotted curve). Plots are in logarithm scale.

Fig. 2. Estimation of separating matrix B (6) thanks to DESM algorithm with three sources. Projections of estimated rows (red curves) and generalised eigenvectors $\phi_1(t)$ with $t \in \Theta$ (blue points) on (x_1, x_2), (x_2, x_3) et (x_1, x_3). Generalised eigenvectors are multiplied by $1/\lambda_1(t)$.

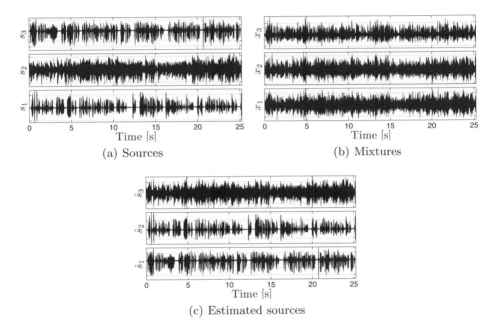

(a) Sources

(b) Mixtures

(c) Estimated sources

Fig. 3. Illustration of DESM algorithm

In the second experiment (Fig. 4), the performance of the proposed algorithm was estimated. To evaluate the estimation of the rows of the separating matrix, we use the performance index defined as

$$PI = \sum_{i \in \mathcal{S}} \sum_{j} \left| \frac{C_{i,j}}{\max_k |C_{i,k}|} \right| - 1, \text{ with } C = B A, \qquad (8)$$

where $\mathcal{S}$ denotes the set of speech sources. So the smaller the performance index is, the better the extraction is. In these experiments, only two sources are speech sources, all the other sources are musical signal. Figure 4 shows the median performance index versus the number of sources. The performance achieved by

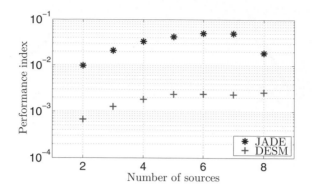

Fig. 4. Performance index

the proposed DESM algorithm are compared with the performance provided by the JADE algorithm [5]. As one can see, the DESM algorithm compares favourably with the JADE algorithm, even with numerous sources.

5   Conclusions and Perspectives

In this paper, a new algorithm denoted DESM (Direction Estimation of Separating Matrix) is proposed to extract the sources with non active periods from a linear instantaneous mixture. The detection of these inactive periods allows to estimate the separating matrix which is then used to extract these sources when they are active. The proposed algorithm was tested with different configurations and shown to be efficient at a low computational cost. Even if in this study, the purpose was to extract speech sources, the DESM algorithm can be used in a more general context (*i.e.* to extract any "sparse" sources). In perspective, this methods could be used with convolutive mixtures in the frequency domain. However, this leads to the classical permutation problem [4] which could be fixed by one of the numerous methods proposed in the literature.

References

1. Abrard, F., Deville, Y.: A time-frequency blind signal separation method applicable to underdetermined mixtures of dependent sources. Signal Processing 85(7), 1389–1403 (2005)
2. Arberet, S., Gribonval, R., Bimbot, F.: A robust method to count and locate audio sources in a stereophonic linear instantaneous mixture. In: Proc. ICA, Charleston, USA, pp. 536–543 (March 2006)
3. Aubrey, A., Rivet, B., Hicks, Y., Girin, L., Chambers, J., Jutten, C.: Two novel visual voice activity detectors based on appearance models and retinal filltering. In: Proc. European Signal Processing Conference (EUSIPCO), Poznan, Poland, pp. 2409–2413 (September 2007)

4. Cardoso, J.-F.: Blind signal separation: statistical principles. Proceedings of the IEEE 86(10), 2009–2025 (1998)
5. Cardoso, J.-F., Souloumiac, A.: Blind beamforming for non Gaussian signals. IEE Proceedings-F 140(6), 362–370 (1993)
6. Comon, P.: Independent component analysis, a new concept? Signal Processing 36(3), 287–314 (1994)
7. Desobry, F., Févotte, C.: Kernel PCA based estimation of the mixing matrix in linear instantaneous mixtures of sparse sources. In: Proc. IEEE Int. Conf. Acoustics, Speech, and Signal Processing (ICASSP), Toulouse, France, vol. 5, pp. 669–672 (May 2006)
8. Gribonval, R., Lesage, S.: A survey of Sparse Component Analysis for Blind Source Separation: principles, perspectives, and new challenges. In: Proc. European Symposium on Artificial Neural Networks (ESANN), Bruges, pp. 323–330 (April 2006)
9. Muller, K.-R., Mika, S., Ratsch, G., Tsuda, K., Scholkopf, B.: An introduction to kernel-based learning algorithms. IEEE Transactions on Neural Networks 12(2), 181–201 (2001)
10. Parra, L., Spence, C.: Convolutive blind separation of non stationary sources. IEEE Transactions on Speech and Audio Processing 8(3), 320–327 (2000)
11. Rivet, B., Girin, L., Jutten, C.: Mixing audiovisual speech processing and blind source separation for the extraction of speech signals from convolutive mixtures. IEEE Transactions on Speech and Audio Processing 15(1), 96–108 (2007)
12. Rivet, B., Girin, L., Jutten, C.: Visual voice activity detection as a help for speech source separation from convolutive mixtures. Speech Communication 49(7-8), 667–677 (2007)
13. Servière, C., Pham, D.-T.A.: A novel method for permutation correction in frequency-domain in blind separation of speech mixtures. In: Puntonet, C.G., Prieto, A.G. (eds.) ICA 2004. LNCS, vol. 3195, pp. 807–815. Springer, Heidelberg (2004)
14. Sodoyer, D., Girin, L., Jutten, C., Schwartz, J.-L.: Developing an audio-visual speech source separation algorithm. Speech Communication 44(1–4), 113–125 (2004)
15. Souloumiac, A.: Blind source detection and separation using second order non-stationarity. In: Proc. IEEE Int. Conf. Acoustics, Speech, and Signal Processing (ICASSP), Detroit, USA, vol. 3, pp. 1912–1915 (May 1995)
16. Wang, W., Cosker, D., Hicks, Y., Sanei, S., Chambers, J.A.: Video assisted speech source separation. In: Proc. IEEE Int. Conf. Acoustics, Speech, and Signal Processing (ICASSP), Philadelphia, USA (March 2005)
17. Yilmaz, Ö., Rickard, S.: Blind Separation of Speech Mixtures via Time-Frequency Masking. IEEE Transactions on Signal Processing 52(7), 1830–1847 (2004)

Blind Separation of Noisy Mixtures of Non-stationary Sources Using Spectral Decorrelation

Hicham Saylani[1,2], Shahram Hosseini[1], and Yannick Deville[1]

[1] Laboratoire d'Astrophysique de Toulouse-Tarbes
Université de Toulouse, CNRS
14 Av. Edouard Belin, 31400 Toulouse, France
[2] Laboratoire des Systèmes de Télécommunications et Ingénierie de la Décision,
Faculté des Sciences, Université Ibn Tofaïl,
BP. 133, 14000 Kénitra, Maroc
{hsaylani,shosseini,ydeville}@ast.obs-mip.fr

Abstract. In this paper, we propose a new approach for blind separation of noisy, over-determined, linear instantaneous mixtures of non-stationary sources. This approach is an extension of a new method based on spectral decorrelation that we have recently proposed. Contrary to classical second-order blind source separation (BSS) algorithms, our proposed approach only requires the non-stationary sources and the stationary noise signals to be *instantaneously* mutually uncorrelated. Thanks to this assumption, it works even if the noise signals are auto-correlated. The simulation results show the much better performance of our approach in comparison to some classical BSS algorithms.

1 Introduction

This paper deals with blind separation of noisy, over-determined, linear instantaneous mixtures of non-stationary sources, in the frequency domain. We therefore consider M noisy mixtures $x_i(t)$ $(i = 1, ..., M)$ of N real, non-stationary, discrete-time sources $s_j(t)$ $(j = 1, ..., N)$, $N < M$, described by the following equation

$$\mathbf{x}(t) = \mathbf{A}\mathbf{s}(t) + \mathbf{n}(t), \tag{1}$$

where $\mathbf{A}$ is a real mixing matrix, of dimension $M \times N$, and $\mathbf{x}(t) = [x_1(t), x_2(t), ..., x_M(t)]^T$, $\mathbf{s}(t) = [s_1(t), s_2(t), ..., s_N(t)]^T$ and $\mathbf{n}(t) = [n_1(t), n_2(t), ..., n_M(t)]^T$ are respectively the observation, source and noise vectors, where ^{T} stands for transpose. Blind Source Separation (BSS) aims at estimating the *pseudo-inverse* of the matrix $\mathbf{A}$, denoted by $\mathbf{A}^+$, provided that the mixing matrix $\mathbf{A}$ is of full rank (equal to N). The BSS approaches based on Independent Component Analysis (ICA) and dealing with noisy mixtures may be split into two principal classes:

1. The approaches based on higher-order statistics assuming that the sources $s_j(t)$ are stationary (possibly auto-correlated), mutually independent and

T. Adali et al. (Eds.): ICA 2009, LNCS 5441, pp. 322–329, 2009.
© Springer-Verlag Berlin Heidelberg 2009

independent from noise signals $n_i(t)$ which are supposed to be stationary, Gaussian, zero-mean, with the same variance σ^2 and such that (see Chapter 15 of [1])

$$\forall (i,j) \in [1,M]^2, \ \forall \ t, \ E[n_i(t)n_j(t)] = \delta(i-j)\sigma^2, \qquad (2)$$

where $\delta(.)$ represents the Kronecker delta.

2. Second-order approaches (e.g. [2],[3],[4]), assuming that
   - The sources $s_j(t)$ are zero-mean, non-stationary "and/or" auto-correlated, mutually uncorrelated i.e.

$$\forall (i,j) \in [1,N]^2, \ i \neq j, \ \forall \ t, \tau, \ E[s_i(t)s_j(t-\tau)] = 0, \qquad (3)$$

and uncorrelated to $n_i(t)$ i.e.

$$\forall (i,j) \in [1,M] \times [1,N], \ \forall \ t, \tau, \ E[n_i(l)s_j(l-\tau)] = 0. \qquad (4)$$

   - The noise signals $n_i(t)$ are stationary, zero-mean, with the same variance σ^2, mutually uncorrelated and temporally uncorrelated i.e.

$$\forall (i,j) \in [1,M]^2, \ \forall \ t, \tau, \ E[n_i(t)n_j(t-\tau)] = \delta(i-j)\delta(\tau)\sigma^2. \qquad (5)$$

The approach proposed in this paper is also a second-order approach but contrary to classical second-order methods mentioned above, it *only* requires the non-stationary sources $s_j(t)$ and the stationary noises $n_i(t)$ to be **instantaneously** mutually uncorrelated. Due to this assumption, it works even if the noise signals $n_i(t)$ are auto-correlated. In other words, we only require Conditions (3), (4) and (5) to be met for $\tau = 0$ which is a much less restrictive assumption. This frequency-domain approach is an extension, to noisy mixtures and to a more general class of source signals, of a new method based on spectral decorrelation that we have recently proposed [5],[6]. A review of that basic method is presented in Section 2. In Section 3, we then describe the extensions proposed in this new work. Simulation results are presented in Section 4, before we conclude in Section 5.

2   Spectral Decorrelation Method

We recently proposed a new BSS method for separating *non-stationary, temporally uncorrelated, mutually uncorrelated, real* sources [5],[6]. This method exploits some interesting properties of random signals in the frequency domain and is based on the following theorem.

Theorem 1. *If $u(t)$ is a temporally uncorrelated, real, zero-mean random signal with a (possibly non-stationary) variance $\gamma(t)$, i.e. $E[u(t_1)u(t_2)] = \gamma(t_1)\delta(t_1 - t_2)$, then its Fourier transform[1] $U(\omega)$ is a wide-sense stationary process with autocorrelation $\Gamma(\nu)$ which is the Fourier transform of $\gamma(t)$, i.e.*[2]

---

[1] The Fourier transform of a discrete-time stochastic process $u(t)$ is a stochastic process $U(\omega)$ defined by $U(\omega) = \sum_{t=-\infty}^{\infty} u(t)e^{-j\omega t}$ [7].

[2] In the following * stands for conjugate and ^{T} for Hermitian transpose.

$$E\left[U(\omega)U^*(\omega - \nu)\right] = \Gamma(\nu) = \sum_{t=-\infty}^{+\infty} \gamma(t)\, e^{-j\nu t}. \tag{6}$$

Moreover, if $u(t)$ is non-stationary with respect to its variance, i.e. if $\gamma(t)$ is not constant, then the process $U(\omega)$ is auto-correlated.

In [5],[6] we studied the case of a determined and noiseless mixture ($M = N$ and $\mathbf{n}(t) = 0$). By mapping the mixtures in the frequency domain, we can write

$$\mathbf{X}(\omega) = \mathbf{AS}(\omega), \tag{7}$$

where $\mathbf{S}(\omega) = [S_1(\omega), S_2(\omega), \cdots, S_N(\omega)]^T$, $\mathbf{X}(\omega) = [X_1(\omega), X_2(\omega), \cdots, X_N(\omega)]^T$; $S_j(\omega)$ and $X_i(\omega)$, $(i, j) \in [1, N]^2$, are respectively the Fourier transforms of $s_j(t)$ and $x_i(t)$. Thus, the frequency-domain observations $X_i(\omega)$ are linear instantaneous mixtures of the frequency-domain sources $S_j(\omega)$. If the real temporal sources $s_j(t)$ are *non-stationary, temporally uncorrelated* and *mutually uncorrelated*, then

1. following Theorem 1, the frequency-domain sources $S_j(\omega)$ are *wide-sense stationary* and *auto-correlated* in the frequency domain,
2. since the property of mutual uncorrelatedness of random signals is kept after computing their Fourier transforms [5],[6], the frequency-domain sources $S_j(\omega)$ are mutually uncorrelated.

Many algorithms have been proposed for separating mixtures of time-domain wide-sense stationary, time-correlated processes (for example *AMUSE* [2] and *SOBI* [3]) and nothing prohibits us from applying them to frequency-domain wide-sense stationary, frequency-correlated processes. For example, a simple algorithm proposed in [5],[6], which may be considered as a modified version of the *AMUSE* algorithm in the frequency domain, consists (in the determined case) in diagonalizing the matrix $\left(E\left[\mathbf{X}(\omega)\mathbf{X}^H(\omega)\right]\right)^{-1} E\left[\mathbf{X}(\omega)\mathbf{X}^H(\omega - \nu)\right]$ for some frequency shift ν meeting the following identifiability condition

$$\frac{E\left[S_i(\omega)S_i^*(\omega - \nu)\right]}{E\left[|S_i(\omega)|^2\right]} \neq \frac{E\left[S_j(\omega)S_j^*(\omega - \nu)\right]}{E\left[|S_j(\omega)|^2\right]}, \quad \forall\, i \neq j. \tag{8}$$

Hence, the separating matrix $\mathbf{A}^{-1}$ is identifiable if and only if the temporal source signals have different normalized variance profiles [5],[6].

Contrary to time-domain classical statistical BSS methods based on non-stationarity of the signals, piecewise stationarity is not required for our above frequency-domain method. However, the temporal uncorrelatedness assumption made in Theorem 1 is too restrictive because a great number of real-world sources are auto-correlated, that is why in [6] we proposed a solution to cope with this problem and to apply the method when the non-stationary sources are auto-correlated. Nevertheless, in [6], like in our previous works [5], we considered determined noiseless mixtures and we supposed that the sources were mutually uncorrelated i.e. that Condition (3) was true for all τ. The approach proposed in the following section is an extension of the above method to noisy over-determined mixtures of non-stationary (possibly auto-correlated) sources with less restrictive hypotheses about mutual correlation of source and noise signals than our previous methods and classical BSS algorithms.

3   Proposed Approach

The new approach introduced in this paper is based on the following theorem and can be applied for separating over-determined noisy mixtures of non-stationary (possibly auto-correlated) sources having different normalized variance profiles when Conditions (3), (4) and (5) are met only for $\tau = 0$.

Theorem 2. *Let $u_p(t)$ $(p = 1, ..., \mathcal{N})$ be $\mathcal{N}$ real, zero-mean and instantaneously mutually uncorrelated random signals i.e.*

$$\forall (p, q) \in [1, \mathcal{N}]^2, p \neq q, \ \forall t, \ E\left[u_p(t)u_q(t)\right] = 0. \tag{9}$$

*Suppose $g(t)$ is a real, zero-mean, stationary, temporally uncorrelated random signal, independent from all signals $u_p(t)$. Then, the signals $u'_p(t)$ defined by $u'_p(t) = g(t)u_p(t)$ are real, zero-mean, **temporally uncorrelated** and **mutually uncorrelated**. Moreover, each new signal $u'_p(t)$ has the same normalized variance profile as the original signal $u_p(t)$.*

Proof: See Appendix.

Therefore, by multiplying Equation (1) by a random signal $g(t)$, which meets the conditions of Theorem 2 (with respect to all signals $s_j(t)$ and $n_i(t)$) and which has a variance σ_g^2, we obtain

$$\mathbf{x}'(t) = \mathbf{A}\mathbf{s}'(t) + \mathbf{n}'(t) \tag{10}$$

where $\mathbf{x}'(t) = g(t)\mathbf{x}(t)$, $\mathbf{s}'(t) = g(t)\mathbf{s}(t)$ and $\mathbf{n}'(t) = g(t)\mathbf{n}(t)$. We assume non-stationary sources $s_j(t)$ and stationary noise signals $n_i(t)$, all of them being real, zero-mean and mutually instantaneously uncorrelated. Thanks to Theorem 2, the new sources $s'_j(t) = g(t)s_j(t)$ are non-stationary, the new noise signals $n'_i(t) = g(t)n_i(t)$ are stationary, and all of them are temporally and mutually uncorrelated, that is for all t and τ:

$$\begin{cases} E\left[\mathbf{s}'(t)\mathbf{s}'^T(t-\tau)\right] \text{ is diagonal} \\ E\left[\mathbf{n}'(t)\mathbf{n}'^T(t-\tau)\right] - \sigma'^2\mathbf{I}_M\delta(\tau) \\ E\left[\mathbf{n}'(t)\mathbf{s}'^T(t-\tau)\right] = \left(E\left[\mathbf{s}'(t)\mathbf{n}'^T(t-\tau)\right]\right)^T = \mathbf{0}_{MN} \end{cases} \tag{11}$$

where $\sigma'^2 = \sigma^2\sigma_g^2$ is the variance of each noise signal $n'_i(t)$, $\mathbf{I}_M$ is the M-dimensional identity matrix and $\mathbf{0}_{MN}$ is the $M \times N$ null matrix. Moreover, by mapping the new mixture of Equation (10) in the frequency domain and denoting $X_i'(\omega)$, $S_j'(\omega)$ and $N_i'(\omega)$ respectively the Fourier transforms of the signals $x_i'(t) = g(t)x_i(t)$, $s_j'(t)$ and $n_i'(t)$ we can write

$$\mathbf{X}'(\omega) = \mathbf{A}\mathbf{S}'(\omega) + \mathbf{N}'(\omega) \tag{12}$$

where $\mathbf{X}'(\omega) = [X_1'(\omega), X_2'(\omega), ..., X_M'(\omega)]^T$, $\mathbf{S}'(\omega) = [S_1'(\omega), S_2'(\omega), ..., S_N'(\omega)]^T$ and $\mathbf{N}'(\omega) = [N_1'(\omega), N_2'(\omega), ..., N_M'(\omega)]^T$. Since the Fourier transform keeps the property of mutual uncorrelatedness, following (11), the covariance matrices $E\left[\mathbf{S}'(\omega)\mathbf{S}'^H(\omega - \nu)\right]$ and $E\left[\mathbf{N}'(\omega)\mathbf{N}'^H(\omega - \nu)\right]$ are also diagonal, and $E\left[\mathbf{S}'(\omega)\mathbf{N}'^H(\omega - \nu)\right]$ and $E\left[\mathbf{N}'(\omega)\mathbf{S}'^H(\omega - \nu)\right]$ are null matrices whatever the frequency shift ν. According to Theorem 1, since non-stationary sources $s_j'(t)$ and stationary noise signals $n_i'(t)$ are real, zero-mean and temporally uncorrelated, then the frequency-domain sources $S_j'(\omega)$ are wide-sense stationary and auto-correlated, and the signals $N_i'(\omega)$ are wide-sense stationary and spectrally uncorrelated. Using (12) we can write :

$$\mathcal{R}_{\mathbf{X}'}(\nu) = E\left[\mathbf{X}'(\omega)\mathbf{X}'^H(\omega - \nu)\right] \tag{13}$$

$$= \mathbf{A}E\left[\mathbf{S}'(\omega)\mathbf{S}'^H(\omega - \nu)\right]\mathbf{A}^H + E\left[\mathbf{N}'(\omega)\mathbf{N}'^H(\omega - \nu)\right] \tag{14}$$

$$= \mathbf{A}\mathcal{R}_{\mathbf{S}'}(\nu)\mathbf{A}^H + E\left[\mathbf{N}'(\omega)\mathbf{N}'^H(\omega - \nu)\right]. \tag{15}$$

Moreover, following Theorem 1

$$\forall i, \quad E\left[N_i'(\omega)N_i'^*(\omega - \nu)\right] = \sum_{t=-\infty}^{+\infty} \sigma'^2 e^{-j\nu t} = 2\pi\sigma'^2 \sum_{k=-\infty}^{+\infty} \delta(\nu - 2k\pi) \tag{16}$$

where δ represents the Dirac delta. According to this result, the covariance of the Fourier Transform of each noise signal $n_i'(t)$ is equal to infinity at $\nu = 2k\pi$. In practice, however, we compute the Discrete Fourier Transform (DFT) of the signals over a finite number of samples. As a result, at $\nu = 2k\pi$ the covariance $E\left[N_i'(\omega)N_i'^*(\omega - \nu)\right]$ has a finite value, denoted by $2\pi\eta$ in the following, which depends on the number of samples. The matrix $E\left[\mathbf{N}'(\omega)\mathbf{N}'^H(\omega - \nu)\right]$ being diagonal, Eq. (15) reads in practice (up to estimation errors)

$$\mathcal{R}_{\mathbf{X}'}(\nu) = \begin{cases} \mathbf{A}\mathcal{R}_{\mathbf{S}'}(\nu)\mathbf{A}^H + 2\pi\eta\mathbf{I}_M & \text{for } \nu = 2k\pi \\ \mathbf{A}\mathcal{R}_{\mathbf{S}'}(\nu)\mathbf{A}^H & \text{for } \nu \neq 2k\pi \end{cases} \tag{17}$$

Since the frequency-domain sources $S_j'(\omega)$ are wide-sense stationary and auto-correlated, Eq. (17) may be used to estimate the separating matrix $\mathbf{A}^+$ using one of the algorithms exploiting autocorrelation of source signals, for example AMUSE [2] or SOBI [3], but in the frequency domain. For example, using the SOBI algorithm in the frequency domain, called SOBI-F in the following, an estimate of the separating matrix $\mathbf{A}^+$, denoted $\widehat{\mathbf{A}}^+$, may be obtained by

$$\widehat{\mathbf{A}}^+ = \Re\{\mathbf{U}^H\mathbf{W}\}, \tag{18}$$

where $\mathbf{W}$ is a whitening matrix of dimension $N \times M$, obtained by eigenvalue decomposition of the matrix $\widehat{\mathcal{R}}_{\mathbf{X}'}(0)$, and $\mathbf{U}$ is a unitary matrix obtained by simultaneous diagonalization of several covariance matrices $\widehat{\mathcal{R}}_{\mathbf{Z}'}(\nu_p)$ $(p = 1, 2, ...)$, where $\mathbf{Z}'(\omega) = \mathbf{W}\mathbf{X}'(\omega)$. Because of working in the frequency domain, the matrix $\mathbf{U}^H\mathbf{W}$ may be complex, that is why $\widehat{\mathbf{A}}^+$ is equal to the real part of this matrix (see [5],[6] for more details). Once the separating matrix $\widehat{\mathbf{A}}^+$ has been estimated, a noisy estimate of the source vector $\mathbf{s}(t)$, denoted $\widehat{\mathbf{s}}_n(t)$, can be obtained using Equation (1) as follows (ignoring permutation and scale indeterminacies)

$$\widehat{\mathbf{s}}_n(t) = \widehat{\mathbf{A}}^+\mathbf{x}(t) \tag{19}$$

$$= \widehat{\mathbf{A}}^+\mathbf{A}\mathbf{s}(t) + \widehat{\mathbf{A}}^+\mathbf{n}(t) \tag{20}$$

$$\simeq \mathbf{s}(t) + \widehat{\mathbf{A}}^+\mathbf{n}(t) \tag{21}$$

4   Simulation Results

We consider three mixtures of two unit-power 20000-sample speech signals, mixed by the following mixing matrix

$$\mathbf{A} = \begin{pmatrix} 1 & 0.9 \\ 0.8 & 1 \\ 0.5 & 0.9 \end{pmatrix}. \tag{22}$$

We compare the performance of our method with those of $AMUSE$ [2], $SOBI$ [3] and $SONS$ [4]. The components of the additive noise vector $\mathbf{n}(t)$ are Gaussian with cross-correlation matrix $\mathcal{R}_{\mathbf{n}}(0) = E[\mathbf{n}(t)\mathbf{n}^T(t)] = \sigma^2\mathbf{I}_3$. We consider both white and colored noise signals. These colored noises are derived from white signals using 32th order FIR filters. In both cases, $10\log_{10}(1/\sigma^2) = 10dB$. The signal $g(t)$ mentioned in Theorem 2 is an i.i.d. uniformly distributed real and zero-mean signal. The results are reported below with and without multiplication of the observations by this signal $g(t)$. The separation performance for each source $s_j(t)$ is measured using the *performance index*, defined by

$$\forall j \in [1, N], \ \ \mathcal{I}_j = max_i 10\log_{10}\left(\frac{b_{ij}^2}{\sum_{k \in [1,N]}^{k \neq j} b_{ik}^2}\right), \tag{23}$$

where b_{ij}, $(i, j) \in [1, N]^2$, are the entries of the N-dimensional square matrix $\mathbf{B} = \widehat{\mathbf{A}}^+\mathbf{A}$, called the *performance matrix*. The global separation performance for N sources is then measured by the *global performance index*, defined by

$$\mathcal{I} = \left(\sum_{j=1}^{N} \mathcal{I}_j\right)/N. \tag{24}$$

Our *SOBI-F* algorithm as well as the *SOBI* and *SONS* algorithms are tested by simultaneously diagonalizing 4 covariance matrices. For the *SONS* algorithm, these matrices are estimated over 200-sample frames[3]. The results with and without an additive noise vector $\mathbf{n}(t)$ are reported in Table 1.

Table 1. Global performance index $\mathcal{I}$ in dB obtained in our simulations

Algorithm	$AMUSE$	$SOBI$	$SONS$	$SOBI$-F without $g(t)$	SOBI-F with $g(t)$
$\mathcal{I}$ (dB), $\mathbf{n}(t) = 0$	33.0	28.2	40.3	39.7	64.3
$\mathcal{I}$ (dB), white $\mathbf{n}(t)$	9.8	17.3	31.8	33.9	51.6
$\mathcal{I}$ (dB), colored $\mathbf{n}(t)$	1.7	1.9	21.1	31.0	51.7

This table deserves the following comments:

- **Noiseless mixtures:** without multiplying the observations by the signal $g(t)$, our *SOBI-F* algorithm has nearly the same performance as *SONS* and outperforms the other two methods. With multiplication by $g(t)$, it is much more efficient than the other algorithms.
- **Noisy mixtures:** our algorithm is more efficient than the other algorithms in all cases, and much better when using multiplication by $g(t)$.
- The *SONS* algorithm is more efficient than *AMUSE* and *SOBI*, and contrary to these two algorithms, it is rather robust to non-whiteness of additive noise.
- Without multiplication by $g(t)$, our algorithm is slightly less efficient with colored noise than with white noise. This result is not surprising because the Fourier transforms of the colored noises $n_i(t)$ are not stationary. It can be seen however that the difference is not significant. Using multiplication by $g(t)$, our algorithm has the same performance with colored noise as with white noise. That is because in the two cases, after multiplication by $g(t)$, the new noise signals $n_i'(t)$ are white.

5   Conclusion and Perspectives

In this paper, we proposed a new BSS approach for over-determined noisy mixtures of non-stationary sources, based on spectral decorrelation method. Our assumptions about source and noise signals are much less restrictive than those made by classical BSS methods. Our simulations confirmed the better performance of our approach using *SOBI-F* algorithm compared to three classical algorithms (*AMUSE*, *SOBI* and *SONS*) for separating speech signals especially with colored noise. More tests using other source and noise signals being only *instantaneously* mutually uncorrelated and a statistical performance test seem however necessary and will be done in the future.

---

[3] The *SONS* algorithm [4] exploits both autocorrelation and non-stationarity of the source signals assuming that they are piecewise stationary.

References

1. Hyvarinen, A., Karhunen, J., Oja, E.: Independent Component Analysis. Wiley, Chichester (2001)
2. Tong, L., Liu, R.-W., Soon, V.C., Huang, Y.-F.: Indeterminacy and identifiability of blind identification. IEEE Trans. Circuits Syst. 38(5), 499–509 (1991)
3. Belouchrani, A., Abed Meraim, K., Cardoso, J.-F., Moulines, E.: A blind source separation technique based on second order statistics. IEEE Trans. on Signal Processing 45, 434–444 (1997)
4. Choi, S., Cichocki, A., Belouchrani, A.: Second order non-stationary source separation. Journal of VLSI Signal Processing 32(1-2), 93–104 (2002)
5. Hosseini, S., Deville, Y.: Blind separation of nonstationary sources by spectral decorrelation. In: Puntonet, C.G., Prieto, A.G. (eds.) ICA 2004. LNCS, vol. 3195, pp. 279–286. Springer, Heidelberg (2004)
6. Hosseini, S., Deville, Y., Saylani, H.: Blind separation of linear instantaneous mixtures of non-stationary signals in the frequency domain. Signal Processing (to appear), http://dx.doi.org/10.1016/j.sigpro.2008.10.024
7. Papoulis, A., Pillai, S.U.: Probability, random variables and stochastic processes, 4th edn. McGraw-Hill, New York (2002)

Appendix: Proof of Theorem 2

Denote $\mathbf{u}(t) = [u_1(t), u_2(t), ..., u_\mathcal{N}(t)]^T$ and $\mathbf{u}'(t) = g(t)\mathbf{u}(t)$.

1. Since $g(t)$ is independent from all the zero-mean signals $u_p(t)$, we can write

$$\forall p \in [1, \mathcal{N}], \ E\left[u'_p(t)\right] = E\left[g(t)u_p(t)\right] = E\left[g(t)\right]E\left[u_p(t)\right] = 0. \quad (25)$$

Hence, the new signals $u'_p(t)$ $(p = 1, ..., \mathcal{N})$ are also zero-mean.

2. Whatever the times t_1 and t_2, we have

$$E\left[\mathbf{u}'(t_1)\mathbf{u}'(t_2)^T\right] = E\left[g(t_1)g(t_2)\mathbf{u}(t_1)\mathbf{u}(t_2)^T\right]. \quad (26)$$

The independence of $g(t)$ from all the signals $u_p(t)$ yields

$$E\left[\mathbf{u}'(t_1)\mathbf{u}'(t_2)^T\right] = E\left[g(t_1)g(t_2)\right]E\left[\mathbf{u}(t_1)\mathbf{u}(t_2)^T\right], \quad (27)$$

and since $g(t)$ is zero-mean, stationary and temporally uncorrelated

$$E\left[\mathbf{u}'(t_1)\mathbf{u}'(t_2)^T\right] = \sigma_g^2\delta(t_1 - t_2)E\left[\mathbf{u}(t_1)\mathbf{u}(t_1)^T\right] \quad (28)$$

where σ_g^2 is the variance of $g(t)$. The signals $u_p(t)$ being zero-mean and instantaneously mutually uncorrelated, the matrices $E\left[\mathbf{u}(t_1)\mathbf{u}(t_1)^T\right]$ and so $E\left[\mathbf{u}'(t_1)\mathbf{u}'(t_2)^T\right]$ are diagonal. As a result, the new zero-mean signals $u'_p(t)$ are mutually uncorrelated. Moreover, according to Eq. (28), the diagonal entries of the matrix $E\left[\mathbf{u}'(t_1)\mathbf{u}'(t_2)^T\right]$ can be written as

$$E\left[u'_p(t_1)u'_p(t_2)\right] = \sigma_g^2\delta(t_1 - t_2)E\left[u_p(t_1)u_p(t_1)\right] = \sigma_g^2\delta(t_1 - t_2)E\left[u_p^2(t_1)\right]. \quad (29)$$

Hence, the new signals $u'_p(t)$ are temporally uncorrelated. Furthermore, by choosing $t_1 = t_2 = t$, Eq. (29) becomes $E\left[u'^2_p(t)\right] = \sigma_g^2E\left[u_p^2(t)\right]$ which means that the new signals $u'_p(t)$ have the same normalized variance profiles as the original signals $u_p(t)$.

Probabilistic Factorization of Non-negative Data with Entropic Co-occurrence Constraints

Paris Smaragdis[1], Madhusudana Shashanka[2],
Bhiksha Raj[3], and Gautham J. Mysore[4],[*]

[1] Adobe Systems Inc.
[2] Mars Incorporated
[3] Mitsubishi Electric Research Laboratories
[4] Stanford University

Abstract. In this paper we present a probabilistic algorithm which factorizes non-negative data. We employ entropic priors to additionally satisfy that user specified pairs of factors in this model will have their cross entropy maximized or minimized. These priors allow us to construct factorization algorithms that result in maximally statistically different factors, something that generic non-negative factorization algorithms cannot explicitly guarantee. We further show how this approach can be used to discover clusters of factors which allow a richer description of data while still effectively performing a low rank analysis.

1 Introduction

With the recent interest in non-negative factorization algorithms we have seen a rich variety of algorithms that can perform this task for a wide range of applications using various models. Empirically it has been observed that the non-negativity constraint in conjunction with the information bottleneck that such a low-rank factorization imposes, often results in data which is often interpreted as somewhat independent. Although this is approximately and qualitatively a correct observation, it is not something that is explicitly enforced in such algorithms and thus more a result of good fortune than planning. Nevertheless this property has proven to be a primary reason for the continued interest in such factorization algorithms. The task of finding independence in non-negative data has been explicitly tackled in the past using non-negative ICA and PCA algorithms [1,2] but such models have not been as easy to manipulate and extend as non-negative factorization models, which resulted in a diminished use of explicit independence optimization for non-negative data.

In this paper we present a non-negative factorization approach that explicitly manipulates the statistical relationships between the estimated factors. We recast the task of non-negative factorization as a probabilistic latent variable decomposition on count/histogram data. Using this abstraction we treat the input data as a multidimensional probability distribution and estimate an additive set of marginal distributions which would approximate it. This approach allows us

---

[*] Work performed while at Adobe Systems Inc.

T. Adali et al. (Eds.): ICA 2009, LNCS 5441, pp. 330–337, 2009.
© Springer-Verlag Berlin Heidelberg 2009

to implicitly satisfy the non-negativity constraint (due to the fact that we estimate the factors as distributions), and at the same time allows for a convenient handle on statistical manipulations. As shown in [3] this approach allows us to use probabilistic approaches to matrix factorization problems and is implicitly tied to a lot of recent work on decompositions and factorizations. In this paper we extend the original PLCA model [4], so that we can manipulate the cross entropy between the estimated marginals. This allows us to extract marginal distributions which are pairwise either similar or dissimilar. We also show how this approach can help in constructing more sophisticated analysis structures by enforcing the creation of related cliques of factors.

2   The PLCA Model

Probabilistic Latent Component Analysis (PLCA) decomposes a multidimensional distribution as a mixture of latent components where each component is given by the product of one-dimensional marginal distributions. Although we will proceed by formulating the PLCA model using a two-dimensional input, this can be easily extended to inputs of arbitrary dimensions which can be seen as non-negative tensors. A given two-dimensional data matrix $\mathbf{V}$ is modeled as the histogram of repeated draws from an underlying two-dimensional distribution $P(x_1, x_2)$. PLCA factorizes this distribution and is formulated as

$$P(x_1, x_2) = \sum_{z \in \mathbf{Z}} P(z)P(x_1|z)P(x_2|z), \tag{1}$$

where z is a latent variable that indexes the latent components and takes values from the set $\mathbf{Z} = \{z_1, z_2, \ldots, z_K\}$. Given $\mathbf{V}$, parameters can be estimated by using the EM algorithm. Maximizing the expected complete data log-likelihood given by

$$\mathcal{L} = \sum_{x_1, x_2} V_{x_1, x_2} \sum_z P(z|x_1, x_2) \log \left[P(z)P(x_1|z)P(x_2|z) \right] \tag{2}$$

yields the following iterative update equations:

$$P(z|x_1, x_2) = \frac{P(z)P(x_1|z)P(x_2|z)}{\sum_{z \in \mathbf{Z}} P(z)P(x_1|z)P(x_2|z)} \tag{3}$$

$$P(x_1|z) = \frac{\sum_{x_2} V_{x_1, x_2} P(z|x_1, x_2)}{\sum_{x_1, x_2} V_{x_1, x_2} P(z|x_1, x_2)}$$

$$P(x_2|z) = \frac{\sum_{x_1} V_{x_1, x_2} P(z|x_1, x_2)}{\sum_{x_1, x_2} V_{x_1, x_2} P(z|x_1, x_2)}$$

$$P(z) = \frac{\sum_{x_1, x_2} V_{x_1, x_2} P(z|x_1, x_2)}{\sum_{z, x_1, x_2} V_{x_1, x_2} P(z|x_1, x_2)}, \tag{4}$$

where equation (3) represents the Expectation step and equations (4) represents the Maximization step of the EM algorithm. As shown in [4], the above formulation can be expressed as a matrix factorization, where $P(x_1, x_2)$ represents a non-negative matrix and $P(x_1|z)$ and $P(x_2|z)$ are the factors along each of the input's dimensions.

3   Imposing Cross-Factor Constraints

In this section we describe how we can manipulate a statistical relationship between two arbitrary sets of marginal distributions in our model. For simplicity we will manipulate the relationship between two sets of marginals as observed along the dimension of x_1. Extending this to other or more dimensions is a trivial extension of the following formulation.

Let the two sets of latent variables be represented by $\mathcal{Z}_1$ and $\mathcal{Z}_2$ where $\mathcal{Z}_1 \cup \mathcal{Z}_2 \subseteq \mathcal{Z}$. To impose our desired constraint, we need to make $P(x_1|z_1)$ and $P(x_1|z_2)$ similar or dissimilar from each other. We can achieve this by modifying the cost function of equation (2) by appending a metric that corresponds to the dissimilarity between the distributions. During estimation, maximizing or minimizing that cost function will result in biasing the estimation towards the desired outcome. One measure we can use to describe the similarity between two distributions is the cross entropy. For two distributions $\mathbf{q}_{z_i} = P(x_1|z_i)$ and $\mathbf{q}_{z_k} = P(x_1|z_k)$, cross entropy is given by

$$H(\mathbf{q}_{z_i}, \mathbf{q}_{z_k}) = -\sum_{x_1} P(x_1|z_i) \log P(x_1|z_k). \tag{5}$$

Appending to the log-likelihood $\mathcal{L}$ cross-entropies $H(\mathbf{q}_{z_i}, \mathbf{q}_{z_k})$ and $H(\mathbf{q}_{z_k}, \mathbf{q}_{z_i})$ for all $z_i \in \mathcal{Z}_1$ and $z_k \in \mathcal{Z}_2$, we obtain the new cost function as[1]

$$\begin{aligned}
Q &= \mathcal{L} + \alpha \sum_{i|z_i \in \mathcal{Z}_1} \sum_{k|z_k \in \mathcal{Z}_2} (H(\mathbf{q}_{z_i}, \mathbf{q}_{z_k}) + H(\mathbf{q}_{z_k}, \mathbf{q}_{z_i})) \\
&= \mathcal{L} - \alpha \sum_{i|z_i \in \mathcal{Z}_1} \sum_{k|z_k \in \mathcal{Z}_2} \sum_t P(x_1|z_k) \log P(x_1|z_i) \\
&\quad - \alpha \sum_{i|z_i \in \mathcal{Z}_1} \sum_{k|z_k \in \mathcal{Z}_2} \sum_t P(x_1|z_i) \log P(x_1|z_k)
\end{aligned}$$

where α is a tunable parameter that controls the extent of regularization. We use the EM algorithm again to estimate all the parameters. The E-step remains the same as given by the equation (3). Since the terms appended to $\mathcal{L}$ in the new cost function does not involve $P(f|z)$ or $P(z)$, the update equations for them remain the same as given by equations (4). Consider the estimation of $\mathbf{q}_{z_i} = P(x_1|z_i)$ for a given value of i. Adding a Lagrange multiplier term and differentiating the new cost function with respect to $P(x_1|z_i)$ and setting it to 0, we obtain

$$\frac{\sum_{x_2} V_{x_1,x_2} P(z_i|x_1,x_2) - \alpha \sum_{z_k} P(x_1|z_k)}{P(x_1|z_i)} - \alpha \sum_{z_k} \log P(x_1|z_k) + \lambda = 0 \tag{6}$$

---

[1] This cost function is equivalent to the log-posterior obtained in a MAP formulation where the exponential of cross-entropy is used as a prior. The prior is given by $P(\mathbf{q}_{z_1}, \mathbf{q}_{z_2}) \propto e^{\alpha H(\mathbf{q}_{z_1}, \mathbf{q}_{z_2})} e^{\alpha H(\mathbf{q}_{z_2}, \mathbf{q}_{z_1})}$ where $z_1 \in \mathcal{Z}_1$ and $z_2 \in \mathcal{Z}_2$.

which implies that

$$\sum_{x_2} V_{x_1,x_2} P(z_i|x_1,x_2) - \alpha \sum_{z_k} P(x_1|z_k) = P(x_1|z_i)\Big(\alpha \sum_{z_k} \log P(x_1|z_k) - \lambda\Big),$$

where λ is the Lagrange multiplier. Treating the term $\log P(x_1|z_k)$ as a constant and utilizing the fact that $\sum_{x_1} P(x_1|z_i) = 1$, we can sum the above equation with respect to x_1 to obtain

$$\sum_{x_1} \sum_{x_2} V_{x_1,x_2} P(z_i|x_1,x_2) - \alpha \sum_{x_1} \sum_{z_k} P(x_1|z_k) = \alpha \sum_{z_k} \log P(x_1|z_k) - \lambda.$$

Utilizing this result in equation (6), we obtain the update equation as[2]

$$P(x_1|z_i) = \frac{\sum_{x_2} V_{x_1,x_2} P(z_i|x_1,x_2) - \alpha \sum_{z_k} P(x_1|z_k)}{\sum_{x_2} \sum_{x_1} V_{x_1,x_2} P(z_i|x_1,x_2) - \alpha \sum_{x_1} \sum_{z_k} P(x_1|z_k)} \tag{7}$$

Since $P(x_1|z_k)$ is treated as a constant during the estimation of $P(x_1|z_i)$ (and similarly $P(x_1|z_i)$ treated as a constant while estimating $P(x_1|z_k)$), the updates for $P(x_1|z_i)$ and $P(x_1|z_k)$ should be alternated between iterations. The above update equation works well when we attempt to perform minimization of the cross-entropy between marginals. It does however present a problem when we attempt to perform cross-entropy maximization while attempting to produce maximally different marginal distributions. To do so we would use a positive α which can potentially result in the outcome of equation 7 to be negative-valued. This is of course an inappropriate estimate for a distribution and would violate its implicit non-negative nature. The easiest way to deal with this problem is to discard any negative valued estimates and replace them with zeroes. In other experiments, more rigorously motivated approaches such as those employed in discriminative training methods [6], which prevent negative probability estimates by employing additive correction terms were observed to result in no appreciable difference in estimates. Finally we note that the estimation equations as presented can be very effective in imposing the cross-entropy prior, sometimes counteracting the fit to the data. In practical situations we found it best to progressively reduce the weight of the prior across the EM iterations.

Examples of cross entropy manipulation: To show the performance of the above estimation rules let us consider a simple non-negative factorization problem. As the input we use the magnitude spectrogram of a drums recording shown in figure 1. In this input we can clearly see the different types of drums in the mixture aided by their distinct spectral and temporal profiles. A factorization algorithm is expected to be able to distinguish between the various drum sounds

---

[2] When α is positive, the update equation for $P(x_1|z_i)$ can also be derived as follows. We construct a lower bound Q' for Q by removing all the $H(\mathbf{q}_{z_i}, \mathbf{q}_{z_k})$ terms from Q. Since $Q' \le Q$, we can estimate parameters by maximizing Q' instead of maximizing Q. Adding a lagrangian to Q' and taking derivatives w.r.t. $P(x_i|z_i)$, it can be easily shown that the resulting update equation is given by equation (7).

for the same reasons. We performed the analysis using the same starting conditions but imposing a cross-entropy prior on the marginals corresponding to the time axis. The results of these analyses are shown in figure 2. The top row

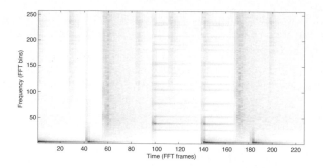

Fig. 1. The drums recording spectrogram we used as an input for the examples in this paper. In it one can clearly see each instrument and a factorization algorithm is expected to discover individual instruments by taking advantage of their distinct spectral and temporal positioning.

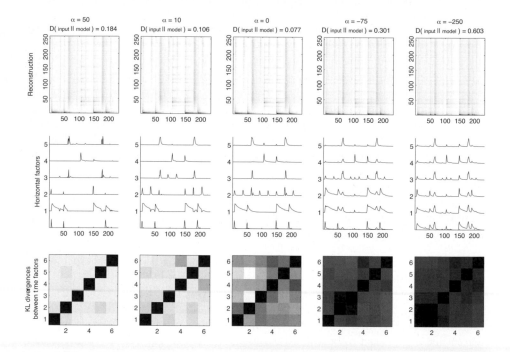

Fig. 2. The results of analyzing the same input with various prior weights. The examples from left to right column show the results with a maximal dissimilarity prior to a maximal similarity prior. The middle column is the case with no prior. The top row shows the reconstruction in each case, the middle row the extracted horizontal marginals and the bottom row their KL divergence.

shows the reconstruction of the input, the middle row shows the extracted time marginals and the bottom row shows their mutual KL divergence (we chose to display the KL divergence since it is a more familiar indicator of relationship between distributions as opposed to the cross entropy). Each column of plots is a different analysis with the prior weight shown in the top. From left to right the prior goes from imposing maximal dissimilarity to maximal similarity. The middle column has no prior imposed. One can clearly see from the plots that for large positive values of the prior's weight we estimate a much more sparse set of marginals, and one where their respective KL divergence is maximal. As the prior weight is moved towards large negative values we gradually observe the discovery of less sparse codes, up to the extreme point where all extracted marginals are very similar.

4   Group-Wise Analysis

A powerful use of the above priors is that of performing a group-based analysis, similar to the concept of the multidimensional independent component analysis [5]. This means factorizing an input with a large number of components which are grouped in a smaller set of cliques of mutually related components. To perform this we need to partition the marginals of an analysis in groups and then use the prior we introduced to request minimal cross-entropy between the marginals in the same groups and maximal cross-entropy between the marginals from different groups. This will result in a collection of marginal groups in which elements of different groups are statistically different, whereas elements in the same group are similar. To illustrate the practical implications of this approach consider the following experiment on the data of figure 1. We partitioned the twelve requested marginals in six groups of two. We performed the analysis with no priors, then with priors forcing the time marginals from separate groups to be different, then with priors forcing time marginals in the same group to be similar, and finally with both types of priors. All simulations were run with the same initial conditions. The results of these analyses are shown in figure 3. Like before each column shows different measures of the same analysis. The left most shows the case where no priors were used, the second one the case where the within group similarity was imposed, the third one where out of group dissimilarity was imposed and the rightmost one where both within group similarity and out of group dissimilarity were imposed. The top row shows the resulting reconstruction, the second row shows the discovered time marginals and the third row shows the KL divergence between the time marginals. Also shown in the titles of the top figures is the KL divergence between the input and the model reconstruction. Occasionally when we impose a prior in the model we observe a slight increase which significs that the model is not representing the input as accurately. Qualitatively this increase is usually fairly minor however and is expected since the model is not optimizing just the fit to the input data anymore. Observing the results we can clearly see that when we impose the within group similarity the extracted marginals belonging to the same group are more similar than otherwise. But by doing so we also implicitly encourage more similarity

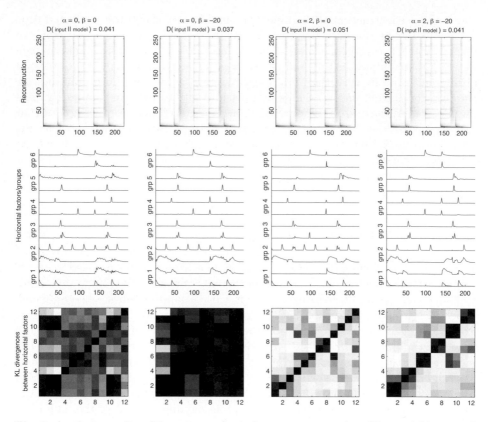

Fig. 3. Analysis results with various values of cross entropy priors. The variables α and β are the out-of-group dissimilarity and in-group similarity prior weights, and they were applied on the horizontal factors only. Each column of plots shows the results of a different analysis with its α and β values shown over the top figures of each column. The top figures show the resulting reconstructions and on their title we show the KL divergence between the input and the reconstruction. The middle figures show the resulting horizontal factors and the bottom figures show the KL divergence values between all factor pairs.

across all marginals since there is nothing to stop two different groups latching on the same instrument. In contrast when we use the out of group dissimilarity prior we see that we get very dissimilar marginals in the outputs, while some of them belonging in the same group happen to have some similarity. Imposing both of the previous priors at the same time results in a more desirable output where each group contains marginals which are dissimilar from marginals of other groups, yet similar to the marginals in the same group. Effectively we see the extracted marginals from the same groups latching on to different parts of the same drum sounds. Also shown at the top of each plot column is the KL divergence between the input and the model. We see that there is no significant deterioration in the fit when imposing these priors. More aggressive values of these priors will result in a worse fit, but an even stronger statistical separation (or not) between the extracted marginals.

In this particular case since the audio stream contains elements with time correlation we used that as the dimension in which the cross-entropy manipulation was performed. This would also be supplemented by using neighboring time samples in order to look for temporal causality as well. In other cases however we might prefer to impose priors on all dimensions as opposed to only one. The estimation process is flexible enough to deal with any number of dimensions and to impose priors that either minimize or maximize cross-entropy on any arbitrary dimension subset.

5   Conclusions

In this paper we presented a non-negative data factorization approach which allows us to discover factors which can be mutually similar or dissimilar. In order to do so we reformulated the nonnegative factorization process as a probabilistic factorization problem and introduced new priors that can minimize or maximize the cross-entropy between any of the discovered marginals. The cross-entropy was shown to be an appropriate measure of similarity which allows us to express arbitrary relationships between any of the estimated marginals. We have shown that using this approach we can perform analysis which is slightly more akin to ICA than NMF by extracting maximally different marginals, and also that we can extract groups of components which contain highly relevant marginals but bear little relation to other groups.

References

1. Plumbley, M.D.: Algorithms for nonnegative independent component analysis. IEEE Transactions on Neural Networks 14(3) (May 2003)
2. Plumbley, M.D., Oja, E.: A "nonnegative PCA" algorithm for independent component analysis. IEEE Transactions on Neural Networks 15(1) (January 2004)
3. Singh, A.P., Gordon, G.J.: A Unified View of Matrix Factorization Models. In: ECML/PKDD 2008 (to appear, 2008)
4. Shashanka, M., Raj, B., Smaragdis, P.: Probabilistic Latent Variable Models as Nonnegative Factorizations. Computational Intelligence and Neuroscience 2008 (2008)
5. Cardoso, J.-F.: Multidimensional independent component analysis. In: Proceedings of the International Workshop on Higher-Order Statistics (1998)
6. Schlüter, R., Macherey, W., Muëller, B., Ney, H.: Comparison of discriminative training criteria and optimization methods for speech recognition. Speech Communication 34, 287–310 (2001)

A Sparsity-Based Method to Solve Permutation Indeterminacy in Frequency-Domain Convolutive Blind Source Separation

Prasad Sudhakar and Rémi Gribonval

METISS Team,
Centre Recherche INRIA Rennes - Bretagne Atlantique, Campus de Beaulieu, 35042
Rennes cedex, France
{firstname.lastname}@irisa.fr
http://www.irisa.fr/metiss

Abstract. Existing methods for frequency-domain estimation of mixing filters in convolutive blind source separation (BSS) suffer from permutation and scaling indeterminacies in sub-bands. However, if the filters are assumed to be sparse in the time domain, it is shown in this paper that the ℓ_1-norm of the filter matrix increases as the sub-band coefficients are permuted. With this motivation, an algorithm is then presented which solves the source permutation indeterminacy, provided there is no scaling indeterminacy in sub-bands. The robustness of the algorithm to noise is also presented.

Keywords: Convolutive BSS, permutation ambiguity, sparsity, ℓ_1-minimization.

1 Introduction

The need to separate source signals from a given set of convolutive mixtures arises in various contexts. The underlying model of having M mixtures $x_m(t)$, $m = 1 \ldots M$ from N source signals $s_n(t), n = 1 \ldots N$, given a discrete time index t, is given by

$$x_m(t) = \sum_{n=1}^{N} \sum_{k=0}^{K-1} a_{mnk} s_n(t - k) + v_m(t) \tag{1}$$

with $v_m(t)$ the noise term. Concisely, it can be written in the matrix notation as

$$\mathbf{x}(t) = \sum_{k=0}^{K-1} \mathbf{A}_k \mathbf{s}(t - k) + \mathbf{v}(t) \tag{2}$$

where $\mathbf{x}(t), \mathbf{v}(t)$ are $m \times 1$ vectors, $\mathbf{A}_k$ is an $M \times N$ matrix which contains the filter coefficients at k^{th} index. The notation $\mathbf{A}_{mn}(t) = a_{mnt}$ will also be used for each mixing filter, which is of length K. The ultimate objective of a BSS system is to recover back the original source signals $s_n, n = 1 \ldots N$ given only the mixtures $x_m(t), m = 1 \ldots M$.

T. Adali et al. (Eds.): ICA 2009, LNCS 5441, pp. 338–345, 2009.

A standard approach is to first estimate the mixing matrix $\mathbf{A}(t) = (\mathbf{A}_{mn}(t))$ $m = 1 \ldots M, n = 1 \ldots N$ and then recover the sources $s_n(t)$. This paper focusses on the estimation of the mixing matrix.

1.1   Permutation Problem Description

Several methods have been proposed by the signal processing community to estimate the mixing matrices in convolutive BSS. Pedersen et. al. present an excellent survey of the existing methods [1]. Broadly, the techniques can be classified into time-domain and frequency-domain techniques. Both approaches have their own advantages and disadvantages. They are summarized in table 3 of [1].

The context of our problem arises in the frequency-domain approach. A survey of these techniques is provided in [2]. The main advantage of frequency-domain techniques is that the convolutive mixture case is transformed (under the narrowband assumption) into complex-valued instantaneous mixture case for each frequency bin:

$$\mathbf{x}(f,t) = \mathbf{A}(f)\mathbf{s}(f,t) + \mathbf{v}(f,t) \tag{3}$$

where $f = 1 \ldots F$ are the sub-band frequencies.

The central task of frequency-domain mixing matrix estimation techniques is to provide an estimate $\hat{\mathbf{A}}(f)$ of $\mathbf{A}(f)$. However, the frequency-domain approach suffers from permutation and of scaling indeterminacies *in each sub-band* f. Specifically, the estimated $\hat{\mathbf{A}}(f)$ is related to the true filter matrix $\mathbf{A}(f)$, for each f in the following form

$$\hat{\mathbf{A}}(f) = \mathbf{A}(f)\mathbf{\Lambda}(f)\mathbf{P}(f) \tag{4}$$

where $\mathbf{P}(f)$ is the frequency-dependent permutation matrix, $\mathbf{\Lambda}(f)$ is a diagonal matrix containing the arbitrary scaling factors.

The frequency-domain methods have to invariably solve the permutation and scaling indeterminacy to eventually estimate $\mathbf{A}(t)$ up to a unique global permutation and scaling $\hat{\mathbf{A}}(t) = \mathbf{A}(t)\mathbf{\Lambda}\mathbf{P}$.

1.2   Existing Approaches to Solve the Described Problem

There are mainly two categories of approaches to solve the permutation indeterminacy in the sub-bands of the estimated mixing filters [1].

The first set of techniques use consistency measures across the frequency sub-bands of the filters to recover the correct permutations, such as inter-frequency smoothness, etc. This category also includes the beamforming approach to identify the direction of arrival of sources and then adjust the permutations [3].

The second set of techniques use the consistency of the spectrum of the recovered signals to achieve the same. The consistency across the spectrum of the recovered signals is applicable for only those signals which have strong correlation across sub-bands, such as speech [4].

Different definitions of consistency have been used to propose different methods. Table 4 in [1] contains a categorization of these approaches.

1.3   Proposed Approach: Sparse Filter Models

Here we propose to use a special type of consistency that can be assumed on the mixing filters: sparsity. That is, the number S of non-negligible coefficients in each filter $\mathbf{A}_{mn}(t)$ is significantly less than its length K. Acoustic impulse responses have few reflection paths relative to its duration, and hence the sparsity of mixing filters is a realistic approximation.

The idea behind our approach is that the permutations in the sub-bands decreases the sparsity of the reconstructed filter matrix $\hat{\mathbf{A}}(t)$. So, one can solve the permutations by maximizing the sparsity.

The sparseness of the filter matrix is measured by its ℓ_0-norm, defined as $||\mathbf{A}(t)||_0 = \sum_{m=1}^{M} \sum_{n=1}^{N} \sum_{t=1}^{K} |\mathbf{A}_{mn}(t)|^0$. Lesser the norm, sparser is the filter matrix. However, the ℓ_1-norm of the filter matrix defined by

$$||\mathbf{A}(t)||_1 = \sum_{m=1}^{M} \sum_{n=1}^{N} \sum_{t=1}^{K} |\mathbf{A}_{mn}(t)| \tag{5}$$

is also a sparsity promoting norm. It is shown by Donoho that working with ℓ_1-norm is as effective as working with ℓ_0-norm while looking for sparse solutions to linear systems [5]. Furthermore ℓ_1-norm being convex, has certain computational advantages over the former and is robust to noise.

We concentrate on the permutation problem and will assume that the estimated filter sub-bands are free from scaling indeterminacy. That is

$$\hat{\mathbf{A}}(f) = \mathbf{A}(f)\mathbf{P}(f), f = 1 \dots F. \tag{6}$$

Hence, when $\hat{\mathbf{A}}(t)$ is reconstructed using $\hat{\mathbf{A}}(f)$, given by Eqn. (6) one can expect an increase in $||\hat{\mathbf{A}}(t)||_1$. In Sec. 2 we experimentally show that this is indeed the case.

Then in Sec. 3 we show that if there is no scaling indeterminacy, a simple algorithm which minimizes the ℓ_1-norm of the filter matrix solves the permutation indeterminacy, even under noisy conditions.

2   Source Permutations and ℓ_1-Norm

This section presents some preliminary experimental study on how the ℓ_1-norm of the filter matrix $\mathbf{A}(t)$ is affected by permutation in the sub-bands.

For experimental purposes, 50 different filter matrices were synthetically created with $N = 5$ sources and $M = 3$ channels, and each filter having a length of $K = 1024$ with $S = 10$ non-zero coefficients. The non-zero coefficients were i.i.d. Gaussian with mean zero and variance one. The locations of non-zero coefficients were selected uniformly at random. Then for each such instance of filter matrix $\mathbf{A}(t)$ the discrete Fourier transform (DFT) $\hat{\mathbf{A}}_{mn}(f)$ was computed for each filter $\mathbf{A}_{mn}(t)$, to obtain the frequency-domain representation $\hat{\mathbf{A}}(f)$ of $\mathbf{A}(t)$. These filters were used in the following two kinds of experiments.

2.1   Random Source Permutations

In a practical scenario, each sub-band can have a random permutation. So, for each filter matrix in the frequency domain, the sources were permuted randomly in an increasing of number of sub-bands (chosen at random), and their ℓ_1-norms were computed. The positive and negative frequency sub-bands were permuted identically.

For one such experimental instance, Fig. 1(a) shows the variation of ℓ_1-norm against increasing number of randomly permuted sub-bands. The circle shows the norm of the true filter matrix. Each star represents the norm after randomly permuting the sub-bands at random locations. Note the gradual increase in the norm as the number of sub-bands being permuted increases. Similar experiments were conducted with combinations of $M = 3, N = 4$ and $M = 2, N = 3$ and $S = 10, 15, 20$, leading to similar observations.

2.2   Sensitivity of ℓ_1-Norm to Permutations

In order to show that even a *single* permutation in only one sub-band can increase the norm, only two sources, chosen at random were permuted in increasing number of sub-bands.

For one such instance, Fig. 1(b) shows a plot of the variation in ℓ_1-norm with the number of sub-bands permuted. The circle in the plot shows the ℓ_1-norm of the true filter matrix. Each star shows the norm after permuting the sources 2 and 3.

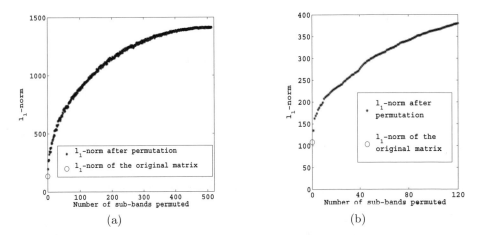

Fig. 1. The variation in the ℓ_1-norm of the filter matrix against the number of sub-bands permuted. (a) All the sources are permuted randomly (b) Only sources 2 and 3 are permuted.

3   Proposed Algorithm

With the inspiration from the previous section, we present an algorithm in this section to solve the permutation indeterminacy.

Assumption: The estimate $\hat{\mathbf{A}}(f)$ of the sub-band coefficients $\mathbf{A}(f)$ are provided by some other independent technique and $\hat{\mathbf{A}}(f) = \mathbf{A}(f)\mathbf{P}(f), f = 1 \ldots F$.

The absence of scaling indeterminacy is not totally a realistic assumption but we feel that this is the first step towards solving the bigger problem.

3.1   Description

We denote the set of all the possible source permutations by $\mathcal{P}$ ($|\mathcal{P}| = N!$) and the inverse discrete Fourier transform by $IDFT$. At each sub-band, every $\mathbf{P} \in \mathcal{P}$ is explored, keeping the other sub-bands fixed, and that permutation is retained which minimizes the ℓ_1-norm. This ensures that ℓ_1-norm of the filter matrix is lowered to the minimum possible extent by aligning that particular sub-band. At the end of one such iteration through all the sub-bands, the norm of the filter matrix would have significantly reduced. However, as the sub-bands were locally examined, the resulting norm may not be the global minimum. Hence, the entire process is iterated until the difference in the norm does not reduce significantly between iterations.

Input: $\hat{\mathbf{A}}, \theta$: The estimated sub-band coefficients of $\mathbf{A}(t)$ and a threshold
Output: $\tilde{\mathbf{A}}$: The sub-band coefficient matrix after solving for the permutations

(1) Initialize $\tilde{\mathbf{A}} \leftarrow \hat{\mathbf{A}}$;
(2) Update all sub-bands;
    foreach $f = 1 : F$ **do**
        $old\tilde{\mathbf{A}} \leftarrow \tilde{\mathbf{A}}$;
        foreach $\mathbf{P} \in \mathcal{P}$ **do**
            $\tilde{\mathbf{A}}(f) \leftarrow \hat{\mathbf{A}}(f)\mathbf{P}$;
            $val(\mathbf{P}) \leftarrow ||IDFT(\tilde{\mathbf{A}}(f'))||_1, f' = 1 \ldots F$;
        end
        $\mathbf{P}(f) \leftarrow \underset{\mathbf{P} \in \mathcal{P}}{\arg\min}\ val(\mathbf{P})$;
        $\tilde{\mathbf{A}}(f) \leftarrow \hat{\mathbf{A}}(f)\mathbf{P}(f)$;
    end
(3) Test if the algorithm should stop;
    if $||\hat{\mathbf{A}}(t)||_1 \geq ||old\hat{\mathbf{A}}(t)||_1 - \theta$ **then** Output $\tilde{\mathbf{A}}$ **else** Go to step (2)

Algorithm 1. Algorithm to solve the permutation indeterminacy by minimizing the ℓ_1-norm of the time domain filter matrix

3.2   Objective of the Algorithm

The aim of the algorithm is to obtain the sub-band matrix which would give the minimum ℓ_1-norm. However, currently we do not have analytical proof about the convergence of the algorithm to the global minimum. Also, the sources will be globally permuted at the end.

3.3   Complexity

A brute force approach to solve the ideal ℓ_1-minimization problem would need $N!^K K \log K$ operations. In our case, each outer iteration needs to inspect $N! \times K$ permutations and at each step, an inverse FFT has a complexity of $K \log K$. Hence, the complexity in each outer iteration is $N! K^2 \log K$. This is still costly because it grows in factorial with the number of sources, but it is tractable for small problem sizes.

4   Experimental Results

In this section we present an illustration of the algorithm presented above.

4.1   The No Noise, No Scaling Case

Firstly, we consider the case where the sub-band coefficients are assumed to be estimated without noise and scaling ambiguity. 20 filter matrices with $N = 3, M = 2, K = 1024$ and $S = 10$ were created, transformed and sub-bands permuted in a similar way as explained in Sec. 2.1 and were the input to the algorithm. The value of θ was set to 0.0001 in all the experiments.

The output was transformed back to time domain to compute the reconstruction error. In all the experiments, the output filters were identical to the true filters up to a global permutation and within a numerical precision in Matlab.

4.2   Effect of Noise

The estimation of $\hat{\mathbf{A}}(f)$ by an actual BSS algorithm invariably involves some level of noise (as well as scaling, which we do not deal with here). Hence, the permutation solving algorithm needs to be robust to certain level of noise. Experiments were conducted by permuting the sub-bands and adding noise to the coefficients:

$$\hat{\mathbf{A}}(f) = \mathbf{A}(f)\mathbf{P}(f) + \mathbf{N}(f) \tag{7}$$

where $\mathbf{N}(f)$ is i.i.d. complex Gaussian with mean zero and variance σ^2.

For illustration, Fig. 2 shows an instance of the reconstructed filter matrix using Algorithm 1 with the input corrupted by additive complex Gaussian noise with $\sigma^2 = 0.2$. Each filter had 10 significant coefficients which have been faithfully recovered, along with some amount of noise.

For a quantitative analysis of the effect of noise, the input SNR was varied between -10 dB and 40 dB in steps of 5 dB and the output SNR was computed. The SNR definitions are given in Eqn. (8). The problem size was $N = 3, M = 2, K = 1024$, and 20 experiments were conducted to obtain each data point.

$$SNR_{in} = 20 \log_{10}\left(\frac{\|\mathbf{A}(t)\|_2}{\|\mathbf{N}(t)\|_2}\right), SNR_{out} = 20 \log_{10}\left(\frac{\|\mathbf{A}(t)\|_2}{\|\mathbf{A}(t) - \tilde{\mathbf{A}}(t)\|_2}\right) \tag{8}$$

The experiments were repeated for $S = 10$ and $S = 25$.

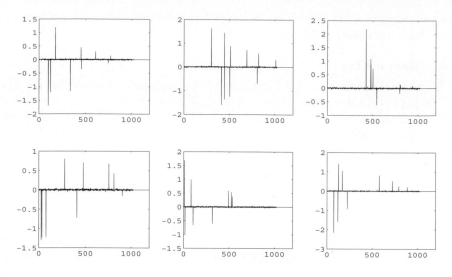

Fig. 2. Reconstructed filter matrix with additive noise in the sub-bands

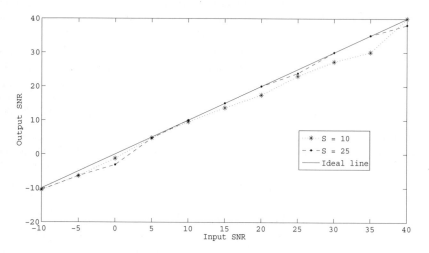

Fig. 3. Output SNR versus input SNR in dB

Figure 3 shows the variation of output SNR (in dB) with input SNR. Due to the absence of scaling, a perfectly reconstructed filter will have the same SNR as the input. The thick line shows the ideal relationship for reference. In the range of 5 to 10 dB input SNR, the curve for $S = 10$ coincides with the ideal line. For $S = 25$, the curve coincides with the ideal line for range of 5 to 20 dB input SNR. At other places, both the curves closely follow the ideal line suggesting perfect reconstruction in most number of experiments.

5   Conclusion and Future Work Direction

Frequency-domain estimation of mixing matrices in convolutive BSS suffers from the indeterminacies of source permutations and scaling of sub-bands. Hence, solving the permutation indeterminacy is an important aspect of such BSS systems. In this paper, it has been shown that in the absence of scaling, the ℓ_1-norm of the filter matrix is very sensitive to permutations in sub-bands. An algorithm has been presented based on the minimization of ℓ_1-norm of the filter matrix to solve for the permutations. Experimental results show that in the absence of scaling, the ℓ_1-minimization principle to solve the permutations performs well even in the presence of noise.

Though the absence of scaling is not a realistic assumption, it can be a first step towards sparsity motivated permutation solving methods. Also, the complexity of the algorithm grows with the $N!$ and K^2, which is expensive even for moderate values for the number of sources N and filter length K. Our future work focusses on replacing the combinatorial optimization step by an efficient convex optimization formulation and devising ℓ_1-norm based methods to solve the permutation problem in presence of arbitrary sub-band scaling.

References

1. Pedersen, M.S., Larsen, J., Kjems, U., Parra, L.C.: A Survey of Convolutive Blind Source Separation. In: Springer Handbook on Speech Processing and Speech Communication (2006)
2. Makino, S., Sawada, H., Mukai, R., Araki, S.: Blind source separation of convolutive mixtures of speech in frequency domain. IEICE Trans. Fundamentals E88-A(7), 1640–1655 (2005)
3. Ikram, M.Z., Morgan, D.R.: A beamforming approach to permutation alignment for multichannel frequency-domain blind source separation. In: Proc. ICASSP 2002, pp. 881–884 (May 2002)
4. Murata, N., Ikeda, S., Ziehe, A.: An approach to blind source separation based on temporal structure of speech signals, RIKEN Brain Science Institute, BSIS Technical reports 98-2 (April 1998)
5. Donoho, D.: For most large underdetermined systems of linear equations, the minimal ell-1 norm near-solution approximates the sparsest near-solution. Communications on Pure and Applied Mathematics 59(7), 907–934 (2006)

Canonical Dual Approach to Binary Factor Analysis

Ke Sun[1], Shikui Tu[1], David Yang Gao[2], and Lei Xu[1]

[1] Department of Computer Science and Engineering,
Chinese University of Hong Kong, Shatin, N.T., Hong Kong, China
{ksun,sktu,lxu}@cse.cuhk.edu.hk
[2] Department of Mathematics, Virginia Polytechnic Institute and State University,
Blacksburg, VA 24061, USA
gao@vt.edu

Abstract. Binary Factor Analysis (BFA) is a typical problem of Independent Component Analysis (ICA) where the signal sources are binary. Parameter learning and model selection in BFA are computationally intractable because of the combinatorial complexity. This paper aims at an efficient approach to BFA. For parameter learning, an unconstrained binary quadratic programming (BQP) is reduced to a canonical dual problem with low computational complexity; for model selection, we adopt the Bayesian Ying-Yang (BYY) framework to make model selection automatically during learning. In the experiments, the proposed approach `cdual` shows superior performance. Another BQP approximation `round` is also good in model selection and is more efficient. Two other methods, `greedy` and `enum`, are more accurate in BQP but fail to compete with `cdual` and `round` in BFA. We conclude that a good optimization is essential in a learning process, but the key task of learning is not simply optimization and an over-accurate optimization may not be preferred.

1 Introduction

Binary Factor Analysis (BFA) explores latent binary structures of data. Unlike in clustering analysis where the observations are scattered around several uncorrelated centers, in BFA the cluster locations are correlated and represented by a binary vector with independent dimensions. From an information theoretic perspective, the observables can be traced to several independent binary random variables as information sources. Research on BFA has been conducted with wide applications. One stream has been focused on analysis of binary data [1] with the aid of Boolean Algebra. The broad variety of binary data, such as social research questionnaires, market basket data and DNA microarray expression profiles, gives this research enormous practical value. Another stream tries to discover binary factors in continuous data [2] [3] [4], taking advantage of the representational capacity of the underlying binary structure. The present work falls in the second category.

A general and difficult problem in BFA is the combinatorial complexity in the inference of a m-bit binary code $y(x)$ or a 2^m-point posterior distribution

T. Adali et al. (Eds.): ICA 2009, LNCS 5441, pp. 346–353, 2009.
© Springer-Verlag Berlin Heidelberg 2009

$p(\boldsymbol{y} \mid \boldsymbol{x})$ for each training sample $\boldsymbol{x}$. Past efforts in tackling this problem include applying the Markov Chain Monte Carlo (MCMC) methods [3], or restricting the model so that the posterior distribution has independent dimensions [2].

Another difficulty in BFA is to determine an appropriate length of the internal binary code $\boldsymbol{y}$. The traditional approach for model selection has to enumerate a set $\mathcal{K}$ and perform maximum likelihood (ML) learning for each candidate length $\dim(\boldsymbol{y}) \in \mathcal{K}$, then the optimal length is selected via minimizing an information criterion [5] [6]. This two-phase approach suffers from excessive computation due to the computational complexity of BFA. In the past decade efforts have also been made to determine a proper model scale during parameter learning. As a general framework for parameter learning and model selection, BYY harmony learning [7] [8] is capable to discard redundant structures during training. The paper [8] investigates machine learning versus optimization from the BYY perspective, where BFA is discussed as a special case. The paper [4] studies BFA under the BYY framework, where $p(\boldsymbol{y} \mid \boldsymbol{x})$ is assumed to be free of structure.

This paper considers the same BFA model as in [4] and [2]. In help of a canonical duality of BQP [9][10], we can compute efficiently for each training sample $\boldsymbol{x}_t$ a binary code $\boldsymbol{y}(\boldsymbol{x}_t)$. As learning proceeds, redundant binary dimensions are identified and discarded with a BYY learning algorithm [8]. A comparison among four BQP methods is presented. The proposed approach cdual is not the best in BQP optimization but presents superior performance in BFA learning. A relax-and-round method round, which is rather rough from an optimization perspective, is also good in model selection and provides a performance even better than the accurate BQP techniques.

The rest of this paper is organized as follows. Section 2 introduces BFA and a BYY learning algorithm. Section 3 imports the canonical duality theory to overcome the BQP computational bottleneck. Section 4 includes an experimental comparison among four BQP methods in BFA learning. Section 5 concludes.

2   Binary Factor Analysis

This paper studies the BFA model

$$q(\boldsymbol{y}) = \prod_{i=1}^{m} \theta_i^{(1+y_i)/2} (1 - \theta_i)^{(1-y_i)/2}, \quad q(\boldsymbol{x} \mid \boldsymbol{y}) = G(\boldsymbol{x} \mid \boldsymbol{A}\boldsymbol{y} + \boldsymbol{c}, \boldsymbol{\Sigma}), \quad (1)$$

where $\boldsymbol{y} \in \{-1, 1\}^m$ is an internal binary code, $0 < \theta_i < 1$, $i = 1, 2, \ldots, m$, $\boldsymbol{x}$ is a continuous observation, $G(\cdot \mid \boldsymbol{\mu}, \boldsymbol{\Psi})$ denotes a Gaussian distribution with mean $\boldsymbol{\mu}$ and covariance $\boldsymbol{\Psi}$, $\boldsymbol{\Sigma}$ is a positive definite diagonal matrix. This model has been studied previously from different perspectives [11] [4] [2].

Within the BYY framework [7], another joint distribution $p(\boldsymbol{x}, \boldsymbol{y})$ describes the observations with $p(\boldsymbol{x})$ and the inference of binary codes with $p(\boldsymbol{y} \mid \boldsymbol{x})$. In this paper, $p(\boldsymbol{x})$ is chosen as $p(\boldsymbol{x}) = \sum_{t=1}^{N} \delta(\boldsymbol{x} - \boldsymbol{x}_t)/N$, where $\delta(.)$ is the Dirac delta function; $p(\boldsymbol{y} \mid \boldsymbol{x}) = \delta(\boldsymbol{y} - \hat{\boldsymbol{y}}(\boldsymbol{x}))$ is assumed to be free of structure, where $\hat{\boldsymbol{y}}(\boldsymbol{x})$ is derived through maximizing the harmony function [7] [8] such that

$$\hat{\boldsymbol{y}}(\boldsymbol{x}) = \underset{\boldsymbol{y} \in \{-1,1\}^m}{\arg\max} \ \log q(\boldsymbol{x}, \boldsymbol{y}) = \underset{\boldsymbol{y} \in \{-1,1\}^m}{\arg\min} \ \left\{ \frac{1}{2}\boldsymbol{y}^T \boldsymbol{Q}_y \boldsymbol{y} - \boldsymbol{f}_y^T(\boldsymbol{x})\boldsymbol{y} \right\}, \qquad (2)$$

$\boldsymbol{Q}_y = \boldsymbol{A}^T \boldsymbol{\Sigma}^{-1} \boldsymbol{A},\ \boldsymbol{f}_y(\boldsymbol{x}) = [\log \boldsymbol{\theta} - \log(\boldsymbol{1} - \boldsymbol{\theta})]\ /2 + \boldsymbol{A}^T \boldsymbol{\Sigma}^{-1}(\boldsymbol{x} - \boldsymbol{c}),\ \boldsymbol{\theta} = (\theta_1, \dots, \theta_m)^T.$
By definition, the harmony function is

$$H(p \,\|\, q) = \int p(\boldsymbol{x})p(\boldsymbol{y}\,|\,\boldsymbol{x}) \log q(\boldsymbol{y})q(\boldsymbol{x}\,|\,\boldsymbol{y})\, d\boldsymbol{y}\, d\boldsymbol{x} = \frac{1}{N}\sum_{t=1}^{N} \widetilde{H}(\boldsymbol{x}_t, \hat{\boldsymbol{y}}(\boldsymbol{x}_t), \boldsymbol{\Theta}),$$

$$\widetilde{H}(\boldsymbol{x}, \boldsymbol{y}, \boldsymbol{\Theta}) = \sum_{i=1}^{m} \left[\frac{1+y_i}{2} \log \theta_i + \frac{1-y_i}{2} \log(1-\theta_i) \right] + \log G\left(\boldsymbol{x} \,|\, \boldsymbol{A}\boldsymbol{y} + \boldsymbol{c}, \boldsymbol{\Sigma} \right). \qquad (3)$$

BYY harmony learning with automatic model selection (BYY-AUTO) is imple-
mented by maximizing $H(p \,\|\, q)$ on the training set. Starting from a large initial
coding length, the gradient flow of $H(p \,\|\, q)$ may either *push θ_i to 0 or 1 when y_i*
turns deterministic, or *push the ratio* $\|\boldsymbol{A}_i\|_2^2 / \sqrt{\boldsymbol{A}_i^T \boldsymbol{\Sigma} \boldsymbol{A}_i}$ *($\boldsymbol{A}_i$ is the i'th column*
of $\boldsymbol{A}$) to 0 when $\boldsymbol{A}_i y_i$ is flooded by noise. In both these cases the i'th bit of
$\boldsymbol{y}$ is identified as redundant and is discarded while the learning proceeds. The
learning process is sketched in Algorithm 1.

Algorithm 1. Free structure BYY-AUTO learning algorithm for BFA

    Input : A set $\{\boldsymbol{x}_t\}_{t=1}^N \subset \Re^n$ of observations
    Output: An estimated binary coding length $\dim(\boldsymbol{y})$; $\boldsymbol{\Theta} = \{\boldsymbol{\theta}, \boldsymbol{A}, \boldsymbol{c}, \boldsymbol{\Sigma}\}$

1 Initialize $\dim(\boldsymbol{y})$ with a large integer m_0; $\boldsymbol{\Theta}_0 = \{\boldsymbol{\theta}_0, \boldsymbol{A}_0, \boldsymbol{c}_0, \boldsymbol{\Sigma}_0\}$;
2 **repeat**
3   Take $\mathcal{X}_e \subset \{\boldsymbol{x}_t\}_{t=1}^N$ sequentially or through a sampling algorithm;
4   Encode $\mathcal{X}_e$ into $\{\hat{\boldsymbol{y}}(\boldsymbol{x}) : \boldsymbol{x} \in \mathcal{X}_e\}$ with a binary encoder;
5   Update $\boldsymbol{\Theta}$ along the gradient flow of $\sum_{\boldsymbol{x} \in \mathcal{X}_e} \widetilde{H}(\boldsymbol{x}, \hat{\boldsymbol{y}}(\boldsymbol{x}), \boldsymbol{\Theta})$;
6   **if** $\theta_i < \epsilon$ *or* $\theta_i > 1 - \epsilon$ *or* $\|\boldsymbol{A}_i\|_2^2 < \delta\sqrt{\boldsymbol{A}_i^T \boldsymbol{\Sigma} \boldsymbol{A}_i}$ **then**
7   Discard the i'th dimension of $\boldsymbol{y}$; update $\boldsymbol{\Theta}$ accordingly;
8 **until** $H(p \,\|\, q)$ *has reached convergence* ;

    In the experiments we fix $\epsilon = 0.1$, $\delta = 2$, $|\mathcal{X}_e| = N$, $m_0 = 2m^\star - 1$, *where*
    $m^\star = \dim(\boldsymbol{y}^\star)$ *is the "true" binary dimension in synthetic data generation.*

---

3 Canonical Dual Approach to Binary Encoding

In BFA, a binary encoder $\hat{\boldsymbol{y}} : \{\boldsymbol{x}_t\}_{t=1}^N \to \{-1,1\}^m$ is usually employed so that
a function can be computed numerically or a mapping, such as $p(\boldsymbol{y}\,|\,\boldsymbol{x}, \boldsymbol{\Theta})$, can
be regularized with the encoding results. In the context here, we need such an
encoding, as in Eq. (2) or line 4 in Algorithm 1, to maximize $H(p \,\|\, q)$ in Eq.
(3) by a gradient-based optimization. Eq. (2) is a BQP that falls in NP-hard.
The BFA-specific formula of $\boldsymbol{Q}_y$ and $\boldsymbol{f}_y(\boldsymbol{x})$ can not make the problem easier.
An approximation is required to avoid the combinatorial complexity.

Both exact and heuristic approaches to BQP have been widely studied in the literature of optimization [12]. Recently Gao et al. have constructed a pair of *canonical dual problems* for BQP [9] [10] with zero duality gap. The solution $\bar{\zeta}$ of the *canonical dual* problem

$$(\mathcal{P}^d): \quad \max_{\zeta>0}\left\{P^d(\zeta) = -\frac{1}{2}\boldsymbol{f}_y^T(\boldsymbol{x})\left[\boldsymbol{Q}_y + diag(\zeta)\right]^{-1}\boldsymbol{f}_y(\boldsymbol{x}) - \frac{1}{2}\boldsymbol{e}^T\zeta\right\}, \quad (4)$$

if exists, will lead to a solution $\hat{\boldsymbol{y}}(\boldsymbol{x}) = (\boldsymbol{Q}_y + diag(\bar{\zeta}))^{-1}\boldsymbol{f}_y(\boldsymbol{x})$ of Eq. (2), where $\boldsymbol{e} = (1, 1, \ldots, 1)^T$. In contrast to the primal BQP, $\mathcal{P}^d$ is a constrained convex optimization problem that can be handled much more efficiently. Algorithm 2 employs a gradient descent to solve BQP through solving $\mathcal{P}^d$ in Eq. (4).

Algorithm 2. $\displaystyle\min_{y\in\{-1,1\}^m}\left\{\frac{1}{2}\boldsymbol{y}^T\boldsymbol{Q}\boldsymbol{y} - \boldsymbol{f}^T\boldsymbol{y}\right\}$ via max its canonical dual

---

1 Normalize $\boldsymbol{Q}^{new} = \boldsymbol{Q}^{old}/tr(\boldsymbol{Q}^{old})$, $\boldsymbol{f}^{new}(\boldsymbol{x}) = \boldsymbol{f}^{old}(\boldsymbol{x})/tr(\boldsymbol{Q}^{old})$;
2 Pre-process $\boldsymbol{Q}^{new} = \boldsymbol{Q}^{old} + \boldsymbol{\Lambda}_Q$, $\boldsymbol{\Lambda}_Q$ is diagonal (optional);
3 Initialize $\zeta = \max(-\boldsymbol{Q}\boldsymbol{e} - \boldsymbol{f}, \boldsymbol{Q}\boldsymbol{e} + \boldsymbol{f} - 2diag(\boldsymbol{Q}))$;
4 **for** *epoch* $\leftarrow$ 1 **to** *max_epochs* **do**
5 $\quad$ $\boldsymbol{y} = (\boldsymbol{Q} + diag(\zeta))^{-1}\boldsymbol{f}$;
6 $\quad$ $\nabla = (\boldsymbol{y} \circ \boldsymbol{y} - \boldsymbol{e})/2$;
7 $\quad$ **if** $\|\nabla\|_\infty < $ *Threshold* **then** break;
8 $\quad$ $\zeta = \zeta + \gamma\nabla$ ($\gamma > 0$ is a small learning rate);
9 Round $\boldsymbol{y}$ to $\{-1,1\}^m$; return $\boldsymbol{y}$.

In the experiments, max_epochs = 50, Threshold = 0.5, γ = 0.02. Step 2 is a classical trick [12] based on $y_i^2 \equiv 1$ but not adopted in the experiments.

Table 1. Algorithms for solving the BQP in Eq. (2)

Name	Description
enum	exhaustively enumerate $\boldsymbol{y} \in \{-1,1\}^m$, which was used in BFA [4]
greedy	the greedy BQP algorithm on page 203 [12]
cdual	the canonical dual approach to BQP (Algorithm 2 in this paper)
round	round $\tilde{\boldsymbol{y}} = \boldsymbol{Q}_y^{-1}\boldsymbol{f}_y(\boldsymbol{x})$ to the nearest binary vector in $\{-1,1\}^m$, which was proposed (Table II, page 836 [11]) for BFA learning under the name "fixed posteriori approximation"

Figure 1 shows the accuracy and efficiency of the BQP algorithms listed in Table 1[1]. **round** is fastest but its performance degenerates greatly as $\dim(\boldsymbol{y})$ increases; **greedy** is most accurate among {round, cdual, greedy} but suffers from $O(\dim^3(\boldsymbol{y}))$ computation [12]; **cdual** is less accurate than **greedy** but is

---

[1] All experiments in this paper are implemented with GNU Octave 3.0.3 on a Intel Core 2 Duo 2.13GHz with 1GB RAM running FreeBSD 7.0.

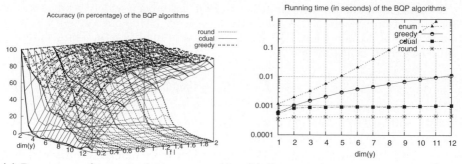

(a) Percentage of correct solutions over a $10\times$ (b) Computation cost in seconds   (note
10 $(\dim(\boldsymbol{y})\times\|\boldsymbol{f}_y(\boldsymbol{x})\|_2)$ grid of configurations the time axis is in log-scale)

Fig. 1. Performance (out of 100 runs) of the BQP algorithms with $tr(\boldsymbol{Q})=1.0$

much more efficient. As $\|\boldsymbol{f}_y(\boldsymbol{x})\|_2$ turns small, round becomes a little more accurate while the error of cdual and greedy raises up significantly.

If $\|\boldsymbol{f}_y(\boldsymbol{x})\|_2$ is small enough such that $\|\boldsymbol{Q}_y^{-1}\boldsymbol{f}_y(\boldsymbol{x})\|_\infty < 1$, then $\triangledown P^d|_{\zeta=0} < \mathbf{0}$. From the convexity of $P^d(\zeta)$ on $\Re^{m+}$, the dual solution $\bar{\zeta} > 0$ does not exist. This explains the failure of cdual on small $\|\boldsymbol{f}_y(\boldsymbol{x})\|_2$. We further consider its impact on BFA learning. In Algorithm 1, dimension i will be deducted if $|\theta_i - 0.5|$ is large enough, hence $\boldsymbol{\theta}$ is in a small neighbourhood around $0.5\boldsymbol{e}$ and $\|\boldsymbol{f}_y(\boldsymbol{x})\|_2^2 = \|\boldsymbol{A}^T\boldsymbol{\Sigma}^{-1}(\boldsymbol{x}-\boldsymbol{c}) + [\log\boldsymbol{\theta} - \log(1-\boldsymbol{\theta})]/2\|_2^2$ is a convex function minimized around $\boldsymbol{x}=\boldsymbol{c}$. A small $\|\boldsymbol{f}_y(\boldsymbol{x})\|_2$ is due to samples lying between the 2^m representative clusters in BFA. cdual is not accurate on these samples.

4   Experiments

This section compares enum, greedy, cdual and round in BFA with synthetic data generated according to Eq. (1). Because of space limitation, we concentrate on the case where $\dim(\boldsymbol{x}) = 10$, $\boldsymbol{y}$ evenly taking values from the 2^{m-1} points $\{\boldsymbol{y}\in\{-1,1\}^m:\mod[\sum_{i=1}^m(y_i+1)/2,2]=0\}^2$, $\boldsymbol{A}=\boldsymbol{Q}\boldsymbol{\Lambda}$, $\boldsymbol{Q}$ is orthogonal, $\boldsymbol{\Lambda}=diag(\lambda_1,\lambda_2,\ldots,\lambda_m)$, λ_i uniformly distributed over the interval $(1,2)$ so that the scale $\|\boldsymbol{A}_i\|_2$ in each binary dimension does not vary too much, and $\boldsymbol{\Sigma}=\sigma^2\boldsymbol{I}$. Three aspects that may affect the learning performance are investigated: the true binary dimension($\dim(\boldsymbol{y}^\star)$), the sample size($N$) and the noise level($\sigma$).

4.1   Binary Matrix Factorization with Fixed Dimension

We fix the binary dimension by skipping line $6\sim 7$ in Algorithm 1 and study its performance on the binary matrix factorization (BMF) $\boldsymbol{X}_{n\times N}=\boldsymbol{A}_{n\times m}\boldsymbol{Y}_{m\times N}$,

---

[2] This is a subset of $\{-1,1\}^m \subset \Re^m$ that can only be well separated with $\geq m$ hyperplanes. Hence the "true" binary dimension is m. It is chosen instead of $\{-1,1\}^m$ to simulate data in real world where not all 2^m binary encodings are valid or observed. A comparison between data generated with this 2^{m-1}-point subset and $\{-1,1\}^m$ is nevertheless interesting but omitted here for saving space.

where A is real and $Y \in \{-1, 1\}^{m \times N}$. Such a factorization differs from the classical BMF in that the matrix to be factored is continuous rather than boolean and is different from [3] in the factorization form. It may be useful in recovering binary signals from continuous observations. For example, a noisy binary image Y can be reverted after rotation or scaling. Figure 2 presents the BMF error and learning time over $\dim(\boldsymbol{y})$. The training error gets an order `round` > `cdual` > `greedy` > `enum` while the running time is in a reverse order. When $\dim(\boldsymbol{y})$ is large, `round` and `cdual` is not as accurate as the others because of their deteriorated BQP accuracy. As a trade-off they are much faster. Training time of `round` appears to be constant over $\dim(\boldsymbol{y})$. `greedy` has avoided the exponential complexity of `enum` but still requires huge computation on a large $\dim(\boldsymbol{y})$. To sum up, `cdual` and `greedy` are recommended for the BMF task.

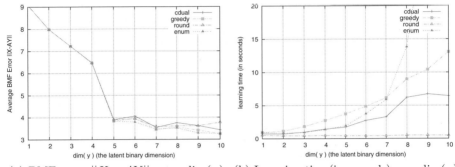

(a) BMF error $\|X - AY\|_2$ over $\dim(\boldsymbol{y})$ (b) Learning time(in seconds) over $\dim(\boldsymbol{y})$

Fig. 2. BMF when $\dim(\boldsymbol{x}) = 10$, $\dim(\boldsymbol{y}^\star) = 5$, $N = 100$, $\sigma = 0.3$, $\sum_{t=1}^{N} \boldsymbol{x}_t = \boldsymbol{0}$

4.2   Model Selection on Synthetic Data

In Algorithm 1, $\dim(\boldsymbol{y})$ is initialized large enough and deducted during learning. Since computational overhead arises when $\dim(\boldsymbol{y})$ is large, how soon this deduction stops determines the learning efficiency. With a true binary dimension $\dim(\boldsymbol{y}^\star) = 5$ and the learner's $\dim(\boldsymbol{y})$ initialized to 9, Figure 3 shows the average dimension deduction curve after 100 independent runs for $\sigma \in \{0.1, 0.3\}$. Two observations are made. (a) the convergence speed of $\dim(\boldsymbol{y})$ is in the same order as the BQP speed in Figure 1(b). Convergence slows down as the noise level increases. (b) `cdual` is robust to noise and yields the best accuracy; `enum` and `greedy` overestimates the model scale; `round` also shows a slight tendency of overestimation. The over/under estimation is controlled by the threshold δ and is related to $|\mathcal{X}_e|$ in Algorithm 1. They are both fixed in this paper for brevity.

Consider one binary dimension $\boldsymbol{y}_i$ with a small $\|A_i\|_2$ and big noise. Maximizing $H(p \| q)$ may further shrink $\|A_i\|_2^2 / \sqrt{A_i^\top \Sigma A_i}$ to achieve model selection. The error of `cdual` on the samples lying among the 2^m representative clusters forms a natural regularization to dimension deduction. An over-accurate binary encoder does not have this type of regularization therefore tends to overestimation. Similar cases may arise in clustering. A carefully designed optimization

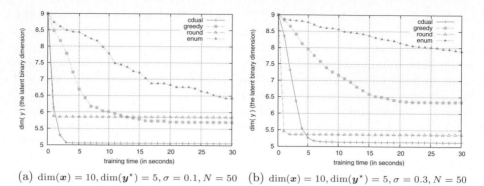

(a) $\dim(\boldsymbol{x}) = 10, \dim(\boldsymbol{y}^{\star}) = 5, \sigma = 0.1, N = 50$ (b) $\dim(\boldsymbol{x}) = 10, \dim(\boldsymbol{y}^{\star}) = 5, \sigma = 0.3, N = 50$

Fig. 3. Dimension deduction over learning time (average of 100 runs)

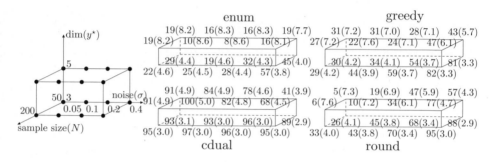

Fig. 4. "$\kappa(\overline{\dim(\boldsymbol{y})})$" out of 100 independent runs for each configuration tetrad $(\dim(\boldsymbol{x}), \dim(\boldsymbol{y}^{\star}), N, \sigma) \in \{10\} \times \{3, 5\} \times \{50, 200\} \times \{0.05, 0.1, 0.2, 0.4\}$, where κ is the number of correctly estimated $\dim(\boldsymbol{y})$ and $\overline{\dim(\boldsymbol{y})}$ is the average $\dim(\boldsymbol{y})$

algorithm which is not so accurate at the cluster boundaries may be good for model selection. As a sub-procedure, an optimization should be customized for learning instead of being isolated and implemented as accurately as possible.

Figure 4 shows the percentage κ of correctly estimated $\dim(\boldsymbol{y})$ and the average resulting $\dim(\boldsymbol{y})$ on a $2 \times 2 \times 4$ configuration grid. According to the experiments κ is sensitive to the threshold δ which is set to 2 here. Generally the performance degrades as N goes small or $\dim(\boldsymbol{y}^{\star})$ goes large. In the batch algorithm where $|\mathcal{X}_e| = N$, the resulting model scale tends to be smaller as σ increases. Therefore κ may increase with σ during overestimation, as in the case of {enum, greedy, round}. cdual is the most robust and shows the best performance in nearly all configurations. round is also good especially when σ is large. Moreover, they outperform greedy and enum considerably in computational cost.

5   Concluding Remarks

The combinatorial complexity in BFA has been avoided through a canonical dual approach and the ML training enumeration in model selection has been avoided by the BYY harmony learning. Among the algorithms investigated, `cdual` provides the best overall performance; `round` is comparably accurate on large noise and is much more efficient; `greedy` and `enum` are both inaccurate and time-consuming. A good optimization is crucial in learning. Learning, however, is not simply optimization. It includes an essential task to select a hierarchy of structures and a proper level in this hierarchy, which is difficult on small sample size and may get promoted with a customized optimization. A more detailed discussion on BFA learning and optimization will follow in subsequent works.

Acknowledgement

The work described in this paper was fully supported by a grant from the Research Grant Council of the Hong Kong SAR (Project No: CUHK4177/07E).

References

1. Keprt, A., Snásel, V.: Binary factor analysis with help of formal concepts. In: Snásel, V., Belohlávek, R. (eds.) CLA. CEUR Workshop Proceedings, CEUR-WS.org., vol. 110, pp. 90–101 (2004)
2. Taylor, G.W., Hinton, G.E., Roweis, S.T.: Modeling human motion using binary latent variables. In: Schölkopf, B., Platt, J., Hoffman, T. (eds.) NIPS, pp. 1345–1352. MIT Press, Cambridge (2007)
3. Meeds, E., Ghahramani, Z., Neal, R.M., Roweis, S.T.: Modeling dyadic data with binary latent factors. In: Schölkopf, B., Platt, J., Hoffman, T. (eds.) NIPS, pp. 977–984. MIT Press, Cambridge (2007)
4. An, Y., Hu, X., Xu, L.: A comparative investigation on model selection in binary factor analysis. Journal of Mathematical Modelling and Algorithms 5 (4), 447–473 (2006)
5. Akaike, H.: A new look at the statistical model identification. IEEE Transactions on Automatic Control 19 (6), 716–723 (1974)
6. Schwarz, G.: Estimating the Dimension of a Model. The Annals of Statistics 6 (2), 461–464 (1978)
7. Xu, L.: A unified learning scheme: Bayesian-Kullback Ying-Yang machines. In: Touretzky, D.S., Mozer, M., Hasselmo, M.E. (eds.) NIPS, pp. 444–450. MIT Press, Cambridge (1995); A preliminary version in Proc. ICONIP 1995, Beijing, pp. 977–988 (1995)
8. Xu, L.: Machine learning problems from optimization perspective. A special issue for CDGO 2007, Journal of Global Optimization (in press, 2008)
9. Gao, D.Y.: Solutions and optimality criteria to box constrained nonconvex minimization problems. Journal of Industrial and Management Optimization 3 (2), 293–304 (2007)
10. Fang, S.C., Gao, D.Y., Shue, R.L., Wu, S.Y.: Canonical Dual Approach for Solving 0-1 Quadratic Programming Problems. Journal of Industrial and Management Optimization 4 (1), 125–142 (2008)
11. Xu, L.: BYY harmony learning, independent state space and generalized APT financial analysis. IEEE Transactions on Neural Networks 12 (4), 822–849 (2001)
12. Merz, P., Freisleben, B.: Greedy and Local Search Heuristics for Unconstrained Binary Quadratic Programming. Journal of Heuristics 8 (2), 197–213 (2002)

Soft Dimension Reduction for ICA by Joint Diagonalization on the Stiefel Manifold

Fabian J. Theis[1], Thomas P. Cason[2], and P.-A. Absil[2]

[1] CMB, Institute of Bioinformatics and Systems Biology,
Helmholtz Zentrum München, Germany, and
MPI for Dynamics and Self-Organization, Göttingen, Germany
fabian.theis@helmholtz-muenchen.de
http://cmb.helmholtz-muenchen.de

[2] Department of Mathematical Engineering, Université catholique de Louvain,
B-1348 Louvain-la-Neuve, Belgium
http://www.inma.ucl.ac.be/~{cason,absil}

Abstract. Joint diagonalization for ICA is often performed on the orthogonal group after a pre-whitening step. Here we assume that we only want to extract a few sources after pre-whitening, and hence work on the Stiefel manifold of p-frames in $\mathbb{R}^n$. The resulting method does not only use second-order statistics to estimate the dimension reduction and is therefore denoted as soft dimension reduction. We employ a trust-region method for minimizing the cost function on the Stiefel manifold. Applications to a toy example and functional MRI data show a higher numerical efficiency, especially when p is much smaller than n, and more robust performance in the presence of strong noise than methods based on pre-whitening.

1 Introduction

The common approach to blind source separation of a set of multivariate data is to first whiten the data and to then search on the more restricted orthogonal group. This has two advantages: (i) instead of optimizing a cost function on n^2, the optimization takes place on a $n(n-1)/2$ dimensional manifold. (ii) During whitening via PCA, the dimension can already be reduced. However a serious drawback of this sometimes called *hard-whitening* technique is that the resulting method is biased towards the data correlation (which is used in the PCA step). Using the empirical correlation estimator, the method perfectly trusts the correlation estimate, whereas it 'mistrusts' any later sample estimates.

In contrast to hard-whitening, *soft-whitening* tries to avoid the bias towards second-order statistics. In algorithms based on joint diagonalization (JD) of a set of source conditions, reviewed for example in [1], this implies using a non-orthogonal JD algorithm [2,3,4,5]. It jointly diagonalizes both the source conditions together with the mixture covariance matrix. Then possible estimation errors in the second-order part do not influence the total error disproportionately high.

T. Adali et al. (Eds.): ICA 2009, LNCS 5441, pp. 354–361, 2009.

Soft-whitening essentially does away with issue (i). In this contribution, we propose a method to deal with issue (ii) instead: The above argument of bias towards correlation with respect to source estimation also applies to the bias with respect to dimension reduction. This can be solved by a subspace extraction algorithm followed by an ICA algorithm, see e.g. [6, 7], which however may lead to an accumulation of errors in the two-step procedure. Hence, we propose the following integrated solution implementing a *soft dimension reduction*: We will first whiten the data, so we will assume (i) and hard-whitening. However we do not reduce the data dimension beforehand. Instead we propose to search for a non-square pseudo-orthogonal matrix (Stiefel matrix) such that it minimizes the JD cost function. An efficient minimization procedure will be proposed in the following. Examples to speech and fMRI data confirm the applicability of the method. In future research a combination of (i) soft-whitening and (ii) soft dimension reduction may be attractive.

2   Joint Diagonalization on the Stiefel Manifold

Let $\mathrm{St}(p, n) = \{Y \in \mathbb{R}^{n \times p} : Y^T Y = I\}$ denote the Stiefel manifold of orthogonal p-frames in $\mathbb{R}^n$ for some $p \leq n$. The JD problem on the Stiefel manifold consists of minimizing the cost function

$$f_{\mathrm{diag}} : \mathrm{St}(p, n) \to \mathbb{R} : Y \mapsto f_{\mathrm{diag}}(Y) = -\sum_{i=1}^{N} \|\mathrm{diag}(Y^T C_i Y)\|_F^2, \tag{1}$$

where $\|\mathrm{diag}(X)\|_F^2$ returns the sum of the squared diagonal elements of X. In the context of ICA, the matrices C_i can for example be cumulant matrices (as in the JADE algorithm [8]) or time-lagged covariance matrices (as in SOBI [9]).

2.1   Diagonal Maximization versus Off-Diagonal Minimization

In the case $p = n$, minimizing f_{diag} is equivalent to minimizing the sum of the squared off-diagonal elements

$$f_{\mathrm{off}} : \mathrm{St}(p, n) \to \mathbb{R} : Y \mapsto f_{\mathrm{off}}(Y) = \sum_{i} \|\mathrm{off}(Y^T C_i Y)\|_F^2.$$

Indeed, $\|\mathrm{off}(Y^T C_i Y)\|_F^2 = \|Y^T C_i Y\|_F^2 - \|\mathrm{diag}(Y^T C_i Y)\|_F^2$ and $\|Y^T C_i Y\|_F^2$ does not depend on $Y \in \mathrm{St}(n, n) = \mathrm{O}(n)$. When $p < n$, we can still observe that if Y_* minimizes f_{diag}, then it minimizes f_{off} over $\{Y_* Q : Q \in \mathrm{O}(p)\} \subset \mathrm{St}(p, n)$; this follows from the same argument applied to the function $Q \mapsto f_{\mathrm{diag}}(Y_* Q)$. Note that minimizing f_{off} is clearly not sufficient for minimizing f_{diag}. As an illustration, consider the case $N - 1$, i.e., there is only one target matrix, C, assumed to be symmetric positive definite with distinct eigenvalues. Then the minimizers of f_{off} are all the matrices $Y \in \mathrm{St}(p, n)$ such that $Y^T C Y$ is diagonal (when $p < n$, there are infinitely many such Y), whereas the optimizers of f_{diag} are $Y_* = \begin{bmatrix} v_1 \ldots v_p \end{bmatrix} \pi$, where $v_1, \ldots, v_p$ are the p dominant eigenvectors of C and π denotes any signed permutation matrix.

2.2   A Trust-Region Method for Minimizing f_{diag}

Minimizing f_{diag} is an optimization problem over the Stiefel manifold. Recently, several methods have been proposed to tackle optimization problems on manifolds; see, e.g., [10, 11, 12] and references therein. In this paper, we use a trust-region approach, which combines favorable global and local convergence properties with a low numerical cost.

In $\mathbb{R}^n$, trust-region methods proceed as follows. At the current iterate x_k, a model m_{x_k} is chosen to approximate a cost function f. The model is "trusted" within a ball of radius Δ_k around x_k, termed the trust region. A candidate for x_{k+1} is selected as an (approximate) solution of the trust-region subproblem, namely, the minimization of m_{x_k} under the trust-region constraint. The new iterate is accepted and the trust-region radius Δ_k updated according to the agreement between the values of f and m_{x_k} at the candidate. We refer to [13] for more information.

The concept of trust-region was generalized to Riemannian manifolds in [14]. On a manifold $\mathcal{M}$, the trust-region subproblem at x_k becomes a subproblem on the tangent space $T_{x_k}\mathcal{M}$. Since the tangent space is a linear space, the classical techniques for solving trust-region subproblems apply as well. The correspondence between the tangent spaces and the manifold $\mathcal{M}$ is specified by a mapping R, called retraction, that is left to the user's discretion but for some rather lenient requirements. The retraction makes it possible to pull back the cost function f on $\mathcal{M}$ to a cost function $\hat{f}_{x_k} = f \circ R_{x_k}$ on $T_{x_k}\mathcal{M}$, where R_{x_k} denotes the restriction of R to $T_{x_k}\mathcal{M}$.

More specifically, the Riemannian trust-region method proceeds along the following steps.

1. Consider the local approximation of the pulled back cost function $\hat{f}_{x_k}$

$$m_{x_k}(\xi) = f(x_k) + \langle \xi, \operatorname{grad} f(x_k) \rangle + \frac{1}{2} \langle \xi, \operatorname{Hess} f(x_k)\,[\xi] \rangle,$$

where $\operatorname{grad} f$ and $\operatorname{Hess} f$ stand for the gradient and the Hessian of f, respectively, and $\langle .,. \rangle$ denotes the scalar product on the tangent space given by the Riemannian metric. Obtain ξ_k by (approximately) solving

$$\min_{\xi \in T_{x_k}\mathcal{M}} m_{x_k}(\xi) \quad \text{s.t. } \|\xi_k\| \le \Delta_k,$$

where Δ_k denotes the radius.

2. Evaluate the quality of the model m_{x_k} through the quotient

$$\rho_k = \frac{\hat{f}_{x_k}(0) - \hat{f}_{x_k}(\xi_k)}{m_{x_k}(0) - m_{x_k}(\xi_k)}.$$

If ρ_k is exceedingly small, then the model is very inaccurate, the trust-region radius is reduced and $x_{k+1} := x_k$. If ρ_k is small but less dramatically so, then $Y_{k+1} = R_{x_k}(\xi_k)$ but the trust-region radius is reduced. Finally, if ρ_k is close to 1, then there is good agreement between the model and the function, and the trust-region radius can be expanded.

Note that trust-region algorithms (much as steepest-descent, Newton, and conjugate gradient algorithms) are *local methods*: they efficiently exploit information from the derivatives of the cost function, but they are not guaranteed to find the global minimizer of the cost function. (This is not dishonorable, as computing the global minimizer is a very hard problem in general.) Nevertheless, they can be shown to converge *globally* (i.e., for every initial point) to stationary points of the cost function; moreover, since they are descent methods, they only converge to minimizers (local or global), except in maliciously-crafted cases. More details on the Riemannian trust-region method can be found in [14, 11]. We also refer to [15] for recent developments.

2.3   Implementing the Riemannian Trust-Region Method

A generic MATLAB code for the Riemannian trust-region method can be obtained from `http://www.scs.fsu.edu/~cbaker/GenRTR/`. The optimization method utilizes MATLAB function handles to access user-provided routines for the objective function, gradient, Hessian, retraction, etc. This allows the encapsulation of a problem into a single driver. In the remainder of this section, we describe the essential elements of the driver that we have created for minimizing f_{diag} (1).

The driver must contain a routine that returns the inner product of two vectors of $T_Y \text{St}(p, n)$, so as to specify the Riemannian structure of $\text{St}(p, n)$. We choose $\langle \xi_1, \xi_2 \rangle = \text{tr}(\xi_1^T \xi_2)$, which makes $\text{St}(p, n)$ a Riemannian submanifold of $\mathbb{R}^{n \times p}$. The retraction is chosen as

$$R_Y : T_Y \text{St}(p, n) \to \text{St}(p, n) : \xi \mapsto R_Y \xi := \text{qf}(Y + \xi)$$

where $\text{qf}(Y)$ denotes the Q factor of the QR decomposition of Y. We further need a formula for the gradient of f_{diag}. We have $\text{grad} f_{\text{diag}}(Y) = P_Y \text{grad} \hat{f}_{\text{diag}}(Y)$, where $\text{grad} \hat{f}_{\text{diag}}(Y) = -\sum_i 4 C_i Y \, \text{ddiag}(Y^T C_i Y)$ is the gradient of

$$\hat{f}_{\text{diag}} : \mathbb{R}^{n \times p} \to \mathbb{R} : Y \mapsto \hat{f}_{\text{diag}}(Y) = -\sum_i \| \text{diag}(Y^T C_i Y) \|_F^2,$$

and where P_Y denotes the orthogonal projection onto $T_Y \text{St}(p, n)$ i.e. $P_Y \xi := \xi - Y \, \text{sym}\left(Y^T \xi\right)$. Finally, the Hessian of f_{diag} is given by

$$\text{Hess} f(Y)[\xi] = \nabla_\xi \text{grad} f(Y)$$

where ∇ is the Riemaniann connection on $\mathcal{M}$ (see [11, Section 5.3.2]). In our case we choose $\nabla_\eta \xi := P_Y(\text{D} \xi(Y)[\eta])$, where Y denotes the foot of η and D the derivative. Therefore, the Hessian of f_{diag} is given by

$$\text{Hess} f_{\text{diag}}(Y)[\xi] = P_Y \text{Dgrad} \hat{f}_{\text{diag}}(Y)[\xi] - \xi \, \text{sym}(Y^T \text{grad} \hat{f}_{\text{diag}}(Y)),$$

where $\text{sym}(A) = (A + A^T)/2$ and $\text{Dgrad} f_{\text{diag}}(Y)$ is

$$\text{Dgrad} f_{\text{diag}}(Y)[\xi] = -\sum_i 4 C_i \begin{pmatrix} \xi \, \text{ddiag}(Y^T C_i Y) \\ + Y \, \text{ddiag}(\xi^T C_i Y) \\ + Y \, \text{ddiag}(Y^T C_i \xi) \end{pmatrix}.$$

3   Simulations

We propose two applications of the above 'JD Stiefel' algorithm in the following. As source conditions, we will follow the SOBI algorithm [9] and calculate lagged auto-covariance matrices.

3.1   Artificial Data

In the first example, we apply the JD Stiefel algorithm to the SOBI cost function in order to separate artificially mixed speech data. $n = 5$ speech sources with 3500 samples were chosen[1]. These were embedded using a matrix chosen with independent normal random elements into a $m = 10$-dimensional mixture space. White Gaussian noise was added with varying strength.

Algorithmically, the noisy mixture data was first whitened. Then either SOBI [9] with dimension reduction was applied or the JD Stiefel algorithm. For both algorithms, $N = 20$ lagged autocovariance matrices were calculated (with lags $2, 4, 6, ..., 40$). The trust-region method was applied with initial/maximal trust-region radius of $1/100$ respectively and maximal 100 outer iterations and additional convergence tolerance. We are interested in the performance of the algorithm when recovering parts of the mixing matrix $\mathbf{A}$. We only recover part of the data in order to simulate the situation of larger dimension than source dimension of interest. This is realized by extracting only a $n' \leq n$ dimensional subspace, either by PCA and SOBI or by the non-squared JD Stiefel algorithm.

In order to measure deviation from perfect recovery, given a projection matrix $\mathbf{W}$, we want the resulting matrix $\mathbf{WA}$ to have only one large number per row. This is measured by Amari's performance index [16] $E(\mathbf{WA})$ generalized to non-square matrices:

$$E(\mathbf{C}) = \sum_{i=1}^{n'} \left(\sum_{j=1}^{n} \frac{|c_{ij}|}{\max_k |\mathbf{c}_{ik}|} - 1 \right)$$

In figure 1(b,c), we show the results for a low and a high noise setting with signal-to-noise ratios of 32.4dB and 4.65dB, respectively. Clearly the JD Stiefel algorithm is able to take advantage of the full dimensionality of the data when searching the correct subspace, so it always considerably outperforms the SOBI algorithm, which only operates on the PCA-dimension-reduced data. Moreover, we see that even in the case of low signal-to-noise ratio (SNR), the JD Stiefel algorithm performs satisfactorily well.

3.2   Recording from Functional MRI

Functional magnetic-resonance imaging (fMRI) can be used to measure brain activity. Multiple MRI scans are taken in various functional conditions; the extracted task-related component reveals information about the task-activated

---

[1] Data set 'acspeech16' from ICALAB http://www.bsp.brain.riken.jp/ICALAB/ICALABSignalProc/benchmarks/

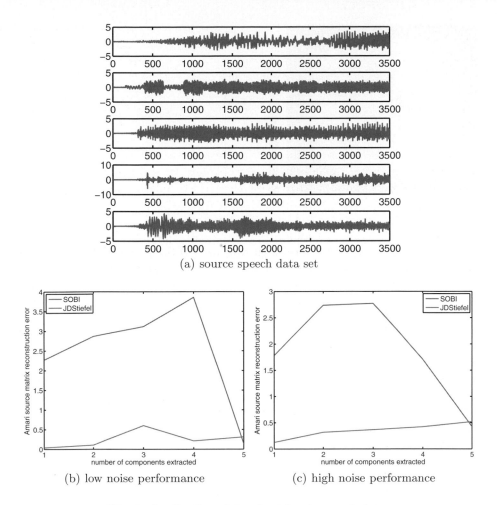

(a) source speech data set

(b) low noise performance   (c) high noise performance

Fig. 1. Application of the algorithm to speech data

brain regions. Classical power-based methods fail to blindly recover the task-related component as it is very small with respect to the total signal, usually around one percent in terms of variance. Hence we propose to use the autoco-variance structure (in this case spatial autocovariances) in combination with the soft dimension reduction to properly identify the task component.

fMRI data with 98 images (TR/TE = 3000/60 msec) were acquired with five periods of rest and five photic simulation periods with rest. Simulation and rest periods comprised 10 repetitions each, i.e. 30s. Resolution was $3 \times 3 \times 4$ mm. The slices were oriented parallel to the calcarine fissure. In order to speed up computation, we reduce the 98 slices to 10 slices by PCA. This corresponds to considering the spatial ICA model $\mathbf{x} = \mathbf{As}$, where each source s_i is a component image.

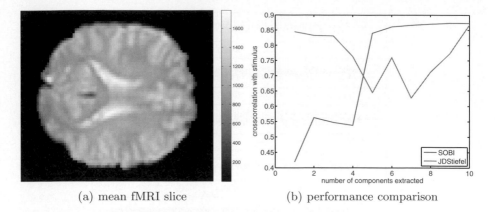

      (a) mean fMRI slice   (b) performance comparison

Fig. 2. Application of the algorithm to data acquired from functional MRI: (a) shows the temporal mean of the $128 \times 128 \times 98$ data set. (b) shows the comparison of JD Stiefel and PCA+SOBI algorithm when recovering the task-related component.

As before, in order to compare the performance of JD Stiefel versus SOBI after PCA (with $N = 100$ lagged autocovariance matrices), we analyze how well the task-related component with the known task vector $\mathbf{v} \in \{0, 1\}^{98}$ is contained in a component by the maximal crosscorrelation of all columns of $\mathbf{A}$ with $\mathbf{v}$. This is motivated by the fact that maximal preservation of the task-related component is key to any dimension reduction method.

We compare the two algorithms for dimension reductions $n' \in \{1, \ldots, 10\}$ in figure 2. The key result is that JD Stiefel already detects the main task component when reducing to only one dimension (crosscorrelation larger than 80%). SOBI is only able to find this when having access to at least 5 dimensions. Then SOBI outperforms JD Stiefel, which is prone to fall in local minima during its search. Multiple restarts and more extended searches should resolve this issue, as the cost functions coincide if $n = m$. More complex analyses of fMRI data using dimension reduction can now be approached, as generalization of e.g. [17].

4   Conclusions

Instead of reducing the dimension of the data and searching for independent components in two distinct steps, we have proposed a two-in-one approach which consists of optimizing the JD cost function on a Stiefel manifold. Numerical experiments on artificially mixed toy and fMRI data are promising. In analogy to soft-whitening, where the correlation estimate is weighed equally with respect to higher-order moments of the data, the proposed method implements a soft dimension reduction strategy, by using both second- and higher-order information of the data. In future work, we propose merging soft-whitening and soft dimension reduction.

Acknowledgements. Partial financial support by the Helmholtz Alliance on Systems Biology (project 'CoReNe') is gratefully acknowledged. This paper presents research results of the Belgian Network DYSCO (Dynamical Systems, Control, and Optimization), funded by the Interuniversity Attraction Poles Programme, initiated by the Belgian State, Science Policy Office. The scientific responsibility rests with its authors.

References

1. Theis, F., Inouye, Y.: On the use of joint diagonalization in blind signal processing. In: Proc. ISCAS 2006, Kos, Greece (2006)
2. Yeredor, A.: Non-orthogonal joint diagonalization in the leastsquares sense with application in blind source separation. IEEE Trans. Signal Processing 50(7), 1545–1553 (2002)
3. Ziehe, A., Laskov, P., Mueller, K.R., Nolte, G.: A linear least-squares algorithm for joint diagonalization. In: Proc. of ICA 2003, Nara, Japan, pp. 469–474 (2003)
4. Pham, D.: Joint approximate diagonalization of positive definite matrices. SIAM Journal on Matrix Anal. and Appl. 22(4), 1136–1152 (2001)
5. Absil, P.A., Gallivan, K.A.: Joint diagonalization on the oblique manifold for independent component analysis. In: Proceedings of the IEEE International Conference on Acoustics, Speech, and Signal Processing (ICASSP), vol. 5, pp. V–945–V–948 (2006)
6. Blanchard, G., Kawanabe, M., Sugiyama, M., Spokoiny, V., Müller, K.: In search of non-gaussian components of a high-dimensional distribution. Journal of Machine Learning Research 7, 247–282 (2006)
7. Kawanabe, M., Theis, F.: Joint low-rank approximation for extracting non-gaussian subspaces. Signal Processing (2007)
8. Cardoso, J.F.: High-order constrasts for independent component analysis. Neural Computation 11(1), 157–192 (1999)
9. Belouchrani, A., Abed-Meraim, K., Cardoso, J.F., Moulines, E.: A blind source separation technique using secton-order statistics. IEEE Trans. Signal Process. 45(2), 434–444 (1997)
10. Edelman, A., Arias, T.A., Smith, S.T.: The geometry of algorithms with orthogonality constraints. SIAM J. Matrix Anal. Appl. 20(2), 303–353 (1998)
11. Absil, P.A., Mahony, R., Sepulchre, R.: Optimization Algorithms on Matrix Manifolds. Princeton University Press, Princeton (2008)
12. Celledoni, E., Fiori, S.: Descent methods for optimization on homogeneous manifolds. Journal of Mathematics and Computers in Simulation 79(4), 1298–1323 (2008)
13. Conn, A.R., Gould, N.I.M., Toint, P.L.: Trust-Region Methods. MPS/SIAM Series on Optimization. Society for Industrial and Applied Mathematics. SIAM, Philadelphia (2000)
14. Absil, P.A., Baker, C.G., Gallivan, K.A.: Trust-region methods on Riemannian manifolds. Found. Comput. Math. 7(3), 303–330 (2007)
15. Baker, C.G., Absil, P.A., Gallivan, K.A.: An implicit trust-region method on Riemannian manifolds. IMA Journal of Numerical Analysis (to appear, 2008)
16. Amari, S., Cichocki, A., Yang, H.: A new learning algorithm for blind signal separation. Advances in Neural Information Processing Systems 8, 757–763 (1996)
17. Keck, I., Theis, F., Gruber, P., Lang, E., Specht, K., Fink, G., Tomé, A., Puntonet, C.: Automated clustering of ICA results for fMRI data analysis. In: Proc. CIMED 2005, Lisbon, Portugal, pp. 211–216 (2005)

Estimating Phase Linearity in the Frequency-Domain ICA Demixing Matrix

Keisuke Toyama[1,2] and Mark D. Plumbley[2]

[1] Sony Corporation, 1-7-1 Konan, Minato-ku, Tokyo, 108-0075, Japan
[2] Queen Mary University of London, Center for Digital Music,
Mile End Road, London, E1 4NS, United Kingdom
keisuke.toyama@jp.sony.com

Abstract. We consider a method for solving the permutation problem in blind source separation (BSS) by the frequency-domain independent component analysis (FD-ICA) by using phase linearity of FD-ICA demixing matrix. However, there is still remaining issue how we can estimate the phase linearity. In this paper, we propose two methods to estimate the linearity of the phase response of the FD-ICA demixing matrix. Our experimental result shows that our new methods can provide better estimation of the phase linearity than our previous method.

1 Introduction

Frequency-domain Independent Component Analysis (FD-ICA) [1] is a blind source separation method for convolutive mixtures, where separation using ICA is performed in the frequency domain, separately for each frequency component. However, successful use of FD-ICA involves solving a permutation problem, since the extracted sources in different frequency bins may be permuted relative to those in other frequency bins. The permutation problem can be tackled by using the direction of arrival (DOA) of each source [2,3], but this can suffer from a spatial aliasing problem above a certain frequency limit. It is possible to deal with this problem by using phase linearity. Sawada et al. have proposed a method to solve this problem by estimating mixing model parameters and fitting to an idealised direct path mixing model, introducing a linear phase assumption on the FD-ICA mixing matrix [4]. We have proposed a method which uses linearity of the phase response of the demixing matrix directly [7,8]. Nesta et al. also have proposed a similar method which uses the phase linearity based on time difference of arrival (TDOA) [6]. However, there is remaining issue how we should estimate the phase linearity. In this paper we propose new methods to estimate the phase linearity based on the DOA method or the Best Fit method.

2 BSS for Convolutive Mixtures

In the time-frequency domain, the observed signals at microphones $X_l(f,t)$ are expressed as

$$X_l(f,t) = \sum_{k=1}^{K} H_{lk}(f)S_k(f,t), \qquad l = 1, ..., L \tag{1}$$

T. Adali et al. (Eds.): ICA 2009, LNCS 5441, pp. 362–370, 2009.

where f represents frequency, t is the frame index, $H_{lk}(f)$ is the frequency response from source k to microphone l, and $S_k(f, t)$ is a time-frequency-domain representation of a source signal. Equation (1) can also be expressed as $\mathbf{X}(f, t) = \mathbf{H}(f)\mathbf{S}(f, t)$ where $\mathbf{X}(f, t) = [X_1(f, t), ..., X_L(f, t)]^T$ is the observed signal vector, $\mathbf{S}(f, t) = [S_1(f, t), ..., S_K(f, t)]^T$ is the source signal vector, and

$$\mathbf{H}(f) = \begin{bmatrix} H_{11}(f) & \cdots & H_{1K}(f) \\ \vdots & \ddots & \vdots \\ H_{L1}(f) & \cdots & H_{LK}(f) \end{bmatrix} \tag{2}$$

is the complex-valued mixing matrix.

In frequency-domain ICA, we perform signal separation from $\mathbf{X}(f, t)$ separately at each frequency f using the complex-valued demixing matrix

$$\mathbf{W}(f) = \begin{bmatrix} W_{11}(f) & \cdots & W_{1L}(f) \\ \vdots & \ddots & \vdots \\ W_{K1}(f) & \cdots & W_{KL}(f) \end{bmatrix} \tag{3}$$

which is adopted so that the reconstructed output signals $\mathbf{Y}(f, t) = [Y_1(f, t), ..., Y_K(f, t)]^T = \mathbf{W}(f)\mathbf{X}(f, t)$ become mutually independent. This can be done using any suitable ICA algorithm, such as the natural gradient approach [1]. Hereafter, we suppose we have two sources ($K = 2$) and two microphones ($L = 2$) for simplicity.

3   Permutation Problem

Since the ICA method has been applied separately at each frequency f, FD-ICA has an ambiguity in the order of the rows of $\mathbf{W}(f)$, such that permuted matrix is also the solution for FD-ICA. This problem is called as the *permutation problem* [1]–[3]. Methods designed to solve the permutation problem include the use of the amplitude correlation between adjacent frequencies [1,3], and the use of the direction of arrival (DOA) [2,3].

In the DOA method, we suppose a signal with frequency f comes from a source in the direction of θ. When the signal $\exp(j2\pi ft)$ is observed at the middle point of the microphones, the observed signals at the microphones are $X_l(f, t) = \exp(j2\pi f [t - d_l \sin(\theta_k(f))/c])$, where d_l is the position of the microphone ($d_1 = -d_2 = D/2$) and c is the speed of sound. The frequency response of the demixing process between the observed signals and the separated signals is expressed by their ratio, $Y_k(f, t)/\exp(j2\pi ft)$. Thus, we can obtain the gain of the frequency response with respect to the direction as

$$\begin{aligned} G_k(\theta_k(f)) &= |Y_k(f, t)/\exp(j2\pi ft)| \\ &= |W_{k1}(f)\exp(-j2\pi f(d_1 \sin(\theta_k(f)))/c) \\ &\quad + W_{k2}(f)\exp(-j2\pi f(d_2 \sin(\theta_k(f)))/c)|. \end{aligned} \tag{4}$$

If $f < c/2D$, the gain $G_k(\theta_k(f))$ has at most one peak and one null point in a half period of $\theta_k(f)$ where $|\theta_k(f)| \leq \pi/2$. The direction where the gain has the unique minimum value (null point) could be regarded as the direction of the unwanted source signal. Therefore, we can solve the permutation problem by comparing the direction of the two sources, $\theta_1(f)$ and $\theta_2(f)$. For more details of this process see [2,3].

However, if $f > c/2D$, the gain $G_k(\theta_k(f))$ has two or more local minimum points so that we cannot uniquely determine the magnitude relationship between $\theta_1(f)$ and $\theta_2(f)$: this problem is called the *spatial aliasing problem*. For example, if the distance between two of microphones is 4 cm and the speed of sound is 343 m/sec, the spatial aliasing problem occurs for $f > 4288$ Hz.

However, by considering phase instead of direction or delay, we can obtain a new insight into this problem allowing us to reduce the spatial aliasing problem. A recent approach [4] to solve the permutation problem with the spatial aliasing problem is to estimate phase and amplitude parameters of an estimated mixing matrix $\hat{\mathbf{A}}(f) = \mathbf{W}^{-1}(f)$ assuming an anechoic direct path model. We have also proposed a method which uses the phase parameters, but for the *demixing* matrix $\mathbf{W}(f)$ directly [7,8].

The direction of arrival $\theta_k(f)$ can be calculated as [3]:

$$\theta_k(f) = \arcsin\left(\frac{(\phi_k(f) - (2n_k(f) + 1)\pi)c}{2\pi f D}\right) \tag{5}$$

where

$$\phi_k(f) = \angle W_{k1}(f) - \angle W_{k2}(f) \tag{6}$$

and $n_k(f)$ is an arbitrary integer to be determined such that $|(\phi_k(f) - (2n_k(f) + 1)\pi)c/2\pi f D| \leq 1$ is satisfied. However, if we plot the phase difference $\phi_k(f)$ itself, this often has an approximate linearity corresponding to constant delay. Thus, the difference could be represented by the following equation:

$$\hat{\phi}_k(f) = a_k f + b_k \tag{7}$$

where $b_k = \pm\pi$ and the equation holds modulo 2π. We know $b_k = \pm\pi$ since the DC component ($f = 0$) does not have phase information so that the two signals at two microphones should have opposite sign to suppress the signal. Our proposed method utilises this linear phase property. To solve the permutation problem, we estimate the a_k in (7) and then we calculate the distance between $\phi_k(f)$ and $\hat{\phi}_k(f)$.

4   Proposed Estimation Methods

In our previous papers [7,8], we have estimated the parameters a_k and b_k by using the method of least squares on low frequency data where the spatial aliasing problem has not occurred ($f < c/2D$). However, the microphone spacing becomes wider, we can use fewer data values and the accuracy of the estimation of those parameters is likely to become worse. To prevent this problem, we propose new two methods to estimate those parameters.

4.1   Method Based on DOA

To estimate the parameter a_k by using the method of least squares, we need reliable data where the permutation problem has not occurred. In a real environment, the DOA method is the most prevalent tool for solving the permutation problem in low frequencies where the spatial aliasing problem has not occurred [2,3]. We therefore first apply the DOA method at low frequencies, then we use the phase linearity property to extend to higher frequencies by the following steps.

[**Step 1**] Solve the permutation problem by using the DOA method in low frequencies where the spatial aliasing problem has not occurred. Set initial loop counter $l := 1$.

[**Step 2**] Estimate a_k in (7) by using the method of least squares, as

$$a_k = \frac{\sum_{f \in \mathcal{F}} f \phi_k(f) - b_k \sum_{f \in \mathcal{F}} f}{\sum_{f \in \mathcal{F}} f^2} \tag{8}$$

$$b_k = \begin{cases} \pi & \text{if } \sum_{f \in \mathcal{F}_{low}} \phi_k(f) > 0 \\ -\pi & \text{otherwise} \end{cases} \tag{9}$$

where $\mathcal{F} = \{f : f_{low} \leq f \leq f_{high}^{(l)}\}$, $\mathcal{F}_{low} = \{f : f_{low} \leq f \leq c/2D\}$, and the frequencies f_{low} and $f_{high}^{(l)}$ are the low and high limits of the frequency range used to estimate a_k. For example, f_{low} is chosen to avoid the effect of low frequencies such as bins 5–20 [3]. $f_{high}^{(l)}$ is calculated at the Step 8. For the first loop, $f_{high}^{(1)} = c/2D$.

[**Step 3**] Estimate the lines $\hat{\phi}_k(f)$ (equation (7)).

[**Step 4**] Wrap the values of $\phi_k(f)$ and $\hat{\phi}_k(f)$ into $-\pi$ to π.

[**Step 5**] Calculate the distance $D_{prop}(f)$ between $\phi_k(f)$ and $\hat{\phi}_k(f)$ using

$$D_{prop}(f) = [E_{11}(f) + E_{22}(f)] - [E_{12}(f) + E_{21}(f)] \tag{10}$$

where

$$E_{ij}(f) = \begin{cases} |\phi_i - \hat{\phi}_j| & \text{if } |\phi_i - \hat{\phi}_j| < \pi \\ 2\pi - |\phi_i - \hat{\phi}_j| & \text{otherwise.} \end{cases} \tag{11}$$

[**Step 6**] Solve the permutation problem by using $D_{prop}(f)$. If $D_{prop}(f) < 0$, consider that a permutation has occurred at the frequency f, whereas if $D_{prop}(f) > 0$, a permutation has not occurred at the frequency f.

[**Step 7**] Calculate the set of "phase wrapping" frequencies $\mathcal{F}_{wrap}$ as

$$\mathcal{F}_{wrap} = \{0 < f \leq f_{max} \cdot \hat{\psi}_k(f) - \pm(2n \mp 1)\pi, \quad k - 1, 2\} \tag{12}$$

where f_{max} is the Nyquist frequency. If no wrapping frequencies exist, $\mathcal{F}_{wrap}$ is a null set.

[**Step 8**] Make the set of the high limit frequencies to estimate the a_k as

$$\{f_{high}^{(l)}\} = \{c/2D\} \cup \mathcal{F}_{wrap} \cup \{f_{\max}\}. \tag{13}$$

The $(l+1)$-th smaller number is used as $f_{high}^{(l)}$ at the Step 2 in the next loop.

[**Step 9**] Unwrap the values of ϕ_k as

$$\phi_k(f) = \phi_k(f) + \text{sign}(a_k)2\pi m_k \tag{14}$$

where m_k is chosen to keep the line continuous.

[**Step 10**] Increase the loop counter $l := l+1$ and update the set of frequencies $\mathcal{F}$ as $\mathcal{F} = \{f : f_{low} \leq f \leq f_{high}^{(l)}\}$, and then repeat from the Step 2. In the final loop when $f_{high}^{(l)} = f_{\max}$, do the Step 2–6 and then stop.

4.2   A "Best Fit" Method

The method which is described in Section 4.1 needs the observed data for which the permutation problem has already been solved using e.g. the DOA method. Thus, the performance of the estimation of the parameters a_k depends on the performance of the initial permutation solver. Here, we propose a new method which does not require these conditions.

To estimate the a_k, we calculate the distance between ϕ_k and assumed linear curves calculated by using all possibility of the value of a_k. We assume the linear curve $\tilde{\phi}(a, f)$ as

$$\tilde{\phi}(a, f) = af + b \tag{15}$$

$$b = \begin{cases} \pi & \text{if } a < 0 \\ -\pi & \text{otherwise} \end{cases} \tag{16}$$

where a is assumed value and the range of the value depends on the location of sources and microphone. Here, we use $\{a : -0.1 < a < 0.1\}$. Next, we calculate the distance $D_{error}(a)$ between $\tilde{\phi}(a, f)$ and $\phi_k(f)$ for all values of a.

$$D_{error}(a) = \sum_f \min(|\tilde{\phi}(a, f) - \phi_1(f)|, |\tilde{\phi}(a, f) - \phi_2(f)|) \tag{17}$$

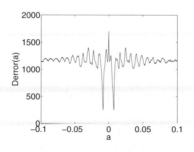

Fig. 1. a and $D_{error}(a)$

The distance $D_{error}(a)$ has two local minimum values (see the Figure 1). These local minimum values should be the estimated value of a_k. To solve the permutation problem, we calculate the distance between $\phi_k(f)$ and $\hat{\phi}_k(f)$ as same as the Step 5 and 6 in Section 4.1.

5   Experiments

To confirm our methods, we performed experiments to separate two speech signals (5 sec of speech at 44.1 kHz) mixed using impulse responses of an anechoic room ($T_{60} = 0$ msec), an echo room ($T_{60} = 300$ msec), a Japanese Tatami floored room ($T_{60} = 600$ msec), and a conference room ($T_{60} = 780$ msec). These impulse responses are supplied by RWCP database. The distance between the two microphones is 2.83 cm, so spatial aliasing would begin at 6060 Hz. For the FD-ICA part, we adopt 2048 as the length of FFT window, and run for 300 iterations. In these experiments, we compared the performance of our methods (method A described in [7], method B described in Section 4.1, and method C described in Section 4.2) to the inter-frequency correlation method [1,3] and the DOA method [2,3]. For the proposed methods, we used 5 as the lowest frequency bin number f_{low}.

Here, we define "correct" permutation data to evaluate the performance of the inter-frequency correlation method and our proposed methods. The "correct" data are obtained by the correlation between the input signal $U_{lk}(f,t) = H_{lk}(f)S_k(f,t)$ observed at microphone and the separated signal $Z_{lk}(f,t) = W_{lk}^{-1}(f)Y_k(f,t)$ which is projected to the microphone by the inverse matrix of the demixing matrix at each frequency [8].

Table 1. Error rate obtained with the inter-frequency correlation method, the DOA method, and the proposed methods

Error rate [%]	Inter-freq. correlation	DOA	Proposed A	Proposed B	Proposed C
Anechoic Room	30.15	0.88	3.41	8.78	8.39
Echoic Room A	6.83	60.59	15.12	4.59	6.44
Japanese Tatami Room	28.78	12.68	41.95	8.68	7.02
Conference Room	34.93	54.44	17.56	9.17	6.63

Table 2. Comparison of average SIR [5] obtained with the inter-frequency correlation method, the DOA method, and the proposed methods. All values are expressed in decibels (dB).

SIR [dB]	Inter-freq. correlation	DOA	Proposed A	Proposed B	Proposed C
Anechoic Room	23.860	22.478	32.725	32.725	32.542
Echoic Room A	17.538	16.037	17.543	17.543	17.549
Japanese Tatami Room	2.346	3.411	3.451	3.451	3.455
Conference Room	0.191	3.234	3.378	3.378	3.406

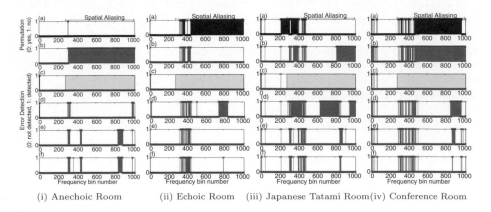

(i) Anechoic Room   (ii) Echoic Room   (iii) Japanese Tatami Room(iv) Conference Room

Fig. 2. Detection of permutation in (i) the anechoic room, (ii) the echoic room, (iii) the Japanese Tatami room, (iv) the conference room; (a) correct detection, (b) detection errors in the inter-frequency correlation method, (c) detection errors in the DOA method, (d) detection errors in proposed method A, (e) detection errors in proposed method B, (f) detection errors in proposed method C

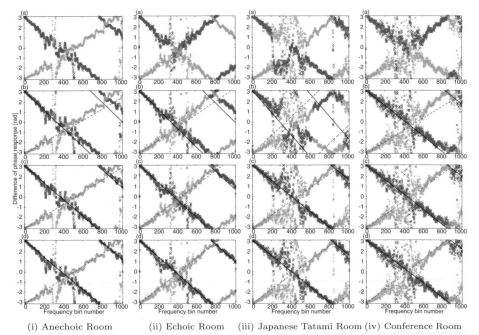

(i) Anechoic Room   (ii) Echoic Room   (iii) Japanese Tatami Room (iv) Conference Room

Fig. 3. The demixing matrix phase response in the same rooms as Fig. 2, showing observed points ϕ_1 ('×'), ϕ_2 ('+'), and estimated lines $\hat{\phi}_1$ (solid line), $\hat{\phi}_2$ (dashed line) from equations (6) and (7). In each environment, (a) shows ϕ_k before solving the permutation problem, (b)–(d) show ϕ_k and $\hat{\phi}_k$ after solving the permutation using method A (b), method B (c), and method C (d).

The results are shown in Figures 2 and 3, and Tables 1 and 2. The methods B and C can solve the permutation better than the inter-frequency correlation method and the DOA method in all the environments. It can be seen from the Figure 3, the estimation of phase linearity is better in methods B and C than for method A which uses only low frequency data to estimate the parameter a_k, especially in the Japanese Tatami Room. It seems that the estimation of the linear curve of method C is slightly better than that of method B. Because the method B relies on the performance of previous permutation solver in each loop for estimating the linear curve, so that the estimation error would be accumulated. On the other hand, method C estimates the linear curve directly from the observed data. Thus, the method C does not suffer from such error essentially. However, if we do not know the direction of sources, we have to calculate all direction for estimating the linear curve.

6   Conclusion

We have proposed two methods which estimate the linearity of the phase response of the demixing matrix to solve the permutation problem. While the permutation errors for methods B and C are significantly reduced from method A, the SIR (Table 2) is only slightly reduced, probably due to most of the energy in these signals being in the lower frequency regions. The proposed methods can estimate the phase linearity better than inter-frequency correlation method, the DOA method, and our previous method in [7,8] especially in long reverberant environments. In future work, we plan to compare the performance of our methods with that of Sawada's method [4] and Nesta's method [6].

Acknowledgement

We used the RWCP Sound Scene Database in Real Acoustical Environments. We would like to thank associate professor Takanobu Nishiura of Ritsumeikan University for giving us useful comments about the database. Mark D. Plumbley is supported by an EPSRC Leadership Fellowship (EP/G007144/1).

References

1. Smaragdis, P.: Blind separation of convolved mixtures in the frequency domain. Neurocomputing 22, 21–34 (1998)
2. Kurita, S., et al.: Evaluation of blind signal separation method using directivity pattern under reverberant conditions. In: Proc. ICASSP 2000, vol. 5, pp. 3140–3143 (2000)
3. Sawada, H., et al.: A robust and precise method for solving the permutation problem of frequency-domain blind source separation. IEEE Transaction. Speech and Audio Proc. 12, 530–538 (2004)
4. Sawada, H., et al.: Grouping separated frequency components by estimating propagation model parameters in frequency-domain blind source separation. IEEE Trans. Audio, Speech, and Language Proc. 15, 1592–1604 (2007)

5. Vincent, E., et al.: Performance measurement in blind audio source separation. IEEE Trans. on Audio, Speech and Language Proc. 14, 1462–1469 (2006)
6. Nesta, F., et al.: A Novel Robust Solution to the Permutation Problem Based on a Joint Multiple TDOA Estimation. In: Proc. IWAENC 2008, paper ID 9023 (2008)
7. Toyama, K., et al.: Using phase linearity to tackle the permutation problem in audio source separation. In: Proc. IWAENC 2008, paper ID 9040 (2008)
8. Toyama, K., et al.: Solving the permutation problem using phase linearity and frequency correlation. In: Proc. ICArn International Workshop 2008, pp. 48–51 (2008)

Nonlinear Blind Source Deconvolution Using Recurrent Prediction-Error Filters and an Artificial Immune System

Cristina Wada[1], Douglas M. Consolaro[1], Rafael Ferrari[2], Ricardo Suyama[2], Romis Attux[1], and Fernando J. Von Zuben[1]

[1] Department of Computer Engineering and Industrial Automation (DCA)
[2] Department of Microwave and Optics (DMO)
School of Electrical and Computer Engineering, University of Campinas (UNICAMP)
Caixa Postal 6101, CEP 13083-970, Campinas – SP – Brazil
{criswada,attux,vonzuben}@dca.fee.unicamp.br,
{rferrari,rsuyama}@dmo.fee.unicamp.br

Abstract. In this work, we propose a general framework for nonlinear prediction-based blind source deconvolution that employs recurrent structures (multi-layer perceptrons and an echo state network) and an immune-inspired optimization tool. The paradigm is tested under different channel models and, in all cases, the presence of feedback loops is shown to be a relevant factor in terms of performance. These results open interesting perspectives for dealing with classical problems such as equalization and blind source separation.

Keywords: Unsupervised Deconvolution, Nonlinear Prediction, Recurrent Neural Networks, Echo State Networks, Artificial Immune Systems.

1 Introduction

The problem of unsupervised (or blind) deconvolution plays a key role in research fields as diverse as digital communications and seismology, and, in a certain sense, may be considered to possess an additional interest from theoretical and practical standpoints due to its relationship with the blind source separation (BSS) problem in its convolutive and undetermined versions [1].

In the classical blind equalization theory, it is known that the problem can be solved using second-order statistics to choose the parameters of linear prediction-error filters (PEFs) [2]. Nevertheless, such approach cannot be adopted when the channel is a mixed phase system: this is indeed a case wherein the difference between uncorrelated and independent signals is essential (analogously to the classical PCA/ICA dilemma [3]). Interestingly, it was shown in [4] and [5] that such limitations are not intrinsically related to the criterion of minimum mean squared prediction error, but to the choice of a linear structure. Through the use of nonlinear prediction techniques, such as neural networks and fuzzy filters, it becomes possible to deconvolve mixed phase channels. This can be explained from two different perspectives: (a) the use of a nonlinear structure increases the flexibility available in

T. Adali et al. (Eds.): ICA 2009, LNCS 5441, pp. 371–378, 2009.

the process of generating adequate input-output mappings; and (b) a nonlinear structure is capable of originating higher-order statistics even if a mean squared error (MSE) criterion is employed (like in nonlinear PCA [3]).

In this work, we propose two extensions of the current state of the nonlinear prediction blind deconvolution paradigm: the use of recurrent neural networks (more specifically, recurrent versions of the multilayer perceptron (MLP) and an echo state network) and of an evolutionary optimization method to choose the parameters of the MLPs (even in the feedforward case). These proposals are analyzed throughout the paper and they are tested under a representative set of channels. The obtained results allow us to reach conclusions that provide a more complete view of the approach and also of its potential implications to blind signal processing tasks.

2   Nonlinear Prediction and Unsupervised Deconvolution

Let $\mathbf{s}(n) = [s(n)\ s(n-1)\ ...\ s(n-K+1)]^T$ be a vector of independent and identically distributed (i.i.d.) samples of an information source and $\mathbf{h}(n) = [h_0\ h_1\ ...\ h_{K-1}]^T$ be the vector representing the channel impulse response. The channel output $x(n)$ is then given by:

$$x(n) = \mathbf{h}(n)^T \mathbf{s}(n) = h_0 s(n) + ... + h_{K-1} s(n - K + 1).$$ (1)

The PEF (in this initial explanation, for the sake of clarity, we consider it to be a feedforward structure) has an input vector $\mathbf{x}(n-1) = [x(n-1)\ x(n-2)\ ...\ x(n-L)]^T$, where L is the number of filter inputs, and attempts to estimate the signal $x(n)$ by minimizing a mean squared error (MSE) cost function:

$$J_{MSE} = E[e(n)^2] = E[(x(n) - \hat{x}(n))^2].$$ (2)

Since the predictor input is the vector $\mathbf{x}(n-1)$, the estimate $\hat{x}(n)$ must be, in accordance with (1), a function of a certain number of past values of the source, $\hat{x}(n) = f(\mathbf{x}(n-1)) = g(s(n-1), s(n-2), ..., s(n-L))$, which means that $\hat{x}(n)$ is independent of $s(n)$. This fact indicates that the best estimate provided by the predictor would be:

$$\hat{x}(n) = h_1 s(n-1) + h_2 s(n-2) + ... + h_{K-1} s(n-K+1),$$ (3)

and, in this situation, the prediction error in the instant n would be $h_0 s(n)$, showing that, in the ideal scenario, the PEF would be able to deconvolve the signal. Although these ideas reveal the potential of solving the deconvolution problem using PEFs, it is well-known that linear predictors are not applicable to the deconvolution of mixed phase systems. In an ideal case, the predictor must be flexible enough to provide a mapping that effectively produces an estimate as close as possible to (3): this fact requires the use of nonlinear predictors.

So in this work, we shall employ three structures: (i) a feedforward MLP, which will be a reference of the standard performance of the nonlinear prediction paradigm; (ii) an MLP with feedback loops, which may be decisive in some cases (e.g. when there are close or coincident channel states [6]); and (iii) an echo state network, which has, in a certain extent, the potential of dynamical processing without a significantly more complex training procedure. Another contribution of this work is an investigation of the applicability of an artificial immune system (AIS) in the adaptation of the connectionist models in (i) and (ii). These efforts lead to a more general formulation of the nonlinear prediction approach, and this is, in fact, our main objective. It is important to remark that this aim, if fulfilled, would have implications beyond the blind deconvolution problem per se, for, as shown in [7], the nonlinear prediction framework can also be applied in a BSS context.

3 MLP-Based Prediction-Error Filters

As stated above, we will employ both a feedforward [8] and a recurrent version of the MLP (which shall be referred to as a *recurrent MLP*). Since we deal with a prediction task, the input layer of a feedforward MLP is composed of the input vector $\mathbf{x}(n\text{-}1)$ and of a bias signal equal to one. We consider networks with a single hidden layer, being the signal $\hat{x}(n)$ estimated by the predictor given by:

$$\hat{x}(n) = \mathbf{w}_o^T \mathbf{u} = \sum_{i=1}^{M} w_{o,i} \tanh\left(w_{i,0} + \sum_{k=1}^{L} w_{i,k} x(n-k) \right), \tag{4}$$

where $w_{o,i}$ represents the synaptic weights of the output neuron, $w_{i,k}$ represents those of the *i-th* neuron of the hidden layer and the vector $\mathbf{u}(n)$ is composed of the outputs of the M neurons of the hidden layer. In (4), we have a predictor with an autoregressive (AR) character.

The structure of the proposed recurrent MLP is similar: however, in the input vector $\mathbf{x}_R(n\text{-}1)$, there are past values of $e(n)$:

$$\mathbf{x}_R(n-1) = \left[x(n-1), \quad ..., \quad x(n-L-r), \quad e(n-1), \quad ..., \quad e(n-r) \right], \tag{5}$$

Applying this vector in the structure of the recurrent MLP, we obtain:

$$\hat{x}(n) = \sum_{i=1}^{M} w_{o,i} \tanh\left(w_{i,0} + \sum_{k=1}^{L-r} w_{i,k} x(n-k) + \sum_{k=L-r+1}^{L} w_{i,k} e(n-k-L+r) \right). \tag{6}$$

A comparison with (4) reveals the addition of a moving-average (MA) parcel to the model. Thus, the recurrent MLP can be conceived in terms of a nonlinear autoregressive moving average (NARMA) model.

In the training of an MLP network, it is well-known that the use of adaptive algorithms built from the derivatives of the cost function is subject to local convergence. In addition to that, in the case of recurrent MLPs, the process is also made more complex due to the constant menace that an unstable solution be reached and to difficulties in carrying out the calculation of derivatives [9]. Having this in mind, we propose the use of an artificial immune system (AIS) [10] that do not require the manipulation of the derivatives of the cost function and are robust to the convergence to poor local minima due to their populational character.

4 Artificial Immune Systems

Artificial immune systems (AISs) are algorithms inspired by the behavior of certain organisms in the presence of antigens – molecules which lead to an immune response [10]. In this work, we have adopted a model based on the clonal selection principle, a version of the CLONALG, first proposed in [11].

According to the clonal selection principle, when pathogens invade the body, the defense cells suffer an intense replication. During cell division, mutations occur in an inverse proportion to the degree of recognition of the pathogens. The higher the affinity relationship (fitness), the lower the mutation suffered by the cells, and vice-versa. Through this mechanism, the cells become more efficient and recognize more quickly a future invasion by the same or by a similar pathogen [10].

In analogy with the immune system, in the adopted AIS each defense cell is represented by a vector whose elements are the parameters of the neural network, while the affinity between the cell and the invader is related to the cost function to be optimized. Since the algorithm is suited to maximization problems and we want to minimize the mean squared prediction error, the fitness is proportional to the inverse of J_{MSE}. The algorithm is summarized below:

1. Randomly initialize a population of N_A cells;
2. Compute the fitness of each cell: $J_{FIT} = (1+J_{MSE})^{-1}$;
3. **While** the *n-th* generation is not attained **do**
   3.1. Create N_C clones for each cell;
   3.2. Keep the original cell and apply a mutation process to each clone following the equation:

$$c' = c + \alpha Y(0,1) \tag{7}$$

$$\alpha = \frac{1}{\beta} \exp(-J_{FIT}) \tag{8}$$

   where c' is the modified clone, $Y(0,1)$ is a random Gaussian variable with zero mean and unit variance and β is the control parameter.
   3.3. Keep in the population only the best solution of each group formed by the individual and its clones;
   3.4. Determine the best individual, which has the highest fitness, among the groups;
   3.5. After each *t* iterations, eliminate *m* elements of the population with the lowest values of fitness and replace them with randomly generated individuals.
4. **End while**

The mutation process is represented in step 3.2, in which the copies of the parameter vector suffer a random modification proportional to the value of α. In this stage, as well as in 3.3 and 3.4, the algorithm establishes an efficient local search mechanism, although there is also a global search potential in the mutation. In 3.5, on the other hand, there is a determinant increase in the global search potential because of the insertion of random individuals in the population, allowing the exploration of novel portions of the search space.

5 MLP Simulation Results

In order to test the proposed technique, we will perform simulations using the MLPs in three scenarios, always with discrete sources formed by i.i.d. binary $\{+1;-1\}$ samples and without noise. The channel models are: $H_1 = 1 - 1z^{-2}$, $H_2 = 0.38 + 0.60z^{-1} + 0.60\,z^{-2} + 0.38z^{-3}$ and $H_3 = -0.8221 + 0.4521z^{-1} + 0.3288z^{-2} + 0.1079z^{-3}$, and, in the simulations, their impulse responses were, without loss of generality, normalized to engender a signal with unit variance. In the first and second channels, there are coincident states [6]: the presence of such states means that the channels cannot be properly equalized by any feedforward filter. In particular, the second channel was chosen also because it has a relatively long impulse response and is proven to be the worst channel of four coefficients in the bit error rate sense [12]. The third channel has no coincident states, but the channel has one zero near the unit circle, which make the task of deconvolution more difficult. Although the third channel can be equalized with a feedforward structure, it is valid to expect that the problem will be simplified by the information brought by the presence of a feedback loop.

In Table 1, we present the values of the relevant parameters of the feedforward and recurrent MLP, and as well as those of the AIS. In order to test the performance of the nonlinear PEF, we divided the available time series into training and test sets with N_{train} and N_{test} samples, respectively. The MSEs will always be those associated with the test phase. The mutation parameter β is increased in some moments when the algorithm reaches a certain generation (indicate in Table 1 by *gen*). This allows the search performed by each individual to become more and more local. All filter and algorithm parameters were chosen with the aid of several preliminary simulations, having always in mind the idea of looking for solutions as efficient and parsimonious as possible.

Table 1. Simulations parameters

	Feedforward MLP			Recurrent MLP		
	H_1	H_2	H_3	H_1	H_2	H_3
M	2	2	8	2	2	2
L	4	2	2	4 (r=2)	5 (r=4)	6 (r=4)
N_{train}	2000	2000	2000	2000	2000	2000
N_{test}	1000	1000	1000	1000	1000	1000
N_A	20	20	25	20	20	20
N_C	4	4	4	4	4	4
gen	60/150	100/200	80/150	80/200	100/200/300	80/200
β	10/50	10/50	10/50	10/50	10/20/50	10/50
T	30	40	40	40	30	40
M	6	7	10	6	6	6

In Table 2, we present the best, worst and average MSE obtained from a set of 10 simulations. A first analysis reveals that the recurrent MLP leads to the best results in all scenarios, always with a number of inputs that is smaller than that of the feedforward MLP. This indicates that the moving average character played a decisive role in the process of equalizing this class of "difficult channels". As a matter of fact, for H_1 and H_2, the presence of coincident states is responsible for the expressive MSE values originated

Table 2. Mean square prediction error

canal	MLP	MSE min	MSE mean	MSE max
H_1	forward	0.1247	0.1335	0.1443
	recurrent	0.0036	0.0058	0.0087
H_2	forward	0.0833	0.0857	0.0895
	recurrent	0.0013	0.0046	0.0082
H_3	forward	0.0260	0.0523	0.0631
	recurrent	0.0033	0.0116	0.0244

by the feedforward structure. As we can see in Figure 1 the "open eye" condition is not attained by the feedforward MLP for the channel H_2 in the minimum MSE realization, while this difficulty is completely overcome by the presence of feedback loops.

The results obtained under the third channel model also favor the performance reached with the recurrent MLP. In this case, there are no coincident states, but, on the other hand, the existence of close states demands a complex mapping that is far from being easily built using a feedforward structure. Such facts are in consonance with the widespread notion that feedback loops are useful when dealing, for instance, with long impulse response channels [13].

Finally, the performance level shown in Table 2 and the relatively uniform distribution verified for the results give strong support to the conclusion that the chosen AIS was able to make an efficient use of the approximation potential brought by the structures. In Figure 2, we have, for a typical trial involving a recurrent MLP and channel H_2, the time evolution of the maximum fitness (solid line) and of the average fitness (dashed line) of the population. The curve of the maximum fitness shows that the technique quickly found a promising solution, whereas the curve of average fitness and its distance from the solid line reveal that a reasonable level of population diversity is maintained. In summary, we conclude that the algorithm achieved its aims, searching for good solutions and avoiding, in recurrent structures, the menace of instability and the difficulties related to cost function manipulation.

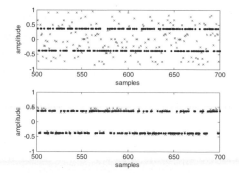

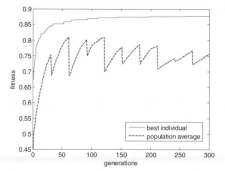

Fig. 1. Best performance of the feedforward (top) and the recurrent (bottom) MLP for H_2. The symbol 'x' corresponds to the prediction error, while '.' to the source signal.

Fig. 2. Maximum fitness and average fitness for recurrent MLPs and channel H_2

6 On the Applicability of Echo State Networks

Echo state networks (ESNs), first proposed in [14], are nonlinear recurrent filtering devices that establish a tradeoff between the dynamical processing capability of feedback structures and the relative simplicity of the training process associated with feedforward structures. The ESN structure we will be concerned with in this work is formed by a layer of fully interconnected neurons (perceptrons with hyperbolic tangent activation functions) followed by a linear output layer. This means that the behavior of the ESN can be described by a pair of equations:

$$\mathbf{u}(n) = \tanh\left[\mathbf{W}^{in}\mathbf{x}(n) + \mathbf{W}\mathbf{u}(n-1)\right]$$
$$y(n) = \mathbf{w}_{out}^{T}\mathbf{u}(n)$$

$$(9)$$

where $\mathbf{W}^{in}$ represents the matrix of the input weights, $\mathbf{W}$ is the matrix of weights of the recurrent layer neurons and $\mathbf{w}_{out}$ is the vector of weights of the output units.

The outputs of the recurrent layer neurons (also known as dynamical reservoir) are called *echo states*. The weights $\mathbf{W}^{in}$ and $\mathbf{W}$ are chosen without the influence of the desired signal: the choice is guided solely by the notion of establishing a repertoire of dynamical patterns that be as diversified as possible in order to facilitate the task of the output layer (which is trained using the desired signal under a classical linear regression framework). Naturally, being the reservoir thus designed, the ESN will not make use of the full potential of the underlying filtering structure. On the other hand, its training process is much simpler than that of a standard recurrent network.

In this work, the study involving the echo state network can be considered preliminary, and will gravitate around a single question: are the feedback loops present in the fixed reservoir enough to ensure the benefits brought by the use of a recurrent predictor? In order to answer that question, we will consider a channel with a zero at −1, which generates coincident states. This channel cannot be properly deconvolved with any feedforward structure [6], which means that, if the ESN be able to equalize it, it can be concluded that its recurrent character made the difference.

We trained an echo state network to play the role of predictor using the methodology discussed in [15]. After some preliminary results, we chose the number of neurons in the reservoir to be equal to 10, the spectral radius to be equal to 0.95 and the matrix $\mathbf{W}^{in}$ to be composed of equiprobable $\{+1;-1\}$ values. Ten simulations were performed with a single-input predictor, and the training and test MSE values were 8.3×10^{-3} and 7.7×10^{-3}, respectively. These values when compared to the MSE of the optimal feedforward predictor [5], which in this case is equal to 0.5, reveal that the ESN played with success the role of nonlinear recurrent predictor in the spirit of the proposal that forms the essence of this work. This certainly calls for a detailed future investigation of the performance of this interesting signal processing solution under a more diversified set of channel models.

7 Conclusions

In this paper, we have proposed a new approach to the problem of nonlinear prediction-based blind source deconvolution that employs recurrent neural networks and an artificial immune system. In all studied scenarios, the presence of feedback loops improve the performance, in particular, when the analyzed channels had coincident states, since the feedforward structures are unable to properly deal with them. The performance of the AIS was also quite satisfactory in terms of MSE, robustness to unstable solutions and global search potential/population diversity. It should also be noticed that the echo state network reached a very satisfactory performance using a training process simpler than the AIS.

As perspectives for future work, we may indicate: extension to cases involving continuous sources, a comparison between the AIS and other optimization tools, and an analysis of the potential of ESN in blind and supervised deconvolution tasks.

References

1. Haykin, S. (ed.): Unsupervised Adaptive Filtering, vol. 2. Wiley, Chichester (2000)
2. Haykin, S.: Adaptive Filter Theory. Prentice Hall, Englewood Cliffs (1997)
3. Hyvärinen, A., Karhunen, J., Oja, E.: Independent Component Analysis. Wiley, Chichester (2001)
4. Cavalcante, C.C., Montalvão Filho, J.R., Dorizzi, B., Mota, J.C.M.: A Neural Predictor for Blind Equalization in Digital Communication. In: Proc. of AS-SPCC, Lake Louise, Canada (2000)
5. Ferrari, R., Suyama, R., Lopes, R.R., Attux, R.R.F., Romano, J.M.T.: An Optimal MMSE Fuzzy Predictor for SISO and MIMO Blind Equalization. In: Proceedings of the IAPR Workshop on Cognitive Information Processing (CIP), Santorini, Greece (2008)
6. Montalvão Filho, J., Dorizzi, B., Mota, J.C.M.: Some Theoretical Limits of Efficiency of Linear and Nonlinear Equalizers. Journal of Communication and Information Systems 14(2), 85–92 (1999)
7. Suyama, R., Duarte, L.T., Ferrari, R., Rangel, L.E.P., Attux, R.R.F., Cavalcante, C.C., Von Zuben, F.J., Romano, J.M.T.: A Nonlinear Prediction Approach to the Blind Separation of Convolutive Mixtures. EURASIP Journal on Applied Signal Processing 2007 (2007)
8. Haykin, S.: Neural Networks: A Comprehensive Foundation. MacMillan, Basingstoke (1994)
9. dos Santos, E.P., Von Zuben, F.J.: Efficient Second-Order Learning Algorithms for Discrete-Time Recurrent Networks. In: Recurrent Neural Networks: Design and Applications. CRC Press, Boca Raton (2000)
10. Castro, L.N.: Fundamentals of Natural Computing: Basic Concepts, Algorithms and Applications. Chapman & Hall, Boca Raton (2006)
11. Castro, L.N., Von Zuben, F.J.: Learning and Optimization Using the Clonal Selection Principle. IEEE Trans. on Evolutionary Computation 6(3), 239–251 (2002)
12. Proakis, J.G.: Digital Communications, 4th edn. McGraw-Hill, New York (2001)
13. Shynk, J.J.: Adaptive IIR filtering. ASSP Magazine, IEEE 6(2), 4–21 (1989)
14. Jaeger, H.: The Echo State Approach to Approach to Analyzing and Training Recurrent Neural Networks (Tech. Rep. n° 148). German National Research Center for Information Technology, Bremen (2001)
15. Ozturk, M.C., Xu, D., Principe, J.C.: Analysis and Design of Echo State Networks. Neural Computation 19, 111–138 (2007)

Arabica: Robust ICA in a Pipeline

Jarkko Ylipaavalniemi and Jyri Soppela

Adaptive Informatics Research Centre,
Department of Information and Computer Science,
Helsinki University of Technology, P.O. Box 5400, FI-02015 TKK, Finland
jarkko.ylipaavalniemi@tkk.fi
http://launchpad.net/arabica

Abstract. We introduce a new robust independent component analysis (ICA) toolbox for neuroinformatics, called "Arabica". The toolbox is designed to be modular and extendable also to other types of data. The robust ICA is the result of extensive research on reliable application of ICA to real-world measurements. The approach itself is based on sampling component estimates from multiple runs of ICA using bootstrapping. The toolbox is fully integrated to a recently developed processing pipeline environment, capable of running on a single machine or in a cluster of servers. Additionally, the toolbox works as a standalone package in Matlab, when the full pipeline is not required. The toolbox is aimed at being useful for both machine learning and neuroinformatics researchers.

Keywords: Toolbox, pipeline, bootstrapping, ICA, fMRI.

1 Introduction

Biomedical signal analysis has been one of the most successful application domains for independent component analysis (ICA) (see, *e.g.*, [1]). In recent years, the neuroscience and neuroinformatics communities have acknowledged the great potential of such exploratory analysis (see, *e.g.*, [2]), particularly under measurement conditions where it is difficult to make valid assumptions or predict the outcome (see, *e.g.*, [3,4,5]). Therefore, researchers in these fields are looking for ways to incorporate ICA into their existing analysis frameworks.

Another field with promising recent development has been building processing pipeline environments that would offer the capabilities of running algorithms and handling data in a cluster of servers, or a grid. These are critical capabilities in modern infrastructures, where large databases can be hosted at a certain facility, algorithms developed in another facility, and possibly a third facility providing computing resources.

However, incorporating ICA seamlessly into existing frameworks is not the only major obstacle. Data-driven analysis of real-world signals as part of bigger analysis pipelines needs to be reliable, both in the sense that the results must be reproducible simply by re-running the whole process, and that analysis of new data can be trusted.

T. Adali et al. (Eds.): ICA 2009, LNCS 5441, pp. 379–386, 2009.

The motivation for our new toolbox is to provide a robust ICA method that is easy to integrate to existing frameworks or use on its own. Also, as a part of such a framework it needs to be capable of running in single machines and cluster servers. The design of the toolbox is fully modular, allowing for the addition of new or alternative parts, such as, algorithms, visualization, or handling of different data formats. Currently, Arabica offers modules to perform robust ICA and handle input, output and visualization of fMRI data.

The name "Arabica" was inspired by the finest species of coffee beans, but also serves as the acronym for "Adaptive Robust Additions to Bagging ICA". Our goal is to setup an open community, based on open source standards, where anyone interested can join in and further develop the toolbox, or integrate it with other existing methods or frameworks. The toolbox community can be found, and the toolbox itself downloaded, at `http://launchpad.net/arabica`.

2   Background

There are many widely used and trusted analysis toolboxes in the neuroinformatics field (see, *e.g.*, [6,7]), but they are mainly designed to be used on their own, as standalone tools. It is very difficult to use parts of one toolbox and combine it with parts of another. Some recent toolboxes even provide ICA (see, *e.g.*, [8,9]), but they are typically designed for a specific purpose and can be difficult to integrate into larger analysis frameworks. Additionally, none of the existing toolboxes offer the level of robustness needed for reliably reproducing the same decomposition when reapplied to the same data.

The robust ICA approach introduced in [10] aims at being reliable and easy to use in practice. By running ICA many times using bootstrapping and random resampling of the data, it allows for analyzing the distribution of the component estimates. The variability information can be used in assessing the reliability and making correct interpretations of the components. One of the main benefits is the knowledge of uncertainty within the components when applying the method to new data.

The recent shift towards cluster or grid computing, to allow for ever larger datasets to be collected and analyzed in a feasible manner has not been taken into account in any of the previous toolboxes. One very promising framework for allowing flexible definition and execution of processing in a cluster of computers is the recently released LONI Pipeline environment ([11], `http://pipeline.loni.ucla.edu`). The environment allows for graphical definition of processing pipelines consisting of inter-connected modules that can be executed in one or many computers. Such an environment is very useful, since it offers the ability to reuse, replace, and modify only certain parts of a big processing pipeline without needing to know every detail of the other parts. This means that an ICA expert can change parameters of an ICA module, or replace a whole module, without being an expert in, *e.g.*, pre-processing of brain measurements. Likewise, a neuroscientist can alter parts of pre- or post-processing without needing to be an expert in ICA. The environment also allows for easy sharing of parts, or even complete, processing pipelines between researchers.

3   Toolbox Design

The toolbox is divided into packages that define one or many modules, using the same module definition as the LONI Pipeline environment. Packages can be installed by the end user or even added at runtime programmatically, allowing easy implementation of new and alternative modules, such as, analysis algorithms, data visualization, and data file handling. Individual modules, or processing flows consisting in many modules, can be automatically exported for use outside of Matlab in the LONI Pipeline environment. Fig. 1 shows an illustration of the control flow during execution of a processing pipeline. The toolbox provides an automatic wrapper for modules defined inside Matlab that allows easy and transparent integration of Matlab functions as parts of a processing pipeline.

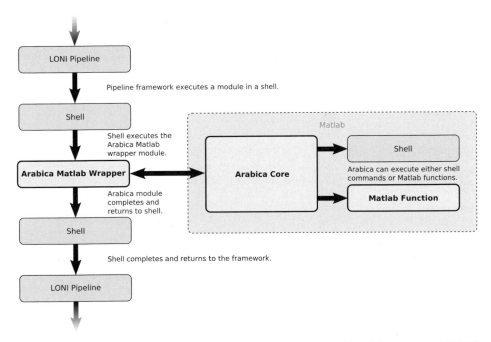

Fig. 1. Designed flow of execution control when running Arabica. The standard LONI Pipeline process on the left, where the pipeline client or server executes module code on the command-line shell in the localhost or in a grid server cluster. The Arabica toolbox provides a wrapper module that launches Matlab with the Arabica framework. Inside Matlab, execution is governed by the Arabica Core package that allows a module to consist in both shell commands and normal Matlab functions.

4   Illustrative Usage

This section demonstrates the use of the toolbox in the analysis of real multi-subject fMRI measurements. Fig. 2 shows the processing pipeline in the LONI

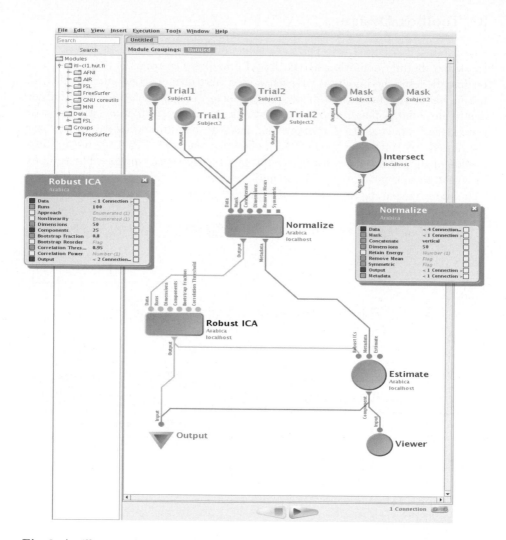

Fig. 2. An illustrative example of a typical Arabica process in the LONI Pipeline client program, when processing multi-subject fMRI data. The interface shows a library of predefined processing modules on the left and an editable graph of the processing pipeline on the right. The popups show selected parameters for two of the Arabica modules. The illustrated process starts from the top with the dataset consisting of functional measurements of two subjects in two trials, followed by the individual skull stripping volume masks. In this simplified illustration, the datasets are already assumed to be suitably pre-processed. The datasets are fed to the Arabica Normalize module which allows for a group analysis of multiple datasets. The module also outputs metadata that is needed in the later stages of the pipeline. The normalized group data is the input to the Arabica Robust ICA module that produces the clustered IC estimates. The last steps in the example process output and visualize the reliable ICs of the individual subjects.

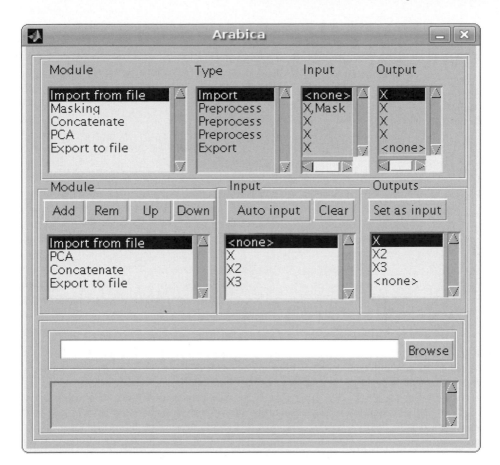

Fig. 3. An illustrative example of a typical Arabica standalone user interface running inside Matlab. It offers a simple way of managing a single execution flow through processing modules. The upper list displays the possible modules to be added to the flow. The lower list the actual process flow and allows management of inputs and outputs of the modules.

Pipeline client program. The figure only illustrate small parts of the data and the processing pipeline. The full pipeline would consist of more subjects and more trials, where each of the datasets may require individual pre-processing, such as, filtering, motion compensation and co-registering to a standard template.

In the example, the inputs are already considered to be pre-processed using some standard fMRI tools and are represented in the pipeline only as single input nodes. The pre-processing could be done, *e.g.*, as a separate processing pipeline. Similarly, the final output and visualization only illustrates the use of a single Estimate module that constructs the reliable component mixing vectors and signals. In practice, the outputs would depend on what kind of information the user is interested in, *e.g.*, separate estimation nodes may be needed to allow for interpretations of how the ICs apply on both individual and group level.

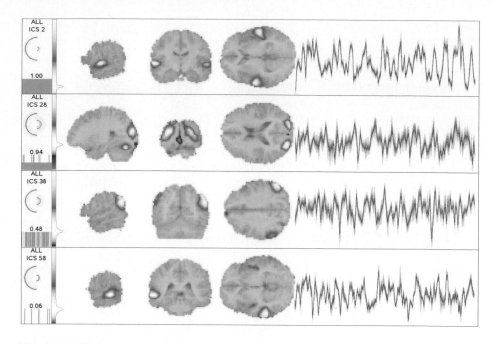

Fig. 4. An illustrative example of results acquired with Arabica Robust ICA. For each of the 4 components, the visualization shows details of robustness, on the left, the spatially independent activation maps, in the middle, and, on the right, the corresponding activation time courses, *i.e.*, mixing vectors. The robustness details include: the IC number; the discrimination rings depicting the distributions of inter-cluster (left arc) and intra-cluster (right arc) distances of all the estimates; and the occurrence probability with a histogram over the 100 runs. The spatial patterns are shown along three different axes with the volume histogram embedded into the colorbar, which shows the skewness and non-gaussianity of the component. Finally, the estimated mean time courses are shown surrounded by the percentiles of the distributions of all estimates, instead of just the typical variance based error-bars. Note particularly, the ability of the method to reliably separate the two last components even though they have very small occurrance probabilities.

Considering the illustrative example further, the first Arabica module to be executed is the Normalize module, which masks and whitens each input dataset individually, and then outputs a concatenated single data matrix suitable ICA.

The normalizing module can be considered as consisting of many sub-modules that can also be controlled individually. Fig. 3 shows the internal structure of the module inside the standalone user interface in Matlab. The interface can also be used without the LONI Pipeline environment. However, unlike the LONI client program, the Matlab interface only supports a linear execution flow at the moment.

The Arabica toolbox includes a visualizer interface that allows the results of the robust ICA method to be plotted in an easy to interpret fashion. Fig. 4 illustrates possible results from the Arabica Robust ICA module. The figure shows 4 typical spatially independent components reliably identified from a multi-subject fMRI study. The analysis of robustness of the ICs and mixing vector variability applies to any type of data. However, visualization of the signal itself is data dependent, e.g., in this case the activation pattern as spatial volumes. The modularity of the toolbox allows for different visualizers to be added for other types of data. Naturally, modules that output the resulting signals in a standard format, loadable in external visualization tools, are possible as well.

For a real application of the robust ICA see, e.g., the study in [12], which also highlights the fact that a typical analysis may also require post-processing after the ICA. The analysis in that study used further clustering and canonical component analysis (CCA) as refinement methods within each set of estimates belonging to a reliable component. The modularity of the Arabica toolbox would allow such methods to be easily defined as modules in Arabica and in the LONI Pipeline environment.

5   Discussion

The robust ICA approach allows for easy and reliable use of ICA. It provides information on the distribution of the component estimates that is required for correct interpretation of the results. One of the main benefits is the knowledge of uncertainty within the components when applying the method to new data.

Arabica provides a toolbox that is fully integrated into an existing processing environment. It works in a cluster of servers or a single machine, and can be used as a part in the LONI Pipeline or as a standalone package in Matlab. The current development version of Arabica offers basic modules to perform robust ICA as a part of larger processing, and provides further modules for input, output and visualization of fMRI data. The toolbox is currently implemented completely with bash shell scripts and Matlab code, but depending on the needs of the community future work may include, e.g., integration of modules implemented as Python code.

Recent developments, such as our Arabica toolbox, offer the ICA community an opportunity to build something useful on a large scale. We hope to gain a wide range of users in machine learning, neuroinformatics, and possibly other fields. There is a clear vision for the future development of many parts of the toolbox, but we invite anyone interested to join the community in developing the core toolbox or new modules that would be useful for them and others.

References

1. Hyvarinen, A., Karhunen, J., Oja, E.: Independent Component Analysis, 1st edn. Wiley-Interscience, New York (2001)
2. McKeown, M.J., Makeig, S., Brown, G.G., Jung, T.P., Kindermann, S.S., Bell, A.J., Sejnowski, T.J.: Analysis of fMRI data by blind separation into independent spatial components. Human Brain Mapping 6(3), 160–188 (1998)

3. Bartels, A., Zeki, S.: The chronoarchitecture of the human brain — natural viewing conditions reveal a time-based anatomy of the brain. NeuroImage 22(1), 419–433 (2004)
4. Damoiseaux, J.S., Rombouts, S.A.R.B., Barkhof, F., Scheltens, P., Stam, C.J., Smith, S.M., Beckmann, C.F.: Consistent resting-state networks across healthy subjects. Proceedings of the National Academy of Sciences 103(37), 13848–13853 (2006)
5. Mantini, D., Perrucci, M.G., Gratta, C.D., Romani, G.L., Corbetta, M.: Electrophysiological signatures of resting state networks in the human brain. Proceedings of the National Academy of Sciences 104(32), 13170–13175 (2007)
6. SPM: MATLAB™Package (1999)
7. Smith, S.M., Jenkinson, M., Woolrich, M.W., Beckmann, C.F., Behrens, T.E.J., Johansen-Berg, H., Bannister, P.R., Luca, M.D., Drobnjak, I., Flitney, D.E., Niazy, R., Saunders, J., Vickers, J., Zhang, Y., Stefano, N.D., Brady, J.M., Matthews, P.M.: Advances in functional and structural MR image analysis and implementation as FSL. NeuroImage 23(S1)(suppl.1), 208–219 (2004)
8. Calhoun, V.D., Adali, T., Pearlson, G.D., Pekar, J.J.: A method for making group inferences from functional MRI data using independent component analysis. Human Brain Mapping 14(3), 140–151 (2001)
9. Beckmann, C.F., Smith, S.M.: Probabilistic Independent Component Analysis for unctional Magnetic Resonance Imaging. IEEE Transactions on Medical Imaging 23(2), 137–152 (2004)
10. Ylipaavalniemi, J., Vigário, R.: Analyzing consistency of independent components: An fMRI illustration. NeuroImage 39(1), 169–180 (2008)
11. Rex, D.E., Ma, J.Q., Toga, A.W.: The LONI Pipeline Processing Environment. NeuroImage 19(3), 1033–1048 (2003), http://pipeline.loni.ucla.edu
12. Ylipaavalniemi, J., Vigário, R.: Subspaces of spatially varying independent components in fMRI. In: Davies, M.E., James, C.J., Abdallah, S.A., Plumbley, M.D. (eds.) ICA 2007. LNCS, vol. 4666, pp. 665–672. Springer, Heidelberg (2007)

Estimating Hidden Influences in Metabolic and Gene Regulatory Networks

Florian Blöchl[1] and Fabian J. Theis[1,2]

[1] Institute for Bioinformatics and Systems Biology, Helmholtz Zentrum München
Ingolstädter Landstrasse 1, 85764 Neuherberg, Germany
{florian.bloechl,fabian.theis}@helmholtz-muenchen.de
[2] Max-Planck-Institute for Dynamics and Self-Organization, Göttingen, Germany

Abstract. We address the applicability of blind source separation (BSS) methods for the estimation of hidden influences in biological dynamic systems such as metabolic or gene regulatory networks. In simple processes obeying mass action kinetics, we find the emergence of linear mixture models. More complex situations as well as hidden influences in regulatory systems with sigmoidal input functions however lead to new classes of BSS problems.

The field of independent component analysis (ICA) as solution to BSS problems has been in the focus of rather intense research during the past decade [1]. With the nowadays available robust algorithms in particular for the linear case, more and more people turn towards model generalizations and applications of ICA. One area of application of ICA and machine learning in general has been bioinformatics [2], which mostly deals with the analysis of large-scale high-throughput data sets from genomics. With the basic methods being robustly established, a trend in this field is to deal with smaller-scale fine-grained models closely integrating information from experiments. This systems-biology ansatz is increasingly bringing forth concise explanations for biological phenomena. In this contribution, we will list a few possible application areas of BSS in systems biology. A key ingredient for the system modeling is the detailed description of the system dynamics; we will consider mass action and Hill kinetics. We then address the question of how to model unknown sources (latent variables) that can be inferred from the observations. In a few situations this can be done even in a linear mixing system, in more general setting we will have to consider extensions of standard BSS models. We will denote measured time-courses by g and hidden influences by h, dropping their explicit time dependence.

1 Mass Action Kinetics

The modeling of a macroscopic chemical system such as a metabolic network is commonly simplified by neglecting the discrete nature of the participating reactants and their reactions. Therefore one introduces continuous concentrations as well as continuous reaction rates linked by a system of *ordinary differential*

T. Adali et al. (Eds.): ICA 2009, LNCS 5441, pp. 387–394, 2009.
© Springer-Verlag Berlin Heidelberg 2009

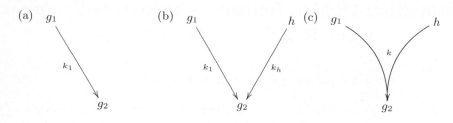

Fig. 1. Reactions of order one and two: (a) shows a reaction, where g_1 reacts to g_2 with rate k_1, see Eq. (1). In (b) we add a hidden influence h, which also contributes to the g_2 concentration. Figure (c) depicts a situation, where g_1 and h react to g_2, see (4).

equations (ODE), the so-called *rate equations*. These rate laws are governed by the law of mass action, which says that the instantaneous rate of a reaction is proportional to the concentration of each reactant, raised to the power of its stoichiometry. The largest degree occurring in this equation is called the *order of the reaction*, which rarely exceeds two [3]. Figure 1 shows the elementary reactions of order one and two, which we will take as starting point for the following discussion.

1.1 First-Order Mass Action Kinetics

For the rate equations of the direct conversion $g_1 \xrightarrow{k_1} g_2$ in Figure 1a, we find

$$\dot{g}_1 = -\tau_1 g_1 - k_1 g_1, \qquad \dot{g}_2 = -\tau_2 g_2 + k_1 g_1, \qquad (1)$$

where the reaction runs with rate k_1. We additionally introduce decay terms quantifying the loss of reactants due to degradation with time constants $\tau_{1/2}$. If we now allow one hidden influence h that produces g_2 in a first-order reaction (Fig. 1b), we have to change the rate law for g_2 to $\dot{g}_2 = -\tau_2 g_2 + k_1 g_1 + k_h h$. In this situation — provided that we know both the decay rates and g_1 — we can directly calculate the time-course of $k_h h$ via $k_h h = \dot{g}_2 + \tau_2 g_2 - k_1 g_1$. Obviously, we cannot determine the scale of the hidden influence and its reaction rate k_h observing only g_1 and g_2.

However, if g_1 and h are assumed to be uncorrelated or independent (e.g. because they stem from different biological processes), we can even estimate k_1 from the observed time-courses: we simply minimize the absolute correlation between g_1 and $k_h h$, which has a unique solution in this case. Simulations with various parameter sets and shapes of hidden influences show that k_1 can be estimated up to an absolute error that depends on the size of the (in practice non-vanishing) correlation between g_1 and h. For instance, with the hidden influences in Figure 2, randomly sampled $k_1 \in (0.1, 1)$ can be estimated with an absolute error of 0.005, averaged over 100 runs with random decay rates in $(0.1, 1)$.

In a general reaction network of this kind with N species g_i, let $g_j \xrightarrow{k_{ij}} g_i$ denote the reaction with rates k_{ij} vanishing if no reaction occurs. We may write the rate law for any of the reactants in a system with n hidden influences as

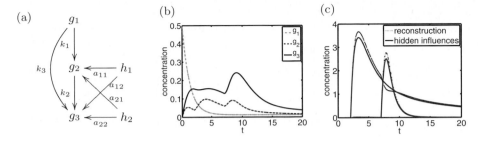

Fig. 2. (a) A coherent feed-forward loop with two hidden influences. (b) Simulated time-courses of the measured reactants for $k_1 = 0.3$, $k_2 = 0.7$, $k_3 = 0.4$ and degradation rates $\tau_1 = 0.1$, $\tau_2 = 0.6$, and $\tau_3 = 0.5$. (c) The two simulated hidden influences and their reconstructions using FastICA. We use the solutions of $\dot{h}_1 = (2(t-1))^{-1} - h_1$ for $t > 2$ and $2\dot{h}_2 - (1 + (t-7)^3)^{-1} - 3h_2$ for $t > 7$. For the mixing matrix we choose $\boldsymbol{A} = (0.6, 0.4; 0, 0.6)$, where we find SNR of 22 and 36 dB.

$$\dot{g}_i + \tau_i g_i - \sum_{l=1}^{N} (k_{il} g_l - k_{li} g_i) = \sum_{j=1}^{n} a_{ij} h_j. \tag{2}$$

Denoting the left hand side of this equation by x_i and using matrix notation, we arrive at the common linear blind-source separation problem $\boldsymbol{x} = \boldsymbol{Ah}$. This problem can be solved by various techniques, if we assume that the hidden influences fulfill certain conditions like decorrelation (PCA [4]), statistical independence (ICA [1]) or non-negativity, possibly combined with sparseness (NMF [5,6]). Moreover, these properties can be used to estimate reaction rates analogously to the upper example. If at least one reactant g_u is known not to be affected by the hidden ones, we may estimate the rates $\boldsymbol{k}$ occurring in its rate law as $\hat{\boldsymbol{k}} = \operatorname{argmin}_{\boldsymbol{k}} |\operatorname{corr}(x_u; g_u)|$. This estimate has — possibly many — indeterminacies depending on the network topology.

Example: A feed-forward loop. A frequently occurring motif in metabolic networks is the coherent feed-forward loop [7], as shown in Figure 2a. We study the first-order coupling of two statistically independent hidden influences $h_{1/2}$ to this system, which for example may correspond to a separation of overlapping metabolic pathways. We write the rate equations for the depicted problem as linear mixing model

$$\dot{g}_2 + \tau_2 g_2 + k_1 g_1 - k_2 g_2 = \sum a_{1j} h_j \tag{3}$$

$$\dot{g}_3 + \tau_3 g_3 + k_2 g_2 + k_3 g_1 = \sum a_{2j} h_j.$$

Hence, if the time-courses of the reactants are measured and the reaction rates and time constants are known, we can estimate $\dot{g}_i$ and reconstruct $h_{1/2}$ by assuming approximate statistical independence. In our simulations we perform ICA using the FastICA algorithm [8]. With this, in 100 simulations (rate parameters given in Figure 2) with random positive mixing coefficients, we reconstruct the hidden influences with a mean signal-to-noise ratio of 25 ± 9 dB.

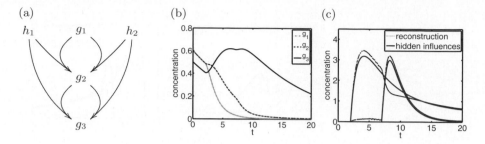

Fig. 3. Second-order mass action kinetics: a three node cascade with 2 hidden influences (a) and a simulated time course (b). All time constants equal 0.01; we choose a mixing matrix $\boldsymbol{A} = (0.8, 0; 0.2, 0.5)$ and the hidden influences from Figure 2. With FastICA we can reconstruct them with a SNR of 14 dB and 27 dB, plotted in (c).

1.2   Second-Order Mass Action Kinetics

The easiest and analytically solvable example for second-order mass action kinetics with a hidden influence h is a reaction $g_1 + h \xrightarrow{k} g_2$, as shown in Figure 1c. The corresponding ODE system contains only degradation terms and the product term representing the reaction:

$$\dot{g}_1 = -\tau_1 g_1 - kg_1 h\,, \qquad \dot{g}_2 = -\tau_2 g_2 + kg_1 h\,. \tag{4}$$

Hence, if we measure the time-courses of g_1 and g_2 and also know their decay rates we can determine $kh = (\dot{g}_2 + \tau_2 g_2)/g_1$, given that $g_1 \neq 0$. If additionally the reaction rate k is known, we can extract the hidden influence. Otherwise this constant and with it the scale of h remains an indeterminacy of this problem, even if e.g. we assume independence of the two reactants.

Now consider a cascade $g_1 \longrightarrow g_2 \longrightarrow g_3$ of second-order reactions. Here we could try to estimate the time-courses of two unobserved reaction partners $h_{1/2}$ that may take part in both reactions — e.g. two enzymes that have similar functions but are regulated by different processes. The rate equations for the first and the third reactant again lead to a linear mixing model $(\dot{g}_l + \tau_l g_l)/g_l = \sum_{m=1}^2 a_{lk} h_k$ for $l = 1, 3$. As before, this can be easily solved by ICA. Figure 3 shows a simulated example.

However, these procedures only work in the discussed simple cases. If, for instance, there exists a direct second-order reaction $g_1 \longrightarrow g_3$ in the three node cascade or if we have a larger cascade with more than two influences, we arrive at a mixing model with time-dependent mixture coefficients.

For a general network of second-order reactions $r : g_{r_a} + h_j \longrightarrow g_{r_b}$ with n hidden influences we find rate equations $\dot{g}_i = -\tau_i g_i - \sum_{r_a=i} \sum_{j=1}^n a_{r_a j} g_i h_j + \sum_{r_b=i} \sum_{j=1}^n a_{jr_a} g_{r_a} h_j + \sum (\text{reactions of } g_i \text{ without hidden influences})$. We therefore will have to solve the BSS problem

$$x_i = \sum_{j=1}^n \left(-\sum_{r_a=i} a_{r_a j} g_i + \sum_{r_b=i} a_{jr_a} g_{r_a} \right) h_j\,, \tag{5}$$

where we know the time-courses of all g. Hence, the estimation of hidden influences in such systems leads to a class of linear mixing models with time-dependent coefficients.

In many reactions, especially enzymatic ones, one observes processes of complex formation and dissociation like $g_1 + g_2 \longleftrightarrow C \longrightarrow g_3 + g_2$. This can be modeled by a detailed second-order mass action system. If however the concentration of g_2 (the enzyme) is much lower than the concentration of g_1 (the substrate), the dynamics can be approximated by *Michaelis-Menten kinetics* (cf. [3]). However, this type of reaction is formally a special case of Hill kinetics, which will be discussed in the following.

2 Gene Regulatory Networks

Molecular interactions on a genetic level are known to show a switch-like behavior. Motivated by the analysis of a promotor binding model (see e.g. [3]), they are usually described by activating $H^{+}_{n,k}(g) = g^n(g^n + k^n)^{-1}$ and inhibiting *Hill functions* $H^{-}_{n,k}(g) = (g^n + k^n)^{-1}$. Here the *Hill coefficient* n is a measure for the cooperativity of the interaction. The *threshold parameter* k corresponds to the concentration at half maximum activation.

A negative feedback loop modeled with Hill kinetics. As showcase for the issues we face in estimating hidden influences in gene regulation, imagine the mutual inhibition of two genes shown in Figure 4a, a bistable motif that is found in many developmental processes:

$$\dot{g}_1 = -\tau_1 g_1 + H^{-}_{n_2,k_2}(g_2) , \qquad \dot{g}_2 = -\tau_2 g_2 + H^{-}_{n_1,k_1}(g_1) . \tag{6}$$

Again we can measure the time-courses of both genes and know all parameters in the rate equations (6). If we now allow two hidden influences h_1 and h_2, we first have to specify the logical operations that couple them to the system. The translation of the potentially complex 'molecular computations' taking place in a gene's promotor region into differential equations is still a field of active research. However, purely additive as well as purely multiplicative coupling of all incoming regulations are widely-used approaches. This corresponds to a combination of the inputs by Boolean *or* and *and* logic, respectively. Both logics can only be transformed to the common linear mixing model, if any hidden influence couples to measured genes with the same Hill function.

In this simplified situation we can proceed analogously to the last section and extract at least the Hill-transformed hidden influences, as demonstrated in Figure 4. In the case of multiplicative coupling, we have rate equations $\dot{g}_1 = -\tau_1 g_1 + H_{n_2,k_2}(g_2) \prod_{j=1}^{2} H^{a_{1j}}_{n_{hj},k_{hj}}(h_j)$ and symmetrically for g_2. Here, after bringing the degradation term to the left hand side, we take the logarithm and subtract the regulation of the measured gene, arriving at

$$\log(\dot{g}_1 + \tau_1 g_1) - \log(H_{n_2,k_2}(g_2)) = \sum_{j=1}^{2} a_{1j} \log\left(H_{n_{hj},k_{hj}}(h_j)\right) . \tag{7}$$

Hence, in this situation we can determine the $H_{n_{hj},k_{hj}}(h_j)$ up to an exponent.

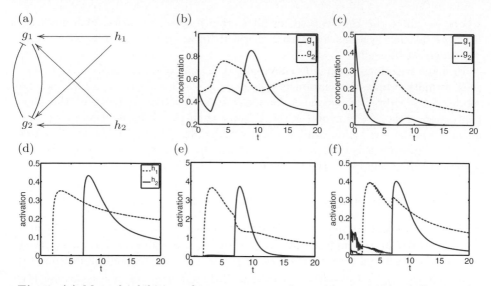

Fig. 4. (a) Mutual inhibition of two genes g_1 and g_2 with two hidden influences h_1 and h_2. We add h_1 and h_2 as defined in Figure 2 as activators with mixing matrix $(1, 1; 1, 0)$ to the system. This leads to completely different time-courses of g_1 and g_2, depending on whether we use additive (b) or multiplicative (c) logic, despite using the same parameters (we take all n and τ equal to one, k_1 to k_4 are 0.4, 0.2, 0.4, 0.3) and initial conditions. Under the assumption of equal coupling, the Hill-transformed hidden influences (d) can be reconstructed fairly well by FastICA: for *or* logic (e) we get SNR of 18 dB and 34 dB, for *and* logic ((f), sign corrected before exponentiating) SNR of 11 dB in both cases.

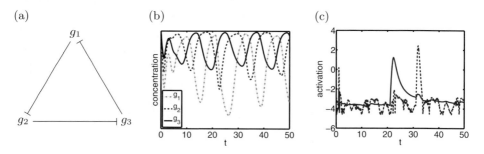

Fig. 5. The repressilator (a), a three-node motif giving stable oscillations and (b) a simulated time-course with our two hidden influences (with a time shift of 20, then oscillations without the h_l would be stable), coupling to g_1 and g_2 with mixing matrix $A = (20, -1; 10, 0)$. We choose $a = 1$, the same for all degradation rates, all time delays are 5, and the non-vanishing interaction weights are $20, 15, 10$. A reconstruction of the transformed hidden influences (c) with FastICA.

2.1   Gene Regulatory Networks with Hill Kinetics

In general regulatory networks, a participating gene as well as a hidden influence can act both as inhibitor $I^-(g)$ and activator $I^+(g)$ to a gene g. Additionally, the Hill parameters may vary in every interaction. For additive coupling, this situation leads to ODEs of the form

$$\dot{g}_i = -\tau_i g_i + \sum_{\sigma=\pm} \left(\sum_{g_j \in I^\sigma(g_i)} H^\sigma_{n_{ji},k_{ji}}(g_j) + \sum_{h_l \in I^\sigma(g_i)} a_{il} H^\sigma_{n_{li},k_{li}}(h_l) \right). \quad (8)$$

If *and* logic is used, we have to write products and exponents instead of the weighted sums of Hill functions, which after logarithmic transformation is converted to weighted sums again. With this, when rearranging the rate equations two new mixing models arise:

$$x_i = \sum_{j \in I^-(g_i)} \frac{a_{ij}}{k_{ij}^{n_j} + h_i^{n_j}} + \sum_{j \in I^+(g_i)} \frac{a_{ij} h_j^{n_j}}{k_{ij}^{n_j} + h_j^{n_j}} \qquad \text{(for \emph{or} logic)} \qquad (9)$$

$$x_i = \sum_j a_{ij} \log\left(k_{ij}^{n_j} + h_j^{n_j}\right) - \sum_{j \in I^+(g_i)} a_{ij} \log\left(h_j^{n_j}\right) \quad \text{(for \emph{and} logic)} \quad (10)$$

The use of different functions for activation and inhibition is a consequence of the non-negativity of Hill functions and their combinations. These models cannot be solved by linear BSS.

2.2 Continuous-Time Recurrent Neural Networks (CTRNN)

A recent approach to gene expression data analysis [9] uses generalized CTRNN as abstract dynamical models of regulatory systems, leading to ODEs of the form

$$\dot{g}_i(t) = \tau_i \left(-g_i(t) + \sum_l W_{li} \sigma(g_l(t - \Delta_l) - k_l) + I_i(t) \right). \quad (11)$$

Here τ_i denotes the degradation rate, I_i an external input and k_l are thresholds. Interactions are incorporated via the activation function $\sigma(x) = (1 + e^{-ax})^{-1}$ and additively connected by a real weight matrix W. The delay constants Δ_l account for the time delay due to gene induction, transcription and translation. Of course, this approach has the advantage of a single function for both inhibition and activation, but on the other hand we loose the biological model and direct interpretability. However, for the estimation of hidden influences h_j in such an ODE system we obtain the mixing model

$$x_i = \sum_j W_{ji} \sigma(h_j(t - \Delta_j) - k_j). \quad (12)$$

Here the time delays will remain an indeterminacy. In the case of equal coupling, this again reduces to a problem solvable using ICA. In Figure 5 we discuss an example for this, two hidden influences to the so-called repressilator [10]. If, which is more realistic, we assume interaction-specific delay times Δ_{ij}, we can use convolutive ICA [1]. However, linear BSS is unable to estimate additional parameters such as the k_j and exponents in σ.

3   Conclusions

With the availability of more and more quantitative data the estimation of hidden influences in biological systems is an upcoming challenge and crucial for the interpretation of many experiments. For simple processes in mass action kinetics and first-order reaction networks we showed that this task leads to a linear BSS problem that can be solved e.g. using ICA. In the more delicate case of second-order reactions and regulatory interactions, where we have to deal with sigmoidal input functions that may be connected by different logical operations, new mixing models arise. The analysis of these models and the estimation of the occurring parameters in nonlinear situations necessitates a treatment within a Bayesian framework, similar to e.g. [11]. In practice, we will be confronted with networks involving reactions and interaction of more than one of the discussed types, leading to a variety of hybrid models and corresponding likelihood functions. Moreover we will only observe a fraction of the known network elements, so there exists an added layer of model inference. These involved learning problems may be solved by extensions of BSS or Bayesian methods.

Acknowledgements. Partial financial support by the Helmholtz Alliance on Systems Biology (project 'CoReNe') is gratefully acknowledged. The authors thank Dominik Wittmann and Daniel Schmidl for helpful discussions.

References

1. Hyvärinen, A., Karhunen, J., Oja, E.: Independent Component Analysis. John Wiley & Sons, New York (2001)
2. Schachtner, R., Lutter, D., Knollmüller, P., Tomé, A.M., Theis, F.J., Schmitz, G., Stetter, M., Vilda, P.G., Lang, E.W.: Knowledge-based gene expression classification via matrix factorization. Bioinformatics 24(15), 1688–1697 (2008)
3. Alon, U.: An Introduction to Systems Biology: Design Principles of Biological Circuits. Chapman & Hall/CRC (2006)
4. Pearson, K.: Principal components analysis. The London, Edinburgh, and Dublin Philosophical Magazine and Journal of Science 6(2), 559 (1901)
5. Lee, D.D., Seung, H.S.: Algorithms for non-negative matrix factorization. In: NIPS, pp. 556–562 (2000)
6. Hoyer, P.O.: Non-negative matrix factorization with sparseness constraints. Journal of Machine Learning Research 5, 1457–1469 (2004)
7. Zhu, D., Qin, Z.S.: Structural comparison of metabolic networks in selected single cell organisms. BMC Bioinformatics 6(1) (2005)
8. Hyvärinen, A.: Fast and robust fixedpoint algorithms for independent component analysis. IEEE Transactions on Neural Networks 10(3), 626–634 (1999)
9. Busch, H., Camacho-Trullio, D., Rogon, Z., Breuhahn, K., Angel, P., Eils, R., Szabowski, A.: Gene network dynamics controlling keratinocyte migration. Molecular Systems Biology 4 (2008)
10. Elowitz, M.B., Leibler, S.: A synthetic oscillatory network of transcriptional regulators. Nature 4(6767), 335–338 (2000)
11. Vyshemirsky, V., Girolami, M.A.: Bayesian ranking of biochemical system models. Bioinformatics 24(6), 833–839 (2008)

Nonnegative Network Component Analysis by Linear Programming for Gene Regulatory Network Reconstruction

Chunqi Chang[1], Zhi Ding[2], and Yeung Sam Hung[1]

[1] Department of Electrical and Electronic Engineering
The University of Hong Kong, Pokfulam Road, Hong Kong
{cqchang,yshung}@eee.hku.hk
[2] Department of Electrical and Computer Engineering
University of California, Davis, CA 95616, USA
zding@ucdavis.edu

Abstract. We consider a systems biology problem of reconstructing gene regulatory network from time-course gene expression microarray data, a special blind source separation problem for which conventional methods cannot be applied. Network component analysis (NCA), which makes use of the structural information of the mixing matrix, is a tailored method for this specific blind source separation problem. In this paper, a new NCA method called nonnegative NCA (nnNCA) is proposed to take into account of the non-negativity constraint on the mixing matrix that is based on a reasonable biological assumption. The nnNCA problem is formulated as a linear programming problem which can be solved effectively. Simulation results on spectroscopy data and experimental results on time-course microarray data of yeast cell cycle demonstrate the effectiveness and anti-noise robustness of the proposed nnNCA method.

1 Introduction

Gene regulatory network reconstruction is an important research problem in systems biology where structure and dynamics of cellular functions are of interest. Since gene regulatory network reveals the underlying inter-dependency and cause-and-effect relationship between various cellular functions, it has become one of the key areas of interest in systems biology.

Gaining a quantitative understanding of gene regulation is of vital importance in modern biology. In general, the problem relates to how and where a particular gene is expressed, often under combinatorial control of regulatory proteins known as transcription factors (TF). The dynamics of gene expression levels, i.e. the mRNA concentrations, in a cell can be measured simultaneously by microarray technology for all genes in the genome in a form of multi-channel time-course signal. However, the dynamics of the regulatory signal, i.e., the transcription factor activities (TFA), cannot be measured by the current technology. In addition, the control strength of a regulatory transcription factor to a gene is another important aspect in the gene regulatory network, which is unfortunately also unknown.

T. Adali et al. (Eds.): ICA 2009, LNCS 5441, pp. 395–402, 2009.
© Springer-Verlag Berlin Heidelberg 2009

In order to understand the entire gene regulatory network, we need to reconstruct the transcription factor activities and the matrix of control strengths from the gene expression measurements. This is a highly challenging inverse problem, especially because microarray data are always extremely noisy.

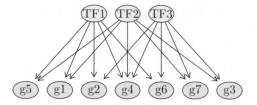

Fig. 1. Gene regulatory network

A conceptual gene regulatory network with 7 genes (g1 ... g7) and 3 transcription factors (TF1, TF2, TF3) is illustrated by Fig 1. In general, gene regulation processes are dynamic and nonlinear. It is assumed that the time scale of change of transcription factor activities (TFA) is much greater than that of gene expression. Therefore, mRNA levels at most time are in a quasi-steady state, and thus at this quasi-steady state the dynamic model becomes approximately instantaneous. In addition, the nonlinear dependence of gene expression on the TFAs is approximately log-linear [1]. Therefore, when the gene expressions are expressed as log-ratios, the model becomes $\mathbf{X} = \mathbf{AS} + \boldsymbol{\Gamma}$, where $\mathbf{X}, \mathbf{A}, \mathbf{S}$, and $\boldsymbol{\Gamma}$ are gene expression, connectivity matrix, TFAs, and noise, respectively.

Under the above instantaneous linear model, gene regulatory network reconstruction is a blind source separation problem, and independent component analysis (ICA) [2,3] had been applied to solve the problem. However, in this special blind source separation problem, the source signals are dependent in general. Therefore, the networks inferred from ICA are not accurate and do not conform to the realistic network structure which is of known sparse structure. There are some other approaches that work on dependent sources [4,5], but the underlying assumptions do not readily apply to gene regulatory network.

It was noted in the pioneering work of [6] that if the sparse network structure is known and satisfies some conditions, then the network can be uniquely reconstructed if there is no noise, and a method called network component analysis (NCA) was proposed to find a connectivity matrix (conformable to the known structure) and a set of transcription factor activities that best fit the model by using alternating least squares (ALS). The original ALS approach to NCA suffers from drawbacks of instability, inefficiency, and local convergence. Tikhonov regularization has been proposed to overcome the problem of instability [7], but on the other hand it is computationally even more inefficient. Then in [8,9] we proposed a more efficient and more effective method called FastNCA that successfully overcomes all the three drawbacks. FastNCA provides a closed-form solution to estimate the connectivity matrix and TFAs through fitting the model by a series of subspace projections.

Although NCA is by far one of the most effective approaches to gene regulatory network reconstruction, existing algorithms, however, lack accuracy and consistency. This motivates us to improve NCA and develop more accurate and robust network reconstruction methods by incorporating some prior information; thereby greatly enhance our ability to accurately reconstruct the networks. Specifically, in this paper we will assume that the entries of the connectivity matrix $\mathbf{A}$ are all nonnegative, and develop a linear programming method to solve the NCA problem with non-negativity constraints on the connectivity matrix.

2   Nonnegative Network Component Analysis

Recall the instantaneous linear gene regulation model mentioned in Section 1.

$$\mathbf{X} = \mathbf{AS} + \mathbf{\Gamma}. \tag{1}$$

Our aim is to estimate the connectivity matrix $\mathbf{A}$ and the TFAs $\mathbf{S}$ from the time-course microarray data $\mathbf{X}$.

Assume that in the network we have N genes and M transcription factors, and the length of time series is K. Then the dimension of $\mathbf{X}$ and $\mathbf{\Gamma}$ is $N \times K$, the dimension of $\mathbf{A}$ is $N \times M$, and the dimension of $\mathbf{S}$ is $M \times K$.

As proved in [6,9], this inverse problem has a unique solution (up to scaling ambiguity of the TFAs) in the noise-less case if the following *NCA criteria* are satisfied:

(i) $\mathbf{A}$ has full column rank;
(ii) when any one column of $\mathbf{A}$ is removed together with the rows corresponding to the nonzero entries of this column, the resulting sub-matrix has full column rank;
(iii) $\mathbf{S}$ has full row rank.

The blind source separation problem or inverse problem of estimating $\mathbf{A}$ and $\mathbf{S}$ from $\mathbf{X}$ based on Eq. (1) and the three NCA criteria is called network component analysis (NCA).

Though the three NCA criteria are enough to estimate the connectivity matrix when there is no noise in the model, additional constraints, if can be used in the algorithm, will certainly yield a more robust estimate in practice when noise is inevitable.

According to [10], we have the biological knowledge that most likely a transcription factor will have the same effect (either positive or negative) on all its regulated genes. This knowledge means that the entries within the same column of $\mathbf{A}$ should have the same sign, and by moving the sign to the corresponding row of $\mathbf{S}$ we can assume that all non-zero entries of $\mathbf{A}$ are positive, i.e., all entries of $\mathbf{A}$ are nonnegative. In other words, if a transcription factor regulates the genes negatively, then we can simply multiply its transcription factor activity (TFA) by -1 and this sign-inversed TFA will regulate the genes positively.

We call the network component analysis problem under the additional nonnegativity constraint on the connectivity matrix $\mathbf{A}$ nonnegative network component analysis (nnNCA).

3   A Linear Programming Approach to nnNCA

If there is no noise in the NCA model Eq. (1), i.e., $\mathbf{X} = \mathbf{AS}$, then the range of $\mathbf{X}$ is equal to the range of $\mathbf{A}$ since $\mathbf{A}$ is of full column rank and $\mathbf{S}$ is of full row rank. Because $\mathbf{X}$ is known to us, we can get the orthonormal basis matrix of the range space of $\mathbf{X}$, denoted by $\bar{\mathbf{X}} = \text{orth}\{\mathbf{X}\}$, and we have

$$\bar{\mathbf{X}} = \text{orth}\{\mathbf{X}\} = \text{orth}\{\mathbf{A}\} \ . \tag{2}$$

With $\bar{\mathbf{X}}$ known, we can get further its orthogonal complement

$$\mathbf{C} = \bar{\mathbf{X}}^{\perp} \tag{3}$$

such that

$$\mathbf{C}^{T}\bar{\mathbf{X}} = \mathbf{0} \ . \tag{4}$$

From Eq. (2) and (4), we have

$$\mathbf{C}^{T}\mathbf{A} = \mathbf{0} \ . \tag{5}$$

Since $\mathbf{C}$ can be estimated from the known time-course microarray data $\mathbf{X}$, the connectivity matrix $\mathbf{A}$ can be obtained by solving the systems of equation consisting of Eq. (5) and the NCA criterion (ii) described in Section 2.

In general, if there is noise in the model Eq. (1), singular value decomposition (SVD) [11] will be applied to $\mathbf{X}$ to obtain a robust estimation of $\mathbf{C}$. We write $\mathbf{X}$ in the standard SVD form as follows

$$\mathbf{X} = \mathbf{U\Sigma V}^{T} \ . \tag{6}$$

Partition $\mathbf{U}$, the matrix of left singular vectors of $\mathbf{X}$, as

$$\mathbf{U} = [\mathbf{U}_S \quad \mathbf{U}_N] \ , \tag{7}$$

where $\mathbf{U}_S$ contains the first M columns of $\mathbf{U}$ and $\mathbf{U}_N$ contains the remaining $N - M$ columns of $\mathbf{U}$. Here we state again that N is the number of genes and M is the number of transcription factors. The matrices $\mathbf{U}_S$ and $\mathbf{U}_N$ are called the signal subspace and noise subspace of $\mathbf{X}$, respectively. Then we get a robust estimate of $\mathbf{C}$ as

$$\mathbf{C} = \mathbf{U}_N \ . \tag{8}$$

Note that in the noisy case with $\mathbf{C}$ estimated by Eq. (8), Eq. 5 does not hold in general. To estimate $\mathbf{A}$ in such case, instead of solving a system of equations as in the noiseless case, we minimize all entries of $\mathbf{C}^{T}\mathbf{A}$ with the constraints imposed by the NCA criteria and non-negativity of $\mathbf{A}$ by solving the following constrained optimization problem via linear programming.

Let $\mathbf{A} = [\mathbf{a}_1, \ldots, \mathbf{a}_M]$ and $\mathbf{C} = [\mathbf{c}_1, \ldots, \mathbf{c}_{N-M}]$, where $\mathbf{a}_i$ is the ith column of $\mathbf{A}$ and $\mathbf{c}_i$ is the ith column of $\mathbf{C}$. In addition, we denote I as the indices where the entries of $\mathbf{A}$ are zeros, and J as the indices where the entries of $\mathbf{A}$ are nonzero (positive). Then, we can estimate the connectivity matrix $\mathbf{A}$ by solving the following linear programming problem

$$\min t \quad \text{s.t.} \quad -t < \mathbf{c}_i^T \mathbf{a}_j < t, \quad \mathbf{A}(I) = 0, \quad \mathbf{A}(J) > 0, \quad \sum_{n=1}^{N} a_{n,j} = L_j \ , \quad (9)$$

for $i = 1, \ldots, N - M$ and $j = 1, \ldots, M$, where L_j is the number of nonzero entries of $\mathbf{a}_j$, $\mathbf{A}(I) = 0$ means the entries of $\mathbf{A}$ indexed by I are zero, and $\mathbf{A}(J) > 0$ means that the entries of $\mathbf{A}$ indexed by J are positive. The last constraint $\sum_{n=1}^{N} a_{n,j} = L_j$ is imposed to avoid the trivial solution $t = 0$ with $\mathbf{A} = \mathbf{0}$, and the right-hand-side L_j is chosen to make the solution conformable to the normalization strategy adopted in [6,9].

For real biological systems, the connectivity matrix $\mathbf{A}$ may be very sparse. Problem (9) can be simplified by considering only the non-zero (positive) entries of $\mathbf{A}$. Denote $\tilde{\mathbf{a}}_j$ as the vector of nonzero entries of the jth column of $\mathbf{A}$, and $\tilde{\mathbf{c}}_{i,j}$ as the vector of entries of the ith column of $\mathbf{C}$ corresponding to the nonzero entries of the jth column of $\mathbf{A}$. Then problem (9) can be simplified as

$$\min t \quad \text{s.t.} \quad -t < \tilde{\mathbf{c}}_{i,j}^T \tilde{\mathbf{a}}_j < t, \quad \tilde{\mathbf{a}}_j > 0, \quad \sum_{n=1}^{L_j} \tilde{a}_{n,j} = L_j \ . \quad (10)$$

Problem (10) is a linear programming problem [12,13], and can be solved by standard algorithms implemented in many mature linear programming software packages. In this paper we use the GLPK package [14] to solve the linear programming problem.

4   Results

4.1   Simulation Results

To test the proposed linear programming based nnNCA algorithm, we use the simulation data described in [6]. In this simulation data, the conceptual gene regulatory network shown in Fig 1 is simulated by the mixing of spectroscopy signals. Three transcription factors are simulated by three kinds of hemoglobins, and the expression of seven genes are simulated by the spectroscopy of the mixed hemoglobins with different mixing ratios that are conformable to the structure of the network in Fig 1. The length of each measured spectroscopy signal is 321. It has been shown in [6,8] that conventional blind source separation methods, based on either higher-order statistics or second-order statistics, cannot recover the true pure spectroscopy signal of the three hemoglobins, while the NCA methods, either by ALS algorithm or FastNCA, can extract the source signals perfectly.

To demonstrate the effectiveness of the nnNCA method, we apply it to the spectroscopy mixtures, and the estimated source signals are shown in the middle column of Fig 2. It is found that the results of nnNCA in this case is almost exactly the same as that of FastNCA [9], with negligible difference of the order of numerical calculation error. Then independent white Gaussian noises are added to the mixing mixture to the level of SNR=5dB (signal to noise ratio). Both FastNCA and nnNCA are applied to the noisy data, and a great number of Monte Carlo runs are performed. Though nnNCA is not guaranteed to always perform better than FastNCA, the overall performance of nnNCA is superior to FastNCA. The estimated source signals by nnNCA and FastNCA from a typical Monte Carlo run are compared in Fig 2. The results demonstrate that the inclusion of the constraints on the positivity of the entries of the connectivity matrix **A** makes the NCA method more robust to measurement noise.

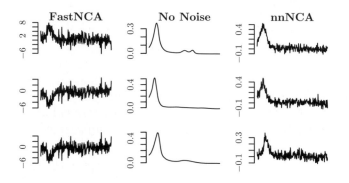

Fig. 2. Simulation results, SNR=5 dB

4.2   Experimental Results

To test the improved robustness to noise of the proposed nnNCA over conventional NCA methods for the analysis of real biological networks, we apply it to analyze the time-course microarray data of yeast cell cycle in [15], and compare the results with that of FastNCA. In this study, there are three experiments with different synchronization methods represented by alpha, cdc15, and elutriation. The time-course microarray data contains 6178 genes and 56 time points. In this analysis, we are interested in the 11 transcription factors that are known to regulate the expression of genes that are involved in the cell cycle process. In order to apply NCA to recover the TFAs of these 11 transcription factors, we need to work on a sub-network that contains the 11 TFs. For this purpose we construct a network that contains only these 11 TFs and those genes that are regulated by only these 11 TFs, based on the network topology information inferred from the ChIP-chip experiment in [16]. Both nnNCA and FastNCA are then applied to this network and the estimated TFAs are displayed in Fig 3 shoulder to shoulder for the ease of comparison, where the curves in black are for FastNCA and the curves in blue are for nnNCA.

The experiment "alpha" contains 2 cell cycles, "cdc15" contains 3 cycles, and "elutriation" contains 1 cycle. The names of the TFs are shown on the right-hand side of the figure. Since these 11 TFs regulate the cell cycle process, it is expected that the TFA of them should be cyclic, and we know that the gene expression signal of many of them are not cyclic at all [6,9]. We observe that the results of these two method are similar in general, and there is no case that the

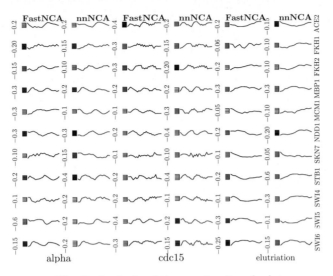

Fig. 3. Analysis of the yeast cell cycle data

estimated TFA by nnNCA is less cyclic than that by FastNCA, while in some cases, such as FKH1 for all 3 experiments and MCM1 for "alpha", the results of nnNCA are significantly more cyclic than that by FastNCA. These demonstrate the superior robustness of nnNCA for the analysis of real biological networks.

5   Discussion and Conclusion

Gene regulatory network reconstruction is an inverse problem similar to blind source separation, but conventional blind source separation methods cannot be applied because the source signals are dependent in general. Network component analysis (NCA) is a suitable source separation method for this specific problem. In this paper a new NCA method, nnNCA, is developed that incorporates a reasonable biological knowledge. It is demonstrated by both simulation and experimental results that nnNCA is more robust. The linear programming based algorithm is also very fast, slightly slower than but comparable to FastNCA, and much faster than the original ALS based NCA. The developed method may also find its applications in some other similar signal processing problems.

Acknowledgements

This work is supported by the University of Hong Kong Seed Funding for Basic Research.

References

1. Savageau, M.A.: Biochemical Systems Analysis: A Study of Function and Design in Molecular Biology. Addison-Wesley, Reading (1976)
2. Liebermeister, W.: Linear modes of gene expression determined by independent component analysis. Bioinformatics 18(1), 51–60 (2002)

3. Lee, S.I., Batzoglou, S.: Application of independent component analysis to microarrays. Genome Biology 4(11), 76 (2003)
4. Abrard, F., Deville, Y.: Blind separation of dependent sources using the "time-frequency ratio of mixtures" approach. In: Proceedings of Seventh International Symposium on Signal Processing and Its Applications, pp. 81–84 (July 2003)
5. Chang, C.Q., Ren, J., Fung, P., Hung, Y., Shen, J., Chan, F.: Novel sparse component analysis approach to free radical EPR spectra decomposition. Journal of Magnetic Resonance 175(2), 242–255 (2005)
6. Liao, J.C., Boscolo, R., Yang, Y.L., Tran, L.M., Sabatti, C., Roychowdhury, V.P.: Network component analysis: Reconstruction of regulatory signals in biological systems. Proceedings of the National Academy of Sciences of the United States of America 100(26), 15522–15527 (2003)
7. Tran, L.M., Brynildsen, M.P., Kao, K.C., Suen, J.K., Liao, J.C.: gnca: A framework for determining transcription factor activity based on transcriptome: identifiability and numerical implementation. Metabolic Engineering 7, 128–141 (2005)
8. Chang, C.Q., Ding, Z., Hung, Y.S., Fung, P.C.W.: Fast network component analysis for gene regulation networks. In: Proc. 2007 IEEE International Workshop on Machine Learning for Signal Processing, Thesaloniki, Greece (2007)
9. Chang, C.Q., Ding, Z., Hung, Y.S., Fung, P.C.W.: Fast network component analysis (FastNCA) for gene regulatory network reconstruction from microarray data. Bioinformatics 24, 1349 (2008)
10. Alon, U.: An Introduction to Systems Biology: Design Principles of Biological Circuits. Chapman & Hall/CRC (2007)
11. Golub, G.H., van Loan, C.F.: Matrix Computation, 3rd edn. The Johns Hopkins University Press (1996)
12. Dantzig, G.: Linear Programming and Extensions. Princeton Univ. Pr., Princeton (1963)
13. Luenberger, D.: Introduction to Linear and Nonlinear Programming. Addison-Wesley Pub. Co., Reading (1973)
14. GNU: GLPK (GNU Linear Programming Kit) [web page and software] (2008), http://www.gnu.org/software/glpk/glpk.html
15. Spellman, P.T., Sherlock, G., Zhang, M.Q., Iyer, V.R., Anders, K., Eisen, M.B., Brown, P.O., Botstein, D., Futcher, B.: Comprehensive identification of cell cycle-regulated genes of the yeast saccharomyces cerevisiae by microarray hybridization. Molecular Biology of the Cell 9(12), 3273–3297 (1998)
16. Lee, T.I., Rinaldi, N.J., Robert, F., Odom, D.T., Bar-Joseph, Z., Gerber, G.K., Hannett, N.M., Harbison, C.T., Thompson, C.M., Simon, I., Zeitlinger, J., Jennings, E.G., Murray, H.L., Gordon, D.B., Ren, B., Wyrick, J.J., Tagne, J.B., Volkert, T.L., Fraenkel, E., Gifford, D.K., Young, R.A.: Transcriptional regulatory networks in saccharomyces cerevisiae. Science 298(5594), 799–804 (2002)

ICA-Based Method for Quantifying EEG Event-Related Desynchronization

Danton Diego Ferreira[1,2], Antônio Maurício Ferreira Leite Miranda de Sá[1],
Augusto Santiago Cerqueira[2], and José Manoel de Seixas[1]

[1] Federal University of Rio de Janeiro/COPPE, Rio de Janeiro, Brazil
[2] Federal University of Juiz de Fora, Minas Gerais, Brazil
danton@lps.ufrj.br, amflms@peb.ufrj.br,
augusto.santiago@ufjf.edu.br, seixas@lps.ufrj.br

Abstract. A new technique for investigating synchronism in the electroencephalogram (EEG) during intermittent photic stimulation (IPS), based on the Independent Component Analysis (ICA) is proposed. It was tested with simulation EEG data and the coherence estimation-based methods were used for performance evaluation. The application of ICA-based allows a suitable investigation about event-related desynchronization (ERD) studies, since it can separate the phase-locked and time-locked spectra to the external stimulation.

Keywords: EEG, ICA, IPS, ERD, ERS, ERP.

1 Introduction

The electroencephalogram (EEG) is a complex signal representing the brain electrical activity that is recorded in the scalp. It results from the temporal and spatial summation of different rhythmic activities that may reflect distinct neurophysiologic mechanisms [1]. Such activities are frequently classified according to frequency bands (e.g. alpha, beta, theta and delta rhythms) [1], and hence the EEG analysis is more conveniently carried out in the frequency domain. It is well known that external stimulation may change the EEG spectrum and several methods have been developed to analyze and detect such changes [2]-[8]. The coherence function has become a standard tool in the quantitative analysis of the EEG, to measure the degree of synchronization between the signals obtained from different cortical areas [2]. It is analogous to correlations coefficient, but in the frequency domain.

Intermittent photic stimulation (IPS) in conjunction with coherence estimates have been proposed by some authors [3]-[6], since the stimulus may reduce the variability of mental states [6] as well as lead to distinct EEG patterns in groups with similar background activities [3]-[5]. However, intermittent stimulation may also change the ongoing EEG in a time-locked manner while not phase-locked. In such case, the changes are not synchronized to the external signal, but do occur during the stimulation process. This influence is often referred as event-related synchronization (ERS) and desynchronization (ERD) [7], depending on whether the stimulation leads to an increase or decrease in the power spectrum.

T. Adali et al. (Eds.): ICA 2009, LNCS 5441, pp. 403–410, 2009.

In ERS (ERD) studies, quantifying the degree of synchronization (desynchronization) in the ongoing EEG activity is an important task. The classical approach [7] consists of first bandpass filtering each event-related trial, squaring next the amplitude samples (to obtain power samples for each trial) and then averaging the pre-processed set of trials. Finally, this time evolution of an instantaneous band power is compared with that of the EEG without stimulation (e.g. the spontaneous activity). However, phased-locked (e.g. ERP) and non-phased-locked activities cannot be separated when they are within the same frequency band.

In [8], the partial coherence (i.e. the coherence between two signals after removal of the linear contribution from a group of other signals) removing the stimuli was proposed for quantifying the similarity between two EEG activities that are not phase-locked to the stimulating signal. It should allow a more suitable evaluation of the ongoing multichannel EEG relationship. Thus, this partial coherence estimate, together with simple coherence might be useful in order to quantify the relationship between cortical activities that would not be completely synchronized to the stimulating signal, but that could reflect time-locked spectral changes [8]. On the other hand, in [9], a statistical frequency-domain signal processing technique was proposed for separating the ongoing EEG activity spectrum from that of the ERP during external rhythmic stimulation. The methodology was based on the coherence estimate between the stimulation signal and the EEG and took into account only the data during stimulation.

In this context, the problem could be viewed as a Blind Source Separation (BSS) problem and the Independent Component Analysis (ICA) [10] could be used. ICA is a technique that, for a given linear mixture of source variables (observed variables), estimates the source variables based on the statistical independence of them.

In this paper, the ICA technique is used to remove the contribution from the stimulation signal from EEG signals and for separating the ongoing EEG activity spectrum from that of the ERP during external rhythmic stimulation. The results of the proposed method are then compared with the ones based in the coherence estimation.

2 Independent Component Analysis

The main goal of the ICA technique is to express a set of random variables as linear combinations of statistically independent component variables [10]. The basic model is represented by (1).

$$\mathbf{o}[k] = \mathbf{A}\mathbf{s}[k] \tag{1}$$

where $\mathbf{o}[k] = [x_1[k], x_2[k], ..., x_M[k]]^T$ is the observed data vector at sample k, $\mathbf{s}[k] = [s_1[k], s_2[k], ..., s_M[k]]^T$ is the statistically mutually independent component vector at sample k and $\mathbf{A}$ is a $M \times N$ scalar matrix which is called mixing matrix.

In the model, only the random variables $o_i[k]$ are known and the mixing matrix should be estimated using only $o[k]$. The goal of the ICA is to find a separation matrix $\mathbf{W}$ that makes the outputs as independent as possible:

$$\mathbf{e}[k] - \mathbf{W}o[k] \tag{2}$$

where $\mathbf{e}[k] = [e_1[k], e_2[k], ..., e_M[k]]^T$ provides the estimations of the independent components $s_i[k]$.

For the estimation of data model, a suitable objective function should be chosen for optimization. In this paper, the ICA is based on the maximization of the nongaussianity. The nongaussianity can be measured by kurtosis or negentropy [11].

The FastICA algorithm [12], used in this work, is based on a fixed-point iteration scheme. It basically consists of two steps, the preprocessing step and the FastICA algorithm itself. The preprocessing consists of centering and whitening data. The centering step is done by subtracting the mean of the observed data x_i. Therefore, the result of this step is a zero mean data. A whitening step is used to remove the correlation between the observed data. The whitening reduces the number of parameters to be estimated.

The FastICA finds a direction in which $e = \mathbf{w}^T\mathbf{z}$ has the maximal nongaussianity, where $\mathbf{z}$ is the mixture vector after the whitening step. The final vector $\mathbf{w}$ gives one of the independent components as the linear combination $\mathbf{w}^T\mathbf{z}$. The directions of the independent components are orthogonal in the whitened space, therefore, the second independent component can be found as the orthogonal direction of $\mathbf{w}$ corresponding to the first estimated independent component. For more dimensions, we need to rerun the algorithm, always constraining the current $\mathbf{w}$ to be orthogonal to the previously estimated vectors $\mathbf{w}$.

3   Coherence-Based Methods

The (magnitude-squared) coherence estimate between the discrete-time signals $y_1[k]$ and $y_2[k]$ may be obtained based on the approach of dividing them into M independent segments and performing the Fourier Transforms of each data segment. This yields to the following estimator [13]:

$$\hat{\gamma}^2_{y_1 y_2}(f) = \frac{\left|\sum_{i=1}^{M} Y_{1i}^*(f)Y_{2i}(f)\right|^2}{\sum_{i=1}^{M}|Y_{1i}(f)|^2 \cdot \sum_{i=1}^{M}|Y_{2i}(f)|^2}, \tag{3}$$

where $Y_{1i}(f)$ and $Y_{2i}(f)$ are, respectively, the i-th window Fourier transform of $y_1[k]$ and $y_2[k]$ at frequency f, "^" and "*" superscript denotes, respectively, an estimation and the conjugate.

Considering the linear model of Figure 1 with a periodic input signal ($x[k]$), the partial coherence estimate expression between the output signals removing the contribution from the input, can be obtained as [8]:

$$\hat{\kappa}^2_{y1y2\bullet}(f) = \frac{\left|\sum_{i=1}^{M} Y_{1i}^*(f)Y_{2i}(f) - \frac{1}{M}\sum_{i=1}^{M} Y_{1i}^*(f)\sum_{i=1}^{M} Y_{2i}(f)\right|^2}{\left[\sum_{i=1}^{M}|Y_{1i}(f)|^2 - \frac{1}{M}\left|\sum_{i=1}^{M} Y_{1i}(f)\right|^2\right] \cdot \left[\sum_{i=1}^{M}|Y_{2i}(f)|^2 - \frac{1}{M}\left|\sum_{i=1}^{M} Y_{2i}(f)\right|^2\right]} \tag{4}$$

where "*" superscript denotes complex conjugate. Such expression can be obtained by simplifying the general matrix expression provided in [14] for the particular case when $x[k]$ is periodic. It has been developed in [2] for estimating the coherence between two Electroencephalographic (EEG) signals during sensory stimulation after removal of this latter. It hence aims at measuring the similarity between the background EEG activity, which is not accounted for by the stimulation.

Critical values for $\hat{\gamma}^2_{y_1y_2}(f)$ are provided in [15] and for $\hat{\kappa}^2_{y1y2\bullet}(f)$ can be obtained with the multivariate extension of the invariance of coherence statistics when one signal is Gaussian and coherence is zero [16] as:

$$\hat{\kappa}^2_{y1y2\bullet} crit = beta_{crit}(1, M-2) \tag{5}$$

where $beta_{crit}(1,M-2)$ is the critical value of the beta distribution with parameters $p=1$ and $q=M-2$ for a given significance level.

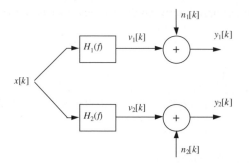

Fig. 1. Linear model used in deriving $\hat{\gamma}^2_{y_1y_2}(f)$ and $\hat{\kappa}^2_{y1y2\bullet}(f)$. $x[k]$ is the stimulus, $v_1[k]$ and $v_2[k]$ are the responses [output of the filters $H_1(f)$ and $H_2(f)$], and $n_1[k]$ and $n_2[k]$ are the contributions of background activity to the measured EEG signals $y_1[k]$ and $y_2[k]$

For the model of Fig. 1, the auto-spectra of signals $v_i[k]$ (evoked response) and $n_i[k]$ (ongoing EEG activity), for i = 1, 2, may be expressed, since the stimulation signal $x[k]$ is periodic, as [8]:

$$\hat{S}_{v_iv_i}(f) = \hat{S}_{y_iy_i}(f) \cdot \hat{\kappa}^2_{y_i}(f) \text{ and } \hat{S}_{n_in_i}(f) = \hat{S}_{y_iy_i}(f)[1 - \hat{\kappa}^2_{y_i}(f)] \tag{6}$$

where $\hat{S}_{y_iy_i}(f) = \dfrac{1}{MT}\sum_{i=1}^{M}|Y_i(f)|^2$ is the auto-spectra of $y_i[k]$ and T is the window duration in seconds. $\hat{\kappa}^2_{y_i}(f)$ is the simplified coherence between $x[k]$ (periodic) and $y_i[k]$:

$$\hat{\kappa}^2_y(f) = \frac{\left|\sum_{j=1}^{M}X^*(f)Y_{ji}(f)\right|^2}{\sum_{j=1}^{M}|X(f)|^2 \cdot \sum_{j=1}^{M}|Y_{ji}(f)|^2} = \frac{|X(f)|^2\left|\sum_{j=1}^{M}Y_{ji}(f)\right|^2}{M|X(f)|^2 \cdot \sum_{j=1}^{M}|Y_{ji}(f)|^2} = \frac{\left|\sum_{j=1}^{M}Y_{ji}(f)\right|^2}{M\sum_{j=1}^{M}|Y_{ji}(f)|^2} \tag{7}$$

In (7) it is implicitly assumed that the M windows are chosen as to contain an integer number of fundamental periods of signal $x[k]$. In such a case the Fourier Transform of the input signal will have the same value in all windows leading to the simplification above.

4   Results

In this section, the proposed method will be evaluated from simulation EEG data. The basic idea of the proposed method is first to apply the ICA on the EEG signals aiming at the removal of the stimulation from them.

For the model of Fig. 1, $x[k]$ was generated as an unit impulse train and both $H_1(f)$ and $H_2(f)$ were set constant $(=a)$. The background noise terms $n_1[k]$ and $n_2[k]$ were obtained as the sum of two parcels - an identical activity (band-filtered white noise within (9-13Hz)) and another uncorrelated activity (unit-variance Gaussian signals).

In order to remove the contribution of the stimulation signal ($x[k]$) from $y_1[k]$ and $y_2[k]$, modeled according to Fig. 1, FastICA algorithm was applied in the mixture matrix consisting of $x[k]$, $y_1[k]$ and $y_2[k]$. The estimated source spectrum by ICA are shown in Fig. 2a, 2c and 2e, respectively from $y_1[k]$, $y_2[k]$ and $x[k]$. Figures 2b, 2d and 2f show the spectrum of $y_1[k]$ and $y_2[k]$ without stimulation and $x[k]$, respectively. It is important to note the similarity between the estimated spectrum and the original ones.

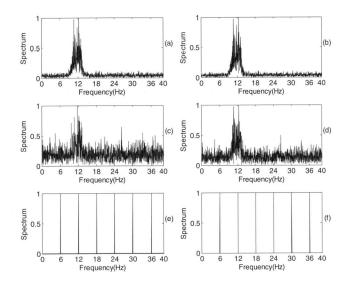

Fig. 2. Normalized spectrum of estimated sources by ICA in (a), (c) and (e); and normalized spectrum of signals y1[k] and y2[k] without stimulation and x[k]

Another way to analyze the ICA-based method for the removal of the contribution from the stimulation is to estimate the simple coherence between the estimated sources by ICA and compare with the partial coherence estimation between the signals $y_1[k]$ and $y_2[k]$ removing the stimulation $x[k]$. This comparison is shown in Fig. 3a. Fig. 3b shows the simple coherence estimation (dashed line) between $y_1[k]$ and $y_2[k]$ and simple coherence estimation (solid line) between estimated sources by ICA referring to $y_1[k]$ and $y_2[k]$. In all these spectral estimations, the window length was set equal to 2 s, leading to $M=12$ data segments in each estimation. This lets to a critical value ($\alpha=5\%$) of 0.259 for $\hat{\kappa}^2_{y1y2\bullet}(f)$, according to (5). The coherence critical value was found to be 0.283 (from [17]). It can be seen that the ICA-based method clearly removes the peaks due to the periodic input signal and preserves the wide peak that occurs within 9-13 Hz, which is due to the nonsynchronized simulated background activity on both signals.

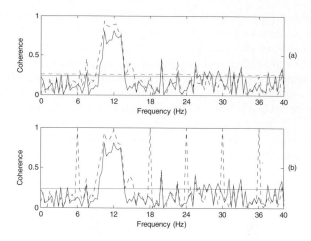

Fig. 3. (a) Simple coherence estimation $\hat{\gamma}^2_{y_1y_2}(f)$ (solid line) between ICA estimated sources refering to y1[k] and y2[k] and the partial coherence estimation $\hat{\kappa}^2_{y_1y_2\bullet}(f)$ (dashed line) between the simulated signals y1[k] and y2[k]. (b) The coherence estimation $\hat{\gamma}^2_{y_1y_2}(f)$ (dashed line) between y1[k] and y2[k] and the coherence estimation $\hat{\gamma}^2_{y_1y_2}(f)$ (solid line) between ICA estimations refering to y1[k] and y2[k]. The signals y1[k] and y2[k] were obtained with an 6-Hz impulse train stimulation (x[k]), according to the model of Fig. 1. The respective critical values are indicated in horizontal lines.

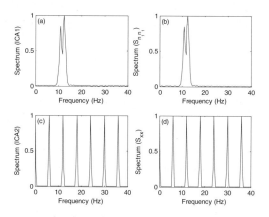

Fig. 4. Normalized spectrum of estimated sources by ICA in (a) and (c), and normalized spectrum of original signals $n_1[k]$ and $x[n]$ in (b) and (d), respectively

Next, FastICA algorithm was applied in the mixture matrix consisting of $x[k]$ and $y_1[k]$. The goal of this application is to obtain the EEG background ($n_1[k]$) estimation. The estimated source spectrums by ICA are shown in Fig. 4a and 4c. Figures 4b and 4d show the spectrum of $n_1[k]$ and $x[k]$, respectively. It is important to note the similarity between the estimated spectrum and the original ones.

Figure 5 illustrates the estimated spectrum of signal $n_1[k]$ using equation (6). It's easy to see that this technique based in coherence can not to estimate the EEG background ($n_1[k]$) very well.

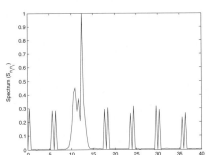

Fig. 5. Estimative of normalized spectrum of $n_1[k]$ obtained by equation (6)

5 Conclusions

First, the ICA was applied on EEG signals in order to remove the stimulation from the acquired signals. The coherence between the derivations were obtained and compared with the one resulted from the partial coherence. Both methods show similar results, but with the ICA method the EEG signals without the contribution of stimulation cold be estimated, which is not achieved by the use of partial coherence. Next, the EEG background ($n_1[k]$) was estimated by ICA and compared ones obtained by coherence-based method. The results showed that the ICA method won best performance. The application of this methodology (ICA-based) allows a suitable investigation of the effect of stimulation in the ongoing activity, as well as separating phase-locked and time-locked spectra to the external stimulation. Thus, they could be useful in the ERD/ERS studies. Additionally, these results were obtained using a linear model (see Fig. 1) and, for nonlinear models, the coherence-based method could produce erroneous results. For future studies, the ICA-based method here proposed will be performed together with the Second-Order Blind Inference (SOBI) method using real EEG data.

References

1. Niedermeyer, E.: The Normal EEG of the Waking Adult. In: Niedermeyer, E., Lopes da Silva, F. (eds.) Electroencephalography - Basic Principles, Clinical Applications, and Related Fields, 3rd edn., ch. 9. Williams &Wilkins, New York (1993)
2. Nunez, P.L., Srinivasan, R., et al.: EEG coherency: I statistics, reference electrode, volume conduction, Laplacians, cortical imaging, and interpretation at multiple scales. Electroenceph. Clin. Neurophysiol. 103, 49–515 (1997)
3. Wada, Y., Nanbu, Y., Kicuchi, Y., Koshino, Y., Hshimoto, T.: Aberant functional organization in shizophrenia: Analysis of EEG coherence during rest and photic stimulation iin drug-naïve patients. Neuropsychology 38, 63–69 (1998)
4. Wada, Y., Nanbu, Y., Koshino, Yamaguchi, N.Y., IIshimoto, T.: Reduced interhemispheric EEG coherence in Alzheimer disease: Analysis during rest and photic stimulation. Alzheimer Dis. Assoc. Disord. 12(3), 175–181 (1998)
5. Jin, Y., Sandman, C.A., Wu, J.C., Bernat, J., Potkin, S.G.: Topographic analysis of EEG photic driving in normal and schizophrenic subjects. Clin. Electroenchephal. 26(2), 102–107 (1995)

6. Kaiser, J., Gruzelier, J.H.: Timing of puberty and EEG coherence during photic stimulation. J. Psychophysiol. 21, 135–149 (1996)
7. Pfurtscheller, G., Lopes da Silva, F.H.: Event-related EEG/MEG synchronization and desynchronization: basic principles. Clin. Neurophysiol. 110(11), 1842–1857 (1999)
8. Miranda de Sá, A.M.F.L., Infantosi, A.F.C.: Evaluating the relationship of non-phase locked activities in the electroencephalogram during intermittent stimulation: a partial coherence-based approach. Med. Bio. Eng. Comput. 45, 635–642 (2007)
9. Infantosi, A.F.C., Miranda de Sá, A.M.F.L.: A coherence-based technique for separating phase-locked from non-phase-locked power spectrum estimates during intermittent stimulation. Journal of neuroscience methods 156(1-2), 267–274 (2006)
10. Hyvärinen, A., Karhunen, J., Oja, E.: Independent Component Analysis. John Wiley & Sons, Chichester (2001)
11. Comon, P.: Independent component analysis – a new concept? Signal Processing 36, 287–314 (1994)
12. Hyvärinen, A.: Fast and robust fixed-point algorithms for independent component analysis. IEEE Transactions on Neural Networks 10(3), 626–634 (1999)
13. Bendat, J., Piersol, A.: Random Data Analysis and Measurement Procedures, 3rd edn. Wiley-Interscience, New York (2000)
14. Otnes, R.A., Enochson, L.: Applied Time Series Analysis. Basic Techniques, vol. 1, pp. 374–376. John Wiley and Sons, New York (1978)
15. Miranda de Sá, A.M.F.L., Iinfantosi, A.F.C., Simpson, D.M.: A statistical technique for measuring synchronism between cortical regions in the EEG during rhythmic stimulation. IEEE Trans. Biom. Eng. 48, 1211–1215 (2001)
16. Nuttall, A.H.: Invariance of distribution of coherence estimate to second-channel statistics. IEEE Trans. Acoust. Speech, Signal Processing, ASSP-29, 120–122 (1981)
17. Miranda de Sá, A.M.F.L.: A note on the sampling distribution of coherence estimate for the detection of periodic signals. IEEE Signal Processing Letters 11, 225–323 (2004)

Localization of Generalized Discharge Sources in Patients with Juvenile Myoclonic Epilepsy Using Independent Components Analysis

Sidcley P. Gomes, Patrícia da Silva Sousa-Carvalho*, Enio Aguiar,
Anderson B. Silva, Fábio M. Braga, and Allan Kardec Barros

UFMA - Federal University of Maranhão, São Luís, Brazil

Abstract. One important information for the classification of epilepsy
is the spatial localization of the discharges source. Juvenile myoclonic
epilepsy (JME) is an idiopathic generalized epilepsy (IGE) that typically presents generalized tonic-clonic, myoclonic, or absence seizures, or
a combination of these. In typical cases of JME, the seizures are usually
bilateral and symmetric, and EEG shows generalized interictal epileptiform discharges and a generalized seizure pattern that also is bilaterally
synchronous. Despite of the generalized pattern of this type of epilepsy,
there are some electroencephalographic and clinical features that suggest a frontal origin for the discharges. In this work, EEG recordings
were analyzed in order to find evidences for this frontal origin in JME.
The analysis of the signals was based on independent component analysis (ICA) for separating epileptiform discharges from artifacts and other
brain sources; then the discharge components were used to spatially localize its source. In the six patients the dipole sources were localized
mainly in the frontal region, what suggests an important participation
of the frontal lobe for this kind of epilepsy.

Keywords: Juvenile myoclonic epilepsy, electroencephalography,
independent component analysis, generalized discharges, dipole source
localization.

1 Introduction

Juvenile myoclonic epilepsy (JME) is an idiopathic generalized syndrome that
appear typically around puberty and it is characterized by myoclonic jerks, absence and generalized tonic-clonic seizures. The electroencephalographic (EEG)
recording in patients with this type of epilepsy usually shows generalized epileptiform discharges such as generalized spikes, polyspikes, spike-wave complexes,
or combinations of these [1]. Due its generalized character on the EEG recording,
the origin of discharges have been under debate. However, a few reports concern
clinical or EEG focality or asymmetry or both in patients with JME [2].

---

* Child neurologist and neurophysiologist of São Domingos Hospital, São Luís, Brazil
and professor of Federal University of Maranhão, São Luís, Brazil.

T. Adali et al. (Eds.): ICA 2009, LNCS 5441, pp. 411–418, 2009.

Generalized spike-wave (GSWC) and polyspike-wave complexes (PSWC), with fronto-central accentuation are the typical EEG pattern shown in JME [3]. Arzimanoglou et al [4] attended in some patients with JME that, on the EEG recording, there is an onset and maximal voltage in the frontocentral regions, then spreading to the parietal, temporal, and occipital regions, suggesting thus a possible frontal origin for JME. The problem is that to analyze epilepsy, principally generalized, only through the voltage of raw EEG is not reliable, since there are many artifacts that contaminate the EEG recording, and consequently tend to change the electrical potential of the discharges sources [5].

Brain topographic maps and source analysis with dipole modeling have been used to study spatial distribution of GSWC and PSWC in JME. Santigo-Rodrigues et al. [6] applied dipole modeling and brain distributed analysis found that pre-frontal medial current sources corresponding to spikes and many diffuse sources in cortical regions corresponding to wave components of PSWC in JME.

Bianchi et al. [7] used mutual synchronization among EEG channels in patients with idiopathic generalized seizures, Juvenile Absence Epilepsy (JAE) and JME. In JAE patients, a leading role of the left frontal lobe was evidenced in correspondence to the seizure, while in JME patients a leading region was not identified. Thus, the location of focuses for JME remains an open issue.

Currently, a new approach has been proposed for preprocessing of EEG, the Independent Components Analysis (ICA) [8]. This method, originally proposed to solve the blind source separation problem, has been applied for several kinds of signals (e.g. voice, music, electrocardiography and other signals). In the case of EEG, studies have demonstrated that ICA is an efficient tool in the separating artifacts from raw EEG[9]. Further, recently, Kobayashi et al [10] demonstrated that ICA can separate two spikes that occurred at approximately the same time, but that had different waveforms and spatial distributions.

This finding have induced some researchers to use ICA not only to separate artifacts from EEG, but also try to separate epileptiform discharges waveforms present on the EEG recordings in patients with epilepsy [11], [12]. Using this approach, they successfully localized discharge sources in Creutzfeldt-Jakob disease [13], primary versus secondary bilateral synchrony [11]. This way we propose in this study, the dipole source localization of epileptiform discharges source in patients with JME using ICA to process the EEG signals.

2 Methods

2.1 Patients

Six patients with JME were included in this study, raging from age from 15-38 yrs, being two males and four females. EEG data were collected on a local data after approval of the hospital ethical committee. It was sampled at 128 Hz. The electrodes impedance was always less than 5 $k\Omega$, and they were placed using the 10/20-electrode International System.

2.2   Data Processing and Analysis

We analyzed separated episodes of spike-and-wave activity in all the six patients. The first three showed ten episodes each, and the last three presented eight, four and three respectively. For each patient the activities were concatenated giving the total time of recording of 13 s for two patients and 31, 33, 34 and 35 s for the others. On fig. 1(a) it is shown ten seconds of EEG presenting three concatenated discharges of patient one.

Nineteen EEG electrodes were placed on the scalp, creating a total of 18 channels. All the channels were referenced to lead CZ. Periods of discharges from the raw digital EEG were determined by a visual inspection of a specialized doctor for each subject. We applied a extended version of Independent Component Analysis called extended infomax ICA algorithm, that separates sources that have either supergaussian or subgaussian distributions, allowing line noise, which is subgaussian, to be focused efficiently into a single source channel and removed from the data [15].

2.3   Dipole Localization

The recorded EEG consists in the mixture of different activities. In other words, each electrode registers different cortical dipole sources added to artifacts, such as: eye blinks, ocular movement, muscular artifacts and others. The great challenge of researchers is finding the origin of this cortical generator through EEG. In order to solve this inverse problem, it is necessary to build a model for the head as being a volume conductor and a model for the cortical sources, usually as dipole sources, as described in [16]. We assumed a spherical head model with conductivities (mhos/m) of 0.33 for the scalp, 0.0042 for the skull, 1.00 for cerebrospinal fluid, and 0.33 for the brain. The radii of the spheres were standardized to 85 mm for the scalp, 79 mm for the skull. 72 mm for the cerebrospinal fluid, and 71 mm for the brain. Goodness of fit was estimated in terms of residual variance (R.V), i.e. the percentage of the spike variance that could not be explained by the model. The head model used is already implemented in the EEGlab toolbox for Matlab (The Mathworks, Natick, MA, USA) developed by Delorme and Makeig [14].

However, since the EEG is a record of various different brain signals in addition to artifacts such as, eyes blink, ocular movement, muscular artifact, the application of source localization in this model to raw EEG data could erroneously localize sources. For this reason many methods have been proposed with the goal of filtering or separating the brain originated sources from these artifacts. A method currently much used in the separation of the artifacts from interest signals is independent components analysis, and we proposed applying this algorithm in the solution of this research

2.4   Independent Component Analysis

Independent Component Analysis (ICA) is a method that finds underlying factors or components from multivariate data. It is assumed that linear mixtures of the independent components form input data records, and there is no need for detailed models of either the dynamics or the spatial structure of the separated components [5]. Even if two processes occur simultaneously, ICA can be used to separate them. Whether we consider the EEG record like a linear combination of several statistically independent sources (like artifacts or other cortical sources) then ICA is able to separate these sources. Recently, Kobayashi et al [10] demonstrated that ICA worked for the separation of two spikes that occurred at approximately the same time, but that had different waveforms spatial distributions. In the same way, ICA could be a reliable method for separating generalized spike-and-wave complexes into patients with JME.

The basic ICA model is:

$$\mathbf{X} = \mathbf{AS} \tag{1}$$

where $\mathbf{X} = [\mathbf{x_1}, \mathbf{x_2}, ..., \mathbf{x_M}]^{\mathbf{T}}$ corresponds to M scalar valued vectors of observations, while $\mathbf{S} = [\mathbf{s_1}, \mathbf{s_2}, ..., \mathbf{s_N}]^{\mathbf{T}}$ are N sources that are linearly mixed by the $M \times N$ matrix $\mathbf{A}$. Assuming that the sources in $\mathbf{S}$ are statistically independent it is possible to estimate the original sources $\mathbf{S}$ and the mixing matrix $\mathbf{A}$ using only $\mathbf{X}$. There are several methods to estimate $\mathbf{S}$ and $\mathbf{A}$; in this work we used the Extended Infomax algorithm [15]. The Infomax algorithm tries to minimize the mutual information among the sources through high order statistical moments. However, it has been showed that the Infomax is limited to separating sources with only supergaussian distributions, but some kinds of noises have subgaussian distributions. So the Extended Infomax algorithm was proposed with the goal of separating sources with both supergaussian and subgaussian distributions.

Here we assume that the EEG data is a mixture of several temporally and spatially independent sources, being the number sources equal to the number of sensors. It is this property of being temporally insensitive that allows the EEG to be divided into sections, so that many epochs can be concatenated together to look for features constant across discharges.

To localize any component, we reconstructed the data using only this component (i.e., setting all columns in the estimated mixing matrix to zero, except the appropriate column associated with this component). The projected component data has the same size of the original data and each row is a single electrode, as in the original data, and is scaled in the original data units (e.g. μV).

3   Results

From the EEG recordings, the selected discharge epochs were concatenated in a single matrix for each patient. ICA algorithm was applied on these data. Fig. 1(a) shows ten seconds of the EEG row with three concatenated discharges selected from patient one. On Fig. 1(b) we see the 18 independent components separated by the ICA algorithm. The components 2, 3, 6 and 9 show characteristics of epileptiform activities. For instance, at 4.5 s and 5.5 s, we found spike-and-wave patterns, typical of epileptiform discharges, while the other independent components are likely to be artifacts and brain underlying activities.

The EEG signal was reconstructed using each independent component that accounted for epileptiform activity separately. For each reconstructed EEG data was selected about 0.2 seconds of discharges (0.1 s before and 0.1 s after the maximum negative amplitude), and mapped in the scalp topography map to estimate the origin of these sources. Fig. 2 shows the distributions of potentials of the independent components 2, 3, 6 and 9 of patient one, within the head. This scalp topography map reveals largely overlapping of components in the frontal regions. The dark values on the head figures show greater contribution (either positive or negative) of a channel to the given component spatial distribution.

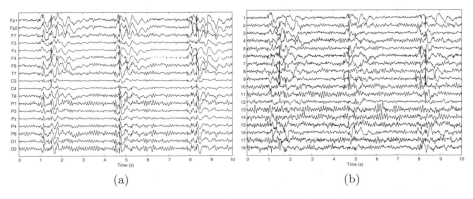

(a)   (b)

Fig. 1. Here, we show three epochs exhibiting epileptiform discharges and the corresponding independent sources. In (a) the concatenated 18 channel EEG data of patient one and (b) the resulting separated sources using ICA.

The source localization was made on the standard 3D MRI head model, Fig. 3. The component 2 showed bilateral activity on the frontal lobes, thus a second dipole needed to be added to show the bilateral activity on this region(represented in the Fig. 3 by blue dipole). Similarly the components 3, 6 and 9 (represented by green, red and pink dipole), that had singular dipole models, was localized on the frontal region in the MRI image on Fig. 3.

On Fig. 2 we see the potential distribution on the scalp maps of the three discharge components that were selected from patient two. Two components are localized on the frontal regions and one on the central region. On Fig. 3 it is shown these components on the head model. The central component had dipole fitting with residual variance (RV) of 22%. If the best-fitting single or dual equivalent estimated dipole has more than 15% residual variance from the spherical forward-model scalp projection (over all scalp electrodes), the component is not further analyzed. Components with equivalent dipole(s) located outside of the model brain volume were also excluded from analysis [17].

For the six patients analyzed, there were about 17 independent components that could be visually accounted as discharge components. In average, about 70.59% (12 of 17) of the independent components had dipole sources within the frontal region, 11.77% (2 of 17) of the components within central regions and 17.65% (3 of 17) had dipole fitting located outside the brain model or had RV

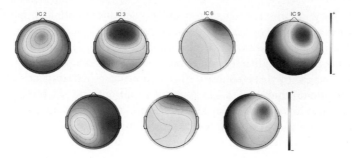

Fig. 2. Scalp topography maps of the independent components of patients one (up) and two (down). It was taken 0.1 s before and 0.1 s after the negative maximum potential of arbitrary discharges. There is an overlapping of components in the frontal regions for both patients, but for patient two there is one component in the central region. Darker values (red or blue) on the maps show greater contribution (either positive or negative) of a channel to the given component spatial distribution. The component two of patient one displays clear symmetrical activity.

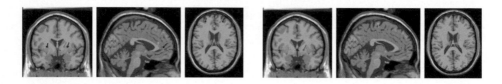

Fig. 3. Localization and orientation of the dipoles in the standard 3D MRI head model, referent to selected independent components of patient one (left) and two (right). The orientations of all dipoles are preferentially radial for patient one, but for patient two the orientation of the green and blue dipoles is radial, while the orientation of red one is tangential. The central component of patient two was not accounted for dipole model fitting since its RV was more than 15%.

more than 15%. About 85.71% (12 of 14) of the fitted dipoles had unilateral symmetry and 14.29% (2 of 14) had bilateral symmetry. On Table 1, we have a summary of each patient's components and dipole analysis.

4   Discussion

It is important in this evaluate whether ICA can be used to separate epileptiform discharges in patients with JME. ICA was processed to GSWD data from EEG recordings. The components that visually accounted as GSWD were selected and reconstructed to be fitted in the head model. The results shown that few independent components (1 to 4) was necessary to represent discharge activities. Similarly, Rodin et al [18] analyzed dipole sources of brain activities, and showed that GSWD of absence seizures could be modeled adequately using three or four equivalent regional dipole sources.

Our results showed that there is a large overlapping of the estimated independent components that accounted as epileptiform discharges in the frontal regions, being few components localized in extrafrontal regions. Analyzing Table 1, almost all components are unilateral with radial orientation. The presence of one patient with bilateral dipole suggest that discharges may spread via the corpus callosum from frontal regions to other regions [11]. These factors suggest an important participation of the frontal regions for this kind of epilepsy, corroborating with the physiological investigations of Santiago et al. [6].

Table 1. Characteristics of the independent components that accounted as epileptiform discharges for each patient

Patients	1	2	3	4	5	6	Total
n. of components	4	2	2	1	2	3	14
Orientation							
Radial	4	2	2	1	1	2	12
Tangential	-	-	-	-	1	1	2
Location							
Frontal	4	2	2	1	1	2	12
Extra-frontal	-	-	-	-	1	1	2
Laterality							
Unilateral	3	2	2	1	1	2	12
Bilateral	1	-	-	-	-	1	2
							Mean
mean RV	3.04%	3.37%	3.19%	7.30%	9.37%	5.96%	5.37%

However, as proposed by Jung et al. [11], a more precise estimative of brain sources location should be made using more electrodes. We attempted to use more signal inputs on bipolar reference, but without success, since independent components were badly estimated, perhaps due the redundant information in the channels of disposable data. Also, despite of a small number of patients our study suggests that ICA models could be useful to analyze EEG records of JME. More reliable and precise investigations concerning the participation of JME discharges in the frontal regions should be made using more patients. Other localization methods should be used to corroborate our hypotheses.

References

1. Commission on Classification and Terminology of the International League Against Epilepsy. Proposal for revised classification of epilepsies and epileptic syndromes. Epilepsia 30, 389–99 (1989)
2. Lancman, M.E., Asconape, J.J., Kiffin Penry, J.: Clinical and EEG asymmetries in juvenile myoclonic epilepsy. Epilepsia 35, 302–306 (1994)
3. Janz, D., Durner, M.: Myoclonic epilepsy. In: Engel Jr., J., Pedley, T.A. (eds.) Epilepsy: A Comprehensive Textbook, pp. 2389–2400. Lippinicott-Raven, Philadelphia (1998)

4. Arzimanoglou, A., Guerrini, R., Aicardi, J.: Epilepsies with predominantly my-oclonic seizures. In: Arzimanoglou, A., Guerrini, R., Aicardi, J. (eds.) Aicardi's epilepsy in children, pp. 58–80. Lippincott Williams & Wilkins, Philadelphia (2004)
5. Makeig, S., Jung, T.P., Bell, A.J., Ghahremani, D., Sejnowisk, T.J.: Blind separa-tion of auditory event-related brain responses into independent components. Proc. Nat. Acad. Sci. USA 94, 10979–10984 (1997)
6. Santigo-Rodriguez, E., Harmony, T., Fernandez-Bouzas, A., Hernandez-Balderas, A., Marrinez-Lopes, M., Graef, a., Garcia, J.C., Fernandez, T.: Source analysis of polyspike and wave complexes in juvenile myoclonic epilepsy. Seizure 11, 320–324 (2002)
7. Bianchi, A.M., Panzica, F., Tinello, F., Franceschetti, S., Cerutti, S., Baselli, G.: Analysis of multichannel eeg synchronization before and during generalized epilep-tic seizures. In: Conference on Neural Engineering, Capri Island, March 20-22 (2003)
8. Comon, P.: Independent component analysis-A new concept? Signal Processing 36, 287–314 (1994)
9. Makeig, S., Bell, A.J., Jung, T.P., Sejnowski, T.J.: Independent component analysis of electroencephalographic data. In: Touretzky, D., Mozer, M., Hasselmo, M. (eds.) Advances in neural information processing systems, vol. 8, pp. 145–151. MIT Press, Cambridge (1996)
10. Kobayashi, K., James, J.J., Nakahori, T., Akiyama, T., Gotman, J.: Isolation of epileptiform discharges from unaveraged EEG by independent component analysis. Clin. Neurophysiol. 112, 1755–1763 (2001)
11. Jung, K.Y., Kimb, J.M., Kimc, D.W., Chung, C.S.: Independent component anal-ysis of generalized spike-and-wave discharges: primary versus secondary bilateral synchrony. Clinical Neurophysiology 116, 913–919 (2005)
12. McKeown, M.J., Humphries, C., lragui, V., Sejnowski, T.J.: Spatially Fixed Pat-terns Account for the Spike and Wave Features in Absence Seizures. Brain Topog-raphy 12(2) (1999)
13. Jung, K.-Y., Seo, D.-W., Na, D.L., Chung, C.-S., Lee Il, K., Oh, K., Im, C.-H., Jung, H.-K.: Source localization of periodic sharp wave complexes using independent component analysis in sporadic Creutzfeldt-Jakob disease. Source localization of periodic sharp wave complexes using independent component analysis in sporadic Creutzfeldt-Jakob disease. Brain Res. 1143, 228–237 (2007)
14. Delorme, A., Makeig, S.: EEGLAB: an open source toolbox for analysis of single-trial EEG dynamics including independent component analysis. J. Neurosci. Meth-ods 134, 9–21 (2004)
15. Lee, T., Girolami, M., Sejnowski, T.: Independent component analysis using an extended infomax algorithm for mixed sub-gaussian and super-gaussian sources. Neural Computation (in press, 1998)
16. Scherg, M., Ebersole, J.S.: Models of brain sources. Brain Topogr. 5, 419–423 (1993)
17. Onton, J., Delorme, A., Makeig, S.: Frontal midline EEG dynamics during working memory. NeuroImage 27(2), 341–356 (2005)
18. Rodin, E.A., Rodin, M.K., Thompson, J.A.: Source analysis of generalized spike-wave complexes. Brain Topogr. 7, 113–119 (1994)

Iterative Subspace Decomposition for Ocular Artifact Removal from EEG Recordings

Cédric Gouy-Pailler[1,*], Reza Sameni[2], Marco Congedo[1], and Christian Jutten[1]

[1] GIPSA-lab, DIS/CNRS/INPG-UJF-Stendhal
Domaine Universitaire, BP 46, 38402 Saint Martin d'Hères Cedex, France
[2] College of Electrical and Computer Engineering, Shiraz University, Shiraz, Iran
cedric.gouypailler@gipsa-lab.inpg.fr

Abstract. In this study, we present a method to remove ocular artifacts from electroencephalographic (EEG) recordings. This method is based on the detection of the EOG activation periods from a reference EOG channel, definition of covariance matrices containing the nonstationary information of the EOG, and applying generalized eigenvalue decomposition (GEVD) onto these matrices to rank the components in order of resemblance with the EOG. An iterative procedure is further proposed to remove the EOG components in a deflation fashion.

1 Introduction

Electroencephalography (EEG) is a widely used technique for analyzing and interpreting human cerebral activity. EEG signals are usually interpreted by means of spectral and topographical measures that reflect global activity of the brain network. However, EEG measures are always contaminated by non-cerebral signals, which may disturb the interpretation of the brain activity. This issue has become a recurrent problem, for example in Brain-Computer Interfaces (BCI), where it has been proved to decrease classification error rates [11].

Ocular artifacts generally occur during blinking or saccades of the eye, and are featured by high amplitude transient artifacts that defect the EEG. They are best recorded by an Electrooculogram (EOG) electrode or a pair of electrodes located close to the eyes. The high amplitude peaks are not seen on all channels, but mainly (and almost exclusively) on the fronto-paretal channels in combination with the occipital electrodes. These peaks are considered as one of the most considerable artifacts in EEG studies [7].

A common way of removing these artifacts is to apply independent component analysis (ICA) on multichannel EEG recordings and to remove the components which show a maximal correlation with a reference EOG channel [6,5]. However, it is not always possible to associate the components extracted by ICA to the EOG in an automatic and unsupervised manner. Moreover, EEG recordings can

---

* This work was supported by Égide, the French Ministry of Defense (Délégation Générale pour l'Armement – DGA), the Open-ViBE project, and the European program of Cooperation in the Field of Scientific and Technical Research.

T. Adali et al. (Eds.): ICA 2009, LNCS 5441, pp. 419–426, 2009.

be rather noisy, and since ICA is based on a measure of independence (and not a measure of signal "cleanness"), the noise in the input channels can be amplified in the output, which again makes the detection of the true EOG component rather difficult. Lastly, most ICA methods are blind to Gaussian noise and as a consequence spread the Gaussian noise among the components.

Another common way of removing EOG artifacts from the EEG is to use a subtraction-based approach [8]. Here, the idea is to use the clean EOG recordings to remove ocular signals from EEG by a simple subtraction of a scaled EOG. Yet there is no evidence that EOG recordings are free of EEG. Thus by subtracting EOG we can also remove EEG signals of interest.

In recent works we have shown the applicability of Generalized Eigenvalue Decomposition (GEVD) for separating pseudo-periodic maternal ECG from fetal ECG signals recorded from the maternal abdomen [9]. In that work, one of the advantages of GEVD over other source separation techniques was the ability of ranking the extracted components in order of periodicity, which provided a means of automatic and unsupervised ECG decomposition and filtering. In this work, we develop a similar idea based on GEVD for the automatic detection and removal of EOG artifacts from multichannel EEG recordings.

The remainder of this paper is organized as follows: in section 2, we present a general nonlinear framework to decompose signals into independent subspaces using GEVD; the results of this method are presented in section 3 over simulated and real signals. The last section is devoted to conclusion and perspectives.

2   Method

2.1   Linear Transform

Suppose that we have an array of N EEG channels $\mathbf{x}(t)$, and a reference EOG channel denoted by $\text{EOG}(t)$. Due to the spiky nature of the EOG, it is possible to detect the onset and offset times of the EOG artifact from the reference EOG channel. To do so, we define $\text{E}(t)$, the averaged power of the EOG signal (or, alternatively the variance) within a sliding window of length w around t, as

$$\text{E}(t) \doteq \frac{1}{w} \sum_{\tau=-w/2}^{w/2} \text{EOG}(t-\tau)^2 \ . \tag{1}$$

Using this definition, an EOG is detected whenever $\text{E}(t)$ passes some predefined threshold th. The *active periods* of the EOG may therefore be defined as

$$t_a \doteq \{t | \text{E}(t) > th\} \ . \tag{2}$$

As we will note later, the procedure of finding the offsets and onsets of the EOG does not need to be perfect, and the results can be further improved in a recursive procedure.

We now seek linear transforms of the multichannel recordings $\mathbf{x}(t)$, that maximally resemble the EOG (in the sense that the power of the extracted signals are concentrated during the active time periods). Therefore, if we denote the linear mixture as $y(t) = \mathbf{b}^T\mathbf{x}(t)$, we can try to maximize the cost function

$$\zeta(\mathbf{b}) = \frac{\mathbb{E}_{t_a}\{y^2(t_a)\}}{\mathbb{E}_t\{y^2(t)\}} \quad, \tag{3}$$

where $\mathbb{E}_t\{\cdot\}$ represents averaging over t. Here, the idea is to find linear mixtures of the input signals, with a maximal energy during the EOG activation time t_a, while minimizing the global component energy. Equation (3) may be rewritten as

$$\zeta(\mathbf{b}) = \frac{\mathbf{b}^T\mathbb{E}_{t_a}\{\mathbf{x}(t_a)\mathbf{x}(t_a)^T\}\mathbf{b}}{\mathbf{b}^T\mathbb{E}_t\{\mathbf{x}(t)\mathbf{x}(t)^T\}\mathbf{b}} \quad, \tag{4}$$

which is in the form of the Rayleigh Quotient [10,3], and may be solved by the joint diagonalization of two covariance matrices: the covariance matrix of the EEG channels over the whole dataset, and the covariance matrix of the data during the active periods of the EOG, respectively defined as

$$C_x \doteq \mathbb{E}_t\{\mathbf{x}(t)\mathbf{x}(t)^T\} \quad, \tag{5}$$

$$A_x \doteq \mathbb{E}_{t_a}\{\mathbf{x}(t_a)\mathbf{x}(t_a)^T\} \quad. \tag{6}$$

The intuition behind this method is to achieve decorrelated components that are at the same time *globally* and *locally* decorrelated. We already have a sense about the global decorrelation, which is achieved by sphering C_x. In addition, the diagonalization of A_x assures that the achieved components are also locally decorrelated over the active EOG epochs, too. This assures that the later extracted components have no redundancy up to second order statistics.

The matrix that jointly diagonalizes the matrix pair (A_x, C_x) is in fact the solution to the GEVD problem written as

$$\begin{cases} UA_xU^T = \Lambda \\ UC_xU^T = I \end{cases}, \tag{7}$$

where Λ is a diagonal matrix containing the generalized eigenvalues on its diagonal in descending order, and U is the matrix containing the generalized eigenvectors on its columns[1].

The decomposed signals may now be found by

$$\mathbf{y}(t) = U^T\mathbf{x}(t) \quad, \tag{8}$$

where the elements of $\mathbf{y}(t)$ correspond to a linear transformation of the original data $\mathbf{x}(t)$ that are ranked according to their resemblance with the EOG activation epochs. This means that $y_1(t)$ contains the most information regarding

---

[1] Note that, contrary to the eigenvalues of symmetric matrices that are mutually orthogonal, generalized eigenvectors, i.e., the columns of U, are not generally orthogonal to each other; but following (7) they are "C_x-orthogonal" [10], p. 344.

the EOG while $y_N(t)$ is the least resembling the EOG. Another interpretation for the method is that $y_1(t)$ is the most dominant component of the EOG while having the least contribution in the overall energy of the signals (corresponding to the largest eigenvalue of Λ). On the other hand $y_N(t)$ is the main non-EOG component, with the least EOG contamination (corresponding to the smallest eigenvalue of A_x, i.e., having the least contribution of the active EOG epochs).

2.2 Signal/Noise Separation

Up to now, the components have been ranked according to their resemblance with the EOG through a linear transformation. The next step is to remove the ocular artifacts from the most contaminated components using a linear or nonlinear transformation. The result of this noise removal, denoted by $\mathbf{z} \doteq (z_1, \cdots, z_N)^T$, can be expressed as

$$\forall i \in [1, ..., N] \quad z_i(t) = y_i(t) - f_i[y_i(t)] \ , \tag{9}$$

The denoising transform must be carefully chosen to remove ocular artifacts from the most contaminated signals, while preserving the non-EOG components. A simple but efficient choice is

$$f : \begin{cases} f_i(u) = u \ , & i \in [1, ..., M] \\ f_i(u) = 0 \ , & i \in [M+1, ..., N] \end{cases} . \tag{10}$$

This transform removes the first M ($M \ll N$) components contaminated by ocular artifacts and keeps the remaining $N - M$ components unchanged. Therefore, in order to find the EOG free components, we can eliminate the first few components of $\mathbf{y}(t)$ and transform back the rest of the components using the inverse of the matrix U. The number of eliminated components M depends on the number of expected dimensions of the EOG subspace. Note that due to the elimination of the first M components, the rank of the multichannel signals are reduced to $N - M$.

2.3 Iterative Improvements

So far, we have removed the most dominant EOG components through a combination of a linear transformation, denoising, and back-projection. The results may be further improved by repeating the upper mentioned method in a recursive procedure. To do so, we can use $y_1(t)$, the component which most resembles the EOG, to re-estimate the onset and offsets of the EOG and its activation epochs using a smaller energy threshold th, and to recalculate C_x, A_x, and the other steps of the algorithm in each iteration. By repeating this procedure in several iterations, a better estimate of the EOG will be achieved. This iterative extension of the algorithm is of special interest for the cases in which a good EOG reference is not available. Therefore, we start with a coarse EOG onset and offset estimate; but we improve this estimate in the next iterations.

3   Experiments

Simulated Data. In order to test the method, we generate artificial EEG contaminated by ocular artifacts. The signals are generated as follows:

$$\mathbf{x}(t) = \mathbf{EEG}(t) + \beta \left(k_1 \cdots k_N\right)^T \text{EOG}(t) \ , \tag{11}$$

where (1) $\mathbf{EEG}(t)$ is generated using a multivariate autoregressive filter (Yule-Walker order $p = 8$, cf. [1] for details), trained using $N = 4$ channels of artifact-free EEG data, (2) EOG(t) results from the convolution of a typical blink segment, extracted from real EOG data, and an ensemble of Dirac impulses having a Poisson distribution with parameter $\lambda = 0.2\,Hz$, (3) β is a scale parameter to adjust the signal-to-noise ratio (SNR) and (4) $(k_1, \cdots, k_N)^T$ is a vector of gain under a linear hypothesis between EOG and EEG, for all i, $k_i \sim \mathcal{N}(1, 0.3)$. A multivariate autoregressive filter has been preferred over a multi-univariate autoregressive filter, because the multivariate filter naturally takes into account spatial correlations of the data and learns those correlations from the real data.

Our approach is compared with a classical ICA approach for EOG removal, based on FastICA [4]. The input of the algorithm consists of the N EEG channels $\mathbf{x}(t)$ and the EOG channel EOG(t). FastICA seeks the most independent $N+1$ components. We then remove the component which is the most correlated to the EOG, by setting it to zero. Components are then back-projected onto the sensor space.

The method proposed in this paper and FastICA, abbreviated GEVD and ICA, respectively, are evaluated by comparing the first N components of the back-projected signals and the EEG generated by the multivariate autoregressive filter. The idea behind this evaluation procedure is that a perfect denoising would lead to perfect recovery of the multivariate autoregressive process EEG(t), free of any EOG contamination. We therefore define the error signals as a function of the denoised processes resulting from ICA and GEVD ($\mathbf{x}_{\text{ICA}}$ and $\mathbf{x}_{\text{GEVD}}$, respectively) as

$$\epsilon_{\text{ICA}}(t) = \mathbf{x}_{\text{ICA}} - \text{EEG}(t) \tag{12}$$

$$\epsilon_{\text{GEVD}}(t) = \mathbf{x}_{\text{GEVD}} - \text{EEG}(t) \ , \tag{13}$$

from which we can define the following performance index (in decibels):

$$\mathcal{Q} = 10 \log \left(\frac{1}{N} \sum_{i=1}^{N} \frac{\mathbb{V}(\epsilon_{\text{GEVD}_i}(t))}{\mathbb{V}(\epsilon_{\text{ICA}_i}(t))} \right)$$

where $\mathbb{V}(\cdot)$ denotes the variance operator. This index is built such that a positive value indicates that ICA outperforms GEVD, while a negative index indicates that GEVD outperforms ICA.

Figure 1 shows the results obtained for simulated data. Two signal lengths are considered (50000 and 100000 points). Two hundred simulations are done for each box-plot. The SNR tuned by β, is varying between -10dB and 30dB.

These values are chosen by considering that a blinking artifact can have an amplitude of up to 100 times stronger than the EEG. Such a condition would yield an SNR of between -10dB to 0dB, depending on the blinking rate.

In the GEVD method, only the first component given by GEVD is removed. This choice is due to the rather simple method used to generate the EOG and EEG mixture. Moreover, in this case the iterative EOG improvement method is not used, since the EOG onsets and offsets are already known. We highlight the fact that in our simulation scenario, the simulation condition for the ICA method is better than it is for real signals, since we provide it with the clean EOG channel, directly.

Our method based on GEVD is shown to clearly outperform a classical ICA approach for a large set of SNRs (t-test highly significant). This effect is much more significant for low SNRs, showing that our method is of particular interest for bad conditionings of the signals.

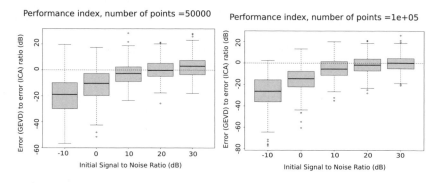

Fig. 1. Method comparison: for different values of initial signal to noise ratios between EEG and EOG, we performed 200 simulations. The performance index is positive when ICA performs better than GEVD whereas it is negative otherwise.

Real Data. We also evaluate our algorithm using a BCI experiment dataset[2] [2]. The data consists of 22 EEG channels and three EOG channels. The inter-electrode distance is about 3.5 cm. The signals are sampled with 250 Hz and bandpass filtered between 0.5 and 100 Hz. An additional 50 Hz notch filter is used to suppress power-line noise.

The back-projected signals after removing the first EOG component are depicted in Fig. 2. In this example, due to the quality of one of the EOG channel, no improvements are achieved by iterating the method. This figure only shows two of the 22 denoised EEG channels. The first one, Fz, is known to be highly contaminated by EOG artifacts because it is close to the eyes. Fig. 2 shows that the EOG component is removed from this channel and also that the EEG signal outside blink contamination is kept perfectly unchanged. Channel Cz, known to be less contaminated than Fz is also drawn (Fig. 2 middle). While not as clear

---

[2] BCI competition IV, dataset 2a, subject A05 [2].

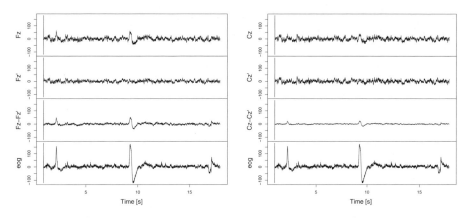

Fig. 2. Denoising of two specific channels, Fz and Cz electrodes (left and right, respectively). From top to bottom, the signals are (1) initial EEG, (2) denoised EEG, (3) residuals, and (4) EOG recordings.

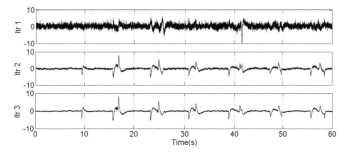

Fig. 3. Iterative improvements of the estimation of EOG. We start from a noisy EEG channel (up). The active periods is then determined according to the energy of the noisy EEG channel. We then select the first component to iterate the method: the active periods of the $(i + 1)^{th}$ iteration are determined using the energy of the first component extracted in the ith iteration.

as in the first presented channel, we can see that the most important part of the EOG contamination is removed from the channel.

To illustrate the interest of the iterative method, we then show that due to the high amplitude features of the EOG contamination, we can even use some EEG channels to evaluate the active periods of EOG. This is of particular interest when no EOG channel is provided. Such a situation is illustrated in Fig. 3 where we have started with a noisy EEG reference and improved this reference after two iterations of GEVD.

4   Conclusion

We presented an automatic method to remove ocular artifacts from EEG measurements. Our algorithm outperformed a classical approach based on independent component analysis of EEG data. This method is a special case of a more general framework designed to generally decompose multivariate signals into independent subspaces. This flexible framework allows to choose different criteria for determining the optimal linear transform. In this paper, the nonstationarity of the EOG was used; but in other applications other prior information such as the periodicity or spectral contrast may be used.

In future works, further improvements can be achieved by replacing our simple component nulling function by some wavelet-based denoising method applied on the first few channels. In that case, it is expected that we separate the EOG from the EEG, without reducing the rank of the EEG signals.

References

1. Anderson, C.W., Stolz, E.A., Shamsunder, S.: Multivariate autoregressive models for classification of spontaneous electroencephalographic signals during mental tasks. IEEE Trans. Biomed. Eng. 45(3), 277–286 (1998)
2. Brunner, C., Leeb, R., Müller-Putz, G.R., Schlögl, A., Pfurtscheller, G.: BCI competion 2008 – Graz dataset A (July 2008),
   http://ida.first.fraunhofer.de/projects/bci/competition_iv
3. Golub, G., van Loan, C.: Matrix Computations, 3rd edn. The Johns Hopkins University Press (1996)
4. Hyvärinen, A., Karhunen, J., Oja, E.: Independent Component Analysis. Wiley Interscience, Hoboken (2001)
5. Ille, N., Berg, P., Scherg, M.: Artifact correction of the ongoing EEG using spatial filters based on artifact and brain signal topographies. J. Clin. Neurophysiol. 19(2), 113–124 (2002)
6. Jung, T., Humphries, C., Lee, T., McKeown, M., Iragui, V., Makeig, S., Sejnowski, T.: Removing electroencephalographic artifacts by blind source separation. J. Psychophysiol. 37, 163–178 (2000)
7. Koles, Z.J.: The quantitative extraction and topographic mapping of the abnormal components in the clinical EEG. Electroencephalogr. Clin. Neurophysiol. 79, 440–447 (1991)
8. Romero, S., Mañanas, M.A., Barbanoj, M.J.: A comparative study of automatic techniques for ocular artifact reduction in spontaneous EEG signals based on clinical target variables: A simulation case. Comput. Biol. Med. (January 2008)
9. Sameni, R., Jutten, C., Shamsollahi, M.B.: Multichannel electrocardiogram decomposition using periodic component analysis. IEEE Trans. Biomed. Eng. 55(8), 1935–1940 (2008)
10. Strang, G.: Linear Algebra and Its Applications, 3rd edn. Brooks/Cole (1988)
11. Thulasidas, M., Guan, C., Ranganatha, S., Wu, J., Zhu, X., Xu, W.: Effect of ocular artifact removal in brain computer interface accuracy. In: Conf. Proc. IEEE Eng. Med. Biol. Soc., vol. 6, pp. 4385–4388 (2004)

Denoising Single Trial Event Related Magnetoencephalographic Recordings

Elina Karp[1], Lauri Parkkonen[2], and Ricardo Vigário[1]

[1] Adaptive Informatics Research Centre
Helsinki University of Technology, P.O. Box 5400, FI-02015 TKK, Finland
[2] Brain Research Unit, Low Temperature Laboratory
Helsinki University of Technology, P.O. Box 5100, FI-02015 TKK, Finland

Abstract. Functional brain mapping is often performed by analysing neuronal responses evoked by external stimulation. Assuming constant brain responses to repeated identical stimuli, averaging across trials is usually applied to improve the typically poor signal-to-noise ratio. However, since wave shape and latency vary from trial to trial, information is lost when averaging. In this work, trial-to-trial jitter in visually evoked magnetoencephalograms (MEG) was estimated and compensated for, improving the characterisation of neuronal responses. A denoising source separation (DSS) algorithm including a template based denoising strategy was applied. Independent component analysis (ICA) was used to compute a seed necessary for the template construction. The results are physiologically plausible and indicate a clear improvement compared to the classical averaging method.

Keywords: Single trial, event related, denoising, magnetoencephalography (MEG), independent component analysis (ICA), denoising source separation (DSS).

1 Introduction

The standard way of analysing brain responses to external stimulation is averaging across trials, i.e. single stimulus-response entities. The averaging is based on the assumption that the brain produces identical responses to identical stimuli while ongoing brain activity and noise varies. Thus, the averaging boosts the responses but suppresses the rest. However, there is evidence that brain responses exhibit inter-trial variations [1]. Therefore, averaging leads to loss and distortion of information. Robust quantification of the variations should bring more accurate insight into the functioning of the brain.

A Bayesian algorithm for the analysis of single trial event related brain activity has been proposed [2]. The algorithm is initialised with the average of the channel capturing the most of stimulus related activity. However, such an initialisation biases the source estimates towards a particular region in the brain. Another algorithmic solution based on maximum likelihood estimation with modelling of the ongoing activity has been introduced as well [3].

T. Adali et al. (Eds.): ICA 2009, LNCS 5441, pp. 427–434, 2009.

In this paper, a denoising source separation (DSS) algorithm including a template based denoising strategy was applied to magnetoencephalographic (MEG) measurements. Independent component analysis (ICA) was used to compute a seed necessary for the template construction. In addition, the trial-to-trial jitter in brain responses was estimated and compensated for. The suitability of ICA in the MEG analysis has been reported [4,5]. However, better separation results can be achieved by the DSS algorithm used in this study. Compared to the Bayesian approach, the DSS algorithm has the advantage of being computationally simpler and using a more generic seed. Furthermore, the use of ICA as a seed generator results in multiple principled templates allowing for a significantly wider range of source search.

2   Data

The experimental data consisted of MEG measurements of brain responses to visual stimuli. In MEG, the magnetic fields produced by the electric activity of neurons are measured on the head surface. MEG is thus a non-invasive method for monitoring a living brain exhibiting good time resolution. In addition, MEG is a good starting point for inverse problem solving, i.e. the localisation of the neuronal sources, due to its fair spatial resolution. However, in event related studies, because of the weakness of the evoked phenomena compared to the ongoing brain activity, the signals have a poor signal-to-noise ratio.

The measurements were conducted in the magnetically shielded room in the Brain Research Unit of the Low Temperature Laboratory at Helsinki University of Technology. The measuring device was a 306-channel MEG system (Elekta Neuromag Oy, Helsinki, Finland) comprising 204 planar gradiometers and 102 magnetometers in a helmet shaped sensor array. Gradiometer data from a single subject was used for the analysis. 106 visual stimuli, consisting of meaningful four-letter words embedded in dynamic noise, were presented at intervals of 990 ms. For details see [6]. The signals were pre-processed by a FIR filter with passband from 1 Hz to 45 Hz and down-sampled from 600 Hz to 300 Hz.

3   Methods

A DSS algorithm with a template based denoising strategy was used for the analysis of the MEG recordings. In the template construction, an initial estimate of underlying sources was needed. Therefore, independent components were estimated by ICA from the whole data set and used as seeds for the DSS algorithm. In addition, an estimation and compensation of trial-to-trial jitter in brain response latencies was included.

3.1   Independent Component Analysis

ICA [7] is a statistical and computational method for revealing hidden factors underlying multidimensional data. It is based on the assumption of statistical independence of the source signals and one of the most widely used methods of solving the blind source separation problem.

In this work, FastICA [8,9] was used for estimating independent components. In order to reduce sensitivity to outliers, hyperbolic tangent was chosen as the nonlinearity. Symmetric orthogonalisation was applied to avoid accumulation of estimation errors.

When ICA estimation is performed, the resulting independent components tend to differ slightly from run to run [10]. The variability can originate from several sources: Real data may not fully fulfil the theoretical assumption of statistical independence. Despite dimension reduction, noise is usually still present although the noiselessness assumption is commonly made. In addition, varying the initialisations of ICA algorithms may result in different estimates. ARABICA toolbox in MATLAB^{TM} [11] was used in order to obtain robust estimates of the independent components: FastICA was run multiple times in a bootstrapping manner and the resulting estimates were clustered based on correlation. The obtained cluster centroids were averaged across the trials and those exhibiting stimulus related activity were chosen for further analysis.

3.2   Denoising Source Separation

DSS [12] is a framework for a variety of source separation algorithms, constructed around different denoising strategies. Whitening of data results in rotational invariance in the sense that no separation of the sources can be considered superior to the others without additional information. In figure 1 a schematic illustration of a DSS algorithm is depicted. In DSS, a denoising function $f(\mathbf{S})$ breaks the symmetry produced by whitening and highlights directions of particular interest resulting thus in a more appropriate separation than blind methods. The approach also enables embedding of prior knowledge in denoising procedures to guide the separation.

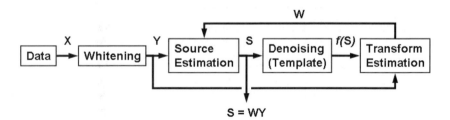

Fig. 1. A denoising source separation algorithm

In this work a template based denoising was applied. Therein the template serves as a target for the algorithm, which searches for a linear transformation ($\mathbf{W}$) that maps the whitened data ($\mathbf{Y}$) as close to the template ($f(\mathbf{S})$) as possible. After each iteration the template is updated based on the emerging characteristics of the data.

In the template construction, the periodicity of the stimulus presentation and the assumption that the brain responds similarly to each stimulus were exploited. In order to get a first robust estimate of the responses, an independent component was averaged over all trials. The average was then placed around each stimulus on an otherwise zero valued signal.

The template was renewed after each iteration and the initial response estimate, i.e. the averaged independent component, was replaced by the average of the strongest denoised source estimate. In addition, the trial-to-trial jitter in the brain responses was estimated and compensated for. Since DSS computes a linear transformation, the estimated brain responses emerge at the very instants found in the original data. Therefore, the jitter can be estimated after each iteration of the DSS algorithm by computing a sliding correlation between the denoised source estimate and its average. In order to compensate for the jitter during the template construction, replicas of the denoised source average were placed at the instants of the highest correlation rather than those of the stimulus presentation.

3.3   Multitemplate Approach

While whitening the data, the dimension was reduced from the original 204 to 50. ICA was computed on the raw data and four independent components exhibiting most prominent stimulus related activity were selected by visual inspection of the averages. The chosen components were used to construct one template each, which were then fed into the DSS algorithm one at a time. From the parallel DSS algorithms, one estimated denoised source per template was obtained. The source estimate with the highest correlation with its corresponding template was considered to be the global outcome of one DSS round. Between the rounds, the chosen source estimate was removed from the whitened data, leaving only directions orthogonal to it for further analysis. The rounds followed one another until the data was exhausted and no interesting signals could be found anymore. The time window used for all averaging was 100 ms before and 800 ms after either the stimulus or the jitter corrected instant. The procedure is illustrated in figure 2.

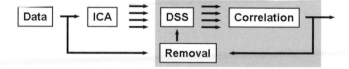

Fig. 2. Multitemplate denoising source separation

4   Results

Illustrative examples of averaged MEG recordings with and without relation to stimuli are shown in figure 3 MEG(a). Corresponding evoked field (EF) image plots [5] are shown in figures 3 MEG(b–e). EF image plots are constructed by stacking vertically single grey coded trial traces sorted by the peak latency and smoothing across neighbouring trials by moving average.

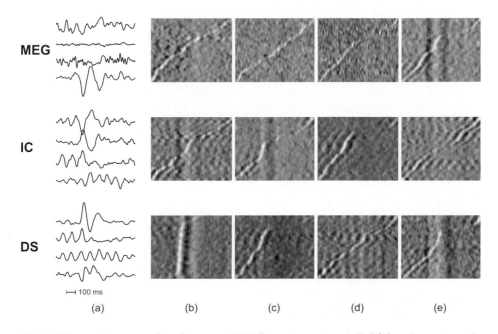

Fig. 3. Illustrative examples of averaged MEG measurements MEG(a) and corresponding EF image plots MEG(b–e) together with obtained independent components IC(a) and denoised sources DS(a) as well as corresponding EF image plots IC(b–e) and DS(b–e). Independent components are sorted by descending amplitude. Denoised sources appear in the order of acquisition.

In the independent components, depicted in figure 3 IC together with their EF image plots, some interesting features start to emerge. Yet, the characteristics of the signals are fuzzy and the time courses unclear. On the contrary, the denoised sources in figure 3 DS(a) exhibit distinct features and clear time courses. The first source shows a strong, transient response to the visual stimuli. Particularly interesting is, however, the emergence of periodic signals: the third source exhibits continuous periodicity fitting the α-range, while in the second one this periodic behaviour ceases after 200–300 ms after the stimulus presentation.

Comparison of the EF image plots of the denoised signals (figure 3 DS(b–e)) and the independent components (figure 3 IC(b–e)) renders jitter compensation visible. E.g. in figure 3 DS(d) the periodic behaviour is approximately in phase

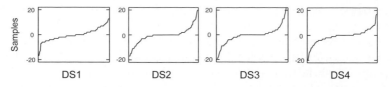

Fig. 4. Estimated trial-to-trial jitter, in samples, sorted in ascending order. In DS3, two outliers with values 34 and 39 were omitted from the figure but used in the template construction.

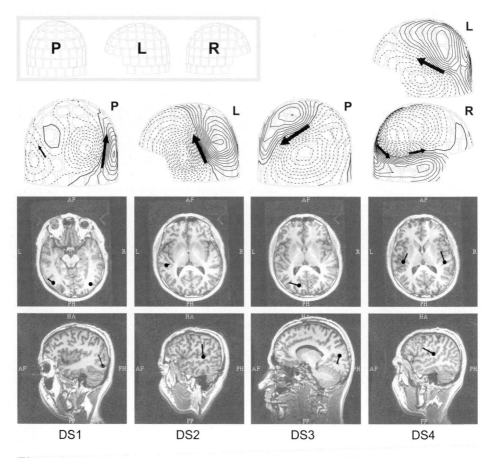

Fig. 5. Field patterns associated with the denoised sources (DS) and the locations of corresponding equivalent current dipoles. P stands for posterior, R for right and L for left view.

compared to the curly activity in the corresponding independent component in figure 3 IC(e). Jitter compensation for each denoised source is shown in figure 4.

From the EF image plots it is clear that the data contains significant trial-to-trial variations. Thus, essential features are lost during averaging as can be seen by comparing the averages and the corresponding EF image plots. The EF image plot of the second independent component (figure 3 IC(c)) shows that the main stimulus locked response, although steady across trials, is not even the strongest signal in the component. It can, however, be detected by DSS because the implicit assumption of constant brain responses is embedded in the template construction (see figure 3 DS(b)).

In addition to the denoised sources, DSS also estimates the corresponding mixing matrix. Each column of the matrix contains a mapping between one denoised source and the measured data. Thus, the mapping can be considered as the magnetic field pattern generated by the denoised source. The field patterns of the denoised sources in figure 3 DS are shown in figure 5. Clearly dipolar structure of the patterns is in accordance with classical models of focal neuronal activity. One exception is the denoised source 3 which a simple dipole model is insufficient to fully explain. However, its periodic nature does not even suggest focal activity.

To investigate further the physiological plausibility of the results, the neuronal sources producing such field patterns were localised by fitting current dipoles through inverse problem solving [13]. The fitted dipoles are shown in figure 5 superimposed on transversal and lateral MRI slices. The dipole locations are physiologically plausible and in agreement with the results of the previous analysis of the same data [6]. Furthermore, the occipital locations of the dipoles associated with the denoised sources one and three suggest processing of visual stimuli. The goodness-of-fit [13] was over 80% for all seven dipoles.

5   Conclusions

The results clearly indicate benefits of single trial analysis. The classical assumption of a constant brain response to a particular stimulus type is embedded in the analysis whereas trial-to-trial latencies were allowed to vary. Because of the variation of the latencies, a considerable amount of information is usually lost in averaging. In addition, the invariance in the responses is assumed rather than imposed and thus changes in wave shape across trials can be observed in single trial analysis (see figure 3 DS(b–e)) providing an additional source of error when averaging.

DSS produces physiologically plausible results. The sources exhibit reasonable dynamics and conceivable latencies. They produce simple dipolar like field patterns with realistic neuronal locations. Moreover, DSS discovered unexpected rhythmic components suggesting the need for further physiological research. In addition, the trial-to-trial jitter in the brain response latencies can be tracked.

The DSS algorithm included the replication of the averaged signal during the template construction. Loosening this implicit assumption of the similar shape of the individual responses could produce interesting results, too. Extending the research to more subjects could bring additional insight into the robustness of the proposed analysis method as well as physiological interpretation of its outcomes.

Acknowledgements

This study was funded by the Academy of Finland through its Centres of Excellence Programme 2006–2011. The data was collected within EU's large-scale facility Neuro-BIRCH III. The authors wish to thank Dr Jaakko Särelä for invaluable discussions on the subject of this paper.

References

1. Truccolo, W.A., Ding, M., Knuth, K.H., Nakamura, R., Bressler, S.L.: Trial-to-Trial Variability of Cortical Evoked Responses: Implications for the Analysis of Functional Connectivity. Clin. Neurophysiol. 113(2), 206–226 (2002)
2. Knuth, K.H., Shah, A.S., Truccolo, W.A., Ding, M., Bressler, S.L., Schroeder, C.E.: Differentially Variable Component Analysis: Identifying Multiple Evoked Components Using Trial-to-Trial Variability. J. Neurophysiol. 95(5), 3257–3276 (2006)
3. Wang, X., Chen, Y., Ding, M.: Estimating Granger Causality after Stimulus Onset: A Cautionary Note. NeuroImage 41(3), 767–776 (2008)
4. Vigário, R., Särelä, J., Jousmäki, V., Hämäläinen, M., Oja, E.: Independent Component Approach to the Analysis of EEG and MEG Recordings. IEEE Trans. Biomed. Eng. 47(5), 589–593 (2000)
5. Onton, J., Westerfield, M., Townsend, J., Makeig, S.: Imaging Human EEG Dynamics Using Independent Component Analysis. Neurosci. Biobehav. Rev. 30(6), 808–822 (2006)
6. Sorrentino, A., Parkkonen, L., Piana, M., Massone, A.M., Narici, L., Carozzo, S., Riani, M., Sannita, W.G.: Modulation of Brain and Behavioural Responses to Cognitive Visual Stimuli with Varying Signal-to-Noise Ratios. Clin. Neurophysiol. 117(5), 1098–1105 (2006)
7. Hyvärinen, A., Karhunen, J., Oja, E.: Independent Component Analysis. Wiley, New York (2001)
8. Hyvärinen, A.: Fast and Robust Fixed–Point Algorithms for Independent Component Analysis. IEEE Trans. Neural Netw. 10(3), 626–634 (1999)
9. FastICA: MATLAB™ toolbox (1998),
   http://www.cis.hut.fi/projects/ica/fastica/
10. Ylipaavalniemi, J., Vigário, R.: Analyzing Consistency of Independent Components: An fMRI Illustration. NeuroImage 39(1), 169–180 (2008)
11. Ylipaavalniemi, J., Soppela, J.: Robust ICA in a Pipeline. In: ICA 2009 (submitted, 2009)
12. Särelä, J., Valpola, H.: Denoising Sourse Separation. Journal of Machine Learning Research 6, 233–272 (2005)
13. Hämäläinen, M., Hari, R., Ilmoniemi, R.J., Knuutila, J., Lounasmaa, O.V.: Magnetoencephalography — Theory, Insrumentation, and Applications to Noninvasive Studies of the Working Human Brain. Rev. Mod. Phys. 65(2), 413–497 (1993)

Eye Movement Quantification in Functional MRI Data by Spatial Independent Component Analysis

Ingo R. Keck[1], Volker Fischer[2], Carlos G. Puntonet[3], and Elmar W. Lang[1]

[1] Institute of Biophysics, University of Regensburg, 93040 Regensburg, Germany
{Ingo.Keck,Elmar.Lang}@biologie.uni-regensburg.de
[2] Institute of Psychology, University of Regensburg, 93040 Regensburg, Germany
Volker.Fischer@psychologie.uni-regensburg.de
[3] Department of Architecture and Technology of Computers, University of Granada, 18071 Granada, Spain
carlos@atc.ugr.es

Abstract. We present a novel analysis to quantify the eye movement from functional magnetic resonance (fMRI) data without the aid of dedicated eye-trackers. We show that by using the parts of the functional image data that contain the motion artefacts of the subject's eyes and by applying spatial independent component analysis to it, it is possible to detect eye movement and to estimate the point of gaze at the time of the data acquisition.

1 Introduction

Functional magnetic resonance imaging (fMRI) is a valuable technique to investigate the activity of the brain. But in many experimental settings the position of the gaze of the subject's eyes is also of interest (e.g. to verify task compliance during the experiment). However, the use of a dedicated eye-tracker is either impossible because of lack of space in the scanner or it is inconvenient because of the discomfort for the subject under investigation and the additional effort in the set up and configuration.

At the same time eye motion artefacts are a common problem in the analysis of fMRI data sets. The variance of the voxels which represent activities related with eye movements can even be higher than the variance induced by brain activities in the regions of interest. Due to the sensitivity of many independent component analysis (ICA) algorithms to outliers it is a common pre-processing step to mask out signals from the eyes for the application of ICA to fMRI data sets. [1]

These motion related artefacts thus might be used to detect the position of the eyes, hence one could avoid problems arising from an application of dedicated eye-trackers inside the scanner. Beauchamp showed in [2] that the *existence* of eye movements can be deduced from the variance of fMRI signals stemming from the eyes. However, he was not able to extract additional information concerning

T. Adali et al. (Eds.): ICA 2009, LNCS 5441, pp. 435–442, 2009.

the actual position or movement of the eyes. Tregellas et al [3] estimated the artefacts resulting from eye movements to reduce the error for motion estimation of the whole head. They used the local changes in voxel intensity to estimate the eye movement within a six dimensional rigid body model. However, they were unable to show that their model correctly represented the actual eye movement. Instead they reported many difficulties, ranging from high noise levels to lack of complete directional information.

In the next sections we will show that applying a spatial ICA to a proper fMRI data-set of an eye saccade experiment, we are able to yield information on the position of the point of gaze at the time of image acquisition.

2   fMRI and Eye Movement

Functional magnetic resonance imaging (fMRI) of the brain relies on a special effect called *Blood Oxygenation Level Dependent* (BOLD) contrast to detect changes in the blood flow that are correlated to increased activity in the brain. [4] The BOLD contrast is based on the detection of local changes in the magnetic susceptibility that affect the signal in T2*-sensitive pulse sequences. However, a large part of the observed signal still derives from the physical structure of the brain. Head movement artefacts in the fMRI signal are well understood and various techniques have been developed to minimise the error that is induced by these movements [5, 6, 7, 8]. These methods typically are based on a rigid body motion estimation of the head and thus should not affect artefacts resulting from eye movements located in the spatial region of the eye in the image.

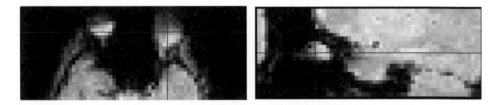

Fig. 1. Functional Magnetic Resonance Imaging data from subject 2 showing a slice through the eye of the subject (left view: from above, right view: from the side of the head). The vitreous body of the eye and the optic nerve appear as high intensity voxels surrounded by low intensity voxels. Artefacts from the sharp change in magnetic susceptibility between tissue and air can be seen extending from the eyes to the air in front of them.

The vitreous humor of the eye consists mainly of water and appears with high intensity on fMRI slices which traverse it. While the normal scan frequency (TR) of an fMRI experiment is within the range of seconds, saccadic eye movements happen on a far more rapid time scale, typically below 100 ms. Therefore it is obvious that fMRI experiments are far too slow to observe a saccadic eye movement directly. However, it should be possible to detect motion artefacts related to different positions of the vitreous humor of the eye, the optic nerve, the surrounding muscles and the aqueous environment inside the vitreous body. The main effect should show up as large changes in local intensity as a voxel outside the eye with low level intensity changes to the high intensity level within the eye (see figure 1).

3   Spatial ICA Applied to a Saccadic Eye Movement Experiment

In the next sections we describe a fMRI experiment and an Independent Component Analysis (ICA) based method to detect the position of the eyes' focal point.

3.1   Experimental Setup

Two healthy volunteers (2 males) with normal vision participated in a simple visual tracking paradigm. Subjects had to look at a fixation cross (0.4° diameter), which was presented at the central fixation point and had to perform a saccade, when the cross jumped to a quadrant (11° horizontal and 7° vertical offset). After each fixation period of 2 seconds, the cross appeared for 7 seconds at one of the four quadrants. A pause of 1 second was included and for each quadrant 16 trials were performed. Both subject reported no difficulties in following the task as expected.

Stimuli were back-projected via an LCD video projector (JVC, DLA-G20, Yokohama, Japan, 72 Hz, 800×600 resolution) onto a translucent circular screen (approximately 30° diameter), placed inside the scanner bore at a distance of 63 cm from the observer. Stimulus presentation was controlled via Presentation Software (Neurobehavioral Systems, Albany, CA, www.neurobs.com).

Imaging was performed using a 3-Tesla MR head scanner (Siemens Allegra, Erlangen, Germany). We continuously acquired a series of functional volumes of 20 ascending axial slices using a standard T2* weighted echo-planar imaging (EPI) sequence (TR= 2000 ms; TE= 30 ms; flip angle=90°; 64 × 64-matrices; in-plane resolution: 3 × 3 mm; slice thickness: 3.75 mm).

For the analysis the data was converted from DICOM to the NII format using SPM [9] and an image registration to the first image of each session was performed to minimise the influence of head movement. No slice timing compensation was applied.

3.2   Spatial ICA Based Eye Movement Quantification

To detect the components related to eye movement artefacts, we applied a spatial Independent Component Analysis (sICA) to the fMRI data of the saccadic eye movement experiment. We expected the movement artefacts to be spatially independent as the spatial position of the artefacts for vertical and horizontal movement should be localised in different voxel positions. The time courses of the components, however, should be statistically dependent due to the selected focal points in the four quadrants.

In spatial ICA, $\mathbf{X}$ contains in its rows the observed fMRI images, $\mathbf{A}$ denotes the mixing matrix and $\mathbf{S}$ will contain the estimated sources:

$$\mathbf{X} = \mathbf{A}\mathbf{S} \tag{1}$$

Before applying ICA, we reduced the data to a rectangular part of the image which contained both eyes, optic nerve and surrounding tissue of the subject. The 3D data of each image was reorganised as a column vector and from these vectors $\mathbf{X}$ was constructed. The ICA then estimates the matrix of independent components $\mathbf{S}$ and the mixing matrix $\mathbf{A}$ which contains the estimated time-courses of the independent components.

Using the SPM toolbox MFBOX [10] as front-end, we applied the JADE algorithm [11] to the problem. For an optimal estimation of the number of sources in the data-set, we applied JADE with increasing number of estimated sources and used the correlation coefficient between the time-courses of the experiment and the time-courses of the resulting independent components to identify the best match.

3.3   Single-slice Analysis

To identify slices where information on eye movements is contained, we applied the spatial ICA separately to each horizontal slice of the recorded image data while varying the maximal number of components from 1 to 30. Then, after normalising each column $\boldsymbol{a_i}$ of each resulting $\mathbf{A}$ to mean 0 and variance 1 and performing the same normalisation on each relevant time-course of horizontal or vertical movement $\boldsymbol{y}_{\mathrm{mov}}(t)$ from the experimental setting, we calculated the correlation coefficient

$$c_{\mathrm{coeff}} = \boldsymbol{a}_i^T \boldsymbol{y}_{\mathrm{mov}} \tag{2}$$

between each independent component's time-course $\boldsymbol{a}_i(t)$ and the relevant time-course of horizontal or vertical movement $\boldsymbol{y}_{\mathrm{mov}}(t)$ from the experimental setting. Figure 2 shows the results for both subjects, where the correlation coefficient is given as the coloured box (ranging from 0 (black) to 1 (white)) depending on the number of the tested image slice and the maximal number of independent components used for the ICA.

We also tested for the possible influence of a time lag in the movement time-course to compensate for the time differences between slice acquisition times. However, the strongest correlations in all slices were found at zero time lag.

As can seen from this result, the information of the eye position is contained in all slices that pass through the eye to some extent. The ICA is able to recover this information to various degrees with best results for slice 13 and slice 12 for subject 1 and subject 2 respectively. These are the slices that contain the optic nerve, which corresponds to the observations made by Trellegas et al in [3].

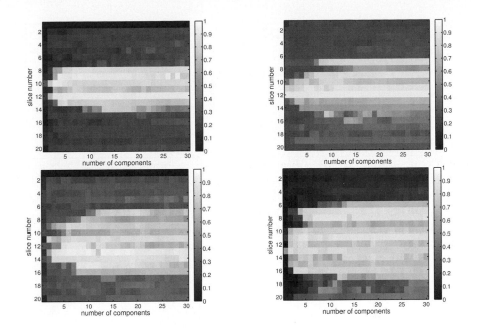

Fig. 2. The correlation coefficients for the time-courses of horizontal (top row) and vertical (bottom row) movement with the time-course of the best correlated independent component, as function of the number of horizontal slice and maximal number of components used in the ICA. Data-sets: subject 1 (left side), subject 2 (right side). The eyes of subject 1 occupy voxels in slice 7 to 16, while for subject 2 these voxels are in slice 8 to 15.

It is also interesting to note that the results of JADE in some slices seem to depend on the given number of maximal independent components. While for obvious reasons the correct number of independent sources in each slice should be at least 2, we are not able to estimate the exact number.

3.4   Multi-slice Analysis

In the next step we used the whole 3D data from each subject within a rectangular region containing both eyes for the spatial ICA to extract the independent components of eye movement. Figure 3 shows the correlation coefficients for the best correlated independent components. In both subjects JADE was able to identify both the horizontal and vertical movements with correlation coefficients well above 0.9.

In figure 4 we plot the time-course of the component related to horizontal movement against the time-course of the component related to vertical movement,

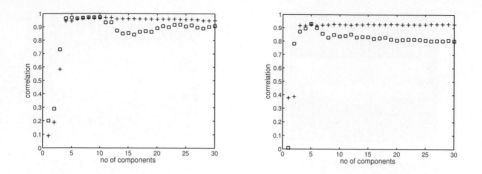

Fig. 3. The correlation coefficients for the time-courses of horizontal (crosses) and vertical (boxes) movement with the time-course of the best correlated independent component, given for the maximal number of components used in the ICA. Data-sets: subject 1 (left), subject 2 (right).

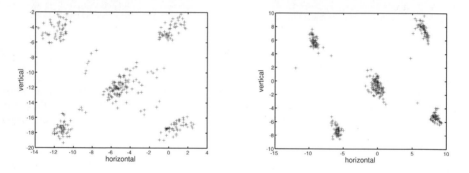

Fig. 4. Scatterplots of the columns a_i of the mixing matrix for the components related to horizontal movement against vertical movement. All five fixation points of the experiment are clearly distinguishable. Data-sets: subject 1 (left), subject 2 (right).

thus resulting in the spatial localisation of the focus point of the eye during image acquisition. For both subjects the five fixation points (four quadrants and central fixation) are clearly visible.

The spatial localisation of the independent components related to the movement follow the form typical for motion artefacts in fMRI. For subject 1 they are located within the vitreous body, while for subject 2 they are located at the position of the optic nerve. This difference, possibly due to anatomical differences in the subjects, could explain why constructing an universal rigid body model in [3] failed in estimating the exact eye position. ICA does not depend on a physical or anatomical model and thus will adapt automatically to each subject.

As further example of how well sICA is performing we refer to figure 5 where we plotted the expected time-courses for horizontal and vertical eye movement

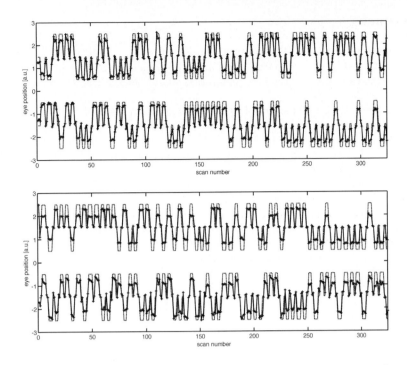

Fig. 5. The time-courses for eye movement (top: horizontal, bottom: vertical) as taken from the experiment (dotted) plotted together with the time-courses estimated by the spatial ICA (crosses, line) for subject 1 (first plot, above) and subject 2 (second plot)

from the experiment together with the estimated time-courses from the spatial ICA.

4   Conclusion

We have shown that it is possible to use spatial ICA to detect the position of the fixation of the human eye in an fMRI experiment without the use of a dedicated eye-tracker. While this result still needs further refinement in the actual procedure of application in fMRI experiments, we were able to show very promising results that may render unnecessary the time consuming and uncomfortable use of a dedicated eye-tracker for the purpose of task conformance compliance testing.

Acknowledgments

This work is supported by the German Ministry for Education and Research (BMBF 01GW0761) within the project *perceptual learning and brain plasticity* and by the project *NAPOLEON* of the Spanish TEC Ministry Program.

References

1. Calhoun, V., Adali, T., Hansen, L.K., Larsen, J., Pekar, J.: ICA of functional MRI data: An overview. In: Fourth International Symposium on Independent Component Analysis and Blind Source Separation (ICA 2003), Nara, Japan, pp. 281–288 (April 2003)
2. Beauchamp, M.S.: Detection of eye movements from fMRI data. Magentic Resonance in Medicine 49, 376 (2003)
3. Trellegas, J.R., Tanabe, J.L., Miller, D.E., Freeman, R.: Monitoring eye movements during fMRI tasks with echo planar images. Human Brain Mapping 17, 237–243 (2002)
4. Frackowiak, R.S.J., Friston, K.J., Frith, C.D., Dolan, R.J., Mazziotta, J.C.: Human Brain Function. Academic Press, San Diego (1997)
5. Ashburner, J., Andersson, J., Friston, K.: High-dimensional nonlinear image registration using symmetric priors. NeuroImage 9, 619–628 (1999)
6. Ashburner, J., Friston, K.: Rigid body registration. In: Frackowiak, R., Friston, K., Frith, C., Dolan, R., Friston, K., Price, C., Zeki, S., Ashburner, J., Penny, W. (eds.) Human Brain Function, 2nd edn. Academic Press, London (2003)
7. Friston, K., Williams, S., Howard, R., Frackowiak, R., Turner, R.: Movement-related effects in fMRI time-series. Magnetic Resonance in Medicine 35, 346–355 (1996)
8. Cox, R., Jesmanowicz, A.: Real-time 3D image registration for functional MRI. Magnetic resonance in medicine 42(6), 1014–1018 (1999)
9. SPM 5: Statistical parametric mapping (September 2008), http://www.fil.ion.ucl.ac.uk/spm/software/spm5/
10. MFBOX: Model-free spm toolbox for blind source separation (September 2008), http://mfbox.sourceforge.net
11. Cardoso, J.F., Souloumiac, A.: Blind beamforming for non gaussian signals. IEE Proceedings - F 140(6), 362–370 (1993)

Non-independent BSS: A Model for Evoked MEG Signals with Controllable Dependencies

Florian Kohl[1,2], Gerd Wübbeler[2], Dorothea Kolossa[1], Clemens Elster[2],
Markus Bär[2], and Reinhold Orglmeister[1]

[1] Berlin University of Technology, EMSP Group, Berlin, Germany
[2] Physikalisch-Technische Bundesanstalt (PTB), Berlin, Germany
{florian.kohl,reinhold.orglmeister}@tu-berlin.de,
d.kolossa@ee.tu-berlin.de,
{gerd.wuebbeler,clemens.elster,markus.baer}@ptb.de

Abstract. ICA is often employed for the analysis of MEG stimulus experiments. However, the assumption of independence for evoked source signals may not be valid. We present a synthetic model for stimulus evoked MEG data which can be used for the assessment and the development of BSS methods in this context. Specifically, the signal shapes as well as the degree of signal dependency are gradually adjustable. We illustrate the use of the synthetic model by applying ICA and independent subspace analysis (ISA) to data generated by this model. For evoked MEG data, we show that ICA may fail and that even results that appear physiologically meaningful, can turn out to be wrong. Our results further suggest that ISA via grouping ICA results is a promising approach to identify subspaces of dependent MEG source signals.

1   Introduction

Stimulus experiments are a common way to investigate brain functioning. An evoked brain response consists of a neuronal current, whose location and time dynamics are of interest. For this, magnetoencephalography (MEG) may be used to record the resulting magnetic fields non-invasively. However, each MEG channel gives access only to a superposition of fields, generated by many neuronal currents, active at the same time instant. Independent component analysis (ICA) methods [1,2] are frequently used in order to decompose the signal mixture. They have been successfully applied to remove artifacts and interferers [3]. However, this framework relies on the assumption that all source signals are statistically independent. When ICA is applied to data from a stimulus experiment, different source signals are triggered by the same stimulus and the independence assumption may no longer hold. For example, evoked signals may have a similar activation and termination time leading to a correlation of their energies.

Recently, ICA has been extended to independent subspace analysis (ISA) or multi-dimensional ICA, where sources are allowed to be dependent within a group as long as different groups are mutually independent [4,5,6,7,8,9,10,11]. In [4], Cardoso proposed that ICA followed by a grouping algorithm may be sufficient to find dependent subspaces from the data. Szabó et al. have given a sufficient condition for this conjecture in [8]. Nevertheless, plain ICA methods are often applied to MEG data consisting of

T. Adali et al. (Eds.): ICA 2009, LNCS 5441, pp. 443–450, 2009.
© Springer-Verlag Berlin Heidelberg 2009

possibly dependent source signals. Here, we present a model for synthetic MEG data in order to develop and assess BSS methods. The model allows generating subspaces of various dimensionalities with gradually adjustable dependencies using a multitude of signals. Applied to this data, ICA shows severe shortcomings, while ISA via grouping ICA results seems a promising method to reveal the subspaces of dependent source signals.

2 SSEJR – A Model for Stimulus Evoked MEG

In the following the Synthetic Stimulus Evoked Jittered Response (SSEJR) Model is presented. The core functionality of this model is a simple mechanism that gradually changes the degree of dependency. Specifically, the signal dependency is controlled by a single parameter, while subspaces remain mutually independent. Against a sweep through values of the dependence parameter, the reliability of algorithms using independence (ICA), relaxed independence (ISA) or no independence (non-independent BSS) assumptions may be evaluated.

2.1 Generation of Dependent Source Signal Subspaces

In this work a setting with two independent subspaces is used. The first has 2 signal dimensions (2D) and the second 3 signal dimensions (3D). Representatively, the generation of dependent source signals for the 2D subspace is explained. For this, two time domain signals are created consisting of 100 epochs assuming 1 kHz sampling rate. Each epoch starts with an assumed stimulus followed by a time response, one for each source, that has a source specific form and latency. The response time duration is about 200 ms. A random stimulus is modeled by randomly drawing the epoch length between 800 ms to 1200 ms, such that the average epoch length is 1000 ms. As an example,

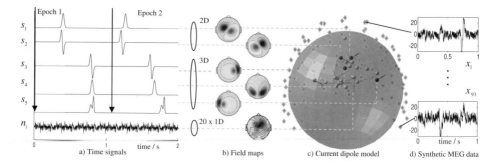

Fig. 1. Synthetic MEG data. (a) 2 epochs of weakly dependent source signals are generated by using the SSEJR model with a high jitter value. Lowering the jitter, all epochs will eventually have responses as depicted in the first epoch leading to two independent signal subspaces (2D and 3D) with highly dependent source signals. The last row depicts one representative AR noise process. (b) Associated field maps. (c) Sphere model of human head with source (black) and noise (gray) dipoles. Sensor coordinates (diamonds) correspond to PTB 93-channel MEG [13]. (d) 2 typical sensor observations exemplify the generated MEG data.

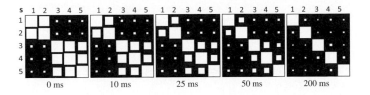

Fig. 2. Hinton diagrams with estimates of pairwise normalized mutual information between 5 SSEJR source signals at different jitter levels. The 2D+3D subspace setting is clearly visible.

one source may have a Gaussian shaped response with 300 ms latency while the other source may have a temporally differentiated Gaussian shaped response with 290 ms latency. This setting is depicted in the first epoch of Fig. 1 a). When all epochs are equal to epoch 1, the resulting signals in the 2D subspace are highly dependent. In order to lower the dependencies, the source specific latencies are changed in each epoch by two normally distributed random numbers with zero mean and standard deviation σ, which leads to a jitter in the latencies. In the following, σ is referred to as the jitter parameter. It is the key parameter of the model controlling gradually the degree of dependence. The higher the jitter the lower the dependency between the signals and vice versa. We refer to this model as the Synthetic Stimulus Evoked Jittered Response (SSEJR) model. Any subspace of any size may be modeled likewise. Subspaces are made mutually independent by choosing all source specific latencies such that signals from different subspaces do not overlap in time. This work's setting uses 5 signals in 2D and 3D subspaces, as depicted in Fig. 1. The dependence of a set of SSEJR generated signals s_i, $i = 1, ..., 5$, will be evaluated in terms of their normalized pairwise mutual information, which may be expressed as

$$I_n(s_i, s_j) = \frac{I(s_i, s_j)}{\sqrt{H(s_i)H(s_j)}}, \text{ with } I(s_i, s_j) = \int p(s_i, s_j) \log \frac{p(s_i, s_j)}{p(s_i)p(s_j)} ds_i ds_j. \quad (1)$$

$H(s_i)$ is the entropy of s_i. Note that the marginal distributions $p(s_i)$ and $p(s_j)$ of any pair of signals within a subspace remain unchanged with varied jitter. The joint distribution $p(s_i, s_j)$ approaches the product of the marginals with increasing the jitter as more sample combinations occur. Hence the mutual information decreases with increasing jitter. Fig. 2 shows the normalized pairwise mutual information between 5 simulated SSEJR source signals for different jitter values. The Hinton diagrams show a block diagonal structure with small off-diagonal elements indicating almost mutually independent subspaces. Elements associated with source signals within each subspace decrease gradually with increasing jitter. This confirms the expected behaviour of the SSEJR model.

2.2 Generating Various Evoked Brain Signals

Fixed signal forms as shown in Fig. 1 a) may not be general enough. As a matter of fact, the use of different signal forms influences the performance of methods under test [13]. Ideally, one would like to consider a large variety of signal forms.

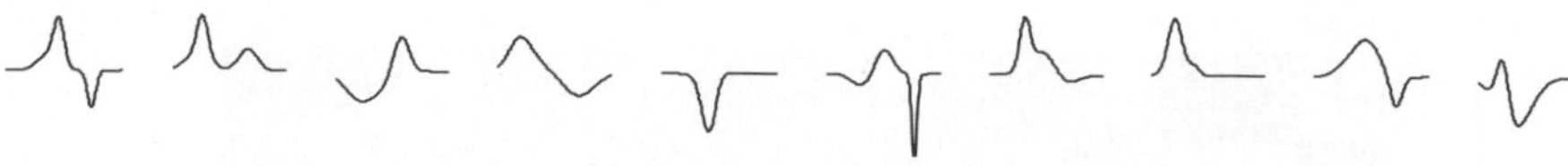

Fig. 3. 10 examples of randomly generated evoked responses using a mixture of 3 Gaussian shaped time signals

Inspired by Gaussian mixture models, time domain signals with various shapes are generated as follows. In the d-th iteration, for a fixed jitter level, the i-th source signal is given by

$$s_i^d(t) = \sum_{k=1}^{3} a_k \exp(-\frac{(t - m_k)^2}{s_k^2}), \qquad (2)$$

where t is the time index with a support of 200 ms, and a_k, m_k and s_k are uniformly distributed random numbers with $-1 < a_k < 1$, $-50 < m_k < 50$ and $5 < s_k < 50$, respectively. Using Equ. 2, a large variety of simulated evoked signals is generated at random. A draw of 10 responses is depicted in Fig. 3.

2.3   Synthetic MEG Data

A homogeneous conducting sphere model serves as an approximation of the human head. The neuronal currents in the brain are modeled by equivalent current dipoles. In this work 25 such dipoles are used, 5 representing evoked sources and 20 independent interfering noise processes. Each such noise dipole follows a 7-th order AR process activated by Laplacian distributed white noise with AR coefficients that reflect prestimulus MEG data. Magnetic fields are calculated at PTB 93-channel MEG sensor coordinates [13], using the quasi-stationary Maxwell equations for dipole currents and emerging volume currents [14]. Sensor noise is introduced with an average SNR of 30 dB. Following the linear ICA model $\mathbf{x} = \mathbf{As}$, $\mathbf{A}$ is the mixing matrix emerging from the contribution of each source to each sensor. The superposition of all magnetic fields gives the observed MEG data $\mathbf{x}$.

For ICA and non-independent BSS the goal is to infer the unmixing matrix $\mathbf{W} = \mathbf{PDA}^{-1}$, where $\mathbf{D}$ is diagonal and $\mathbf{P}$ a permutation matrix. ISA aims at separating only the subspaces with an arbitrary linear combination of rows of $\mathbf{W}$ associated with the source signals within each subspace.

3   Performance of ICA for SSEJR Data

In this section the robustness of ICA against a violation of the independence assumption is investigated. We tested JADE, Extended Infomax and FastICA as sample distribution processing methods [1,2]. The application of methods exploiting the time structure, such as SOBI or TDSEP [1,2], are deferred to future work. The performance of PCA, as a preprocessing step to ICA, is given as well.

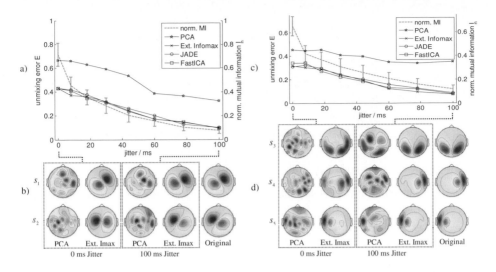

Fig. 4. Performance of ICA (a,c) and associated recovered field maps (b,d) for the 2D subspace (a,b) and 3D subspace (c,d), respectively. For convience, the estimated normalized mutual information $I_n(s_1, s_2)$ (a) and $I_n(s_3, s_4)$ (c) are depicted, too.

3.1 Performance Metric

A normalized version of Amari's performance index [2] gives the unmixing error E, which may be expressed as

$$E(G) = \frac{1}{2m(m-1)} \left[\sum_j \left(\sum_i \frac{G_{ij}}{\max_k G_{kj}} - 1 \right) + \sum_i \left(\sum_j \frac{G_{ij}}{\max_k G_{ik}} - 1 \right) \right], \qquad (3)$$

where the $m \times m$ matrix $\mathbf{G} = \mathbf{WA}$ is the product of the estimated unmixing matrix $\mathbf{W}$ with the simulated mixing matrix $\mathbf{A}$. E is always between 0 and 1 and equal to 0 iff $\mathbf{G} = \mathbf{PD}$, i.e. unmixing is perfect.

3.2 Simulation Setup

The SSEJR model is used with 20 iterations per jitter value, hence considering the results over 20 different signal forms for each source. The simulated MEG data $\mathbf{x}$ is whitened and reduced to 25 dimensions prior to the application of ICA. For the configuration of 20 interferers and 5 SSEJR source signals belonging to a 2D and a 3D subspace, the unmixing error is calculated based on the matrix $\mathbf{G}_k$ of the same dimensions as the k-th subspace. In simulations, the sequence of the 25 simulated signals is known in advance. Hence, we may extract $\mathbf{G}_k$ by taking only elements from $\mathbf{G}$ that are associated with the SSEJR source signals in the k-th subspace. For example, for the 2D subspace, the first two elements of two rows in $\mathbf{G}$ with best signal to interference ratio give the 2×2 matrix $\mathbf{G}_1$. This matrix contains the information to what extent the 2 SSEJR source signals in the 2D subspace have been separated.

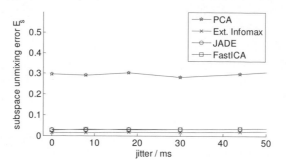

Fig. 5. Performance of ICA to unmix the 2D and 3D subspace containing dependent SSEJR source signals

3.3 Results

The median of the unmixing error and the recovered field maps for the 2D and 3D subspaces are depicted in Fig. 4. A histogram based estimator of mutual information [12] illustrates the functionality of the SSEJR model. The mutual information decreases gradually with increasing jitter. As expected, PCA fails to decompose the signals for all jitter values. In contrast, all tested ICA methods show good performance for almost independent signals at 100 ms jitter, i.e. E drops below 0.2. However, when lowering the jitter value, all ICA methods loose performance. Strikingly, their unmixing error closely follows the estimated mutual information. For highly dependent source signals separation fails completely. In the MEG community, the recovered field maps are of interest, too. Fig. 4 b), d) depict maps for independent and dependent signals. For independent signals at 100 ms all patterns have been recovered by ICA. For dependent signals false pattern emerge, however. Most critically, for closely spaced dipoles, as in the 2D subspace, false recovered maps show a dipolar structure, cf. Fig. 4 b). This finding stresses that even physiologically meaningful results recovered by ICA do not proove their correctness.

4   Grouping the ICA Results – An Approach to ISA

It has been conjectured that ICA can separate mutually independent subspaces of dependent source signals [4]. Based on this conjecture, we grouped the ICA estimated source signals according to their residual dependencies. More precisely, the histogram based estimated normalized pairwise mutual information [12] is used to build an adjacency matrix. To this matrix, a tree-based grouping algorithm is applied to find the underlying subspace structure. Note that other researchers followed the same conjecture but used different techniques to build the adjacency matrix [7,10,11].

4.1 Simulation Setup

The simulation setup follows Subsec. 3.2. The subspace unmixing error E_s is evaluated using Bach's performance index E_s [6]. It is always between 0 and 1 and equals zero

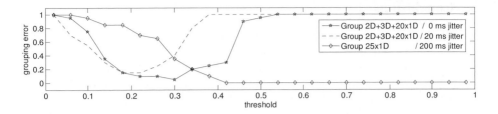

Fig. 6. Grouping error against the threshold. Grouping is perfromed by a tree-based algorithm utilizing the residual mutual information of extended infomax ICA results.

iff the m x m matrix **G** is a permuted block diagonal matrix, which means that dependent signals in a subspace are possibly mixed, while the different subspaces have been unmixed.

4.2 Grouping Method

The histogram based estimate of normalized pairwise mutual information of ICA recovered source signals is used to construct an adjacency matrix. Subsequently, a maximum weight spanning tree is learned. By cutting edges with weights inferior to a threshold, an estimate to the number of subspaces and their associated dependent source signals is obtained [6]. The threshold is varied from 0 to 1, i.e. from cutting no edges to cutting all edges of the tree. It has to be chosen such that dependent and independent source signals are correctly grouped. In the limit, choosing the threshold equal to one will always result in finding only 1D subspaces.

4.3 Results

Fig. 5 depicts the subspace unmixing error E_s against dependency levels ranging from 0 ms to 50 ms jitter. The ICA methods under test show an error close to zero. This strongly supports the conjecture that ICA may be used to unmix mutual independently SSEJR subspaces. Representatively, the tree-based grouping results for extended Infomax are given in Fig. 6. Three graphs show the median grouping error over 20 iterations at 0 ms, 20 ms and 200 ms jitter. The first two jitter values aim at detecting the subspace setting 2D+3D+20x1D, the latter the setting 25x1D. For almost independent source signals, at 200 ms jitter, correct grouping is obtained choosing a threshold larger than 0.42. In order to detect the correct subspaces at jitter values below 20 ms, the threshold has to be below 0.35. Hence, there is a trade off between detecting the correct group of almost independent source signals and dependent source signals. The grouping success for SSEJR signals at 0 ms, 20 ms and 200 ms jitter is 95%, 60% and 65% for a threshold of 0.3. A threshold of 0.18 yields rates of 85%, 85% and 15% and a threshold of 0.42, 70%, 0% and 100%. Note that a grouping is judged successful only if the whole subspace decomposition was found correctly. Hence, finding subspaces based on postprocessing ICA results seems possible for SSEJR data. Still, the performance is sensitive to the choice of threshold, which is a target for future improvements.

5 Conclusions

This work considers source dependencies in MEG stimulus experiments. The SSEJR model is introduced for the generation of synthetic stimulus evoked MEG signals. Various signals in subspaces of various size with gradually adjustable dependencies are the key features. We applied ICA and ISA to data generated by this model. For evoked MEG data, our results suggest that ICA may fail. Most critically, even results that appear physiologically meaningful, can be significantly wrong. Our results further suggest that ISA via postprocessing ICA results may be a promising approach to identify subspaces of dependent MEG source signals.

Acknowledgment

Financial support of the Deutsche Forschungsgemeinschaft DFG (Grant OR 99/4-1) is gratefully acknowledged.

References

1. Hyvärinen, A., Karhunen, J., Oja, E.: Independent component analysis. Wiley, Chichester (2001)
2. Cichocki, A., Amari, S.: Adaptive blind signal and image processing. Wiley, Chichester (2002)
3. Vigario, R., Jousmäki, V., Hämäläinen, M., Hari, R., Oja, E.: Independent component analysis for identification of artifacts in magnetoencephalographic recordings. In: Proc. NIPS, vol. 10, pp. 229–235 (1997)
4. Cardoso, J.F.: Multidimensional independent component analysis. In: Proc. ICASSP, vol. 4, pp. 1941–1944 (1998)
5. Hyvärinen, A., Hoyer, P.O.: Emergence of phase and shift invariant features by decomposition of natural images into independent feature subspaces. Neural Comput. 12, 1705–1720 (2000)
6. Bach, F.R., Jordan, M.I.: Finding clusters in independent component analysis. In: Proc. ICA, vol. 4, pp. 891–896 (2003)
7. Theis, F.J.: Towards a general independent subspace analysis. In: Proc. NIPS, vol. 19, pp. 1361–1368 (2006)
8. Szabó, Z., Pószos, B., Lörincz, A.: Separation theorem for k-independent subspace analysis with sufficient conditions, Tech. Rep., Eötvös Lorand University, Budapest (2006)
9. Póczos, B., Lőrincz, A.: Independent subspace analysis using k-nearest neighborhood distances. In: Duch, W., Kacprzyk, J., Oja, E., Zadrożny, S. (eds.) ICANN 2005. LNCS, vol. 3697, pp. 163–168. Springer, Heidelberg (2005)
10. Póczos, B., Lörincz, A.: Non-combinatorial estimation of independent autoregressive sources. Neurocomp. Lett. 69, 2416–2419 (2006)
11. Meinecke, F., Ziehe, A., Kawanabe, M., Müller, K.R.: A resampling approach to estimate the stability of one-dimensional or multidimensional independent components. IEEE Trans. Biomed. Eng. 49, 1514–1525 (2002)
12. Kraskov, A., Stögbauer, H., Grassberger, P.: Estimating mutual information. Phys. Rev. E 69 (2004)
13. Kohl, F., Wübbeler, G., Sander, T., Trahms, L., Kolossa, D., Orglmeister, R., Elster, C., Bär, M.: Performance of ICA for dependent sources using synthetic stimulus evoked MEG data. In: Proc. DGBMT-Workshop Biosignalverarbeitung, pp. 32–35 (2008)
14. Sarvas, J.: Basic mathematical and electromagnetic concepts of the biomagnetic inverse problem. Phys. Med. Biol. 32, 11–22 (1987)

Atrial Activity Extraction Based on Statistical and Spectral Features

Raúl Llinares, Jorge Igual, Addisson Salazar, Julio Miró-Borrás, and Arturo Serrano

Departamento de Comunicaciones, Universidad Politénica de Valencia,
Plaza Ferrándiz i Carbonell s/n,
03804 Alcoy (Alicante), Spain
{rllinares,jigual,asalazar,jmirob}@dcom.upv.es,
arsercar@upvnet.upv.es

Abstract. Atrial fibrillation is the most common human arrhythmia. The analysis of the associated atrial activity provides features of clinical relevance. Previously, the extraction of the atrial signal is necessary. We follow the semi Blind Source Extraction S-BSE approach to solve the problem. The proposed algorithm satisfies the prior knowledge about the atrial signal: its statistical properties and its spectral content. The introduction of this prior information allows obtaining a new algorithm with the following advantages: it allows the extraction of only the atrial component and it improves the quality of the recovered atrial signal in terms of spectral concentration as we show in the results.

Keywords: Atrial Tachyarrhythmia, Power Spectrum, Kurtosis, Independent Component Analysis, Blind Source Extraction, Electrocardiogram.

1 Introduction

Atrial fibrillation (AF) is a kind of tachycardia. A normal, steady heart rhythm typically beats 60-80 times a minute. In cases of atrial fibrillation, the rate of atrial impulses can range from 300-600 beats per minute. The analysis of these very fast, irregular fibrillatory waves (F-waves) can be used for the identification of underlying AF mechanisms and prediction of therapy efficacy [1]. In particular, the fibrillatory rate has primary importance in AF spontaneous behavior [2], response to therapy [3] or cadioversion [4].

The proper analysis of atrial fibrillation requires the extraction of the atrial activity from the recordings, i.e., the cancellation of the ventricular activity. Taking into account the independence between the atrial and ventricular activity, ICA algorithms based on HOS and SOS have been applied directly to the standard ECG recordings, obtaining the different sources: the atrial rhythm, the ventricular activity and other artefacts such as noise or breathing components [5], [6], [7]. Other algorithms not exploiting the decoupling of the activities have also been proposed to obtain the F-waves [8], obtaining similar performance results when all of them are compared [9].

All these ICA algorithms are focused in a linear transformation of the ECG in order to obtain a set of independent components, not only the atrial rhythm. It means

T. Adali et al. (Eds.): ICA 2009, LNCS 5441, pp. 451–458, 2009.

that they are applied directly without considering the problem of the extraction, not separation, of the only interesting atrial component; in the case of the standard 12-lead ECG recordings, these are decomposed as the linear combination of at most 12 as independent as possible components. As a consequence, depending on the algorithm, the atrial activity is recovered in different positions. All of these solutions require a postprocessing step to distinguish which one is the atrial activity from the set of recovered sources. The identification can be carried out using a kurtosis-based reordering of the separated signals followed by spectral analysis of the subgaussian sources [5] or searching for the source with a peak in the range of 3-10 Hz from the set of final candidates [6].

2 Semi-Blind Source Extraction of the Atrial Rhythm

2.1 Review Stage

None of the proposed solutions explained in the previous Section includes the specific characteristics of the target signal, i.e., the F-waves, in the separation process. In [10], a prior knowledge about the sources is used in the separation process of the atrial activity from ECG recordings. The statistical distribution of the ventricular and atrial activities, supergaussian and subgaussian respectively, is introduced in the maximum likelihood ICA approach [11]. Nevertheless, it needs the post-processing step to confirm which one is the correct atrial activity. Some recent algorithms introduce the spectral content information in the separation step, avoiding the postprocessing identification stage [12], [13]. But none of the previous algorithms takes into account, simultaneously, the statistical and spectral features of the atrial activity to extract the signal.

2.2 Atrial Activity Features

The statistical analysis of the atrial activity reveals that it presents a Gaussian or subgaussian behavior (it depends on the patient and the state of the disease). It can be modeled as a saw-tooth signal consisting of a sinusoid with several harmonics [8]. In this case, the kurtosis values are close to zero.

On the other hand, atrial activity signal typically exhibits a narrowband spectrum with a main frequency between 3.5–9 Hz [6]. The main peak in the band determines the most common fibrillatory rate of nearby endocardial sites [14]. None of the other signals involved in the ECG has the same spectral characteristics. We show in Fig. 1 the normalized spectra of four typical sources mixed in an ECG recording with AF (atrial, ventricular, power line interference and breathing sources); as we can see the atrial signal is concentrated in the aforementioned range of frequencies.

These are the main characteristics of this signal and they will be our prior knowledge about the atrial activity. Hence, we will try to estimate a signal with minimum kurtosis and simultaneously with spectral content as close as possible to the spectral known properties of the atrial signal.

The mathematical formulation of the power spectral density information about the atrial waveform consists on the maximization of the of the relative power contribution to the relevant frequency band:

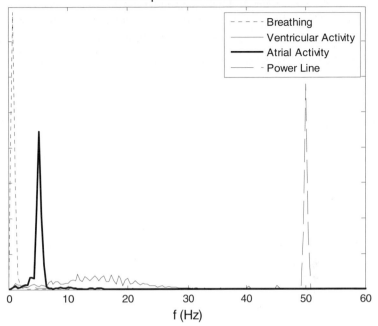

Fig. 1. Superposed power spectra of four sources mixed in the ECG (atrial activity, ventricular activity, 50 Hz power line interference and breathing)

$$
\max\left(\frac{\int_{f_1}^{f_2} \phi_y(f)df}{\int_0^{\infty} \phi_y(f)df} \right)
\tag{1}
$$

where $\phi_y(f)$ is the power spectrum of the recovered signal $y(t)$.

It is maximum for the atrial activity $y_A(t)$; the interval of integration $[f_1, f_2]$ depends on the prior information about the patient and the criterion to fix the bandwidth. One option is the full range of frequencies 3.5-9 Hz, i.e., there is no prior knowledge about the patient or the kind of AF.

2.3 Algorithm for Atrial Activity Extraction Using Prior Information

The atrial fibrillation mixture model reads:

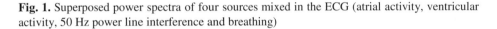

$$
\mathbf{x}(t) = \mathbf{A}\mathbf{s}(t)
\tag{2}
$$

where **x** is the mixtures vector, i.e., the ECG recording, **A** the mixing matrix and **s** the sources vector. In atrial fibrillation episodes, the ventricular and atrial activities are the statistically independent sources, in addition to artifacts and noise sources.

We will assume that the recorded signals are first whitened by PCA in order to reduce the dimensions of the problem, to exhaust the use of the second order statistics and to assure that all the sources have the same variance (this simplifies (1) to the numerator): $\mathbf{z}(t) = \mathbf{V}\mathbf{x}(t)$, where $\mathbf{V}$ is the KxM whitening matrix, so $E\left[\mathbf{z}(t)\mathbf{z}(t)^T\right] = \mathbf{I}$.

The Blind Source Extraction BSE of the atrial component $y_A(t)$ is:

$$y_A(t) = \mathbf{w}^T \mathbf{z}(t) \tag{3}$$

with the restriction $\mathbf{w}^T \mathbf{w} = 1$ to keep the unit variance constraint. In order to estimate $y_A(t)$, i.e., the recovering $Kx1$ unit norm $\mathbf{w}$ vector, we will use the combination of the cost functions, resulting in the following optimization problem:

$$\arg\max_{\mathbf{w}}\left(-k(y_A) + \int_{f_1}^{f_2} \phi_{y_A}(f)df\right), \quad subject\ to\ \mathbf{w}^T \mathbf{w} = 1 \tag{4}$$

where $k(y_A)$ is the kurtosis of y_A. The integral in (4) is calculated numerically and the frequencies normalized by the sampling frequency. The method used to estimate the power spectrum $\phi_{y_A}(e^{j\omega})$ is the classical periodogram due to its simplicity. Then, we can express (4) as function of the unknown vector $\mathbf{w}$ and the whitened observations $\mathbf{z}$:

$$\arg\max_{\mathbf{w}}\left(-\left(E\left[(\mathbf{w}^T\mathbf{z})^4\right] - 3E\left[(\mathbf{w}^T\mathbf{z})^2\right]^2\right) + \sum_{i=I_1}^{I_2} \mathbf{w}^T \mathbf{Z}(e^{j\omega_i})\mathbf{Z}^H(e^{j\omega_i})\mathbf{w}\right) \tag{5}$$

$$subject\ to\ \mathbf{w}^T \mathbf{w} = 1$$

where $\mathbf{Z}(e^{j\omega})$ is the column vector with the Fourier transforms of $z_i(t)$ and $[I_1, I_2]$ the corresponding interval of digital frequencies.

To solve (4), we use a gradient based algorithm that, after every iteration, enforces the constraint $\mathbf{w}^T \mathbf{w} = 1$ dividing by its norm, obtaining the following updating rule:

$$\mathbf{w} \leftarrow \mathbf{w}_{old} - \mu\left(E\left[\mathbf{z}\left(\mathbf{w}_{old}^T \mathbf{z}\right)^3\right] + \left(\mathbf{C}\mathbf{w}_{old} + \mathbf{C}^T \mathbf{w}_{old}\right)\right),$$

$$\mathbf{C} = \sum_{i=I_1}^{I_2} \mathbf{Z}(e^{j\omega_i})\mathbf{Z}^H(e^{j\omega_i}) \tag{6}$$

$$\mathbf{w}_{new} \leftarrow \mathbf{w} / \|\mathbf{w}\|$$

where μ is the learning step.

3 Results

3.1 Database

The data correspond to the "PTB Diagnostic ECG Database" from the MIT database Physionet [15]. There is a total of fifteen patients with atrial fibrillation.

Each record includes 15 simultaneously measured signals: the conventional 12 leads (I, II, III, VR, VL, VF, V1, V2, V3, V4, V5, V6) together with the 3 Frank lead ECGs (VX, VY, VZ). Each signal is digitized at 1000 samples per second, with 16 bit resolution over a range of ± 16.384 mV. In our simulations, we have used the 15 leads during 10 seconds for every patient. The data were downsampled by a factor of 5 and filtered with a 0.5-40 Hz bandwidth filter; this preprocessing does not alter the results.

3.2 Algorithms

We applied the proposed algorithm (we will refer it as KSC – Kurtosis and Spectral Concentration) and two ICA algorithms, one based on high order statistics (FastICA [16]) and other based on second order statistics (SOBI [17]).

All the necessary parameters for the algorithms are experimentally adjusted; for KSC, the learning step has the form $\dfrac{\mu}{\mu+k}$ (k is the iteration variable); for FastICA, the "pow3" non-linearity (kurtosis approximation) is used; for SOBI, the value of time-lags is 55.

3.3 Quality Measurements

The measurement of the quality of the estimation of the residual ECG, i.e., the extraction of the F-waves, is not easy because of the real atrial signal is unknown by definition; it is a hidden or unobserved variable that is only unveiled through the ECG recordings mixed with the rest of biological and non biological signals.

We will use a performance index that measures the quality of the extraction in a quantitative way: the Spectral Concentration (SC). It is defined such as [6]:

$$SC = \frac{\int_{f_p-1}^{f_p+1} P_A(f)df}{\int_0^\infty P_A(f)df} \tag{7}$$

where $P_A(f_i)$ is the power spectrum of the extracted atrial signal and f_p the main peak frequency. The estimation of the spectra is carried out with the modified periodogram using the Welch-WOSA method with the following parameters: a Hamming window of 4096 points length, a 50% overlapping between adjacent windowed sections and a Fast Fourier Transform (FFT) with 8192 points.

SC is a measure of the relative power contained in the narrowband around the peak frequency. It means that, when we compare different algorithms applied to the same patient, a higher SC value indicates a better extraction of the atrial F-waves.

In Table 1, we show the SC obtained from the different algorithms. The quality of the estimated atrial signal is similar in terms of spectral concentration for KSC and SOBI (a little bit better for the KSC algorithm) and worse for FastICA. We must remember that the $[f_1, f_2]$ interval corresponds to the full range of the atrial component, so this prior is only used for the purpose of focus the KSC algorithm in the search for the atrial signal instead of any source.

The real advantage of KSC with respect to SOBI and FastICA is that it allows the extraction of the atrial signal, i.e., we avoid to recover all the sources and after the identification step of the atrial signal, as ICA separation algorithms require.

The mean of the kurtosis obtained by the different methods are -0.0171±0.3736 for KSC, 0.9962±1.2550 for FastICA and 0.4653±0.9500 for SOBI. These values confirm the hypothesis of subgaussian behavior of the atrial activity. In Fig. 2 we show the atrial signal and the associated histogram (with the value of the kurtosis) corresponding to patient #9.

Table 1. Spectral Concentration obtanied by the different methods and patients

Patient	KSC	FastICA	SOBI
#1	0.5363	0.4260	0.5326
#2	0.5420	0.3286	0.5259
#3	0.5898	0.3442	0.5958
#4	0.6835	0.6193	0.6849
#5	0.5933	0.3202	0.5840
#6	0.4337	0.2775	0.4211
#7	0.6240	0.4282	0.6185
#8	0.4567	0.3236	0.4419
#9	0.8238	0.7704	0.8255
#10	0.4049	0.1846	0.3319
#11	0.5413	0.3554	0.5391
#12	0.6423	0.4929	0.6463
#13	0.6075	0.5444	0.6050
#14	0.4448	0.2225	0.4274
#15	0.4947	0.2531	0.4867
Mean	0.5612	0.3927	0.5511
Std. Dev.	0.1100	0.1586	0.1226

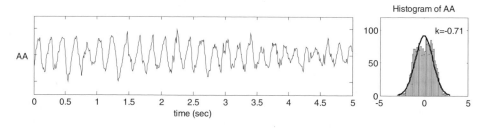

Fig. 2. Five seconds of atrial activity and associated histogram (with kurtosis value) of patient #9

4 Conclusions

We have presented an algorithm for the extraction of the atrial rhythm in atrial fibrillation based on the prior statistical and spectral knowledge of the atrial signal. We have show that it is able to extract the signal of interest, instead of recovering all the sources superposed in the ECG. It allows the simplification of the extraction of the atrial signal but at the same time the use of all the information contained in all the leads. We have improved the results obtained by ICA algorithms, following a Semi –Blind approach to solve the problem, i.e., adapting the algorithm to the specific application.

Acknowledgments. This work has been supported by Spanish Administration and the FEDER Programme of the European Community under grants TEC 2005-01820, TEC 2008-02975/TEC, and *Universidad Politécnica de Valencia* under the Programme of Support to Research and Development.

References

1. Bollmann, A., Husser, D., Mainardi, L., Lombardi, F., Langley, P., Murray, A., Rieta, J.J., Millet, J., Bertil Olsson, S., Stridh, M., Sörnmo, L.: Analysis of surface electrocardiograms in atrial fibrillation: techniques, research, and clinical applications. Europace 8(11), 911–926 (2006)

2. Asano, Y., Saito, J., Matsumoto, K., et al.: On the mechanism of termination and perpetuation of atrial fibrillation. Am. J. Cardiol. 69, 1033–1338 (1992)

3. Stambler, B.S., Wood, M.A., Ellenbogen, K.A.: Antiarrhythmic actions of intravenous ibutilide compared with procainamide during human atrial flutter and fibrillation: electrophysiological determinants of enhanced conversion efficacy. Circulation 96, 4298–4306 (1997)

4. Manios, E.G., Kanoupakis, E.M., Chlouverakis, G.I., et al.: Changes in atrial electrical properties following cardioversion of chronic atrial fibrillation: relation with recurrence. Cardiovasc. Res. 47, 244–253 (2000)

5. Rieta, J.J., Castells, F., Sanchez, C., Zarzoso, V.: Atrial activity extraction for atrial analysis using blind source separation. IEEE Trans. Biomed. Eng. 51(7), 1176–1186 (2004)

6. Castells, F., Rieta, J.J., Millet, J., Zarzoso, V.: Spatiotemporal Blind Source Separation approach to atrial activity estimation in atrial tachyarrhythmias. IEEE Trans. Biomed. Eng. 52(2), 258–267 (2005)

7. Llinares, R., Igual, J., Millet, J., Guillem, M.: Independent component analysis of body surface potential mapping recordings with atrial fibrillation. In: International Joint Conference on Neural Networks, pp. 4595–4601 (2006)

8. Stridh, M., Sornmo, L.: Spatiotemporal QRST cancellation techniques for analysis of atrial fibrillation. IEEE Trans. Biomedical Engineering 48, 105–111 (2001)

9. Langley, P., Bourke, J.P., Murray, A.: Frequency analysis of atrial fibrillation. Comput. Cadiol., 65–68 (2000)

10. Castells, F., Igual, J., Millet, J., Rieta, J.J.: Atrial activity extraction from atrial fibrillation episodes based on maximum likelihood source separation. Signal Processing 85, 523–535 (2005)

11. Pham, D.T., Garat, P., Jutten, C.: Separation of a mixture of independent sources through a maximum likelihood approach. In: EUSIPCO, pp. 771–774 (1992)

12. Igual, J., Llinares, R., Camacho, A.: Source separation with priors on the power spectrum of the sources. In: ESSAN, pp. 331–336 (2006)

13. Phlypo, R., Zarzoso, V., Comon, P., D'Asseler, Y., Lemahieu, I.: Extraction of atrial activity from the ECG by spectrally constrained ICA based on kurtosis sign. In: Davies, M.E., James, C.J., Abdallah, S.A., Plumbley, M.D. (eds.) ICA 2007. LNCS, vol. 4666, pp. 641–648. Springer, Heidelberg (2007)

14. Bollmann, A., Lombardi, F.: Electrocardiology of atrial fibrillation. IEEE Engineering in Medicine and Biology Magazine 25(6), 15–23 (2006)

15. Goldberger, A.L., Amaral, L.A.N., Glass, L., Hausdorff, J.M., Ivanov, P.C., Mark, R.G., Mietus, J.E., Moody, G.B., Peng, C.K., Stanley, H.E.: Physio Bank, PhysioToolkit, and Physio Net: Components of a New Research Resource for Complex Physiologic Signals. Circulation 101(23), 215–220 (2000)
16. Hyvarinen, A.: Fast and robust fixed point algorithms for independent component analysis. IEEE Transactions on Neural Networks 10(3), 626–634 (1999)
17. Belouchrani, A., Kabed Meraim, K., Cardoso, J.F., Moulines, E.: A blind source separation technique based on second order statistics. IEEE Trans. on Signal Processing 45(2), 434–444 (1997)

Blind Source Separation of Sparse Overcomplete Mixtures and Application to Neural Recordings

Michal Natora[1], Felix Franke[2], Matthias Munk[3], and Klaus Obermayer[1,2]

[1] Institute for Software Engineering and Theoretical Computer Science,
Berlin Institute of Technology, Germany
{natora,ff,oby}@cs.tu-berlin.de
[2] Bernstein Center for Computational Neuroscience, Berlin, Germany
[3] Max Planck Institute for Biological Cybernetics, Tübingen, Germany
{matthias.munk}@tuebingen.mpg.de

Abstract. We present a method which allows for the blind source separation of sparse overcomplete mixtures. In this method, linear filters are used to find a new representation of the data and to enhance the signal-to-noise ratio. Further, "Deconfusion", a method similar to the independent component analysis, decorrelates the filter outputs. In particular, the method was developed to extract neural activity signals from extracellular recordings. In this sense, the method can be viewed as a combined spike detection and classification algorithm. We compare the performance of our method to those of existing spike sorting algorithms, and also apply it to recordings from real experiments with macaque monkeys.

1 Introduction

In order to understand higher cortical brain functions, an analysis of the simultaneous activity of a large number of individual neurons is essential. One common way to acquire the necessary amount of neural activity data is to use acute extracellular recordings, either with electrodes or, more recently, with multi electrodes (e.g. tetrode arrays). However, the recorded data does not directly provide the isolated activity of single neurons, but a mixture of neural activity from many neurons additionally corrupted by noise. The signal of the neurons is represented by spikes, which have a length of up to 4 ms and an occurrence frequency of up to 350 Hz. In order to maximize the information yield, one aims at recording from as many neurons as possible; the number of recording channels in acute recordings, however, is mostly limited to 4 (in tetrodes). Thus, these recordings represent a sparse and overcomplete mixtures of neural signals.

The task of so called "spike sorting" algorithms is to reconstruct the single neural signals (i.e., spike trains) from these recordings. There are several reasons to favor *realtime online* sorting over *offline* sorting, although more methods are available in the latter category. For example, realtime online spike sorting techniques are particularly desired for conducting "closed loop" experiments and for brain interface devices. The approaches in realtime online sorting (see

T. Adali et al. (Eds.): ICA 2009, LNCS 5441, pp. 459–466, 2009.

[1] and the references therein) are cluster-based and have mainly the following drawbacks: 1) They do not resolve overlapping spikes, 2) they do not perform well on data with a low signal-to-noise ratio (SNR), 3) they are not able to adapt to nonstationarities of the data as caused by tissue drifts.

An approach based on blind source separation (BSS) techniques and addressing primarily problems 1) and 3) was presented in [2]. Therein, independent component analysis (ICA) was applied on multichannel data recorded by tetrodes. Later, the method was adopted to data recorded by dodecatrodes (12 channels) [3]. However, both approaches had to deal with several new problems: Amongst others, time delays across the channels were not considered, biologically meaningless independent components had to be discarded manually, and different neural signals with similar channel distribution could not be classified correctly. The most severe problem, though, is the fact that the method can not process data containing neural activity from a greater number of neurons than recording channels (overcompleteness).

A possible solution to the overcompleteness was provided in [4]. Therein, the ICA was combined with a conventional clustering technique (k-means clustering) and an aggregation procedure. This approach works only offline and is computationally very expensive, making a real time implementation infeasible. Moreover, the other just mentioned problems persist.

In the following, we present an online realtime algorithm based on the BSS idea, which successfully handles all of the problems 1)-3), but also avoids the discussed drawbacks of the methods in [2,4]. We compare the performance of our algorithm to other common spike sorting techniques and demonstrate its capabilities on real data from extracellular recordings made in the prefrontal cortex of awake behaving macaque monkeys. In order to avoid repetition, we directly formulate the method in the framework of spike sorting.

2   Method

2.1   Generative Model

We assume an explicit model for the neural data recorded extracellularly. The assumptions made are:

1. Each neuron generates a unique spike waveform $\boldsymbol{\xi}^i$ (called template), which is constant over time windows.
2. All time series $\boldsymbol{v}^i$ of spike times of neuron i (called spike trains) are statistically independent of the noise $\boldsymbol{\eta}$. Furthermore, these quantities sum up linearly.
3. The noise statistics is entirely captured by a covariance matrix $\boldsymbol{C}$.

These assumptions are reasonable and underlie, explicitly or implicitly, most spike sorting techniques [5]. Consequently, the measured signal $x_{k,t}$ on channel k at time t can be expressed as:

$$x_{k,t} = \sum_{i=1}^{M} \sum_{\tau} v_{t-\tau}^i \xi_{k,\tau}^i + \eta_{k,t} \qquad\qquad k = 1, ..., N \qquad (1)$$

The measured data is a convolution of the templates with the corresponding intrinsic spike trains corrupted by colored Gaussian noise.

2.2 Calculation of Linear Filters

Spike sorting is achieved when the intrinsic spike train v_t^i is reconstructed from the measured data. Since, according to the model (1), the data was generated by a convolution of the intrinsic spike trains with fixed templates, the intuitive procedure would be to apply a deconvolution on the matrix $\boldsymbol{X}$ (where $(\boldsymbol{X})_{k,t} := x_{k,t}$) in order to retrieve v_t^i. However, for an exact deconvolution a filter with an infinite impulse response is necessary. In general, such a filter is not stable and would amplify noise [6]. Nevertheless, a noise robust approximation for an exact deconvolution can be achieved with finite impulse response filters, to which we will refer as linear filter.

In brief, a set of filters $\{\boldsymbol{f}^1, \ldots, \boldsymbol{f}^M\}$ is desired, such that each filter $\boldsymbol{f}^i$ has a well defined response of 1 to its matching template $\boldsymbol{\xi}^i$ at shift 0 (i.e. $\boldsymbol{\xi}^{i\top} \cdot \boldsymbol{f}^i = 1$), but minimal response to the rest of the data. This means that the spikes of neuron i are the signal to detect for filter $\boldsymbol{f}^i$, but will be treated as noise by filter $\boldsymbol{f}^{j\neq i}$.

Incorporating these conditions leads to a constrained optimization problem, to which the solution are the desired filters. A major advantage is the fact that the mentioned optimization problem can be solved analytically. In particular, the filters are given by the following expression:

$$\boldsymbol{f}^i = \frac{\boldsymbol{R}^{-1}\boldsymbol{\xi}^i}{\boldsymbol{\xi}^{i\top}\boldsymbol{R}^{-1}\boldsymbol{\xi}^i} \qquad\qquad i = 1, \ldots, M \qquad (2)$$

where $\boldsymbol{\xi}^i := \left(\xi_{1,1}^i \cdots \xi_{1,T}^i \cdots \xi_{N,1}^i \cdots \xi_{N,T}^i\right)^\top$ ($\boldsymbol{f}^i$ is defined analogously) and $\boldsymbol{R}$ is the multichannel data covariance matrix. Linear filters maximize the signal-to-noise ratio and minimize the sum of false negative and false positive detections, and are therefore optimal in this sense [7].

2.3 Filtering the Data

Once the filters are calculated, they are cross-correlated with the measured signal, i.e. $\sum_{k,\tau} x_{k,\tau+t} f_{k,\tau}^i =: y_t^i$. Note that we do not have to preprocess the data with a whitening filter, but the filters can be applied directly to $\boldsymbol{X}$, because the noise statistics is already captured in the matrix $\boldsymbol{R}$.

From a different point of view, the filtering just changed the representation of the templates. While in the original space the i-th template was represented by $\boldsymbol{\xi}^i$, its representation in the filter output space is given by the vectors $\boldsymbol{\xi}^i \star \boldsymbol{f}^j$, $j = 1, \ldots, M$, where $\left(\boldsymbol{\xi}^i \star \boldsymbol{f}^j\right)_t := \sum_{k,\tau} \xi_{k,t+\tau}^i f_{k,t}^j$ is the cross-correlation function. This interpretation of filtering will be useful in the next section.

2.4 Deconfusion

The linear filters derived in the previous section should suppress all signal components except their corresponding template with zero shift. Thus, the filter response to all templates (and their shifted variants) has to be minimal. Ignoring the fact that noise has to be suppressed as well, this leads already to $(2T-1) \cdot M$ minimization constraints; a number which is regularly larger than the number of free variables of a filter which is $T \cdot N$. Furthermore, if the SNR is low, the data covariance matrix $\boldsymbol{R}$ is dominated by the noise covariance matrix $\boldsymbol{C}$. Hence, the lower the SNR, the less spikes from non-matching neurons a filter will suppress. Simple positive thresholding in the filter output is, therefore, not the optimal way to sort spikes. This is because only the filter output of filter $\boldsymbol{f}^i$ is taken into account for the detection of spikes from neuron i, but the filter response of filter $\boldsymbol{f}^{j \neq i}$ to template $\boldsymbol{\xi}^i$ is ignored. The idea is to use this additional information for improved sorting.

Since the detection and classification of the spikes is based on the detection of high positive peaks in the filter output (by construction), all values below zero in the filter output do not carry any valuable information. Therefore, we ignore all values below zero by applying a half-wave rectification function $I(x)$ to the filter output $\boldsymbol{Y}$, where $I(x) := x \cdot H(x)$, $H(x)$ being the Heaviside function.

In the next step, $I(\boldsymbol{Y})$ is considered as a linear mixture of different sources. Every source is the intrinsic spike train $\boldsymbol{v}^i$ of a neuron. However, it is not guaranteed that the maximal response of filter $\boldsymbol{f}^i$ to spikes from neuron j will be at a shift of 0, i.e., when the filter and the template overlap entirely. This leads to the following model for the rectified filter output:

$$I(y_t^i) = \sum_j (\boldsymbol{A})_{i,j} \, v_{t+\tau_{i,j}}^j \tag{3}$$

with $\boldsymbol{A}$ being the mixture matrix, and $\tau_{i,j}$ being the associated shifts, i.e., $(\boldsymbol{A})_{i,i} = 1$ and $\tau_{i,i} = 0 \; \forall i$ by construction. We want to reconstruct the sources $\boldsymbol{v}^i$ by solving the corresponding inverse problem:

$$v_t^i \approx z_t^i = \sum_j (\boldsymbol{W})_{i,j} \, I(y_{t-\tau_{j,i}}^j) \tag{4}$$

with $\boldsymbol{W} = \boldsymbol{A}^{-1}$. Here, the relation to ICA becomes clear, since this is a similar inverse problem ICA solves. In contrast to ICA, we do not have to estimate $\boldsymbol{W}$ and $\tau_{i,j}$ from the data, but can calculate them directly from the responses (i.e., cross-correlation functions) of all filters to all templates.

The mixing matrix $\boldsymbol{A}$ is given by the maxima of these responses and $\tau_{i,j}$ by their respective shifts:

$$(\boldsymbol{A})_{i,j} = \max_\tau \left\{ \left(\boldsymbol{\xi}^i \star \boldsymbol{f}^j \right)_\tau \right\} \qquad \tau_{i,j} - \operatorname{argmax}_\tau \left\{ \left(\boldsymbol{\xi}^i \star \boldsymbol{f}^j \right)_\tau \right\} \tag{5}$$

We will refer to the procedure summarized in (4) as "Deconfusion". After Deconfusion the false responses of filters to non-matching templates are suppressed, and hence the filter outputs are de-correlated. In principle, it is possible that the inverse problem is not exactly solvable, if the shifts are not consistent, i.e., if the following equation is violated:

$$\tau_{j_1,k} - \tau_{j_1,i} = \tau_{j_2,k} - \tau_{j_2,i} \qquad \forall i, j_1, j_2, k \tag{6}$$

For arbitrary templates and data covariance structures inconsistent shift may occur, however, with templates from real experiments we did not observe this to be a concern.

In the final step, thresholding is applied to every output channel z^i. All local maxima after a threshold crossing are identified as spiking times of neuron i. In this sense, spike detection and spike classification are performed simultaneously. All found spikes are used to re-estimate the templates. Consequently, the linear filters and the Deconfusion are adjusted as well, allowing the method to adapt to non-stationary templates and to a varying noise structure.

2.5   Initialization

All of the analysis done in the precedent sections was based on the assumption of known initial templates. Hence, before applying our method to the data, one needs an initialization phase during which the templates are found. In principle, any supervised or unsupervised learning method can be applied.

In our experiments we used an energy based spike detection method (as in [1]) and a Gaussian mixture model in combination with the expectation-maximization algorithm for clustering. The number of clusters was determined by the Bayesian inference criterion. The clustering was done on the first 6 principle components of the detected, pre-whitened spikes.

We want to emphasize that the initialization phase is only necessary at the beginning of a recording session. Once the initial templates are estimated, the main algorithm runs online. Usually we used an initialization phase of about 30 s in our real recordings.

3   Evaluation

3.1   Comparison to Existing Spike Sorting Methods

The performance of our method was compared to two popular spike sorting algorithms, of which the implementation is publicly available. *OSort* is an online sorting method based on the Euclidean distance between the estimated cluster means and the pre-whitened spikes [1]. On the other hand, *WaveClus* is an offline method using superparagmetic clustering of wavelet coefficients [8]. Note that neither *OSort* nor *WaveClus* are formulated for tetrode data, so a comparative evaluation with these methods on real data (see Sec. 3.2) would be difficult.

The artificially generated data simulates a single channel recording of 50 s length and a sample frequency of 25 kHz containing activity from two neurons with firing frequencies of 10 Hz each. The two used templates were extracted manually from real recordings beforehand and scaled to equal maximal height. The overcompleteness and the equal heights constraint do not allow for the use of the ICA method described in [3,4] (see Sec. 1 for a more detailed discussion). The noise was generated by an ARMA model approximating a noise characteristic of real recordings [9].

We evaluated the performance on two different settings, in which either the noise level or the amount of overlapping spikes was varied. Our method yields even better results than the offline method, especially in the case of low SNR and in the presence of many overlapping spikes; see Fig. 1.

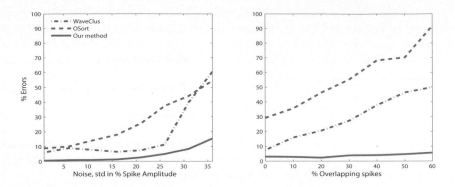

Fig. 1. Average performances over 10 simulations of different spike sorting methods. The parameters in the rival methods were set according to their references. The error is defined as the sum of false positive detections and false negative detections times 100, divided by the total number of inserted spikes. *Left*: Performance for different noise levels. The noise level is varied by changing the noise standard deviation in respect to the maximal height of the template. *Right*: Performance for different amounts of overlapping spikes; the noise level was set to 21%. The amount of temporal overlap between the two templates is drawn from a standard uniform distribution.

3.2   Application to Real Data

The sorting capability of the algorithm was also evaluated on extracellular recordings of in vivo experiments. Activity was recorded while a macaque monkey performed a visual short-term memory tasks; approximatively 2000 trials per session.

We recorded from the ventral prefrontal cortex with an array of 16 individually movable micro-tetrodes. Data were sampled at 32 kHz and bandpass filtered between 0.5 kHz and 10 kHz. Since the tetrodes are inserted anew before every experiment (acute recordings), tissue drifts occur due to tissue relaxation, which leads to constantly varying spike shapes and noise levels. For the initialization phase we used the first 7 trials of the recordings. Our method was executed in the exactly same way as described in Sec. 2. The 7 trials used for the initialization were also processed with our method in order to improve the sorting quality.

Two well-established tests to quantitatively asses the sorting quality of a method performing on real data are the interspike interval distribution and the projection test [1,5]. We evaluated our sorting with both tests (see Fig. 2, for illustration only the results obtained from one tetrode are shown) and conclude that our method is well suited for the use in real experiments.

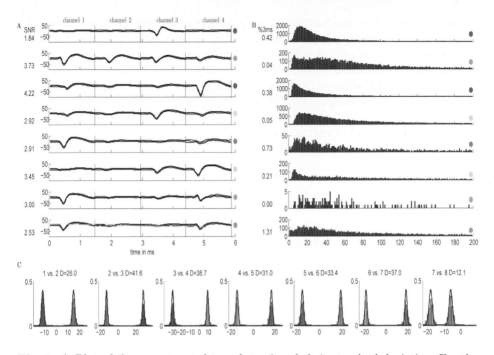

Fig. 2. *A*: Plot of the concatenated templates $\boldsymbol{\xi}$ and their standard deviation. For the averaging all detected spikes from trials 50–90 were used. The vertical lines indicate the concatenation points, while the colored dots on the right serve as a label of the corresponding neuron. On the left, the estimated SNR values are given. *B*: Histogram of the interspike interval distribution with a bin size of 1 ms. The numbers on the left represent the percentage of spikes with an inter-spike interval of less than 3 ms. A small value means that the refractory period of the neurons is respected; thus, the found sources correspond to putative neurons. *C*: Projection test of the found clusters. The solid line represents a Gaussian distribution whose mean is the corresponding template and whose variance is 1. A good fit indicates that the assumptions in Sec. 2.1 are appropriate. The D value is the distance in standard deviations between the means, indicating the separation quality of the clusters.

4   Conclusion

We proposed a blind source separation method for sparse overcomplete mixtures. In the first step, linear filters were applied to the data, which transformed the overcomplete separation problem into a complete one. Signals are represented in the filter output as high positive peaks allowing to ignore all negative values (half-wave rectification) and making following linear transformations nonredundant. A source separation technique, which we called "Deconfusion", was performed on the rectified filter output providing simultaneous detection and classification of the signals. Although this last step is similar to the ICA approach, the parameters of the Deconfusion do not have to be estimated from the data by assuming some statistical properties about it, but can be directly calculated from the cross-correlation functions of the filters and templates. We used this method as a spike sorting algorithm in order to retrieve spike trains from extracellular recordings. Since the linear filters maximize the SNR and the filtering and Deconfusion are linear operations, our method performs well on noisy data and resolves overlapping spikes. Simulations revealed that our method indeed outperforms other existing spike sorting algorithms. Moreover, our method operates online and adapts to non-stationaries present in the data. Therefore, the method is well suited for the analysis of acute recordings, which was also demonstrated in this contribution.

Acknowledgments

This research was supported by the Federal Ministry of Education and Research (BMBF) with the grants 01GQ0743 and 01GQ0410. We thank Sven Dähne for technical support.

References

1. Rutishauser, U., Schuman, E.M., Mamelak, A.N.: Online detection and sorting of extracellularly recorded action potentials in human medial temporal lobe recordings, in vivo. J. Neurosci. Methods 154(1-2), 204–224 (2006)
2. Takahashi, S., Sakurai, Y., Tsukuda, M., Anzai, Y.: Classification of neural activities using independent component analysis. Neurocomputing 49, 289–298 (2002)
3. Takahashi, S., Sakurai, Y.: Real-time and automatic sorting of multi-neuronal activity for sub-millisecond interactions in vivo. Neuroscience 134(1), 301–315 (2005)
4. Takahashi, S., Anzai, Y., Sakurai, Y.: A new approach to spike sorting for multi-neuronal activities recorded with a tetrode–how ICA can be practical. Neurosci. Res. 46(3), 265–272 (2003)
5. Pouzat, C., Mazor, O., Laurent, G.: Using noise signature to optimize spike-sorting and to assess neuronal classification quality. J. Neurosci. Methods 122(1), 43–57 (2002)
6. Robinson, E.A., Treitel, S.: Geophysical Signal Analysis. Prentice-Hall, Englewood Cliffs (1980)
7. Melvin, W.: A STAP overview. IEEE Aerospace and Electronic Systems Magazine 19(1), 19–35 (2004)
8. Quiroga, R.Q., Nadasdy, Z., Ben-Shaul, Y.: Unsupervised spike detection and sorting with wavelets and superparamagnetic clustering. Neural Comput. 16(8), 1661–1687 (2004)
9. Hayes, M.H.: Statistical Digital Signal Processing and Modeling. Wiley, New York (1996)

BSS-Based Feature Extraction for Skin Lesion Image Classification

Hans Georg Stockmeier[1], Heidi Bäcker[2], Wolfgang Bäumler[2],
and Elmar W. Lang[1]

[1] Institute for Biophysics and Physical Biochemistry,
University of Regensburg, Germany
[2] Department of Dermatology, Regensburg University Hospital, Germany
hans.stockmeier@biologie.uni-regensburg.de

Abstract. We apply filter-based image classification methods to skin
lesion images obtained by two different recording systems. The task is to
distinguish different malignant and benign diseases. This is done by extracting features form fluorescence images by applying adaptively learnt
or predefined filters and applying a standard classification algorithm to
the filter outputs. Several methods for filter bank creation such as ICA,
PCA, NMF and Gabor filters are compared.

1 Introduction and Motivation

For many modern medical imaging methods there exist computational solutions
to help the physician in the interpretation of the recorded material. In dermatology there are software solutions commercially available, which proved to
be useful in the discrimination of malignant melanoma from harmless nevi [1].
Horsch et al. describe an algorithm which proposes a diagnosis based on heuristic
criteria derived from the well-known "ABCD-rule". This rule originally refers to
the features "asymmetry", "border", "color" and "diameter". This feature set was
later expanded by Horsch et al., and is widely accepted among dermatologists.
The algorithm achieved a rate of success of 91% in a study of 749 cases [2].

In contrast to only detecting and delineating lesions, photodynamic diagnosis
(PDD) [3] based on fluorescence images is a very recent approach in actually
diagnosing skin lesions. Thus is almost no clinical experience available which
could give information about which features were typical and discriminative for
the diseases under consideration and therefore could be useful for classification.
Hence appropriate features need to be extracted from the data adaptively. This
approach thus necessitates the design of filter-based classification methods. It
relies completely on features that are either entirely independent of the image
material (e.g. pre-calculated Gabor filter banks) or are obtained by applying
unsupervised learning algorithms. In both cases the filter-based approach does
not need previous knowledge about discriminative features of the image classes.

The only assumption taken is that the texture of the images contains information which allows us to determine their class. This assumption seems reasonable
on physiological grounds and justifies the use of filter-based methods. All of the

T. Adali et al. (Eds.): ICA 2009, LNCS 5441, pp. 467–474, 2009.

considered diseases change the layered structure of the skin and the cohesion and growth of the keratinocytes in a typical way. The latter cells form the majority of cells in the epidermis. These structural changes give the lesions a typical appearance on the surface of the skin which is used by dermatologists to distinguish the different diseases. It seems plausible that these physiological changes also cause typical, disease-specific changes in the fluorescence patterns observed during fluorescence diagnosis.

If doubts remain, in clinical practice, of course, a biopsy is performed and the lesion in question is judged on the cellular level by histology. This is the 'gold standard' in diagnostic dermatological methods to definitely determine the type and especially the malignancy of suspicious tissue but implies of course additional costs and additional stress for the patient both of which non-invasive methods as the proposed one could help to avoid.

2   Filter-Based Image Classification

There is a considerable amount of methods available for the analysis of images with filters. Two tasks are common: the detection of certain features within an image (segmentation) and the assignment of an image to a group (classification). For both tasks the methods used in the feature extraction stage are very similar.

The first step in feature extraction is the calculation of the filter bank. Several methods are possible here which will be addressed later. The second step is the filtering of the images to determine the filter outputs. This filtering is in fact a convolution. Denoting image intensities at location (x, y) with $I(x, y)$ and the j-th filter with $w_j(x, y)$, the related filter response is denoted as $r_j(x, y)$ (boundary effects are neglected here):

$$r_j(x, y) = \sum_{(k,l)} w_j(k, l) I(x + k, y + l)$$

As a third step the actual features for classification (or segmentation) are calculated from these filter responses. In the case of segmentation it is the goal to find regions in an image that meet certain prescribed conditions. This is in general a match between one of the filters and the image to at least a certain degree. For the classification task it mostly suffices to detect the presence or absence of features anywhere in an image, their exact position in the image being of lesser (or no) importance.

One possible method to do this detection is to reduce the filter response to a single value which measures the presence of textural features and their degree of similarity with the corresponding filter. Very common for this task is the 'energy' of a filter response which is basically $E_j = \sum_{(x,y)} (r_j(x, y))^2$ [4][5]. Other functions are also possible and will be addressed later.

When filtering with a whole set of filters $w = (w_1, ..., w_n)$, a vector of 'energy' feature values $E = (E_1, ..., E_n)$ is obtained from an image. This is done for every image in a training set. So each image $I^q(x, y)$ is represented by a feature vector E^q. The training and classification procedures themselves are straight forward then. The feature vectors of all images of the training set are presented to a classifier together with their class labels for training. In our case this was a two-layered perceptron with ten neurons in the hidden layer and two neurons in the output layer in order to distinguish two diseases at a time. When a satisfactory training error is reached, the trained system can be used to classify an image with unknown class label. Filtering this test image $I^t(x, y)$ with the filter bank w results in a feature vector E^t which is presented to the classifier to determine the class label of $I^t(x, y)$.

A good estimation of the performance of the trained classifier can be obtained with leave-one-out cross-validation. This is training the classifier with the whole set of available images with known class labels except one. The latter is used to test the classification result. The procedure is repeated for every single image in the data set and the generalization error estimated as the fraction of wrongly classified images.

2.1   Filtering Methods for Filter Bank Construction

ICA. The procedure to extract filters by ICA is well known and described in numerous publications [6][7]. The same number of patches of size $n \times n$ is extracted at random positions in every image. These patches are considered to be the data vectors $x(t)$ in the ICA-model $x(t) = As(t)$. The components of the data vector $x(t)$ are given by the pixel positions in the patches: $x(t) = (x(1, 1, t), ..., x(n, n, t))$. Application of an ICA algorithm – in our case fastICA – yields the underlying independent sources (or causes) $s(t)$ and the decomposition $W = A^{-1}$. The rows of $W = A^{-1}$ represent the desired ICA filters w_j which produce filter responses as statistically independent as possible for the same input $x(t)$:

$$r_j(t) = s_j(t) = \sum_{(k,l)} w_j(k, l)x(k, l, t)$$

Due to the independence of the $r_j(t)$, the ICA filters w_j cover a range of different features in the image material. The whole set of filters w describes the most important structural elements in the images in the sense that it is able to represent every patch-sized image section with a small error. This was shown to be the case for photographs of the natural surroundings of mammals (like trees, grass, landscape and so on) where the ICA filters resemble two-dimensional Gabor-functions [8][9]. Le Borgne et al. reported similar observations on ICA filters tuned to four different classes of images ("cities", "open", "indoor", "closed") whose Gabor-fits were clearly distinct in center frequencies and orientation angle. Moreover they found that the energies of the filter responses of filters learnt from just one class indeed sufficed to determine the class label of an image with high probability [11]. Several authors investigated in more depth the usability of this ICA filter based approach for image classification and texture segmentation tasks [5][4] and proved its applicability to real world problems, e.g. cell counting [12], [13].

PCA. Principal Component Analysis is a well-known method for source decorrelation and can be used for filter set calculation completely analogously to ICA. Besides that PCA can also be used to limit the size of the filter bank to m filters. This is done by keeping only the filters yielding the responses with the m highest variances in the bank and omitting the rest. This kind of dimension reduction was also utilized as a preprocessing step for the calculation of the ICA filter banks with fastICA. In both cases m was 30.

Gabor filters. Gabor filters are calculated from the two-dimensional Gabor-function $h(x, y)$:

$$h(x', y') = A \exp\left(-(x'/\sqrt{2}\sigma_x)^2 - (y'/\sqrt{2}\sigma_y)^2\right) \cos(2\pi f x' + \phi)$$

$$x' = (x - x_0)\cos\theta + (y - y_0)\sin\theta$$
$$y' = (x - x_0)\sin\theta + (y - y_0)\cos\theta$$

The parameter θ denotes the orientation of the filter which was successively set to 0°, 30°, 60°, 90°, 120° and 150°. The parameters σ_x and σ_y denote the widths of the Gaussian along the orientation θ and perpendicular to it, and were set to a fixed ratio of $\sigma_x/\sigma_y = 2$. σ_x was calculated in dependence of the spatial frequency f of the plane wave equal to $\sigma_x = 1/\pi f$. f was chosen to be $\sqrt{2}/2^n$ with $n = 2...6$. These parameter values are very similar to that given in [14] and result in a 'dyadic Gabor filter bank' of 36 filters. The size of the filters was fixed for all frequencies and orientations to 20 pixels.

NMF. Non-negative matrix factorization aims to find an approximate factorization $V \approx WH$ of a non-negative data matrix V into non-negative factors W and H. W and H are supposed to provide parts-based linear representations of V. NMF can be reformulated in terms of the filtering model by identifying V with the image patches $x(t)$ and H with the coefficients $r_j(t)$. The rows of W correspond to the filters. NMF can then be used to learn filters from image material. Again, these are supposed to catch the most important structural information in the image [15]. Consequently NMF filter banks can also be used for image classification. This is an application of NMF which to our knowledge has not been studied yet.

NMFsc. NMF decomposes the data matrix V according to the error criterion $||V - WH||^2$ which is to be minimized. This decomposition is not unique and does not always produce a sparse representation of the data. Such a sparse representation is desirable because it encodes much of the data using only few concurrently 'active' filter components and is therefore supposed to give an encoding which can be interpreted more easily. It is similar to one obtained by ICA with the difference that NMFsc is aiming directly at a sparse decomposition while in ICA this is rather a side-effect and due to the fact that independent components mostly yield a super-gaussian distribution and are therefore sparsely distributed. Hoyer [15] included an additional sparseness criterion in the NMF algorithm which enables the user to explicitly control the sparseness of W and H:

$$s(x) = \frac{\sqrt{\dim(x)} - \left(\sum |x_i|\right) / \sqrt{\sum x_i^2}}{\sqrt{\dim(x)} - 1}$$

In our study we chose a sparseness of 0.99 for the coefficients H and left W unconstrained. The dimension of H, which equals the number of filters, was set to 30 while the dimension of V was given by the size of the patches which again was 10x10 pixels. The number of patches $x(t)$ for the unsupervised training of PCA, ICA, NMF and NMFsc filters was 3000 from every single image in the training set.

2.2   Filter Response Evaluation

There were several methods proposed to measure if and to what amount typical structures, to which a filter is tuned, are present in an image. The most common one is to calculate the 'energy' $E_j = \sum_{(x,y)} \left(r_j(x,y)\right)^2$ of a filter response image which is the sum of the squared gray values of all its pixels as previously mentioned.

Since ICA and also NMFsc filter responses are in general sparsely distributed, it seems reasonable to use sparseness measures to detect if a filter is well tuned to textural structures in an image. Several of such measures were proposed e.g. in [14]. We used the kurtosis $K_j = \sum_{(x,y)} \left(r_j(x,y)\right)^4 - 3$ which is the 4th order statistical moment of the filter response distribution. It was computed after normalization of the variance of $r_j(x,y)$ to unity.

The kurtosis is also a measure of the deviation of a distribution from the Gaussian distribution. Therefore it seemed interesting to also try the negentropy as evaluation function which is also a measure for the above mentioned deviation. Hyvärinen et al. give an estimation for the negentropy which is more robust and less sensitive to outliers than the kurtosis and which is used in the fastICA algorithm [16]. It is basically an estimation of the entropy and leads to the evaluation criterion $N_j = \sum_{(x,y)} \log\left(\cosh\left(r_j(x,y)\right)\right)$.

3   Image Acquisition Methods and Images Available

The first image ensemble was recorded with a fluorescence diagnosis prototype which consisted of a commercially available light source called 'D-Light' (Karl Storz GmbH & Co KG, Tuttlingen, Germany) emitting between 380nm and 450nm and an 8bit CCD-camera-prototype (PCO AG, Kelheim, Germany) with a resolution of 735x582 pixels and equipped with an edge filter which was permissive for wavelengths above 610nm.

The second image ensemble was recorded with an integrated fluorescence diagnosis system called 'Dyaderm' (Biocam GmbH, Regensburg, Germany). Fluorescence was excited with the integrated flash lamp emitting light in the spectral range of 380nm to 450nm at a repetition rate of about 10Hz and recorded with an integrated CCD-camera with a resolution of 320x240 pixels at 12 bits depth.

Table 1 gives an overview of the number of recorded skin lesions organized by disease. In the 'prototype' image ensemble actinic keratoses and Bowen's disease

Table 1. Numbers of Images Available

'Prototype' ensemble	lesions	patients	'Dyaderm' ensemble	lesions	patients
basal cell carcinomas	45	31	basal cell carcinomas	61	18
precancerous lesions	42	28	actinic keratoses	51	21
psoriasis lesions	41	2	Bowen's disease lesions	21	5
total	131	61	total	133	34

lesions were combined in one class called 'precancerous lesions'. This is because both are precursors to the squamous cell carcinoma of the skin albeit in different stages of development. The differences between these diseases are in many cases subtle and so lesions are often clinically hard to differentiate. A distinction is nevertheless of interest to the dermatologist and therefore both classes were separately registered for the 'Dyaderm' ensemble of images. Psoriasis is mostly easily distinguished from other diseases by practitioners and so was included in the first image ensemble mainly to test the system's performance. Besides that exists a possible application in form of mass screening of psoriasis lesions with the goal to detect single cancerous or precancerous lesions among the psoriasis lesions which often occur in great numbers or cover whole parts of the body. However, since only ten psoriasis images were available for the 'Dyaderm' ensemble these were omitted.

4   Results and Conlusions

Table 2 displays the results for the experiments performed. Each experiment consisted of the task to distinguish two classes of skin lesions. For each of the classes a separate filter bank was learnt for the ICA, PCA, NMF and NMFsc methods. So for each of these experiments 60 filters per filter bank were available. The Gabor filter bank consisted of 36 distinct filters. Each of the experiments was repeated five times. Table 2 gives the mean values of the respective classification results.

First of all the evaluation function seems to be of some importance for the quality of the classification. Entropy estimation and energy criterion are clearly superior to kurtosis in almost all experiments. Because both performed equally

Table 2. Rates of Correct Classification in %

Filtering Method	ICA			PCA			NMF	NMFsc	Gabor
Evaluation Crit.	Ener.	Kurt.	Entr.	Ener.	Kurt.	Entr.	Entr.	Entr.	Entr.
Prototype									
Prec. − Basal.	55.9	54.8	55.0	51.4	60.5	50.7	50.3	51.2	56.4
Psor. − Basal.	85.6	72.1	88.6	87.0	68.6	85.8	74.0	84.0	88.6
Psor. − Prec.	87.9	70.0	87.6	87.9	65.7	85.5	68.3	83.3	89.9
Dyaderm									
Act.K. − Basal.	67.2	53.3	65.9	65.7	57.1	64.5	62.2	−	57.6
Act.K. − Bowen's	70.4	65.4	72.6	68.7	66.3	69.6	72.6	−	59.1
Basal. − Bowen's	76.7	74.1	79.5	76.3	71.0	76.4	74.9	−	66.9

well we relied on the entropy estimation criterion only for the Gabor and NMF experiments.

The performance among the filtering methods was rather similar with the exception of the Gabor filter set and NMF (without sparseness constraints). The latter achieved considerable less accurate results in all experiments. In contrast, NMF with sparseness constraints proved very good considering the classification rate but was extremely demanding regarding computation times. NMFsc exceeded the calculation times of fastICA (both implemented in Matlab) by a factor of 20 to 25 and took six to eight days for a single run of one experiment on a 2.8GHz Pentium 4. Experiments on the 'Dyaderm' image ensemble with NMFsc were therefore omitted so far.

Gabor filtering performed very well compared to the unsupervised learning methods in the case of the 'Prototype' image ensemble but rather poor with the 'Dyaderm' ensemble. One possible explanation could be the fact that the image ensembles were recorded with different camera resolutions. So in one case the Gabor filters would cover the distinctive textural features with their preselected range of orientations and frequencies but not in the other case with a lower camera resolution. ICA, presumably adapting to the right scale of these structures by learning, would perform well in both cases. This assumption is supported by some preliminary results we obtained from experiments where the images were linearly scaled before training and classification and which show that the classification rate is dependent of the scaling factor.

Overall the classification rates achieved are not yet satisfying. It also has to be stated that the number of images which were available does not allow for statistically reliable conclusions. But regarding the facts that classification rates better than 90% are considered to be feasible for clinical practice [1] or even sufficient for commercial success [2] in the task of distinguishing malignant melanoma from melanocytic nevi and that the tasks in the 'Dyaderm' ensemble are interesting for dermatologists and were classified with a reasonably good rate of success, we are hopeful that filter based classification can contribute to the field of dermatological image recognition research. Accordingly we plan future work along several lines. These are the combination of several evaluation functions and several different filtering methods in one classifier, the test of different filter sizes and filter numbers together with different scaling factors as well as the detection of the most discriminative features (filters) and feature boosting.

References

1. Perrinaud, A., Gaide, O., French, L.E., Saurat, J.-H., Marghoob, A.A., Braun, R.P.: Can automated dermoscopy image analysis instruments provide added benefit for the dermatologist? A study comparing the results of three systems. Br. J. Derm. 157, 926–933 (2007)
2. Horsch, A., Stolz, W., Neiß, A., Abmayr, W., Pompl, R., Bernklau, A., Bunk, W., Dersch, D., Gläßl, A., Schiffner, R., Morfill, G.: Improving Early Recognition of Malignant Melanomas by Digital Image Analysis in Dermatoscopy. In: Conference on Medical Informatics in Europe, pp. 531–535. IOS Press, Amsterdam (1997)

3. Bäumler, W., Abels, C., Szeimies, R.-M.: Fluorescence Diagnosis and Photodynamic Therapy in Dermatology. Med. Laser Appl. 18, 47–56 (2003)
4. Jenssen, R., Eltoft, T.: Independent Component Analysis for Texture Segmentation. Pattern Recognition 36(10), 2301–2315 (2003)
5. Labbi, A., Bosch, H., Pellegrini, C.: Image Categorization using Independent Component Analysis. In: Proc. of the Workshop on Biologically-inspired Machine Learning, ECCAI Advanced Course on Artificial Intelligence ACAI 1999 (1999)
6. Hoyer, P., Hyvärinen, A.: Independent Component Analysis Applied to Feature Extraction from Colour and Stereo Images. Network: Computation in Neural Systems 11(3), 191–210 (2000)
7. Bell, A., Sejnowski, T.: The Independent Components of Natural Scenes are Edge Filters. Vision Research 37(23), 3327–3338 (1997)
8. van Hateren, J., van der Schaaf, A.: Independent Component Filters of Natural Images Compared with Simple Cells in Primary Visual Cortex. Proc. R. Soc. Lond. B 265, 359–366 (1998)
9. Willmore, B., Watters, P., Tolhurst, D.: A Comparison of Natural-Image-based Models of Simple-Cell Coding. Perception 29(9), 1017–1040 (2000)
10. Ziegaus, C., Lang, E.W.: A Comparative Study of ICA Filter Structures Learnt from Natural and Urban Images. In: Mira, J., Prieto, A.G. (eds.) IWANN 2001. LNCS, vol. 2085, pp. 295–302. Springer, Heidelberg (2001)
11. Le Borgne, H., Guerin-Dugue, A.: Sparse-Dispersed Coding and Images Discrimination with Independent Component Analysis. In: Lee, T., Jung, T., Makeig, S., Sejnowski, T. (eds.) Third International Conference on Independent Component Analysis and Blind Signal Separation, San Diego (2001)
12. Kämpfe, T., Nattkemper, T.W., Ritter, H.: Combining independent component analysis and self-organizing maps for cell image classification. In: Radig, B., Florczyk, S. (eds.) DAGM 2001. LNCS, vol. 2191, pp. 262–268. Springer, Heidelberg (2001)
13. Theis, F., Kohl, Z., Stockmeier, H., Lang, E.: Automated Counting of labelled Cells in Rodent Brain Section Images. In: Proc. BIOMED 2004, pp. 209–212. Acta Press, Canada (2004)
14. Randen, T., Husøy, J.: Filtering for Texture Classification: a Comparative Study. IEEE Transactions on Pattern Analysis and Machine Intelligence 21(4), 291–310 (1999)
15. Hoyer, P.: Non-negative matrix factorization with sparseness constraints. Journal of Machine Learning Research 5, 1457–1469 (2004)
16. Hyvärinen, A., Karhunen, J., Oja, E.: Independent Component Analysis. Wiley & Sons, New York (2001)

Spike Detection and Sorting: Combining Algebraic Differentiations with ICA

Zoran Tiganj[1] and Mamadou Mboup[1,2]

[1] Projet ALIEN, INRIA Futurs,
Parc Scientifique de la Haute Borne,
40, avenue Halley Bt.A, Park Plaza, 59650 Villeneuve d'Ascq, France
[2] UFR Mathématiques et Informatique (CRIP5),
Université Paris Descartes,
45 rue des Saints-Péres, 75270 Paris cedex 06, France

Abstract. A new method for action potentials detection is proposed. The method is based on a numerical differentiation, as recently introduced from operational calculus. We show that it has good performance as compared to existing methods. We also combine the proposed method with ICA in order to obtain spike sorting.

1 Introduction

Decoding the neural information is one of the most important problems in neuroscience. As it is well known, this information is conveyed by the spike train of electrical discharges, called action potential. Action potentials are generated when the equilibrium of electrical charges across the axonal membrane of a neuron is broken [1]. Decoding of the communication between neurons is an extremely challenging problem. One of the difficulties stems from the impossibility, in general, to record the activity of a single neuron but only a mixture activity of all neurons in a measured region. Imperative requirement for the neural information decoding is the ability of action potential detection and sorting (finding out which action potentials are fired by the same neuron) [2], [3], [4], [5].

To make the detection and sorting easier, recording systems usually consist of several electrodes. Each electrode receives the action potentials from all the surrounding neurons. The contribution from a single neuron depends on its distance from the electrode and on the type of the tissue that the action potentials go through. It is shown that ICA [6], [7] has a strong potential in neural signals detection and sorting [8], [9], even though it is well known that neurons are not independent sources – they are communicating with each other and the breaking of the equilibrium of electrical charges is caused by that communication.

In this paper we show how a new algebraic technique for numerical differentiation [10] can lead to a very good performances in neural spike detection. Further on, we use a combination of the technique with ICA to perform spike sorting.

In a neural recording settings, the number of neurons significantly exceeds that of the intracranial electrodes. We assume a measurement configuration where the

T. Adali et al. (Eds.): ICA 2009, LNCS 5441, pp. 475–482, 2009.

distances between the neurons and the electrodes are such that the contribution of only few closest neurons will have a strong influence on the electrodes. The spiking activity of the remaining neurons is considered as the background noise [11].

The neural signal is first represented using a local piecewise polynomial model. The occurrence of an action potential (a spike) is represented by a discontinuity point in the model. Now, such singularities are fully characterized and easy to handle by the differential algebra and operational calculus approach to parameter estimation, introduced in [12] (see also [10] and [13]). Following this approach, we give in section 2 an explicit characterization of the spiking instants, as solutions of a polynomial equation, whose coefficients are composed of a short time window iterated integrals of the signal observation. Using classical numerical integration, we implement these coefficients as outputs of FIR filters. The filters are, as a matter of fact, numerical differentiators and moreover, they inherit the very important low pass property of the iterated integrals. The proposed method is compared in section 3 with two among the most successful methods in the literature. The simulation is performed using synthetic signal. We also combine our detection approach with ICA (see section 4). Benefit of such approach is not only a spike detection, but also a spike sorting, done by ICA. Some concluding remarks are pointed out in section 5.

2   Characterization of the Spike Instants

In this section we give an explicit characterization of the spiking instants. More precisely we show that if an action potential occurs at time instant $t^\star$, then this may be described as a solution of a polynomial equation.

2.1   Mathematical Model

To begin, the noisy observation spike train $y(t)$ is represented by the piecewise regular model

$$y(t) = \sum_{i=1}^{K} \chi_{[t_{i-1}, t_i]}(t) p_i(t - t_{i-1}) + n(t), \tag{1}$$

where $\chi_{[t_{i-1}, t_i]}(t)$ is the characteristic function of the interval $[t_{i-1}, t_i]$, the t_i are the spiking instants and $n(t)$ is the noise corruption. In the sequel, we denote by $x(t) = y(t) - n(t)$ the unobserved noise-free signal. Each $p_i(t)$ is assumed to be a polynomial. Set $t_0 = 0$. Let T be given such that there is at most one discontinuity point in each interval $I_\tau^T = (\tau - T, \tau)$, $\tau \geq T$. Let set $x_\tau(t) = x(\tau - t)$, $t \in [0, T]$ for the restriction of the signal in I_τ^T and redefine the discontinuity point, say t_τ, relatively to I_τ^T with:

 – $t_\tau = 0$ if $x_\tau(t)$ is smooth
 – $0 < t_\tau \leq T$ otherwise

Now, the Nth order derivative of $x_\tau(t)$ (in the sense of distributions theory [14]) satisfies

$$\frac{d^N}{dt^N} x_\tau(t) = [x_\tau^{(N)}](t) + \sum_{k=1}^{N} \left(\mu_{N-k}^0 \delta(t)^{(k-1)} + \mu_{N-k} \delta(t - t_\tau)^{(k-1)} \right) \tag{2}$$

where:

- $\mu_k^0 = x^{(k)}(0+) - x^{(k)}(0-) = x^{(k)}(0)$
- μ_k is the jump of the kth order derivative at the point t_τ

$$\mu_k = x^{(k)}(t_{\tau+}) - x^{(k)}(t_{\tau-})$$

with:

- $\mu_0 = \mu_1 = ... = \mu_{N-1} = 0$ if there is no spike (action potential)
- $\mu_k \neq 0, \ 0 \leq t_\tau \leq T$ if there is a spike (action potential)

where the notation $f^{(k)}$ means k^{th} order derivative of f.
- $[x_\tau^{(N)}]$ represents the regular part of the Nth order derivative of the signal.

The action potential detection problem is now casted into the estimation of the location of the signal discontinuities, viz spike locations, t_τ.

We assume that the degree of each $p_i(t)$ in (1) does not exceed $N - 1$. The regular part $[x_\tau^{(N)}]$ then vanishes. Notice that different choices for N lead to different descriptions of t_τ. In the sequel we set $N = 2$. Equation (2) then reduces to

$$\frac{d^2}{dt^2} x_\tau(t) = \sum_{k=1}^{2} \left(\mu_{2-k}^0 \delta(t)^{(k-1)} + \mu_{2-k} \delta(t - t_\tau)^{(k-1)} \right) \tag{3}$$

2.2   Finding the Firing Instants

In order to solve the equation (3) for t_τ, the problem is transferred into the operational domain (it is possible to solve the equation in the time domain too, but it is easier to handle expression like $\delta(t)$ using operational calculus) where equation (3) reads as:

$$s^2 \hat{x}_\tau(s) - s x_\tau(0) - \dot{x}_\tau(0) = \mu_1 e^{-t_\tau s} + s \mu_0 e^{-t_\tau s} \tag{4}$$

Unknown parameters μ_0, μ_1 and initial conditions $x_\tau(0)$ and $\dot{x}_\tau(0)$ are irrelevant for our purpose. We consider them as undesired perturbations which are easy to eliminate by a Nth order differentiation with respect to s. This results in:

$$t_\tau^2 (s^2 \hat{x}_\tau)^{(2)} + 2 t_\tau (s^2 \hat{x}_\tau)^{(3)} + (s^2 \hat{x}_\tau)^{(4)} = 0 \tag{5}$$

Multiplication by s^k, $k > 0$ in the operational domain implies k order derivative in the time domain. Now differentiation is known to be difficult and ill-conditioned. On the other hand division by s^ν, $\nu > 0$ corresponds in the time domain to the νth order iterated integral

$$\frac{\hat{u}}{s^\nu} \longrightarrow f(t) = \frac{1}{(\nu - 1)!} \int_0^t (t - \lambda)^{\nu-1} u(\lambda) d\lambda.$$

Let us then divide equation (5) by s^{ν}, $\nu > 2$. The resulting equation becomes in the time domain:

$$t_{\tau}^2 \int_0^t (t-\lambda)^{\nu-1}\lambda^2 x_{\tau}^{(2)}(\lambda)d\lambda - 2t_{\tau}\int_0^t (t-\lambda)^{\nu-1}\lambda^3 x_{\tau}^{(2)}(\lambda)d\lambda$$
$$+ \int_0^t (t-\lambda)^{\nu-1}\lambda^4 x_{\tau}^{(2)}(\lambda)d\lambda = 0. \quad (6)$$

In the sequel we fix t to $t = T$. Next, we integrate by parts and we replace the unobserved signal $x_{\tau}(t)$ by its noisy observation counterpart $y_{\tau}(t) = y(\tau - t) = x(\tau - t) + n(\tau - t)$. This leads to the explicit characterization of $\hat{t}_{\tau}$:

$$\hat{t}_{\tau}^2 \int_0^T \left[(T-\lambda)^{\nu-1}\lambda^2\right]^{(2)} y(\tau-\lambda)d\lambda - 2\hat{t}_{\tau}\int_0^T \left[(T-\lambda)^{\nu-1}\lambda^3\right]^{(2)} y(\tau-\lambda)d\lambda$$
$$+ \int_0^T \left[(T-\lambda)^{\nu-1}\lambda^4\right]^{(2)} y(\tau-\lambda)d\lambda = 0 \quad (7)$$

Let us define:

$$h_g(\lambda) = \begin{cases} \left[(T-\lambda)^{\nu-1}\lambda^{g+1}\right]^{(2)} & 0 \le \lambda < T \\ 0 & \text{otherwise} \end{cases} \quad (8)$$

where $g = 1, 2, 3$. After inserting $h_g(\lambda)$ back in equation (7) we obtain

$$\hat{t}_{\tau}^2 \int_0^T h_1(\lambda)y(\tau-\lambda)d\lambda - 2\hat{t}_{\tau}\int_0^T h_2(\lambda)y(\tau-\lambda)d\lambda + \int_0^T h_3(\lambda)y(\tau-\lambda)d\lambda = 0 \quad (9)$$

The coefficients of this equation can be easily implemented as FIR filters.

Only discrete samples of the observation are available. We assume a regular sampling with the sampling period T_s. Let $\tau = nT_s$ and $T = MT_s$. Using a numerical integration method with abscissas mT_s and weight W_m, $m = 0, ..., M$, we obtain the discrete approximation of equation (9) given by

$$\hat{t}_{\tau}^2 \sum_{m=0}^M h_{1,m}y_{n-m} - 2\hat{t}_{\tau}\sum_{m=0}^M h_{2,m}y_{n-m} + \sum_{m=0}^M h_{3,m}y_{n-m} = 0 \quad (10)$$

where: $h_{g,m} = W_m h_g(mT_s)$ and $y_m = y(mT_s)$. Trapezoidal rule for numerical integration defines the parameters W_m:

$$W_0 = W_M = \frac{T_s}{2}, \; W_i = 1, \; i = 1, 2, ..., M - 1$$

3   Spike Detection Algorithm

Recall that if the input observation x_{τ} is smooth, then all the three coefficients of the second order equation (6) are equal to 0 and we have $t_{\tau} = 0$. Otherwise, the spiking instant satisfies $0 \leqslant t_\tau \leqslant T$. In the discrete time domain each of these coefficients can be derived as the input observation x_τ filtered with one of the three filters $\{h_{g,m}\}$, $g = 1, 2, 3$ from equation (10). In the presence of a noise, $y_\tau = x_\tau + n_\tau$, each filter's output will be different from 0, wether x_τ is smooth or not. But still, the values of the filters outputs will be highly correlated with the smoothness of the noise-free observation x_τ (see [10] where it is shown that the filters behave as a numerical differentiators). Therefore we use a product of the three filter's outputs as the final spike detection decision parameter. In this way, the occurrences of the spikes are highly emphasized with respect to the noise.

Comparison of the proposed detection method with two methods among the most accurate ones in the literature is hereafter presented. These are the method based on a combination of median filter, matched filter and nonlinear energy operator (NEO) [4] and the method based on wavelet transformation [5]. The spike train observation data are simulated[1] as in [4]. The value of the parameter T, that defines the filter window length, is set to be 20% above the simulated spike length ($T = 100$ samples). Figure 1 displays the receiver operating characteristic (ROC) curves for the different methods. These results show that the proposed method compares favorably with respect to the others.

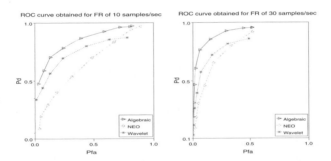

Fig. 1. ROC curves obtained by using the proposed method (Algebraic), NEO, and wavelet based method. Figures for two different firing rates (FR) are shown.

4   Combination with ICA

In this section, we use a more elaborated neural model for the simulation. This is described first. The description is followed by the implementation of the algebraic differentiation filters with ICA. Finally the simulation results are presented.

4.1   Creating the Model

Signals for the simulation are created from cell attached recording performed simultaneously on one of the Purkinje cells[2] (figure 2(a)). Two seconds activity,

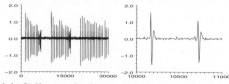
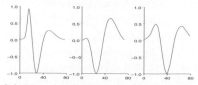

(a) Cell attached recording on one of Purkinje cells (left: 30000 recorded samples; right: zoom with 1000 samples)

(b) Three action potential templates

Fig. 2. Cell attached recording and action potential templates

with 15kHz sampling frequency (30000 samples in total), of 1000 artificial neurons are simulated. For this, spikes extracted from the recording were aligned into three clusters by using a classical K-means algorithm. Three different templates (figure 2(b)) are obtained by the centroids of each cluster.

For each neuron, we simulate first the firing instants as Poisson distributed with a refractory period of 6.7ms. To each location, we then associate randomly one of the above spike templates. The neurons are spatially randomly distributed in the form of a cube, with minimal and maximal mutual distances. Four electrodes are placed on one side of the cube. We assume an isotropic medium (constant conductivity of the tissue), so that the contribution of each neuron is determined only by its distance from each electrode.

4.2   Spike Detection and Sorting

Superposition of the distant neurons activity occurs as a noise on the electrodes – original action potential shape waveforms are destroyed. Only the action potentials coming from the neurons that are close to the electrodes are recorded as spike shape waveforms. Next, signals from the four electrodes are applied to each of the three filters in order to emphasize action potential shape waveforms. The product of the filters outputs at each electrode may be used as a decision function. This was done in the previous section. Here we use the filters outputs as inputs for ICA in order to perform both, spike detection and sorting. We apply ICA on each set of the four outputs from each of the three filters. The goal is to reconstruct the activity of one neuron per electrode (the closest one). FastICA MATLAB toolbox[3] is used for ICA implementation. ICA gives three results for each of the three filters. The final detection and sorting result is given as a product of these three ICA outputs.

4.3   Results

To gauge the effects of the filters, we also consider the same settings, but with the filters replaced by the identity: we apply ICA directly on the signals from

---

[3] The toolbox is developed at Laboratory of Information and Computer Science in the Helsinki University of Technology and it is available on `http://www.cis.hut.fi/projects/ica/fastica/`

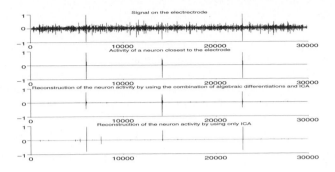

Fig. 3. Results of spike detection and sorting on one electrode with 30000 samples when only ICA is applied and when the combinations of algebraic differentiations and ICA is applied

the electrodes and then raise each ICA output to the power 3. Figure 3 compares the two normalized results obtained with and without the filters, on one sample realization of the reconstructed activity of one neuron. The figure shows that the amplitudes of the spikes that are not fired by the neuron closest to the observed electrode are attenuated more with the proposed method than when only ICA is used. This is confirmed in Table 1 that displays the ratio of the mean amplitude of the correctly detected action potentials (15 in total) over the mean amplitude of the falsely detected ones (80 in total). At a given electrode, a correct detection is meant when the detected action potential stems from the closest neuron. Otherwise, the detection is qualified as false. The first row of the table concerns the combination of the algebraic differentiations and ICA while the second row is for ICA alone. The different columns correspond to the four electrodes.

Table 1. Comparison of action potentials (AP) sorting when ICA is combined with the algebraic differentiation filters and when only ICA is used

	Averaged normalized amplitudes of correctly detected AP / Averaged normalized amplitudes of falsely detected AP			
	Neuron 1	Neuron 2	Neuron 3	Neuron 4
AD and ICA	29.45	48.97	29.76	86.31
ICA	19.72	46.80	15.41	78.12

5   Conclusion

We have proposed a piecewise regular model for the spike train where the occurrence of a spike (transient) is represented by a discontinuity points. Using operational calculus and basic algebraic manipulations we give an explicit characterization of these discontinuity points. The characterization provides a spike detection method which compares favorably with the existing methods. By combining this approach with ICA we obtain not only good results for detection, but also for spike sorting.

References

1. Rieke, F., Warland, D., de Ruyter van Steveninck, R., Bialek, W.: Spikes: Exploring the neural code. MIT Press, Cambridge (1999)
2. Lewicki, M.S.: A review of methods for spike sorting: the detection and classification of neural action potentials. Network: Comput. Neural Syst. (1998)
3. Delescluse, M., Pouzat, C.: Efficient spike-sorting of multi-state neurons using inter-spike intervals information. J. Neurosci. Methods 150(1), 16–29 (2006)
4. Land, B., Spence, A.: Spike (action potential) detection and sorting. Journal Club 25 (April 2002)
5. Nenadic, Z., Burdick, J.W.: Spike detection using the continious wavelet transform. IEEE Trans. Biomed. Eng. 52(1), 74–87 (2005)
6. Cardoso, J.F.: Blind signal separation: statistical principles. Proceedings of the IEEE 10(9), 2009–2025 (1998)
7. Comon, P.: Independent component analysis, A new concept? Sig. Proc. 36 (1994)
8. Mamlouka, A.M., Sharp, H., Menne, K.M.L., Hofmann, U.G., Martinetz, T.: Unsupervised spike sorting with ICA and its evaluation using GENESIS simulations. Neurocomputing (2005)
9. Brown, G.D., Yamada, S., Sejnowski, T.J.: Independent component analysis at the neural cocktail party. Trends in Neurosciences (2001)
10. Mboup, M., Join, C., Fliess, M.: Numerical differentiation with annihilators in noisy environment. Numerical Algorithms (to appear, 2008)
11. Gerstner, W., Kistler, W.: Spiking Neuron Models. Single neurons, Populations, Plasticity. Cambridge University Press, Cambridge (2002)
12. Fliess, M., Sira-Ramirez, H.: An algebraic framework for linear identification ESAIM Contr. Opt. Calc. Variat. 9 (2003)
13. Mboup, M.: Parameter estimation for signals described by differential equations, Applicable Analysis (to appear, 2008), doi:10.1080/00036810802555441
14. Schwartz, L.: Theorie des distributions, 3rd edn. Hermann (1998)
15. Pouzat, C.: Methods and Models in Neurophysics. Les Houches 2003 Summer School, pp. 729–786. Elsevier, Amsterdam (2005),
   http://fr.arxiv.org/abs/q-bio.QM/0405012

Segmentation of Natural and Man-Made Structures by Independent Component Analysis

André Cavalcante[1], Fausto Lucena[1], Allan Kardec Barros[2],
Yoshinori Takeuchi[1], and Noboru Ohnishi[1]

[1] Depto. of Media Science, School of Information Science, Nagoya University,
Furo-cho, Chikusa-ku, Nagoya, 464-8603, Japan
[2] Laboratory for Biological Information Processing, Universidade Federal do
Maranhão, Avenida dos Portugueses, São Luís, MA, 65080-040, Brazil
andre@ohnishi.m.is.nagoya-u.ac.jp, lucena@ohnishi.m.is.nagoya-u.ac.jp,
allan@ufma.br, takeuchi@ohnishi.m.is.nagoya-u.ac.jp,
ohnishi@is.nagoya-u.ac.jp

Abstract. Multi-scale processing is one of the main issues in the segmentation of natural and man-made structures in real worlds scenes. In this work, we use independent component analysis (ICA) to learn sets of multi-scale features specialized for natural and man-made structures, respectively. Then, we use the learned features to represent images according to a simple linear generative model. Finally, we separate each group of structures by analyzing the error of representation for each set of features. The features learned by ICA reflected both second and higher-order statistical information of each dataset. The average time consumed in the segmentation was 3 milliseconds by image block. The system was validated using scenes from different image databases.

1   Introduction

Even for humans subjects, the segmentation of natural and man-made structures in real-world scenes at different ranges is not a trivial task. Many works have focused on classical edge detection to segment the geometry of man-made structures at both ground and aerial level [1][2]. Indeed, contour detection along with texture analysis can be useful for such applications. However, considering scenes at long range and involving complex structures, edge detection can be very noisy. Furthermore, geometry analysis normally involves the use of pre-defined shape models, which might not fit to diverse set of structures.

On the other hand, second order statistics have also been used to classify scenes into natural or man-made [3]. Interestingly, it has been shown that the power spectra information can be easily used to differentiate natural and man-made scenes even at different ranges. The drawback of such technique is that it requires the use of the whole scene to estimate a consistent spectral signature, which is not suitable for a block-level classification.

Another interesting approach is the use of a probability model which capture local dependencies to label image sites [4]. Also, a set of features learned from

T. Adali et al. (Eds.): ICA 2009, LNCS 5441, pp. 483–490, 2009.
© Springer-Verlag Berlin Heidelberg 2009

multi-scale analysis is used. However, according to the authors, such technique is not feasible to close range and fails to differentiate image blocks that contains sky from those of man-made images that do not contain lines or edges.

Therefore, the purpose of this work is to propose a system to segregate natural and man-made structures at different ranges. Firstly, we use independent component analysis (ICA) to learn features adapted to the statistical characteristics of each image group. Secondly, we use the learned features to represent image blocks according to a linear generative model. Finally, we carry out a block-level classification by analyzing the error of representation for each set of features.

2   Methods

Let us divide an image into a set of m non-overlapping blocks indicated by the vector $\mathbf{y} = [y_1, y_2, \ldots, y_m]^\mathrm{T}$. Then, assume that each block y_i can be generated as a linear combination of vectors from a stochastically learned subspace $\Phi = [\phi_1, \phi_2, \ldots, \phi_n]^\mathrm{T}$, where ϕ_i is also called basis function. This process might be written as

$$\hat{y}_i = w_1\phi_1 + w_2\phi_2 + \ldots + w_n\phi_n, \tag{1}$$

where $\hat{y}_i$ is the reconstructed version of block y_i and each component w_i is the projection coefficient on the ith base function.

Also, let us define the error ε_i for the projection of the image block y_i onto the subspace Φ as

$$\varepsilon_i = y_i - \hat{y}_i. \tag{2}$$

Now, let us assume that the subspaces Φ_{Nat} and Φ_{Man} represent underlying bases of natural and man-made images, respectively. Then, we can estimate errors $\varepsilon_{i_{Nat}}$ and $\varepsilon_{i_{Man}}$ by projecting an image block y_i onto the those two subspaces. Finally, to segregate natural and man-made structures within a real world image, we use the following hypothesis:

$$\text{If} \quad y_i \quad \text{is} \quad \text{natural}, \quad \text{then} \quad \mathrm{E}\{\varepsilon_{i_{Nat}}^2\} < \mathrm{E}\{\varepsilon_{i_{Man}}^2\}. \tag{3}$$

$$\text{If} \quad y_i \quad \text{is} \quad \text{man} - \text{made}, \quad \text{then} \quad \mathrm{E}\{\varepsilon_{i_{Nat}}^2\} > \mathrm{E}\{\varepsilon_{i_{Man}}^2\}. \tag{4}$$

To learn the subspaces Φ_{Nat} and Φ_{Man}, and estimate the projection coefficients, we used the system shown in Figure 1. The following subsections provide a full explanation of the structure of our model.

Learning. Let $\mathbf{x} = [x_1, x_2, \ldots, x_m]^\mathrm{T}$ be a set of observations taken from the same data class and written as in Eq. 1. Using $\mathbf{x}$ as a training input, ICA learns basis functions ϕ_i for the data class so that the set of variables which composes vector $\mathbf{a} = [a_1, a_2, \ldots, a_n]^\mathrm{T}$ are mutually statistically independent.

$$\mathbf{x} = \mathbf{a}^\mathrm{T}\Phi. \tag{5}$$

To achieve statistically independence, ICA algorithms work with higher-order statistics which point out directions where data is maximally independent. Here, we used the FastICA algorithm [5].

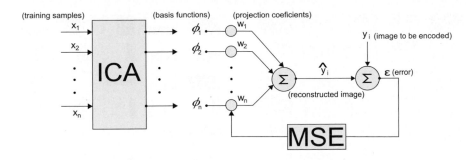

Fig. 1. The system consist of two phases: *learning* and *projection*. In the *learning* phase, we use independent component analysis (ICA) to *learn* subspaces specialized for natural and man-made images. In the *projection* phase, we estimate the projections coefficients through a minimum mean square error (MSE) estimation.

Projection. The projection phase consists of finding out the image representation for the subspace Φ. Such representation is given by the projection vector $\mathbf{w} = [w_1, w_2, \ldots, w_n]^T$ where each coefficient w_i shows how the ith base function is activated inside the image. In our model, the projection vector is found out through minimum mean square error (MSE) estimation.

Hence, for a given image block y_i, the projection vector is estimated such that it minimizes the MSE between the reconstructed block $\hat{y}_i$ and the original one. For this estimation method, the solution vector is given by

$$\mathbf{w} = E[\Phi\Phi^T]^{-1}E[\Phi y_i]. \tag{6}$$

The term $E[\Phi\Phi^T]^{-1}$ in Eq. 6 holds information about the angles between every two basis functions of Φ and is necessary to achieve the minimum error once that there is no constraint about orthogonality for the ICA basis.

3   Results

In the learning phase, we have used 15 natural and 15 man-made images from McGill Calibrated Image Database [6]. This database contains over 850 real-world images of 576 x 768 pixels from natural and urban environments. In both learning and projection phase, images were processed in gray scale not requiring color information.

The natural ensemble included open and close-up views of forests, fields, mountainous landscapes and also natural objects such as trees and flowers. Here, image blocks containing are also considered natural. This way, only images in the natural training ensemble contained sky. The man-made set included urban scenes with restriction for natural structures. For both groups, images in close, medium, and long range were used.

To form the learning input, blocks of 16 x 16 pixels were randomly selected. Overlapping was permitted. For each block, two learning samples were generated. The first sample was obtained performing a line-wise reading of the pixels inside

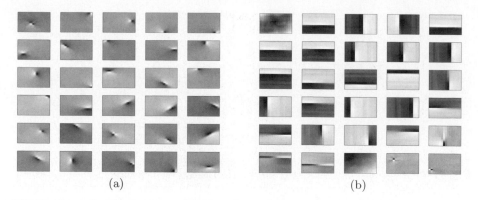

Fig. 2. Example of subspaces. (a) Φ_{Nat} learned from natural and (b) Φ_{Man} learned from man-made images.

the block and second one, through a column-wise reading. We have extracted 100.000 samples from each image group. Figure 2 shows examples of subspaces Φ_{Nat} and Φ_{Man} learned from natural and man-made images.

In the projection phase, have selected several images (not previously used for training) that contains both natural and man-made structures. The selected scenes are shown in Figures 3(a), 3(c), and 3(e). These scenes include a diverse type of structures in close, medium and long range. Also, sky blocks are presented. Here, we used 50-dimensional subspaces. The average time taken was 3 milliseconds by image block on Matlab in a 2.4 Ghz Intel processor. Figures 3(b), 3(d), and 3(f) shows the results of the segmentation by using the hypothesis defined in Equations 3 and 4. Here, the image blocks that contains man-made structures are highlighted by a cross '+'.

To validate the system performance and also to test generality of the features learned by ICA, we have also used image from different image databases. Figure 4(a) and 4(c) shows images from the Labelme database [7] and Hateren database [8], respectively. The segmentation results are shown in Figures 4(b) and 4(d).

4   Discussion

The system proposed here to segregate natural and man-made structures in real-world scenes is based on a quite intuitive idea: structures should be best represented by a set of features which are adapted to their statistical proprieties. This way, firstly, let us analyze the subspaces Φ_{Nat} and Φ_{Man} learned by ICA.

From Figure 2(a), one can see that the basis functions of Φ_{Nat} are Gabor-like in accordance with results of classical works on the efficient coding of natural images. On the other hand, the basis functions learned from man-made images resembles bars and step edges. These two different subspaces may reflect important characteristics of their respective classes.

First, the Gabor-like basis of Φ_{Nat} present no preference for specific orientations as suggested by the almost isotropic power spectra of natural images

Fig. 3. Segmentation of natural and man-made structures in real world scenes. Figures (a), (c) and (e) show the original images. The respective results of the segmentation are shown in Figures (b), (d) and (f). The image blocks that contains man-made structures are highlighted by a cross '+'. All unmarked blocks were classified as natural.

[3]. In case of Φ_{Man} however, the bar and step edge-like basis appear only at vertical and horizontal orientations. This propriety match the power spectra of man-made images which is strongly dominated by those two orientations.

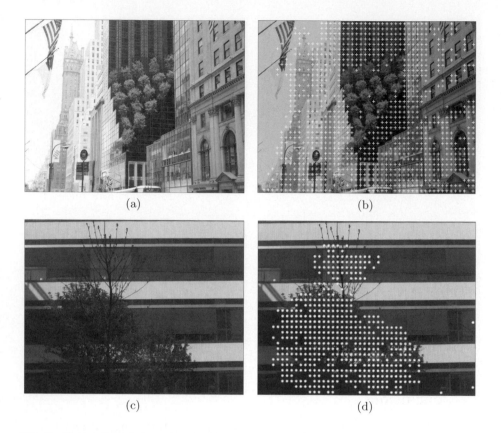

Fig. 4. Generality of the subspaces Φ_{Nat} and Φ_{Man} for other image database. Figures (a) and (c) show the original images from Labelme database [7] and Hateren database [8]. Figures (b) and (d) show the segmented images. For the first scene, the blocks which contains man-made structures are highlighted by a cross '+'. In second scene, natural structures are highlighted by a star '*'.

More importantly, these subspaces may reflect phase information of each image class. In fact, the ICA learning is based on higher-order statistics which are thought to extract phase information from the input data [9]. In this regard, natural and man-made images are likely to exhibit phase spectra with different characteristics. For instance, it has been reported that contrary to natural scenes, edges found in man-made scenes present high degrees of coherence [10]. This means that in man-made images, the pixel arrangement across the length of an edge normally present one orientation. According to [11], a coherent or a step edge produce a global alignment of frequency components in the phase spectra. On the other hand, the orientation of edges in natural scenes continuously turns and twists producing local or phase alignment only in frequency bands. This way, it is reasonable to expect that the step-edge-like basis of Φ_{Man} can capture global phase alignments and the Gabor-like basis of Φ_{Nat}, which are band-pass, can capture local phase alignments.

Since, the subspaces Φ_{Nat} and Φ_{Man} learned by ICA seem to be well-adapted to represent both second and higher order statistical information of their respective classes, let us now analyze the detection results in several real-world scenes shown in Figure 3. The first scene (Figure 3(a),3(b)) shows natural and man-made structures at both medium and long range. The image also contains sky blocks. Analyzing the original and the processed image, we can notice that the system has a higher accuracy in medium range than in long range. Notice that in Figure 3(b), the man-made structures in long range, especially roofs, are more often misclassified as a natural texture. For this scene, almost all sky blocks were correctly classified as a natural structure.

The second scene in Figure 3(a),(b) present both types of structures at close range. The accuracy of the detection appears to be constant from medium to close range. For this scene, most of the misclassified blocks were from grass and textures from the superficies of vehicles. The third scene present a natural structure at medium range and man-made structures at medium and long range. Also, sky blocks are present. Although many sky blocks are misclassified, it is noticeable that the majority is still corrected classified as natural.

In Figure 4, we test the generality of the subspaces Φ_{Nat} and Φ_{Man}. The system works with similar accuracy for a different database even though using features learned from a previous one. The accuracy of this method can be improved by using local dependencies over image sites or a lateral inhibition process. Indeed, such strategies may reduce the quantity of isolated blocks especially in sky areas.

5   Conclusions

Our system consists of two phases: learning and projection. In the learning phase, we used independent component analysis (ICA) to learn subspaces which can represent the underlying basis of natural and man-made images. We have found that the basis functions of man-made images consists of sharp discontinuities resembling bars and step edges in contrary to the Gabor-like basis functions of natural images. According to these features, both subspaces appear to be well-adapted to the second and higher-order statistics of their respective classes. In the projection phase, the system calculate the error of projection of a real-world scene onto those two subspaces. Then, each image block is classified as natural or man-made according to the subspace that generates the lower error. We have found that the system have a higher accuracy at medium and close range. The system also can distinguish sky blocks from close and medium range man-made blocks.

References

1. Krishnamachari, S., Chellappa, R.: Delineating buildings by grouping lines with mrfs. IEEE Trans. on Pat. Anal. Mach. Intell. 5(1), 164–168 (1996)
2. Mayer, H.: Automatic object extraction from aerial imagerya survey focusing on buildings. Computer Vision and Image Understanding 74(2), 138–149 (1999)

3. Torralba, A., Oliva, A.: Statistics of natural image categories. Network: Comput. Neural Syst. 14, 391–412 (2003)
4. Kumar, S., Hebert, M.: Man-Made Structure Detection in Natural Images using a Causal Multiscale Random Field. In: IEEE International Conference on Computer Vision and Pattern Recognition (CVPR), vol. 1, pp. 119–126 (2003)
5. Hyvarinen, A., Karhunen, J., Oja, E.: Independent Component Analysis. John Wiley and Sons, New York (2001)
6. Olmos, A., Kingdom, F.A.: McGill Calibrated Colour Image Database (2004), http://tabby.vision.mcgill.ca
7. Russell, B.C., Torralba, A., Murphy, K.P., Freeman, W.T.: LabelMe: a database and web-based tool for image annotation. International Journal of Computer Vision 77(1-3), 157–173 (2008)
8. van Hateren, J.H., van der Schaaf, A.: Independent component filters of natural images compared with simple cells in primary visual cortex. Proc. R. Soc. Lond. B 265, 359–366 (1998)
9. Field, D.J.: What is the goal of sensory coding? Neural Computation 6, 559–601 (1994)
10. Vailaya, A., Jain, A.K., Zhang, H.J.: On image classification: City images vs. landscapes. Pattern Recognition 31, 1921–1936 (1998)
11. Olshausen, B.A., Field, D.J.: Natural image statistics and efficient coding. Network 7, 333–339 (1998)

Topographic Independent Component Analysis Based on Fractal and Morphology Applied to Texture Segmentation

Klaus Fabian Côco, Evandro Ottoni Teatini Salles, and Mário Sarcinelli-Filho

Graduate Program on Electrical Engineering, Federal University of Espirito Santo
Av. Fernando Ferrari, 514 29075-910 Vitoria, ES, Brazil
{klaus,evandro,mario.sarcinelli}@ele.ufes.br

Abstract. The topographic independent component analysis (TICA) is a technique for texture segmentation in which the image base is obtained from the mixture matrix of the model through a bank of statistical filters. The use of energy as the topographic criterion in connection with the TICA filter bank has been explored in the literature, with good results. In such context, this paper proposes the use of energy plus a morphologic fractal descriptor as a new topographic criterion to be used in connection with the TICA filter bank. The new approach, called TICA fractal multi-scale (TICAFS) approach, results in a meaningful reduction of the segmentation error and/or in a meaningful reduction in the number of filters, when compared to the TICA energy (TICAE) approach.

Keywords: TICA, Fractals, Morphology, Texture, Natural Images.

1 Introduction

Models taking into account high-order statistics, like Independent Component Analysis (ICA) and fractal analysis, have been adopted to perform texture segmentation. ICA performs blind source separation, using high-order statistics to obtain a linear system of statistically independent components, while statistic fractals are characterized by the property of statistical self-similarity [1].

A variation of the ICA technique, the Topographic Independent Component Analysis (TICA), allows getting a more precise model of the human vision system. Developed by Hyvarinen and co-authors [2], the TICA approach is a particularization of the Independent Subspace Analysis (ISA) approach [3] allowing approximating the model of the simple V1 cells of the visual cortex and the complex cells of the human vision system as well [3]. As TICA inserts a spatial organization in the ICA model, which can be seen as a topographic organization of the V1 cells of the visual cortex [2], it establishes a relationship of residual dependency in the structure of the cells of the visual cortex [2], trying to find subspaces with invariant characteristics, in terms of high-order statistics [3].

Textures are characterized by two basic properties, namely homogeneity and localness of the representation, which means that the visual perception of textures can be addressed by analyzing the statistical behavior of the image in a

T. Adali et al. (Eds.): ICA 2009, LNCS 5441, pp. 491–498, 2009.
© Springer-Verlag Berlin Heidelberg 2009

window of limited dimension [4]. This can be perceived through using statistical filters which map the image values in each neighborhood in a subspace of relevant perceptivity, thus reducing the size of the representation while preserving the structural information [4]. On the other hand, the natural appearance of the fractals is a strong evidence that they capture perceivable relevant structures of the natural surfaces [5]. In such context, which is proposed in this work is just to change the residual dependency relationship of the TICA model, in order to improve the sensitivity of the bank of statistical filters in the regions with "visible" singularities, in terms of high-order statistics, using fractals.

2   Background

Natural images are well represented by non-Gaussian distributions, extracted through TICA [2], and by statistical fractals [5]. A merging of these two mathematical models is the essence of the method here addressed.

The TICA model is similar to the ICA one, the difference being the algorithm used to get the model. Considering the ICA approach based on the maximum likelihood estimation, the adaptation for the TICA method is expressed as [6]

ICA Case:

$$\log L(W) = \sum_{t=1}^{T} \sum_{i=1}^{n} \log p_i \left(\boldsymbol{w}_i^T \boldsymbol{x}(t) \right) + T \log |\det W|, \tag{1}$$

TICA Case (energy criterion - TICAE approach):

$$\log L(W) = \sum_{t=1}^{T} \sum_{j=1}^{n} G \left(\sum_{i=1}^{n} h(i,j) \left(\boldsymbol{w}_i^T \boldsymbol{x}(t) \right)^2 \right) + T \log |\det W|, \tag{2}$$

where G is a nonlinear monotonic transformation of real positive numbers, $h(i,j)$ is the neighborhood function, $\left(\boldsymbol{w}_i^T \boldsymbol{x}(t) \right)^2$ is the energy of the high-order non-negative independent components, $\boldsymbol{x}(t)$ is a mixture component of the ICA model and $\boldsymbol{w}_i^T$ is a component of the inverse mixture matrix [6]. The properties of the TICA model are described in [2], where its statistical order is defined as

$$\sigma_i = \varphi \left(\sum_{k-1}^{n} h(i,k) u_k \right), \tag{3}$$

with $\varphi(.)$ being another nonlinear monotonic transformation of positive real numbers, $h(i,k)$ being the neighborhood function once more, and u_k being the non-negative independent components of high statistical order used to generate the statistical components of the model.

In this work the fractal dimension is adopted as the high-order statistical component of the TICA model. It is able to statistically describe a same phenomenon in an image in different scales, thus providing a more efficient tool to analyze textures [7]. Such ability is explained, at least in part, by the fact that the image singularities are preserved, which means that the fractal dimension in the border regions of a texture is always lower than the fractal dimension of the texture as a whole [5]. Thus, using the fractal dimension it is possible to determine the border lines separating regions of different textures, as it is shown in the analysis of the experimental results (Section 5).

The fractal model used in the experiments here reported is adapted from the texture descriptor called Local Multi-fractal Morphologic Exponent (LMME) [7]. How to obtain this adapted descriptor is now considered.

An $N \times N$ image is considered as a 3-D surface defined as the set of triples $\{i, j, f(i, j); i, j = 1, ..., N\}$. For a given scale ε and a structuring element Y_ε, it is defined another set of triples, $\{i_{\varepsilon k}, j_{\varepsilon k}, \beta \varepsilon\}$, where $k = 1, ..., P_\varepsilon$, P_ε being the number of elements in Y_ε, and β is a shape factor, which defines the shape of the structuring element. The dilatation of the image with Y_ε, at the pixel (i, j), is calculated as

$$f_\varepsilon(i, j) = \max_{k=1,...,P_\varepsilon} \{f(i + i_{\varepsilon k}, j + j_{\varepsilon k})\} + \beta \varepsilon. \tag{4}$$

A natural local measure is defined in a window of dimension $W \times W$ as

$$\mu_\varepsilon(i, j) = \frac{|f_\varepsilon(i, j) - f(i, j)|}{\sum_{i,j}^W |f_\varepsilon(i, j) - f(i, j)|}, \tag{5}$$

so that the measure of order q in the scale ε is given by

$$I(q, \varepsilon) = \sum_{i,j}^W \mu_\varepsilon(i, j)^q, \tag{6}$$

the LMME corresponding to the window is

$$L_q = \left(\frac{1}{1-q}\right) \lim_{\varepsilon \to 0} \left(\frac{\ln I(q, \varepsilon)}{\ln N_{/\varepsilon}}\right), \tag{7}$$

and, finally, the LMME of the grid is given by

$$L_q(i, j) = \left(\frac{1}{1-q}\right) \lim_{\varepsilon \to 0} \left(\frac{\ln \mu_\varepsilon(i, j)}{\ln N_{/\varepsilon}}\right). \tag{8}$$

The region of the neighborhood delimited by the structuring element, in a given scale, can be seen as a "click" to the theory of Markov fields. The successive dilatations of the structuring element correspond to successive dilatations of the "click", which corresponds to the expansion of the region under consideration in different scales, which will compose the Markovian fields. Such fields, by their turn, are correlated by a single measure obtained from the successive ones. The measure correlating the successive fields is the fractal descriptor LMME, L_q, which can be obtained from the gradient of the line interpolating the points of $\ln(I(q, \varepsilon)) \times \ln(1/\varepsilon)$, for different scales [7].

3   The Proposed Model

It is known that the ICA model is an approximation of the simple cells in V1 [6] and that the TICA model performs a spatial organization of such cells, thus approximating the behavior of the complex cells in V1 [2]. As the human beings are able to observe and distinguish self-similar characteristics in images like the carpet of Sierpinsk [1], there is evidence that the human visual system interprets simple self-similar structures.

Taking into account that the simple cells in V1 process just local statistical information and that the complex cells perform the statistical organization of the information coming from the V1 cells, a model for the self-similar statistical organization is here proposed as the topographic criterion for the TICA.

By its turn, the fractal descriptor LMME provides a good measure of high-order statistical self-similarity to be used in connection with the TICA model. Instead of the function u_k (see (3)), the fractal descriptor LMME is here adopted to represent the non-negative high-order independent components used to generate the statistical dependency. Actually, it is adopted the function $R(\mathbf{s}_i)$ ($\mathbf{s}_i$ is the independent component $\mathbf{s}_i = \boldsymbol{\omega}_i^T \boldsymbol{x}(t)$ (see (9)), which is based on the LMME descriptor of a region of the image. This way, the TICA model using the descriptor LMME here proposed, hereinafter referred to as TICAFS (TICA Fractal with multiple scales - see (8)), is

$$\log L(W) = \sum_{t=1}^{T} \sum_{j=1}^{n} G \left(\sum_{i=1}^{n} h(i,j) R\left(\boldsymbol{w}_i^T \boldsymbol{x}(t) \right) \right) + T \log |\det W|, \qquad (9)$$

which is based on Eq. (2). The $R(\mathbf{s}_i)$ function corresponds to the application of Eq.(8) in each grid of the neighborhood function, similar to the energy calculation in the TICAE model. By its turn, $\boldsymbol{x}(t)$ are parts of the original image randomly selected.

4   Experiments

In order to check the effectiveness of the proposed method, two stochastic textures with linear Markovian dependency were generated, which are different only in the variance. The method adopted to generate such samples was the Pearson system of distributions [8]. Such textures were mixed to compose the mosaic shown in Figure 1-a, which will be used to test the method here proposed for segmentation. Two other tests were also performed, now using Brodatz's textures [9] (the D9, D22, D55 and D103 textures), to compose the test images of Figures 1-b and 1-c. The texture D9 is a non-periodic one, while the textures D22, D55 and D103 are periodic.

For each image of Figure 1, a set of samples of the mixture, $X(t)$, was built with 2000 windows, $\boldsymbol{x}(t)$, of dimension 8×8 pixels (that determines the filter-bank length - 64 filters - see [10]), which were randomly selected to avoid any temporal correlation. Each window comprises a neighborhood of the TICA model. For

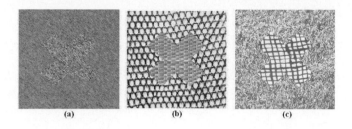

Fig. 1. (a)Test-image with stochastic textures (b)Test-image with the textures D55 (*the inner one*) and D22 (*the outer one*) (c)Test-image with the textures D103 (*the inner one*) and D9 (*the outer one*)

each pixel of it there is a structuring element whose dimension is lower than the dimension of the window, which is centered in such pixel. To calculate the LMME spectrum (Eq. (8)), the scales considered for the structuring element were 2, 3, 4 and 5, the value adopted for β was 3 and the value adopted for q was 1. To calculate the energy, the dimension of the structuring element was fixed in 5.

The segmentation was implemented by adopting the algorithms proposed in [11] and using the k-means algorithm for clustering.

5   Results

Figure 2 shows the classification error resulting from the segmentation of the stochastic textures of Figure 1-a (the TICAE method does not separate the image regions), while Figures 3 and 4 show the same error when segmenting the textures of Figures 1-b and 1-c, respectively. In all cases, different lengths were adopted for the bank of filters (see [11]), as shown in the graphics, for both the TICA energy (TICAE) method and the TICA with fractal descriptor using multiple scale factors (TICAFS) method here proposed.

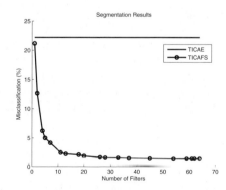

Fig. 2. Classification error for the textures of Figure 1-a

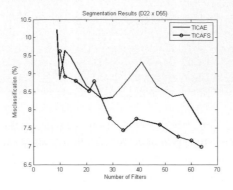

Fig. 3. Classification error for the textures of Figure 1-b

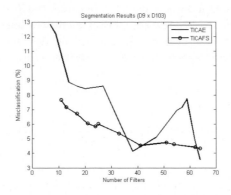

Fig. 4. Classification error for the textures of Figure 1-c

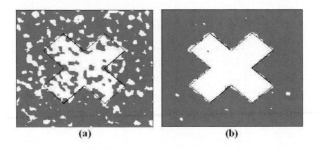

Fig. 5. Results of texture classification considering the mosaic of Figure 1-a for the (a)TICAE method and (b)TICAFS method here proposed, for a bank of 62 filters

Finally, Figures 5-a and 5-b show the result of the classification considering the textures of the mosaic of Figure 1-a, for the same two techniques. The black line delimits the borderline separating the areas corresponding to the two textures.

As one can see from the figures presented, the classification error is much lower when using the proposed TICAFS method, considering images with high order statistics (like that in Figure 1-a). Other advantage of the proposed method, in comparison with the TICAE method, is that the same classification error can be obtained with a bank containing much less filters, as it can be seen from Figures 3 and 4. Another meaningful result is that the border of the region characterized by a certain texture is much better defined when using the TICAFS technique here proposed (see Figure 5).

In addition, it should be stressed that the method here proposed, as all ICA-based methods, is a non-supervised one, thus not demanding any previous training.

6   Conclusion

The fractal multi-scale analysis in the topographic criterion of the TICA technique has shown to be more robust than the traditional model using energy. The use of the morphologic fractal spectrum gives to the TICA model information about the high-order statistics of the textures, what allows a more effective separation of the statistically independent components.

The value of the fractal spectrum LMME is obtained over multi-scale Markovian fields in the neighborhood region of the TICA model, thus determining a representative measure of a stochastic process, which yielded to a better result, considering the synthetic image of Figure 1-a.

The quality of the texture classification by using the TICA model depends on the length, on the disposition and on the approach adopted over the Markovian fields in the neighborhood region. Thus, it is natural that a same texture be classified with different approximation rates for distinct approaches of the neighborhood model. The behavior of the model here proposed is justified by the higher sensitivity of the fractal descriptor regarding the singularities of the textures. On the other hand, the compromise between the generalization and the specification internal to the texture was enough to guarantee a good classification in spite of the noisy representation inherent to the model of fractal surfaces.

Therefore, the conclusion is that the proposal here presented improves the sensitivity of the neighborhood function in the border region of contiguous textures, thus providing a more robust classification, which is validated by the results presented.

References

1. Turner, M.J., Blackledge, J.M., Andrews, P.R.: Fractal Geometry in Digital Imaging. Academic Press, San Diego (1998)
2. Hyvarinen, A., Hoyer, P.O., Inki, M.: Topographic independent component analysis. Neural Computation 13, 1527–1558 (2001)
3. Hyvarinen, A., Hoyer, P.O.: Emergence of phase and shift invariant features by decomposition of natural images into independent feature subspaces. Neural Computation 12, 1705–1720 (2000)

4. Manduchi, R., Portilla, J.: Independent component analysis of textures. In: Proceedings of the International Conference on Computer Vision, Kerkyra, Greece, pp. 1054–1060 (1999)
5. Pentland, A.P.: Fractal based description of natural scenes. IEEE Transactions on Pattern Analysis and Machine Intelligence 6, 661–674 (1984)
6. Hyvarinen, A., Karhunen, J., Oja, E.: Independent Component Analysis. John Wiley and Sons, New York (2001)
7. Xia, Y., Feng, D., Zhao, R.: Morphology-based multifractal estimation for texture segmentation. IEEE Transactions on Image Processing 15, 614–623 (2006)
8. The Matworks, Inc.: Matlab: The language for technical computing, http://www.mathworks.com
9. Brodatz, P.: Texture: a Photograph Album for Artists and Designers. Dover, New York (1956)
10. Jenssen, R., Eltoft, T.: Ica filter bank for segmentation of textured images. In: Proc. Int'l. Workshop on Independent Component Analysis and Blind Signal Separation (ICA 2003), vol. 1, pp. 827–832 (2003)
11. Unser, M., Edem, M.: Nonlinear operators for improving texture segmentation based on features extracted by spatial filtering. IEEE Transactions on Systems, Man and Cybernetics 20, 804–815 (1990)

Image Source Separation Using Color Channel Dependencies[*]

Koray Kayabol[1,**], Ercan E. Kuruoglu[1], and Bulent Sankur[2]

[1] ISTI, CNR, via G. Moruzzi 1, 56124, Pisa, Italy
koray.kayabol@isti.cnr.it, ercan.kuruoglu@isti.cnr.it
[2] Bogazici University, Bebek, 80815, Istanbul, Turkey
bulent.sankur@boun.edu.tr

Abstract. We investigate the problem of source separation in images in the Bayesian framework using the color channel dependencies. As a case in point we consider the source separation of color images which have dependence between its components. A Markov Random Field (MRF) is used for modeling of the inter and intra-source local correlations. We resort to Gibbs sampling algorithm for obtaining the MAP estimate of the sources since non-Gaussian priors are adopted. We test the performance of the proposed method both on synthetic color texture mixtures and a realistic color scene captured with a spurious reflection.

1 Introduction

The problem of blind image separation is encountered in various applications such as document image restoration, astrophysical component separation, analysis of functional MRI (Magnetic Resonance Imaging) and removal of spurious reflections. Most of the previous studies for image source separation have assumed that the source signals are statistically independent. There are cases, however, where this assumption does not hold anymore. For example, the RGB components of a color image are known to be strongly correlated. We conjecture that a source separation algorithm that explicitly takes into consideration this prior information can perform better.

In this paper, we propose a Bayesian approach for the image source separation problem which takes the color channel dependencies into consideration using coupled MRF model. We limit our study to the separation of color image components, which are dependent, and an independent reflection source. The more general problem of source separation from multi-band observations, with interdependencies between the bands, is out of scope of this study. The application scenario is a color image corrupted by spurious reflections. Notice that reflection

---

[*] This work was supported by CNR-TUBITAK joint project No. 104E101. Partial support has also been given by the Italian Space Agency (ASI), under project COFIS.

[**] This work has been supported in part by the "Abdus Salam" International Centre for Theoretical Physics, TRIL Program, through a specific operational agreement with CNR-ISTI, Italy.

T. Adali et al. (Eds.): ICA 2009, LNCS 5441, pp. 499–506, 2009.

removal problem was addressed in [1] where at least two observed images were required. These had to be obtained in different lighting or under different polarization conditions [1]. In contrast our method can remove reflections by using only one observed color image.

The Bayesian frameset allows for under-determined linear mixture models, where the number of observations is smaller than the number of sources. For example, consider a color scene behind a transparent surface photographed with camera, but with an undesired achromatic reflection on the surface. We assume that reflected images are generally achromatic. In this case, the number of sources is four, with three dependent (RGB) components and the fourth one, the reflection, independent from them. Notice that the RGB components are considered as the three sub-sources of a single color source, which is itself mixed with an achromatic image. Since the source separation works with a single observed mixture image, the problem is severely under-determined. However, this under-determined problem is converted into a better posed one, by supplying the missing information via a constrained coupled MRF source model.

There are some recent papers that address the dependent source separation problem. For astrophysical images, Bedini et al. [2] proposed a method for correlated component separation, and Caiafa et al. [3] developed a technique for non-independent components using entropic measures. Gencaga et al. [4] proposed a method for separation of non-stationary mixtures of dependent sources. Hyvarinen and Hurri [5] and Kawanabe and Muller [6] developed separation methods for sources that have variance dependencies. There are also studies which investigate the frequency domain [7] and time-frequency [8] dependencies. Except for [2] and [3], these methods are all developed for dependent 1D signal separation applications.

We overcome the difficulty of separation of such under-determined and sparse models by using spatio-chromatic coupled MRF source models. The coupled MRF model takes both the inter- and the intra-channel dependencies into account. To obtain the MAP estimate of the sources we used Gibbs sampling [9], which is a fully Bayesian algorithm. The algorithm used in this study is an extension of the algorithm in [10].

2   Problem Definition in the Bayesian Framework

In this study, we limit our study to the mixture case of a trichromatic color image and an achromatic component, which results in a single observed color image. The three components of the observed image are $y_j(n), j \in \{r, g, b\}$, where n indexes the pixels. Thus in the parlance of source signal separation one has $L = 4$ sources, $s_i(n), i \in \{r, g, b, m\}$ and $K = 3, y_j(n), j \in \{r, g, b\}$ observations. The linear mixing model is given as:

$$\begin{bmatrix} y_r(n) \\ y_g(n) \\ y_b(n) \end{bmatrix} = \begin{bmatrix} a_{r,r} & 0 & 0 & a_{r,m} \\ 0 & a_{g,g} & 0 & a_{g,m} \\ 0 & 0 & a_{b,b} & a_{b,m} \end{bmatrix} \begin{bmatrix} s_r(n) \\ s_g(n) \\ s_b(n) \\ s_m(n) \end{bmatrix} + V \qquad (1)$$

where V is a zero-mean Gaussian noise vector with $\Sigma = \text{diag}\{\sigma_r^2, \sigma_g^2, \sigma_b^2\}$ covariance matrix. Noise terms are independent identically distributed (iid) in every pixel of the three component images.

When we formulate the BSS problem in the Bayesian framework, the joint posterior density of unknowns $\mathbf{s}$ and $\mathbf{A}$ can be written as:

$$p(\mathbf{s}_{r,g,b,m}, \mathbf{A}|\mathbf{y}_{r,g,b}) \propto p(\mathbf{y}_{r,g,b}|\mathbf{s}_{r,g,b,m}, \mathbf{A})p(\mathbf{s}_{r,g,b,m}, \mathbf{A}) \qquad (2)$$

where $\mathbf{s}$ and $\mathbf{y}$ are lexicographically ordered vector representation of source and observation images, respectively. The unknown sources and the elements of the mixing matrix must be found by using their joint posterior density in (2).

Since the joint solution of (2) is not tractable, we must separate the problem into more manageable parts. According to the source and observation models, we assume that the mixing matrix and the sources are mutually independent, and also the achromatic source is independent from the color image. But the components of the color image are dependent. Using the Bayes rule, conditional densities for sources and mixing matrix can be written as:

$$p(\mathbf{s}_{r,g,b}|\mathbf{s}_m, \mathbf{y}_{r,g,b}, \mathbf{A}) \propto p(\mathbf{y}_{r,g,b}|\mathbf{s}_{r,g,b,m}, \mathbf{A})p(\mathbf{s}_{r,g,b}) \qquad (3)$$

$$p(\mathbf{s}_m|\mathbf{s}_{r,g,b}, \mathbf{y}_{r,g,b}, \mathbf{A}) \propto p(\mathbf{y}_{r,g,b}|\mathbf{s}_{r,g,b,m}, \mathbf{A})p(\mathbf{s}_m) \qquad (4)$$

$$p(\mathbf{A}|\mathbf{s}_{r,g,b,m}, \mathbf{y}_{r,g,b}) \propto p(\mathbf{y}_{r,g,b}|\mathbf{s}_{r,g,b,m}, \mathbf{A})p(\mathbf{A}) \qquad (5)$$

One can use the maximum-a-posteriori (MAP) estimate by alternating variable approach, such as Iterated Conditional Mode. Difficulties may arise in MAP estimation due to non-Gaussian prior densities since they disturb the convexity of the MAP estimate. So we resort to Markov Chain Monte Carlo-based (MCMC) numerical Bayesian methods such as Gibbs sampling.

We use Gibbs sampler to break down the multivariate sampling problem into a set of univariate ones [9]. This iterative procedure continues until the samples converge to those that would have been obtained by sampling the joint density. Our random sampling scheme from source images is hybrid wherever we cannot use direct sampling methods, in other words Metropolis steps are embedded within Gibbs sampling. The normalization term of the Gibbs distribution, namely the partition function, is intractable. Since the posterior of a pixel is formed by the product of a Gaussian likelihood and a Gibbs prior, the posterior is also intractable and direct sampling is not possible. Therefore we resort to Metropolis steps.

After sampling all the images with Metropolis method, the next step is drawing samples from the mixing matrix. When all of the unknowns are sampled, one iteration of the Gibbs sampling algorithm is completed. The Gibbs sampling modified by embedded Metropolis steps, as used in this study, is given in Table 1.

Table 1. Gibbs sampling algorithm

for all source image, $l = r, g, b, m$

    for all pixels, $n = 1 : N$
        Using Metropolis method
        $s_l^{t+1}(n) \longleftarrow \text{sample}_{s_l(n)} \left\{ p(s_l(n)|\mathbf{s}_{l\setminus(n)}^t, \mathbf{s}_{\{r,g,b,m\}\setminus l}^t, \mathbf{y}_{r,g,b}, \mathbf{A}^t) \right\}$

    for all elements of mixing matrix, $(k, l) = (1, 1) : (K, L)$

    $a_{k,l}^{t+1} \longleftarrow \text{sample}_{a_{k,l}} \left\{ p(a_{k,l}|\mathbf{A}_{-(k,l)}^t, \mathbf{y}_{r,g,b}, \mathbf{s}_{r,g,b,m}^t) \right\}$

3   Observation Model

Since the observation noise is assumed to be iid zero-mean Gaussian at each pixel, the likelihood is also Gaussian such that

$$p(\mathbf{y}_{r,g,b}|\mathbf{s}_{r,g,b,m}, \mathbf{A}) = \prod_{k\in\{r,g,b\}} \mathcal{N}(\mathbf{y}_k|\bar{\mathbf{y}}_k, \sigma_k^2) \qquad (6)$$

where $\mathcal{N}(\mathbf{y}_k|\bar{\mathbf{y}}_k, \sigma_k^2)$ represents a Gaussian density with mean $\bar{\mathbf{y}}_k$ and variance σ_k^2 and

$$\bar{\mathbf{y}}_k = \sum_{l=1}^{L} a_{k,l}\mathbf{s}_l, \qquad L = 4. \qquad (7)$$

The prior distributions of elements of $\mathbf{A}$ are chosen as a non-negative uniform distribution due to the lack of any information to the contrary.

$p(a_{k,l}) = [u(a_{k,l}) - u(a_{k,l} - A_{max})]/A_{max}$ where A_{max} is the maximum allowable value of the mixing matrix and $u(.)$ is the unit step function. Using this prior and likelihood in (6), the posterior density of $a_{k,l}$ is formed as

$$p(a_{k,l}|\mathbf{y}_{r,g,b}, \mathbf{s}_{r,g,b,m}, \mathbf{A}_{-a_{k,l}}, \sigma_{1:K}^2) \propto \mathcal{N}(a_{k,l}|\mu_{k,l}, \gamma_{k,l})[u(a_{k,l}) - u(a_{k,l} - A_{max})]. \qquad (8)$$

The mean $\mu_{k,l}$ and the variance σ_k^2 are calculated only for the nonzero elements of the mixing matrix in (1) such that

$$\mu_{k,l} = \frac{1}{\mathbf{s}_l^T \mathbf{s}_l} \mathbf{s}_l^T (\mathbf{y}_k - \sum_{i\in\{r,g,b,m\}, i\neq l} a_{k,i}\mathbf{s}_i) \qquad (9)$$

and $\gamma_{k,l} = \sigma_k^2/\mathbf{s}_l^T \mathbf{s}_l$ and $l \in \{r, g, b, m\}$ and $k \in \{r, g, b\}$.

4   Source Model

Two source models have been used in this study; one of them is for achromatic source and the other is for the color source.

4.1   Achromatic Source Model

For independent achromatic source, we assume that it is modelled as MRFs and that its density $p(\mathbf{s}_m)$ is chosen as Gibbs distribution with possibly non-convex energy potential function. The energy function of Gibbs distribution is

$$U(\mathbf{s}_m) = \frac{1}{2} \sum_{\{n,q\} \in \mathcal{C}} \beta_m \rho_m(s_m(n) - s_m(q)) \tag{10}$$

where $\rho_m(.)$ is the non-convex potential function and $\mathcal{C}$ is the entire clique set. The Gibbs distribution is given as

$$p(\mathbf{s}_m) = \frac{1}{Z(\beta_m)} e^{-U(\mathbf{s}_m)} \tag{11}$$

where $Z(\beta_m)$ and β_m are the partition function and the parameter of MRF, respectively.

In this work we opt to describe pixel differences in terms of iid Cauchy density. Notice that while pixels are dependent, the pixel differences can often be modeled as an iid process [10]. Then the clique potential of the Gibbs density under Cauchy assumption becomes:

$$\rho_m(s_m(n) - s_m(q)) = \ln \left[1 + \frac{(s_m(n) - s_m(q)))^2}{\delta_m} \right] \tag{12}$$

where δ_m is the scale parameter of Cauchy distribution and is called also as threshold parameter of the regularization function. In this study, both of the β_m and δ_m parameters are assumed homogeneous over the MRF.

4.2   Color Source Model

For the prior of the color image, we use multivariate Cauchy density. The clique potential of dependent color component $\mathbf{s}_c(n) = [s_r(n) \quad s_g(n) \quad s_b(n)]^T$ can be written as

$$\rho_c(\mathbf{s}_c(n) - \mathbf{s}_c(q)) = \frac{5}{2} \ln \left[1 + [\mathbf{s}_c(n) - \mathbf{s}_c(q)]^T \Delta^{-1} [\mathbf{s}_c(n) - \mathbf{s}_c(q)] \right]. \tag{13}$$

where Δ is 3×3 symmetric matrix which defines the correlation between color channels. Again we want to point out that these correlations are between pixel differences, which implies that the component edge images are mutually correlated. Our model does not assume any correlation between the pixel values directly, so that $\rho_c(.)$ is a function of pixel differences $[\mathbf{s}_c(n) - \mathbf{s}_c(q)]$. only. The choice of a non-convex clique potential function helps to preserve the edges. The matrix Δ can be expressed explicitly as

$$\Delta = \begin{bmatrix} \delta_{r,r} & \delta_{r,g} & \delta_{r,b} \\ \delta_{r,g} & \delta_{g,g} & \delta_{g,b} \\ \delta_{r,b} & \delta_{g,b} & \delta_{b,b} \end{bmatrix} \tag{14}$$

and the elements of this matrix are defined by the user.

5   Simulation Results

We illustrate the performance of the proposed dependent color source separation algorithm with two examples: The first example consists of a synthetic mixture of texture images, while the second one is real mixture case actually taken with a camera. We use the Peak Signal-to-Inference Ratio (PSIR) as performance indicator. Furthermore, we assume that the correlation coefficients of the color components in (14) as well as the variances of the noise terms in (1) are known or user defined.

5.1   Synthetic Mixture Case

The mixture consists of a color texture image whose components are dependent and another gray-valued texture image independent of the color image. Sample images of original textures and their mixtures are shown in the first and second columns of Fig. 1, respectively. The true color version of the original color image and mixtures (observation) can be seen in Fig. 2. The mixing matrix was chosen as

$$\mathbf{A} = \begin{bmatrix} 1.0 & 0 & 0 & 0.4 \\ 0 & 0.7 & 0 & 0.6 \\ 0 & 0 & 0.5 & 0.8 \end{bmatrix}. \tag{15}$$

These results are compared with those obtained under independent color channels assumption. The PSIR (dB) results obtained under dependence and independence assumptions are shown in Table 2. To obtain the results for the

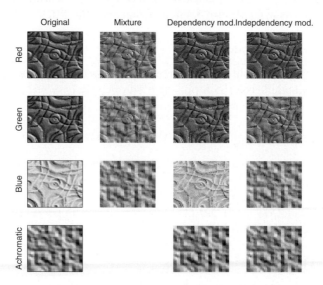

Fig. 1. Simulation experiments with noiseless texture images. First column: original images; second column: images mixed with **A** in (15); third column: sources estimated with dependence assumption; fourth column: sources estimated with independence assumption.

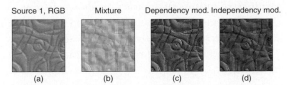

Fig. 2. True color images: (a) original color texture, (b) true color mixture (observation), (c) and (d) estimated color texture under dependence and independence assumption, respectively

Table 2. PSIR (dB) results of the separated sources under dependence and independence assumptions. While source 1 has red, green and blue components, source image 2 consists of only one gray-valued component.

	red	green	blue	achromatic
Dependence assumption	48.98	37.69	36.39	38.55
Independence assumption	48.30	31.83	25.75	34.76

Fig. 3. Removal of reflection image mixed to a color image. (a) Observed actual mixed image, (b) Color image with mixed reflection removed, (c) Estimated achromatic reflection image.

independence assumption, the source model and the algorithm in [10] was used. Dependence assumption yields better results as compared to the independence assumption. For example, the blue component is not separated under independence assumption.

5.2   Real Mixture Case

For a realistic application, we used a color image taken with digital camera and corrupted by a reflection as in Fig. 3 (a)[1]. The scene is organized such that a toy is standing behind a transparent CD box and a reflection occurs on the box surface. The reflection is assumed as an achromatic source while the scene behind is a color image source. We were justified in using the proposed linear mixing model in (1) since the reflection image is both transparent and additive. However, as would occur in real life situations, the mixing is not stationary, which means that the mixing matrix changes from pixel-to-pixel. Although the mixing is not stationary since the amount of reflection has been changing over the surface,

---

[1] The authors would like to thank to Alpay Kucuk for taking the photograph used in Sect. 5.2.

the separation results are quite satisfactory as shown in Fig. 3 (b) and (c). The entries of the Δ matrix has been manually tuned to optimal values by trial and error. The color image of the toy scene is well separated with the reflection almost removed while the achromatic reflection component still contains some vestige of the color image.

6   Conclusion

In this study, we have proposed statistical models for blind separation of dependent source images, and have shown that taking into account the prior information on the dependence of color components results in higher performance as compared to the independence model. The proposed model can find application in color document analysis, restoration of ancient documents and polarized image applications. In a follow-up study we will extend the algorithm for automatic estimation of MRF parameters and noise variances.

References

1. Bronstein, A.M., Bronstein, M.M., Zibulevsky, M., Zeevi, Y.Y.: Sparse ICA for blind separation of transmitted and reflected images. Int. J. Imaging Science Technology 15, 84–91 (2005)
2. Bedini, L., Herranz, D., Salerno, E., Baccigalupi, C., Kuruoglu, E.E., Tonazzini, A.: Separation of correlated astrophysical sources using multiple-lag data covariance matrices. EURASIP Journal on Applied Signal Processing 15, 2400–2412 (2005)
3. Caiafa, C.F., Kuruoglu, E.E., Proto, A.N.: A minimax entropy method for blind separation of dependent components in astrophysical images. In: AIP-Proceedings of MaxEnt 2006, pp. 81–88 (2006)
4. Gencaga, D., Kuruoglu, E.E., Ertuzun, A.: Bayesian separation of non-stationary mixtures of dependent Gaussian sources. In: AIP-Proceedings of MaxEnt 2005 (2005)
5. Hyvarinen, A., Hurri, J.: Blind separation of sources that have spatiotemporal variance dependenies. Signal Process. 84, 247–254 (2004)
6. Kawanabe, M., Muller, K.-R.: Estimating functions for blind separation when sources have variance dependenies. Journal Machine Learning 6, 453–482 (2005)
7. Zhang, K., Chan, L.-W.: An adaptive method for subband decomposition ICA. Neural Computation 18, 191–223 (2006)
8. Abrard, F., Deville, Y.: A time-frequency blind signal separation method applicable to underdetermined mixtures of dependent sources. Signal Process. 85, 1389–1403 (2005)
9. Gilks, W.R., Richardson, S., Spiegalhalter, D.J.: Markov Chain Monte Carlo in Practice. Chapman & Hall, London (1996)
10. Kayabol, K., Kuruoglu, E.E., Sankur, B.: Bayesian separation of images modelled with MRFs using MCMC. IEEE Trans. Image Process. (accepted, 2008)

Learning Natural Image Structure with a Horizontal Product Model

Urs Köster[1,2], Jussi T. Lindgren[1,2], Michael Gutmann[1,2],
and Aapo Hyvärinen[1,2,3]

[1] Department of Computer Science
[2] Helsinki Institute for Information Technology
[3] Department of Mathematics and Statistics University of Helsinki, Finland

Abstract. We present a novel extension to Independent Component Analysis (ICA), where the data is generated as the product of two sub-models, each of which follow an ICA model, and which combine in a horizontal fashion. This is in contrast to previous nonlinear extensions to ICA which were based on a hierarchy of layers. We apply the product model to natural image patches and report the emergence of localized masks in the additional network layer, while the Gabor features that are obtained in the primary layer change their tuning properties and become less localized. As an interpretation we suggest that the model learns to separate the localization of image features from other properties, since identity and position of a feature are plausibly independent. We also show that the horizontal model can be interpreted as an overcomplete model where the features are no longer independent.

1   Introduction

The study of natural images statistics has recently received a great deal of attention in machine learning as well as in computational neuroscience for its wide applicability from machine vision to the understanding of cortical processing. There is now a large body of evidence suggesting that neural visual systems are adapted to the statistics of the input [1,2], where the timescale of adaptation can range from evolutionary scale to the scale of seconds. Hence, visual mechanisms reflect the statistical structure of the visual data. For example the features obtained by applying Independent Component Analysis (ICA) to natural images have very similar properties to those of Simple Cells in mammalian primary visual cortex[3,4,5].

In ICA, the observed data vector $\mathbf{x}$ is assumed to be generated as a linear superposition of features, $\mathbf{x} = \mathbf{As}$, where the distribution of the sources is usually assumed to be a known supergaussian probability density function (pdf). Due to the assumption that sources are independent, we can write $p(\mathbf{s}) = \prod_i p_i(s_i)$ or for the log-probability $\log p(\mathbf{s}) = \sum_i \log p_i(s_i)$. If the mixing matrix $\mathbf{A}$ is invertible and has inverse $\mathbf{W}$, consisting of vectors $\mathbf{w}_i$ we can make a transformation of density to obtain the pdf for the data as $\log p(\mathbf{x}) = \sum_i \log p_i(\mathbf{w}_i^T \mathbf{x}) - \log |\det \mathbf{W}|$. This model can easily be optimized with maximum likelihood.

T. Adali et al. (Eds.): ICA 2009, LNCS 5441, pp. 507–514, 2009.

A weakness of ICA is, that as an inherently linear model, it is not able to recover independent sources from data with complex, nonlinear dependencies such as most natural signals. Therefore attempts have been made [6,7] to extend ICA to model more general densities. Taking these ideas in a different direction, here we try to nonlinearly extend the ICA model to include two classes of sources, which are mixed independently to reflect different aspects of an observed data vector. The two parts are then combined nonlinearly to produce the actual observed data vector. For modelling natural image patches this means that we independently sample from submodels $\mathbf{x}_l$ and $\mathbf{x}_r$, and the actual observed image patch $\mathbf{x}$ is obtained as $\mathbf{x} = \mathbf{x}_l \odot \mathbf{x}_r$, where $\odot$ denotes elementwise multiplication.

This kind of model can be interpreted as taking the basic principle from a linear superposition model such as ICA but generalizing it to a nonlinear superposition of different "sources", where the sources themselves are now generated as ICA-like linear superpositions. As a general example of this idea, one visual subsystem could specialize in 'what' there is in a particular scene, whereas another would code for 'where' in the scene the stimulus is located. These two are plausibly independent in general, but obviously cannot be captured by independent sources in a linear model.

2   Methods

2.1   The Proposed Model

We define the generative model for the data as follows:

$$\mathbf{x} = \mathbf{x}_l \odot \mathbf{x}_r = \mathbf{As} \odot (\mathbf{Bt} + c) \tag{1}$$

where $\mathbf{x}_l = \mathbf{As}$ is the "classical" ICA or sparse coding part and $\mathbf{x}_r = \mathbf{Bt} + c$ codes for aspects of data that cannot be captured by the linear ICA model. The $\odot$ indicates elementwise multiplication, so each pixel is defined by the product of two independent parts. The matrix $\mathbf{A}$ is square and invertible whereas $\mathbf{B}$ is undercomplete, with significantly fewer columns (features) than $\mathbf{A}$. The vectors $\mathbf{s}$ and $\mathbf{t}$ are the independent components of the two subimages. We require both $\mathbf{B}$ and $\mathbf{t}$ to be non-negative to ensure that that $\mathbf{x}_r$ is always positive. c is a small constant that is added for numerical stability, and it was set to $c = 0.1$ for all experiments. The model can be written more succinctly as

$$\mathbf{x} = \mathbb{D}(\mathbf{Bt} + c)\mathbf{As} \tag{2}$$

where $\mathbb{D}$ indicates diagonalization of the vector.

2.2   Maximum Likelihood Optimization

As $\mathbf{A}$ is assumed to be invertible, we can solve for the components $\mathbf{s}$ as

$$\mathbf{s} = \mathbf{A}^{-1}\mathbb{D}(\mathbf{Bt} + c)^{-1}\mathbf{x} = \mathbf{W}\mathbb{D}(\mathbf{Bt} + c)^{-1}\mathbf{x} \tag{3}$$

where we define the filter matrix $\mathbf{W} = \mathbf{A}^{-1}$ to be the inverse of the feature matrix $\mathbf{A}$. Now we define a pdf on $\mathbf{s}$ following the ICA model

$$p(\mathbf{s}) = \prod_i \exp\left(g(s_i)\right) = \frac{4\sqrt{3}}{\pi} \prod_i \frac{1}{\cosh^2(\pi/\sqrt{12}s_i)} \tag{4}$$

where the function $g(\mathbf{s})$ defines the normalized log-pdf, which we choose to be the logistic distribution. Now we transform the density to obtain the probability distribution for $\mathbf{x}$ as

$$\log p(\mathbf{x}|\mathbf{W}, \mathbf{B}, \mathbf{t}) = \sum_i g(\mathbf{w}_i^T \mathbb{D}(\mathbf{Bt}+c)^{-1}\mathbf{x}) + \log|\det \mathbf{W}| - \sum_i \log|\mathbf{b}_i^T \mathbf{t} + c| \tag{5}$$

where the extra terms due to the normalization constant are given by the determinant of the Jacobian of the matrix $\mathbf{W}\mathbb{D}(\mathbf{Bt})^{-1}$. From this we get the log-likelihood of the parameters for a sample of data vectors of size T. We choose a flat prior for $\mathbf{A}$ and $\mathbf{B}$ and a Laplacian prior for $\mathbf{t}$, so the log-likelihood for one data sample becomes:

$$\log p(\mathbf{W}, \mathbf{B}, \mathbf{t}|\mathbf{x}) = \sum_i g(\mathbf{w}_i^T \mathbb{D}(\mathbf{Bt}+c)^{-1}\mathbf{x}) + \log|\det \mathbf{W}| - \sum_i \log|f(\mathbf{b}_i^T \mathbf{t})| - |\mathbf{t}|_1$$

$$\tag{6}$$

This can now be optimized by taking gradients of the sample expectation w.r.t. both the weight matrices and the components $\mathbf{t}$.

3   Identifiability with Artificial Data

To create random data following the model, we sample from a logistic distribution for $\mathbf{s}$ and from an exponential distribution for $\mathbf{t}$. The mixing matrices are also generated randomly, with the restriction that the matrices have to be well-conditioned for the algorithm to converge. We arbitrarily constrained the condition number of $\mathbf{A}$ and $\mathbf{B}$ to be no larger than ten. Furthermore, $\mathbf{B}$ is constrained to be non-negative, following the model definition. The independent mixtures $\mathbf{x}_l = \mathbf{As}$ and $\mathbf{x}_r = \mathbf{Bt} + c$ are multiplied elementwise to obtain data following the model distribution. We generated 20,000 samples with a dimensionality of 60, and with 4 and 8 features in $\mathbf{B}$. Then, we attempted to estimate the model parameters $\mathbf{A}$ and $\mathbf{B}$ from the data. Like in ICA, the order and the sign of the components cannot be determined, so given the true mixing matrix $\tilde{\mathbf{A}}$ we expect the product $\tilde{\mathbf{A}}\mathbf{A}^{-1}$ to be a permuted diagonal matrix with random sign. Similarly, for the second part of the model we expect $\tilde{\mathbf{B}}\mathbf{B}^\dagger$ to be a permuted identity matrix. Here the pseudo-inverse is used, since $\mathbf{B}$ is not a square matrix.

The results for our experiments on artificial data are given in Fig. 1. Up to some noise, both $\mathbf{A}$ and $\mathbf{B}$ are correctly identified. The product $\tilde{\mathbf{A}}\mathbf{A}^{-1}$ shows that the vectors in $\mathbf{A}$ and $\tilde{\mathbf{A}}$ are identical up to randomly flipped signs, but the order of the vectors is randomized. Since the vectors in $\mathbf{B}$ are constrained to be non-negative, there is no sign indeterminacy but only the order of the vectors is shuffled. This shows that the parameters of the proposed model can be identified for a range of different sizes of $\mathbf{B}$.

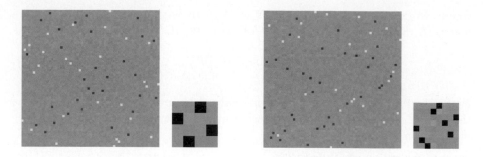

Fig. 1. Both parameter matrices **A** and **B** can be identified up to order and sign indeterminacies. We show the product of the true and the inverse of the estimated matrices, which results in permuted diagonal matrices. In the plots we code 0 as gray, 1 as black and -1 as pure white. The two plots on the left are for data generated with 4 vectors in **B**, on the right there are 8 vectors. The larger plots show $\tilde{\mathbf{A}}\mathbf{A}^{-1}$, the smaller ones $\tilde{\mathbf{B}}\mathbf{B}^{\dagger}$, the product of the true parameter matrix and the pseudoinverse of the estimated matrix.

4   Experiments on Natural Images

4.1   Preprocessing

Experiments were performed on natural image patches sampled from natural images available in P. O. Hoyer's ImageICA package[1]. We used 20,000 patches of size 16×16 pixels for all experiments and performed zero phase whitening on the data [8]. The dimensionality was not reduced, and the DC-component was retained. We discarded 20% of the patches with the lowest variance, which correspond to blank image regions and do not significantly affect the gradient.

We performed experiments with **B** having a varying number of features between 2, 4, 8 and 16. The estimation was started with **A** initialized randomly, and **B** to a matrix of all ones divided by the number of elements. The hidden variables **t** were also initialized randomly, but each vector **t** was then normalized to unit L_1-norm. This had the effect that, with $c = 0.1$, each pixel of $\mathbf{x}_r$ was close to one initially and not influencing the $\mathbf{x}_l$ part of the model. The estimation was then started by learning only the matrix **A** for $\mathbf{x}_l$ with a stepsize of 0.1, until visual inspection showed that it had converged to an ICA basis set characterized by Gabor-like receptive fields. After this initialization, **A**, **B** and **t** were estimated simultaneously. The stepsize for **t** was chosen to be 1, while the stepsizes for **A** and **B** were both 0.1. The non-negativity of the components $\mathbf{x}_r$ was ensured by forcing both **B** and **t** to be non-negative after every update step.

4.2   Separation into Gabors and Localized Masks

Since the novel model presented here is a generalization of ICA, and in fact feature matrix **A** is initialized with an ICA basis, it should not be surprising that

---

[1] Available at http://www.cis.hut.fi/projects/ica/imageica/

the "independent components" recovered by the model are Gabor-like filters that are localized, oriented and band-pass, as shown in Fig. 2(a) for different numbers of columns in **B**. However there are important differences that emerge once the modulation due to the **Bt** component is taken into account. While for a small number of columns in **B** the features look like the Gabor filters familiar from the classical ICA model, they become less localized as the number increases. In all cases the filters in **B** learn to perform a localized modulation, that dampens some of the image to create a patch with blank areas. The vectors in **B** evenly tile all of the pixel space, but selectively boost or mask regions of individual patches. This is shown in Fig. 2(b).

4.3   Dependence of Tuning Properties on the Number of Filters

To investigate the change in appearance of the features in **A**, we parametrized them with a least-squares fit to Gabor functions, i.e. Sinusoids with a Gaussian envelope. We then analyzed the tuning statistics of the Gabors in terms of frequency and size. As we show in Fig. 3, there is a significant change in aspect ratio and modulation (number of zero crossings of the sinusoid) of the Gabors as the number of filters in **B** is increased.

5   Discussion

5.1   Separation of Structure and Position

The most striking aspect of the results is that with an increasing number of vectors in **B**, the appearance of the features starts to differ significantly from the Gabor-like features that are obtained by most other ICA or Sparse Coding models. All features become less localized, and especially the highest frequency features, which tend to be very localized in the classical ICA model, loose all localization and cover the whole image patch. In an ICA model, this would clearly be less than optimal because most natural images patches have only localized structure. In the nonlinear model the situations is quite different though: Depending on the structure of $\mathbf{x}_r$, the localization properties can be recovered by "masking off" the part of the reconstruction $\mathbf{x}_l$ that does not contribute to the total image patch $\mathbf{x} = \mathbf{x}_l \odot \mathbf{x}_r$ that is being coded. This is conceivable since most image patches have blank regions and only localized structure such as edges or textured objects. Rather than having a set of features that can code for arbitrary image patches, it is advantageous to independently specify the region of the image patch that contains structure, and the kind of structure. Our new model can be viewed as accomplishing this by coding image structure in $\mathbf{x}_l$ and location in $\mathbf{x}_r$. By having the ICA reconstruction $\mathbf{x}_l = \mathbf{As}$ matched just to the structure, and discarding most localization information from the basis **A**, a representation with higher likelihood can be achieved. The additional part of the model $\mathbf{x}_r$ then simply masks off *where* in the image patch the particular structure occurs, leaving the rest of the image patch blank. In this way, is possible to encode a particular image patch with fewer basis functions than with classical ICA, since

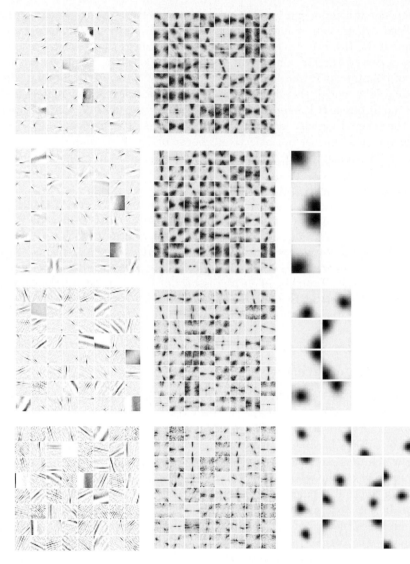

Fig. 2. Comparison of the features obtained with ICA (top) and the new product model (bottom three rows). We show a subset of 64 randomly chosen feature vectors of **A** in the first column, and their Fourier power in the second column. For the product model the with 4, 8 and 16 secondary features, the vectors of **B** are shown in the third column. While **A** converges to the familiar ICA features, **B** produces localized masks. As the number of features in **B** increases, the Gabors in **A** spread out to cover more of the image patch, this is particularly evident for 16 columns in **B**. Intuitively, this can be explained as a masking, where combining one feature from **A** with different masks from **B** can produce new Gabors in various positions. The Fourier transforms show how the features become more localized in Fourier space as the number of vectors of **B** increases, but also helps to identify the unlocalized highest frequency features as aliasing artifacts: All the Fourier power should be confined to a cicle around the origin, the four corners are artifacts due to the rectangular sampling grid.

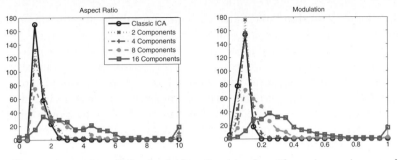

Fig. 3. Change in tuning of the basis functions in **A** with an increasing number of local fields in **B**. The aspect ratio increases for more fields, i.e. the Gabors become more elongated, filling most of the patch rather than just a small portion. The number of sidelobes of the Gabors also increases, making the basis functions less localized and more similar to a Fourier basis.

the features can become more specialized for orientation and frequency, while the localization in preserved in the second part of the model.

Along these lines, it is also possible to view the novel model as an implicitly overcomplete version of ICA. By multiplying each of the n features in **A** with each of the m features in **B**, a new set of features of size mn is obtained. For a large number of secondary features, e.g. $m = 16$ the vectors in **A** are close to sinusoids and the vectors of **B** are nearly Gaussian. Each of the sinusoids is masked with Gaussians at different positions, which corresponds to constructing a new set of Gabor features. It is important to note that the weights of the new features will no longer be independent, since the "mask" $\mathbf{x}_r$ chosen for one of the features in **A** will also be applied to each other features.

5.2   Relation to Contrast Gain Control

One of the initial motivations for the way the model was specified, in particular the nonnegativity of $\mathbf{x}_r$, was that the secondary features would perform divisive Contrast Gain Control (CGC) on the image patches. This can be easily seen by rewriting 1 as

$$\frac{\mathbf{x}}{\mathbf{Bt} + c} = \mathbf{As} \tag{7}$$

where with slight abuse of notation the fraction is taken to be elementwise. Models of divisive normalization in various ways are abundant in the literature [9] and are motivated from the observation that natural images are not stationary, and the statistics vary considerably from one image region to another [10].

However, in preliminary experiments (results not shown) we could not confirm a significant reduction in energy dependencies in our model compared to the classical ICA model.

6   Conclusion

We have presented an extension of ICA with a second layer, where, in contrast to previous work, the layers are horizontal rather than hierarchical. After showing the identifiability on simulated data, we have applied the novel model to natural images. We report the emergence of localized "masks" in the additional layer, while the Gabor-like features in the primary layer become less localized than in classical ICA. As a possible interpretation we suggest that the model learns to separately code for the structure and position of features in image patches. This gives the features an implicit position invariance, with one feature in **A** being able to code for many different positions conditional on **B**. This is a powerful principle which is outside the scope of linear models but may be of great importance in neural visual systems.

Acknowedgements. We wish to thank David C.J. Senne for comments on the manuscript. Urs Köster is kindly supported by a Scholarship from the *Alfried Krupp von Bohlen und Halbach-foundatation.*

References

1. Olshausen, B.A., Field, D.J.: Sparse coding with an overcomplete basis set: A strategy employed by V1? Vision Research 37(23), 3311–3325 (1997)
2. van Hateren, J.H., van der Schaaf, A.: Independent component filters of natural images compared with simple cells in primary visual cortex. Proc. R. Soc. Lond. B 265, 359–366 (1998)
3. Jutten, C., Herault, J.: Blind separation of sources part i: An adaptive algorithm based on neuromimetic architecture. Signal Processing 24, 1–10 (1991)
4. Comon, P.: Independent component analysis – a new concept? Signal Processing 36, 287–314 (1994)
5. Hyvärinen, A., Karhunen, J., Oja, E.: Independent Component Analysis. Wiley, Chichester (2001)
6. Hyvärinen, A., Hoyer, P.: Emergence of phase and shift invariant features by decomposition of natural images into independent feature subspaces. Neural Computation 12(7), 1705–1720 (2000)
7. Hyvärinen, A., Hoyer, P., Inki, M.: Topographic independent component analysis. Neural Computation (2001)
8. Atick, J.J.: Could information theory provide an ecological theory of sensory processing? Network: Computation in neural systems 3, 213–251 (1992)
9. Schwartz, O., Simoncelli, E.P.: Natural signal statistics and sensory gain control. Nature Neuroscience 4(8), 819–825 (2001)
10. Karklin, Y., Lewicki, M.S.: A hierarchical bayesian model for learning non-linear statistical regularities in non-stationary natural signals. Neural Computation 17(2), 397–423 (2005)

Estimating Markov Random Field Potentials for Natural Images

Urs Köster[1,2], Jussi T. Lindgren[1,2], and Aapo Hyvärinen[1,2,3]

[1] Department of Computer Science, University of Helsinki, Finland
[2] Helsinki Institute for Information Technology, Finland
[3] Department of Mathematics and Statistics, University of Helsinki, Finland

Abstract. Markov Random Field (MRF) models with potentials learned from the data have recently received attention for learning the low-level structure of natural images. A MRF provides a principled model for whole images, unlike ICA, which can in practice be estimated for small patches only. However, learning the filters in an MRF paradigm has been problematic in the past since it required computationally expensive Monte Carlo methods. Here, we show how MRF potentials can be estimated using Score Matching (SM). With this estimation method we can learn filters of size 12×12 pixels, considerably larger than traditional "hand-crafted" MRF potentials. We analyze the tuning properties of the filters in comparison to ICA filters, and show that the optimal MRF potentials are similar to the filters from an overcomplete ICA model.

1 Introduction

Probabilistic models of natural images are useful in a wide variety of applications, such as denoising and inpainting[1], novel view synthesis[2], texture modelling [3], and in modelling the early visual system [4]. Such models can also provide controllable test stimuli for experiments in neurophysiology and psychophysics.

Two approaches that have received significant interest with relation to image modelling are Markov Random Fields (MRF, e.g. [5]) and Independent Component Analysis (ICA [6], in images context see e.g. [4]). Traditionally, in the MRF framework the model parameters have been selected by hand (e.g. [3]) rather than learned, whereas in the ICA approach the model parameters are learned from the data. Only recently Roth and Black have shown that MRF performance can be improved by fitting the model parameters to natural image data [1].

In ICA, the observed data vector $\mathbf{x}$ is assumed to be generated as a linear superposition of features, $\mathbf{x} = A\mathbf{s}$, where the distribution of the sources is usually assumed to be a known supergaussian probability density function (pdf). Due to the assumption that sources are independent, we can write $p(\mathbf{s}) = \prod_i p_i(s_i)$ or for the log-probability $\log p(\mathbf{s}) = \sum_i \log p_i(s_i)$. If the mixing matrix A is invertible and has inverse W, consisting of vectors $\mathbf{w}_i$, we can make a transformation of density to obtain the pdf for the data as $\log p(\mathbf{x}) = \sum_i \log p_i(\mathbf{w}_i^T \mathbf{x}) + \log |\det W|$. This model can easily be estimated with maximum likelihood.

T. Adali et al. (Eds.): ICA 2009, LNCS 5441, pp. 515–522, 2009.

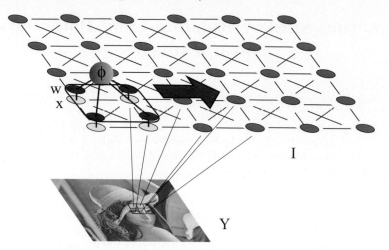

Fig. 1. Sketch of a Markov Random Field: The MRF has maximal cliques of size 2×2 pixels; one clique $\mathbf{x}_i$ is highlighted. Each unit of the field is associated to a pixel of the underlying image Y. The potential energy V for each clique is defined as a function of the inner product of the image patch corresponding to the clique with a bank of filters of the same size as the clique, $V(\mathbf{x}) = \phi(\mathbf{w}^T\mathbf{x})$. This is visualized by the filter vector $\mathbf{w}$ that is scanned over the whole image, and the product is computed with each clique. In general there will be several filters in a filter matrix W, but for visualization purposes only one is shown.

On the other hand, a MRF is a graphical model that is defined as a 2-D lattice of units with undirected links, as illustrated in Fig. 1. The maximal cliques formed by these connections play an important role as the potential energy of the field is given as a function of these cliques. The key property of a MRF is conditional independence, so the state of each unit on the field depends only on those units it is directly linked to and the unit is independent of all other units in the field. While ICA is limited to modelling small image patches, the MRF provides a principled model for *whole* images of arbitrary size, even if the clique size is limited.

The paper is structured as follows: In Section 2 we present the MRF model and how it can be estimated with Score Matching. In Section 3 we discuss the application of the model to natural images, show the filters that are obtained, and analyze the properties of the filters compared to ICA models. Finally we discuss the implications of this work in Section 4.

2   The MRF Model and Estimation

In contrast to ICA, where filter responses are computed by a simple inner product, the energy (i.e. the negative logarithm of the non-normalized pdf) of a MRF is given by a convolution of the image I with potential functions U_k

$$V(\mathbf{I}, \theta) = \sum_{k,x,y} \phi\left(\mathbf{U}_k * \mathbf{I}\right) = \sum_{k,x,y} \phi\left(\sum_{x',y'} \mathbf{U}_{k,x',y'}\mathbf{I}_{x-x',y-y'}\right) \tag{1}$$

where the convolution (denoted by $*$) runs over pixels indices x and y. The elementwise nonlinearity ϕ gives the energy of the cliques of the field, which are simply summed up to obtain the energy V of the field. As it is customary in ICA to work on whitened data, we insert a whitening filter $\mathbf{Q}$ in the convolution so it becomes $V(\mathbf{I}, \theta) = \sum_{k,x,y} \phi(\mathbf{U}_k * \mathbf{Q} * \mathbf{I})$. The whitening filter can be absorbed into the image, which corresponds to estimating a model for white data, but it can also be viewed as a part of the potential function when the model is applied to non-whitened data. It is important to use a whitening filter rather than an arbitrary whitening matrix for this to hold.

The unnormalized probability of the model is given by the exponential of the negative energy, and must be normalized by the partition function Z. Since Z cannot be computed in closed form, we estimate the model using Score Matching [7], which works on the non-normalized distribution. To estimate the model with Score Matching we need to compute the Score function $\Psi_j = \frac{\partial V}{\partial \mathbf{I}_j}$, the Score Matching objective function J and its derivatives w.r.t. the parameter vectors.

$$J = \sum_j \left(\frac{1}{2} \Psi_j^2 + \Psi_j' \right), \qquad \frac{\partial J}{\partial \mathbf{w}_k} = \sum_j \left(\Psi_j \frac{\partial \Psi_j}{\partial \mathbf{w}_k} + \frac{\partial \Psi_j'}{\partial \mathbf{w}_k} \right) \qquad (2)$$

For further analysis it is convenient to rewrite the convolution as a discrete sum of inner products. We rewrite the convolution $\mathbf{I} * \mathbf{U}_k = \mathbf{X}\mathbf{w}_k$ where $\mathbf{X}$ is a matrix containing vectorized patches $\mathbf{x}_i$ from the image, and $\mathbf{w}_k$ is a vectorized filter. Similarly we write $\mathbf{X}_j$ as a subset of $\mathbf{X}$ containing only those patches which include the image pixel I_j. Thus the energy becomes

$$V(I, \theta) = \sum_{k,i} \phi \left(\mathbf{w}_k^T \mathbf{x}_i \right) \qquad (3)$$

Where the sum over i is over the patches contained in the matrix $\mathbf{X}$. Using this notation we can compute the score function w.r.t. to the image pixels $\mathbf{I}_j$

$$\Psi_j = \frac{\partial V}{\partial \mathbf{I}_j} = \frac{\partial}{\partial \mathbf{I}_j} \sum_{k,x,y} \phi \left(\mathbf{U}_k * I \right) = \sum_k \check{\mathbf{w}}_k^T \phi' \left(\mathbf{X}_j \mathbf{w}_k \right) \qquad (4)$$

$$\Psi_j' = \frac{\partial^2 V}{\partial \mathbf{I}_j^2} = \sum_k (\mathbf{w}_k \odot \mathbf{w}_k)^T \phi'' \left(\mathbf{X}_j \mathbf{w}_k \right) \qquad (5)$$

We denote elementwise multiplication of vectors by $\odot$, and $\check{\mathbf{w}}$ indicates reversal of the order of elements in a vector. It is important to note that in order to avoid border effects, the index j does not run over *all* image pixels, but only those that lie in the central region of the image so it can be reached by all pixels in the filter $\mathbf{w}$. The gradient of the objective function is now easily obtained from the gradients

$$\frac{\partial \Psi_j}{\partial \mathbf{w}_k} = \check{\phi}' \left(\mathbf{X}_j \mathbf{w}_k \right) + \mathbf{X}_j \left[\check{\mathbf{w}}_k \cup \phi'' \left(\mathbf{X}_j \mathbf{w}_k \right) \right] \qquad (6)$$

$$\frac{\partial \Psi_j'}{\partial \mathbf{w}_k} = 2\check{\mathbf{w}}_k \phi'' \left(\mathbf{X}_j \mathbf{w}_k \right) + \mathbf{X}_j \left[\check{\mathbf{w}}_k \odot \check{\mathbf{w}}_k \odot \phi''' \left(\mathbf{X}_j \mathbf{w}_k \right) \right] \qquad (7)$$

Thus the Score Matching objective can easily be optimized by gradient descent.

3   Experiments on Natural Images

3.1   Methods

We performed experiments on natural images from P.O. Hoyer's ImageICA package[1]. For the ICA and overcomplete ICA experiments we randomly sampled 20,000 image patches of 8×8 and 12×12 pixels size. For the MRF, we sampled 5,000 larger "images" of size 25×25 and 36×36 for use with filters of size 8×8 and 12×12 respectively. Now since the main advantage of the MRF model over ICA is that it can be applied to arbitrarily large images, it may seem surprising that we use images that are not significantly larger than the patches ordinarily used in ICA. However, what is important for estimating the model is the size of the filters relative to the images. In particular, since we use only the valid region of the convolution, only the central pixels of the image contribute to the objective function. Thus the full range of dependencies is captured, and the filters should be identical if they were estimated with larger images.

We used less samples for the MRF than for ICA since each of the images is effectively generating more data points due to the convolution. In preprocessing we removed the DC value of the images and normalized them to unit variance. After sampling, we whitened the image vectors with a zero phase whitening filter [8]. We did not reduce the dimensionality with PCA as it is customary in ICA models, since this would destroy the structure that we wish to capture with the MRF. Therefore the highest frequencies containing aliasing artifacts due to the rectangular sampling grid will be boosted, which has to be taken into account in the analysis of the results.

We performed experiments on a complete ICA model with 144 filters, and a 16 times overcomplete ICA model with 2304 filters. The MRF had 144 filters, and all three models were estimated with Score Matching. The filters were initialized randomly and estimated by gradient descent, in case of the MRF a stochastic gradient with a batch size of 20 was used. The experiments were repeated 10 times with different random seed for the sampling of image patches and initialization of the weight matrix. All the filters were normalized to unit norm, which is necessary to prevent filters from going to zero in the overcomplete ICA and MRF models. Convergence was determined by optical inspection of the filters. Because the estimation of ICA with Score Matching is not widely used, we also estimated the complete ICA model with FastICA, to control for differences that are due to the estimation method.

3.2   Results

In most classical MRF work, the potentials that were used were of rather small size such as 3×3 pixels and typically chosen to be directional derivatives. Thus it is perhaps not surprising that the larger MRF filters we estimated are very similar to ICA filters in appearance, being localized Gabor-like filters with tuning for spatial frequency, phase and orientation. This is shown in Fig. 2, where ICA and MRF filters are compared directly for different image patch sizes.

---

[1] Available at http://www.cs.helsinki.fi/u/phoyer/imageica.tar.gz

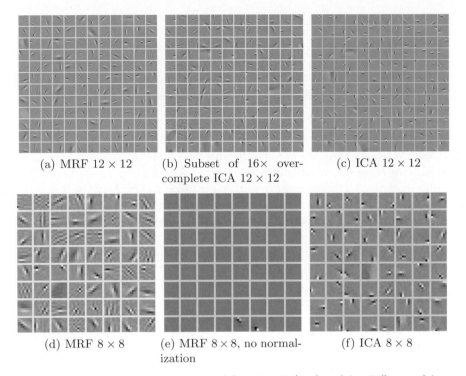

(a) MRF 12 × 12   (b) Subset of 16× over-complete ICA 12 × 12   (c) ICA 12 × 12

(d) MRF 8 × 8   (e) MRF 8 × 8, no normal-ization   (f) ICA 8 × 8

Fig. 2. Comparison of the filters obtained for 12 × 12 (top) and 8 × 8 (bottom) image patches. The complete and overcomplete ICA models shows the well-known Gabor like filters, and the MRF potentials are very similar, sharing the properties of localization and tuning for spatial frequency, phase and orientation. While for the ICA model it is not necessary to normalize the filters, it is interesting to note that in the MRF case almost all the filters go to zero unless the norms of the vectors **w** are constrained to be unity.

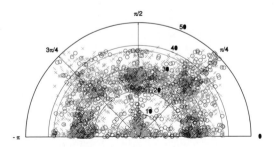

Fig. 3. Polar plot of frequency vs. orientation for 12 × 12 image patches for ICA (circles) and MRF (crosses). The orientations are not uniformly distributed, with filters preferring to be aligned along the pixels, horizontal or vertical, and at 45 degrees to these directions. Due to the rectangular sampling grid, the maximum frequency is higher along the diagonal, which may also account for the non-uniformity. Usually this problem is alleviated by dimensionality reduction amounting to high-pass filtering, which is not easily possible with the MRF model.

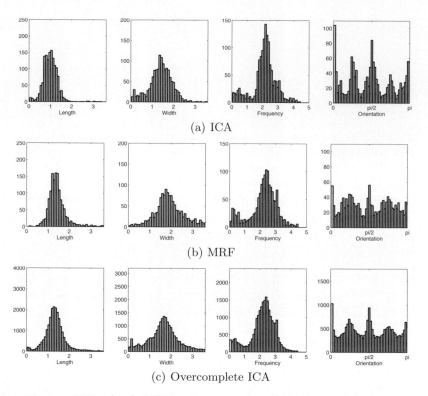

(a) ICA

(b) MRF

(c) Overcomplete ICA

Fig. 4. Tuning of ICA (top), MRF and overcomplete ICA (bottom) for 12×12 image patches. We show the size (length and width in pixels) of the Gaussian envelope of the Gabors we fit, and the distribution of frequencies (rad per pixel). Additionally, we show the distribution of orientations, which is clearly not uniform in both cases.

To analyze the similarity between the two models further, we fit Gabor function to the filters so we can analyze their tuning properties. We used a least squares fit adapted from [9] to parametrize the filters in terms of length and width, frequency, phase and orientation. In Fig. 3 we show a polar plot plotting orientation against frequency.

In Fig. 4 we show histograms of the size and frequency distribution for the three models. The complete ICA model produces very localized filters which cover a relatively narrow band of frequencies. Both overcomplete ICA and the MRF give slightly larger filters with a slightly broadened distribution of frequencies. While the distributions for the MRF and ICA are somewhat different, it is important to note that the filters for overcomplete ICA are also slightly different and in some respects more similar to the MRF (e.g. somewhat larger filters with less peaked frequency tuning). This may suggest that there are no fundamental differences between the filters obtained from the two models.

We performed t-tests to quantify the statistical significance of the difference in mean length, width and frequency of the filters between the four models,

FastICA, ICA and overcomplete ICA estimated with Score Matching and the MRF. Only in the comparison between the MRF and overcomplete ICA, there was no sufficient difference in the tuning properties to reject the null hypothesis at the Bonferroni corrected threshold of 0.017. It is interesting to note that estimating the same ICA model with two different estimation methods produces a larger difference in the filters, than the difference between overcomplete ICA and the MRF estimated with Score Matching.

4   Discussion

Estimating optimal MRF potentials from natural images has previously been attempted by Roth and Black [1], making use of Contrastive Divergence (CD) [10]. We would like to point out that the filters obtained in the work of Roth and Black have a very different appearance, being disjoint and distributed over the whole image patch rather than the coherent and smooth Gabors that we obtain. The patch size used by those authors was considerably smaller (3×3 and 5×5, which may forces features to spread out more to capture the longer range dependencies of natural images. In addition, it is conceivable that with the particular Monte Carlo method used by the authors, a different local optimum is found or the method encountered some other problems.

It is possible to view the MRF model as a highly overcomplete ICA model with some additional constraints. In particular, the convolution in (1) can be interpreted as keeping the image fixed, and multiplying it with the filters in different positions.The resulting "filters" would be shifted copies of the original filters at different positions in the image and padded with zeros. Thus, while the model is highly overcomplete, non of the filters model the whole image, but only regions. If we assume natural images to be stationary having copies of the filters at different positions does not have an effect, and the main difference to ICA would be that the size of the filters is restricted to be much smaller than the image. This makes it quite obvious that optimal MRF filters should not be vastly different from ICA filters. It would be interesting to investigate if there are systematic differences in the sets of filters, and how they tile the parameter space of positions, orientations etc. In particular, while an ICA basis may contain nearly identical filters in different positions, this should not be the case with the MRF model. Therefore, if one were to attempt to use ICA filters in place of MRF potentials for e.g. a denoising task, one would face the problem of selecting the correct subset of an ICA basis to form a set of near-optimal MRF potentials.

To conclude, we have show that it is possible to learn the filters used in a non-Gaussian Markov Random Field model. The learning is based on score matching and leads to Gabor-like filters. This gives a well-defined probabilistic model of whole images instead of just small patches.

Acknowledgements. We wish to thank Jascha Sohl-Dickstein for helpful discussions. Urs Köster is supported by a Scholarship from the *Alfried Krupp von Bohlen und Halbach-foundatation.*

References

1. Roth, S., Black, M.: Fields of experts: A framework for learning image priors. In: CVPR, vol. 2, pp. 860–867 (2005)
2. Woodford, O., Reid, I., Torr, P., Fitzgibbon, A.: Fields of experts for image-based rendering. In: Proceedings British Machine Vision Conference (2006)
3. Zhu, S., Wu, Y., Mumford, D.: FRAME: Filters, random field and maximum entropy – towards a unified theory for texture modeling. International Journal of Computer Vision 27(2), 1–20 (1998)
4. van Hateren, J.H., van der Schaaf, A.: Independent component filters of natural images compared with simple cells in primary visual cortex. Proc. R. Soc. Lond. B 265, 359–366 (1998)
5. Li, S.Z.: Markov Random Field modelling in image analysis, 2nd edn. Springer, Heidelberg (2001)
6. Comon, P.: Independent component analysis – a new concept? Signal Processing 36, 287–314 (1994)
7. Hyvärinen, A.: Estimation of non-normalized statistical models using score matching. Journal of Machine Learning Research 6, 695–709 (2005)
8. Atick, J.: Could information theory provide an ecological theory of sensory processing? Network: Computation in neural systems 3, 213–251 (1992)
9. Hyvärinen, A., Hoyer, P., Inki, M.: Topographic independent component analysis. Neural Computation (2001)
10. Hinton, G.E.: Training products of experts by minimizing contrastive divergence. Neural Comput. 14(8), 1771–1800 (2002)

Blind Source Separation with *Spatio-Spectral* Sparsity Constraints – Application to Hyperspectral Data Analysis

Yassir Moudden[1], Jerome Bobin[1,2], Jean-Luc Starck[1], and Jalal Fadili[3]

[1] DSM /IRFU/SEDI, CEA/Saclay, F-91191 Gif-sur-Yvette, France
[2] Department of Applied & Computational Mathematics , California Institute of Technology, Pasadena, California 91125
[3] GREYC CNRS UMR 6072, Image Processing Group, ENSICAEN 14050, Caen Cedex, France
yassir.moudden@cea.fr, jerome.bobin@cea.fr, jstarck@cea.fr,
jalal.fadili@greyc.ensicaen.fr

Abstract. Recently, sparsity and morphological diversity have emerged as a new and effective source of diversity for Blind Source Separation giving rise to novel methods such as Generalized Morphological Component Analysis. The latter takes advantage of the very sparse representation of structured data in large overcomplete dictionaries, to separate sources based on their morphology. Building on GMCA, the purpose of this contribution is to describe a new algorithm for *hyperspectral* data processing. It assumes the collected data exhibits sparse spectral signatures in addition to sparse spatial morphologies, in specified dictionaries of spectral and spatial waveforms. Numerical experiments are reported which demonstrate the validity of the proposed extension.

Keywords: Morphological diversity, GMCA, MCA, sparsity, BSS, wavelets, curvelets, hyperspectral data, multichannel data.

1 Introduction

Generalized Morphological Component Analysis (GMCA) is a recent algorithm for multichannel data analysis and blind source separation introduced in [3], which relies on sparsity and morphological diversity to disentangle the source signals from observed mixtures. Indeed, sparsity has been recognized recently as an effective source of diversity for BSS [15]. The successes of GMCA in many generic multichannel data processing applications, including BSS, color image restoration and inpainting [4,3], strongly motivated research to extend its applicability. In particular, there are instances where one is urged by *prior* knowledge to set additional constraints on the estimated parameters (*e.g.* equality constraints, positivity). Building on GMCA, the purpose of this contribution is to describe a new algorithm for so-called *hyperspectral* data processing. Hyperspectral imaging systems collect data in a large number (up to several hundreds) of contiguous spectral intervals so that it makes sense to consider *e.g.* the regularity of some measurement from one channel to its neighbors. Typically, the

T. Adali et al. (Eds.): ICA 2009, LNCS 5441, pp. 523–531, 2009.

spectral signatures of the structures in the data may be known *a priori* to have a sparse representation in some specified possibly redundant dictionary of template *spectral* waveforms.

In what follows, regardless of other definitions or models living in other scientific communities, the term *hyperspectral* will be used generically to identify multichannel data with the above two specific properties *i.e.* that the number of channels is large and that these achieve a *regular* if not *uniform* sampling of some additional and meaningful physical index (*e.g.* wavelength, space, time) which we refer to as the *spectral* dimension. We further assume that hyperspectral data is structured *a priori* according to the simple yet common instantaneous linear mixture model given as follows:

$$\mathbf{X} = \mathbf{AS} + \mathbf{N} = \sum_k \mathbf{X}_k + \mathbf{N} = \sum_k a^k s_k + \mathbf{N} \tag{1}$$

where the measurements $\mathbf{X} \in \mathbb{R}^{m,t}$ are a mixture of contributions from various objects $\mathbf{X}_k$ with different statistical and *spatio-spectral* properties. These induce *inter-* and *intra-* sensor structures or signal coherence in space and across sensors which we classically assume to be well represented by rank one matrices, product of a spectral signature $a^k \in \mathbb{R}^{m,1}$ and a spatial density profile $s_k \in \mathbb{R}^{1,t}$. Here, s_k and a^k are respectively the kth row and column of $\mathbf{S} \in \mathbb{R}^{n,t}$ and $\mathbf{A} \in \mathbb{R}^{m,n}$. The $m \times t$ random matrix $\mathbf{N}$ is included to account for Gaussian instrumental noise assumed, for simplicity, to be uncorrelated inter- and intra- channel with variance σ^2. The image from the pth channel is formally represented here as the pth row of $\mathbf{X}$, x_p.

We describe next the proposed modified GMCA algorithm for hyperspectral data processing when it is known *a priori* that the underlying objects of interest $\mathbf{X}_k = a^k s_k$ exhibit sparse spectral signatures and sparse spatial morphologies in known dictionaries of spectral and spatial waveforms. Accounting for this prior requires a modified objective function which is discussed in section 2. The modified GMCA algorithm this entails is then given in section 3. Finally, numerical experiments in section 4 demonstrate the efficiency of the proposed method.

2   Objective Function

Initially, the GMCA algorithm aims at solving the following non-convex problem:

$$\min_{\mathbf{A},\mathbf{S}} \sum_k \lambda_k \|\nu_k\|_1 + \frac{1}{2\sigma^2} \left\| \mathbf{X} - \sum_k a^k s_k \right\|^2 \quad \text{with } s_k = \nu_k \mathbf{\Phi} \tag{2}$$

which is readily derived as a MAP estimation of the model parameters $\mathbf{A}$ and $\mathbf{S}$ where the ℓ_1 penalty terms imposing sparsity come from Laplacian priors on the sparse representation ν_k of s_k in $\mathbf{\Phi}$. Interestingly, the treatment of $\mathbf{A}$ and $\mathbf{S}$ in the above is asymmetric. This is a common feature of the great majority of BSS methods which invoke a uniform *improper* prior distribution for the spectral parameters $\mathbf{A}$. Truly, $\mathbf{A}$ and $\mathbf{S}$ often have different roles in the model and very different sizes. However, dealing with so-called *hyperspectral* data, assuming that the spectral signatures a^k also have sparse representations γ^k in spectral dictionary $\mathbf{\Psi}$, this asymmetry is no longer so obvious. Also, a well known property of the linear mixture model (1) is its *scale and permutation invariance*. A consequence is that unless *a priori* specified otherwise, information on the separate scales of a^k and s_k is lost du to the multiplicative mixing, and only a joint scale parameter for a^k, s_k can be estimated. This loss of information needs to be translated into a *practical* prior on $\mathbf{X}_k = a^k s_k = \mathbf{\Psi}\gamma^k \nu_k \mathbf{\Phi}$. Unfortunately, deriving the distribution of the product of two independent random vectors γ^k and ν_k based on their marginal densities is notoriously cumbersome. We propose instead that the following p_π is a good and practical candidate *joint sparse prior* for γ^k and ν_k after the loss of information induced by multiplication:

$$p_\pi(\gamma^k, \nu_k) = p_\pi(\gamma^k \nu_k, 1) \propto \exp(-\lambda_k \|\gamma^k \nu_k\|_1) \propto \exp(-\lambda_k \sum_{i,j} |\gamma_i^k \nu_k^j|) \quad (3)$$

where $\gamma_i^k = \gamma^k \nu_k$ is the ith entry in γ^k and ν_k^j is the jth entry in ν_k. Note that the proposed distribution has the nice property, for subsequent derivations, that the conditional distributions of γ^k given ν_k and of ν_k given γ^k are both Laplacian distributions. Finally, inserting the latter prior distribution in a Bayesian MAP estimator leads to the following minimization problem:

$$\min_{\{\gamma^k, \nu_k\}} \frac{1}{2\sigma^2} \left\| \mathbf{X} - \sum_k \mathbf{\Psi}\gamma^k \nu_k \mathbf{\Phi} \right\|^2 + \sum_k \lambda_k \|\gamma^k \nu_k\|_1 \quad (4)$$

Interestingly, this can be expressed slightly differently as follows:

$$\min_{\{\alpha_k\}} \frac{1}{2\sigma^2} \left\| \mathbf{X} - \sum_k \mathbf{X}_k \right\|^2 + \sum_k \lambda_k \|\alpha_k\|_1 \text{ with } \mathbf{X}_k = \mathbf{\Phi}\alpha_k \mathbf{\Psi} \text{ and } \forall k, \text{rank}(\mathbf{X}_k) \leq 1 \quad (5)$$

thus uncovering a nice interpretation of our problem as that of approximating the data $\mathbf{X}$ by a sum of rank one matrices $\mathbf{X}_k$ which are sparse in the specified dictionary of rank one matrices [14]. This is the usual ℓ_1 minimization problem [7] but with the additional constraint that the $\mathbf{X}_k$ are all rank one at most. The latter constraint is enforced here mechanically through a proper parametric representation of $\mathbf{X}_k = a^k s_k$ or $\alpha_k = \gamma^k \nu_k$.

We note that rescaling the columns of $\mathbf{A} \leftarrow \rho\mathbf{A}$ while applying the proper inverse scaling to the rows of $\mathbf{S} \leftarrow 1/\rho\mathbf{S}$, leaves both the quadratic measure of fit and the ℓ_1 sparsity measure in equation (4) unaltered. Although renormalizing is still worthwhile numerically, it is no longer dictated by the lack of scale invariance of the objective function and the need to stay away from trivial solutions, as in GMCΛ.

There have been previous reports of a symmetric treatment of $\mathbf{A}$ and $\mathbf{S}$ for BSS [12,8] however in the noiselesscase. We also note that very recently, the objective function (5) was proposed in [10] however for dictionary learning oriented applications. The algorithm used in [10] is however very different from the method proposed here which benefits from all the good properties of GMCA, notably its speed and robustness which come along the iterative thresholding with a decreasing threshold.

3 GMCA Algorithm for *Hyperspectral* Data

For the sake of simplicity, consider now that the multichannel dictionary $\Omega = \Psi \otimes \Phi$ reduces to a single orthonormal basis, tensor product of orthonormal bases Ψ and Φ of respectively spectral and spatial waveforms. In this case, the minimization problem (4) is best formulated in coefficient space as follows:

$$\min_{\{\gamma_k, \nu_k\}} \frac{1}{2\sigma^2} \|\alpha - \gamma\nu\|^2 + \sum_{k=1}^{n} \lambda_k \|\gamma^k \nu_k\|_1 \tag{6}$$

where the columns of γ are γ^k, the rows of ν are ν_k and $\alpha = \Psi^T \mathbf{X} \Phi^T$ is the coefficient matrix of data $\mathbf{X}$ in Ω. Thus, we are seeking a decomposition of matrix α into a sum of sparse rank one matrices $\alpha_k = \gamma^k \nu_k$.

Unfortunately, there is no obvious closed form solutions to problem (6). We propose instead a numerical approach by means of a block-coordinate relaxation iterative algorithm, alternately minimizing with respect to γ and ν. Indeed, thanks to the chosen prior, for fixed γ (resp. ν), the *marginal* minimization problem over ν (resp. γ) is convex. Inspired by [6,5,13], we obtain the following system of update rules, akin to a Projected Landweber algorithm [1]:

$$\begin{cases} \nu^{(+)} = \Delta_\eta \left(\nu^{(-)} + \mathbf{R}_\nu \left(\alpha - \gamma\nu^{(-)} \right) \right) \\ \gamma^{(+)} = \Delta_\zeta \left(\gamma^{(-)} + \left(\alpha - \gamma^{(-)}\nu \right) \mathbf{R}_\gamma \right) \end{cases} \tag{7}$$

where $\mathbf{R}_\nu$ and $\mathbf{R}_\gamma$ are appropriate relaxation matrices for the iterations to be non-expansive. Assume left invertibility of $\mathbf{A}$ and right invertibility of $\mathbf{S}$. Then, taking $\mathbf{R}_\nu = \left(\gamma^T \gamma \right)^{-1} \gamma^T$ and $\mathbf{R}_\gamma = \nu^T \left(\nu\nu^T \right)^{-1}$, the above are rewritten as follows:

$$\nu^{(+)} = \Delta_\eta \left(\left(\gamma^T \gamma \right)^{-1} \gamma^T \alpha \right) \tag{8}$$

$$\gamma^{(+)} = \Delta_\zeta \left(\alpha\nu^T \left(\nu\nu^T \right)^{-1} \right) \tag{9}$$

where vector η has length n and entries $\eta[k] = \lambda_k \|\gamma^k\|_1 / \|\gamma^k\|_2^2$, while ζ has length m and entries $\zeta[k] = \lambda_k \|\nu_k\|_1 / \|\nu_k\|_2^2$. The multichannel soft-thresholding operator Δ_η acts on each row k of ν with threshold $\eta[k]$ and Δ_ζ acts on each column k of γ with threshold $\zeta[k]$. Equations (8) and (9) rules are easily interpreted as thresholded alternate least squares solutions. Finally, in the spirit of the fast GMCA algorithm [4,3], it is proposed that a solution to problem (6) can be approached efficiently using the following symmetric iterative thresholding scheme with a progressively decreasing threshold, which we refer to as hypGMCA:

1. Set the number of iterations $I_{\max}$ and initial thresholds $\lambda_k^{(0)}$
2. Transform the data $\mathbf{X}$ into α
3. While $\lambda_k^{(h)}$ are higher than a given lower bound $\lambda_{\min}$,
   – Update ν assuming γ is fixed using equation (8).
   – Update γ assuming ν is fixed using equation (9) .
   – Decrease the thresholds $\lambda_k^{(h)}$.
5. Transform back γ and ν to estimate $\mathbf{A}$ and $\mathbf{S}$.

The *salient to fine* process is also the core of hypGMCA. With the threshold successively decreasing towards zero along iterations, the current sparse approximations for γ and ν are progressively refined by including finer structures spatially and spectrally, alternatingly. The final threshold should vanish in the *noiseless* case or it may be set to a multiple of the noise standard deviation as in common detection or denoising methods. A discussion of different possible thresholding strategies is given in [2]. When non-unitary or redundant transforms are used, the above is no longer strictly valid. Nevertheless, simple shrinkage still gives satisfactory results in practice.

Fig. 1. Image data set used in the experiments

4   Numerical Experiments

In this section, we report on two toy BSS experiments in 1D and 2D to compare GMCA to its extension hypGMCA. First we consider synthetic 2D data consisting of $m = 128$ mixtures of $n = 5$ image sources. The sources were drawn at random from a set of 128×128 structured images shown on Figure 1. The spectra were generated as sparse process in some orthogonal wavelet domain given *a priori*. The wavelet coefficients of the spectra were sampled from a Laplacian probability density with scale parameter $\mu = 1$. Finally, white Gaussian noise with variance σ^2 was added to the pixels of the synthetic mixture data in the

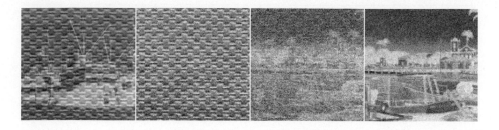

Fig. 2. Four 128×128 mixtures out of the 128 channels. The SNR is equal to 20dB.

different channels. Figure 2 displays four typical noisy simulated mixture data
with SNR = 20dB. A visual inspection of figure 3 allows a first qualitative assessment of the improved source recovery provided by correctly accounting for
a priori spatial as well as spectral sparsity. The top images were obtained with
GMCA while the bottom images, were obtained with hypGMCA. In all cases,
both methods were run in the curvelet domain [11] with the same number of
iterations. The graph on figure 4 gives more quantitative results. It traces the
evolution of the mixing matrix criterion $\mathcal{C}_{\mathbf{A}} = \|\mathbf{I}_n - \mathbf{P}\tilde{\mathbf{A}}^\dagger \mathbf{A}\|_1$ as a function of the
SNR which was varied from 0 to 40dB, where P serves to reduce the scale and
permutation indeterminacy of the mixing model and $\tilde{\mathbf{A}}^\dagger$ is the pseudo-inverse
of the estimated mixing matrix. In simulation, the true source and spectral matrices are known and so that $\mathbf{P}$ can be computed easily. The mixing matrix
criterion is then strictly positive unless the mixing matrix is correctly estimated
up to scale and permutation. Finally, as we expected since it benefits from the
added *a priori* spectral sparsity constraint it enforces, the proposed hypGMCA
is clearly more robust to noise.

Fig. 3. Top row: Estimated sources using the original GMCA algorithm. **Bottom
row:** Estimated sources using the new hypGMCA.

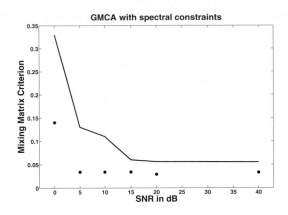

Fig. 4. Evolution of the mixing matrix criterion $\mathcal{C}_A$ **as a function of the SNR in dB.** *Solid line:* recovery results with GMCA. $\bullet$: recovery results with hypGMCA.

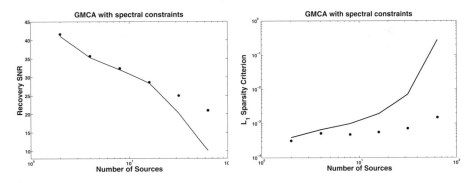

Fig. 5. Abscissa: Number of sources. **Ordinate - left:** Recovery SNR. **Right:** sparsity-based criterion $\mathcal{C}_{\ell_1}$. *Solid line:* recovery results with GMCA. $\bullet$: recovery results with hypGMCA.

In a second experiment, GMCA and hypGMCA are compared as the number n of sources is increased while the numbers of samples t and channels m are kept constant. Increasing the number of sources makes the separation task more difficult. We consider now 1D synthetic source processes **S** generated from *i.i.d.* Laplacian probability density distributions with scale parameter $\mu = 1$. The Dirac basis was taken as the dictionary of spatial waveforms Φ. The entries of the mixing matrix are also drawn from *i.i.d.* Laplacian distributions with scale parameter $\mu = 1$ and the Dirac basis was also taken as dictionary of spectral waveforms Ψ. The data are not contaminated by noise. The number of samples is $t = 2048$ and the number of channels is $m = 128$. Figure 5 depicts the comparisons between GMCA and its extension to the hyperspectral setting. Each point of this figure has been computed as the mean over 100 trials. The panel on the left of Figure 5 features the evolution of the recovery SNR when the number of sources varies from 2 to 64. At lower n, the *spatiospectral* sparsity constraint only slightly enhances the source separation. However, as n becomes larger than 15 the spectral sparsity constraint clearly enhances the recovery results. For instance, when $n = 64$, the GMCA algorithm with the spectral sparsity constraint outperforms the original GMCA by up to 12dB. The right of Figure 5 shows the behavior of both algorithms in terms of $\mathcal{C}_{\ell_1} = \sum_{i=1}^n \left\| a^i s_i - \tilde{a}^i \tilde{s}_i \right\|_1 / \sum_{i=1}^n \left\| a^i s_i \right\|_1$. As expected, accounting for spectral sparsity yields sparser results. Furthermore, as the number of sources increases, the deviation between the aforementioned methods becomes wider.

5   Conclusion

We described a new algorithm, hypGMCA, for blind source separation in the case where it is known *a priori* that the spatial and spectral features in the data have sparse representations in known dictionaries of template waveforms. The proposed method relies on an iterative thresholding procedure with a progressively decreasing threshold. This alone gives the method true robustness to noise. As expected, taking into account the additional prior knowledge of spectral sparsity leads to enhanced performance. It was illustrated by numerical experiments that spatiospectral sparsity yields robustness to noise contamination, as well as statbility when the dimensionality of the problem increases. Current work is on enforcing other prior constraints such as positivity and on applications of the proposed method. In fact promising results have been obtained on real hyperspectral data from Mars Observer previously studied in [9].

References

1. Blumensath, T., Davies, M.: Iterative thresholding for sparse approximations. Journal of Fourier Analysis and Applications (submitted, 2007)
2. Bobin, J., Starck, J.-L., Fadili, J., Moudden, Y., Donoho, D.L.: Morphological Component Analysis: An Adaptive Thresholding Strategy. IEEE Trans. On Image Processing 16(11), 2675 (2007)
3. Bobin, J., Starck, J.-L., Fadili, M.J., Moudden, Y.: Sparsity and morphological diversity in blind source separation. IEEE Transactions on Image Processing 16(11), 2662 (2007)
4. Bobin, J., Starck, J.-L., Moudden, Y., Fadili, J.: Blind Source Separation: the Sparsity Revolution. Advances in Imaging and Electron Physics (2008)
5. Chaux, C., Combettes, P.L., Pesquet, J.-C., Wajs, V.R.: A variational formulation for frame based inverse problems. inverse problems 23, 1495–1518 (2007)
6. Combettes, P.L., Wajs, V.R.: Signal recovery by proximal forward-backward splitting. SIAM Journal on Multiscale Modeling and Simulation 4(4), 1168–1200 (2005)
7. Donoho, D.L., Elad, M.: Optimally sparse representation in general (non-orthogonal) dictionaries via ℓ^1 minimization. Proc. Nat. Aca. Sci. 100, 2197–2202 (2003)
8. Hyvarinen, A., Karthikesh, R.: Imposing sparsity on the mixing matrix in independent component analysis. Neurocomputing 49, 151–162 (2002)

 9. Moussaoui, S., Hauksdottir, H., Schmidt, F., Jutten, C., Chanussot, J., Brie, D., Douté, S., Benediktsson, J.A.: On the decomposition of Mars hyperspectral data by ICA and Bayesian positive source separation. Neurocomputing (in press, 2008)
10. Rubinstein, R., Zibulevsky, M., Elad, M.: Sparsity, take 2: Accelerating sparse-coding techniques using sparse dictionaries (submitted, 2008)
11. Starck, J.-L., Candès, E.J., Donoho, D.L.: The curvelet transform for image denoising. IEEE Transactions on Image Processing 11(6), 670–684 (2002)
12. Stone, J.V., Porrill, J., Porter, N.R., Wilkinson, I.W.: Spatiotemporal Independent Component Analysis of Event-Related fMRI Data Using Skewed Probability Density Functions. NeuroImage 15(2), 407–421 (2002)
13. Tseng, P.: Convergence of a block coordinate descent method for nondifferentiable minimizations. J. of Optim. Theory and Appl. 109(3), 457–494 (2001)
14. Zhang, Z., Zha, H., Simon, H.: Low-rank approximations with sparse factors I: Basic algorithms and error analysis. SIAM Journal on Matrix Analysis and Applications 23(3), 706–727 (2002)
15. Zibulevsky, M., Pearlmutter, B.A.: Blind source separation by sparse decomposition in a signal dictionary. Neural Computation 13, 863–882 (2001)

Image Coding and Compression with Sparse 3D Discrete Cosine Transform

Hamid Palangi[1], Aboozar Ghafari[1], Massoud Babaie-Zadeh[1],
and Christian Jutten[2,*]

[1] Electrical Engineering Department and Advanced Communications Research
Institute (ACRI), Sharif University of Technology, Tehran, Iran
[2] GIPSA-lab, Grenoble, France
hmd_palangi@yahoo.com, aboozar412@yahoo.com
mbzadeh@yahoo.com, Christian.Jutten@inpg.fr

Abstract. In this paper, an algorithm for image coding based on a
sparse 3-dimensional Discrete Cosine Transform (3D DCT) is studied.
The algorithm is essentially a method for achieving a sufficiently sparse
representation using 3D DCT. The experimental results obtained by the
algorithm are compared to the 2D DCT (used in JPEG standard) and
wavelet db9/7 (used in JPEG2000 standard). It is experimentally shown
that the algorithm, that only uses DCT but in 3 dimensions, outperforms
the DCT used in JPEG standard and achieves comparable results (but
still less than) the wavelet transform.

Keywords: Sparse image coding, 3 dimensional DCT, wavelet transform.

1 Introduction

In data compression reducing or removing redundancy or irrelevancy in the data
is of great importance. An image can be lossy compressed by removing irrelevant
information even if the original image does not have any redundancy [1]. The
JPEG standard [2] which is based on Discrete Cosine Transform (DCT) [3],
is widely used for both lossy and lossless image compression, especially in web
pages. However, the use of the DCT on 8×8 blocks of pixels results sometimes in
a reconstructed image that contains blocking effects (especially when the JPEG
parameters are set for large compression ratios). Consequently, JPEG2000 was
proposed based on Discrete Wavelet Transform (DWT) [4,5] which provides more
compression ratios than JPEG for comparable values of Peak Signal-to-Noise
Ratio (PSNR).

* This work has been partially funded by Iran National Science Foundation (INSF)
under contract number 86/994, by Iran Telecommunication Research Center (ITRC)
and also by center for International Research and Collaboration (ISMO) and French
embassy in Tehran in the framework of a GundiShapour program. We would also
like to thank National Elite Foundation of Iran for their financial supports.

T. Adali et al. (Eds.): ICA 2009, LNCS 5441, pp. 532–539, 2009.

Compression systems are typically based on the assumption that the signal can be well approximated by a linear combination of a few basis elements in the transform domain. In other words, the signal is sparsely represented in the transform domain, and hence by preserving a few high magnitude transform coefficients that convey most of information of the signal and discarding the rest, the signal can be effectively estimated. The sparsity of representation depends on the type of the transform used and also the signal properties. In fact the great variety in natural images makes impossible for any fixed 2D transform to achieve good sparsity for all cases [1]. Thus, the commonly used orthogonal transforms can achieve sparse representations only for particular image patterns.

In this article an image coding strategy based on an enhanced sparse representation in transform domain is studied which is based on a recently proposed approach [6] for image denoising. Based on this approach an enhanced sparse representation can be achieved by grouping similar 2D fragments of input image (blocks) into 3D data arrays. *We have used this approach with a 3D DCT transform for image coding purposes.* The procedure includes three steps: 3D DCT transformation of a 3D array, shrinkage of the transform domain coefficients, and inverse 3D DCT transformation. Due to the similarity between blocks in a 3D array, the 3D DCT transform can achieve a highly sparse representation. Experimental results demonstrate that it achieves outstanding performance in terms of both PSNR and sparsity.

The paper is organized as follows. The next section describes the main idea and discusses its effectiveness. The algorithm is then stated in Section 3. Finally, Section 4 provides some experimental results of algorithm and its comparison with DWT.

2   The Basic Idea

The basic idea of this article is achieving an enhanced sparse representation by grouping similar 2D fragments of the input image into 3D arrays, and then using a 3D DCT transformation to transform 3D arrays. In fact this idea has been introduced in [6] for image denoising and has been shown to outperform state of the art denoising algorithms [6]. Then, in this article, we consider applying an approximately similar idea for image compression and study its performance.

A simple justification for the effectiveness of the proposed idea is as follow [6]:

- Assume that the grouping is done, i.e. similar blocks are placed in groups and a 2D DCT transformation is used for each group.
- In each group we have similar blocks and hence after transformation we will have the same number of *high-magnitude* coefficients for each block in a group, say α high-magnitude coefficients for each block.
- Assuming n blocks in each group, we will have $n\alpha$ high-magnitude coefficients in that group. In other words this group can be represented by $n\alpha$ coefficients.
- Now we should perform a 1D DCT transform on the third dimension (along each row) of each group.

- Components of this row are similar (because only similar blocks are in this group), i.e. there is a kind of similarity for all members of the row.
- As an example, after using 1D DCT transform the first or second coefficients of this row will be high-magnitude (because of the compaction property of DCT transform). This means that the whole group can be represented by α or 2α coefficients instead of $n\alpha$ coefficients (i.e. a much more sparse representation).

3   The Algorithm

Based on the main idea of the previous section, the algorithm is as follows:

- Grouping:
    1. Block input image to 8×8 fragments with one pixel overlap
    2. Save blocks in Y.
- while Y is not empty
    for i=1,...,*Number Of Fragments:*
        1. *Choose one block as a reference block (Y_r).*
        2. *Calculate* $d(Y_r, Y_i) = \frac{\|Y_r - Y_i\|_2^2}{N^2}$ *were Y_i is the ith block.*
        3. *if $d(Y_r, Y_i) \leq$ Threshold Distance*
            - *Assign Y_i to a group.*
            - *Remove Y_i from Y.*
        Save resulted group in a 3D array named Group Array
- 3D DCT
    1. for every group of *Group Array*
        Perform a 2D DCT on that group
    2. *Perform a 1D DCT on the third dimension of Group Array*
- Shrinkage
    1. if *Transform Domain Coefficients $\leq$ Hard Threshold*
        Discard that coefficient.
- Calculate inverse 3D DCT transform
- Place each decoded block in its original position.

Remark 1. For image blocking we have used (as suggested in [6]) blocks with one pixel overlap to increase PSNR and also overcome the blocking effects resulted from image blocking.

Remark 2. Grouping can be realized by various techniques; e.g., K-means clustering, self-organizing maps, fuzzy clustering, vector quantization and others. A complete overview of these approaches can be found in [7]. A much simpler and effective grouping of mutually similar signal fragments can be realized by matching as discussed in [6]. In matching we want to find blocks which are similar to a reference block. It needs a search between all blocks to find blocks similar to a given reference block. The fragments whose distance from the reference block is smaller than a grouping threshold are stacked in a group. Any image fragment can be used as a reference block and thus a group can be constructed for it. The similarity between image fragments is typically computed as the inverse of some distance measure. Hence, a smaller distance implies higher similarity. In particular, we use the same distance proposed in [6] which is defined below as a measure of dissimilarity.

$$d(Y_r, Y_i) = \frac{\|Y_r - Y_i\|_2^2}{N^2} \tag{1}$$

In (1), Y_r is the reference block from which the distance of the ith block (Y_i) is calculated. N is the size of the chosen blocks (for all our simulations 8×8 blocks are used, that is $N = 8$). This distance can also be computed in the transform domain; i.e., we can do the grouping after 2D transformation (*transform domain grouping*) and then perform a 1D DCT on third dimension along the rows. This idea was tested and the changes in PSNR were in the order of 10^{-2} with the same sparsity.

Remark 3. Note that in the 3D DCT transformation at first a 2D DCT transform is applied on groups and then a 1D DCT transform is performed on the third dimension, which is on the rows of every group. Both of the used DCT transformations are complete DCT transforms.

Remark 4. In the shrinkage we have used a hard thresholding methodology; i.e., we have simply discarded those coefficients in the transform domain whose magnitude is less than some fixed threshold.

Remark 5. Obviously a straightforward implementation of this algorithm is highly computationally demanding. In order to realize a practical and efficient algorithm, some constraints should be considered. For example to reduce the number of processed blocks we can only use a limited number of reference blocks by choosing reference blocks between every N_1 blocks. In this way we will have *(Total Number Of Blocks)*$/N_1$ reference blocks. A complete set of such ideas to reduce the computational complexity and increase the speed of the algorithm can be found in [6].

4   Simulation Results

In this section, we study the performance of the presented approach and compare it with 2D DCT and wavelet transform for image compression. The wavelet transform that we have used in this comparison is db9/7 which is used in JPEG2000 standard [8]. This wavelet transform is also used in *FBI* fingerprint database [9]. The images which have been used for all simulations are 441×358 *Tracy* and *Barbara* images. Our criterion to measure sparsity is simply the ℓ^0 norm, that is, the number of nonzero coefficients. The simulation results are as presented in Table 1 (note that all transforms mentioned in this table are complete). In this table d_G stands for distance used for grouping and th_C stands for the hard threshold used to shrinkage the coefficients.

As it can be seen in Table 1, generally the performance of 3D DCT is better than 2D DCT (an improvement about $2dB$ in PSNR with the same degree of sparsity). Note that in the last row of the table for Tracy image the results of 2D DCT and 3D DCT are very close to each other. The reason is that in this case the distance threshold used for grouping is very high (245) and therefore we don't have an exact grouping; i.e., similarity of the third dimension is not high and this yields weak results with 3D DCT. Generally it can be deduced from the table that *with more precise grouping we will have better results but only in terms of PSNR. If we want to achieve high sparsity at the same time, we would need some sort of balance between the number of nonzero elements (ℓ^0 norm as a criterion to measure sparsity) and PSNR.* This result was expected because the main idea was based on the similarity between blocks and if this similarity increases then the similarity that exists in the third dimension of every array will increase and therefore more compaction can be achieved. The best result from 3D DCT idea has been shown in bold in the table. It should also be noted that generally 3D DCT results are weaker than results obtained by wavelet transform in terms of PSNR with the same sparsity for Tracy image. It is interesting to note the results for Barbara image. In this case results of 3D DCT are closer to (or even better than) those of the Wavelet transform.

Although the complexity of wavelet transforms depends on the size of filters and the use of floating point vs integer filters, wavelet transforms are generally more computationally complex than the current block- based DCT transforms [10].

Table 1. 3D DCT Versus 2D DCT and Wavelet db9/7

$TestImage$	d_G	th_C	ℓ^0 Norm	PSNR in dB		
				2D DCT	4 Level Wavelet db 9/7	3D DCT
	10	50	3327	32.8294	38.2536	36.7798
	10	30	4703	36.3101	39.8067	38.2879
Tracy	**10**	**20**	**6175**	**37.9671**	**40.9493**	**39.3012**
	20	20	5608	37.3384	40.5386	39.0774
	50	20	5189	36.8697	40.1906	38.7131
	245	20	6009	37.8331	40.8519	38.1795
	10	50	9027	27.5271	29.4975	29.3795
	10	30	15068	30.3920	32.0041	32.2399
Barbara	**10**	**20**	**22138**	**33.9195**	**34.5035**	**34.6991**
	20	20	21758	33.6757	34.5024	34.6826
	50	20	21205	33.5548	34.4704	34.6386
	245	20	20273	33.1230	34.1353	33.9469

Figures 1 and 3 show the original images and their decoded versions using wavelet db9/7, DCT and 3D DCT for the bold rows of Table 1. As it can be seen from these figures, the blocking effect when 2D DCT is used is clearly visible. But when 3D DCT is used there is almost no blocking effect. In Figs. 2 and 4, a comparison between the performances of these three transforms is made for both test images.

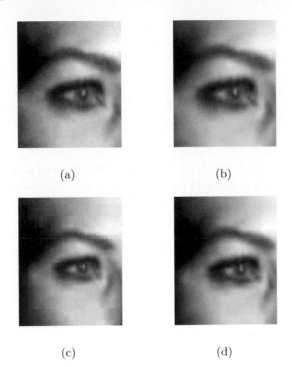

Fig. 1. The zoomed results of using various transforms for *Tracy* test image (a) The original image (b) Decoded image after compression using wavelet db9/7 (c) Decoded image after compression using 2D DCT (d) Decoded Image after compression using 3D DCT

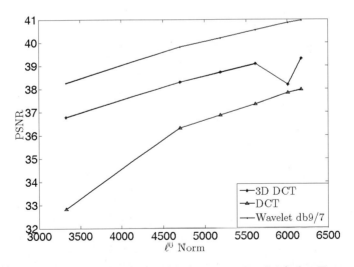

Fig. 2. Comparison between DCT, 3D DCT and wavelet db9/7 for *Tracy* test image

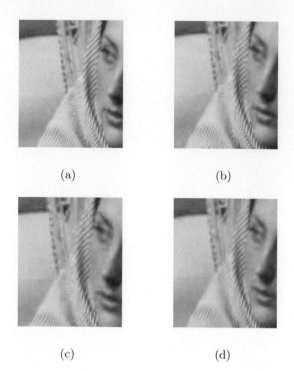

(a)   (b)

(c)   (d)

Fig. 3. The zoomed results of using various transforms for *Babara* test image. (a) The original image (b) Decoded image after compression using wavelet db9/7 (c) Decoded image after compression using 2D DCT (d) Decoded Image after compression using 3D DCT.

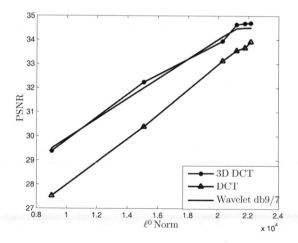

Fig. 4. Comparison between DCT, 3D DCT and wavelet db9/7 for *Barbara* test image

5   Conclusions

In this article the idea of a recently proposed approach for image denoising was studied to be used for image compression. This idea is based on 3D DCT transform to enhance the sparsity of the coefficients. Our simulations show that the usage of this idea enhances the results compared to 2D DCT transform (used in JPEG), and gives the results comparable (but still below) what is obtained using wavelet transform (used in JPEG2000).

References

1. Salomon, D.: Data Compression, The Complete Reference, 3rd edn. Springer, New York (2004)
2. Wallace, G.K.: The JPEG still Picture Compression Standard. Commun. ACM 34(4), 30–44 (1991)
3. Rao, K., Yip, P.: Discrete Cosine Transform, Algorithms, Advantages, Applications. Academic, New York (1990)
4. Taubman, D., Marcellin, M.: JPEG 2000: Image Compression Fundamentals, Standards and Practice. Kluwer, Norwell (2002)
5. Topiwala, P.N.: Wavelet Image and Video Compression. Kluwer, Norwell (1998)
6. Dabov, K., Foi, A., Katkovnik, V., Egiazarian, K.: Image Denoising by Sparse 3D Transform-Domain Collaborative Filtering. IEEE Transactions on Image Processing 16(8), 2080–2095 (2007)
7. Jain, A.K., Murty, M.N., Flynn, P.J.: Data Clustering: A Review. ACM Computing Surveys 31(3), 264–323 (1999)
8. Antonini, M., Barlaud, M., Mathieu, P., Daubechies, I.: Image Coding using Wavelet Transforms. IEEE Transactions on Image Processing 1(2), 205–220 (1992)
9. Wickerhauser, M.V.: Adapted wavelet analysis from theory to software algorithms. AK Peters, Ltd., Wellesley (1994)
10. McDowell, D.: Why Do We Care About JPEG 2000? Image Science and Technology Reporter 14(3) (1999)

Bayesian Non-negative Matrix Factorization

Mikkel N. Schmidt[1], Ole Winther[2], and Lars Kai Hansen[2]

[1] University of Cambridge, Department of Engineering
mns@imm.dtu.dk
[2] Technical University of Denmark, DTU Informatics
{owi,lkh}@imm.dtu.dk

Abstract. We present a Bayesian treatment of non-negative matrix factorization (NMF), based on a normal likelihood and exponential priors, and derive an efficient Gibbs sampler to approximate the posterior density of the NMF factors. On a chemical brain imaging data set, we show that this improves interpretability by providing uncertainty estimates. We discuss how the Gibbs sampler can be used for model order selection by estimating the marginal likelihood, and compare with the Bayesian information criterion. For computing the maximum a posteriori estimate we present an iterated conditional modes algorithm that rivals existing state-of-the-art NMF algorithms on an image feature extraction problem.

1   Introduction

Non-negative matrix factorization (NMF) [1,2] has recently received much attention as an unsupervised learning method for finding meaningful and physically interpretable latent variable decompositions. The constraint of non-negativity is natural for a wide range of natural signals, such as pixel intensities, amplitude spectra, and occurence counts. NMF has found widespread application in many areas, and has for example been used in environmetrics [1] and chemometrics [3] to find underlying explanatory sources in series of chemical concentration measurements; in image processing [2] to find useful features in image databases; in text processing [4] to find groups of words that constitute latent topics in sets of documents; and in audio processing [5] to separate mixtures of audio sources.

In this paper, we discuss NMF in a Bayesian framework. Most NMF algorithms can be seen as computing a maximum likelihood (ML) or maximum a posteriori (MAP) estimate of the non-negative factorizing matrices under some assumptions on the distribution of the data and the factors. Here, we derive an efficient Markov chain Monte Carlo (MCMC) method for estimating their posterior density, based on a Gibbs sampling procedure. This gives not only an estimate of the factors, but also an estimate of their marginal posterior density, which is valuable for interpreting the factorization, computing uncertainty estimates, etc. This work is related to the Bayesian spectral decomposition (BSD) method of Ochs et al. [6], that uses a (computationally expensive) atomic point-mass prior and is implemented in a modified commercial MCMC toolbox; and to the Bayesian non-negative source separation method of Moussaoui et al. [7]

T. Adali et al. (Eds.): ICA 2009, LNCS 5441, pp. 540–547, 2009.

that incorporates a hybrid Gibbs-Metropolis-Hastings sampling procedure. The contributions of this paper are three-fold: 1) We present a fast and direct Gibbs sampling procedure for the NMF problem that, compared with BSD, reduces computation time by more than an order of magnitude on the same data. 2) We present a marginal-likelihood estimation-method based on the Gibbs sampler, which leads to a novel model order selection method for NMF. 3) We propose an iterated conditional modes algorithm for computing the MAP estimate of the Bayesian NMF, and show that this algorithm rivals current state-of-the-art NMF algorithms. Matlab implementations of the presented algorithms are available at http://www.mikkelschmidt.dk/ica2009.

2   Bayesian Non-negative Matrix Factorization

The non-negative matrix factorization problem can be stated as $\boldsymbol{X} = \boldsymbol{AB} + \boldsymbol{E}$, where $\boldsymbol{X} \in \mathbb{R}^{I \times J}$ is a data matrix that is factorized as the product of two element-wise non-negative matrices, $\boldsymbol{A} \in \mathbb{R}_+^{I \times N}$ and $\boldsymbol{B} \in \mathbb{R}_+^{N \times J}$ ($\mathbb{R}_+$ denotes the non-negative reals), and $\boldsymbol{E} \in \mathbb{R}^{I \times J}$ is a residual matrix. In the Bayesian framework, we state our knowledge of the distribution of the residual in terms of a likelihood function, and the parameters in terms of prior densities. The priors are chosen in accordance with our beliefs about the distribution of the parameters; however, to allow efficient inference in the model it is desirable to choose prior densities with a convenient parametric form. In this paper, we choose a normal likelihood and exponential priors, which are suitable for a wide range of problems, while permitting an efficient Gibbs sampling procedure. We assume that the residuals, $E_{i,j}$, are i.i.d. zero mean normal with variance σ^2, which gives rise to the likelihood,

$$p(\boldsymbol{X}|\boldsymbol{\theta}) = \prod_{i,j} \mathcal{N}\left(X_{i,j}; (\boldsymbol{AB})_{i,j}, \sigma^2\right), \tag{1}$$

where $\boldsymbol{\theta} = \{\boldsymbol{A}, \boldsymbol{B}, \sigma^2\}$ denotes all parameters in the model and $\mathcal{N}(x; \mu, \sigma^2) = (2\pi\sigma^2)^{-1/2} \exp(-(x - \mu)^2/(2\sigma^2))$ is the normal density. We assume $\boldsymbol{A}$ and $\boldsymbol{B}$ are independently exponentially distributed with scales $\alpha_{i,n}$ and $\beta_{n,j}$,

$$p(\boldsymbol{A}) = \prod_{i,n} \mathcal{E}\left(A_{i,n}; \alpha_{i,n}\right), \qquad p(\boldsymbol{B}) = \prod_{n,j} \mathcal{E}\left(B_{n,j}; \beta_{n,j}\right), \tag{2}$$

where $\mathcal{E}(x; \lambda) = \lambda \exp(-\lambda x) u(x)$ is the exponential density, and $u(x)$ is the unit step function. The prior for the noise variance is chosen as an inverse gamma density with shape k and scale θ,

$$p(\sigma^2) = \mathcal{G}^{-1}\left(\sigma^2; k, \theta\right) = \frac{\theta^k}{\Gamma(k)} (\sigma^2)^{-k-1} \exp\left(-\frac{\theta}{\sigma^2}\right). \tag{3}$$

Using Bayes rule, the posterior can now be written as the product of Equations (1–3). The posterior can be maximized to yield an estimate of $\boldsymbol{A}$ and $\boldsymbol{B}$; however, we are interested in estimating the marginal density of the factors, and because we cannot directly compute marginals by integrating the posterior, we proceed in the next section by deriving an MCMC sampling method.

2.1   Gibbs Sampling

In Gibbs sampling, a sequence of samples is drawn from the conditional posterior densities of the model parameters, that converges to a sample from the joint posterior. To derive the Gibbs sampler for the Bayesian NMF problem, we first consider the conditional densities of A and B which are proportional to a normal multiplied by an exponential, i.e., a rectified normal density, which we denote by $\mathcal{R}\left(x; \mu, \sigma^2, \lambda\right) \propto \mathcal{N}(x; \mu, \sigma^2)\mathcal{E}(x; \lambda)$. The conditional density of $A_{i,n}$ is

$$p(A_{i,n}|X, A_{\backslash(i,n)}, B, \sigma^2) = \mathcal{R}\left(A_{i,n}; \mu_{A_{i,n}}, \sigma^2_{A_{i,n}}, \alpha_{i,n}\right), \tag{4}$$

$$\mu_{A_{i,n}} = \frac{\sum_j \left(X_{i,j} - \sum_{n' \neq n} A_{i,n'} B_{n',j}\right) B_{n,j}}{\sum_j B^2_{n,j}}, \qquad \sigma^2_{A_{i,n}} = \frac{\sigma^2}{\sum_j B^2_{n,j}}, \tag{5}$$

where $A_{\backslash(i,n)}$ denotes all elements of A except $A_{i,n}$, and due to symmetry, the similar expression for $B_{n,j}$ can easily be derived. The conditional density of σ^2 is an inverse-gamma

$$p(\sigma^2|X, A, B) = \mathcal{G}^{-1}(\sigma^2; k_{\sigma^2}, \theta_{\sigma^2}) \tag{6}$$

$$k_{\sigma^2} = \frac{IJ}{2} + 1 + k, \quad \theta_{\sigma^2} = \frac{1}{2}\sum_{i,j}(X - AB)^2_{i,j} + \theta. \tag{7}$$

The posterior can now be approximated by sequentially sampling from these conditional densities.

A few remarks on the implementation: Since the elements in each column of A (row of B) are conditionally independent, we can samples an entire column of A (row of B) simultaneously. When $I \times J \gg (I + J) \times N$ it advantageous to implement equations (5) and (7) in a way that avoids explicitly computing large matrix products of size $I \times J$. The bulk of the computation is the comprised of the matrix products $XB^\top$ and $A^\top X$ that can be precomputed in each iteration. Based on this, an efficient NMF Gibbs sampler is given as Algorithm 1, where R and G^{-1} denotes drawing a random sample from the rectified normal and inverse-gamma densities, and the notation $A_{:,\backslash n}$ is used to denote the submatrix of A that consists of all colums except the n'th.

2.2   Estimating the Marginal Likelihood

An important problem in NMF is to choose the number of factors, N, for a given data set (when this is not evident from the nature of the data). In the Bayesian framework, model selection can be performed by evaluating the marginal likelihood, $p(X)$, which involves an intractable integral over the posterior. Several methods exist for estimating the marginal likelihood, including annealed importance sampling, bridge sampling, path sampling, and Chib's method [8]. The latter is of particular interest here since it requires only posterior draws, and can thus be implemented directly using the described Gibbs sampler.

Chib's method is based on the relation $p(X) = \frac{p(X|\theta)p(\theta)}{p(\theta|X)}$. The numerator can easily be evaluated for any θ, so the problem is to evaluate the denominator, i.e., the the posterior density at the point $\boldsymbol{\theta}$. If the parameters are segmented into K blocks, $\{\boldsymbol{\theta}_1, \ldots, \boldsymbol{\theta}_K\}$, using the chain rule we may write the denominator as the product of K terms, $p(\boldsymbol{\theta}|\boldsymbol{X}) = p(\boldsymbol{\theta}_1|\boldsymbol{X}) \times p(\boldsymbol{\theta}_2|\boldsymbol{\theta}_1, \boldsymbol{X}) \times \cdots \times p(\boldsymbol{\theta}_K|\boldsymbol{\theta}_1, \ldots, \boldsymbol{\theta}_{K-1}, \boldsymbol{X})$. If these K parameter blocks are chosen such that they are amenable to Gibbs sampling, each term can be approximated by averaging over the conditional density $p(\boldsymbol{\theta}_k|\boldsymbol{\theta}_1, \ldots, \boldsymbol{\theta}_{k-1}, \boldsymbol{X}) \approx \frac{1}{M} \sum_{m=1}^{M} p(\boldsymbol{\theta}_k|\boldsymbol{\theta}_1, \ldots, \boldsymbol{\theta}_{k-1}, \boldsymbol{\theta}_{k+1}^{(m)}, \ldots, \boldsymbol{\theta}_K^{(m)}, \boldsymbol{X})$, where $\boldsymbol{\theta}_{k+1}^{(m)}, \ldots, \boldsymbol{\theta}_K^{(m)}$ are Gibbs samples from $p(\boldsymbol{\theta}_{k+1}, \ldots, \boldsymbol{\theta}_K|\boldsymbol{\theta}_1, \ldots, \boldsymbol{\theta}_{k-1}, \boldsymbol{X})$. Thus, the marginal likelihood can be evaluated by K runs of the Gibbs sampler. This procedure is valid for any value of $\boldsymbol{\theta}$, but the estimation is most accurate when $\boldsymbol{\theta}$ is chosen as a high density point, e.g., the posterior mode.

In the NMF problem, the columns of $\boldsymbol{A}$ and rows of $\boldsymbol{B}$ can be used as the parameter blocks, and the marginal likelihood can thus be estimated by $2N$ runs of the Gibbs sampler, which makes this method attractive especially for NMF models with a small number of components.

2.3   An Iterated Conditional Modes Algorithm

With the conditional densities of the parameters in the NMF model in place, an efficient iterated conditional modes (ICM) [9] algorithm can be derived for computing the MAP estimate. In this approach, we iterate over the parameters of the model, but instead of drawing random samples from the conditionals, as in the Gibbs sampler, we set each parameter equal to the conditional mode, and after a number of iterations the algorithm converges to a local maximum of the joint posterior density. This forms a block coordinate ascent type algorithm with the benefit that the optimum is computed for each block of parameters in each iteration. Since the modes of the conditional densities have closed-form expressions, the ICM algorithm has a low computational cost per iteration. The ICM NMF is given as Algorithm 2. In the algorithm, P_+ sets negative elements of its argument to zero.

3   Experimental Evaluations

3.1   Analysis of Chemical Shift Brain Imaging Data

We demonstrate the proposed bayesian NMF method on a chemical shift imaging (CSI) data set [6], that consists of 369-dimensional spectra measured at 256 positions in a human head. The measured spectra are mixtures of different underlying spectra, and the NMF decomposition finds these as the columns of $\boldsymbol{A}$ and the corresponding active positions in the head as the rows of $\boldsymbol{B}$. Ochs et al. [6] demonstrate that the data is well described by two components that correspond to brain tissue and muscle tissue, and present a bilinear MCMC method that provides physically meaningful results but takes several hours to compute. Sajda et al. [10,3] demonstrate, on the same data set, that a constrained NMF method provides meaningful results, and Schmidt and Laurberg [11] extend the NMF approach by including advanced prior densities.

In our experiments, the priors for A and B were chosen as $\alpha_{i,n} = 1$ and $\beta_{n,j} = 10^{-6}$ to match the overall amplitude of the data. For the noise variance we used an uninformative prior, $k = \theta = 0$. We sampled $40,000$ points from the posterior and discarded the first half to allow for the sampler to burn in. The computations took 80 seconds on a 2 Ghz Pentium 4 computer. We then computed the mean and the 5'th and 95'th percentile of the marginal distributions of the resolved spectra. For comparison, we computed the MAP estimate using the ICM algorithm, that provides a solutions almost identical to the results of Sajda et al. [3]. The results are shown in Figure 1.

Uncertainty in NMF can be caused by noise in the data, but since NMF in general is not unique multiple competing solutions may also be reflected in the posterior. Also, because of the bilinear structure of NMF, uncertainties may be correlated between A and B, which can not be seen in plots of marginal distributions, but can be assesed through further analysis of the posterior density.

3.2   NMF Model Order Selection

To demonstrate the proposed model order selection method, we generated a data matrix by multiplying two random unit mean i.i.d. exponential distributed matrices with $I = 100$, $J = 20$, and $N = 3$, and added unit variance zero mean i.i.d. normal noise. Using Chib's method, we computed the marginal likelihood for model orders between 1 and 5. We generated $20,000$ samples per parameter block, and discarded the first half to allow burn-in; several runs of the algorithm

---

Algorithm 1. Gibbs sampler

> **for** $m = 1$ to M **do**
>> $C = BB^\top, D = XB^\top$
>> **for** $n = 1$ to N **do**
>>> $A_{:,n} \leftarrow \mathrm{R}\left(a_n, \frac{\sigma^2}{C_{n,n}}, \alpha_{:,n}\right)$
>>
>> **end**
>> $\sigma^2 \leftarrow$
>> $\mathrm{G}^{-1}\left(\frac{IJ}{2} + k + 1, \chi + \theta + \xi\right)$
>> $E = A^\top A, F = A^\top X$
>> **for** $n = 1$ to N **do**
>>> $B_{n,:} \leftarrow \mathrm{R}\left(b_n, \frac{\sigma^2}{E_{n,n}}, \beta_{n,:}\right)$
>>
>> **end**
>> $A^{(m)} \leftarrow A, \quad B^{(m)} \leftarrow B$
>
> **end**
> **Output.** $\left\{A^{(m)}, B^{(m)}\right\}_{m=1}^{M}$

Algorithm 2. ICM

> **repeat**
>> $C = BB^\top, D = XB^\top$
>> **for** $n = 1$ to N **do**
>>> $A_{:,n} = P_+(a_n)$
>>
>> **end**
>> $\sigma^2 = \dfrac{\theta + \chi + \xi}{\frac{IJ}{2} + k + 1}$
>> $E = A^\top A, F = A^\top X$
>> **for** $n = 1$ to N **do**
>>> $B_{n,:} = P_+(b_n)$
>>
>> **end**
>
> **until** convergence
> **Output.** A, B

Definitions: $\chi = \dfrac{1}{2}\sum_{i,j} X_{i,j}^2$, $\xi = \dfrac{1}{2}\sum_{i,n} A_{i,n}\left(AC - 2D\right)_{i,n}$

$$a_n = \frac{D_{:,n} - A_{:,\backslash n}C_{\backslash n,n} - \alpha_{:,n}\sigma^2}{C_{n,n}}, \quad b_n = \frac{F_{n,:} - E_{n,\backslash n}B_{\backslash n,:} - \beta_{n,:}\sigma^2}{E_{n,n}}$$

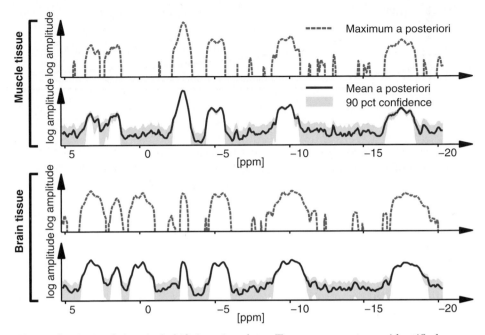

Fig. 1. Analysis of chemical shift imaging data. Two components are identified corresponding to (top) muscle and (bottom) brain tissue. MAP estimation provides a point estimate of the components whereas Gibbs sampling gives full posterior marginals, that provide an uncertainty estimate on the spectra, and leads to better interpretation of the results. For example, the confidence intervals show, that many of the low amplitude peaks in the MAP spectra may not be significant.

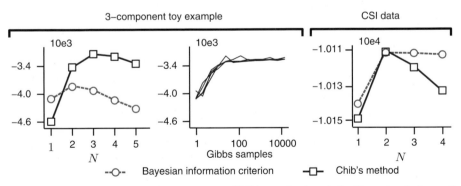

Fig. 2. NMF model order selection using Chib's method and the Bayesian information criterion (BIC). Left: A three-component toy example demonstrates that the method finds the correct number of components where BIC fails due to the small sample size. Center: Several runs of the algorithm suggest that the estimate of the marginal likelihood is stable after a few thousand Gibbs samples. Right: Analysis of the chemical shift imaging data confirms that it contains two spectral components. In this experiment, Chib's method and BIC were in agreement.

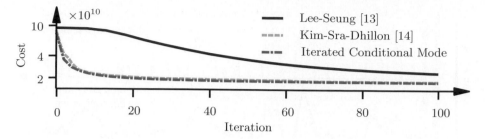

Fig. 3. Convergence rate for three least squares NMF algorithms on an image feature extraction problem. The ICM algorithm converges much faster than the Lee-Seung algorithm and with similar rate per iteration as the Kim-Sra-Dhillon algorithm.

suggested that this was sufficient for the sampler to stabilize. For comparison, we computed the Baysian information criterion (BIC), where the number of effective parameters was chosen as the number of non-zero parameters in the MAP estimate. The marginal likelihood attained its maximum at the correct model order, $N = 3$, whereas BIC favored a simpler model, $N = 2$. The reason why BIC fails in this case is the small number of samples in the data set, and our results suggest that Chib's method is more robust. Next, we applied the marginal likelihood estimation technique to the CSI data set described in the previous section, and here Chib's method and BIC agreed in confirming that the data contains two spectral components. The results are shown in Figure 2.

3.3   Image Feature Extraction

To compare our ICM algorithm with other methods, we computed an $N = 32$ components factorization of a database of images. We used the cropped UMIST Face Database [12], which consists of 564 grayscale images of size 92×112, that were vectorized to form a data matrix of size $I = 10304 \times J = 564$. To be able to directly compare with existing least squares NMF methods, we used a flat prior, $\alpha_{i,n} = \beta_{n,j} = k = \theta = 0$. We compared with two state-of-the-art methods: Lee and Seung's multiplicative update algorithm [13] and Kim, Sra, and Dhillon's fast Newton algorithm (FNMA[I]) [14].

The results (see Figure 3) show that the ICM algorithm converges much faster than the Lee-Seung algorithm and with approximately the same rate per iteration as the Kim-Sra-Dhillon algorithm. Since all three algorithms are dominated by the computation of the same matrix products they have approximately the same computational cost per iteration.

4   Conclusions

We have taken a Bayesian approach to NMF and presented a fast MCMC sampling procedure for approximating the posterior density, and we have showed that this can be valuable for the interpretation of the non-negative factors recovered in NMF. The sampling procedure can also directly be used to estimate the marginal likelihood, which is useful for model order selection. Finally, we have presented an iterated conditional modes algorithm for computing the MAP estimate, that rivals existing state-of-the-art NMF algorithms.

Acknowledgments. We thank Paul Sajda and Truman Brown for making the chemical shift brain imaging data set available to us.

References

1. Paatero, P., Tapper, U.: Positive matrix factorization: A nonnegative factor model with optimal utilization of error-estimates of data values. Environmetrics 5(2), 111–126 (1994)
2. Lee, D.D., Seung, H.S.: Learning the parts of objects by non-negative matrix factorization. Nature 401(6755), 788–791 (1999)
3. Sajda, P., Du, S., Parra, L.: Recovery of constituent spectra using non-negative matrix factorization. In: Wavelets: Applications in Signal and Image Processing, Proceedings of SPIE, vol. 5207, pp. 321–331 (2003)
4. Gaussier, E., Goutte, C.: Relation between PLSA and NMF and implication. In: Proceedings of the International SIGIR Conference on Research and Development in Information Retrieval, pp. 601–602 (2005)
5. Schmidt, M.N., Olsson, R.K.: Single-channel speech separation using sparse non-negative matrix factorization. In: ISCA International Conference on Spoken Language Processing, (INTERSPEECH) (2006)
6. Ochs, M.F., Stoyanova, R.S., Arias-Mendoza, F., Brown, T.R.: A new method for spectral decomposition using a bilinear bayesian approach. Journal of Magnetic Resonance 137, 161–176 (1999)
7. Moussaoui, S., Brie, D., Mohammad-Djafari, A., Carteret, C.: Separation of non-negative mixture of non-negative sources using a Bayesian approach and MCMC sampling. IEEE Transactions on Signal Processing 54(11), 4133–4145 (2006)
8. Chib, S.: Marginal likelihood from the Gibbs output. Journal of the American Statistical Association 90(432), 1313–1321 (1995)
9. Besag, J.: On the statistical analysis of dirty pictures. Journal of the Royal Statistical Society 48(3), 259–302 (1986)
10. Sajda, P., Du, S., Brown, T.R., Stoyanova, R., Shungu, D.C., Mao, X., Parra, L.C.: Nonnegative matrix factorization for rapid recovery of constituent spectra in magnetic resonance chemical shift imaging of the brain. IEEE Transactions on Medical Imaging 23(12), 1453–1465 (2004)
11. Schmidt, M.N., Laurberg, H.: Nonnegative matrix factorization with gaussian process priors. Computational Intelligence and Neuroscience (February 2008)
12. Graham, D.B., Allinson, N.M.: Characterizing virtual eigensignatures for general purpose face recognition. In: Face Recognition: From Theory to Applications. Computer and Systems Sciences, vol. 163, pp. 446–456 (1998)
13. Lee, D.D., Seung, H.S.: Algorithms for non-negative matrix factorization. In: Advances in Neural Information Processing Systems (NIPS), pp. 556–562 (2000)
14. Kim, D., Sra, S., Dhillon, I.S.: Fast Newton-type methods for the least squares nonnegative matrix approximation problem. In: Proceedings of SIAM Conference on Data Mining (2007)

Blind Decomposition of Spectral Imaging Microscopy: A Study on Artificial and Real Test Data

Fabian J. Theis[1], Richard Neher[2], and Andre Zeug[3]

[1] CMB, Institute of Bioinformatics and Systems Biology,
Helmholtz Zentrum München, Germany, and
MPI for Dynamics and Self-Organization, Göttingen, Germany
fabian.theis@helmholtz-muenchen.de
http://cmb.helmholtz-muenchen.de

[2] Kavli Institute for Theoretical Physics, University of California, USA

[3] Department of Cellular Neurophysiology, Hannover Medical School, Germany

Abstract. Recently, we have proposed a blind source separation algorithm to separate dyes in multiply labeled fluorescence microscopy images. Applying the algorithm, we are able to successfully extract the dye distributions from the images. It thereby solves an often challenging problem since the recorded emission spectra of fluorescent dyes are environment and instrument specific. The separation algorithm is based on nonnegative matrix factorization in a Poisson noise model and works well on many samples. In some cases, however, additional cost function terms such as sparseness enhancement are necessary to arrive at a satisfactory decomposition.

In this contribution we analyze the algorithm on two very well controlled real data sets. In the first case, known sources are artificially mixed in varying mixing conditions. In the second case, fluorescent beads are used to generate well behaved mixing situations. In both cases we can successfully extract the original sources. We discuss how the separation is influenced by the weight of the additional cost function terms, thereby illustrating that BSS can be be vastly improved by invoking qualitative knowledge about the nature of the sources.

Blind source separation (BSS) and independent component analysis in particular have become well established and are studied from applied and theoretical perspectives. However, while possible applications are numerous, real-world examples of successful applications of BSS are surprisingly rare. They include audio source separation [1], applications to data from neuroscience such as EEG [2] and fMRI [5] and analysis of astrophysical data sets. Typically whole groups have devoted a significant amount of time to the detailed development of a successful strategy for applying ICA.

Recently [6] we have proposed a novel application of BSS algorithms[1]: We applied BSS to multiply labeled fluorescence microscopy images, in which the

---

[1] Accepted manuscript available at
http://wwwuser.gwdg.de/~azeug/paper/Zeug2008.pdf

T. Adali et al. (Eds.): ICA 2009, LNCS 5441, pp. 548–556, 2009.

sources are the individual labels. Their distributions are the quantities of interest and have to be extracted from the images. We have developed a non-negative matrix factorization (NMF) algorithm to detect and separate spectrally distinct components of multiply labeled fluorescence images. It operates on spectrally resolved images and delivers both the emission spectra of the identified components and images of their abundances. We tested the proposed method using biological samples labeled with up to 4 spectrally overlapping fluorescent labels. In most cases, NMF accurately decomposed the images into the contributions of individual dyes. However, the solutions are not unique, when spectra overlap strongly or else when images are diffuse in their structure. To arrive at satisfactory results in such cases, we extended NMF to incorporate preexisting qualitative knowledge about spectra and label distributions. For this an additional sparseness constraint denoted as segregation bias was introduced.

In this contribution we will shortly present the algorithm as variant of NMF, which instead of Gaussian noise assumes a Poisson noise model more adequate for the microscopy environment. We then present a detailed study on both 'semi toy' data, i.e. data generated on the basis of experimental data, and controlled real data. The segregation bias is not required in the toy examples but turns out to be crucial for real data. To illustrate the nature of the ambiguity and the effect of the segregation bias, we introduce a visualization of the very common case of three sources (dyes).

1   The PoissonNMF Algorithm

Given image data $\mathbf{X} \in \mathbb{R}^{N \times n}$, N being the number of pixels and n the number of observed images i.e. the number of spectral wave lengths, consider the factorization problem $\mathbf{X} \approx \mathbf{AS}$. Here $\mathbf{A} \in \mathbb{R}^{N \times r}$ denotes the r concentration maps and $\mathbf{S} \in \mathbb{R}^{r \times n}$ the dye spectra.

We search for a factorization as above with both $\mathbf{A}$ and $\mathbf{S}$ non-negative. In our case, a sensible noise model is to assume that at each pixel i in image j, we independently observe x_{ij} photons observing a Poisson statistic $P(x_{ij}) = \exp(-\lambda_{ij})\lambda_{ij}^{x_{ij}}/x_{ij}!$ with rate parameter $\lambda_{ij} = (\mathbf{AS})_{ij} = \sum_{k=1}^{r} a_{ik}s_{kj}$ given by the noiseless model. The joint likelihood can therefore be written as

$$L(\mathbf{X}|\mathbf{A}, \mathbf{S}) = \prod_{ij} P(x_{ij}) = \prod_{ij} \exp(-\lambda_{ij})\frac{\lambda_{ij}^{x_{ij}}}{x_{ij}!}.$$

Hence, determining the parameters $\mathbf{A}$ and $\mathbf{S}$ by maximum likelihood implies minimizing the negative log likelihood $f_0(\mathbf{A}, \mathbf{S}) := -\ln L(\mathbf{X}|\mathbf{A}, \mathbf{S}) = \sum_{ij}(\lambda_{ij} - x_{ij}\ln\lambda_{ij} + \ln x_{ij}!)$. In practice, we include a sparseness term denoted segregation bias with weight λ, favoring solutions with high spectral overlap and segregated labels: $f(\mathbf{A}, \mathbf{S}) = f_0(\mathbf{A}, \mathbf{S}) + \lambda\sum_i(\sum_j|a_{ij}|)(\sum_j a_{ij}^2)^{-1/2}$. Here sparseness is measured by the mean ratio of 1- over 2-norm of the rows of $\mathbf{A}$. Taking derivatives yields a gradient descent method, which by appropriate choice of the update rates can be rewritten as multiplicative updates rules [6]:

$$a_{pq} \leftarrow a_{pq} \frac{1}{\sum_j s_{qj}} \left(\sum_j \frac{s_{qj} x_{pj}}{(\mathbf{AS})_{pj}} + \lambda \left(\frac{1}{(\sum_j a_{pj}^2)^{1/2}} - \frac{a_{pq} \sum_j a_{pj}}{(\sum_j a_{pj}^2)^{3/2}} \right) \right),$$

$$s_{pq} \leftarrow s_{pq} \frac{1}{\sum_j a_{jp}} \sum_j \frac{a_{jp} x_{jq}}{(\mathbf{AS})_{jq}}.$$

This update rule is similar to the one derived by [3,4] using the Kullback-Leibler divergence. Alternative minimization methods have been discussed for general NMF algorithms, however many algorithms use such multiplicative update rules simply due to the fact of easy preservation of non-negativity.

A key requirement for a useful algorithm is a direct assessment of the quality of a separation. To enable such quality control for 3 or more source, we visualize the extracted sources by constructing a so-called *simplex-projection* of the mixtures along the estimated source spectra $\mathbf{S} \in \mathbb{R}^n$. We project a series of values $\mathbf{x} \in \mathbb{R}^n$ onto an affine 2-dimensional plane $H \subset \mathbb{R}^n$ that is being spanned by three 'principal spectral vectors' $\mathbf{s}_1, \mathbf{s}_2, \mathbf{s}_3 \in \mathbb{R}^n$. The projection onto the simplicial cone spanned by $\mathbf{s}_1, \mathbf{s}_2, \mathbf{s}_3$ can be visualized in two dimension by intersection it with the plane spanned by the unit vectors. Then the mixture data lies in a triangle spanned by the three dye spectra $\mathbf{s}_i$, see e.g. figure 1(b). The details of this projection are given as supplement of [6]. For more than three sources, this projection can be applied using subset of the sources.

2   Application to Artificially Mixed Data

We will first show the results of poissonNMF when applied to the following test data: Using a confocal laser scanning microscope (cLSM, Zeiss LSM 510 META) equipped with a spectroscopic detection unit, we acquired $xy\lambda$ image stacks of three dye solutions, i.e. Alexa Fluor 488 (A 488), Alexa Fluor 546 (A 546, both Invitrogen), and ethidium bromide (EtBr). Two acquisition parameter sets have been used, one with 8 λ-channels of 21.4nm spectral width and the other with 16 λ-channels of 10.7nm spectral width.

Spatially non-overlapping sources. This sequentially recorded dataset was rearranged to a 4d data stack ($x \times y \times$ dye $\times \lambda$), see figure 1(a). A trivial mixture was generated where each pixel contains one dye only, which corresponds to mixing matrix $\mathbf{A} = \mathbf{I}$. By applying the poissonNMF algorithm, we were able to accurately retrieve the spectra of the individual dyes, shown in figure 1(b). The reference spectra were obtained from the separate measurements of each dye before concatenation. The simplex projection of the dataset $\mathbf{X}$ using the obtained spectra $\mathbf{S} = (s_{A488}, s_{A546}, s_{EtBr})$ is shown in figure 1(c). The contour plot representing the distribution of $\mathbf{X}$ shows three well distinct pixel clusters in the appropriate region of the principle vectors ($s_{A488}, s_{A546}, s_{EtBr}$). The size of the individual distribution around one principle spectral vector is a measure for the noise of the data $\mathbf{X}$. The spectra found by the presented algorithm nicely represent the distributions but slightly deviate from the individual central positions. This minor off-center position is one the scale of the noise in the dataset.

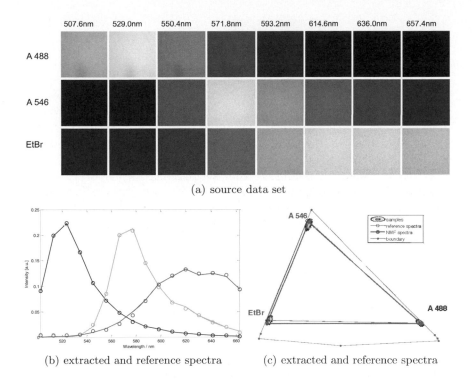

(a) source data set

(b) extracted and reference spectra   (c) extracted and reference spectra

Fig. 1. Non-overlapping mixture: the data is concentrated at the corners of the NMF simplex. (a) sequentially acquired test data set of fluorescent dye solution. (b) extracted source spectra (o) of A488 (blue), A546 (green) and EtBr (red) with respect to their reference spectra (line). (c) data projection (contour plot) onto the simplex plane defined by the three NMF spectra (red dots), which deviate slightly from the reference spectra (blue dots), with a mean signal-to-noise ratio (SNR) of 32.9 dB.

The simplex projection also shows the boundary of the region on the affine hyperplane, where all components of the data are positive. Indeed, no data points were found in the negative region of the surface. They are, however, very close to the boundary, because the spectra are almost zero at some wavelength.

Despite of the significant overlap of the three spectra, NMF performed well on the data shown in figure 1 and the additional segregation bias was required. The algorithm converged after a few (3–5) iterations where for the presented dataset of $\sim 10^6$ pixels each iteration cycle takes about 2 seconds on a standard Laptop (2 MHz CPU) without further speed optimization. The following iterations resulted in only slight changes of the spectra, which are practically not visible but still lead to an additional 20% decrease of the cost function. The presented results have been obtained within 300 iterations, which roughly took 6 minutes including pre and post processing. At the final iterations the cost function changed by less than 0.01 percent. Due to the highly overdetermined structure of the data set no significant differences have been found between the data set with 8 and 16 channels—except for the doubled computational cost for the doubled data size.

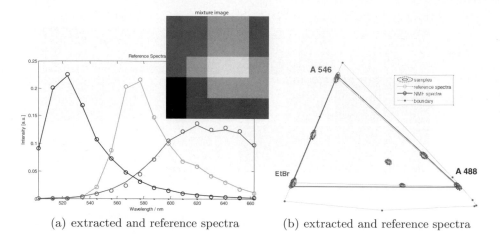

(a) extracted and reference spectra   (b) extracted and reference spectra

Fig. 2. Discrete spatial overlap: the data is artificially mixed on three different levels, see inset of (a). (a) itself shows the extracted source spectra (o) of A488 (blue), A546 (green) and EtBr (red) with respect to their reference spectra (line). (b) data projection (contour plot) onto the simplex plane defined by the three NMF spectra (red dots), which deviate slightly from the reference spectra (blue dots) as can be seen from a high mean SNR of 34.4 dB.

Mixture with discrete spatial overlap. In a second application we generate a patch of partially overlapping spectra by simply adding integer combinations of the three single dye data to an artificial dataset. Here, the pixel show a mixture of none, one, two or three dye spectra (figure 2). The resulting mixture image is shown as inset in figure 2(a). Again the algorithm delivers precise spectra after only a few iterations. According to the reduced size of the dataset ($3.3 \cdot 10^5$ data points) the computation time was one third of the calculation described before. The spectra are similar to that in figure 1(b). Here the superposition of dyes can be nicely found in the projection: Data points are not only found at the corners of the simplex, but also at the middle of the sides for data points with two dyes, and one at the center of the triangle when all three dyes are present at a pixel. The estimated spectral basis again nicely encloses all data points in the projection. A small fraction of datapoints to lie in the negative space of the spectral basis, but the distance to the positive domain in less than the noise in the data and data points that appear as if they had negative contributions are consistent with that contribution being zero.

Sources with continuous spatial overlap. Realistic conditions are simulated in the third example. Here we introduced a gradient to the dye sources to generate a fractional superposition of all dye contributions (figure 3). We took care that A546 (green) does never occurs by itself; this can be seen by comparing the simplex projections of figures 1–3). Due to the inhomogenity of the dye distributions, the spectra show a stronger deviation from the reference spectra. The algorithm does not converge as fast as in the previous examples, because the

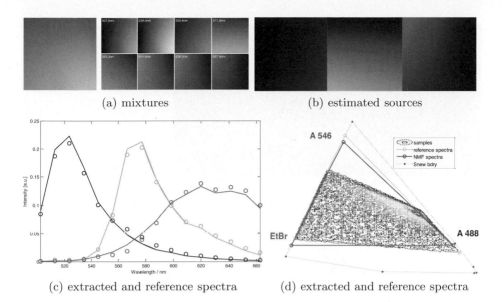

Fig. 3. Continuous spatial overlap: (a) a gradient was added in the artificially superimposed dyes. (b) dye distribution of A488 (blue), A546 (green) and EtBr (red) obtained from NMF analysis. (c) extracted source spectra (o) of A488 (blue), A546 (green) and EtBr (red) with respect to their reference spectra (line). (d) data projection (contour plot) onto the simplex plane defined by the three NMF spectra (red dots), which only slightly deviate from the reference spectra (blue dots), mean SNR 27.3 dB.

main contribution to the gradient comes from data points in the periphery of the triangle, of which there are far fewer here. After 10 iterations the spectra did not reach their final shape. The cost value reaches its 20% limit after 60 iterations. After 300 iterations the resulting dye distribution was acceptable, but continuous to show deviations most obvious in the simplex projection. To find a more precise decomposition the algorithm requires more iteration steps (or sparseness must be introduced to the algorithm).

The simplex projection nicely visualizes the principle of the presented algorithm. NMF will find spectra vectors such that the projected triangle (red in figure 1(c), 2(c), 3(d)) optimally encloses all data points $\mathbf{X}$ (contour plot). Due to the non negativity of spectral superposition only data points inside the triangle can be described with this set of spectral vectors. Thus the spectral vectors always have a slight tendency to larger triangles, especially in the case of datasets with higher noise level i.e. with enlarged contours in the regions of the principle spectra. To force the algorithm to tighter triangles a sparseness enhancing term was introduced, which results in smaller triangles. With this term, the algorithm weighs data points outside the triangle against smaller triangle, such that a number of data points can lie outside the triangle.

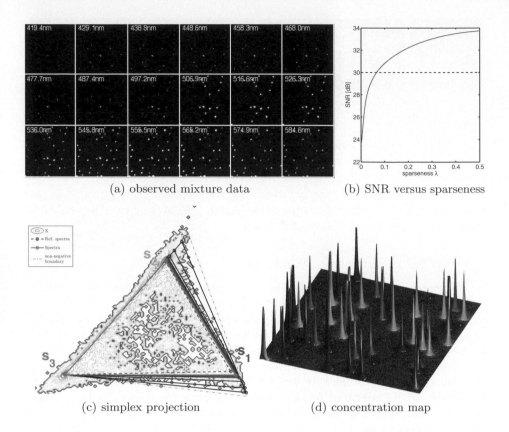

(a) observed mixture data

(b) SNR versus sparseness

(c) simplex projection

(d) concentration map

Fig. 4. Bead data set: (a) shows the observed mixture data set consisting of beads of three different colors, measured in 16 channels. The extracted source SNRs are given in (b) and are visualized via the simplex projection in (c). (d) gives a surface plot of the concentration map with false colors encoding the three different dye sources.

3   Application to Bead Data

In a real experiment we examined a standard bead slide with a distribution of three kinds of subresolution, single color beads using a spectra resolving cLSM (Zeiss LSM 710). 18 lambda channels have been recorded simultaneously for a 3d image stack of $1024 \times 1024 \times 100$ voxels. The resulting 3.5GB of data have been analyzed with poissonNMF (figure 4). Presented in figure 4(a) is a λ-stack of z-projections of a zoomed region of the data. The brightness of the three different types of beads differs significantly. Moreover, even beads of one type do not have a single brightness level. Due to clustering effects, beads frequently aggregate so that spots can be found with a superposition of two or occasionally three spectra, although the latter case is very unlikely due to steric conditions.

PoissonNMF was applied with different strengths of sparseness enhancement (dark red to bright red with increasing sparseness). Due to the size of the dataset slightly more computational power is required than is available from standard personal computers. However only a fraction of the data contains significant information and the relevant subset can be extracted by thresholding. The simplex projection was calculated for the spectral basis at highest sparseness. In the simplex plane projection in figure 4(c) we see that the data is concentrated on the bounding cone of the nonnegative mixing model. With increasing sparseness the triangle of the spectral basis converges to the center of the three dye distributions. We observed this stays more or less constant even when increasing the sparseness parameter. Indeed when increasing λ the mean SNR of the concentration images is above 30 dB already after 15% of the tested sparseness values, figure 4(b). This shows the robustness of the method with respect to the choice of sparseness parameter and underscores that the inclusion of such a bias is not a fudge parameter that can be arbitrarily adjusted to obtain the desired result. Instead, by adding the segregation bias to the data, we exploit qualitative knowledge about the structure of the data.

4   Conclusions

We have studied a previously proposed NMF algorithm [6] in the application context of spectral microscopy. A key requirement for the broad applicability of such a separation algorithm is that it is well behaved in controlled examples. Here we have analyzed artificial mixtures of real sources with varying concentration overlap and real data with close-to point activity in the concentration maps (beads). In all the examples we see good separation performance. Moreover the necessary use of sparseness in the case of realistic data is acceptable in the sense that the additional sparseness parameter λ does not seem to strongly disturb separation performance in a wide range of values — which can be determined from the simplex projections. In future work we will extend and study the algorithm in the case of higher-dimensional data sets and multiple excitation wave lengths, which leads to nonnegative tensor factorization.

Acknowledgements. FT gratefully acknowledges partial financial support by the Helmholtz Alliance on Systems Biology (project 'CoReNe'). RAN acknowledges financial support by the National Science Foundation under grant no. NSF PHY05-51164.

References

1. Araki, S., Mukai, R., Makino, S., Nishikawa, T., Saruwatari, H.: The fundamental limitation of frequency domain blind source separation for convolutive mixtures of speech. IEEE Transactions on Speech and Audio Processing 11(2), 109–116 (2003)
2. Delorme, A., Makeig, S.: Eeglab: an open source toolbox for analysis of single-trial eeg dynamics. Journal of Neuroscience Methods 134, 9–21 (2004)
3. Lee, D.D., Seung, H.S.: Learning the parts of objects by non-negative matrix factorization. Nature 40, 788–791 (1999)

4. Lee, D.D., Seung, H.S.: Algorithms for non-negative matrix factorization. In: Proc. NIPS 2000, vol. 13, pp. 556–562. MIT Press, Cambridge (2001)
5. McKeown, M.J., Makeig, S., Brown, G.G., Jung, T.P., Bell, A.J., Kindermann, S.S., Sejnowski, T.: Analysis of fMRI data by blind separation into independent spatial components. Human Brain Mapping 6, 160–188 (1998)
6. Neher, R.A., Mitkovski, M., Kirchhoff, F., Neher, E., Theis, F.J., Zeug, A.: Blind source separation techniques for the decomposition of multiply labeled fluorescence images. Biophysical Journal (accepted, 2008)

Image Denoising Using Sparse Representations

SeyyedMajid Valiollahzadeh[1,*], Hamed Firouzi[1], Massoud Babaie-Zadeh[1],
and Christian Jutten[2]

[1] Department of Electrical Engineering, Sharif University of Technology,
Tehran, Iran
[2] GIPSA-lab, Grenoble, France

Abstract. The problem of removing white zero-mean Gaussian noise
from an image is an interesting inverse problem to be investigated in this
paper through sparse and redundant representations. However, finding
the sparsest possible solution in the noise scenario was of great debate
among the researchers. In this paper we make use of new approach to
solve this problem and show that it is comparable with the state-of-art
denoising approaches.

1 Introduction

Being a simple inverse problem, the denoising is a challenging task and basically
addresses the problem of estimating a signal from the noisy measured version
available from that. A very common assumption is that the present noise is additive zero-mean white Gaussian with standard deviation σ. Many solutions have
been proposed for this problem based on different ideas, such as spatial adaptive filters, diffusion enhancement [1], statistical modeling [2], transfer domain
methods [3], [4], order statistics [5] and yet many more. Among these methods, methods based on with sparse and redundant representations has recently
attracted lots of attentions [8]. Many researchers have reported that such representations are highly effective and promising toward this stated problem [8].
Pioneered by Donoho [5], sparse representations firstly examined with unitary
wavelet dictionaries leading to the well-known shrinkage algorithm [5]. A major
motivation of using overcomplete representations is mainly to obtain translation-
invariant property [6]. In this respect, several multiresolutional and directional
redundant transforms are introduced and applied to denoising such as curvelets,
contourlets, wedgelets, bandlets and the steerable wavelet [5] [8].

The aim of all such transforms is to provide a redundant sparse decomposition of the signal. In parallel, beside providing a suitable redundant transform,
representation of a signal with these transforms is also of high value, since such
a representation is not necessarily unique. Several methods are then proposed to

---

* This work has been partially supported by Iran National Science Foundation (INSF)
under contract number 86/994 and also by ISMO and French embassy in Iran in the
framework of a Gundi-Shapour collaboration program.

T. Adali et al. (Eds.): ICA 2009, LNCS 5441, pp. 557–564, 2009.

find the best possible representation of a signal from a redundant, overcomplete dictionary obtained by these transforms, namely Matching Pursuit(MP), Basis Pursuit(BP), FOCUSS, and Smoothed ℓ^0- Norm (SL0) [7]. All these approaches basically try to find the sparsest possible solution among all the possible representations a signal can obtain. As an alternative point of view to obtain the sparse representation, example-based dictionary learning of K-SVD which is introduced by Aharon, *et. al.* [8] attempts to find the sparse dictionary over the image blocks. When using the Bayesian approach to address this inverse problem with the prior of sparsity and redundancy on the image, it is the dictionary to be used that we target as the learned set of parameters. Instead of the deployment of a pre-chosen set of basis functions like the curvelet or contourlet, this process of dictionary learning can be done through examples, a corpus of blocks taken from a high-quality set of images and even blocks from the corrupted image itself. This idea of learning a dictionary that yields sparse representations for a set of training image blocks has been studied in a sequence of works [8] and specifically the one using K-SVD has shown to outperform in both providing the sparse representation and capability of denoising. While the work reported here is based on the same idea of sparsity and redundancy concepts, a different method is used to solve the sparsest possible solution in presence of noise. An example-based dictionary learning such as K-SVD along with here used technique can provide better solutions in estimation of the original clean signal.

The paper is organized as follows. In section 2, we briefly present modeling of the scenario in decomposing a signal on an overcomplete dictionary in the presence of noise. In section 3 we discuss this algorithm in the real image denoising task. At the end we conclude and give a general overview to future's work.

2   Finding the Sparse Representation in Presence of Noise

Consider the problem of estimation of $\mathbf{x}$ from the observed signal

$$\mathbf{y} = \mathbf{x} + \mathbf{n}$$

where $\mathbf{n}$ denotes the observation noise. Assume that $\mathbf{x}$ has a sparse representation over the dictionary $\boldsymbol{\Phi}$, i.e. $\mathbf{x} = \boldsymbol{\Phi\alpha}$ with a small $\|\boldsymbol{\alpha}\|_0^0$ (the number of nonzero elements of a vector) and also assume that a good estimation on the energy of the present noise, $\|\mathbf{n}\|_2^2 \leq \epsilon^2$ is provided.

The sparsest representation we are looking for, is simply

$$P_0: \qquad \min \|\boldsymbol{\alpha}\|_0^0 \qquad \text{subject to} \qquad \|\mathbf{y} - \boldsymbol{\Phi\alpha}\|_2^2 \leq \epsilon^2 \qquad (1)$$

Note that the above-stated problem rarely has a unique solution [11], since once the sparsest solution is found, many feasible variants of it sharing the same support can be built. Since the above-stated problem is highly nonconvex and hard to deal with, many researchers pursue a strategy of convexification with

---

- Initialization: let $\boldsymbol{\alpha} = \lambda(\mathbf{I} + \boldsymbol{\Phi}^T\boldsymbol{\Phi})^{-1}\boldsymbol{\Phi}^T\mathbf{y}$
  (This is equivalent to the solution when the σ tends to be infinity)
  i.e.:

  $$\text{argmin}_{\boldsymbol{\alpha}} \qquad \|\boldsymbol{\alpha}\|_2^2 + \lambda\|\mathbf{y} - \boldsymbol{\Phi}\boldsymbol{\alpha}\|_2^2$$

- Choose a suitable decreasing sequence for σ, $[\sigma_1 \ldots \sigma_J]$.
- for $n = 1, \ldots, J$:
  1. Let $\sigma = \sigma_n$.
  2. find $\boldsymbol{\alpha}_\sigma^{opt} = \text{argmin}_{\boldsymbol{\alpha}} \qquad (m - F_\sigma(\boldsymbol{\alpha})) + \lambda\|\mathbf{y} - \boldsymbol{\Phi}\boldsymbol{\alpha}\|_2^2$
     using any kind of optimization tool ,
     say steepest decent with fixed number of iterations
- Final answer is $\boldsymbol{\alpha} = \boldsymbol{\alpha}^{opt}$.

Fig. 1. Algorithm for finding the sparse coefficients in presence of noise

replacing ℓ^0 norm with ℓ^1- norm. so simply try to solve the following problem instead:

$$P_1: \qquad \min\|\boldsymbol{\alpha}\|_1 \qquad \text{subject to} \qquad \|\mathbf{y} - \boldsymbol{\Phi}\boldsymbol{\alpha}\|_2^2 \leq \epsilon^2 \qquad (2)$$

where $\|\boldsymbol{\alpha}\|_1 = \sum \alpha_i$ is the ℓ^1-norm of $\boldsymbol{\alpha}$. Note that the replacing ℓ^0-norm by other convex cost functions such as ℓ^1-norm is only asymptotic and the equivalence does not always hold [9]. Hereafter, motivated by the recently stated work of Mohimani, *et al.* [7] we seek to find the sparsest possible answer without such a replacement and instead, attempt to relax the replacing ℓ^0- norm by a continuous, differentiable cost function, say $F_\sigma(\boldsymbol{\alpha}) = \sum_i \exp(-\alpha_i^2/2\sigma^2)$.

This function tends to count the number of zero elements of a vector. So, as stated in [7] the above problem can be converted to:

$$P_0: \qquad \min_{\boldsymbol{\alpha}}(m - F_\sigma(\boldsymbol{\alpha})) \qquad \text{subject to} \qquad \|\mathbf{y} - \boldsymbol{\Phi}\boldsymbol{\alpha}\|_2^2 \leq \epsilon^2 \qquad (3)$$

The above optimization task can be converted to optimizing the Lagrangian:

$$P_0: \qquad \min_{\boldsymbol{\alpha}}(m - F_\sigma(\boldsymbol{\alpha})) + \lambda\|\mathbf{y} - \boldsymbol{\Phi}\boldsymbol{\alpha}\|_2^2 \qquad (4)$$

So that the constraint becomes a penalty and the parameter λ is dependent on ϵ. Solution toward this problem was recently proposed in [12] and it is shown that for a proper choice of λ, these two problems are equivalent. The σ parameter determines the smoothness of the approximated cost function. By gradual decrease in this parameter it is highly probable to skip trapping in local minimum. The overall algorithm which is used through this paper is shown in Fig. 1 is a slight modification of the same idea presented in [12].

Once the sparsest solution of (3) has been found with the stated algorithm summarized in Fig. 1, we can retrieve the approximate image by $\hat{\mathbf{x}} = \boldsymbol{\Phi}\hat{\boldsymbol{\alpha}}$.

3 Image Denoising

The problem of estimation of X from an observed noisy version of it under the sparsity prior has two essential issues: first, to find a dictionary $\boldsymbol{\Phi}$ which permits a sparse representation regarding the fact that the sample are noisy and second to find the coefficients of this sparse representation. The second phase was what explained so far. As it was shown by Aharon [8], *et. al.*, the K-SVD learning is a very efficient strategy which leads to satisfactory results. This method along with all other types of dictionary learning fails to act properly [8] when the size of dictionary grows. Beside that, the computational complexity and thus time needed for training will grow awesome.

When we are dealing with larger size images we are still eager to apply this method but as stated it is computationally costly and both dictionary learning and optimization to find the coefficients of sparse representation are sometimes intractable. To overcome this difficulty, an image with size $\sqrt{N} \times \sqrt{N}$ is divided to blocks of size of $\sqrt{n} \times \sqrt{n}$. These blocks are chosen highly overlapped for two reasons: first, to avoid blockiness and second to have better estimate in noise removal process. Then a dictionary is tried to be found over these blocks and all these blocks are cleaned with algorithm of Fig. 1. Let $\mathbf{L}_{ij}$ be a matrix representing each block to be located in (ij)-th position of the image. $\mathbf{L}_{ij}$ is a matrix of size $n \times N$ which provides the location information of all possible blocks of the images. So in this respect, the noise removal process changes to:

$$\{\hat{\mathbf{X}}, \hat{\boldsymbol{\alpha}}\} = \operatorname{argmin}_{\mathbf{X},\boldsymbol{\alpha}} \lambda \|\mathbf{Y} - \mathbf{X}\|_2^2 + \sum_{ij} \gamma \|\boldsymbol{\alpha}_{ij}\|_0^0 + \sum_{ij} \|\boldsymbol{\Phi}\boldsymbol{\alpha} - \mathbf{L}_{ij}\mathbf{X}\|_2^2 \quad (5)$$

in which $\mathbf{X}$ is the original image to be estimated and the $\mathbf{Y}$ is the observed available noisy version of it. This equation is similar to (1) with this slight difference that local analysis was taken into account and a linear combination of ℓ^0-norm and ℓ^2-norm of all sparse representation and error between the original signal and the reconstructed one tried to be minimized. In this process, visible artifacts may occur due to blocking phenomenon. To avoid this, we choose the blocks with overlap and at the end average the results in order to prevent blockiness artifact. After determining all the approximated coefficients, we estimate the original image by solving the following equation:

$$\hat{\mathbf{X}} = \operatorname{argmin}_{\mathbf{X}} \lambda \|\mathbf{Y} - \mathbf{X}\|_2^2 + \sum_{ij} \|\boldsymbol{\Phi}\boldsymbol{\alpha} - \mathbf{L}_{ij}\mathbf{X}\|_2^2 \quad (6)$$

This quadratic equation has the solution:

$$\hat{\mathbf{X}} = (\lambda \mathbf{I} + \sum_{ij} \mathbf{L}_{ij}^T \mathbf{L}_{ij})^{-1} (\lambda \mathbf{Y} + \sum_{ij} \mathbf{L}_{ij}^T \boldsymbol{\Phi}\hat{\boldsymbol{\alpha}})^{-1} \quad (7)$$

This estimated modified image can be interpreted as a relaxed averaging between the noisy observed image with the cleaned estimated one. The summarized overall algorithm is shown is Fig. 2.

- Goal: denoise a given image $\mathbf{Y}$ from additive white Gaussian noise with variance of $\|\mathbf{n}\|_2^2$
- parameters:

  n block-size ,k dictionary, λ Lagrangian multiplier.the task is to optimize

  $$\{\hat{\mathbf{X}}, \hat{\boldsymbol{\alpha}}\} = \operatorname{argmin}_{\mathbf{X}, \alpha} \lambda \|\mathbf{Y} - \mathbf{X}\|_2^2 + \sum_{ij} \gamma \|\alpha_{ij}\|_0^0 + \sum_{ij} \|\boldsymbol{\Phi}\alpha - \mathbf{L}_{ij}\mathbf{X}\|_2^2$$

- train a dictionary $\boldsymbol{\Phi}$ of size $n \times k$ using K-SVD.

- find the sparse noisy coefficients of $\boldsymbol{\alpha}$ using algorithm stated in Fig. 1.

- Final estimation is $\hat{\mathbf{X}} = (\lambda \mathbf{I} + \sum_{ij} \mathbf{L}_{ij}^T \mathbf{L}_{ij})^{-1} (\lambda \mathbf{Y} + \sum_{ij} \mathbf{L}_{ij}^T \boldsymbol{\Phi}\hat{\boldsymbol{\alpha}})^{-1}$.

Fig. 2. The final denoising algorithm

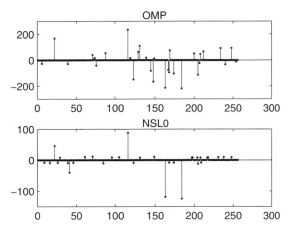

Fig. 3. Coefficients of a sample block represented with OMP above and in bottom. The latter, leads to the same result or sparsely superior one.

4   Experimental Results

In this work, the underlying dictionary was trained with the K-SVD method and once the learning is done, the image blocks was represented sparsely via Fig. 1. The algorithm of section 2 was used for such a representation. The overall denoising method explained above was examined with numerous test images mainly of size 256×256 and 512×512 with different noise levels. Blocks of size 8×8 was driven by the synthesis noisy image and a dictionary of size 64×256 was learned through this blocks using K-SVD method. Then we applied the algorithm of Fig. 1 to represent each block on the provided dictionary, while the similar approach done by Aharon [8] make use of Orthogonal Matching Pursuit (OMP) [10] for this stage. The tested images are all the same ones as those used in the denoising experiments reported in [8], in order to enable a fair comparison. Table 1 summarizes the denoising results in the same database of

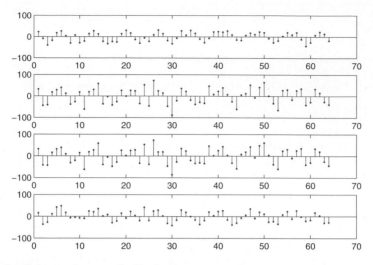

Fig. 4. Coefficients of a sample block. From top to bottom: the original clean signal, the signal corrupted with additive white Gaussian noise of $\|\mathbf{n}\|_2 = 20$, recovered block via OMP and the recovered block with the algorithm of Fig. 1.

Fig. 5. From left to right: original image, noisy image with zero-mean white gaussian noise of $\|\mathbf{n}\|_2 = 20$, the cleaned image via sparse representation described

images. In a quite large experiments we found sparser solution and better quality of representations. Every result reported is an average over 5 experiments, having different realizations of the noise. To show a comparison in sparsity yielded with different methods coefficients in representations of a sample block with OMP and the stated algorithm was depicted in Fig. 3. The quite same results is valid for other blocks as well.

The denoised blocks were averaged, as described in Fig.2 .In Fig. 5 the results of the overall algorithm for the image "Barbara" for $\|\mathbf{n}\|_2 = 20$ is shown. By refereing to Table 1, as it is seen, when the level of noise grows, our approach outperforms K-SVD with OMP and we can conclude the mentioned algorithm is suitably designed for noisy cased with known energy.

Also a comparison was done with other types of sparse coding phase such as FOCUSS and SL0 [8] and yet the proposed algorithm outperforms them. A

Table 1. Summary of denoising PSNR results. In each column the bottom is corresponding to our approach and the above is corresponding to the K-SVD with OMP. the bold one corresponds with better response.

$\frac{\sigma}{PSNR}$	LENA	BARBARA	BOAT	Fgrpt	House	Peppers	Average	σ_{PSNR}
2/42.11	**43.58**	**43.67**	**43.14**	**42.99**	**44.47**	**43.33**	**44.47**	**43.33**
	42.11	42.38	42.17	41.85	42.92	42.51	42.92	42.51
5/34.15	**38.60**	**38.08**	**37.22**	**36.65**	**39.37**	**37.78**	**39.37**	**37.78**
	38.18	37.41	36.68	36.17	38.25	37.08	38.25	37.08
10/28.13	**35.47**	34.42	**33.64**	**32.39**	**35.98**	34.28	35.98	**34.28**
	35.42	**34.51**	33.62	32.31	35.60	**34.53**	35.60	**34.53**
15/24.61	33.70	32.36	31.73	30.06	34.32	32.22	34.32	32.22
	33.91	**32.79**	**32.13**	**30.258**	**34.40**	**32.79**	**34.40**	**32.79**
20/22.11	32.38	30.83	30.36	28.47	33.20	30.82	33.20	30.82
	33.46	**32.01**	**31.29**	**29.16**	**34.19**	**31.58**	**34.19**	**31.58**
25/20.17	31.32	29.60	29.28	27.26	32.15	29.73	32.15	29.73
	32.72	**31.01**	**30.46**	**28.90**	**33.61**	**30.83**	**33.61**	**30.83**
50/14.15	27.79	25.47	25.95	23.24	27.95	26.13	27.95	26.13
	28.98	**26.93**	**27.30**	**24.43**	**28.69**	**27.70**	**28.69**	**27.70**
75/10.63	25.80	23.01	23.98	19.97	25.22	23.69	25.22	23.69
	26.93	**24.71**	**25.33**	**21.69**	**26.83**	**24.28**	**26.83**	**24.28**
100/8.13	24.46	21.89	22.81	18.30	23.71	21.75	23.71	21.75
	26.32	**23.55**	**24.36**	**22.19**	**25.08**	**23.14**	**25.08**	**23.14**

sample comparison has been done in Fig. 4. In this experiment after providing the dictionary, the sparse representation coefficients are found with different approaches. The coefficients of the original clean signal, the signal corrupted with additive white gaussian noise of $v\|\mathbf{n}\|_2 = 20$, recovered block via OMP and the recovered signal via Fig. 1 is depicted in Fig. 4 and as it can be seen the our recovered signal resembles more to the original signal.

5   Discussions and Conclusions

In this paper a simple algorithm for denoising application of an image was presented leading to state-of-the-art performance, equivalent to and sometimes surpassing recently published leading alternatives. It is basically on the basis of sparse representation of an image in the presence of noise. The stated algorithm considers local approach, splits the noisy observed image to several blocks and learns a dictionary over these blocks and attempts to find the best possible sparse representation of each block with this dictionary. In order to find the cleaned image some averaging is needed to avoid the blocking effect in boundaries. Experimental results show satisfactory recovering of the image. Future theoretical work on the general behavior of this algorithm is on our further research agenda.

References

1. Gilboa, G., Sochen, N., Zeevi, Y.: Forward-and-Backward Diffusion Processes for Adaptive Image Enhancement and Denoising. IEEE Trans. on Image Processing 11(7), 689–703 (2002)
2. Mihcak, M.K., Kozintsev, I., Ramchandran, K., Moulin, P.: Low complexity image denoising based on statistical modeling of wavelet coefficients. IEEE Signal Processing Letters 6(12), 300–303 (1999)
3. Grace Chang, S., Yu, B., Vetterli, M.: Adaptive wavelet thresholding for image denoising and compression. IEEE Trans. on Image Processing 9(9), 1532–1546 (2000)
4. Starck, J., Candes, E., Donoho, D.L.: The curvelet transform for image denoising. IEEE Trans on Image Processing 11(6), 670–684 (2002)
5. Donoho, D.L.: De-noising by soft thresholding. IEEE Trans. Inf. Theory 41(3), 613–627 (1995)
6. Coifman, R., Donoho, D.L.: Translation invariant de-noising. In: Wavelets and Statistics. Lecture Notes in Statistics, pp. 125–150. Springer, New York (1995)
7. Mohimani, H., Babaei-Zadeh, M., Jutten, C.: A Fast approach for overcomplete sparse decomposition based on smoothed L0 norm. IEEE Trans on Signal Processing (to appear)
8. Aharon, M., Elad, M., Bruckstein, A.: K-SVD: An Algorithm for Designing Overcomplete Dictionaries for Sparse Representation. IEEE Trans. on Signal Processing 54(11) (November 2006)
9. Candes, E.J., Wakin, M.B., Boyd, S.: Enhancing Sparsity by Reweighted l1 Minimization. Journal of Fourier Analysis and Applications, special issue on sparsity (October 2008)
10. Davis, G., Mallat, S., Zhang, Z.: Adaptive time-frequency approximations with matching pursuits, Tech. Rep. TR1994-657 (1994)
11. Donoho, D.L., Elad, M., Temlyakov, V.N.: Stable recovery of sparse overcomplete representations in the presence of noise. IEEE Transactions on Information Theory 52(1), 6–18 (2006)
12. Firouzi, H., Farivar, M., Babaie-Zadeh, M., Jutten, C.: Approximate Sparse decomposition Based on Smoothed L-0 norm. In: ICASSP 2009 (2009), http://www.arxiv.org

Target Speech Enhancement in Presence of Jammer and Diffuse Background Noise

Jani Even, Hiroshi Saruwatari, and Kiyohiro Shikano

Graduate School of Information Science, Nara Institute of Science and Technology
8916-5, Takayama-cho, Ikoma, Nara, Japan
{even,sawatari,shikano}@is.naist.jp
http://spalab.naist.jp/

Abstract. This paper presents a method for enhancing a target speech in the presence of a jammer and background diffuse noise. The method is based on frequency domain blind signal separation (FD-BSS). In particular, the permutation resolution is done using both the direction of arrival (DOA) information contained in the estimated filters and some statistical features computed on the estimated signals. This enables the separation of the target speech, the jammer and the diffuse background noise which is not possible if using only the DOA or the statistical features. Since in presence of diffuse noise, FD-BSS cannot provide a good estimate of the target speech a channel wise modified Wiener filter is proposed as post processing to further enhance the target speech.

Keywords: Acoustical blind signal separation, permutation resolution, nonlinear post filter.

1 Introduction

The cocktail party problem is one of the well known applications of blind signal separation (BSS), the goal is to separate the speeches of several persons talking at the same time by means of multiple microphones (also referred to as BSS for convolutive mixture, see review paper [1]). In particular, the frequency domain approach (FD-BSS) is of great practical interest because of its lower computational cost [2]. However, a specific problem of FD-BSS is the so called *permutation indeterminacy* that requires the addition of a permutation resolution method to achieve the separation.

Since most of the research has been focused on the cocktail party problem another application of FD-BSS has been overlooked: the extraction of a close target speech signal from a diffuse background noise. This situation is of great interest since it describes the simple human/machine hands-free speech interface. The user interacting with the machine, target speech, is close to the microphone array whereas other sources are at a larger distance and create the diffuse background noise. The permutation methods developed for the cocktail party problem, that exploit the direction of arrival (DOA) of the speeches [3], are not a reliable solution in the presence of diffuse background noise. Thus an approach exploiting

T. Adali et al. (Eds.): ICA 2009, LNCS 5441, pp. 565–572, 2009.

the statistical discrepancy between the target speech and the diffuse background noise was proposed to overcome this problem [4]. This method performs well as long as the user is the only point source close to the microphone array. If another speech source, referred to as *jammer* in the remainder, is close enough to the microphone array to be considered as a point source the performance degrades dramatically.

In this paper we propose a permutation resolution method combining the statistical information used in [4] with the DOA information. This approach solves the problem of extracting the target speech in the presence of both diffuse background noise and jammer. Another issue is that in presence of diffuse background noise, FD-BSS gives a better estimate of the diffuse background noise than of the point sources [5]. Consequently the target speech is usually further enhanced by some nonlinear post processing exploiting the noise estimate given by FD-BSS (for example spectral subtraction [5] or Wiener filtering [6]). Similarly, in our case, the estimates of the target speech and the jammer are still highly contaminated by the diffuse background noise whereas the diffuse background noise is accurately estimated. Consequently, we apply a modified Wiener filter that exploits the target speech and diffuse background noise estimates in order to further enhance the target speech after the FD-BSS.

Some experimental results show that the proposed method is able to suppress efficiently the diffuse background noise and the jammer with moderate distortion of the target speech estimate.

2   Preliminaries: FD-BSS for Cocktail Party

In the cocktail party problem, the goal of FD-BSS is to recover some speeches when only convolutive mixtures of these speeches are observed. Performing the separation in the frequency domain replaces the time domain convolutive mixture by several simpler instantaneous mixtures in the frequency domain. The frequency domain model of the mixture is obtain by applying a short time Fourier transform (STFT with a F points analysis frame) to the received signals. The observed signal at the fth frequency bin is $X(f,t) = A(f)S(f,t)$, where the $n \times n$ matrix $A(f)$ represents the instantaneous mixture and $S(f,t) = [s_1(f,t), \ldots, s_n(f,t)]^T$ is the emitted signal at the fth frequency bin (t denotes the frame index and f the frequency bin).

In the fth frequency bin, the estimates $Y(f,t) = [y_1(f,t), \ldots, y_n(f,t)]^T$ are obtained by applying an unmixing matrices $B(f)$ to the observed signals

$$Y(f,t) = B(f)X(f,t) = B(f)A(f)S(f,t).$$

If the components of $S(f,t)$ are statistically independent it is possible to recover them up to scale and permutation indeterminacy by finding the unmixing matrix $B(f)$ that gives an estimate $Y(f,t)$ with statistically independent components. For such unmixing matrix we have $Y(f,t) = P(f)\Lambda(f)S(f,t)$ where $P(f)$ is a $n \times n$ permutation matrix and $\Lambda(f)$ is a diagonal $n \times n$ matrix ([7]) .

Because of the permutation indeterminacy $P(f)$, before transforming back the signals to the time domain, it is necessary to match the components belonging to the same signal across all the frequency bins. This is done by applying a *permutation resolution method*. For the cocktail party problem, speech signal can be approximately considered as a point source and FD-BSS is equivalent to a set of adaptive null beamformers each having its null toward different speakers [8]. Then it is possible to resolve the permutation by determining the position of these nulls using the directivity pattern of the matrices $B(f)$ [8]. A fast method exploiting the DOA information that does not requires the estimation of the directivity patterns was proposed in [3].

After resolving the permutation, the estimated speeches are still filtered by an indeterminate filter because of the scaling indeterminacy $\Lambda(f)$. A solution is to *project back* the estimated speeches to the microphone array [9]. The projection back of the ith estimate is a n component signal defined by

$$Z(f,t) = B(f)^{-1}D_i Y(f,t)$$

where D_i is a matrix having only one non null entry $d_{ii} = 1$. If we assume perfect separation $B(f)A(f) = P(f)\Lambda(f)$ and the estimated speech is $s_j(f,t)$ then $P(f)$ is such that $P(f)^{-1}D_i P(f) = D_1$ and $Z(f,t) = A(f,t)^{(:,j)}s_j(f,t)$ where $A(f,t)^{(:,j)}$ is the jth column of $A(f,t)$. Namely Z(f,t) is equal to the contribution of the jth speech at the microphone array.

3   Proposed Method

3.1   Problem Formulation

The simple model of the human/robot hands-free speech interface is very different from the cocktail party model. The user is assumed to be close to the microphone array and thus is modeled as a point source. But the other sources are seen as a diffuse background noise as they are far from the microphone array and the environment is reverberant. In particular the diffuse background noise has no clear DOA preventing the use of a permutation resolution method based solely on DOA.

In [4], we proposed a permutation resolution method that exploits the statistical discrepancies between the diffuse background noise and the target speech. This method works very well for the simple human/robot hands-free speech interface. But for a more complicated model where the target speech and several other speech sources, the jammers, stand out of the diffuse background noise the performance are very poor. To treat this problem, we propose to improve the method in [4] by exploiting the DOA information of some of the separated components.

3.2   Permutation Resolution

In the time domain, the distribution of the speech signal amplitude is often modeled by a Laplacian distribution because the speech is a non stationary signal having activity and non activity parts (silence). On the contrary, the diffuse background noise is composed of the superposition of many sounds consequently its amplitude has a distribution that is close to the Gaussian distribution. After the STFT, in each of the frequency bins, we can also observe that the modulus of the speech signal has a spikier distribution than that of the diffuse background noise. This statistical discrepancy between speech and diffuse background noise is the key of the permutation resolution method presented in [4]. In each frequency bins, after convergence of the separation matrices $B(f)$, the permutation resolution is performed using statistical features computed on the components of $Y(f,t)$ (several methods are presented in [4]).

In this paper, we use the scale parameter $\alpha_i(f)$ of the Laplacian distribution that fits the distribution of the modulus. The maximum likelihood estimate of this parameter is $\alpha_i(f) = (\mathcal{E}\{|y_i(f,t)|\})^{-1}$. In [4], the component with the largest parameter was selected as the target speech. In presence of several jammers, we have several speeches thus we have several estimated components $y_i(f,t)$ with large $\alpha_i(f)$ (see Fig. 1(c) obtained for a target in front of the array and one jammer; each symbol $\circ$, $\times$, $\square$ or $\triangledown$ corresponds to a component). We first estimate the number of speeches standing out of the diffuse background noise by determining how many components have larger $\alpha_i(f)$. Here we simply computed the means of the $\alpha_i(f)$ for the middle range frequency where the contrast between speech and diffuse noise is the highest and applied a simple thresholding (see Fig. 1(d) $\square$ and $\triangledown$).

After selecting the number m of speech components, we estimate the DOA of the m components with higher $\alpha_i(f)$, $i.e.$ we use the spatial information contained in the separation matrices $B(f)$. This reduces the DOA information of $B(f)$ (in Fig. 1(b)) to that of the selected speech components only (in Fig. 1(e)). Clustering the DOAs in Fig. 1(b) results in much permutation errors whereas we are able to cluster easily the ones in Fig. 1(e) (see the clustered DOAs in Fig. 1(f)).

Finally in the plane $\{\alpha_i(f), \theta_i(f)\}$, Fig. 1(d), we can see the clustering of the noise ($\square$ and $\triangledown$), the jammer ($\circ$) and the target ($\times$). The noise components have low $\alpha_i(f)$ and broad range for $\theta_i(f)$ whereas the speech and jammer components have narrow range $\theta_i(f)$ and large $\alpha_i(f)$. The discrimination between the jammers and target requires some additional knowledge (for example the target is assumed to be the closest to the front direction). *Note: Using direct clustering of the plane $\{\alpha_i(f), \theta_i(f)\}$ in Fig. 1(a) is another possible approach.*

3.3   Channel-Wise Wiener Post Filter

After resolving the permutation and projecting back the estimated signals to the microphone array we have (dropping frequency and frame indexes)

$$\widehat{X}_S \approx X_S + X_N, \quad \widehat{X}_J \approx X_J + X_N, \quad \widehat{X}_N \approx X_N,$$

where $X_S(f,t)$, $X_J(f,t)$ and $X_N(f,t)$ are the true values at the microphone array of the target, the jammer and the noise. The reason is that FD-BSS can only efficiently cancel $n-1$ point sources when using a n microphone array [8] (the proposed method can theoretically treats up to $n-2$ jammers).

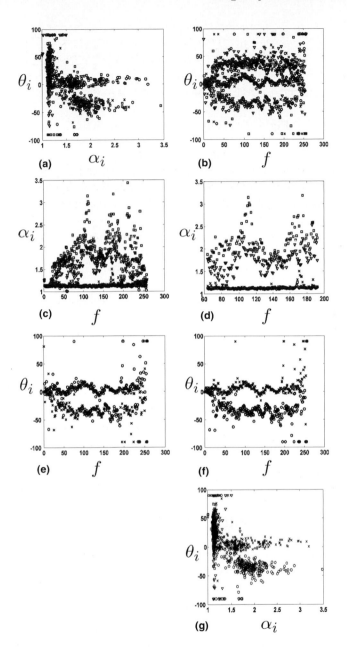

Fig. 1. Statistical index $\alpha_i(f)$ versus DOA $\theta_i(f)$ before clustering (a), all DOA $\theta_i(f)$ versus frequency (b), statistical index $\alpha_i(f)$ versus frequency (c), statistical index $\alpha_i(f)$ versus middle frequency range after speech selection (d), DOA of speeches versus frequency before clustering (e), DOA of speeches versus frequency after clustering (f) and statistical index $\alpha_i(f)$ versus DOA $\theta_i(f)$ after clustering (g)

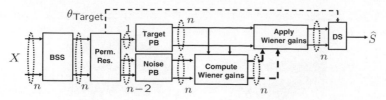

Fig. 2. Overview of the proposed method

The jammer estimate is discarded whereas the speech and noise estimates are used to compute the gain of a Wiener filter for further processing the target speech (see Fig. 2). The modified Wiener filtering is applied on each of the components of the projection back of the estimated target. The Wiener gain for the ith component is

$$G^{(i)}(f,t) = \frac{|\widehat{X}_S^{(i)}(f,t)|^2 \circledast K(f,t)}{|\widehat{X}_S^{(i)}(f,t)|^2 \circledast K(f,t) + \gamma|\widehat{X}_N^{(i)}(f,t)|^2 \circledast K(f,t)}$$

where the subscript (i) denotes the ith component, $\circledast$ denotes the 2D convolution, $K(f,t)$ is a small kernel (spectral smoothing) and γ is a parameter controlling the noise reduction. The ith component of the filtered target speech is

$$\widehat{S}^{(i)}(f,t) = \sqrt{G^{(i)}(f,t)|\widehat{X}_S^{(i)}(f,t)|^2} \frac{\widehat{X}_S^{(i)}(f,t)}{|\widehat{X}_S^{(i)}(f,t)|}.$$

finally the speech estimate is obtained by applying a delay and sum (DS) beamformer in the direction θ of the target speech (the target DOA is estimated during the permutation resolution step but requires some *a priori* knowledge to discriminate between jammer and target) $\widehat{S}(f,t) = \sum_{i=1}^{n} G_{\mathrm{DS}\theta}^{(i)}(f,t)\widehat{S}^{(i)}(f,t)$ where $G_{\mathrm{DS}\theta}^{(i)}(f,t)$ the gain of the DS beamformer at the ith microphone.

4   Experimental Results

We used a four microphone array with inter-microphone spacing of 2.15cm. The target speaker is in front at one meter of the microphone array whereas the jammer is at two meters with an angle of $-40°$. The diffuse noise source is created by a vacuum cleaner at two meters from the array at an angle of $40°$. The room reverberation time is T60 = 200ms. The SNR of the target speech to the jammer is 0dB. We performed simulations with different level of diffuse noise 5dB, 10dB and 15dB SNR. For each case we used 100 different utterances for a female target speaker and five different utterances for a male jammer (the signals are from a database of Japanese utterances at 16kHz). All results are averaged values computed on these 500 trials.

The short time Fourier transform uses a 512 point hanning window with 50% overlap and pre-emphasis (a first order high pass filter $z_p = 0.97$). Speech separation is performed by 600 iterations of the FD-BSS method with adaptation step of 0.3 divided by two every 200 iterations. The nonlinear functions in the update rule are estimated from the data (adapted from [10]). The spectral smoothing uses a Gaussian kernel $K(f, t)$ of size 9×9.

First we checked that the proposed permutation resolution method solves the problem encountered in [4] by comparing the quality of the target and noise estimates. The estimation quality is measured in term of noise reduction rate (NRR) defined as the difference of the SNR before and after processing, *i.e.* a positive value indicates improvement [8]. In Fig. 3, we can see that the speech estimate obtained with the proposed method (LAP-DOA) is better than that of the method in [4] (LAP) and also outperforms the delay and sum beamformer (DS). DS refers to a simple delay and sum beamformer in the target direction (not to the proposed post processing). The quality of the noise estimation remains the same.

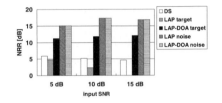

Fig. 3. NRR for target and noise estimation with proposed permutation solver (LAP-DOA) and permutation solver in [4] (LAP) and target estimate with DS beamformer

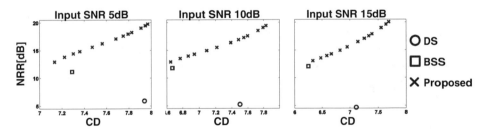

Fig. 4. Noise rate reduction (NRR) versus Cepstral distance (CD) for DS beamformer (○), BSS (□) and the proposed method (×)

Then to see the improvement of the post-processing, we both considered the NRR and the cepstral distance (CD) between the estimated signal and the contribution of the target speech at the microphone array. The CD measures the amount of nonlinear distortion introduced by the post processing stage. Figure 4 shows the NRR versus the CD at different SNRs (input SNR: between speeches and noise). The simple delay and sum beamformer (*circ*), the BSS target speech estimate (□) and the proposed method (×) with parameter $\gamma \in \{0.1, 0.5, 1, 1.5, 3, 5, 10, 15, 20, 25, 50, 75, 100\}$ are compared. For the proposed method NRR and CD are increasing with γ. For a CD of the same order as the DS, the proposed method has an NRR improvement over BSS of 8.34dB, 5.1dB and 4.4dB at 5dB, 10dB and 15dB of SNR respectively ($\gamma = 75$, $\gamma = 20$ and $\gamma = 10$).

5   Conclusions

In this paper we proposed a permutation resolution method efficient for the human/machine hands-free interface as it can deal with diffuse background noise and jammer in a blind manner. Moreover we show how to further enhance the target speech by applying a modified Wiener filter that exploits the good estimation of the diffuse background noise.

References

1. Pedersen, M., Larsen, J., Kjems, U., Parra, L.: A Survey of Convolutive Blind Source Separation Methods. Springer, Heidelberg (2007)
2. Takahashi, Y., Saruwatari, H., Shikano, K.: Real-time implementation of blind spatial subtraction array for hands-free robot spoken dialogue system. In: 2008 IEEE/RSJ International Conference on Intelligent Robots and Systems, Nice, France, pp. 1687–1692 (2008)
3. Sawada, H., Mukai, R., Araki, S., Makino, S.: A robust and precise method for solving the permutation problem of frequency-domain blind source separation. IEEE Trans. Speech and Audio Processing 12, 530–538 (2004)
4. Even, J., Saruwatari, H., Shikano, K.: An improved permutation solver for blind signal separation based front-ends in robot audition. In: 2008 IEEE/RSJ International Conference on Intelligent Robots and Systems, Nice, France, pp. 2172–2177 (2008)
5. Takahashi, Y., Takatani, T., Saruwatari, H., Shikano, K.: Blind spatial subtraction array with independent component analysys for hands-free speech recognition. In: International Work Shop on Acoustic Echo and Noise Control (IWAENC) (CD-ROM) (2006)
6. Takahashi, Y., Osako, K., Saruwatari, H., Shikano, K.: Blind source extraction for hands-free speech recognition based on wiener filtering and ica-based noise estimation. In: Joint Workshop on Hands-free Speech Communication and Microphone Arrays (HSMCA), pp. 164–167 (2008)
7. Comon, P.: Independent component analysis, a new concept? Signal Processing 36, 287–314 (1994)
8. Saruwatari, H., Kurita, S., Takeda, K., Itakura, F., Nishikawa, T., Shikano, K.: Blind source separation combining independent component analysis and beamforming. EURASIP Journal on Applied Signal Processing 2003(11), 1135–1146 (2003)
9. Murata, N., Ikeda, S., Zieh, A.: An approach to blind source separation based on temporal structure of speech signals. Neurocomputing 41(1-4), 1–24 (2001)
10. Vlassis, N., Motomura, Y.: Efficient source adaptivity in independent component analysis. IEEE Trans. Neural Networks 12(3), 559–566 (2001)

Constructing Time-Frequency Dictionaries for Source Separation via Time-Frequency Masking and Source Localisation

Ruairí de Fréin[1], Scott T. Rickard[1], and Barak A. Pearlmutter[2]

[1] Complex & Adaptive Systems Laboratory, University College Dublin, Ireland
[2] Hamilton Institute, National University of Ireland Maynooth, Co. Kildare, Ireland
rdefrein@ee.ucd.ie, scott.rickard@ucd.ie, barak@cs.nuim.ie

Abstract. We describe a new localisation and source separation algorithm which is based upon the accurate construction of time-frequency spatial signatures. We present a technique for constructing time-frequency spatial signatures with the required accuracy. This algorithm for multi-channel source separation and localisation allows arbitrary placement of microphones yet achieves good performance. We demonstrate the efficacy of the technique using source location estimates and compare estimated time-frequency masks with the ideal 0 dB mask.

1 Introduction

Speech is sparse in the time-frequency (T-F) domain, a property which has been exploited for Blind Source Separation (BSS), [4]. Related assumptions, namely the log-max [5] or Windowed Disjoint Orthogonality (WDO) assumption [8] in various transform domains are exploited for decompositions of financial data [6] and images [9]. Localisation can be performed in the time or frequency domain when the technique relies on a sparse representation in a dictionary of pre-computed transfer functions [1,2]. We discuss the challenges involved in constructing a T-F dictionary of spatial signatures for source localisation in T-F in Section 3. A typical office contains recording devices, such as mobile phones, MP3 players, PDAs, hearing aids, and computers all equipped with (largely unused) microphones. Consider a dedicated teleconferencing room, with an arbitrary number of inexpensive microphones. The source location, detected using this sensor array, is used to automatically identify the speaker or indicate the position of the speaker in the room. Our goal is to perform localisation and separation using multiple observations from arbitrarily placed sensors. The T-F domain lends itself to this problem as speech typically has increased WDO [8] and sparsity [7] in the T-F domain than in the time domain or frequency domain.

In an anechoic environment, a continuous time source signal $s_j(t)$ is attenuated and delayed as it propagates the direct path to sensor x_i. The attenuation and delay effect on the jth source received at the ith sensor is (a_{ji}, δ_{ji}), consequently $\hat{s}_{ji}(t) = a_{ji}s_j(t - \delta_{ji})$. I mixture signals are observed, $x_i(t)$, at physical locations x_i, where $h_{ji}(t)$ is the continuous time transfer function from source to sensor.

T. Adali et al. (Eds.): ICA 2009, LNCS 5441, pp. 573–580, 2009.
© Springer-Verlag Berlin Heidelberg 2009

The source is constrained to lie at one of P arbitrarily placed grid points. We consider a synthetic scenario where the sensors are placed arbitrarily in a 2m × 2m × 2m room in Section 4 and observe the signals,

$$x_i(t) = \sum_{j=1}^{J} \hat{s}_{ji}(t) = h_{1i}(t) \star s_1(t) + h_{2i}(t) \star s_2(t) + \cdots + h_{Ji}(t) \star s_J(t). \quad (1)$$

2   Fractional Delay of Discrete Signals

A continuous time signal $s(t)$ is denoted by $s[n] = s(nT)$ in the discrete time domain where T is the sampling period and $n = 0, 1, 2, \ldots$. A continuous time signal delayed by $\delta \in \mathbb{R}$ seconds is denoted by $s(t - \delta)$. A discrete signal, $s[n]$, can be delayed by an integer, d, number of samples giving $s[n-d]$ or by rounding down, $\lfloor d \rfloor$, if $d = \delta/T$ is non-integer yielding $s[n - \lfloor d \rfloor]$. A source signal $s(t)$ is delayed by δ seconds propagating to a sensor x_i in an ideal anechoic teleconferencing room. Constraining the source physical locations such that the signals can only be delayed by an integer number of samples in the discrete time domain when propagating to each sensor limits the possible source locations. Alternatively, rounding down (denoted by $\lfloor d \rfloor$) introduces error. Although sources are constrained to lie on a grid, this grid can be refined and a space of interest more densely populated to locate arbitrarily placed sources. Non-integer sample delay,

$$s^\delta[n] = s(nT - \delta), \quad (2)$$

can be computed using sinc interpolation, given that the signal is bandlimited and sampled at a sufficiently high sampling rate,

$$s^\delta[n] = \sum_{n=-\infty}^{\infty} s[n]\mathrm{sinc}(nT - \delta). \quad (3)$$

In practice a finite length approximation of the sinc function leads to error in the estimate of $s^\delta[n]$. A non-integer sample delay of a bandlimited signal sampled above the Nyquist rate can also be determined by multiplying the Discrete Fourier Transform of $s[n]$, $\mathrm{DFT}\{s[n]\} = S[k] = \sum_{n=0}^{N-1} s[n]W^{kn}$ where $k = 0, 1, \ldots, N - 1$ and $W = e^{-j\frac{2\pi}{N}}$, by a linear phase term W^{kd}. This corresponds to a circular shift of the signal by $d = \delta/T \in \mathbb{R}$ samples. We define the zero-padding function $\mathrm{ZP}\,(b, s[n], e)$ which appends b and e zeros to the beginning and end of the signal respectively. The inverse-pad function $\mathrm{IP}\,(b, s[n], e)$ removes b and e samples from the beginning and end of the signal. Zero-padding by $\lceil d \rceil$, where $\lceil \cdot \rceil$ is the ceiling function, taking the DFT, multiplying by the linear phase term, taking the IDFT and inverse-padding gives the desired result. Defining $\mathrm{IDFT}\{S[k]\} = \frac{1}{N}\sum_{k=0}^{N-1} S[k]W^{-kn}$ to be the inverse DFT, the frequency domain method in (Eqn. 4) is the benchmark method we shall use for the remainder of this work.

$$s^\delta[n] = \mathrm{IP}\,\left(\lceil d \rceil, \mathrm{IDFT}\{\mathrm{DFT}\{\mathrm{ZP}\,(\lceil d \rceil, s[n], 0)\}W^{kd}\}, \lfloor d \rfloor\right) \quad (4)$$

The contributions of this paper are the formulation of the T-F spatial signatures problem, the construction of a practical solution via Algorithm 1 and the synchronized STFT, and a demonstration of the efficacy of the technique by implementing a source localisation algorithm in the discrete T-F domain.

3   Time-Frequency Spatial Signatures

$S[k, m]$ is the Short-Time-Fourier-Transform (STFT) of $s[n]$,

$$\text{STFT}\{s[n]\} = S[k, m] = \sum_{n=mR}^{N-1-mR} s[n]w_a[n - mR]W^{k(n-mR)} \quad (5)$$

positioned at sample mR where $w_a[n]$ is the analysis window function and R is the number of window hop-size samples. [k,m] are the discrete frequency and time indices respectively. N is the FFT size. The STFT is inverted using the synthesis window $w_s[n]$ and Over-Lap and Add (OLA) re-synthesis. A discrete source signal $s_j[n]$ is delayed by $d \in \mathbb{R}$ samples (neglecting attenuation effects) as it propagates to sensor x_i yielding $s_j^\delta[n]$.

Problem. *Construct a T-F spatial signature $H^\delta[k, m]$ so that,*

$$S_j^\delta[k, m] = S_j^0[k, m]H^\delta[k, m] \quad \text{for } j = 1, \dots, J \text{ and } \forall |\delta|/T < \Delta. \quad (6)$$

$S_j^\delta[k, m]$ and $S_j^0[k, m]$ are the synchronized STFT (sSTFT) of $s_j^\delta[n]$ and $s_j[n]$ respectively. We define the sSTFT so that the analysis window is centered on the same portion of the signal for the signal delayed by $|\delta|/T < \Delta$ samples, and the windowed segment of $s^\delta[n]$ is a circular shift of the windowed segment of $s[n]$. Δ is a user defined upper-bound on the range of delays under consideration specified by the maximum expected propagation distance.

The sSTFT is an alternative T-F analysis to the typical STFT method in (Eq. 5). It has the property that if a single source is active in T-F bin $[k, m]$, the T-F representation of the source at each x_i is calculated such that $w_a[n]$ is aligned with the received delayed source $s_j^\delta[n] \quad \forall |\delta|/T < \Delta$.

Definition 1. For a delay $|\delta|/T < N/4$ samples, we define the sSTFT of $s_j^\delta[n]$,

$$S_j^\delta[k, m] = \sum_{n-mR}^{N-1-mR} s_j^\delta[n]w_{az}^\delta[n - mR]W^{k(n-mR)}, \quad (7)$$

where the window hop-size is $R = N/4$ samples. We define analysis and synthesis windows which are non-zero for $N/2$ samples and zero-padded by $N/2$ zeros,

$$w_{az}[n] = \text{ZP}\left(N/4, w_a[n], N/4\right), \quad w_{sz}[n] = \text{ZP}\left(N/4, w_s[n], N/4\right) \quad (8)$$

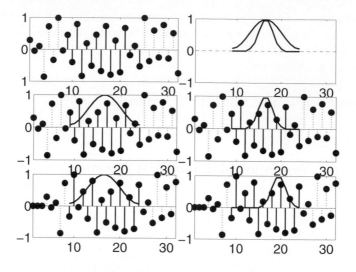

Fig. 1. The STFT (Eq.5) using $w_a[n]$ (which is not zero-padded) does not have the circular shift property in col. 1. The sSTFT (Eq.7) using $w_{az}[n]$ (which is zero-padded) allows the analysis to be synchronized with delays up to $N/4$ samples in col. 2.

to form a pair of windows of length N. Firstly, $w_{az}[n]$ allows the signal $s[n]$ to be shifted by $|\delta|/T < N/4$ samples such that the circular shift property still holds for each local windowed version of $s[n]$. Secondly, $w_{az}[n]$, is aligned with the source signal by delaying it by $d = \delta/T$ samples so that the same samples of $s[n]$ and $s^\delta[n]$ are multiplied by the same samples of $w_{az}^\delta[n]$. We define,

$$w_{az}^\delta[n] = \text{IDFT}\{\text{DFT}\{w_{az}\}[n]W^{kd}\} \text{ and } w_{sz}^\delta[n] = \text{IDFT}\{\text{DFT}\{w_{sz}\}W^{kd}\} \quad (9)$$

Fig. 1 illustrates the difference between the STFT and the sSTFT using a signal, $s[n]$, of length 32 samples in column 1 row 1. We index each subplot using the figure number and then row and column index in parenthesis, e.g. Fig. 1(1,1). Samples 9-24 are analyzed using a 16-point FFT and 16-point Hamming window. The Hamming window is superimposed on the signal in Fig. 1(2,1). Samples 9-24 supported by the window, $w_a[n]$, have linear stalks. Samples 1-8 and 25-32 are denoted by dotted stalks. $s[n]$ is delayed by 3 samples in Fig. 1(3,1). Different portions of the signal $s[n]$ and its delayed counter-part are analyzed in Fig. 1(2,1) and Fig. 1(3,1) using the STFT. Samples common to both windowed signals are scaled differently due to the shifted version of the signal in Fig. 1(2,1) and Fig. 1(3,1) relative to the window. The Fourier transform circular shift property does not hold for the local windowed signal Fig. 1(2,1) and its delayed windowed version Fig. 1(3,1). Analyzing $s[n]$ as specified by (Eq.7) using the zero-padded window $w_{az}^0[n]$ (Eq. 8) with $\delta = 0$ in Fig. 1(2,2) and analyzing $s^\delta[n]$ with $w_{az}^\delta[n]$ (Eq. 9) in Fig. 1(3,2) preserves the Fourier transform circular shift property. $w_{az}^0[n]$ is pre-zero-padded and post-zero-padded by $N/4$ samples. $w_{az}^0[n]$ is non-zero for $N/2$ samples. A circular shift of $w_{az}^0[n]$ by $d < N/4$ samples yielding,

$w^\delta[n]$, times $s^\delta[n]$ results in a windowed signal which is a circular shift of the original windowed signal in Fig.1(2,2). This facilitates the construction of T-F spatial signatures using the Fourier transform circular shift property. Due to the analysis window alignment of the sSTFT, the re-synthesis is not dependent on the analyzed signal $s[n]$. The synthesis window $w_{sz}^\delta[n]$ corresponding to $w_{az}^\delta[n]$ is used to re-synthesize any signal accurately using OLA. Synchronized T-F spatial signatures independent of the signal to be delayed and unbiased by wrap-around and windowing can be estimated using Alg. 1 and sSTFT.

Algorithm 1. Synchronized T-F Spatial Signatures

$s[n]$ delayed by $\delta/T < \frac{N}{4}$ samples, resulting in $s^\delta[n]$, is calculated by element-wise multiplication with the matrix $H_\delta[k,m]$ in discrete T-F given the $s[n]$ is analyzed using (Eq. 7) yielding $S^0[k,m]$ using the analysis window w_{az}^0 in (Eq. 8), e.g. $S^\delta[k,m] = S^0[k,m]H_\delta[k,m]$. Using the sample indices q, g

$$H_\delta[k,m] = \frac{\sum_{q=\frac{mN}{4}}^{Q-1-\frac{mN}{4}} \operatorname{sinc}(qT - \frac{mN}{4} - \delta)w_{az}^\delta[q - \frac{mN}{4}]W^{k(q-\frac{mN}{4})}}{\sum_{g=\frac{mN}{4}}^{G-1-\frac{mN}{4}} \operatorname{sinc}(gT - \frac{mN}{4})w_{az}[g - \frac{mN}{4}]W^{k(g-\frac{mN}{4})}}. \tag{10}$$

We form $\boldsymbol{x}[k,m] = [X_1[k,m], \ldots, X_i[k,m], \ldots, X_I[k,m]]^T \in \mathbb{C}^{I\times1}$ a vector of the observations at each sensor for each T-F point $[k,m]$ using sSTFT. We construct a T-F spatial signatures matrix for each $[k,m]$, $\boldsymbol{D}[k,m] \in \mathbb{C}^{I\times P}$, using Alg. 1 and attenuating each $H_{pi}[k,m]$ using a_{ji}. $\boldsymbol{D}[k,m]$ gives the transfer function, $H_{pi}[k,m]$, for every location, p, in the grid relative to each sensor, x_i, for $[k,m]$.

$$\boldsymbol{D}[k,m] = \begin{matrix} \uparrow \\ \text{Sensor } i \\ \downarrow \end{matrix} \overset{\longleftarrow \text{Location } p \longrightarrow}{\begin{pmatrix} H_{11}[k,m] & \ldots & H_{p1}[k,m] & \ldots & H_{P1}[k,m] \\ \vdots & \ldots & \vdots & \ldots & \vdots \\ H_{1i}[k,m] & \ldots & H_{pi}[k,m] & \ldots & H_{Pi}[k,m] \\ \vdots & \ldots & \vdots & \ldots & \vdots \\ H_{1I}[k,m] & \ldots & H_{pI}[k,m] & \ldots & H_{PI}[k,m] \end{pmatrix}} \tag{11}$$

The J sources $[\mathrm{s}_1, \ldots \mathrm{s}_j \ldots, \mathrm{s}_J]$ are constrained to lie on a subset of the P grid points. We locate a source by estimating the vector $\boldsymbol{c}[k,m] \in \mathbb{C}^{P\times1}$ which explains the sensor observations in the most parsimonious manner given $\boldsymbol{D}[k,m]$.

$$\boldsymbol{x}[k,m] = \boldsymbol{D}[k,m]\boldsymbol{c}[k,m]. \tag{12}$$

4   Source Localisation and Separation Simulations

We solve each subsystem $[k,m]$ in (Eq. 12) independently. This approach lends itself to real-time parallel implementation on dedicated processors. Assuming WDO in T-F, a single source is active in $[k,m]$. The element of the estimated

$c[k, m]$ with the most energy, for example $p = 25$ in Fig. 2(b), indicates the position of the source. For example in Fig. 2(b) a mixed $L_1 + \lambda L_2^2$ objective function prejudiced in favor of the sparsity constraint, L_1-norm yields a solution with significant energy at one grid location $p = 25$ in comparison with a solution with an emphasis on the L_2^2-norm. The weight λ is a data dependent heuristic. Masking the mixture at x_i, $X_i[k, m]$, using $c_p[k, m]$ as an indicator for each $[k, m]$ for each p of interest, separates the source at p. We consider a mixed objective

$$\min_{\mathbf{c}} \underbrace{\frac{1}{2} \sum_{i=1}^{P} |\mathbf{c}_p|}_{E_1} + \lambda \underbrace{||\mathbf{Dc} - \mathbf{x}||_2^2}_{E_2}, \text{ and define } \tilde{\boldsymbol{D}} = \begin{pmatrix} \text{Re}\{\hat{\boldsymbol{D}}\} & -\text{Im}\{\hat{\boldsymbol{D}}\} \\ \text{Im}\{\hat{\boldsymbol{D}}\} & \text{Re}\{\hat{\boldsymbol{D}}\} \end{pmatrix}. \quad (13)$$

We use an iteratively re-weighted least squares approach in the spirit of [3],

$$\min_{\mathbf{c}} \hat{E}(\mathbf{c}) = \hat{E}_1 + \lambda E_2, \ \hat{E}_1 \equiv \frac{1}{2} \sum_{p=1}^{P} \alpha_p^k |\mathbf{c}_p^k|^2, \ \boldsymbol{\alpha} \in \mathbb{R}^{P \times 1} \text{ and } \boldsymbol{\Lambda} = \text{diag}(\boldsymbol{\alpha}). \quad (14)$$

Using a modified objective (Eq. 14) we solve $\frac{\partial \hat{E}(\boldsymbol{c})}{\partial \boldsymbol{c}^k} = \mathbf{0}$ for $\boldsymbol{c}^k$, where k denotes the iteration index and $(\cdot)^H$ denotes the conjugate transpose operation.

$$\left(\boldsymbol{\Lambda} + \lambda \tilde{\boldsymbol{D}}^T \tilde{\boldsymbol{D}}\right) \begin{pmatrix} \text{Re}\{\boldsymbol{c}\} \\ \text{Im}\{\boldsymbol{c}\} \end{pmatrix} = \lambda \tilde{\boldsymbol{D}}^T \begin{pmatrix} \text{Re}\{\boldsymbol{x}\} \\ \text{Im}\{\boldsymbol{x}\} \end{pmatrix} = (\boldsymbol{\Lambda} + \lambda \boldsymbol{D}^H \boldsymbol{D})\boldsymbol{c} = \lambda \boldsymbol{D}^H \boldsymbol{x} \quad (15)$$

Solving (15) iteratively and setting $\alpha_p^{k+1} = 1/|\boldsymbol{c}_p^k|$ yields a fixed point solution.

In the first experiment we show that construction of accurate T-F spatial signatures, $H_{ji}(k, m)$, is crucial for a sparse solution. Wrap-around and windowing effects inherent in unsynchronized analysis using STFT of (Eq. 5) introduce error and occlude the true solution. We estimate $S^\delta[k, m]$ using (Eq. 7) to analyze $s[n]$. We delay $s[n]$ in T-F using Alg. 1 and re-synthesize using OLA with $w_{sz}^\delta[n]$ and hop-size of $N/4$ (method 1). We compare method 1 with a STFT approach (method 2) e.g., $s[n - d] \approx \text{ISTFT}(\text{STFT}(s[n], w_a, N, N/2)W^{kd}, w_s, N, N/2)$. Method 2 performs analysis of $s[n]$ using the STFT (Eq. 5) with a window $w_a[n]$ of N samples and hop-size $N/2$ (as in Fig. 1 column 1). A linear phase term W^{kd} shifts each frame of the signal in T-F. The resulting signal is re-synthesized using the inverse STFT with a synthesis window $w_s[n]$ of length N and overlap $N/2$ using OLA re-synthesis. SNR is defined as $20 * \log(||s_B^\delta[n]||_2)/||s_B^\delta[n] - s_{est}^\delta[n]||_2)$, where $s_{est}^\delta[n]$ is the estimate of the delayed signal and the benchmark, $s_B^\delta[n]$, is computed using (Eq. 4). We analyze speech from the TIMIT database, sampled at 16kHz, with the parameters $N = 2048$ and $R = N/4$ and $R = N/2$ for method 1 and 2 respectively. Fig. 2(a) illustrates the window wrap-around and windowing effects on the delayed source using the sSTFT (method 1) and STFT (method 2). The sSTFT method exhibits sub-sample dips in SNR due to the numerical instability of the truncated delayed sinc function yet degrades gracefully as a function of delay and achieves an estimate > 40dB when delayed by 511 samples. The SNR of method 2 in Fig. 2(a) decreases rapidly as a function of delay in samples due to windowing effects. These inaccurate T-F spatial signatures are unsuitable for localization via sparse representations.

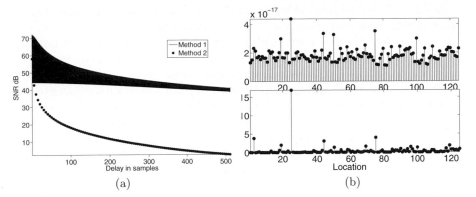

Fig. 2. Fig. 2(a) illustrates the effects of window misalignment and wrap-around using the STFT compared to sSTFT analysis. Row 2 Fig. 2(b) shows that a sparse solution reveals the source location (position 25) compared to a more dense solution (Row 1).

Table 1. Average Localisation and Separation Performance

Source		PSR			WDO			WDO (0dB)			Loc %	
p	3	5	6	3	5	6	3	5	6	3	5	6
s_1 25	0.89	0.60	0.86	0.88	0.60	0.83	0.92	0.87	0.85	25.0	8.0	6.3
s_2 97	0.90	0.69	0.67	0.89	0.68	0.65	0.92	0.82	0.75	47.0	14.0	11.0
s_3 105	0.80	0.58	0.67	0.79	0.57	0.65	0.87	0.79	0.69	17.0	5.0	5.0
s_4 5	−	0.86	0.70	−	0.85	0.68	−	0.79	0.73	−	24.0	3.0
s_5 62	−	0.94	0.90	−	0.92	0.89	−	0.90	0.86	−	43.0	54.0
s_6 112	−	−	0.77	−	−	0.74	−	−	0.78	−	−	18.0

In the second experiment we perform source localisation, using the sSTFT approach, in a synthetic 2m × 2m × 2m room with grid points every 50cm. We tune (Eq.13) over a range of λ. A sparse solution decreases the Euclidean distance between our signal estimate and the original signal as the trade-off between the reconstruction penalty E_2 and sparsity penalty E_1 is adjusted in (Eq.13). Mixtures of 1 to 6 speakers are generated from the TIMIT database by assigning them randomly to grid-points $\{25, 97, 105, 5, 62, 112\}$. We perform multiple experiments with different initial conditions to test the accuracy of the localisation experiment. We use 10 microphones in each experiment. We choose candidate locations—based on the signal power at each location—by analyzing a subset of the T-F points with the optimal analysis window for each grid point p, e.g. w_{az}^p. Localisation and separation is performed using these optimal windows. The mean results for the $3, 5, 6$ speaker cases are tabulated in Table 1. The T-F mask metrics PSR and WDO introduced in [8] are used for comparison with the 0dB ideal mask defined in [8] to gauge separation performance. Our technique achieves performance comparable with the 0dB mask. A significant percentage of the signal energy is located at the correct source positions.

With regard to existing T-F separation algorithms, relying on relative measurements between multiple observations of the mixture could introduce bias if the analysis is unsynchronized using STFT (Eq. 5). The DUET algorithm [8] estimates relative attenuations and delays between two closely spaced sensors using the ratios of the STFT of the observations at sensor x_1 and x_2. A bias has been noted in attenuation and delay estimates in [8] where the delay between 2 sensors ranges from -5 to 5 samples and attenuation from -0.15 to 0.15. It is clear from Fig. 1 that using (Eq. 5) to compute $X_1[k, m]$ and $X_2[k, m]$ and taking the ratio of these two channels in T-F, introduces error as the mixture at x_2 is shifted relative to the window and the mixture $x_1[n]$. The circular shift property is not satisfied. The delay estimate is only approximate $\delta[k, m] \approx -\frac{1}{2\pi k} \angle \frac{X_2[k,m]}{X_1[k,m]}$. The samples of the windowed signal $x_1[n]$ common to the windowed $x_2[n]$ are scaled differently due to window misalignment for analysis and so the relative attenuation $\alpha[k, m] \approx |\frac{X_2[k,m]}{X_1[k,m]}|$ is approximate. To conclude, the synchronized STFT method combined with the algorithm for T-F spatial signature construction facilitates the implementation of a global signal shift as a circular shift with in the support of each frame of the analyzed signal. We have presented a sparse source localisation technique using synthetic experiments to motivate this approach. Incorporating sSTFT into a existing techniques, such as DUET and its extensions will lead to gains in accuracy of the mixing parameter estimation. We will extend this approach to the echoic case in future work.

References

1. Malioutov, D.M., Çetin, M., Willsky, A.S.: A sparse signal reconstruction perspective for source localization with sensor arrays. IEEE Transactions on Signal Processing 53, 3010–3022 (2005)
2. Model, D., Zibulevsky, M.: Signal Reconstruction in Sensor Arrays using Sparse Representations. Signal Processing 86, 624–638 (2006)
3. Gorodnitsky, I.F., Rao, B.D.: Sparse signal reconstruction from limited data using FOCUSS: A re-weighted minimum norm algorithm. IEEE Transactions on Signal Processing 45, 600–616 (1997)
4. Zibulevsky, M., Pearlmutter, B.A.: Blind source separation by sparse decomposition in a signal dictionary. Neural Computation 13, 863–882 (2001)
5. Reddy, A.M., Raj, B.: Soft Mask Methods for Single-Channel Speaker Separation. IEEE Transactions on Audio, Speech, and Language Processing 15, 1766–1776 (2007)
6. de Frein, R., Drakakis, K., Rickard, S.T.: Portfolio diversification using subspace factorizations. In: Proceedings of the Annual Conference on Information Sciences and Systems, pp. 1075 1080 (2008)
7. Rickard, S.T.: Sparse sources are separated sources. In: Proceedings of the 16th Annual European Signal Processing Conference, Florence, Italy (2006)
8. Yilmaz, O., Rickard, S.T.: Blind separation of speech mixtures via time-frequency masking. IEEE Transactions on Signal Processing 52, 1830–1847 (2004)
9. Hoyer, P.O.: Non-negative sparse coding. In: Proceedings of the IEEE Workshop on Neural Networks for Signal Processing, pp. 557 565 (2002)

Time Frequency Masking Strategy for Blind Source Separation of Acoustic Signals Based on Optimally-Modified LOG-Spectral Amplitude Estimator

Eugen Hoffmann, Dorothea Kolossa, and Reinhold Orglmeister

Berlin University of Technology, Electronics and Medical Signalprocessing Group,
Einsteinufer 17, 10587 Berlin, Germany
{Eugen.Hoffmann.1,Reinhold.Orglmeister}@tu-berlin.de,
d.kolossa@ee.tu-berlin.de

Abstract. The problem of *Blind Source Separation* (BSS) of convolved acoustic signals is of great interest for many classes of applications such as in-car speech recognition, hands-free telephony or hearing devices. The quality of solutions of ICA algorithms can be improved by applying *time-frequency masking*. In this paper, a number of time-frequency masking algorithms are compared and a post-processing algorithm is presented that improves the quality of the results of ICA algorithms by applying a modified speech enhancement technique. The proposed method is based on a combination of "classical" time-frequency masking methods and an extended Ephraim-Malah filter. The algorithms have been tested on real-room speech mixtures with a reverberation time of 130 - 159 ms, where a SIR-improvement of up to 23dB has been obtained, which was 11dB above ICA performance for the same dataset.

1 Introduction

Blind source separation (BSS) is a technique of recovering the source signals using only observed mixtures when the mixing process is unknown. In recent years, the problem has been widely studied and many methods have been proposed [11,14]. However, in the majority of cases, the estimated source signals are still corrupted by remaining interferences. The quality of the recovered source signals can be improved by applying post-processing on the ICA outputs [14]. There exist a number of algorithms for calculating time-frequency masks from the estimated direction of arrival of the separated signals [15,19], from the estimates of the local SNRs [9], from the cosine distance between a sample vector and the basis vector corresponding to the target [16], or from an approximation of the remaining interferences [7]. The main idea of the proposed approach is to combine the information contained in the time-frequency masks from the algorithms above with a better speech enhancement technique in order to improve the separation performance. For this propose, a method for the post processing stage, based on the optimally-modified Log-Spectral Amplitude (OM-LSA) estimator by Cohen [4], is adapted for the multichannel case. The proposed method is a further development of the approach presented in [8].

T. Adali et al. (Eds.): ICA 2009, LNCS 5441, pp. 581–588, 2009.

A further goal of this paper is a comparative study of different time frequency masking algorithms to show the effectiveness of different approaches on the same datasets.

2   Proposed Method

In this section, the applied method for blind source separation is presented. First the time-domain signals are converted into frequency-domain time-series signals using the Short-Time Fourier Transform (STFT), on which the JADE-Algorithm is applied, to compute the unmixing filter matrices $\mathbf{W}(\Omega)$. At the next step, the permutation problem is treated, so the unmixing filter matrices can be corrected by multiplication with the permutation matrix $\mathbf{P}(\Omega)$ as described already in [1,17]. At the post-processing stage of the algorithm, a speech enhancement algorithm is applied to improve the quality of the extracted signals and to minimize those crosstalk components, which were not eliminated by the ICA algorithm.

2.1   ICA

Acoustic signal mixtures in reverberant environments can be described by

$$\mathbf{x}(t) = \mathbf{A} * \mathbf{s}(t), \tag{1}$$

where $\mathbf{s}(t)$, $\mathbf{x}(t)$ and $\mathbf{A}$ denote the the vector of source signals, the vector of mixed signals and a mixing matrix containing the impulse responses between the sources and the sensors and $*$ denotes the convolution operator. Transforming (1) into the frequency domain reduces the convolutions to multiplications

$$\mathbf{X}(\Omega, \tau) \approx \mathbf{A}(\Omega)\mathbf{S}(\Omega, \tau), \tag{2}$$

where Ω is the normalized angular frequency, τ represents the frame index, $\mathbf{A}(\Omega)$ is the mixing system in the frequency domain, $\mathbf{S}(\Omega, \tau) = [S_1(\Omega, \tau), \ldots, S_N(\Omega, \tau)]$ represents the source signals, and $\mathbf{X}(\Omega, \tau) = [X_1(\Omega, \tau), \ldots, X_N(\Omega, \tau)]$ denotes the observed signals. So for each frequency bin an instantaneous ICA problem has to be solved. For this purpose the JADE-algorithm has been used, which is based on joint diagonalization of the most significant cumulant matrices of higher order [2].

2.2   Permutation Correction

The filter matrices calculated by ICA can be randomly permuted. To solve the permutation problem, the phase differences in the estimated unmixing filter matrices are used [1,16]. For this purpose we normalize the estimated mixing matrix $\hat{\mathbf{A}}(\Omega))$ on the first row, so the normalized mixing matrix can be written as

$$\hat{\mathbf{A}}(\Omega) = \begin{bmatrix} 1 & \cdots & 1 \\ \cdots & \cdots & \cdots \\ \hat{a}_{j1}e^{-j\Omega\delta_{j1}} & \cdots & \hat{a}_{jn}e^{-j\Omega\delta_{jn}} \\ \cdots & \cdots & \cdots \end{bmatrix}. \tag{3}$$

To correct the permutations, the phase differences δ_i are compared and the columns of the mixing filter matrix are sorted, so $\delta_i < \delta_k$ for all $i < k$.

2.3   Time-Frequency Masking

In this section, a brief review of existing time-frequency masking algorithms will be given. Each approach calculates a mask in the range $0 \leq \mathbf{M}(\Omega, \tau) \leq 1$.

Amplitude Based Mask Estimation. The time-frequency mask is calculated by comparing the estimates of the local SNRs of the output signal $Y_i(\Omega, \tau)$ with a threshold T [9]

$$M_i(\Omega, \tau) = \Psi\left(\log\left(|Y_i(\Omega, \tau)|^2\right) - \max_{\forall j \neq i} \log\left(|Y_j(\Omega, \tau)|^2\right) - \frac{T}{10}\right) \tag{4}$$

and a sigmoid nonlinearity Ψ defined by

$$\Psi(x) = \frac{1}{1 + \exp(-x)}. \tag{5}$$

The threshold T was set to 3dB, with higher thresholds leading to better SNR gains but in some test cases to musical noise.

Phase Angle Based Mask Estimation. The algorithm proposed in [16] considers closeness of the phase angle $\theta_i(\Omega, \tau)$ between a column of the mixing matrix $\mathbf{a}_i(\Omega)$ and the observed signal $\mathbf{X}(\Omega, \tau)$ calculated in the space transformed by a whitening matrix $V(\Omega) = \mathbf{R}^{-1/2}(\Omega)$, $\mathbf{R}(\Omega) = \langle \mathbf{X}(\Omega, \tau)\mathbf{X}(\Omega, \tau)^H \rangle$. The phase angle is given by

$$\vartheta_i(\Omega, \tau) = \arccos \frac{\left|\mathbf{b}_i^H(\Omega)\mathbf{Z}(\Omega, \tau)\right|}{\|\mathbf{b}_i(\Omega)\| \, \|\mathbf{Z}(\Omega, \tau)\|}, \tag{6}$$

where $\mathbf{Z}(\Omega, \tau) = V(\Omega)\mathbf{X}(\Omega, \tau)$ are whitened samples and $\mathbf{b}_i(\Omega) = V(\Omega)\mathbf{a}_i(\Omega)$ is the basis vector i in the whitened space. Then the mask is calculated by

$$M_i(\Omega, \tau) = \frac{1}{1 + \exp(g(\vartheta_i(\Omega, \tau) - \vartheta_T))} \tag{7}$$

where ϑ_T and g are parameters specifying the transition point and its steepness, respectively.

Interference Based Mask Estimation. The mask is estimated by

$$M_i(\Omega, \tau) = \frac{1}{1 + \exp(g(\tilde{\mathbf{S}}_i(\Omega, \tau) - \lambda_s))} \times \left(1 - \frac{1}{1 + \exp(g(\tilde{\mathbf{N}}_i(\Omega, \tau) - \lambda_n))}\right) \tag{8}$$

where λ_s, λ_n and g are parameters specifying the threshold points and the steepness of the sigmoid function and $\tilde{\mathbf{S}}_i(\Omega, \tau)$ and $\tilde{\mathbf{N}}_i(\Omega, \tau)$ are speech and noise dominance measures given by

$$\tilde{\mathfrak{S}}_i(\Omega, \iota, R_\Omega, R_\tau) - \frac{\left\|\Phi(\Omega, \tau, R_\Omega, R_\tau)(Y_i(\Omega, \tau) - \sum_{m \neq i} Y_m(\Omega, \tau))\right\|}{\left\|\Phi(\Omega, \tau, R_\Omega, R_\tau) \sum_{m \neq i} Y_m(\Omega, \tau)\right\|} \tag{9}$$

and

$$\tilde{\mathbf{N}}_i(\Omega, \tau, R_\Omega, R_\tau) = \frac{\left\|\Phi(\Omega, \tau, R_\Omega, R_\tau)(Y_i(\Omega, \tau) - \sum_{m \neq i} Y_m(\Omega, \tau))\right\|}{\left\|\Phi(\Omega, \tau, R_\Omega, R_\tau)Y_i(\Omega, \tau)\right\|}. \tag{10}$$

Here, $\|\cdot\|$ denotes the Euclidean norm operator and

$$\Phi(\Omega, \tau, R_\Omega, R_\tau) = \begin{cases} \mathcal{W}(\Omega - \Omega_0, \tau - \tau_0, R_\Omega, R_\tau), & |\Omega - \Omega_0| \leq R_\Omega, \\ & |\tau - \tau_0| \leq R_\tau \\ 0, & \text{otherwise} \end{cases} \quad (11)$$

utilizes a two dimensional window function $\mathcal{W}(\Omega - \Omega_0, \tau - \tau_0, R_\Omega, R_\tau)$ of the size $R_\Omega \times R_\tau$ (e.g. a two dimensional Hanning window) [7].

Power Ratio Based Mask Estimation. Power ratio was proposed in [17] and was originally used for detection of permutations.

$$powRatio_i(\Omega, \tau) = \frac{\left\|\mathbf{a}_i^H(\Omega)Y_i(\Omega, \tau)\right\|^2}{\sum_{k=1}^{N} \|\mathbf{a}_k(\Omega)Y_k(\Omega, \tau)\|^2} \quad (12)$$

where $\mathbf{a}_i(\Omega)$ is the i-th column of the estimated mixing matrix and $Y_i(\Omega, \tau)$ is the un-mixed signal i. Since this measure is in the range $0 \leq powRatio_i(\Omega, \tau) \leq 1$ and represents the dominance of the i-th separated source in the observation, it is possible to use the power ratio for masking purposes, so $M_i(\Omega, \tau) = powRatio_i(\Omega, \tau)$.

2.4   Speech Enhancement

In the following, an algorithm is described to minimize the remaining noise components based on a speech enhancement technique. For this purpose, the following signal model is assumed

$$\mathbf{Y}(\Omega, \tau) = \mathbf{Y}_c(\Omega, \tau) + \mathbf{Y}_n(\Omega, \tau), \quad (13)$$

where the clean signal $\mathbf{Y}_c(\Omega, \tau)$ is corrupted by a noise component $\mathbf{Y}_n(\Omega, \tau)$, the remaining sum of the interfering signals and the background noise. To improve the quality of the separated signals, an optimally modified log-spectral amplitude (OM-LSA) speech enhancement technique has been used. The estimated signals are obtained by

$$\tilde{\mathbf{Y}}(\Omega, \tau) = \mathbf{G}(\Omega, \tau)\mathbf{Y}(\Omega, \tau), \quad (14)$$

where $\mathbf{G}(\Omega, \tau)$ is the amplitude estimator gain. It is calculated by

$$\mathbf{G}(\Omega, \tau) = \mathbf{G}_{SE}(\Omega, \tau)^{p(\Omega,\tau)}\mathbf{G}_{min}^{(1-p(\Omega,\tau))}, \quad (15)$$

where $\mathbf{G}_{min}$ is a spectral floor constant, $\mathbf{G}_{SE}$ the gain of the speech enhancement method and $p(\Omega, \tau)$ is the speech presence probability [3,6]. The speech presence probability is used to modify the spectral gain function. Since the probability functions are not known, the masks from section 2.3 are used at this point to approximate the speech presence probability. For the calculation of the gain $\mathbf{G}_{SE}(\Omega, \tau)$ in (15), different speech enhancement algorithms can be used. In [8] the method by McAulay and Malpass [12] has been used. In this paper it is suggested to use the log spectral amplitude estimator (LSA) as proposed by Ephraim and Malah [5].

For our algorithm, we defined the a posteriori $\gamma_i(\Omega, \tau)$ and a priori SNR $\xi_i(\Omega, \tau)$ by

$$\gamma_i(\Omega, \tau) = \frac{|Y_i(\Omega, \tau)|^2}{\lambda_D(\Omega, \tau)} \quad (16)$$

and

$$\xi_i(\Omega, \tau) = \alpha\xi_i(\Omega, \tau - 1) + (1 - \alpha)\frac{\lambda_X(\Omega, \tau)}{\lambda_D(\Omega, \tau)}, \tag{17}$$

where α is a smoothing parameter that controls the trade-off between the noise reduction and the transient distortionis [6], $Y_i(\Omega, \tau)$ is the i-th ICA-output, $\lambda_D(\Omega, \tau)$ is the noise power

$$\lambda_D(\Omega, \tau) = \alpha_D\lambda_D(\Omega, \tau - 1) + (1 - \alpha_D)D_i^2(\Omega, \tau) \tag{18}$$

with the noise estimate $D_i(\Omega, \tau)$ given by

$$D_i(\Omega, \tau) = |Y_i(\Omega, \tau)|(1 - p_i(\Omega, \tau)), \tag{19}$$

and $\lambda_X(\Omega, \tau)$ is the approximate clean signal power

$$\lambda_X(\Omega, \tau) = (|Y_i(\Omega, \tau)| \, p_i(\Omega, \tau))^2. \tag{20}$$

With these parameters, the log spectral amplitude estimator is given by:

$$\mathbf{G}_{SE}(\Omega, \tau) = \frac{\xi(\Omega, \tau)}{1 + \xi(\Omega, \tau)} \exp\left(\int_{t=v(\Omega, \tau)}^{\infty} \frac{e^{-t}}{t} dt\right) \tag{21}$$

with $\xi(\Omega, \tau)$ denoting the local a priori SNR and $v(\Omega, \tau)$

$$v(\Omega, \tau) = \left(\frac{\xi(\Omega, \tau)}{1 + \xi(\Omega, \tau)}\right)\gamma(\Omega, \tau). \tag{22}$$

3   Experiments and Results

3.1   Recording Conditions

For the evaluation of the proposed approaches, two different sets of recordings were used. The first data set was created at TU Berlin. In these recordings different audio files from the TIDigits database [10] were used and mixtures with up to four speakers were recorded. The distance L_i between loudspeakers and microphones were varied between 0.9 and 2 m. The second dataset was recorded in a meeting room under mildly reverberant conditions [18]. Here, also mixtures with up to four speakers are represented. The experimental setup is shown schematically in Figure 1. The experimental conditions are summarized in Table 1.

3.2   Parameter Settings

The algorithm was tested on both recordings, which were first transformed to the frequency domain at a resolution of $N_{FFT} = 1024$. For calculating the spectrogram, the signals were divided into overlapping frames with a Hanning window and an overlap of $3/4 \cdot N_{FFT}$ and the STFT was then calculated.

Source-Signal i

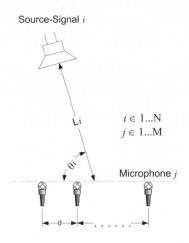

$i \in 1...N$
$j \in 1...M$

Microphone j

Fig. 1. Experimental Setup

Table 1. Mixture description

Mixture	Mix. 1 TU Berlin	Mix. 2 TU Berlin	Mix. 3 TU Berlin	Mix. 4 NTT Labs [18]	Mix. 5 NTT Labs [18]	Mix. 6 NTT Labs [18]
Reverberation time T_R	159 ms	159 ms	159 ms	130 ms	130 ms	130 ms
Distance between two sensors d	3 cm	3 cm	3 cm	4 cm	4 cm	4 cm
Sampling rate f_S	11 kHz	11 kHz	11 kHz	8 kHz	8 kHz	8 kHz
Number of speakers N	2	3	4	2	3	4
Distance between speaker i and array center	$L_1 = L_2 =$ 0.9 m	$L_1 = L_2 =$ $L_3 = 0.9$ m	$L_1 = L_2 =$ $L_3 = L_4 =$ $= 0.9$ m	$L_1 = L_2 =$ 1.2 m	$L_1 = L_2 =$ $L_3 = 1.2$ m	$L_1 = L_2 =$ $L_3 = L_4 =$ $= 1.2$ m
Angular position of the speaker i (as shown in Figure 1)	$\theta_1 = 50°$ $\theta_2 = 115°$	$\theta_1 = 30°$ $\theta_2 = 80°$ $\theta_3 = 135°$	$\theta_1 = 25°$ $\theta_2 = 80°$ $\theta_3 = 130°$ $\theta_4 = 155°$	$\theta_1 = 30°$ $\theta_2 = 70°$	$\theta_1 = 30°$ $\theta_2 = 70°$ $\theta_3 = 110°$	$\theta_1 = 30°$ $\theta_2 = 70°$ $\theta_3 = 110°$ $\theta_4 = 150°$

3.3   Performance Measurement

For calculation of the effectiveness of the proposed algorithm, the signal to interference ratio (SIR) was used as a measure of the separation performance and the signal to distortion ratio (SDR) as a measure of the signal quality.

$$SIR_i = 10 \log_{10} \frac{\sum_n y_{is_i}^2(n)}{\sum_{j \neq i} \sum_n y_{is_j}^2(n)} \qquad (23)$$

$$SDR_i = 10 \log_{10} \frac{\sum_n x_{ks_i}^2(n)}{\sum_n (x_{ks_i}(n) - \alpha y_{is_i}(n - D))^2}, \qquad (24)$$

where y_{i,s_j} is the i-th separated signal with only the s_j source active, and x_{k,s_j} is the observation obtained by microphone k when only s_j is active. α and D are parameters for phase and amplitude chosen to optimally compensate the difference between y_{i,s_j} and x_{k,s_j}.

Table 2. Experimental results (mean value of output SIR/SDR in [dB])

Mixture	Mix. 1 TU Berlin	Mix. 2 TU Berlin	Mix. 3 TU Berlin	Mix. 4 NTT Labs	Mix. 5 NTT Labs	Mix. 6 NTT Labs
ICA only	6.8/8.2	5.1/4.3	3.2/1.6	12.2/12.7	7.1/7.7	4.5/4.4
ICA and TF-Mask from [9]	12.0/6.5	7.6/4.0	5.4/1.4	17.5/10.2	9.1/7.1	5.2/4.4
ICA and proposed method using [9]	16.1/5.4	10.4/3.3	9.0/0.9	21.0/7.8	10.6/5.6	5.8/3.5
ICA and TF-Mask from [16]	11.5/5.5	13.1/2.9	11.8/0.2	19.3/8.8	12.5/4.8	9.1/1.8
ICA and proposed method using [16]	12.6/4.8	15.7/2.1	14.3/-0.2	20.5/7.7	13.6/3.9	11.4/1.2
ICA and TF-Mask from [7]	14.3/5.7	13.1/2.6	10.9/0.3	19.0/9.3	16.4/4.1	12.9/1.5
ICA and proposed method using [7]	16.0/5.4	14.8/2.4	12.7/0.1	20.2/8.4	17.5/3.5	13.8/1.4
ICA and TF-Mask using Power Ratio	11.0/7.2	9.5/3.6	7.7/0.9	15.8/10.8	11.1/5.9	6.9/2.7
ICA and proposed method using Power Ratio	14.6/6.2	12.6/2.7	10.1/0.4	17.7/8.9	13.6/4.6	7.8/2.0
ICA and TF-Mask combination of [9] and [16]	13.9/5.0	13.7/2.8	12.6/0.1	21.9/8.3	12.4/4.7	8.7/2.1
ICA and proposed method using a combination of [9] and [16]	18.2/4.0	17.6/2.1	15.9/-0.0	**25.8/6.4**	13.1/3.7	9.6/1.5
ICA and TF-Mask combination of [7] and [16]	15.9/4.5	16.9/2.2	15.3/-0.3	23.0/7.9	17.7/3.7	14.9/1.3
ICA and proposed method using a combination of [7] and [16]	**18.3/3.9**	**20.9/1.8**	**18.3/-0.6**	25.0/6.8	**19.9/2.9**	**17.4/1.0**
ICA and TF-Mask combination of Power Ratio and [16]	13.9/5.2	14.5/2.6	13.3/0.3	21.6/8.4	13.5/4.4	9.3/1.8
ICA and proposed method using a combination of Power Ratio and [16]	17.8/4.2	19.1/1.9	16.2/-0.2	24.6/6.9	16.6/3.2	11.8/1.1

3.4   Experimental Results

In this section the experimental results of the signal separation will be compared. All the mixtures from Table 1 were separated with the JADE algorithm and subsequently the time frequency masking from Sections 2.3-2.4 was performed using parameter settings as shown in Section 3.2. For each result the performance is calculated using Eq. (23)-(24). Table 2 shows the results of the applied methods.

4   Conclusions

In this paper, an approach for the post processing of ICA outputs for speech mixtures has been presented. It uses the combination of an optimally modified log-spectral amplitude (OM-LSA) speech enhancement technique and information from different time frequency masking algorithms for approximation of speech probabilities. The proposed algorithm has been tested on reverberant speech mixtures with $RT_{60} \approx 130$ - 159 ms. As can be seen in Table 2, the best results were achieved by applying a combination of the algorithms presented in [7] and [16] for probability calculation. Still, the results of the proposed method depend on the results of the preceding BSS algorithm. Thus, given a low ICA performance (in terms of SIR,SDR), a stronger signal distortion should be expected, which is the case for all tested masking algorithms. This tendency can be seen especially in Mixture 3 in Table 2.

With the proposed method an SIR-improvement of up to 23dB has been obtained. This is 11dB above ICA performance for the same dataset.

References

1. Baumann, W., Kolossa, D., Orglmeister, R.: Maximum Likelihood Permutation Correction for Convolutive Source Separation. In: ICA 2003, pp. 373–378 (2003)
2. Cardoso, J.-F.: High Order Contrasts for Independent Component Analysis. Neural Computation 11, 157–192 (1999)
3. Cohen, I.: On Speech Enhancement under Signal Presence Uncertainty. In: International Conference on Acoustic and Speech Signal Processing, pp. 167–170 (May 2001)
4. Cohen, I.: Optimal Speech Enhancement Under Signal Presence Uncertainty Using Log-Spectral Amplitude Estimator. IEEE Signal Processing Letters 9(4) (April 2002)
5. Ephraim, Y., Malah, D.: Speech Enhancement using a Minimum Mean Square Error Log-Spectral Amplitude Estimator. IEEE Trans. Acoust., Speech, Signal Processing ASSP-33, 443–445 (1985)
6. Ephraim, Y., Cohen, I.: Recent Advancements in Speech Enhancement. In: The Electrical Engineering Handbook. CRC Press, Boca Raton (2006)
7. Hoffmann, E., Kolossa, D., Orglmeister, R.: A batch algorithm for blind source separation of acoustic signals using ICA and time-frequency masking. In: Davies, M.E., James, C.J., Abdallah, S.A., Plumbley, M.D. (eds.) ICA 2007. LNCS, vol. 4666, pp. 480–487. Springer, Heidelberg (2007)
8. Hoffmann, E., Kolossa, D., Orglmeister, R.: A Soft Masking Strategy based on Multichannel Speech Probability Estimation for Source Separation and Robust Speech Recognition. In: 2007 IEEE Workshop on Applications of Signal Processing to Audio and Acoustics, New Paltz, NY (2007)
9. Kolossa, D., Orglmeister, R.: Nonlinear postprocessing for blind speech separation. In: Puntonet, C.G., Prieto, A.G. (eds.) ICA 2004. LNCS, vol. 3195, pp. 832–839. Springer, Heidelberg (2004)
10. Leonard, R.G.: A Database for Speaker-Independent Digit Recognition. In: Proc. ICASSP 1984, vol. 3, p. 42.11 (1984)
11. Mansour, A., Kawamoto, M.: ICA Papers Classified According to their Applications and Performances. IEICA Trans. Fundamentals E86-A(3), 620–633 (2003)
12. McAulay, R.J., Malpass, M.L.: Speech Enhancement using a Soft-Decision Noise Suppression Filter. IEEE Trans. ASSP-28, 137–145 (April 1980)
13. Mukai, R., Araki, S., Sawada, H., Makino, S.: Removal of Residual Cross-talk Components in Blind Source Separation using LMS Filters. In: NNSP 2002, pp. 435–444 (September 2002)
14. Pedersen, M.S., Larsen, J., Parra, L.C.: A Survey of Convolutive Blind Source Separation Methods. In: Springer Handbook on Speech Processing and Speech Communication (2007)
15. Roman, N., Wang, D.L., Brown, G.J.: Speech segregation based on sound localization. JASA 114(4), 2236–2252 (2003)
16. Sawada, H., Araki, S., Mukai, R., Makino, S.: Blind Extraction of a Dominant Source from Mixtures of Many Sources using ICA and Time-Frequency Masking. In: ISCAS 2005, pp. 5882–5885 (May 2005)
17. Sawada, H., Araki, S., Makino, S.: Measuring Dependence of Bin-wise Separated Signals for Permutation Alignment in Frequency-domain BSS. In: ISCAS 2007, pp. 3247–3250 (2007)
18. Sawada, H.:
    http://www.kecl.ntt.co.jp/icl/signal/sawada/demo/bss2to4/index.html
19. Yilmaz, O., Rickard, S.: Blind Separation of Speech Mixtures via Time-Frequency Masking. IEEE Trans. Signal Process. 52(7), 1830–1847 (2004)

Solving the Acoustic Echo Cancellation Problem in Double-Talk Scenario Using Non-Gaussianity of the Near-End Signal

Intae Lee[1], Charles E. Kinney[2], Bowon Lee[3], and Antonius A.C.M. Kalker[3]

[1] Institute for Neural Computation, University of California, San Diego
[2] Mechanical and Aerospace Engineering, University of California, San Diego
[3] Hewlett-Packard Laboratories, Palo Alto
intelli@ucsd.edu

Abstract. The acoustic echo cancellation (AEC) problem in the double-talk scenario and the blind source separation (BSS) problem resemble each other in that, for both problems, mixed signals are given and the objectives are to remove unwanted signals from the mixed signals. As many BSS algorithms have utilized the non-Gaussianity of the source signals to solve the separation problem, the super-Gaussianity of the near-end speech signal can be utilized to perform AEC in the double-talk scenario. Here, we propose a maximum likelihood (ML) approach using a super-Gaussian source prior to solve the double-talk-scenario AEC problem and compare the algorithm with minimizing mean squared error (MSE). The simulation results and analysis support the efficiency of the proposed method.

1 Introduction

In hands-free teleconferencing where microphones and loudspeakers are located in the same room, acoustic echo is unavoidable. Echo refers to the sound that is heard again after being reflected. In a teleconferencing system as in Fig. 1, the voice of the far-end speech (FES) is played by the near-end loudspeakers and is inevitably captured by the near-end microphones, and thus it is sent back to the far-end loudspeakers to become echo. Also, the echo can be reflected again to produce a chain of echoes in the far-end room. During a conversation, such echoes are mostly annoying and they usually disturb the participants in their conversation by letting the talkers hear their own voice back. Hence, in a teleconferencing system, it is essential to prevent the FES from being reflected.

Other than physically blocking the unwanted signal from entering the microphone, a common approach to this problem is to algorithmically remove the unwanted signal. While sound is played in the near-end loudspeaker, the signal can be sampled by the system (which we will call the reference signal) and can be used to remove itself that is filtered and added into the microphone signal. This kind of problems are called acoustic echo cancellation (AEC) problems and are basically system identification problems where the input signals and the

T. Adali et al. (Eds.): ICA 2009, LNCS 5441, pp. 589–596, 2009.
© Springer-Verlag Berlin Heidelberg 2009

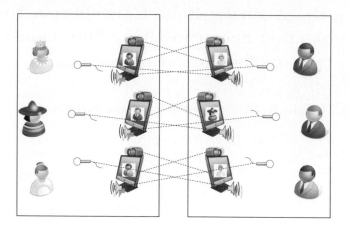

Fig. 1. Configuration of a duplex videoconferencing system

corresponding output signals of the system are given. Here in the problem, the system to be identified is the convolution of the room impulse response and the impulse response of the loudspeaker, that are mostly identified together as one filter.

AEC has been researched for several decades and there have been many successful algorithms proposed where most of them assume that in the microphone signal there is no foreground speech, or only noise, added to the convolved reference signal. In this case, popular objective functions are such ones as mean squared error (MSE), least square, or weighted least square. (For more details, please see [1] and the references therein). However, in the microphone signal, local speech can easily be mixed with the convolved reference signal. The AEC problem in such a double-talk scenario is more challenging for system identification than the AEC problem in single-talk scenario.

In order to solve this problem, we take advantage of the non-Gaussian property of the echo-canceled near-end speech signal. And thus, a maximum likelihood (ML) approach using super-Gaussian source priors is proposed. This kind of approach closely resembles the methods of independent component analysis (ICA) or blind source separation (BSS) since non-Gaussianity has been one of the most important key factors in those algorithms. In this paper, we will show that maximizing non-Gaussianity is also very efficient, and superior to minimizing MSE, in solving the double-talk-scenario AEC problem.

2   Notations

In this section, the notations are defined. It should be noted that the signals including the filters whose notations we define here are the frequency-domain counterparts. Scalars (including complex-valued ones), vectors, and matrices will be denoted with normal lowercase, bold lowercase, and bold uppercase letters, respectively. Also, all vectors will be column vectors.

- $(\cdot)^{\mathrm{T}}$: the transpose operator.
- $(\cdot)^{\mathrm{H}}$: the Hermitian transpose operator.
- $|\cdot|$: the absolute value of a number or variable.
- $\mathrm{P}_{\Theta}(z|\Theta)$: a probability distribution family with respect to parameter(s) Θ.
- m: the number of microphones, the near-end talkers, the loudspeakers, or the far-end talkers, which are all the same.
- N: the number of frames in short-time Fourier domain.
- $\tilde{\mathrm{E}}[\cdot]$: the expectation with respect to the empirical distribution, i.e. $\frac{1}{N}\sum_{n=1}^{N}\cdot$.
- $s_i^f[n]$: the short-time Fourier transformed (STFT-ed) i-th NES, i.e. the speech of the i-th near-end talker in f-th frequency bin.
- $r_i^f[n]$: the STFT-ed i-th reference signal, i.e. the signal sampled from the i-th loudspeaker in f-th frequency bin.
- $x_i^f[n]$: the STFT-ed i-th microphone signal in f-th frequency bin.
- a_{ij}^f: the STFT-ed transfer filter from s_j to the x_i in f-th frequency bin. Here we assume that the time-domain filter is time-invariant and it should be noted that the "short time" includes the whole filter length such that no frame variable is applicable to this filter.
- h_{ij}^f: the STFT-ed transfer filter from r_j to the x_i in f-th frequency bin. Again, we assume that the time-domain filter is time-invariant and it should be noted that the "short time" includes the whole filter length such that no frame variable is applicable to this filter.

Hence, their relation is

$$x_i^f[n] = \sum_{j=1}^{m} a_{ij}^f s_j^f[n] + \sum_{j=1}^{m} h_{ij}^f r_j^f[n], \qquad i = 1, \cdots, m. \qquad (1)$$

Note that in such (rough) STFT-domain expressions, there is some error that comes from the difference between the actual linear convolution and the circular convolution in the STFT domain. This problem will be ignored in this paper by assuming that the discrete Fourier transform (DFT) size is big enough to keep this error negligible.

The (STFT-ed) parameters and signals that have to be learned are defined as the following.

- $\hat{h}_{ij}$: the estimate of h_{ij}.
- $\hat{\mathbf{h}}_i^f$: the i-th column of $(\hat{\mathbf{H}}^f)^{\mathrm{H}}$.
- $y_i^f[n]$: the echo-canceled $x_i^f[n]$ with respect to $\hat{h}_{ij}$ $(j = 1, \cdots, m)$, i.e.

$$\begin{aligned} y_i^f[n] &= x_i^f[n] - \sum_{j=1}^{m} \hat{h}_{ij}^f r_j^f[n] \\ &= x_i^f[n] - (\hat{\mathbf{h}}_i^f)^{\mathrm{H}} \mathbf{r}^f[n], \qquad i = 1, \cdots, m, \end{aligned} \qquad (2)$$

and the ideal goal is

$$y_i^f[n] = \sum_{j=1}^{m} a_{ij}^f s_j^f[n], \qquad i = 1, \cdots, m. \qquad (3)$$

3   Frequency-Domain Acoustic Echo Cancellation from the Maximum Likelihood Perspective

The lengths of the filters to be learned depend on the reverberation times of the room. The time-domain filters have counterparts in the T-F domain and thus they can be learned in both domains. The approaches in both domains have been compared in [2] for BSS where some of the comparisons also apply to AEC. The main advantage of the T-F-domain approach is that it better handles longer filter length by decomposing the problem into smaller problems since convolution in the time domain (roughly) decomposes into bin-wise multiplication in the T-F domain, and thus each filter component can be learned separately in the T-F domain which typically results in reducing the amount of computation and the convergence time. The algorithms that we propose are also T-F-domain approaches.

3.1   Objective Function

ML Objective functions are derived using a group of symmetric and unimodal probability density functions (PDFs). The PDFs are expressed in the following form where z denotes a complex-valued dummy variable. Note that this is equivalent to having two-dimensional joint PDFs where the two real-valued variables are the real and the imaginary parts of z.

$$P_{(p,\sigma)}(z|(p,\sigma)) = \frac{1}{Z_{p,\sigma}} e^{-\frac{1}{\sigma}|z|^p}, \tag{4}$$

where $Z_{p,\sigma}$ is the normalization constant,

$$Z_{p,\sigma} = \frac{2\pi\sigma^{\frac{2}{p}}\Gamma(\frac{2}{p})}{p}, \tag{5}$$

and $\Gamma(\cdot)$ is the gamma function,

$$\Gamma(z) = \int_0^\infty t^{z-1}e^{-t}dt. \tag{6}$$

The positive parameters p and σ control the super-Gaussianity, which is characterized by peakedness about mean and heavy tail, and the variance of the PDF, respectively. Note that this PDF model is an extension of the generalized Gaussian density (GGD) model to two dimensions and that the model satisfies circularity which, in signal processing, is a common assumption for complex-valued signals.

In Fig. 2, the PDFs are drawn with respect to some p values. It can be seen that as p becomes smaller, the PDF increases in super-Gaussianity. For the short time frequency components of white Gaussian noise, the PDF with $p = 2$ is a good match since the PDF becomes a two-dimensional independent Gaussian distribution. For the short-time frequency components of speech, p being close

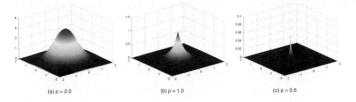

(a) p = 2.0   (b) p = 1.0   (c) p = 0.5

Fig. 2. The PDF model in (4) with respect to different p values, (a) $p = 2.0$, (b) $p = 1.0$, and (c) $p = 0.5$. The σ values are chosen such that the PDFs have unit variance.

to 1 is a good match. Hence in the scenario where local speech is present, the p in the PDF should be close to 1.

Using the two-dimensional GGD, ML approach is taken with respect to x_i^f. The probability distribution family of x_i^f is denoted as

$$P_{\hat{\mathbf{h}}_i^f}\left(x_i^f[n]|\hat{\mathbf{h}}_i^f\right) = \frac{1}{Z_{p,\sigma}} e^{-\frac{1}{\sigma}\left|x_i^f[n]-(\hat{\mathbf{h}}_i^f)^H\mathbf{r}^f[n]\right|^p}, \tag{7}$$

with the parameters $\hat{\mathbf{h}}_i^f \ (= [h_{i1}^f,\cdots,h_{im}^f]^H)$. Note that the unknown parameters to be learned are not p or σ but $\hat{\mathbf{h}}_i^f$ and that p and σ will be fixed a priori in the algorithm.

The log-likelihood function is then

$$\log\left(\prod_{n=1}^{N} P_{\hat{\mathbf{h}}_i^f}\left(x_i^f[n]|\hat{\mathbf{h}}_i^f\right)\right) = \sum_{n=1}^{N}\log P_{\hat{\mathbf{h}}_i^f}\left(x_i^f[n]|\hat{\mathbf{h}}_i^f\right) \tag{8}$$

$$= -\log Z_{p,\sigma} - \frac{1}{\sigma}\sum_{n=1}^{N}\left|y_i^f[n]\right|^p. \tag{9}$$

3.2 Algorithm

As it can be seen from (9), in each frequency bin, maximizing likelihood is equivalent to minimizing the p-power of $y_i^f[n]$, i.e.

$$C_{AEC} = \tilde{E}\left[\left|y_i^f[n]\right|^p\right]$$

$$= \tilde{E}\left[\left|x_i^f[n] - (\hat{\mathbf{h}}_i^f)^H\mathbf{r}^f[n]\right|^p\right]. \tag{10}$$

Please note that when $p = 2$, the corresponding cost function is equivalent to MSE. Also, note that the result of optimizing the objective function is not affected by the value of σ.

The gradient is

$$\frac{\partial C_{AEC}}{\partial(\hat{h}_{ij}^f)^*} = -\frac{p}{2}\tilde{E}\left[(y_i^f[n])^*\left|y_i^f[n]\right|^{p-2}(r_j^f[n])^*\right], \tag{11}$$

and the Hessian is

$$\frac{\partial^2 C_{AEC}}{\partial \hat{h}_{ij}^f \partial (\hat{h}_{ij}^f)^*} = \frac{p^2}{4} \tilde{E}\left[|y_i^f[n]|^{p-2} |r_j^f[n]|^2\right]. \tag{12}$$

In order to avoid matrix inversion in the algorithm, the Newton's update is derived with respect to each $\hat{h}_{ij}^f$ instead of the vector $\mathbf{h}_i^f$. Then, the Newton's update with an additional learning rate $\mu (\leq 1)$ which helps the algorithm to avoid diverging is

$$\hat{h}_{ij}^f \leftarrow \hat{h}_{ij}^f - \mu \left(\frac{\partial^2 C_{AEC}}{\partial \hat{h}_{ij}^f \partial (\hat{h}_{ij}^f)^*}\right)^{-1} \frac{\partial C_{AEC}}{\partial (\hat{h}_{ij}^f)^*} \tag{13}$$

$$= \hat{h}_{ij}^f + \mu \frac{2\tilde{E}\left[y_i^f[n]|y_i^f[n]|^{p-2}(r_j^f[n])^*\right]}{p\tilde{E}\left[|y_i^f[n]|^{p-2}|r_j^f[n]|^2\right]}. \tag{14}$$

For the details on the Newton's update in the complex domain, please look at [3,4].

4   Experiments

For simpler analysis of the proposed AEC algorithm, experiments will be constrained to cases where the number of foreground sounds and the number of reference signals are 1. In addition, since the algorithm is dealing with each microphone signal separately, it will be assumed that m equals 1.

In order to have better understanding of the proposed AEC algorithm, two cases have been tested with various p values; that is, cases when the local sound, or the time-domain counterpart of $\{s_i^1[n], \cdots, s_i^F[n]\}$, is not only speech but also Gaussian noise have been tested for the p values from 0.2 to 2.5 being incremented by 0.1. Please note that the test of using MSE as the objective function is included in this experiment since the proposed objective function is equivalent to MSE when $p = 2$.

The following parameters have been chosen for the experiments:

- sampling rate: 8 kHz
- length of the time-domain signals: 12 seconds
- FFT size: 4096 taps ($\approx$ 500 ms)
- window: Hanning window
- window shift size: 1024 taps
- filter length of the time-domain room impulse responses: 1600 taps (= 200 ms)

As the reference signal, clean speech signal has been used. Room impulse responses were obtained by an image method [5] and then convolved with the time-domain signals. It was assumed that no convolution takes place in the loudspeaker. Before mixing, the powers of the signals have been matched to be equal.

Performance was measured by the cross-correlation between the known target signal (i.e. the perfectly echo-canceled signal) and the learned echo-canceled signal. Both of the signals are normalized such that their autocorrelations (with no shift) are 1 and thus the maximum performance to be achieved is also 1 which is the case when they exactly equal.

4.1   The Results When Local Sound Is Either Speech or Noise

The performance has been tested for various p values in the case when the local sound is speech. The results are shown in Fig. 3(a). As it can be seen, the performance is best for p being near 1 or slightly smaller and monotonically degrades as the value of p increases. For comparison, the performance has also been tested in the case when the local sound is white Gaussian noise. The results are shown in Fig. 3(b). In this case, the performance is best for p being near 2.

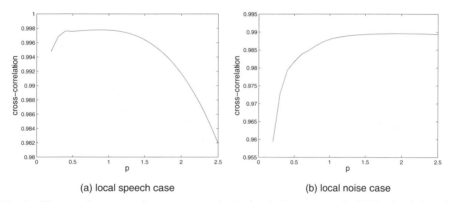

(a) local speech case   (b) local noise case

Fig. 3. The performance (in cross-correlation) of the proposed AEC algorithm for various p values in cases when the local sound is either (a) speech or (b) noise

4.2   Maximum Likelihood p Values of the Local Sound Data

The results imply that the fixed p value that results in the best performance of the AEC algorithm depends on the local sound signal. In order to have a better understanding of their relationship, the ML parameters, p and σ, of the PDF family (4) that correspond to the convolved local sound data in the previous experiments have been learned in each frequency bin by using the MATLAB function "fminbnd()" iteratively. Please note that the unknown parameters to be learned are p and σ now and no filter coefficients are to be learned. The results are shown in Fig. 4 for p values only since the σ values have no effect on the proposed AEC algorithm as it was to be seen in (10). Fig. 4(a) and Fig. 4(b) are the histograms of the 2049 (which is the number of frequency bins that were used) ML p values with respect to the local speech and the local noise, respectively. The optimum p values of the local speech are distributed around less than 1 while the ones of the local noise are distributed around 2 or larger.

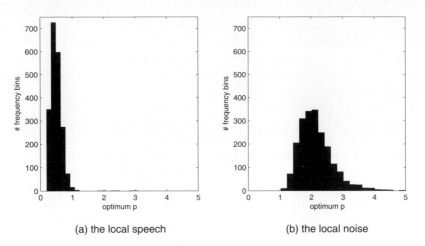

(a) the local speech   (b) the local noise

Fig. 4. The histograms of the bin-wise ML p values of the PDF family in (4) with respect to (a) the local speech and (b) the local noise

5   Conclusions

The p value that results in the best performance of the proposed AEC algorithm tends to be around the ML p values of the two-dimensional GGD family with respect to the local signal in the simulated data. Hence, it is shown, in double-talk scenario, ML approach with super-Gaussian source priors ($p < 2$) provides better AEC result than the approach of using MSE ($p = 2$). Convergence results of the proposed algorithm will be shown in a longer version of this draft.

References

1. Benesty, J., Gänsler, T., Morgan, D.R., Sondhi, M.M., Gay, S.L.: Advances in Network and Acoustic Echo Cancellation. Springer, Berlin (2001)
2. Nishikawa, T., Saruwatari, H., Shikano, K.: Blind source separation of acoustic signals based on multistage ICA combining frequency-domain ICA and time-domain ICA. IEICE Trans. Fundamentals (4), 846–858 (2003)
3. Yan, G., Fan, H.: A newton-like algorithm for complex variables with applications in blind equalization. IEEE Trans. on Signal Processing 48(2), 553–556 (2000)
4. Lee, I., Kim, T., Lee, T.W.: Fast fixed-point independent vector analysis algorithms for convolutive blind source separation. Signal Processing 87(8), 1859–1871 (2007)
5. Allen, J.B., Berkley, D.A.: Image method for efficiently simulating small room acoustics. J. Acoust. Soc. Amer. 65, 943–950 (1979)

An ICA-Based Method for Blind Source Separation in Sparse Domains

Everton Z. Nadalin[1,3], Ricardo Suyama[2,3], and Romis Attux[1,3]

[1] Departament of Computer Engineering and Industrial Automation
School of Electrical and Computer Engineering – University of Campinas (Unicamp) – Caixa
Postal 6101 – CEP 13083-970 – Campinas – SP – Brazil
[2] Departament of Microwave and Optics
School of Electrical and Computer Engineering – University of Campinas (Unicamp) – Caixa
Postal 6101 – CEP 13083-970 – Campinas – SP – Brazil
[3] Lab. of Signal Processing for Communications (DSPCom)
School of Electrical and Computer Engineering – University of Campinas (Unicamp) – Caixa
Postal 6101 – CEP 13083-970 – Campinas – SP – Brazil
{nadalin,attux}@dca.fee.unicamp.br, rsuyama@dmo.fee.unicamp.br

Abstract. In this work, we propose and analyze a method to solve the problem of underdetermined blind source separation (and identification) that employs the ideas of sparse component analysis (SCA) and independent component analysis (ICA). The main rationale of the approach is to allow the possibility of reaching a method that is more robust with respect to the degree of sparseness of the involved signals and more effective in the use of information brought by multiple sensors. The ICA-based solution is tested with the aid of three representative scenarios and its performance is compared with that of one of the soundest SCA techniques available, the DEMIXN algorithm.

Keywords: Sparse component analysis, independent component analysis, blind identification, audio signal processing.

1 Introduction

The interrelated problems of blind source separation (BSS) and blind identification have been studied in the last years under a wide range of applications [1]. An emblematic instance of BSS is the Cocktail Party Problem (CPP), which fits very well the scope of this work. In the CPP, it is assumed that a group of people is gathered in a room and that a number of conversations are simultaneously carried out. It is also considered that a set of sensors (perhaps the pair of ears of a certain guest) is excited with different mixtures of their voices, and that there is a natural aim to separate these signals or to identify the mixing model. If the voices are supposed to be mutually independent, this problem can be treated within the classical independent component analysis (ICA) framework. Indeed, ICA generally leads to good results when the number of information sources is equal to or less than the number of sensors [1].

Things are different when the number of sources is greater than the number of sensors, i.e., in an underdetermined case. Such scenario is associated with the impossibility of perfectly inverting the mixture model, which constitutes a problem *per se*

T. Adali et al. (Eds.): ICA 2009, LNCS 5441, pp. 597–604, 2009.
© Springer-Verlag Berlin Heidelberg 2009

and, in addition to that, poses serious difficulties to the direct use of any measure based on mutual independence.

When it is possible to rely on *a priori* information about the sources, very interesting perspectives of dealing with the underdetermined BSS problem emerge. A clear example of this statement is the potential use of the hypothesis that, in the temporal or in a transformed domain, not all sources are always superposed when the mixture is considered. Several forms of this idea and the multifarious ways of employing it have originated the research field of *sparse component analysis* (SCA) [2]. Notice that this explanation evokes our initial example of the CPP: not all people talk, for instance, at the same time, which means that, *for certain whiles*, an eventual underdetermined mixing model may turn into a perfectly invertible model. This same idea may eventually hold in the context of another variable (e.g. frequency).

In this paper, we explore BSS problems that are composed of mutually independent signals and, moreover, in which the notion of sparseness is relevant. In such context, we propose a novel approach to identify the mixture matrix by applying ICA to "locally well-determined mixtures" in the sparse domain. The proposal opens interesting perspectives from two standpoints: 1) it brings the well-established ICA framework to the context of SCA, which raises relevant points of connection between two important approaches to the BSS problem and 2) it allows an extension of the applicability of existing SCA approaches insofar as the number of sources present in the "well-determined" mixtures is concerned.

2   Sparse Component Analysis

Sparse component analysis (SCA) is a tool employed to solve the blind source separation (BSS) problem under the assumption that the sources are sparse in some domain [2]. For instance, in audio applications, it has been observed that the involved signals typically show a certain degree of sparseness in the time-frequency domain, which means that there are time-frequency regions in which not all the sources contribute to the mixture. The relevance of this fact becomes clear when one considers, for instance, the case of underdetermined mixtures, i.e., when there are more sources than sensors: in such case, with the aid of the *a priori* information concerning sparseness, it is possible to identify the mixing matrix.

The first approaches based on this idea [3, 4, 5] attempted, in general, to identify the main directions in the global scatter plot by clustering all the data together. These methods, however, exhibit poor performances in regions in which there is source overlap. Instead of considering the global scatter plot, some alternative methods use information contained in local scatter plots. In such cases, an interesting possibility is to treat the problem within a principal component analysis (PCA) framework, having in mind the goal of identifying the mixture matrix by seeking regions in which only one source appears [6, 7].

It should be noted that these methods do not employ the idea of independent component analysis (ICA), which is a very sound and useful tool in other BSS applications. In this work, we will analyze the implications of using ICA in sparse contexts and, moreover, we will consider its potential in performing identification even in the presence of more than one source, in contrast, for instance, with solutions like the PCA-based method discussed above, which require the existence of regions with a single active source (which characterizes a single dominant problem [8]). It is also interesting to remark that, by allowing that situations with multiple dominant sources be considered, there arises a promising perspective of employing information brought by additional sensors to improve the identification performance. Having presented these ideas, let us turn our attention to a more detailed exposition of the method.

3 Sparse Signals and ICA

In accordance with the above exposed ideas, if the sources are mutually independent in some domain, it is possible to apply, at least locally, a classical ICA strategy to estimate the directions that form the mixing matrix. This ICA-based method will work properly whenever there be, in some time-frequency regions, a number of sources smaller than or equal to the number of sensors. In other words, due to sparseness, a globally underdetermined problem is turned into a locally standard separation problem. In order to illustrate the potential of the method, let us consider the effects of applying ICA to the vicinity of a point (t, f) in the time-frequency domain under four possible scenarios:

1. **The number of active sources is equal to the number of sensors:** in such case, it is well-known that ICA leads effectively to the identification of the directions associated with the mixing matrix, as illustrated in Fig. 1.
2. **The number of active sources is smaller than the number of sensors:** when the number of sources is smaller than the number of sensors, ICA tends to find the directions associated with all active sources, being the remaining degrees of freedom influenced by other factors like additive noise. It should be noted, however, that the directions not associated with active sources will vary considerably from one time-frequency region to another. Fig. 2 illustrates a situation of this sort, in which there are two sensors and a single active source.
3. **The number of sources is greater than the number of sensors:** this originates a locally underdetermined problem, which means that probably some directions will be estimated with some degree of precision whereas others will not be found. The output of the separating system will not be composed of independent signals, and, in general, not even a perfect uncorrelatedness condition will be reached. In Fig. 3, we have an example with four sources and two sensors, an example in which no direction of the mixing matrix was precisely estimated.
4. **There are no active sources:** In such case, we shall have, once more, available degrees of freedom that cannot be properly employed, which means that the outcome of an ICA method will not possess a definite significance. Analogously to the second scenario, the directions obtained in these data blocks will vary considerably. Fig. 4 illustrates a situation of this nature (with no sources and two sensors).

Therefore, after applying ICA to several different time-frequency regions and discarding the solutions that lead to highly correlated signals (as described in the third scenario), the set of estimates should be more concentrated around the values of the mixing directions, since the estimates that are not associated with sources will tend to be equally distributed among all possible directions.

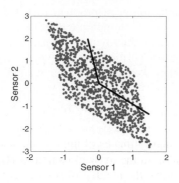

Fig. 1. Two sources and two sensors

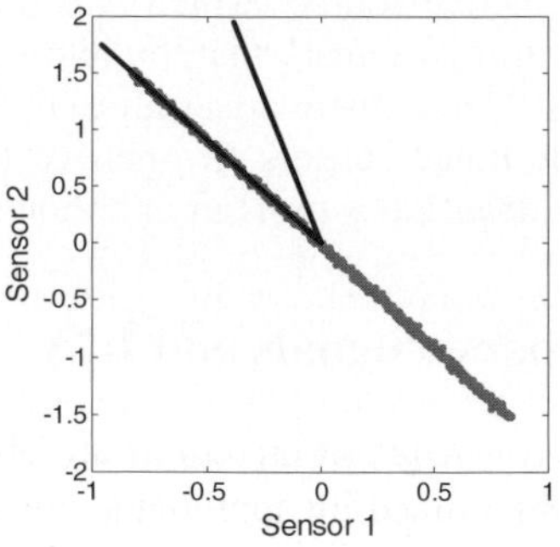

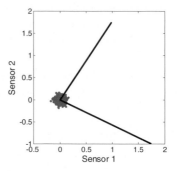

Fig. 2. One source and two sensors

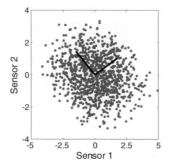

Fig. 3. Four sources and two sensors

Fig. 4. No sources and two sensors

4   Proposed Method

In this work, we reach a sparse time-frequency domain by applying the Short-Time Fourier Transform (STFT) to the signals. This option is not the only possibility, but is an effective and widespread solution [9]. The method also assumes that the total number of sources is known *a priori*, although it is possible that, in future works, this demand be less stringent. In the tests we will carry out in this work, we shall consider exclusively a two-sensor case.

In order to provide a clear and concise explanation of the proposed method, we will present it with the aid of the following algorithmic description.

Step 1 – Obtain the time-frequency representation of each observed signal using the STFT, choosing appropriately the window size $N_{samples}$ and the overlap ratio between consecutive sample windows;

Step 2 – Estimate the ICA for the neighborhood $\Omega_{t,f}$ around each time-frequency point (t, f). The ICA algorithm should provide an estimate of the "local $M \times M$ mixing matrix", where M denotes the number of sensors. From this matrix, it is possible to estimate the mixing directions associated with all sources using the following equation:

$$\theta_i(t,f) = \tan^{-1} \frac{[W]_{2,i}}{[W]_{1,i}}, \qquad i = 1, 2$$

The validity of the obtained estimate depends, as discussed in the previous section, on the existence or absence of more than two active sources. This explains the inclusion of step 2.1.

Step 2.1 – In order to avoid unreliable estimates, ICA solutions with an output correlation coefficient greater than a certain threshold (typically 0.08) were discarded. This policy aims at rejecting directions obtained from undermodeled scenarios.

Step 3 – After all directions are obtained, the final estimates of the mixing directions are determined by applying a clustering algorithm to the mass of estimates $\theta(t,f)$ that were not discarded in step 2.1. As it is assumed that the number of sources is *a priori* known, the well-known k-means algorithm was employed [10]. However, constructive methods and, in fact, any effective tool could have been applied: this is not fundamentally related to the *modus operandi* of the proposal *per se*.

The reader will notice that the method bears a certain degree of structural similarity with other approaches [6,7]: the main difference is exactly the use of ICA and some refinements conceived to aid the estimation process. This difference, notwithstanding, may open very relevant perspectives: we will discuss some of them in the following section, wherein the ICA-based method is tested.

5 Simulation Results

As mentioned in the previous section, we employ, in this work, a time-frequency representation given by the Short-Time Fourier Transform, which is evaluated considering a window of 512 samples and half-window overlap.

The ICA method was the one proposed by Comon [12], which is based on the maximization of a kurtosis measure, there being also a pre-whitening step. The main reason for choosing this setup was its interesting tradeoff between performance and convergence rate. As the input data of the ICA algorithm are complex-valued, we have chosen to treat the real and imaginary parts as separate vector elements [real () imag ()]. In all cases, the proposed method was compared with the DEMIXN algorithm [7], which is, in a certain sense, akin to it in some key methodological aspects, such as knowledge of the number of sources, use of clustering etc. To establish proper bases for comparison, the DEMIXN was also used with a window of 512 samples. The experimental scenario used in this work was very similar to that described in [7]: mixtures were built from a number of sources ranging from 2 to 15, being the mixing directions always uniformly distributed. The DEMIXN was run for scenarios of increasing difficulty (with respect to the number of underlying sources) until it was considered that it ceased to work properly. For each number of sources N, twenty runs were performed and the angular mean error (AME) was computed for each of them. The sources were drawn from a set of 200 five-second duration excerpts of Polish voices, all of them sampled at 4 kHz[1].

---

[1] Available at http://mlsp2005.conwiz.dk/index.php@id=30.html

5.1 First Scenario

The first scenario was built in accordance with the conditions described in [7]. In Fig. 5 (a), the spectrogram of a mixture of two voices is shown. Notice the presence of a contrast between darker regions (where the mixture signal is stronger) and lighter ones (where the mixture signal is weaker or absent), which is an indicative of the existence of a certain degree of sparseness.

In Fig. 6, we have the evolution of the AME with respect to the number of sources (for an average of 20 trials, each using different voice samples randomly drawn from the total set). The curves associated with the proposed approach and with DEMIX show that the performance of both methods is compatible (in a scenario that was analyzed in [7]).

In this case, the DEMIXN algorithm was unable to find the entire set of mixing directions in 4% of the cases. This was caused by the algorithm having created two clusters related to the same direction: these cases were not taken into account in the process of calculating the average shown in Fig. 6. The ICA-based method, on the other hand, does not produce this kind of wrong result in any of the tests. As explained in [7], the DEMIXN converges up to 13 sources, but, for a higher number of sources, the algorithm does not find all clusters: the proposed technique, on the other hand, does not seem to be subject to this limitation.

5.2 Second Scenario

In the second scenario, we always include a uniform white noise as one of the sources. This inclusion can be justified as a robustness test, for this kind of source is non-sparse in the time-frequency domain. Particularly, in the situation in which we have more sources than sensors and one of the sources is non-sparse, it is neither possible to apply a classical ICA algorithm nor the typical SCA approaches. Fig. 5(b) shows the spectrogram of the mixture of one voice and the white noise source.

In this situation, the proposed method does not experience a significant performance loss in any of the tests. On the other hand, DEMIXN does not find all directions in 40% of the cases, and, moreover, ceases to converge with 10 sources. Fig. 7 shows that, up to 9 sources, the proposed method and DEMIXN have close performances (in these cases, the latter also reaches acceptable results). The increase of the AMEs can be explained by the fact that the direction of the white noise is harder to find: if we discard this direction, the AME is close to that found in the first case.

5.3 Third Scenario

The third scenario is, in a certain sense, a variation of the second one. Instead of a white noise, which is not a particularly representative model for audio signals, the additional source is changed to one that has a broad spectrum and is present for a long period of time: the sound of a crowd clapping and screaming. Fig. 5 (c) shows the spectrogram of a mixture of this source and one voice.

The results obtained for this case are, in a certain sense, similar to those obtained in the second scenario. The DEMIXN approach, which is strongly based on the assumption that most of the time-frequency regions will have only one active source, fails in the task of finding the mixing directions in more than 10% of the trials. As shown in Fig. 7, this result is slightly better than that obtained for the second scenario, but is very different from the error level obtained in the first scenario.

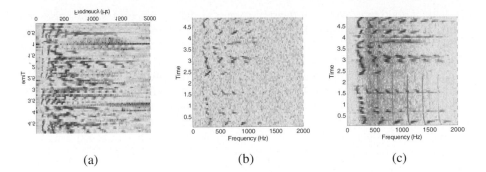

Fig. 5. Spectrogram of a mixture of (a) two voice sources, (b) one voice source and a white noise, and (c) one voice source and one crowd source

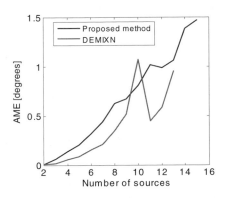

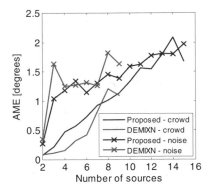

Fig. 6. Average AME of the algorithms in the first scenario

Fig. 7. Average AME of the algorithms in the second (noise source) and third (crowd source) scenarios

Even though there are regions in which only one source is active, the proposed method takes advantage of the ICA algorithm and can estimate the true angles from a richer set of time-frequency regions.

6 Conclusions

In this work, we proposed a new approach to the task of performing sparse component analysis in underdetermined contexts. The method is based on the idea of applying ICA to identify local determined mixtures in the time-frequency domain, which allows the estimation of the mixing directions of the model as a whole. Such strategy is very interesting, since it combines two elements that bring useful *a priori* information: the existence of a certain degree of sparseness and the hypothesis of mutual independence between sources in the transformed domain. This combination is particularly auspicious and valid for audio source separation.

The results show that the ICA-based method was able to provide a performance level compatible with that of the DEMIXN in a scenario similar to one discussed in

[7], and, furthermore, was significantly robust in the presence of white noise or of audio sources with a reduced degree of sparseness. This, in our opinion, is indicative of the validity of conducting a more thorough study concerning the applicability of ICA as a tool for SCA in the context of audio sources.

There are many perspectives for future work, among which we highlight the investigation of more efficient clustering methods - which could remove the need for *a priori* knowledge of the number of sources and improve the performance of the method as a whole -, the use of alternative ICA techniques and the application of different time-frequency representations - particularly via wavelet transforms -, and the analysis of the use of an ICA-based performance index (in step 2.1 of the algorithm) instead of a correlation measure.

Acknowledgments. The authors would like to thank CAPES and FAPESP for the financial support and Diogo C. Soriano for his comments and suggestions.

References

1. Hyvärinen, A., Karhunen, J., Oja, E.: Independent Component Analysis. John Wiley & Sons, New-York (2001)
2. Gribonval, R., Lesage, S.: A survey of Sparse Component Analysis for Blind Source Separation: principles, perspectives, and new challenges. In: ESANN 2006 proceedings - European Symposium on Artificial Neural Networks Bruges, Belgium (2006)
3. Bofill, P., Zibulevsky, M.: Underdetermined blind source separation using sparse representations. Signal Processing 81, 2353–2363 (2001)
4. Yilmaz, O., Rickard, S.: Blind Separation of Speech Mixtures via Time-Frequency Masking. IEEE Transactions on Signal Processing 52, 1830–1847 (2004)
5. Van Hulle, M.M.: Clustering Approach to Square and Non-Square Blind Source Separation. In: Proc. IEEE Neural Networks for Signal Processing IX, pp. 315–323 (1999)
6. Abrard, F., Deville, Y.: Blind separation of dependent sources using the time-frequency ratio of mixtures approach. In: Proc. ISSPA 2003, France (2003)
7. Arberet, S., Gribonval, R., Bimbot, F.: A Robust Method to Count and Locate Audio Sources in a Stereophonic Linear Instantaneous Mixture. In: Rosca, J.P., Erdogmus, D., Príncipe, J.C., Haykin, S. (eds.) ICA 2006. LNCS, vol. 3889, pp. 536–543. Springer, Heidelberg (2006)
8. Noorshams, N., Babaie-Zadeh, M., Jutten, C.: Estimating the Mixing Matrix in Sparse Component Analysis Based on Converting a Multiple Dominant to a Single Dominant Problem. In: Davies, M.E., James, C.J., Abdallah, S.A., Plumbley, M.D. (eds.) ICA 2007. LNCS, vol. 4666, pp. 397–405. Springer, Heidelberg (2007)
9. Rioul, O., Vetterli, M.: Wavelets and Signal Processing. IEEE Signal Processing Magazine 8, 14–38 (1991)
10. Duda, R.O., Hart, P.E., Stork, D.G.: Pattern Classification. John Wiley & Sons, New-York (2001)
11. Comon, P.: Independent component analysis, A new concept? Signal Processing 36, 287–314 (1994)
12. Comon, P., Moreu, E.: Improved contrast dedicated to blind separation in communications. In: Proc. ICASSP, Munich, pp. 3453–3456, April 20-24(1997)

Extension of Sparse, Adaptive Signal Decompositions to Semi-blind Audio Source Separation

Andrew Nesbit[1,*], Emmanuel Vincent[2], and Mark D. Plumbley[1,**]

[1] Queen Mary University of London
School of Electronic Engineering and Computer Science
Mile End Road, London, E1 4NS, United Kingdom
{andrew.nesbit,mark.plumbley}@elec.qmul.ac.uk
[2] METISS Group, IRISA-INRIA
Campus de Beaulieu, 35042 Rennes Cedex, France
emmanuel.vincent@irisa.fr

Abstract. We apply sparse, fast and flexible adaptive lapped orthogonal transforms to underdetermined audio source separation using the time-frequency masking framework. This normally requires the sources to overlap as little as possible in the time-frequency plane.

In this work, we apply our adaptive transform schemes to the semi-blind case, in which the mixing system is already known, but the sources are unknown. By assuming that exactly two sources are active at each time-frequency index, we determine both the adaptive transforms and the estimated source coefficients using ℓ^1 norm minimisation. We show average performance of 12–13 dB SDR on speech and music mixtures, and show that the adaptive transform scheme offers improvements in the order of several tenths of a dB over transforms with constant block length. Comparison with previously studied upper bounds suggests that the potential for future improvements is significant.

1 Introduction

Our goal is to tackle the problem of *audio source separation* for *underdetermined* and *instantaneous* mixtures. Specifically, given an observed two-channel mixture $x(n) = (x_1(n), x_2(n))$, we aim to estimate all $J > 2$ simultanously active sources $s(n) = (s_1(n), \ldots, s_J(n))$, assuming the mixture has been generated thus:

$$x(n) = As(n) \ , \tag{1}$$

where $A = (a_{i,j})$ is a $2 \times J$ matrix with real-valued entries $a_{i,j}$, the mixture and source indices are i and j respectively, and the discrete-time index ranges as $0 \le n < N$.

---

* AN is supported by EPSRC Grant EP/E045235/1.
** MDP is supported by EPSRC Leadership Fellowship EP/G007144/1.

T. Adali et al. (Eds.): ICA 2009, LNCS 5441, pp. 605–612, 2009.

In the *blind* case only $x(n)$ is known. If $s(n)$ remains unknown but A is given, then the problem is called *semi-blind*. If both A and $s(n)$ are known, then we can determine upper performance bounds; this ideal *oracle* estimation case is useful for algorithm benchmarking purposes [10].

Underdetermined audio source separation is typically addressed by *time-frequency* (TF) *masking*, which assumes that we can transform $x(n)$ by a linear, invertible TF transform so that the sources overlap as little as possible [4]. State-of-the-art methods have the potential to yield sparser representations and superior performance compared to non-adaptive transforms with constant block lengths [9, 10]. Such methods include adaptive, dyadic *lapped orthogonal transforms* (LOTs) [6] and adaptive, non-dyadic LOTs, which give better performance in return for higher computational complexity [7]. We recently introduced *MPEG-like* LOTs, which aim for a trade-off between improving computation time, and decreasing artifacts at window boundaries and improving performance, and evaluated them in oracle contexts [8]. In this paper, we extend this previous work by evaluating them in semi-blind contexts.

2   Time-Frequency Masking

Let us denote by $X(m) = (X_1(m), X_2(m))$ the TF transform of $x(n)$, and let $S(m) = (S_1(m), \ldots, S_J(m))$ be the transform of $s(n)$, where $0 \leq m < N$. We assume that exactly two sources are active at each m because this gives better performance than the simpler *binary masking* case which allows only one active source [1, 10]. The set of both source indices contributing to $X(m)$ is denoted by $\mathcal{J}_m = \{j : S_j(m) \neq 0\}$, and is called the *local activity pattern* at m. Given a particular $\mathcal{J}_m$, Equation (1) then reduces to a determined system:

$$X(m) = A_{\mathcal{J}_m} S_{\mathcal{J}_m}(m) \;, \tag{2}$$

where $A_{\mathcal{J}_m}$ is the 2×2 submatrix of A formed by taking columns A_j, and $S_{\mathcal{J}_m}(m)$ is the subvector of $S(m)$ formed by taking elements $S_j(m)$, whenever $j \in \mathcal{J}_m$. Once $\mathcal{J}_m$ has been estimated for each m we estimate the sources according to the following:

$$\begin{cases} \widehat{S}_j(m) = 0 & \text{if } j \notin \mathcal{J}_m \;, \\ \widehat{S}_{\mathcal{J}_m}(m) = A_{\mathcal{J}_m}^{-1} X(m) & \text{otherwise} \;, \end{cases} \tag{3}$$

where $A_{\mathcal{J}_m}^{-1}$ is the inverse of $A_{\mathcal{J}_m}$ [4]. Finally, we recover the estimated source vector in the time domain $\hat{s}(n)$ by using the inverse transform.

The assumption that exactly two sources are active at each m can be modelled probabilistically by assuming that the source coefficients $S_j(m)$ follow a Laplacian prior distribution, independently and identically for all j and m [1]. In the semi-blind case, the maximum a posteriori solution of (2) is then equivalent to minimising the ℓ^1 norm *cost* of the source coefficients [1] given by the following:

$$C(\widehat{S}) = \sum_{m=0}^{N-1} \sum_{j=1}^{J} |\widehat{S}_j(m)| \;. \tag{4}$$

Then for an orthogonal transform, the estimated semi-blind activity patterns are given by

$$\widehat{\mathcal{J}}_m^{\text{sb}} = \arg \min_{\mathcal{J}_m} \sum_{j=1}^{J} |\widehat{S}_j(m)| \ , \tag{5}$$

which depends implicitly on (3).

3   Adaptive Signal Expansions

Let us now describe how to construct an adapted LOT which better fulfills the sparsity assumption of the sources. This entails forming a *partition* of the domain $\{0, \ldots, N-1\}$ of the mixture channels $x_i(n)$, that is,

$$\lambda = \{(n_k, \eta_k)\} \ , \tag{6}$$

such that

$$0 = n_0 < n_1 < \cdots < n_k < \cdots < n_{K-1} = N - 1 \ , \tag{7}$$

where K is the number of partition points. This segments the domain of $x_i(n)$ into adjacent intervals $\mathcal{I}_k = \{n_k, \ldots, n_{k+1} - 1\}$ which should be relatively long over durations which require good frequency resolution, and relatively short over the durations requiring good time resolution. This is achieved by windowing $x_i(n)$ with windows $\beta_k^\lambda(n)$, each of which is supported in $\{n_k - \eta_k, \ldots, n_{k+1} + \eta_{k+1} - 1\}$, thus partly overlapping with its immediately adjacent windows β_{k-1}^λ and β_{k+1}^λ by $2\eta_k$ and $2\eta_{k+1}$ points respectively (see Fig. 1). The *bell parameters*

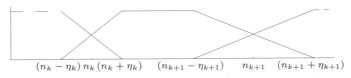

$$(n_k - \eta_k)\, n_k\, (n_k + \eta_k) \quad (n_{k+1} - \eta_{k+1}) \quad n_{k+1} \quad (n_{k+1} + \eta_{k+1})$$

Fig. 1. Schematic representation of window β_k^λ partly overlapping with its adjacent windows β_{k-1}^λ and β_{k+1}^λ

η_k and η_{k+1} determine how quickly β_k^λ rises and falls on its left and side right sides. To avoid 'double overlapping', these are subject to the constraint

$$n_{k+1} - n_k \geq \eta_{k+1} + \eta_k \ . \tag{8}$$

Note that for $\eta_0 = \eta_{K-1} = 0$ appropriate modifications are needed [6].

For every partition λ we form its associated windows according to

$$\beta_k^\lambda(n) = \begin{cases} r\left(\dfrac{n - (n_k - \frac{1}{2})}{\eta_k}\right) & \text{if } n_k - \eta_k \leq n < n_k + \eta_k \ , \\ 1 & \text{if } n_k + \eta_k \leq n < n_{k+1} - \eta_{k+1} \ , \\ r\left(\dfrac{(n_{k+1} - \frac{1}{2}) - n}{\eta_{k+1}}\right) & \text{if } n_{k+1} - \eta_{k+1} \leq n < n_{k+1} + \eta_{k+1} \ , \\ 0 & \text{otherwise} \ , \end{cases} \tag{9}$$

where the *bell function* $r(t)$ satisfies $r^2(t) + r^2(-t) = 1$ for $-1 \leq t \leq 1$, $r(t) = 0$ for $t < -1$ and $r(t) = 1$ for $t > 1$, where t is real-valued, and also satisfies various regularity properties; in practice, we use a sine bell [6].

The *local cosine basis* associated with $\mathcal{I}_k$ is

$$\mathcal{B}_k^\lambda = \left\{ \beta_k^\lambda \sqrt{\frac{2}{n_{k+1} - n_k}} \cos\left[\pi \left(f + \frac{1}{2} \right) \frac{n - (n_k - \frac{1}{2})}{n_{k+1} - n_k} \right] \right\}_{0 \leq f < n_{k+1} - n_k}, \quad (10)$$

where the index m in Sect. 2 is now expressed as $m = (k, f)$, where f indexes the 'frequency'. The basis B^λ spanning the space of signals of length N, for the partition λ, is given by $B^\lambda = \bigcup_{k=0}^{K-1} \mathcal{B}_k^\lambda$. Our aim is to find, of all admissible partitions $\lambda \in \Lambda$, the partition which determines the *best orthogonal basis* (BOB) for representing signals of length N. The set of all candidate bases is called the *library* and is given by $\mathcal{L} = \bigcup_{\lambda \in \Lambda} B^\lambda$.

4   Fast and Flexible Partitioning Schemes

For any additive function C, we can use dynamic programming to determine the BOB which minimises $C(\widehat{S})$ over all $B^\lambda \in \mathcal{L}$ [3,6]. Such algorithms jointly estimate the local activity patterns $\mathcal{J}_m$ according to (5) and find the best orthogonal basis which minimises the ℓ^1 norm given by (4) according to

$$\widehat{\lambda} = \arg\min_{\lambda \in \Lambda} C(\widehat{S}) \ . \tag{11}$$

In previous work [7] we described a *flexible segmentation* (FS) partitioning scheme which admits all possible partitions λ with some 'resolution' L, so that if the signal length N is an integral multiple of L, then each partition point can be written as $n_k = cL$ for $c \geq 0$, and where η_k is subject only to the condition (8). The FS library $\mathcal{L}$ is very large due to a combinatorial explosion between the range of allowed interval lengths, interval onsets and bell parameters, so the computation time is typically very high. To decrease this burden of computational complexity, but still wishing to maintain highly flexible partitioning, we subsequently added some constraints to the FS scheme and introduced the following *MPEG-like* partitioning schemes [8]:

Long-Short (LS). We restrict the range of allowable partitions to admit intervals $\mathcal{I}_k$ of only two lengths, that is, a *long interval* of length L_L and a *short interval* of length $L_S = L$, where L_L is an integral multiple of L_S, and $2\eta_k \in \{L_L, L_S\}$. Apart from this restriction of interval lengths and bell parameters, there are no additional constraints, and LS is otherwise the same as FS.

Window Shapes (WS). This is equivalent to LS with the additional constraint that if $\mathcal{I}_k$ is long, then at most one of η_k and η_{k+1} is short. In other words, the four different window shapes admitted (compared to five in LS) correspond to a long window ($2\eta_k = 2\eta_{k+1} = L_L$), a short window ($2\eta_k =$

$2\eta_{k+1} = L_S$), a long-short *transition window* ($2\eta_k = L_L, 2\eta_{k+1} = L_S$), and a short-long ($2\eta_k = L_S, 2\eta_{k+1} = L_L$) transition window in the MPEG-4 framework.

Onset Times (OT). This is equivalent to LS with the additional constraint if any interval $\mathcal{I}_k$ is long, then n_k must satisfy $n_k = cL_L$ for some integer $c = 0, \ldots, \frac{N}{L_L} - 1$.

WS/OT. This scheme imposes both the WS and OT constraints simultaneously.

WS/OT/Successive Transitions (WS/OT/ST). This scheme imposes the WS/OT constraints in addition to disallowing adjacent transition windows, i.e., a transition window must be adjacent to a long window and a short window. This implements the MPEG-4 windowing scheme [5], with the exception that here, we have more freedom in choosing the bell function $r(t)$.

Clearly, the sizes of the libraries become smaller as we impose more constraints.

5   Experiments and Results

We performed two sets of experiments to test our algorithms. Performance is measured through the *signal to distortion ratio* (SDR) [10],

$$\text{SDR [dB]} = 10 \log_{10} \frac{\sum_{n=0}^{N-1} \sum_{j=1}^{J} (s_j(n))^2}{\sum_{n=0}^{N-1} \sum_{j=1}^{J} (\hat{s}_j(n) - s_j(n))^2} \ . \tag{12}$$

In the first set of experiments, we applied our methods to twenty mixtures in total (ten music mixtures, ten speech mixture), where each mixture each had $J = 3$ sources at a sampling rate of 22.05 kHz, with a resolution of 16 bits per sample, and of length $N = 2^{18}$ (approximately 11.9 s). The sources were mixed according to following mixing matrix:

$$A = \begin{pmatrix} 0.21 \ 0.95 \ 0.64 \\ 0.98 \ 0.32 \ 0.77 \end{pmatrix} \ . \tag{13}$$

For each mixture, we performed semi-blind estimations of $s(n)$ for each of the LS, WS, OT, WS/OT and WS/OT/ST partitioning schemes, with long intervals $L_L = 2^c$, where $c \in \{8, \ldots, 11\}$ (12 ms to 93 ms), and short intervals $L_S = 2^c$, where $n \in \{4, \ldots, 9\}$ (0.73 ms to 23 ms). We exclude all long-short combinations with $L_L \leq L_S$. Results are presented in Table 1 where each entry is the average over the twenty different mixtures corresponding to a particular transform scheme with given block lengths. We also compare the MPEG-like schemes to the baseline *fixed basis* (FB) transform (where $L_L = L_S$ and $2\eta_k = L_L$ for all k) and find that the maximum average SDR is 12.06 dB at $L_L = L_S = 2^{10}$.

For the second set of experiments, we indicate the performance achievable on particular types of mixtures. We applied the best transform scheme as determined by Table 1 (LS) to each instantanous mixture in the *dev1* data set of the *Signal Separation Evaluation Campaign* (SiSEC 2008)[1]. These optimal

---

[1] Available online at http://sisec.wiki.irisa.fr/tiki-index.php

Table 1. Average results for MPEG-like transforms for semi-blind separation on music and speech mixtures (see text). The baseline (fixed basis, FB) transform scheme yields maximum average SDR of 12.06 dB at $L_L = L_S = 2^{10}$.

Scheme	L_L	L_S					
		2^4	2^5	2^6	2^7	2^8	2^9
LS	2^8	10.45	10.50	10.51	10.55	-	-
	2^9	11.72	11.71	11.72	11.72	11.79	-
	2^{10}	12.14	12.10	12.19	12.16	12.23	12.29
	2^{11}	11.70	11.59	11.73	11.77	11.92	**12.34**
WS	2^8	10.45	10.51	10.52	10.55	-	-
	2^9	11.76	11.71	11.74	11.74	11.80	-
	2^{10}	12.16	12.14	12.18	12.16	12.23	**12.28**
	2^{11}	11.62	11.66	11.69	11.75	11.91	12.22
OT	2^8	10.68	10.66	10.65	10.64	-	-
	2^9	11.83	11.83	11.85	11.85	11.83	-
	2^{10}	12.07	12.07	12.07	12.06	12.15	12.19
	2^{11}	11.65	11.56	11.60	11.61	11.86	**12.29**
WS/OT	2^8	10.68	10.67	10.66	10.64	-	-
	2^9	11.84	11.83	11.85	11.85	11.83	-
	2^{10}	12.07	12.07	12.08	12.08	12.16	12.20
	2^{11}	11.62	11.56	11.59	11.61	11.83	**12.29**
WS/OT/ST	2^8	10.69	10.68	10.67	10.64	-	-
	2^9	11.84	11.84	11.85	11.85	11.85	-
	2^{10}	12.05	12.04	12.06	12.08	12.16	12.21
	2^{11}	11.57	11.52	11.53	11.55	11.77	**12.28**

semi-blind results are presented in Table 2; also shown are oracle estimation results, where the L_L and L_S which give best results were determined in previous work [8]. Oracle results are computed by jointly determining the local activity patterns $\mathcal{J}_m$ and the best orthogonal basis $B^\lambda \in \mathcal{L}$ which maximise the SDR given by (12), given knowledge of the reference sources [9].

6   Discussion

For the results in Table 1, the best average SDR is approximately 12.3 dB for each transform scheme. Previous results demonstrated oracle performance of

Table 2. Results for LS scheme for semi-blind and oracle separation on SiSEC 2008 data (see text)

Mixture	J	Semi-blind			Oracle		
		L_L	L_S	Av. SDR [dB]	L_L	L_S	Av. SDR [dB]
3 Female Speakers	3	2^9	2^5	10.35	2^{10}	2^4	24.09
4 Female Speakers	4	2^{11}	2^9	7.04	2^{10}	2^4	18.61
3 Male Speakers	3	2^9	2^9	8.41	2^{10}	2^4	18.56
4 Male Speakers	4	2^{10}	2^9	5.62	2^{10}	2^4	14.37
Music with No Drums	3	2^{10}	2^7	16.33	2^{10}	2^4	34.26
Music with Drums	3	2^9	2^4	11.95	2^{10}	2^4	28.06

23–25 dB, but the differences between the two cases are not suprising; the oracle estimation criterion is the same as the performance measurement criterion (SDR), whereas the semi-blind estimation criterion (ℓ^1 norm) is different.

The greatest variability in average SDR occurs with changing the long interval length L_L. The SDR improvements in the demonstrated range of 1–2 dB may be significant in high fidelity applications. Varying L_S or changing transform scheme has a much smaller effect on performance, in contrast to previous oracle results, where performance naturally decreases as the partitioning schemes get more restrictive and their respective libraries becoming smaller.

In each case in Table 1, the best average SDR is achieved at the *greatest* length for the short intervals ($L_S = 2^9$). In contrast, Table 2 shows indvidual, rather than average, results. Previous oracle results for the LS and WS schemes show that the best average SDR was obtained at the *least* length for the short intervals ($L_S = 2^4$), where we suggested that a library which allows fine-grained placement of the long windows improves performance [8]. The current ℓ_1 criterion does not achieve this, but a semi-blind criterion which admits such fine-grained placement will be a good step towards closing the performance gap between semi-blind and oracle performance. This claim is strengthened by noting that the average SDR improvement yielded by adaptive schemes compared to FB is in the order of 0.3 dB in the semi-blind case, and 1–2 dB in oracle contexts.

7 Conclusions and Further Work

We demonstrated average SDR performance of 12–13 dB on mixtures of music and speech signals by extending our adaptive signal decomposition schemes to the semi-blind case. Table 1 suggests that optimal results are obtained when both L_L and L_S are long, but this requires further investigation. Further work includes extending this technique to the fully blind case. Preliminary experiments on mixing matrix estimation with the SiSEC 2008 data sets using histogram-based methods [2] have shown very promising results, and we intend to incorporate that framework into our adaptive transform schemes.

References

1. Bofill, P., Zibulevsky, M.: Underdetermined blind source separation using sparse representations. Signal Process. 81(11), 2353–2362 (2001)
2. Bofill, P.: Identifying single source data for mixing matrix estimation in instantaneous blind source separation. In: Koutník, J., Kůrková, V., Neruda, R. (eds.) ICANN 2008, Part I. LNCS, vol. 5163, pp. 759–767. Springer, Heidelberg (2008)
3. Huang, Y., Pollak, I., Bouman, C.A., Do, M.N.: Best basis search in lapped dictionaries. IEEE Trans. Signal Process. 54(2), 651–664 (2006)
4. Gribonval, R.: Piecewise linear source separation. In: Proc. SPIE (Wavelets X), vol. 5207, pp. 297–310 (2003)
5. ISO: Information technology—Coding of audio-visual objects—Part 3: Audio (ISO/IEC 14496-3:2005). ISO, Geneva, Switzerland (2005)
6. Mallat, S.: A Wavelet Tour of Signal Processing, 2nd edn. Academic Press, London (1999)
7. Nesbit, A., Plumbley, M.D., Vincent, E.: Oracle evaluation of flexible adaptive transforms for underdetermined audio source separation. In: Proc. ICArn 2008, pp. 17–20 (2008)
8. Nesbit, A., Vincent, E., Plumbley, M.D.: Benchmarking flexible adaptive time-frequency transforms for underdetermined audio source separation. In: ICASSP 2009 (submitted, 2009)
9. Vincent, E., Gribonval, R.: Blind criterion and oracle bound for instantaneous audio source separation using adaptive time-frequency representations. In: Proc. WASPAA 2007, pp. 110–113 (2007)
10. Vincent, E., Gribonval, R., Plumbley, M.D.: Oracle estimators for the benchmarking of source separation algorithms. Signal Process. 87(8), 1933–1950 (2007)

Validity of the Independence Assumption for the Separation of Instantaneous and Convolutive Mixtures of Speech and Music Sources

Matthieu Puigt[1], Emmanuel Vincent[2], and Yannick Deville[1]

[1] Laboratoire d'Astrophysique de Toulouse-Tarbes, Université de Toulouse, CNRS,
14 Av. E. Belin, 31400 Toulouse, France
{mpuigt,ydeville}@ast.obs-mip.fr
[2] IRISA-INRIA, Campus de Beaulieu, 35042 Rennes cedex, France
emmanuel.vincent@irisa.fr

Abstract. In this paper, we study the validity of the assumption that speech source signals exhibit lower dependency and therefore better separability with Independent Component Analysis algorithms than music sources. In particular, we investigate some dependency measures in the temporal and the time-frequency domains, resp. in the framework of instantaneous and convolutive mixtures. Moreover, we test several ICA methods, based on the above dependency measures, on the same source signals. We experimentally show that speech and music sources tend to have the same mean behaviour for excerpt durations above 20 ms, but music signals provide more spread dependency measures and SIR values. Lastly, we experimentally show that Gaussian nonstationary mutual information is better suited to audio signals than mutual information.

1 Introduction

Independent Component Analysis (ICA) [1] is the most investigated class of methods to solve the Blind Source Separation (BSS) problem. Among the applications of BSS, audio processing is one of the major areas of interest. In this paper, we aim to study the behaviour of dependency measures when applied to speech or music source signals, with respect to the length of the signal recordings, for linear instantaneous and frequency-domain convolutive mixtures. Indeed, one generally assumes that, in the "cocktail party" problem, speech sources are independent and are thus separable thanks to ICA while, on the contrary, this is not the case for music sources, because musicians play in a coherent way, thus yielding dependent source signals [2].

A dependency measure, i.e. the mutual information, has been previously studied, only for speech sources, and only in the framework of linear instantaneous mixtures [3]. The authors show that speech source signals are independent (resp. have some dependencies) when source signal excerpts have long duration (resp. short duration). However, the chosen estimator has larger bias and variance than estimators developed more recently [4]. Moreover, [3] only provides mean values of the estimated mutual information and therefore omits the variance of

T. Adali et al. (Eds.): ICA 2009, LNCS 5441, pp. 613–620, 2009.

this measure. One also finds some papers in the literature, e.g. [5], which study the effects of the length of the time-frequency windows on the performance of frequency-domain convolutive methods. However, [5] only investigates the above effects as a function of the length of the impulse response of the mixing filters, for speech signals. It therefore does not show any possible difference of performance with respect to the nature of source signals. Moreover, the authors only measure the source correlation, which does not correspond to the dependency measure used in the tested ICA method.

As a consequence, in Sect. 2, we generalize [3] by computing the statistics (mean values and variance) of two dependency measures of speech *and* music sources (contrary to [3], we not only test the mutual information but also the Gaussian nonstationary mutual information), and by applying ICA algorithms to these signals. In Sect. 3, we extend the above procedure to the framework of frequency-domain convolutive ICA which was studied in [5] in more specific conditions. Conclusions are derived from this investigation in Sect. 4.

2   Independence and ICA Performance in Time Domain

Many ICA techniques achieve source separation by minimizing some dependency measure between the estimated source signals. In this paper, we study two classical measures, i.e. the zero-lag mutual information [1] and the Gaussian nonstationary zero-lag mutual information [6], resp. defined as

$$\mathcal{I}\{s_1, \ldots, s_N\} = -\mathbb{E}\left\{\log \frac{\mathbb{P}_{s_1}(s_1) \ldots \mathbb{P}_{s_N}(s_N)}{\mathbb{P}_{s_1, \ldots, s_N}(s_1, \ldots, s_N)}\right\} \tag{1}$$

and

$$\mathcal{GI}\{s_1, \ldots, s_N\} = \frac{1}{Q}\sum_{q=1}^{Q}\frac{1}{2}\log \frac{\det \operatorname{diag} \widehat{\mathbf{R}}_s(q)}{\det \widehat{\mathbf{R}}_s(q)}, \tag{2}$$

where $\mathbb{E}\{.\}$ stands for expectation, $\mathbb{P}_{s_1, \ldots, s_N}$ and $\mathbb{P}_{s_i}$ ($i \in \{1 \ldots N\}$) are resp. the joint and marginal probability density functions of the sources and Q is the number of disjoint time frames over which the source correlation matrices $\widehat{\mathbf{R}}_s(q)$ ($q \in \{1 \ldots Q\}$) are computed. Note that we do not need to normalize the signals, since the above measures do not depend on their scales. The separation performance may be related to the value of these measures over the true source signals, which is assumed to be near zero. The validity of this assumption depends on the considered mixture. As explained in Sect. 1, while speakers at a "cocktail party" tend to speak freely without attention to distant speakers, musicians often play synchronous sounds at related frequency ratios as specified by the rules of music harmony [2], regardless of the recording duration. Therefore, for each of these types of signals, the dependency between such source signals should intuitively be the same for all durations. One may also expect it to be larger for music than for speech. We are going to study whether these intuitions are true.

We consider the audio BSS dataset [7], which consists of thirty pairs of speech sources and thirty pairs of music sources, sampled at 22.05 kHz. These signals are resp. collected from English audio books read by different speakers and from synchronized multitrack recordings. All pairs of signals are then split into disjoint excerpts of equal durations, from 2^7 samples (5.7 ms) to 2^{18} samples (11.9 s). The mutual information is estimated via the software proposed[1] in [4] and we set the number Q of frames in (2) to $Q = 8$ for computing the Gaussian mutual information. The above dependency measures are computed for each above-defined excerpt.

Figure 1 shows the variations of the estimates of the Gaussian mutual information, for one pair of speech signals, with respect to the index of the excerpt and their time duration. The obtained values have a random behaviour over time. However, when the excerpt length increases, the highest measures of dependency significantly decrease. A similar behaviour has been observed for music sources and for mutual information dependency measures. Therefore, for the sake of brevity, it is not shown in Fig. 1. As a consequence, contrary to [3], we hereafter always study both the mean values and the associated standard deviations (that we consider as a spread measure) of these measures.

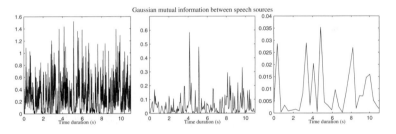

Fig. 1. Variations over time of the Gaussian mutual information between two speech signals in [7], computed for several lengths of test excerpt (*left:* 2^7 samples, *center:* 2^{10} samples, *right:* 2^{13} samples)

Figure 2 thus shows the mean values over all excerpts and all sources of the estimated mutual information and Gaussian mutual information and their above-defined spreads, vs excerpt length. In order to measure the variance of the estimators, this experiment was also conducted for thirty pairs of independent Gaussian white noise signals. These results show that both dependency measures are much larger for audio sources than for independent noise signals, regardless of their excerpt duration. This is partly due to the short-term periodicity of some speech and music sounds [3]. The results also show that both dependency measures span a larger range for music than for speech, but that they are similar for both types of sources on average, except for short durations, i.e. below 20 ms where there are higher for music. In this case, both mean measures of the dependency are high, thus showing that the independence assumption is not fulfilled for speech nor for music, which is in agreement with [2,3]. However, they significantly decrease when the excerpt time duration increases. This phenomenon is

---

[1] Matlab code is accessible at `http://www.klab.caltech.edu/~kraskov/MILCA/`

observed for all pairs of sources, except for one pair of electronically-generated music sources whose mutual information keeps a high value for durations above 2 s due to repetitions of the same note samples over time. Similarly, keeping high dependencies might be observed in a civilized dialogue situation with one speaker being silent when the other speaks and vice-versa. Gaussian mutual information is significantly smaller (resp. slightly smaller or similar) and has a significantly lower (resp. slightly lower or similar) variance than mutual information for durations above (resp. below) 100 ms. These variances indicate that the nonstationary Gaussian source model is more appropriate for audio sources than the stationary non-Gaussian model.

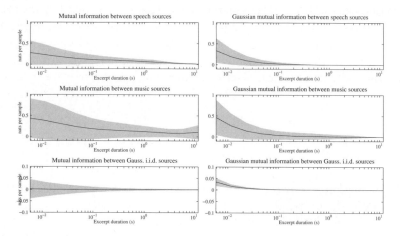

Fig. 2. Mutual information and Gaussian mutual information between the source signals in [7] and between Gaussian i.i.d. signals vs excerpt length. The plain lines and the gray areas resp. denote the mean and the spread (one standard deviation) of the measured values.

In order to show the influence of the dependency measures on the performance of classical ICA methods, we consider the above excerpts, we mix them with the identity matrix and we run the parallel version of FastICA [8][2] and the Pham-Cardoso algorithm [6], since these methods are resp. based on non-Gaussian stationary and Gaussian nonstationary dependency measures. The performance index is the well-known Signal-to-Interference Ratio (SIR) [9]. Figure 3 shows the performance of the above BSS methods, with respect to the excerpt length. Note that in some cases, we found ill-conditioning problems with the Pham-Cardoso algorithm: in the joint-diagonalization procedure of this BSS method, the inverse mixing matrix is estimated in an iterative procedure. We found cases when the determinant of $\widehat{\mathbf{R}}_s(q)$ in (2) is equal to zero, thus making the iterative procedure stop early and the final estimated separation matrices equal to its initial value. Since it is initialized as the identity matrix (which is also our mixing matrix), the

---

[2] FastICA matlab code is accessible at http://www.cis.hut.fi/projects/ica/fastica/

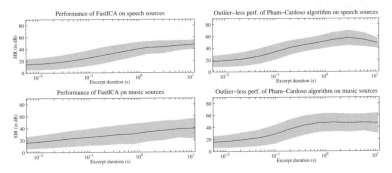

Fig. 3. Performance of ICA methods on the source signals in [7], mixed by the identity matrix. The plain lines correspond to mean SIR (in dB) while the gray areas show the spread.

corresponding SIR is very large, yielding an aberrant value. These outliers are removed in Fig. 3. The performance of FastICA and of the outlier-less version of the Pham-Cardoso algorithm is in agreement with the above dependency measures, except in the case of speech sources separated by the Pham-Cardoso algorithm, since the performance slightly decreases for the longest durations. Indeed, when the size of the excerpts decreases, the dependency between the sources increases and the SIR obtained by the BSS methods significantly decrease. Since the mean dependency measures between sources are high for short-duration excerpts, one could expect the corresponding SIR to be low. Figure 3 shows that it is close to 20 dB, which still yields significant source enhancement. Lastly, except for the shortest music source excerpts, the Pham-Cardoso algorithm yields the highest mean SIR, which confirms the above comments on the dependency measures.

As a consequence of the above analysis, for the sake of brevity, we only study the Gaussian mutual information measure and the corresponding Pham-Cardoso BSS method below, since they are resp. more appropriate for speech and music source signals than the non-Gaussian dependency estimators and the ICA approaches based on these measures.

3   Independence and ICA Performance in Time-Frequency Domain

As explained in Sect. 1, the performance of frequency-domain convolutive BSS has been discussed in [5] by Araki *et al.* with respect to the time-frequency window length of the time-frequency transform and the length of the mixing filter impulse responses. Indeed, they show that the performance of frequency approaches highly depends on the temporal width of the time frequency windows. Here, we extend to the time-frequency domain the procedure proposed in Sect. 2. This work differs from [5] since Araki *et al.* only measured correlation between sources (instead of the dependency measures used in their BSS method) and did not consider the nature of source signals. However, they took into account the length of the impulse responses of the mixing filters, which is not studied here, for the sake of brevity. As stated above, we only consider the Gaussian mutual information, since it is more appropriate for audio sources than the mutual information. We still use the signals [7] tested in the previous section and we compute their short-time Fourier transform (STFT), defined as

$$S_i(t, \omega) = \frac{1}{\sqrt{2\pi}} \int_{-\infty}^{+\infty} s_i(t')h(t' - t)e^{-j\omega t'} dt' \qquad i = 1, 2, \tag{3}$$

where $h(t' - t)$ is a windowing function centered on time t. In these tests, the length of this windowing function geometrically increases from 2^7 to 2^{13} samples. For each STFT window length, for each pair of source signals and for each frequency bin, we compute the Gaussian mutual information. In Fig. 4, we show the mean values and the spread of the considered dependency measure, over the frequency bins, with respect to the length of the windowing function used in (3). Like in Sect. 2, we carry out the same experiment for thirty pairs of independent Gaussian white noises. Contrary to the previous analysis in time domain, here, audio and noise mean dependency measures are of the same order of magnitude. Moreover, the estimated dependency of the sources increases when the size of $h(.)$ increases. This phenomenon, explained in [5], can be summarized as follows: when the STFT window size is high, the number of time-frequency samples in each frequency bin is small and estimating correctly the statistics becomes harder. However, even if all the mean dependency measure values decrease with the STFT size, the ratio between music and speech mean dependency measures (not presented here) significantly increases when the STFT size decreases. This means that music sources present more dependencies than speech for low STFT sizes. Lastly, music and speech sources provide a larger variance than white noise. In particular, the largest variance is obtained with music source signals, which is coherent with the results obtained in Sect. 2.

We then apply the Pham-Cardoso algorithm on each frequency bin of the time-frequency transforms of the sources. In order to analyze the sensitivity of this algorithm to dependency only, we do not generate "real" convolutive mixtures. Indeed, the approximation of convolution by complex-valued multiplication in each frequency bin also affects performance. We avoid measuring this effect by actually generating mixtures from a complex-valued mixing matrix in each bin. Since the considered ICA algorithm is equivariant, its performance does not depend on the value of the chosen matrix. Therefore we simply choose the identity matrix, as in Sect. 2. In order to handle the band-to-band permutation effects, we compute the SIR on the output signals obtained for each frequency bin. Figure 5 shows the mean SIR, over all the frequency bins and the sets of sources, with respect to the STFT window length. The results are in agreement with those in Fig. 4: both classes of signals yield decreasing SIR when the STFT window size increases, which is again in agreement with [5]. For the shortest STFT sizes, the performance obtained with speech sources is higher than for music ones, which is in agreement with the above analysis on dependency measures. Moreover, the variance of the SIR obtained with speech sources is somewhat lower than the one

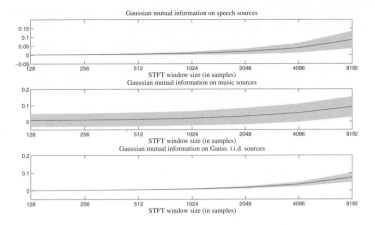

Fig. 4. Gaussian mutual information between the time-frequency transforms of the source signals in [7]. The plain lines correspond to mean value while the gray areas show the spread.

obtained with music sources. Lastly, note that the mean SIR are much higher than those obtained by frequency-domain convolutive ICA algorithms in the literature (which are generally around 10 dB). This is due to the band-to-band permutation problem and to the length of the impulse response of the mixing filters, which have been occulted here.

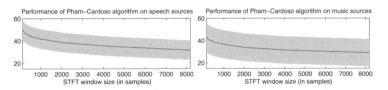

Fig. 5. Performance of the Pham-Cardoso method for the source signals in [7], mixed by the identity matrix in the framework of frequency domain convolutive BSS. The plain lines correspond to the mean SIR (in dB) while the gray areas show the spread.

4   Conclusion

In this paper, we studied the validity of the independence assumption for speech or music source mixtures, with respect to the excerpt size (resp. the STFT window size) in the time (resp. time-frequency) domain. Starting from previous work stating that music sources are dependent [2] for short durations, we looked for statistical differences between speech and music dependency measures in several configurations. We finally showed that these classes of sources almost yield the same mean behaviour for long excerpt durations and high STFT window sizes. However, for both linear instantaneous and convolutive mixture models, the variance of the dependency measures is significantly higher for music than for speech. Moreover, in the time domain, we showed that these classes of sources are separable by ICA for long-enough durations while, for short-time excerpts, the mean dependencies are high, which implies that ICA methods are not appropriate in these cases, which is in agreement with [2,3]. This limitation may be solved thanks to BSS methods not based on independence, e.g. [2,10]. Moreover, even if the independence assumption is met, sparse algorithms often separate much better speech and music sources than classical ICA methods (see e.g. [11]). It could be interesting to compare the performance of sparse approaches with respect to the degree of sparsity of the source signals, for several classes of sources.

References

1. Hyvärinen, A., Karhunen, J., Oja, E.: Independent Component Analysis. Wiley-Interscience, New York (2001)
2. Abrard, F., Deville, Y.: Blind separation of dependent sources using the TIme-Frequency Ratio Of Mixtures approach. In: Proc. Int. Symp. on Signal Processing and its Applications (ISSPA), pp. 81–84 (2003)
3. Smith, D., Lukasiak, J., Burnett, I.S.: An analysis of the limitations of blind signal separation application with speech. Signal Processing 86(2), 353–359 (2006)
4. Kraskov, A., Stögbauer, H., Grassberger, P.: Estimating mutual information. Physical Review E 69(6) (2004); preprint 066138
5. Araki, S., Makino, S., Mukai, R., Nishikawa, T., Saruwatari, H.: The fundamental limitation of frequency domain blind source separation for convolutive mixtures of speech. IEEE Trans. on Speech and Audio Processing 11(2), 109–116 (2003)
6. Pham, D.T., Cardoso, J.-F.: Blind separation of instantaneous mixtures of nonstationary sources. IEEE Trans. on Signal Processing 49(9), 1837–1848 (2001)
7. Vincent, E., Gribonval, R., Plumbley, M.D.: Oracle estimators for the benchmarking of source separation algorithms. Signal Processing 87(8), 1933–1950 (2007)
8. Hyvärinen, A.: Fast and Robust Fixed-Point Algorithms for Independent Component Analysis. IEEE Trans. on Neural Networks 10(3), 626–634 (1999)
9. Mansour, A., Kawamoto, M., Ohnishi, N.: A survey of the performance indexes of ICA algorithms. In: Proc. of Int. Conf. on Modelling, Identification, and Control (2002)
10. Caiafa, C.F., Proto, A.N.: Separation of statistically dependent sources using an L^2-distance non-Gaussianity measure. Signal Processing 86, 3404–3420 (2006)
11. Deville, Y., Puigt, M.: Temporal and time-frequency correlation-based blind source separation methods. Part I: Determined and underdetermined linear instantaneous mixtures 87(3), 374–407 (2007)

Limitations of the Spectrum Masking Technique for Blind Source Separation

Gustavo Fernandes Rodrigues and Hani Camille Yehia

Universidade Federal de Minas Gerais, UFMG
gfernandes@acad.unibh.br,
hani@cefala.org

Abstract. The objective of this study is to analyze the limitations of techniques for blind source separation (BSS) of convolved mixtures based on time-frequency domain binary masking. In this study, initially, the limitations of the spectrum masking technique as a function of the reverberation time of the signals that compose the mixture to be separated are analyzed. When the ideal masks are known, a separation of about 9 dB is obtained for the case of an environment with reverberation time less than 300 ms. From this point on, as the reverberation time increases, the signals that compose the mixture spread over the time-frequency plane, progressively reducing the separation process performance. Finally, a subjective test to measure the intelligibility of the separated voice signals using an ideal binary mask is also proposed for different reverberation times.

Keywords: Blind Source Separation, Time-Frequency Masking, Reverberation, Ideal Binary Mask, Speech Intelligibility.

1 Introduction

The purpose of this work is to analyze the limitations of the spectrum masking technique in the time-frequency domain for blind source separation using mixtures of reverberate sounds. The initial approaches to blind source separation did not consider the presence of delay and reverberation in the mixtures. Although there are systems of blind separation of reverberated voice signals in real environments, the performance of the algorithms still has much to be improved [1]. Recently, some methods based on masking in the field of time-frequency have been proposed [2,3]. The separation is based on non-overlapping spectrogram of sources in the time-frequency domain. The concept of ideal binary mask was first mentioned in [4], as computational auditory scene analysis ($CASA$) and after proposed in [5].

In [5] was presented speech intelligibility tests and showed that estimated masks that are very close to ideal improve speech intelligibility when extracted from multisource mixtures or in reverberant mixtures [6]. Recent studies [7,8] have shown that the application of the ideal binary masks improves the intelligibility of speech masked by one or more interfering sounds. In many studies, it is

T. Adali et al. (Eds.): ICA 2009, LNCS 5441, pp. 621–628, 2009.

assumed that the ideal masks are available. As it is very difficult or practically impossible to estimated the ideal binary masks from real mixtures (due to noise presence or in situations where the source signals are not available), it is very important to analyze the effects of estimation errors on speech intelligibility. Relatively few studies have addressed this question [8]. The result found in [7] shown that the ideal binary mask is a effective technique to improve human speech intelligibility performance in the presence of competing voices. The ideal masking specifies the time-frequency units, where the target signal energy is greater than the interference signal energy. Ideal binary masking retains time-frequency regions of a mixture where the target source is stronger (those assigned to a one in the mask) and eliminates the time-frequency regions where interfering sources are stronger (those assigned to a zero in the mask). The ideal binary mask is generated by checking whether the SNR (signal-to-noise ratio) in each time-frequency unit is greater than $0\ dB$ [7]. In [7] is suggested that the SNR value greater than $-6\ dB$ is a better criterion to estimate the ideal binary mask, at least for improving speech intelligibility. The effects of the local SNR threshold were also analyzed in [8] that show the existence of the plateau region near 100% of the words identified correctly in the ineligibility tests, ranging from -20 to 5 dB. In [9] was derived the theory and the requirements to enable calculations of the ideal binary mask using a directional system without the availability of the source signals. The percentage of correct time-frequency units were measured for three different environments: anechoic, low reverberation time ($400\ ms$) and high reverberation time ($1000\ ms$). In an anechoic room, the percentage of correct time-frequency units is around 95%, when the masker is located at 180°. Using impulse responses from the low and the high reverberant room, the percentage of correct time-frequency units is around 83% and 73%, respectively.

In this study the performance of the ideal binary masks for the voice signals separation is analyzed. The limitations of the spectrum masking techniques, used in blind signals separation, considering different reverberation times are evaluated. Real results obtained by measuring of room impulse response and signal recordings in rooms with different reverberation times are presented and discussed. The voice signals intelligibility separated by a ideal binary mask for the case of convolved mixtures was also analyzed by the application of a subjective test.

2   Analysis of Non-ideal Binary Mask for Source Separation in the Time-Frequency Domain

The performance of the ideal mask for separation of two source signals (with sampling rate of $10000\ kHz$) from an artificial mixture, when errors were introduced in the mask, has been analyzed. The errors were introduced at random, and they were increased gradually ("bit by bit"). Before each increase of one error bit, 1000 different masks were generated. The length frame of the ideal masks had 512 samples ($51.2\ ms$), so the maximum amount of error corresponds to 256 samples. The size of the window (512 samples) was chosen based on performance achieved in the tests shown in [2]. To evaluate the performance of separated sources was used a signal-distortion measure (SDR) given by:

$$SDR = 10log_{10}\left(\frac{\sum_k s_i^2(k)}{\sum_k (s_i(k) - \hat{s}_i(k))^2}\right), \qquad (1)$$

where s_i represents the original source signal i and $\hat{s}_i$ represents the separated source signal i.

In Figure 1, the signal-to-distortion ratio decreases when the number of errors introduced into the ideal mask increases. The performance loss of the separation process is measured as a function of the distance between the ideal mask and the mask effectively used. The results show a performance loss of 3 dB when approximately 10% of the first 256 bits of the mask are inverted. Despite the performance loss of 3 dB, the signals obtained have good intelligibility. These results are similar to those obtained in [8], where the score remained high, near 100% of the words identified correctly in the intelligibility tests and dropped relatively fast thereafter. An error of 10% of the bits in estimating binary masks is acceptable without compromising speech intelligibility [8].

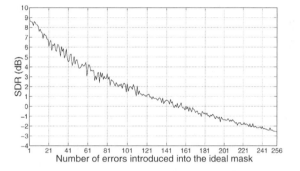

Fig. 1. Mean signal-to-distortion ratio values obtained from the increase of the number of errors introduced into the ideal mask. The ideal masks had frames with 512 samples, so the maximum amount of error is 256 samples.

3   Limitations of Spectrum Masking Methods for Reverberated Signals

The implications caused by the increase of reverberation time in the separation process via masking techniques in the time-frequency domain were studied. Several room impulse responses with different reverberation times were simulated. The convolved mixtures were obtained artificiality. The reverberation time is given in function of room volume and the absorption walls. Spectrograms obtained in different rooms and with same reverberation time can be significantly different. For example, a room of high volume and high absorption walls can have its room reverberation time similar to that of a small volume and low absorption walls. However, preliminary analysis for the above cases showed no difference in performance of separation by ideal masks. The room impulse responses were

Table 1. Reverberation times

Reflections	Reverberation Times (ms)
One wall	33
Two Walls	40
Three Walls	182
Four Walls	290
Five Walls	980
Six Walls	1750
Double Dimensions	3900

simulated using a MatLab software developed by [10]. The dimensions of the simulated room are 6,5 m x 3,25 m x 2,5 m, the sensor position is 2,0 m x 1,75 m x 1,0 m and the source position is 3,25 m x 1,75 m x 1,0 m. The following room impulse responses were simulated: room impulse response in an anechoic room, considering a reflection in just one wall, reflections on two walls, reflections on three walls, reflections on four walls, reflections on five walls, reflection on all walls, and impulse response in a room with double dimensions. The reverberation times for each case mentioned above are shown in Table 1.

The separated signals quality was evaluated for the following situations: i) use of the ideal mask on the source signals; ii) use of the mask, obtained in an anechoic environment by *DUET* algorithm, on the source signals; iii) use of the ideal mask on the signal mixture and iv) use of the mask, obtained in an anechoic environment by *DUET* algorithm, on the signal mixtures. In the first case, the ideal mask was applied directly on each source signal convolved with a room impulse response of a specific source and sensor position. As shown in Figure 2, the mean signal-distortion ratio is 12 dB. Increasing reverberation time up to 300 ms, the signal-distortion ratio decreases 1 dB. In the third case, the results were obtained by applying the ideal mask on the mixed signals and considering different reverberation times, can be observed that the mean signal-distortion ratio in an anechoic room is 9 db and for reverberation times less than 300 ms this measure is 8 dB, but the signals show good intelligibility. In the second and fourth cases, the binary mask was obtained by the *DUET* algorithm for an anechoic room (in this case were used two mixtures to obtain the mask) and applied on the sources and signal mixtures. In these cases, can be verified that for reverberation times less than 300 ms, the results are the same found in the first and third cases. However, if the time reverberation is larger than 300 ms, the signal-distortion ratio show a low performance when compared with the results found in the first and third case.

4   Real Recording in Reverberation Rooms

The reverberation effect for source separation using an ideal mask was analyzed. Speech signals were recorded in two reverberations rooms with reverberation times equal to approximately 100 ms and 1100 ms, respectively. The speakers and microphones positions as well the room dimensions are shown in Figure 4.

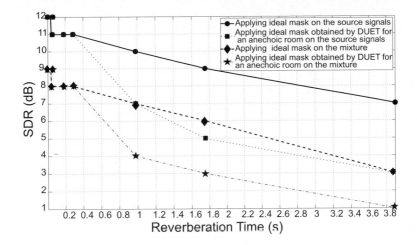

Fig. 2. Signal-distortion ratios of the separated signals using a binary mask for different reverberation times

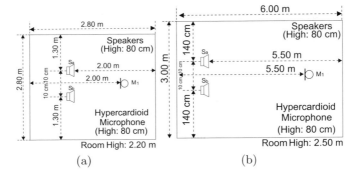

Fig. 3. Room where the signals were recorded. (a) Room 1: studio with reverberation time equal to 100 ms. (b) Room 2: environment with reverberation time equal to 1100 ms.

The signals were recorded with a sampling rate of 44.100 kHz and after the signals were downsampled to 10.000 kHz. The duration time of each recording was 45 seconds. The following signals were used: i) male voice (s_1); ii) female voice (s_2); iii) music - the fifth symphony of Bethoven (s_3) and iv) white noise with a uniform distribution (s_4). The ideal masks were extracted separately for each signal from the recordings. The recordings were made as follows: first, the source signals were recorded separately (to allow the ideal masks extraction) and later, the signals were recorded in order to obtain the mixtures (2x2). The room impulse responses were measured by using a time-stretched pulse - TSP [11]. The TSP signal used was given by [11]:

$$T(k) = \begin{cases} exp(jpk^2), & 0 \le k < N/2, \\ 1, & k = N/2, \\ T^*(N - k), & N/2 < k < N, \end{cases} \qquad (2)$$

where p determines the stretch of the signal. The $IDFT$ (inverse discrete Fourier transform) of $T(k)$ can be obtained by $T^{-1}(k) = 1/T(K)$, as detail in [11]. The reverberation time (T_{60}) was specified using the method proposed by [12].

The signal-distortion measures found in mixtures (2x2) of signals recorded in room with different reverberation times (100 ms and 1100 ms) are shown in table 2. By applying the ideal mask on a mixture with two speech signal voice (male and female voices), the mean SDR decreases about 3 dB in the room with reverberation time equal to 1100 ms. In the other situations proposed, the mean SDR also decreases in the room with reverberation time equal to 1100 ms: about 5 dB for mixture of the signals s_1 (male voice) and s_3 (music) and approximately 8 dB for mixture of the signals s_1 (male voice) and s_4 (white noise).

Table 2. Applying the ideal mask on a mixture of signals recorded in environments with reverberation times equal to 100 and 1100 ms

Room 1: Reverberation time equal to 100 ms		$\hat{s}_a$	$\hat{s}_b$	Mean
Mixture: s_1 (male voice) and s_2 (female voice)	$SDR(dB)$	9	8	9
Mixture: s_1 (male voice) and s_3 (music)	$SDR\ (dB)$	8	21	15
Mixture: s_1 (male voice) and s_4 (white noise)	$SDR\ (dB)$	4	34	19
Room 2: Reverberation time equal to a 1100 ms		$\hat{s}_a$	$\hat{s}_b$	Mean
Mixture: s_1 (male voice) and s_2 (female voice)	$SDR(dB)$	8	4	6
Mixture: s_1 (male voice) and s_3 (music)	$SDR\ (dB)$	8	11	10
Mixture: s_1 (male voice) and s_4 (white noise)	$SDR\ (dB)$	4	18	11

5   Subjective Tests of Intelligibility of the Voice Signals Separated Using an Ideal Masking for Convolved Mixtures

The intelligibility of the voice signals separated using an ideal masking for convolved mixtures are analyzed considering different reverberation times. The database used for the tests is composed of sixteen sentences. The sentences correspond to separate voice signals by applying the ideal mask on mixtures with various reverberation times (anechoic, $33ms$, $40ms$, $182ms$, $290ms$, $980ms$, $1750ms$ and $3900ms$). The tests were carried out by 10 people, being handed a questionnaire to fill during the test with necessary instructions. The tests were conducted individually and each participant heard (by an earphone) the sixteen sentences, the intelligibility of the sentence was analyzed and the phrase perceived was written in the form. The intelligibility of the sentences was expressed by the percentage of correct words defined by the ratio between the number of words correctly perceived and quantity of words in the sentences. The results show that the intelligibility of the signals decreases when the reverberation time increases, as shown in Figure 4. The results are very similar to those obtained in the same test mentioned previously, but using objective measures (signal-to-distortion), as shown in Figure 2. The subjective tests show a high rate of correct words perceived (over 90 %) to separated signals from mixtures with reverberation time less than 300 ms. In [13] the experiments showed the speech intelligibility are degraded when the reverberation time increases. Three environments: anechoic room, living room and conference room were designed to simulate the effect of listening to the sentences in the presence of different levels of reverberation. The reverberation times for the living room and the conference room are equals to 180 ms and 600 ms, respectively. Five people listening three hundred of nonsense sentences spoken conversationally by a male talker. The average percentage of correct words found is 80%, 75% and 67% for the anechoic, living room and conference room, respectively. Other studies must be conducted to find out how much of the drop in intelligibility is caused by the lack of separation, and how much is caused by reverberation time (on clean speech) itself.

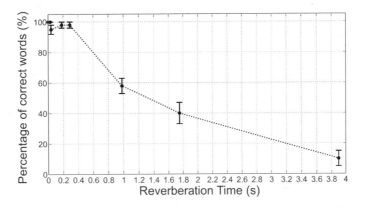

Fig. 4. Subjective tests of intelligibility of voice signals separated by a spectral ideal binary mask considering different reverberation times. The dotted curve shows the average percentage of correct words. The lines at each point represent the standard deviation of the results for each reverberation time.

6   Conclusions

In this work the distortions caused by the use of ideal masks in source separation are analyzed. The results show that the quality and intelligibility of the signals are perfectly acceptable in some situations. When the ideal masks was applied directly on the source signal (original signal), the signal-distorting mean found is 12 dB. In the case, that the ideal mask was applied in a artificially convolved mixture with reverberation time less than 300 ms, the SDR average reduces to 8 dB, but the signal still remains intelligible. In subjective and objectives tests, the results found show that intelligibility of the signals and the SDR ratios are good to mixtures with reverberation time less than 300 ms.

References

1. Araki, S., Mukai, R., Makino, S., Nishikawa, T., Saruwatari, H.: The fundamental limitation of frequency domain blind source separation for convolutive mixtures of speech. IEEE Trans. on Speech and Audio Proc. 11(2), 109–116 (2003)
2. Yilmaz, O., Rickard, S.: Blind separation of speech mixtures via time-frequency masking. IEEE Transactions on Signal Processing 7(52), 1830–1847 (2004)
3. Araki, S., Makino, S., Sawada, H., Mukai, R.: Underdetermined Blind Separation of Convolutive Mixtures of Speech with Directivity Pattern based Mask and ICA. In: Puntonet, C.G., Prieto, A.G. (eds.) ICA 2004. LNCS, vol. 3195, pp. 898–905. Springer, Heidelberg (2004)
4. Hu, G., Wang, D.L.: Speech segregation based on pitch tracking and amplitude modulation. In: Proc. IEEE Workshop on Applications of Signal Processing to Audio and Acoustics, pp. 79–82 (2001)
5. Roman, N., Wang, D.L., Brown, G.: Speech segregation based on sound localization. J. Acoust. Soc. Am. 114, 2236–2252 (2003)
6. Roman, N., Wang, D.L.: Pitch-based monaural segregation of reverberant speech. J. Acoust. Soc. Am. 120, 458–469 (2006)
7. Brungart, D., Chang, P., Simpson, B., Wang, D.L.: Isolating the energetic component of speech-on-speech masking with ideal time-frequency segregation. J. Acoust. Soc. Am. 120, 4007–4018 (2006)
8. Li, N., Loizou, P.C.: Factors influencing intelligibility of ideal binary-masked speech: Implications for noise reduction. J. Acoust. Soc. Am. 123(3), 1673–1682 (2008)
9. Boldt, J.B., Kjems, U., Pedersen, M.S., Lunner, T., Wang, D.L.: Estimation of the ideal binary mask using directional systems. In: Proceedings of the 11th International Workshop on Acoustic Echo and Noise Control - IWAENC (2008)
10. Campbell, D.R., Palomäki, K.J., Brown, G.J.: A MATLAB Simulation of Shoebox Room Acoustics for use in Research and Teaching (2007),
    http://media.paisley.ac.uk/~campbell/Roomsim
11. Suzuki, Y., Asano, F., Kim, H.Y., Sone, T.: An optimum computer-generated pulse signal suitable for the measurement of very long impulse responses. J. Acoust. Soc. Am. 97, 1119–1123 (1995)
12. Schroeder, M.R.: New Method of Measuring Reverberation Time. J. Acoust. Soc. Am. (1965)
13. Payton, K.L., Uchanski, R.M., Braida, L.D.: Intelligibility of conversational and clear speech in noise and reverberation for listeners with normal and impaired hearing. J. Acoust. Soc. Am. 95, 1581–1591 (1965)

Improvement of the Initialization of ICA Time-Frequency Algorithms for Speech Separation

Auxiliadora Sarmiento, Sergio Cruces, and Iván Durán

University of Seville, Camino de los Descubrimientos, Seville, Spain
sarmiento@us.es, sergio@us.es, iduran@us.es

Abstract. The blind separation of speech signals in reverberant environments is a well-known problem for which many algorithms have been developed. In this paper, we propose a novel initialization procedure for those ICA algorithms that work in the time-frequency domain and use the prewhitening of the observations. In comparison with classical initializations, this method allows to reduce drastically the number of permutations. The effectiveness of the proposed technique in realistic scenarios is illustrated by means of simulations.

1 Introduction

In this paper, we address the problem of the blind separation of convolutive speech mixtures assuming the same number of observations and sources. Among the different approaches to solve this problem, we have focused in the one which works in the time-frequency domain. It is well known that the convolutive mixture model in time domain can be approximated as a set of parallel instantaneous mixing problems in the time-frequency domain. Each of these problems can be solved by instantaneous ICA algorithms such as FastICA [1], JADE [2], or algorithms based on the non-stationarity of signals [3,4]; separating the sources for each frequency. However, the permutation ambiguity of the parallel ICA solutions can degrade the estimation of the sources. Various methods have been proposed to align the solutions, for instance, using the assumption of similarity among the envelopes of the source signal waveforms [5] or the estimation of the direction of arrival [6,7].

We propose a new initialization procedure for those ICA algorithms that use the whitening of the observations in the time-frequency domain. This initialization exploits the local continuity of the demixing filter in the frequency domain and our experiments show that it reduces drastically the number of permuted solutions and also improves the quality of the separation with respect to several evaluation criteria. Therefore, it could be used to alleviate the computational burden of the algorithms that solve the permutation problem, although this issue will not be discussed in this work.

The paper is structured as follows. We start presenting the signal model and notation in Sect. 2. In Sect. 3, we describe a classical initialization that exploits

T. Adali et al. (Eds.): ICA 2009, LNCS 5441, pp. 629–636, 2009.
© Springer-Verlag Berlin Heidelberg 2009

the continuity of the mixing system in the frequency domain. Then, in Sect. 4, we present the novel initialization method. Sect. 5 describes how we fix the existing ambiguities, after the separation of the sources. Then, in Sect. 6, we show the results of our simulations. Finally, we summarize the contributions of this article in Sect. 7.

2   Signal Model and Notation

Consider the standard convolutive mixing model of N speech sources, $s_j(n), j = 1, \cdots, N$, in a noiseless situation

$$x_i(n) = \sum_{j=1}^{N} \sum_{k=-\infty}^{\infty} h_{ij}(k)s_j(n-k), \quad i = 1, \cdots, N \ , \tag{1}$$

where $x_i(n), i = 1, \ldots, N$ are the N sensor signals and $h_{ij}(n)$ is the impulse response from source j to microphone i. By applying short-time Fourier transform (STFT), we can convert the convolutive mixture to an instantaneous mixture problem for each frequency f. Let $X_i(f,t)$ and $S_i(f,t)$ be, repectively, the STFT of $x_i(n)$ and $s_i(n)$, and $H_{ij}(f)$ be the frequency response of the channel $h_{ij}(n)$. From (1) we obtain

$$X_i(f,t) = \sum_{j=1}^{N} H_{ij}(f)S_j(f,t), \quad i = 1, \cdots, N \ , \tag{2}$$

which for can be rewritten, in matrix notation, as

$$\mathbf{X}(f,t) = \mathbf{H}(f)\mathbf{S}(f,t) \ , \tag{3}$$

where $\mathbf{X}(f,t) = [X_1(f,t), \ldots, X_N(f,t)]^T$ and $\mathbf{S}(f,t) = [S_1(f,t), \ldots, S_N(f,t)]^T$, are the observation and source vectors for each time-frequency point, respectively, and $\mathbf{H}(f)$ is the frequency response of the the mixing filter whose elements are $H_{ij}(f) = [\mathbf{H}(f)]_{ij} \ \forall i,j$. The vector of outputs or estimated sources $\mathbf{Y}(f,t) = [Y_1(f,t), \ldots, Y_N(f,t)]^T$ can be obtained by pre-multiplying the observations by the separation matrix $\mathbf{B}(f)$, for each frequency f, i.e.

$$\mathbf{Y}(f,t) = \mathbf{B}(f)\mathbf{X}(f,t) \ . \tag{4}$$

There are many ICA algorithms in which the first processing stage consists in whitening the observations in the time-frequency domain as

$$\mathbf{Z}(f,t) = \mathbf{W}(f)\mathbf{X}(f,t) \ , \tag{5}$$

where $\mathbf{W}(f)$ is chosen so as to enforce the covariance of $\mathbf{Z}(f,t)$ to be the identity matrix $\mathbf{C}_Z(f,t) = \mathbf{I}_N$. Since the residual mixture after the whitening

$$\mathbf{Z}(f,t) = \mathbf{U}_*(f)^H \mathbf{S}(f,t) \tag{6}$$

is characterized by the unitary mixing matrix $\mathbf{U}_*(f)$, the separation matrix also inherits this latter property $\mathbf{U}(f)^{-1} = \mathbf{U}(f)^H$, and the outputs are

$$\mathbf{Y}(f,t) = \mathbf{U}(f)^H \mathbf{Z}(f,t) = \mathbf{B}(f)\mathbf{X}(f,t), \tag{7}$$

where the overall separation system decomposes as

$$\mathbf{B}(f) = \mathbf{U}(f)^H \mathbf{W}(f) \ . \tag{8}$$

According to a certain criterion, the optimal separation system $\mathbf{B}_o(f)$ yields the source estimates

$$\mathbf{Y}(f,t) = \mathbf{B}_o(f)\mathbf{X}(f,t) \ . \tag{9}$$

3   Classical Initialization

The idea of exploiting the continuity of the frequency response of the mixing filter and its inverse is not new. Many existing blind source separation algorithms in the time-frequency domain (see for instance [8]), assume that the optimal separating solution $\mathbf{B}_o(f)$, at the frequency f, is close to the optimal solution in contiguous frequencies. Under this assumption, it seems reasonable to initialize the separation system $\mathbf{B}(f)$ from the value of the optimal separation system at the previous frequency, i.e., setting for $f > 0$

$$\mathbf{B}(f) = \mathbf{B}_o(f-1) \ , \tag{10}$$

before optimizing $\mathbf{B}(f)$ with an ICA algorithm. Unfortunately, (10) cannot be directly applied in those algorithms which require the whitening of the observations. As we have seen in (8), these algorithms decompose the separation system as the product of an unitary matrix and the whitening system. Due to the variability of the sources spectra, even at contiguous frequencies, the whitening matrices $\mathbf{W}(f)$ and $\mathbf{W}(f-1)$ are different, so solving directly for $(\mathbf{U}(f))^H = \mathbf{B}_o(f-1)\mathbf{W}^{-1}(f)$, in general, violates unitary assumption of $(\mathbf{U}(f))$.

An alternative method to initialize from previous solutions while avoiding the previously described problem consists in initially preprocessing the observations at frequency f by the optimal separation matrix determined for the previous frequency. This *classical initialization* technique, first, computes the new observations

$$\mathbf{X}^{inic}(f,t) = \mathbf{B}_o(f-1)\mathbf{X}(f,t) \ , \tag{11}$$

and then, determines matrix $\mathbf{W}(f)$ which whitens these new observations. Later, the rotation matrix $(\mathbf{U}(f))^H$ can be obtained by any preferred ICA method. Therefore, this *classical initialization* method, consists in decomposing the overall separation matrix in the following three factors

$$\mathbf{B}(f) = (\mathbf{U}(f))^H \mathbf{W}(f)\mathbf{B}_o(f-1) \ . \tag{12}$$

4   Proposed Initialization

In order to exploit the continuity of the frequency response of the separating filter, we propose to initialize the separation system $\mathbf{B}(f)$ from the its joint closest value to a set of optimal separation systems already computed at nearby frequencies. This leads to the following constrained minimization problem

$$\sum_i \alpha_i \|\mathbf{B}_o(f-i) - \mathbf{B}(f)\|_F^2 \quad s.t. \quad (\mathbf{U}(f))^H \mathbf{U}(f) = \mathbf{I}_N , \tag{13}$$

where $\|\cdot\|_F$ denotes the Frobenius norm. The corresponding Lagrangian function $\mathcal{L}$ is given by

$$\mathcal{L}(\mathbf{U}(f)) = \sum_i \alpha_i \left\|\mathbf{B}_o(f-i) - (\mathbf{U}(f))^H \mathbf{W}(f)\right\|_F^2 - \mathrm{Tr}\left\{\mathbf{\Lambda}\left((\mathbf{U}(f))^H \mathbf{U}(f) - \mathbf{I}\right)\right\} ,$$

where $\mathbf{\Lambda}$ is the Hermitian matrix of multipliers. The minimization of the Lagrangian is obtained solving for $\mathbf{U}(f)$ from the equation

$$\frac{\partial \mathcal{L}}{\partial (\mathbf{U}(f))^*} = \sum_i \alpha_i \left(\mathbf{W}(f)\mathbf{W}(f)^H \mathbf{U}(f) - \mathbf{W}(f)\mathbf{B}_o(f-i)^H\right) - \mathbf{U}(f)\mathbf{\Lambda} = \mathbf{0} .$$

Then, one obtains the desired solution

$$\mathbf{U}(f) = \mathbf{Q}_L \mathbf{Q}_R^H , \tag{14}$$

where $\mathbf{Q}_L$ and $\mathbf{Q}_R$ are, respectively, the left and right singular vectors of the following factorization

$$[\mathbf{Q}_L, \mathbf{D}, \mathbf{Q}_R] = \mathrm{svd}\left(\mathbf{W}(f)\sum_i \alpha_i \mathbf{B}_o(f-i)^H\right) . \tag{15}$$

Since whitening constitutes an essential preprocessing stage of many ICA algorithms, there are potentially a wide number of algorithms that can benefit from our proposed initialization.

5   Avoiding the Indeterminacies

Due to the decoupled nature of the solutions across different frequencies, the correspondence between the true sources and their estimates, in general suffers from scaling, phase and ordering ambiguities. Thus, the vector of source estimates can be approximately modeled as

$$\mathbf{Y}(f,t) \approx \mathbf{P}(f)\mathbf{D}(f)\mathbf{S}(f,t) , \tag{16}$$

where $\mathbf{P}(f)$ is a permutation matrix and $\mathbf{D}(f)$ is a diagonal matrix of complex scalars. In sequel we describe how to avoid these ambiguities. The problem of permutation across frequencies is solved by a modified version of the solution proposed in [9] but using the optimal pairing proposed in [10]. For each possible frequency f_k we set $\mathbf{Y}^{(1)}(f_k,t) = \mathbf{Y}(f_k,t)$, and sequentially correct the ordering of the signals by premultiplying with the estimated permutation matrices $\mathbf{P}^{(l)}(f_k)$ for $f_k = 1, \ldots, n_F$ at iteration l, i.e.

$$\mathbf{Y}^{(l+1)}(f_k,t) = \mathbf{P}^{(l)}(f_k)\mathbf{Y}^{(l)}(f_k,t) \ . \tag{17}$$

Then, provided that we did not achieve the convergence, which occurs when $\mathbf{P}^{(l)}(f) \neq \mathbf{I}$ for any frequency f, we increment the iteration index $l = l + 1$ and restart the previous ordering procedure.

Let us denote $\tilde{Y}_j(f_k,t)$ the logarithmic function

$$\tilde{Y}_j^{(l)}(f_k,t) = \ln\left(\left|Y_j^{(l)}(f_k,t)\right|^2\right)\left\langle \ln\left(\left|Y_j^{(l)}(f_k,t)\right|^2\right)\right\rangle_t, \quad j = 1,\cdots,N \tag{18}$$

and the correlation matrix

$$C_{\tilde{Y}_i\tilde{Y}_j}^{(l)}(f_k) = \sum_{\substack{f=1 \\ f \neq f_k}}^{n_F} \left\langle \tilde{Y}_i^{(l)}(f_k,t)\tilde{Y}_j^{(l)}(f,t)\right\rangle_t \tag{19}$$

between $\tilde{Y}_i^{(l)}(f_k,t)$ and a function of the other estimated sources at other frequencies. The estimation of the permutation at each frequency f_k is obtained, in the case of two sources, after the evaluation of the following statistic

$$d(f_k)^{(l)} = C_{\tilde{Y}_1\tilde{Y}_1}^{(l-1)}(f_k) + C_{\tilde{Y}_2\tilde{Y}_2}^{(l-1)}(f_k) - C_{\tilde{Y}_2\tilde{Y}_1}^{(l-1)}(f_k) - C_{\tilde{Y}_1\tilde{Y}_2}^{(l-1)}(f_k) \ . \tag{20}$$

We set $\mathbf{P}^{(l)}(f_k) = \mathbf{I}$ if $d(f_k)^{(l)} \geq 0$, and we set an antidiagonal permutation otherwise. This procedure can be extended in a straightforward way for more than two sources. After the alignment of the estimated sources, we enforce them to absorb the corresponding diagonal coefficients of the separating system at each frequency. This way we avoid the scaling and phase ambiguities.

6 Simulations

We simulated the 2×2 mixing system of Fig. 1 with the help of the image technique. The Roomsim toolbox[1] was used to determine the channel impulse responses for the configuration of microphones and loudspeakers we propose in Fig. 2. The two original speech sources were randomly chosen from male and female speakers, in a database[2] of 12 individual recordings of 5 s duration and sampled at 10KHz. We computed the STFT using hanning windows of length 1024 samples and 90% overlap.

---

[1] http://media.paisley.ac.uk/not campbell/Roomsim/
[2] http://www.imm.dtu.dk/pubdb/p.php?4400

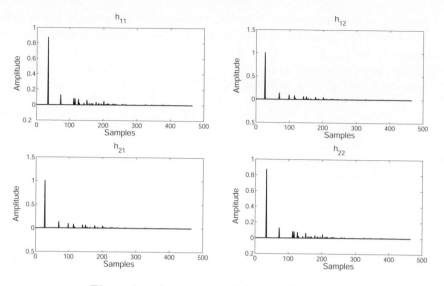

Fig. 1. Impulse response of the considered filter

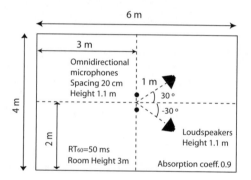

Fig. 2. Microphone and loudspeaker positions for the simulated room recordings

We estimated the separation system $\mathbf{B}_o(f)$ by initializing the Thin-ICA separation algorithm [11] with both the *classical initialization* of Sect. 3 and with the proposed initialization of Sect. 4. After that, we fixed the ambiguities of $\mathbf{B}_o(f)$, and then filtered the observations to obtain the time-domain estimated sources $y_1(t)$ and $y_2(t)$. We first compared both initializations in terms of the number of permutations that were necessary to align the estimated sources in the time-frequency domain. Fig. 3 illustrates this number of permutations for the *classical initialization* and the proposed method to initialize from a set of $k = 1, 3, 5, 7, 9,$ previous separating solutions. One can observe that the proposed initialization reduces significatively the number of permutations with respect to the classical initialization, and that the best result is obtained when the initialization is based on the solution at the previous frequency.

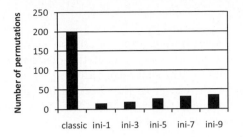

Fig. 3. Comparison of the number of permutations required to align the estimated sources with the classical initialization and the proposed initialization based in the k preceding separation solutions. Results were averaged over 23 different mixtures.

We also compared the initializations in terms of the quality of the estimated sources. According to [12], each estimated source was decomposed into three terms $y_i(t) = s_{target} + e_{interf} + e_{artif}$ which, respectively, represent the target source, the interference from other sources and a last component of artifacts. This decomposition was done by the BSS_EVAL toolbox, where we allowed the time-invariant filtering mode by a 100 samples filter length. Then, we analyzed three performance measures: the Source to Interferences Ratio (SIR), the Sources to Artifacts Ratio (SAR) and the Source to Distortion Ratio (SDR) defined by

$$\text{SIR} = 10 \log_{10} \frac{\|s_{target}\|^2}{\|e_{interf}\|^2}, \qquad \text{SAR} = 10 \log_{10} \frac{\|s_{target} + e_{interf}\|^2}{\|e_{artif}\|^2},$$

$$\text{SDR} = 10 \log_{10} \frac{\|s_{target}\|^2}{\|e_{interf} + e_{artif}\|^2}. \tag{21}$$

Table 1 shows the average performance measures of the different initialization techniques. These results evidence that the SIR is essentially not affected by the initialization method. On the contrary, the proposed initialization improves the SAR and SDR measures up to 8 dB in comparison with the classical method. This suggests that the proposed initialization helps to reduce the artifacts or echoes of the true sources introduced by the ICA algorithm. The best performance is again obtained by the initialization from the previous solution.

Table 1. Comparison of the average SIR, SAR and SDR for different initializations

	classic	$ini-1$	$ini-3$	$ini-5$	$ini-7$	$ini-9$
SIR (dB)	20.16	22.03	22.27	22.10	22.23	21.85
SAR (dB)	0.86	9.32	8.87	7.54	7.44	6.30
SDR (dB)	0.70	8.92	8.53	7.25	7.16	6.06

7   Conclusions

We have proposed a novel initialization operation that can be used by any ICA algorithm that uses a pre-whitening of the observations. This initialization exploits the continuity of the separating filter in the frequency domain, and it compares favorably with respect to the classical initialization which exploits the same property. Our simulation results evidence a remarkable reduction in the number of permuted solutions and an improvement of the source estimates.

References

1. Hyvärinen, A.: Fast and robust fixed-point algorithm for independent component analysis. IEEE Trans. Neural Network 10(3), 626–634 (1999)
2. Cardoso, J.F.: Blind beamforming for non-Gaussian signals. IEE Proceedings-F, 362–370 (December 1993)
3. Pham, D.T., Cardoso, J.F.: Blind separation of instantaneous mixtures of non stationary sources. IEEE Trans. Signal Processing 49(9), 1837–1848 (2001)
4. Belouchrani, A., Abed-Meraim, K., Cardoso, J.F., Moulines, E.: A blind source separation technique using second-order statistics. IEEE Trans. Signal Process. 45(2), 434–444 (1997)
5. Murata, N., Ikeda, S., Ziehe, A.: An approach to blind source separation based on temporal structure of speech signals. Neurocomputing 41(14), 1–24 (2001)
6. Ikram, M.Z., Morgan, D.R.: A beamforming approach to permutation alignment for multichannel frequency-domain blind speech separation. In: Proc. IEEE Int. Conf. on Acoustic, Speech and Signal Processing, pp. 881–884 (2002)
7. Sawada, H., Mukai, R., Araki, S., Makino, S.: A robust and precise method for solving the permutation problem of frequency-domain blind source separation. In: Proc. Int. Conf. on Independent Component Analysis and Blind Source Separation, pp. 505–510 (2003)
8. Anemüller, J., Kollmeier, B.: Adaptive separation of acoustic sources for ane-choic conditions: A constrained frequency domain approach. Speech Communication 39(1-2), 79–95 (2003)
9. Pham, D.T., Servire, C., Boumaraf, H.: Blind separation of convolutive audio mixtures using nonstationarity. In: Proc. of ICA 2003 Conference, Nara, Japan (April 2003)
10. Tichavsky, P., Koldovsky, Z.: Optimal paring of signal components separated by blind techniques. IEEE Signal Processing Letters 11(2), 119–122 (2004)
11. Cruces, S., Cichocki, A., De Lathauwer, L.: Thin QR and SVD factorizations for simultaneous Blind Signal Extraction. In: Proc. of the European Signal Processing Conference (EUSIPCO), Viena, Austria, pp. 217–220 (2004)
12. Fèvotte, C., Gribonval, R., Vincent, E.: BSS_EVAL toolbox user guide, IRISA, Rennes, France, Tech. Rep. 1706, (2005),
   http://www.irisa.fr/metiss/bss_eval/

Audio Watermark Detection Using Undetermined ICA

Jongwon Seok[1] and Hafiz Malik[2]

[1] Department of Information and Communication Engineering,
Changwon National University, South Korea
[2] Department of Electrical and Computer, University of Michigan-Dearborn, USA
jwseok@changwon.ac.kr, {jwseok,hafiz}@umd.umich.edu

Abstract. This paper presents a blind watermark detection scheme for additive watermark embedding model. The proposed estimation-correlation-based watermark detector first estimates the embedded watermark by exploiting non-Gaussian of the real-world audio signal and the mutual independence between the host-signal and the embedded watermark and then a correlation-based detector is used to determine the presence or the absence of the watermark. For watermark estimation, blind source separation (BSS) based on underdetermined independent component analysis (UICA) is used. Low *watermark-to-signal ratio* (WSR) is one to the limitations of blind detection for additive embedding model. The proposed detector uses two-stage processing to improve WSR at the blind detector; first stage removes the audio spectrum from the watermarked audio signal using linear predictive (LP) filtering and the second stage uses resulting residue from the LP filtering stage to estimate the embedded watermark using BSS based on UICA. Simulation results show that the proposed detector performs significantly better than existing estimation-correlation-based detection schemes.

Keywords: Audio Watermark Detection, Blind Source Separation, Mean-Field Approaches, Independent Component Analysis, Linear Predictive Coding.

1 Introduction

The spread spectrum (SS) based watermarking is one of the most representative of blind additive embedding (AE) model which relies on the theory of spread spectrum communication for information embedding and detection. More specifically, in case of SS-based watermarking, the host signal acts as interference at the blind detector (the host signal, x, is not used during watermark detection process) and as the host signal has much higher energy than the watermark therefore host interference deteriorates detection performance at the blind detector (or detectors hereon unless otherwise states). Superior detection performance is one of the desirable features for blind AE model.

Main motivation of this paper is to design a blind detector for SS-based watermarking. Existing detectors for SS-based watermarking schemes are bounded by the host signal interference at the detector. The proposed detector intends to reduce the host signal interference by developing an estimation-correlation-based detection

T. Adali et al. (Eds.): ICA 2009, LNCS 5441, pp. 637–645, 2009.

framework. The proposed detector therefore consists of two stages: 1) watermark estimation stage, and, 2) watermark detection stage. Objective of watermark estimation stage is to estimate embedded watermark which has higher *watermark-to-signal-ratio* (WSR) than the watermarked audio. To accomplish this goal, blind source separation based on the underdetermined ICA (UICA) framework (i.e. ICA for more sources than sensors) is used for watermark estimation. To this end, we model the problem of blind watermark detection for AE as that of BSS for underdetermined mixtures. To ensure better WSR at the watermark estimation stage, the watermarked audio is pre-processed to remove correlation in audio signal using linear predictive (LP) filtering. It has been shown that the received watermarked signal is an underdetermined linear mixture of the underlying independent sources obeying non-Gaussian distributions, therefore BSS based on the UICA framework can be used for the watermark estimation [1]. Similarity measure based on correlation is then used to detect the presence or the absence of the embedded watermark in the estimated watermark. Performance of the proposed watermark detection scheme is evaluated using sound quality assessment material (SQAM) downloaded from [2]. Simulation results for the SQAM dataset show that the proposed scheme performs significantly better than existing estimation-correlation-based detection schemes [2] based on median filtering and Wiener filtering.

2 Motivation

Majority of existing SS-based watermarking schemes [3] use AE model to insert the watermark into the host audio. Mathematically, the SS-based watermark embedding process can be expressed as,

$$y(n) = x(n) + w(n), \ n = 1,2,...,N \tag{1}$$

where $y(n)$ is the watermarked audio signal in the marking space, $x(n)$ the original signal in the marking space, and $w(n)$ the watermark signal. It is reasonable to assume that $x(n)$ and $w(n)$ are zero-mean and independent and identically distributed (i.i.d.) random variables with variance, σ_x^2, and σ_w^2, respectively. It is assumed further that x and w are mutually independent. In the data hiding literature, the embedding model given by Eq. (1) is referred as blind AE, as embedder ignores the host signal information during watermark embedding process.

Adversary attacks or distortion due to signal manipulations, v, can be modeled as an additive channel distortion. Therefore, the watermarked audio signal subjected to adversary attacks or channel distortions,

$$y_n(n) = y(n) + v(n), \tag{2}$$

is processed at the detector to detect the presence or the absence of the embedded watermark. The basic additive embedding and correlation based detection framework is shown in Figure 1.

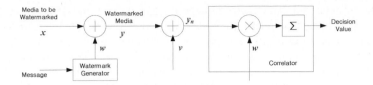

Fig. 1. Semantic diagram of the basic additive embedding and correlation based detection framework

Correlation based detector is commonly used to detect the presence or the absence of the embedded watermark. The decision threshold, d, is obtained by correlating the received watermarked audio and the watermark sequence at the detector is given as,

$$d = \frac{1}{N}\sum_{i=1}^{N} x(i)w(i) + \frac{1}{N}\sum_{i=1}^{N} w^2(i) + \frac{1}{N}\sum_{i=1}^{N} w(i)v(i) \qquad (3)$$

$$= d_x + d_w + d_v$$

where $d_w \ (= \sigma_w^2)$ is the energy of the watermark, $E_x\{d_x\}= 0$, and $E_v\{d_v\}= 0$, where $E_x\{.\}$ denotes expectation over random variable x.

Let us assume no attack scenario, that is, $v = 0 \Rightarrow d = d_x + d_w$. Now according to the central limit theorem d_x can be modeled as a Gaussian random variable, for a sufficiently large N. It is important to mention that detection performance of the correlation-based detector depends on the decision threshold used. Let us assume that watermark detection threshold, $T = d_x /2$. In this case, the probability of a false negative P_{fn} (the watermark is present, but the detector decides otherwise) equals the probability of a false positive P_{fp} (the watermark is not present but the detector decides otherwise), i.e.,

$$P_{fn} = P_{fp} = \frac{1}{2} erfc \left(\frac{d_w}{\sqrt{8}\sigma_x} \right) \qquad (4)$$

where $erfc(\cdot)$ is the complementary error function. Following observations can be made from the Eq. (4):

1. Zero bit error probability is not achievable even in the absence of attack-channel distortion, that is, $v = 0$. In addition, decoding bit error probability is a function of *signal-to-watermark ratio* (SWR = 1/WSR), here $SWR = (\sigma_x / d_w)$. Therefore, the host audio acts as interference at the detector which deteriorates detection performance significantly as, $\sigma_x^2 >> \sigma_w^2$ for most of the watermarking applications [3].

2. The false positive and the false negative probabilities decrease as the strength of watermark d_w increases. However, the strength of the embedded watermark is bounded by the masking threshold the host audio signal which is estimated using human auditory system (HAS) [8].

3. The false positive and the false negative probabilities decrease as the strength of the host audio, σ_x, decreases. This can be achieved by applying whitening before correlation in the detection procedure.

It can be summarized that the detection performance of a blind detector for additive watermarking schemes is inherently bounded by the host-signal interference at the detector. The motivation behind this paper is to design a watermark detector for AE with improved watermark detection performance. Towards this end, the proposed detector uses the theory of ICA by posing watermark estimation as a blind source separation (BSS) problem from an underdetermined mixture of independent sources. The fundamentals of the ICA theory are briefly outlined in the following section followed by the details of the proposed ICA-based detector.

2.1 Independent Component Analysis

The ICA is a statistical framework for estimating underlying hidden factors or components of multivariate statistical data. In the ICA model, the data variables are assumed to be linear or nonlinear mixtures of some unknown latent variables, and the mixing system is also unknown [4]. Moreover, these hidden variables are assumed to be non-Gaussian and mutually independent. The linear-statistical, static ICA generative model considered in this paper which is given as,

$$X^{(i)} = AS^{(i)} + V^{(i)} : i = 1,..., N \tag{5}$$

where $X^{(i)}$ represents N-realizations of m-dimensional random vector known as observation vector, $S^{(i)}$ are the realizations of n_1-dimensional source vector or the hidden variables, $A \in R^{m \times n_1}$ is an unknown matrix also referred as mixing matrix, and $V^{(i)}$ represents realizations of the noise, independent of the underlying sources, $S^{(i)}$.

The mixtures in which number of observations (dimensionality of observation vector, $X^{(i)}$), m, is smaller than the number of sources (dimensionality of source vector $S^{(i)}$), n_1, i.e. $m < n_1$ is referred as underdetermined mixtures. In this paper, we will consider such mixtures to model blind watermark estimation problem using blind source separation (BSS) based on ICA framework. The BSS framework using ICA model tries to find a linear representation of the underlying sources, $S^{(i)} i = 1,...,n_1$ which are statistically independent based on the observation only. In other words, the BSS framework intends to estimate the mixing matrix or the hidden sources $S^{(i)}$ or both from the observation alone. As pointed out in many works such as in [1, 4] (and references therein) that the underdetermined mixtures cannot be linearly inverted due to bound on the rank of mixing matrix, which makes it even more difficult to estimate the underlying sources even if the mixing matrix is known. Therefore, the problem of extracting sources from the observation following underdetermined mixture model is nontrivial [4]. Recently, many researchers have proposed methods to solve BSS for underdetermined mixtures problem [1, 4, 5]. For example, Lathauwer et al in [5] have proposed BSS scheme based on multi-linear analysis and Hojen-Sorensen et. al. in [1] have proposed a statistical framework based on mean-field approaches to solve BSS for underdetermined mixtures.

Before using BSS based on UICA can be used to estimate watermark from the watermarked audio we need to verify, 1) watermarked audio is an underdetermine mixture of independent sources, and 2) the underlying sources obey non-Gaussian distribution. It can be observed from Eq. (1) that the AE model fits into underdetermined linear mixture model, therefore, BSS for underdetermined mixtures can be used to estimate the embedded watermark given that the underlying latent sources (x and w) satisfies non-Gaussianity and independence constraints. This is a realistic assumption, as multimedia data can be modeled using the non-Gaussian distribution [6]. Therefore, if embedded watermark obeys non-Gaussian distribution then BSS based on UICA can be used for watermark estimation from the watermarked signal [6].

3   Watermark Detection

This section provides an overview of the proposed blind watermark detection scheme from the received watermarked audio signal obtained by additive embedding. The proposed watermark detection scheme consists of two stages, 1) watermark estimation stage, and 2) watermark detection stage. It is mentioned earlier that the watermark estimation stage is further divided into two sub-stages 1.a) spectral removal stage, and 1.b) source separation stage.

3.1   Watermarking Estimation

The goal of the watermark embedder is that embedded watermark should survive intentional and unintentional attacks, whereas the goal of the watermark detector is to detect embedded watermark with very low false rates in the presence of an active adversary and signal manipulations. In case of AE model low false rates are hard to achieve even in the absence of attach channel due to strong host interference. For detector performance analysis, existing correlation-based schemes, model the audio signal as a white Gaussian channel. Recent results in audio processing and compression community, however, show that samples of the real audio signals are highly correlated, which can be exploited to improve the detection performance by de-correlating the input audio before detection. The proposed detection scheme achieves this goal by applying whitening or de-correlation before watermark estimation. Simulation results presented in this paper show that the whitening before watermark estimation using ICA improves detection performance significantly. This improvement can be attributed to the fact that whitening actually increases watermark to interference ratio hence yields superior detection performance.

To remove the correlation in the audio signal, an autoregressive modeling named linear predictive coding (LPC) [7] can be used. The LPC method approximates the original audio signal, $x(n)$, as a linear combination of the past p audio samples, such that

$$x(n) \approx a_1 x(n-1) + a_2 x(n-2) + \cdots + a_p x(n-p) \tag{6}$$

where the coefficients $a_1, a_2, \cdots a_p$ are assumed to be constant over the selected audio segment. Rewriting Eq. (6) by including an error term $e(n)$, that is,

$$x(n) = \sum_{i=1}^{p} a_i x(n-i) + e(n) \tag{7}$$

where $e(n)$ is an excitation or residual signal of $x(n)$. Now using Eq. (7), Eq. (1) can be expressed as,

$$y(n) = \sum_{i=1}^{p} a_i x(n-i) + e(n) + w(n) \tag{8}$$

Likewise, watermark audio can also be expressed as,

$$y(n) = \sum_{k=1}^{p} a_k y(n-k) + \widetilde{w}(n) \tag{9}$$

where $\widetilde{w}(n)$ is the residual signal of the watermarked audio signal. We assume that, by the characteristics of the linear predictive analysis, $\widetilde{w}(n)$ has the characteristics of both $e(n)$ and $w(n)$. We can consider the linear combination of the past audio sample as the estimate $\widetilde{y}(n)$, defined as,

$$\widetilde{y}(n) = \sum_{k=1}^{p} a_k y(n-k) \tag{10}$$

Here prediction error, $e(n)$, can be expressed as,

$$e(n) = y(n) - \widetilde{y}(n) = y(n) - \sum_{k=1}^{p} a_k y(n-k) = \widetilde{w}(n) \tag{11}$$

It can be observed from Eq. (11) that the estimate $\widetilde{w}(n)$ with the audio spectrum removed has the characteristics of both the excitation signal of the original audio $x(n)$ and the watermark signal $w(n)$.

This method transforms the non-white watermarked audio signal to a whitened signal by removing the audio spectrum. It can be observed from Fig. 2 that the empirical probability density function (pdf) of a small segment of the watermarked audio signal before LP filtering and the residual or error signal of the watermarked audio signal after LP filtering. The empirical pdf of the watermarked audio signal is clearly not smooth and has large variations due to voiced part. On the other hand, the empirical pdf of the residual signal has a smoother distribution and a smaller variance than the watermarked audio signal.

It is important to mention that the LPC stage also improves WSR which ultimately improves the source separation performance of the BSS used for watermark estimation. This is because, the watermark sequence is i.i.d., so de-correlation stage does not reduce its energy in the residual signal, whereas, de-correlation does reduce audio signal energy hence improving the WSR.

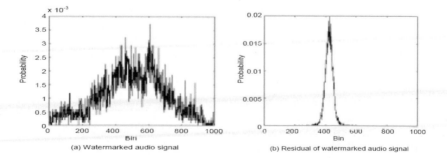

Fig. 2. Empirical probability density functions before and after linear prediction filtering

Residual signal is the then used to estimate the hidden watermark using BSS based on UICA. For watermark estimation, probabilistic ICA method based on mean-field approaches is used. Superior source separation performance is the only motivation behind using of the probabilistic ICA presented in [1]. This is however not the limitation of the proposed scheme, as any of the BSS scheme based on UICA can be used for watermark estimation from the residual signal. Estimated sources are then correlated with the watermark, w, to determine the presence or the absence of the embedded watermark. Fig. 3 shows the block diagram of the proposed audio watermark detection scheme.

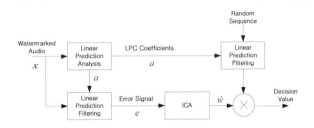

Fig. 3. Block diagram of the proposed watermark detection procedure

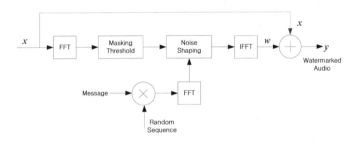

Fig. 4. Watermark embedding procedure

4   Experimental Results

Detection performance of the proposed watermark estimation scheme is evaluated for the watermarked audio clips obtained using Eq. (1). In addition, detection performance of the proposed detector is also compared with existing estimation-correlation based watermark detection schemes based on median filtering, Wiener filtering, and ICA without LP filtering. Detection performance of these four watermark estimation-correlation-based detection schemes is evaluated for watermarked audio obtained using blind AE model discussed in Section 4.1.1. Detection performance is evaluated in terms of the value of correlation coefficient. Here higher correlation value here indicates better detection and vice versa.

4.1 Watermark Embedding and Experimental Setup

4.1.1 Watermark Embedding

The binary message to be embedded is first modulated by a key-dependent random sequence. The watermark is then spectrally shaped in the frequency domain according to masking threshold estimated based on the human auditory system (HAS) ISO/MPEG-1 Audio Layer III model [8]. Motivation here is to design the weighting function that maximizes energy of the embedded watermark subject to a required acceptable distortion. Resulting watermark is then added into the original audio signal in the frequency domain which is then transformed to the time domain to obtain the watermarked audio. A semantic diagram of the audio watermark embedding scheme discussed above is shown in Fig. 4.

4.1.2 Experimental Setup

Simulation results presented in this section are based on the following system settings: 1) 44. kHz sampled and 16 bits resolution audio signals are used as the host audio, 2) 1024-point watermark is then embedded into four consecutive non-overlapping frames, 3) watermarked signal is first segmented into non-overlapping frames of 4096 samples each, then each frame is further segmented into four non-overlapping sub-frames which are then applied to the ICA block to estimate the embedded watermark after LPC filtering. For performance evaluation SQAM downloaded from [2] was used.

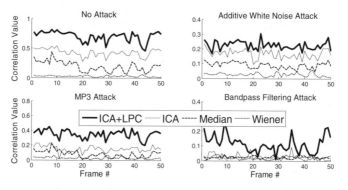

Fig. 5. Robustness performance: No Attack (top-left), AWGN Attack [5% noise power] (top-right), MP3 Compression Attack [128 kbps] (bottom-left), and Bandpass Filtering Attack (bottom-right)

Table 1. Correlation value depending on detection methods using SQAM[2] database

Audio File	ICA	ICA+LPC	MEDIAN	WIENER
Bass47_1	0.449	0.775	0.315	0.046
Gspi35_2	0.519	0.787	0.206	0.056
Harp40_1	0.294	0.582	0.288	0.057
Horn23_1	0.562	0.781	0.329	0.033
Quar48_1	0.411	0.729	0.347	0.054
Sopr44_1	0.499	0.771	0.375	0.055
Trpt21_2	0.480	0.782	0.466	0.062
Vioo10_2	0.493	0.751	0.560	0.067

4.2 Simulation Results

Robustness performance of the proposed the proposed watermark estimation scheme is evaluated for the following attacks scenarios: 1) no adversary attack, 2) additive white Gaussian noise 3) MP3 compression (128 kbps) and 4) Bandpass filtering (2nd-order Butterworth filter with cutoff frequencies 100 and 6000 Hz).

Detection performance of the proposed estimation-correlation based detector scheme and the existing schemes for these attacks is given in Fig 5. It is observed from Fig. 5 that for four attack scenarios the proposed detector outperforms the exiting detectors. In addition, detection performance of the watermark detectors under consideration for SQAM database is given in Table 1. It can be observed from both Fig. 5 and Table 1 that the proposed detector performs significantly better than its counterparts. Improved detection performance of the proposed detector can be attributed to its better host signal interference cancelation capability.

5 Conclusion

In this paper, we described a new framework for estimation-correlation based detection for additive embedding. The proposed blind detection method extracts the embedded watermark signal suppressing the host signal interference at the detector. The proposed framework exploits mutual independence and non-Gaussianity of the audio signal and the embedded watermark to estimate the embedded watermark using BSS based UICA. Experimental results showed that the proposed detection scheme is robust to common signal processing attacks and performs significantly better than existing estimation-correlation based detection schemes.

References

1. Hojen-Sorensen, P., Winther, O., Hansen, L.K.: Mean-Field Approaches to Independent Component Analysis Neural Computation, vol. 14, pp. 889–918. MIT Press Journals, Cambridge (2002)
2. SQAM: http://sound.media.mit.edu/mpeg4/audio/sqam/
3. Cox, I.J., Miller, M.L., Bloom, J.A.: Digital Water-marking. Morgan Kaufmann, San Francisco (2001)
4. Hyvarinen, A., Karhunen, J., Oja, E.: Independent Component Analysis. John Wiley & Sons, Chichester (2001)
5. De Lathauwer, L., Comon, P., De Moor, B., Van-dewalle, J.: ICA Algorithms for 3 Sources and 2 Sensors. In: Proc. IEEE Signal Processing (HOS 1999), pp. 116–120 (1999)
6. Malik, H., Khokhar, A., Ansari, R.: An Improved Detector/Decoder for Spread Spectrum based Water-marking using Independent Component Analysis. In: Fifth Workshop Digital Rights Management (DRM 2005), pp. 171–183 (2004)
7. Atal, B.S., Hanauer, S.L.: Speech Analysis and Synthesis by Linear Prediction of the Speech Wave. J. Acoust. Soc. Am. 50, 637–655 (1971)
8. ISO/IEC IS 11172, Information Technology – Coding of Moving Pictures and Associated Audio for Digital Storage up to about 1.5 Mbits/s

Mixtures of Gamma Priors for Non-negative Matrix Factorization Based Speech Separation

Tuomas Virtanen[1] and Ali Taylan Cemgil[2]

[1] Tampere Univ. of Technology, Korkeakoulunkatu 1, FI-33720 Tampere, Finland
[2] Boğaziçi University, Dept. of Computer Eng., TR-34342 Istanbul, Turkey

Abstract. This paper deals with audio source separation using supervised non-negative matrix factorization (NMF). We propose a prior model based on mixtures of Gamma distributions for each sound class, which hyperparameters are trained given a training corpus. This formulation allows adapting the spectral basis vectors of the sound sources during actual operation, when the exact characteristics of the sources are not known in advance. Simulations were conducted using a random mixture of two speakers. Even without adaptation the mixture model outperformed the basic NMF, and adaptation furher improved slightly the separation quality. Audio demonstrations are available at `www.cs.tut.fi/~tuomasv`.

1 Introduction

Separation of mixtures of sound sources has many applications in the computational analysis of audio, speech enhancement, and noise-robust speech recognition. Particularly, non-negative matrix factorization (NMF) and its extensions have produced good results [1,2,3].

The signal model in non-negative spectrogram factorization approximates the spectrum vector $\mathbf{x}_t$ in frame t as a weighted sum of N basis vectors $\mathbf{b}_n$:

$$\mathbf{x}_t \approx \sum_{n=1}^{N} \mathbf{b}_n g_{n,t}, \tag{1}$$

where $g_{n,t}$ is the gain of the nth component in frame $t = 1, \ldots, T$.

The basis vectors and gains can be estimated by minimizing the error of the approximation (1). In audio signal processing, the divergence

$$\sum_{t=1}^{T} \sum_{f=1}^{F} d(x_{t,f}, \sum_{n} b_{n,f} g_{n,t}) \tag{2}$$

where $d(p, q) = p \log(p/q) - p + q$, has turned out to produce good results [1]. Here $b_{n,f}$ denotes the fth entry of $\mathbf{b}_n$, and f is the frequency index. The same procedure can be derived from a maximum likelihood perspective

$$p(\mathbf{x}_{1:T}|\mathbf{b}_{1:N}, g_{1:T,1:N}) = \sum_{t=1}^{T} \sum_{f=1}^{F} \delta(x_{t,f} - \sum_{n=1}^{N} s_{t,f}^n) \prod_{n=1}^{N} p(s_{t,f}^n|b_{n,f} g_{n,t}) \tag{3}$$

T. Adali et al. (Eds.): ICA 2009, LNCS 5441, pp. 646–653, 2009.

where each spectrogram entry $x_{t,f}$ equals the sum $x_{t,f} = \sum_{n=1}^{N} s_{t,f}^n$ of component spectrograms $s_{t,f}^n$ having Poisson distribution $p(s_{t,f}^n | b_{n,f} g_{n,t}) = Po(s_{t,f}^n; b_{n,f} g_{n,t})$ [5,6]. The divergence (2) can be efficiently minimized using the multiplicative update rules proposed in [7].

When training material of each source in isolation is available, NMF can be used in "supervised mode", i.e., to train class conditional basis spectra of each source in advance [2,3,4]. In the training phase, all the training material of a specific sound class is first concatenated into a single signal, and the spectrogram of the resulting signal is then decomposed into a sum of components using NMF. This results to a class specific set of basis vectors for each source. For the actual separation, the trained basis vector sets of all the source classes are combined and a mixture signal can then be processed using the learned spectra. The previous studies have kept the basis vectors fixed and re-estimated the gains only.

In the real world scenarios, it is either not possible to have training material of a particular target source, or the acoustic conditions in the training and actual operation stages vary. In these situations, adaptive models may be advantageous. One obvious possibility is to train prior distributions $p(\mathbf{b}_n | \Theta)$ instead of fixed parameters $\mathbf{b}_n^*$. Rennie et al. [4] obtained better results in the separation of two speakers by using prior distributions instead of fixed spectra. The computational burden caused by prior distributions can be alleviated if appropriate conjugate priors are chosen, so that one can retain the efficiency of the original NMF algorithm in maximum a posterior (MAP) estimation [5] as explained in Section 2, or in a full Bayesian treatment [6].

This paper discusses the supervised use of NMF where the basis vectors are trained in advance using material where each sound class is present in isolation. We propose here a practical procedure to estimate a Gamma mixture prior model for basis vectors. Section 4 shows simulations using mixtures of two speakers, where the proposed method is shown to outperform the existing ones.

2   Supervised Non-negative Spectrogram Factorization

The characteristics of acoustic sources in real environments are highly variable, hence it is advantageous to have adaptive models that can capture these characteristics. In a probabilistic framework, this can be accomplished by using prior distributions for the basis vectors $\mathbf{b}_n$ instead of fixing them. Formally, in the training phase of supervised non-negative spectrogram factorisation, we ideally wish to estimate class-conditional hyperparameters Θ_c for each source class $c = 1 \ldots C$ by maximising the marginal log-likelihood:

$$\mathcal{L}(\Theta_c) = \int p(s_c | \mathbf{b}_c, g_c) p(\mathbf{b}_c | \Theta_c) p(g_c | \Theta_c) d\mathbf{b}_c dg_c \qquad (4)$$

where s_c is a known spectrogram from a source class c, and $\mathbf{b}_c$ and g_c denote all the basis vector and gains of class c, respectively. Then, the actual separation of mixture spectrogram $\mathbf{x}$ is achieved via computation of

$$p(s_{1:C} | \mathbf{x}, \Theta_{1:C}) = \int \delta(\mathbf{x}_{1:T} - \sum_{c=1}^{C} s_c) \left[\prod_{c=1}^{C} p(s_c | \mathbf{b}_c, g_c) p(\mathbf{b}_c | \Theta_c) p(g_c | \Theta_c) \right] d\mathbf{b}_{1:C} dg_{1:C}$$

However, these integrals can be hard to evaluate and more practical approaches are taken in practice, such as computing MAP estimates for $\mathbf{b}_c$ and g_c.

As a prior for basis vectors $p(\mathbf{b}_c|\Theta_c)$, one can use a Gamma distribution $\mathcal{G}(b_{n,f}; k_{n,f}, \theta_{n,f})$ for each element $b_{n,f}$ of each basis vector $\mathbf{b}_n$. In the MAP framework with the Poisson observation model it results to minimizing the sum of the divergence (2) and the penalty term

$$\sum_{n=1}^{N}\sum_{f=1}^{F}(k_{n,f} - 1)\log(b_{n,f}) - b_{n,f}\theta_{n,f} \tag{5}$$

which is the logarithm of the Gamma distribution [5], up to additive terms which are independent of the basis vector entries.

A typical gain prior $p(g|\Theta_c)$ is an exponential distribution with rate parameter λ which translates to the penalty term $\lambda\sum_{n}\sum_{t} g_{n,t}$. Sparse prior for the gains has been found to improve the separation quality [2].

During separation, when the basis vectors are fixed, the MAP estimation of gains can be obtained by applying iteratively updates

$$\mathbf{r}_t = \mathbf{x}_t ./ \sum_{n}\mathbf{b}_n g_{n,t} \qquad g_{n,t} \leftarrow g_{n,t}\frac{\mathbf{r}^{\mathsf{T}}\mathbf{b}_n}{\mathbf{1}^{\mathsf{T}}\mathbf{b}_n + \lambda}, \quad n = 1,\ldots,N, \tag{6}$$

where ./ denotes element-wise division and $\mathbf{1}$ is a all-one column vector. Similarly, when the gains are fixed, the basis vectors can be updated via

$$b_{n,f} \leftarrow \frac{k_{n,f} - 1 + b_{n,f}\sum_{t}(g_{n,t}x_{t,f}/\sum_{n'} g_{n',t}b_{n',f})}{1/\theta_{n,f} + \sum_{t} g_{n,t}} \tag{7}$$

which is guaranteed to increase the posterior probability of the basis vectors when $k_{n,f} \geq 1$ [5]. It is important to note that under this formalism, the basis is adapted during the actual separation.

3   Training Mixture of Gamma Priors

In single channel source separation, when two or more sources overlap in time and frequency, we have to use redundancy of the sources to achieve good sound source separation. Redundancy in frequency can be used efficiently when basis vectors correspond to entire spectra of sound events instead of just parts of their spectra. Basis vectors corresponding to entire spectra can be trained by restricting only one basis vector to be active in each frame.

We make two assumptions which allow us to train efficiently distributions of basis vectors corresponding to entire spectra: 1) only one basis vector is active in each frame, and 2) the training data can be normalized so that the normalized observations correspond to observed basis vectors. The first assumption can be viewed as an extreme case of sparseness whereas the second cancels out the effect of gains in training basis vector priors. While this is omitting the variation in the Poisson model, we find this procedure virtually identical to the more principled approach where $\mathbf{b}_c$ are considered as latent.

A prior for the basis vectors based on the above assumptions can be trained using a mixture model. In the sequel, we omit the class label c as each class is learned separately. The model for a source class is

$$p(\mathbf{b}|\Theta) = \sum_{n=1}^{N} w_n \prod_{f=1}^{F} \mathcal{G}(b_{n,f}; k_{n,f}, \theta_{n,f}),\tag{8}$$

where k, θ are the shape and scale parameters of individual Gamma distributions and w_n are the prior weights. All the hyperparameters are denoted as $\Theta = (k, \theta, w)$. We do not model the dependencies between frequency bins so that the distribution of a mixture component is the product of its frequency marginals.

3.1   Training Algorithm

The observations are first preprocessed by normalizing each observation vector $\mathbf{b}_t$ so that the norm of $\log(\mathbf{b}_t + \epsilon)$, where ϵ is a small fixed scalar, is unity. The EM algorithm is initialized by running the k-means algorithm with random initial clusters for 10 iterations using the normalized log-spectrum observations to get cluster centroid vectors $\boldsymbol{\mu}_n$. Centroids of linear observations are then calculated as $\mu_{n,f} = e^{\mu_{n,f}} - \epsilon$. From the linear cluster centroids we estimate the initial Gamma distribution parameters as $k_{n,f} = \mu_{n,f}^2$ and $\theta_{n,f} = 1/\mu_{n,f}$. This generates a Gamma distribution having mean $\mu_{n,f}$ and variance 1. Cluster weights are set to $w_n = 1/N$. The iterative estimation procedure is as follows:

1. Evaluate the posterior distribution $z_{n,t}$ that the nth cluster has generated the tth observation as

$$z_{n,t} = \frac{w_n \prod_f \mathcal{G}(b_{t,f}; k_{n,f}, \theta_{n,f})}{\sum_{n'=1}^{N} w_{n'} \prod_f \mathcal{G}(b_{t,f}; k_{n',f}, \theta_{n',f})}\tag{9}$$

2. Re-estimate the mixture weights as

$$w_n = \frac{\sum_{t=1}^{T} z_{n,t}}{\sum_{n'=1}^{N} \sum_{t=1}^{T} z_{n',t}}.\tag{10}$$

3. Re-estimate the shape parameters by solving

$$\log(k_{n,f}) - \psi(k_{n,f}) = \log\left(\frac{\sum_t z_{n,t} b_{t,f}}{\sum_t z_{n,t}}\right) - \sum_t \log(b_{t,f}) z_{n,t}\tag{11}$$

using the Newton-Raphson method, where $\psi(k_{n,f}) = \Gamma'(k_{n,f})/\Gamma(k_{n,f})$ is the digamma function. We used 10 iterations, and the previous estimates of $k_{n,f}$ as initial values.

4. Re-estimate the scale parameters as

$$\theta_{n,f} = \frac{\sum_{t=1}^{T} z_{n,t} b_{t,f}}{k_{n,f} \sum_t z_{n,t}}.\tag{12}$$

The steps 1-4 are repeated for 100 iterations, or until the algorithm converges. In order to prevent too narrow clusters, we found it advantageous to restrict the variance of each cluster above a fixed minimum m after each iteration as follows. The variance of each Gamma distribution is $\mu_{n,f}\theta_{n,f}^2$. For clusters for which the variance is smaller than the minimum limit m, we calculate the ratio $y_{n,f} = m/(\mu_{n,f}\theta_{n,f}^2)$, and then modify the distribution parameters as $\theta_{n,f} \leftarrow \theta_{n,f}y_{n,f}$ and $k_{n,f} \leftarrow k_{n,f}/y_{n,f}$. The above procedure sets the variance of the cluster to m without changing its mean. We also found it advantageous to keep the mixture weights fixed for the first 90 iterations.

3.2   Alternative Gamma Prior Estimation Methods

In addition to the Gamma mixture model, we tried out alternative methods for generating the priors. In general, one can generate a Gamma distribution from fixed basis vectors $b_{n,f}$ obtained with NMF (or by any other algorithm) by selecting arbitrary shape k, and then calculating the scale as $\theta_{n,f} = b_{n,f}/k$. The mean of the resulting distribution equals $b_{n,f}$ and its variance $b_{n,f}^2/k$ scales quadratically with the mean.

In addition to direct training of the Gamma mixture model parameters, we obtained good results by applying a Gaussian mixture model for the log-spectrum observations and then deriving the corresponding Gamma mixture model by matching the moments of each cluster. We calculated the log-spectrum as $\log(\mathbf{b}_t + \epsilon)$ and then trained a Gaussian mixture model, which mean and variance are denoted as $\mu_{n,f}$ and $\sigma_{n,f}^2$, respectively. The mean and variance of the linear observations are $\tilde{\mu}_{n,f} = e^{\mu_{n,f}+\sigma_{n,f}^2/2}$ and $\tilde{\sigma}_{n,f}^2 = (e^{\sigma_{n,f}^2} - 1)e^{2\mu_{n,f}+\sigma_{n,f}}$, respectively. Gamma distributions of linear observations can be generated by matching the mean and variance as $k_{n,f} = \tilde{\mu}_{n,f}^2/\tilde{\sigma}_{n,f}$ and $\theta_{n,f} = \tilde{\sigma}_{n,f}/\tilde{\mu}_{n,f}$.

4   Simulations

We evaluated the performance of the proposed methods in separating signals consisting of two speakers of different genders. We used the Grid corpus [8], which consists of short sentences spoken by 34 speakers. We generated 300 random test signals where three sentences spoken by a male speaker and a female speaker were mixed. Each test signal was generated by concatenating random three sentences of a random male speaker, concatenating random three sentences of a random female speaker, and mixing the signals at equal power level.

The data representation is similar to the one used in [2]: the signals were first filtered with a high-frequency emphasis filter, then windowed into 32 ms frames using a Hamming window with 50 % overlap between adjacent frames. DFT was used to calculate the spectrum of each frame, and the spectra were decimated to 80-band Mel frequency scale by weighting and summing the DFT bins.

In the training phase we learned a model for both genders. We used leave-one-out training where the model of a gender was trained by excluding each test speaker at time from the training data, resulting in 18 male and 16 female models in total. We used only every 10th sentence (in the alphabetical order) of the training data to keep the computation time reasonable. The purpose of the leave-one-out training was to simulate a situation where the exact characteristics of the target and interfering sources were not known in advance.

4.1   Training Algorithms

Four different algorithms were tested in training the priors:

- NMF estimates fixed priors using the sparse NMF algorithm [1] which uses the divergence criterion (2). Sparseness factor which produced approximately the best results was used (the optimal value was different in testing). In order to test basis vector adaptation, Gamma distributions were generated using the procedure explained in Section 3.2 with parameter $k = 0.01$.
- Gamma mixture model was trained using the algorithm in Section 3.1.
- Gaussian mixture model was trained using log-spectrum observations, and the Gamma mixture model was generated as explained in Section 3.2.
- Gaussian mixture model was trained using linear-spectrum observations and the Gamma mixture model was generated by matching the moments.

The above algorithms are denoted as NMF, Gamma, Gaussian-log, and Gaussian-lin, respectively. All the algorithms were tested with 30 and 70 components per speaker. We found that it is advantageous to control the variance of the trained distributions by scaling their parameters as $k_{n,f} = k_{n,f}/q$ and $\theta_{n,f} = \theta_{n,f}q$, which retains the mean of the distribution but scales its variance by q. Value $q = 0.1$ produced approximately the best results. Normalizing each basis vector to unity norm and scaling the distributions accordingly by multiplying the scale parameter was also found to improve the results slightly.

4.2   Testing

In the test phase, the bases of male and female speakers obtained by a particular training algorithm were concatenated. Each of the 300 test signals was processed using sparse NMF by applying the update rules (6) and (7). Sparseness factor λ which produced approximately best result was used.

All the algorithms were tested with fixed and adaptive bases: adaptive bases used the distributions obtained from the training, whereas fixed based were set equal to the mean of each prior distribution. The basis vectors in all the algorithms were initialized with the mean of each prior distribution, and random positive values were used to initialize the gains.

The basis vectors and gains were estimated using each test signal at time. The weighted sum of male basis vectors in frame t is calculated as $\mathbf{m}_t = \sum_{n \in \mathcal{M}} \mathbf{b}_n g_{n,t}$, where $\mathcal{M}$ is the set of male basis vectors. Similarly, the weighted sum $\mathbf{f}_t$ of female basis vectors is calculated using the set of female speaker basis vectors. The male speaker spectrum $\hat{\mathbf{m}}$ in each frame is then reconstructed as

$$\hat{\mathbf{m}}_t = \mathbf{x}_t. * \mathbf{m}_t./(\mathbf{m}_t + \mathbf{f}_t), \tag{13}$$

where $.*$ and denotes element-wise multiplication. Female spectra are obtained as $\mathbf{x}_t - \hat{\mathbf{m}}_t$.

The quality of separation was measured by the signal-to-noise ratio of the separated spectrograms. The SNRs were averaged over both the speakers in all the test signals.

4.3   Results

The average signal-to-noise ratios of each of the tested algorithm are illustrated in Table 1. The Gamma and Gaussian-log methods produce clearly better results than NMF and Gaussian-lin. Adaptive bases increase the performance of Gamma method, but for other methods the effect is small. A larger number of components improves significantly the performance of all the methods except NMF.

Sparseness in testing was found to improve the quality of the separation slightly. Figure 1 illustrates the performance of the Gamma method with different sparseness factors λ. Sparseness improved more clearly the performance of the NMF training, but the results are omitted because of space limitation restrictions. Figure 1 also illustrates the effect of scaling the variances of the distributions. Value $q = 0$ corresponds to fixed priors, and larger values (adaptation) improve the quality slightly up to certain value of q.

All the parameters of the training algorithms were not completely optimized for this application, so final judgment about the relative performance of Gamma and Gauss-log methods cannot be made. However, the results show that these models perform clearly better than NMF in training the basis vectors.

Table 1. Average signal-to-noise ratios of the tested methods in dB, obtained with fixed and adaptive basis vectors and with either 30 or 70 components per source. The best algorithm in each column is highlighted with bold face font.

method	30 components		70 components	
	fixed	adaptive	fixed	adaptive
NMF	5.68	5.71	5.57	5.55
Gamma	**6.55**	**6.73**	6.95	**7.04**
Gaussian-log	6.54	6.56	**7.02**	7.03
Gaussian-lin	3.34	3.27	3.75	3.71

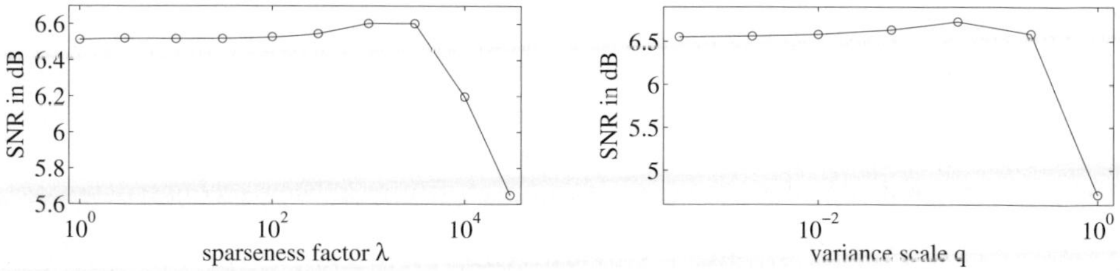

Fig. 1. The effect of the sparseness factor λ (left panel) and the distribution variance scale q (right panel) on the average signal-to-noise ratio

5   Conclusions

We have proposed the use of a Gamma mixture model in representing the basis vector distributions in supervised non-negative matrix factorization based sound source separation. The proposed method is shown to produce better results than previous sparse NMF training. In addition to better separation quality, the method also simplifies the training since there is no need to tune the sparseness factor in NMF. Mixture model training also opens up new possibilities of incorporating hidden state variables in the model, which allow modeling the temporal dependency in the signals.

References

1. Virtanen, T.: Monaural sound source separation by non-negative matrix factorization with temporal continuity and sparseness criteria. IEEE Transactions on Audio, Speech, and Language Processing 15(3) (2007)
2. Schmidt, M.N., Olsson, R.K.: Single-channel speech separation using sparse non-negative matrix factorization. In: Proceedings of the International Conference on Spoken Language Processing, Pittsburgh, USA (2006)
3. Wilson, K.W., Raj, B., Smaragdis, P.: Regularized non-negative matrix factorization with temporal dependencies for speech denoising. In: 9th Annual Conf. of the Int. Speech Communication Association (Interspeech 2008), Brisbane, Australia (2008)
4. Rennie, S.J., Hershey, J.R., Olsen, P.A.: Efficient model-based speech separation and denoising using non-negative subspace analysis. In: Proceedings of IEEE Int. Conf. on Audio, Speech and Signal Processing, Las Vegas, USA (2008)
5. Virtanen, T., Cemgil, A.T., Godsill, S.: Bayesian extensions to non-negative matrix factorisation for audio signal modelling. In: Proceedings of IEEE International Conference on Audio, Speech and Signal Processing, Las Vegas, USA (2008)
6. Cemgil, A.T.: Bayesian inference in non-negative matrix factorisation models. Technical Report CUED/F-INFENG/TR.609, University of Cambridge (July 2008)
7. Lee, D.D., Seung, H.S.: Algorithms for non-negative matrix factorization. In: Proceedings of Neural Information Processing Systems, Denver, USA, pp. 556–562 (2000)
8. Cooke, M.P., Barker, J., Cunningham, S.P., Shao, X.: An audio-visual corpus for speech perception and automatic speech recognition. Journal of the Acoustical Soc. of America 120(5) (2006)

An Implementation of Floral Scent Recognition System Using ICA Combined with Correlation Coefficients

Byeong-Geun Cheon, Yong-Wan Roh, Dong-Ju Kim, and Kwang-Seok Hong

School of Information and Communication Engineering, Sungkyunkwan University,
Suwon 440-746, South Korea
bkorigin@ece.skku.ac.kr, elec1004@skku.ac.kr,
radioguy@korea.com, kshong@skku.ac.kr

Abstract. We propose a floral scent recognition system using ICA combined with correlation coefficients estimated by statistical analysis between floral scent and prescribed reference models. The proposed floral scent recognition system consists of three modules such as floral scent acquisition module using Metal Oxide Semiconductor (MOS) sensor array, entropy-based floral scent detection module, and floral scent recognition module using ICA combined with correlation coefficients. To evaluate the proposed floral scent recognition system, we implemented an individual floral scent recognition system using K-NN with PCA that are generally used in conventional electronic noses. In the experimental results, the proposed system achieves higher performance than traditional odor recognition methods. We confirmed that the proposed system is practical and effective.

Keywords: Independent Component Analysis, Correlation Coefficient, Floral Scent Recognition, Electronic Nose.

1 Introduction

An electronic nose is an appliance consisting of an array of gas sensors coupled to a pattern recognition algorithm [1]. The most common types of sensor utilized electronic nose are Conducting polymer (CP) sensor, Metal oxide semiconductor (MOS) sensor, Surface acoustic wave (SAW) sensor, Quartz crystal microbalance (QCM) sensor, Optical sensor, and Metal oxide semiconductor field effect transistor (MOSFET) sensor [2]. The electronic nose has generated much recent worldwide interest for its potential to solve a wide variety of problems in food and environmental monitoring [3], beverage manufacturing [4], bioprocesses [5] and medical diagnostics [6]. Pattern recognition methods in electronic noses are usually based on parametric algorithms that use PCA (Principal Component Analysis), LDA (Linear Discrimination Analysis), ICA (independent component analysis), HCA (Hierarchical Cluster Analysis), and clustering algorithms such as k-means or non-parametric algorithms that use ANN (Artificial neural networks), GA (Genetic algorithms), and Fuzzy logic and fuzzy rules based algorithms [7].

In this paper, we propose an odor recognition method based on ICA combined with correlation coefficients; its performance is an improvement over traditional odor

T. Adali et al. (Eds.): ICA 2009, LNCS 5441, pp. 654–661, 2009.

recognition systems. We apply this proposal to a practical floral scent recognition system, verifying the potential of this application. The proposed floral scent recognition method includes a selection step to select the two floral scent models that has a maximum similarity. The maximum similarity is decided after calculating the average of the correlation coefficients of the individual sensors between feature vector from the floral scent input point of the floral scent input until the stable region and 12 types of reference models. We performed an individual floral scent recognition experiment using K-NN (K-Nearest Neighbor) with PCA recognition algorithm to evaluate performance of the proposed method.

2   Floral Scent Recognition System

The three steps of the suggested floral scent recognition system using ICA combined with correlation coefficients are: 1) floral scent acquisition step using MOS sensor array, 2) entropy based floral scent detection step, and 3) floral scent recognition step using ICA combined with correlation coefficients. Figure 1 depicts the flow chart of the proposed system. The details of these steps are explained in the following sections.

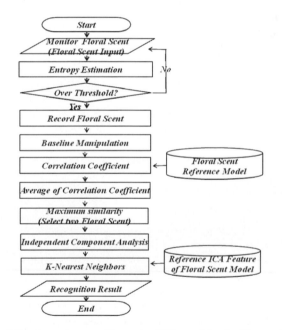

Fig. 1. Flow chart of floral scent recognition using ICA combined with correlation coefficients

2.1   Floral Scent Acquisition Module Using MOS Sensor Array

The MOS sensor has been developed as a chemical sensor. It analyzes specific odors in electrical appliances. The major advantages of MOS sensors are fast response and recovery times; these mainly depend on the temperature and the level of interaction between the sensor and gas. MOS sensors are small and relatively inexpensive to fabricate. They consume lower power compared to other sensors, and can be integrated directly into the measurement circuitry [8].

The proposed floral scent acquisition module has been developed using 16 MOS sensors (Table 1). It includes a heater, MCU (C8051F305) for 24bit AD conversion, data processing, and Bluetooth communication to deliver sensor data.

Table 1. List of MOS(Metal Oxide Semiconductor) sensors

NO	SENSOR	TARGETED GASES	NO	SENSOR	TARGETED GASES
1	TGS800	Air Contaminants	9	TGS2620	Alcohol
2	TGS822	Alcohol	10	MICS2710	Nitrogen
3	TGS825	Ammonia	11	MICS5131	Alcohol
4	TGS826	Ammonia	12	MICS5132	Carbone Dioxide
5	TGS833T	Water Vapors	13	MICS5135	VOCs
6	TGS880	Alcohol	14	SP-53	Ammonia
7	TGS2602	Air Contaminants	15	MQ-3	Alcohol
8	TGS2611	Methane	16	SP3s-AQ2	Air Contaminants

Figure 2 shows the floral scent acquisition module using the MOS sensor array and an example of the acquisition process to scent rose oil.

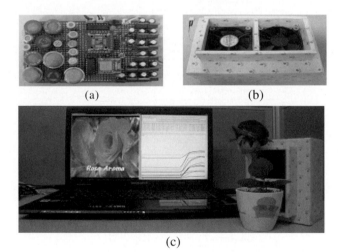

(a)   (b)

(c)

Fig. 2. MOS sensor array (a), exterior of floral scent acquisition module (b) and example of acquisition process for rose oil scent (c)

The measured 16-channel analog sensor signals are converted to digital signals via MCU and delivered to the Notebook via the Bluetooth channel defined by the serial profile. Sensor signal acquisition was performed by firmware programmed under Silicon Laboratories IDE environment at a sampling frequency of 17Hz/sensor. Sensor signals were simultaneously measured for all sensing elements. Figure 3 (a) shows a simultaneously measured example using rose oil scent.

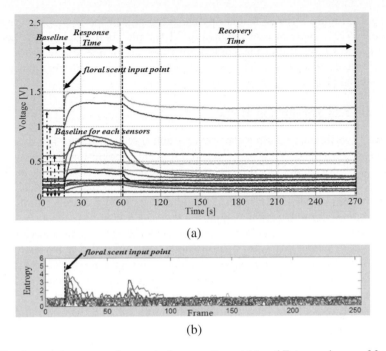

(a)

(b)

Fig. 3. Simultaneously measured example for rose oil scent (a) and Entropy change of frame (b)

2.2 Entropy-Based Floral Scent Detection Module

In the sensor array based electronic nose system, the pre-processing step that includes accurate detection of corresponding feature vector and a revision and conversion phase of detected feature vector is significant [7]. The floral scent detection module is needed, to detect the starting point of the floral scent input region, to extract the feature vector. We propose the floral scent detection module that applies Shannon's entropy. The entropy of the frame in the individual sensor is defined in equation (1). A frame consists of 32 samples and has a 50% overlapped window. If the entropy of the nth frame, i.e. $E(n)$, is more than the threshold value, the current frame is selected for the floral scent input point (by experimentation, the threshold value selected was 1.0).

$$E(n) = -\sum_{l=0}^{L-1} p(l) \times \log_2 p(l)$$
(1)

where parameter p(l) is the probability a specific range from the sensor response will occur in one frame. In equation (2), p(l) divides c(l) with S.

$$p(l) = \frac{c(l)}{S}, \qquad 0 \le l \le L-1$$
(2)

where c(l) is the occurrence frequency of each sample within one frame and S is the number of samples from the corresponding frame. l divides the whole sensor response with range L. Figure 3 (b) is the entropy change graph of the frame in the measured floral scent. 512 samples/sensor, beginning 32 samples before the floral scent input point until 480 samples after the floral scent input point, are selected as the feature vector to execute recognition. The average detection error of this module is by less than 0.2 second [10].

2.3 Floral Scent Recognition Module

Correlation coefficients measure the strength and direction of the linear association between two variables. We propose the floral scent recognition module using ICA combined with correlation coefficients that measure a linear association between the feature vector of floral scent input and reference models. The individual reference models for floral oils were constructed from an average of feature vectors which are selected by floral scent detection module.

The proposed module calculates correlation coefficients of the individual sensor between feature vector from floral scent input point until the stable region and 12 types of reference models in equation (3). Then, this module selects the two floral scent models with the maximum similarity to the calculated average of individual correlation coefficients in equation (4). We use two candidates that are selected the two floral scent models with the maximum similarity for ICA. The recognition result is decided by minimum Euclidean distance between the ICA feature vector of the floral scent input and ICA feature vectors of selected two candidates.

$$r_k = \frac{\sum (X_{ki} - \overline{X}_k)(Y_{ki} - \overline{Y}_k)}{\sqrt{\sum (X_{ki} - \overline{X}_k)^2} \times \sqrt{\sum (Y_{ki} - \overline{Y}_k)^2}} \qquad (3)$$

where k is the sensor channel number and $1 \le k \le K$ ($K=16$) based on the 16 MOS sensors. X_{ki} is the i th feature of the k th sensor from floral scent input and Y_{ki} is the i th feature of the k th sensor from the reference model and $1 \le i \le 512$. $\overline{X}_k$ is the average of the features from the k th sensor and $\overline{Y}_k$ is the features average from the k th sensor.

$$S_m = \frac{\sum_{k=1}^{K} r_k}{K} \qquad (4)$$

where S_m is the individual average of correlation coefficients of the m th reference model and $1 \le m \le 12$ (m is the total number of reference models).

Figure 4 shows the correlation coefficients of individual sensors and the individual average of correlation coefficients between the feature vector of a specific Hyssop scent and 12 types of reference model. Most correlation coefficients from the sensor between specific Hyssop scent and Hyssop reference model appear highly correlated.

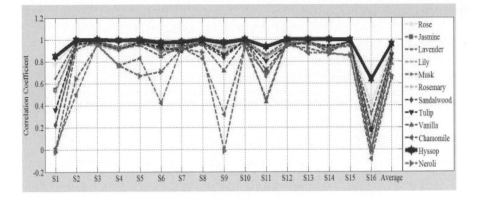

Fig. 4. Comparative chart of correlation coefficients by sensor and average correlation coefficients

In the context of face recognition, Bartlett and Sejnowski first proposed to use ICA for face representation. They presented two different approaches to apply ICA to face recognition [9]. In the first approach (ICA Architecture I), the input face images are considered to be a linear mixture of an unknown set of statistically independent source images. The source images obtained in this architecture are spatially local and sparse in nature. In the second approach (ICA Architecture II), the representation coefficients are assumed to be statistically independent. So the goal of ICA Architecture II is to find statistically independent coefficients for input data. In this architecture, while mixing coefficient vectors are independent, source images tend to have global face appearance, as in the case of PCA.

The proposed module works with the ICA Architecture I for a fair comparison with the PCA-based method. The feature representation of floral scents using ICA Architecture I is summarized as follows [11].

1) PCA is first performed using 12 types of reference models. And then the D eigenvectors associated with the largest Eigen values are selected to consist of matrix E as mentioned in the above section.

2) If we assume that the input floral scent is formed of a row vector of X, and then the PCA coefficients Y can be obtained as $Y = XE$.

3) ICA is performed on E^T, where eigenvectors form the rows of this matrix. The independent basis floral scents in the rows of U are computed as $U = WE^T$.

4) The ICA coefficients matrix $B = \{b_1, b_2, \cdots, b_N\}$ for the linear combination of independent basis floral scent in U is computed as follows. From $Y = XE$, we get $X = YE^T$. Since $U = WE^T$ and the assumption that W is invertible, we get $E^T = W^{-1}U$. Therefore, we obtain the ICA coefficients matrix $B = YW^{-1}$ because of the relation $X = (YW^{-1})U = BU$.

3 Experimental Results

We implemented the floral scent recognition system using ICA combined with correlation coefficients. Floral oils include Rose, Jasmine, Lavender, Lily, Musk, Rosemary, Sandalwood, Tulip, Vanilla, Chamomile, Hyssop, and Neroli.

In the experiments, we used a database of floral scents consisting of 600 (50×12) measured floral scents for 12 types of floral oils. The individual reference model for floral oils was constructed from an average of 50 feature vectors, 512 samples/sensor. These 512 sample/sensor begin 32 samples before the floral scent input point until 480 samples after the floral scent input point. The experiments include individual floral scent recognition systems using PCA, PCA combined correlation coefficients, ICA and proposed method. PCA and ICA were implemented with the 64 variables selected using reference models as a training set to check classification capability. Figure 5 shows the floral scent recognition rate of four different methods. The average recognition rate of PCA and ICA obtained were 71.3% and 74.7% respectively. The average recognition rate of PCA combined with correlation coefficients and ICA combined with correlation coefficients obtained were 92.3% and 94.3% respectively.

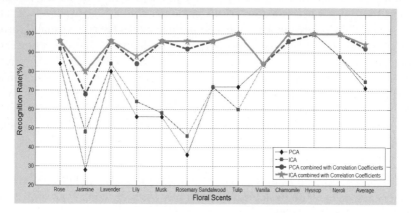

Fig. 5. Comparative chart of four different recognition methods

The floral scent recognition rate using ICA combined with correlation coefficients. For each of the tasks addressed, the percentage recognition obtained was above 80%. Experimental results show that ICA performs approximately 2.7% higher than PCA and ICA combined with correlation coefficients performs approximately 2.0% higher than PCA combined with correlation coefficients.

4 Conclusions

This study proposed an electronic nose system consisting of 16 MOS sensors developed for the floral scent recognition system. We proposed floral scent recognition system using ICA combined with correlation coefficients for 12 floral oils. We implemented individual floral scent recognition system using K-NN with PCA. This is widely used in conventional electronic noses. We evaluated the performance of the proposed system. The proposed system performs better than traditional odor recognition methods. The proposed method improves the average recognition rate more than 2.0%, compared to PCA combined with correlation coefficients. We confirm that the proposed system is practical and effective. Future work in this area includes investigating an analysis of various smells.

Acknowledgments. This work is supported by MKE, Korea under ITRC IITA-2008-(C1090-0801-0046) and the Korea Science and Engineering Foundation (KOSEF) grant funded by the Korea government (MEST) (No. 2008-000-10642-0).

References

1. Gardner, J.W., Bartlett, P.: A brief history of electronic noses. Sensors and Actuators, B 18(19), 211–220 (1994)
2. Arshak, K., Moore, E., Lyons, G.M., Harris, J.: A review of gas sensors employed in electronic nose applications. Sensor Review 24(2), 181–198 (2004)
3. Arnold, C., Harms, M., Goschnick, J.: Air Quality Monitoring and Fire Detection With The Karlsruhe Electronic Micronose KAMINA. IEEE Sensors Journal 2(3) (June 2002)
4. Panigrahi, S., Balasubramanian, S., Gu, H., Logue, C., Marchello, M.: Neural-network-integrated electronic nose system for identification of spoiled beef. LWT 39, 135–145 (2006)
5. Natale, C.D., Macagnano, A.: Human skin odor analysis by means of an electronic nose. Sensors and Actuators B 65 (2000)
6. Di Natale, C., Macagnano, A., Martinelli, E., Paolesse, R., D'Arcangelo, G., Roscioni, C., Finazzi-Agro, A., D'Amico, A.: Lung cancer identification by the analysis of breath by means of an array of non-selective gas sensors. Biosensors and Bioelectronics 18(10), 1209–1218 (2003)
7. Scott, S.M., James, D., Ali, Z.: Data analysis for electronic nose systems. Microchim Acta 156, 183–207 (2007)
8. Pearce, T.C., Schiffman, S.S., Nagle, H.T., Gardner, J.W.: Handbook of machine olfaction-electronic nose technology. Wiley-VCH, Chichester (2003)
9. Bartlett, M.S., Movellan, J.R., Sejnowski, T.J.: Face recognition by independent component analysis. IEEE Trans. Neural Networks 13(6), 1450–1464 (2002)
10. Roh, Y.-W., Cheon, B.-G., Jeon, J.-B., Hong, K.-S.: Rediscovery of the Entropy for Pattern Recognition. In: Speech Communication and Signal Processing Conference (KSCSP 2007), vol. 25(1) (2007)
11. Ekenel, H.K., Sankur, B.: Feature selection in the independent component subspace for face recognition. Pattern Recognition Letters 25, 1377–1388 (2004)

Ion-Selective Electrode Array Based on a Bayesian Nonlinear Source Separation Method

Leonardo Tomazeli Duarte[1,*], Christian Jutten[1], and Saïd Moussaoui[2]

[1] GIPSA-lab, Institut Polytechnique de Grenoble, CNRS, France
Leonardo.duarte@gipsa-lab.inpg.fr, Christian.Jutten@gipsa-lab.inpg.fr
[2] IRCCyN, UMR CNRS 6597, Ecole Centrale Nantes, France
Said.Moussaoui@irccyn.ec-nantes.fr

Abstract. Ion-selective electrodes (ISE) offer a practical approach for estimating ionic activities. Nonetheless, such devices are not selective, i.e., the ISE response can be affected by interfering ions other than the target one. With the aim of overcoming this problem, we propose a Bayesian nonlinear source separation method for processing the data acquired by an ISE array. The Bayesian framework permits us to easily incorporate prior information such as the non-negativity of the sources into the separation method. The effectiveness of our proposal is attested by experiments using artificial and real data.

1 Introduction

Ion-selective electrodes (ISEs) are devices used for measuring the ionic activity, a measure of effective concentration of an ion in aqueous solution [1]. In contrast to more sophisticated analytical techniques, an ISE distinguishes itself because of its low cost and its ease of manipulation. Nonetheless, a well-known problem associated with an ISE regards its lack of selectivity [1,2], i.e., the ISE response can be affected by interfering ions other than the target one.

One possible way to deal with the interference problem relies on the use of an electrode array followed by a signal processing block designed for extracting the relevant information from the acquired data. If, for instance, signal processing blocks based on blind source separation (BSS) techniques are considered [3], then one may skip almost totally[1] the usual calibration stage, which, for an ISE array, is extremely time-demanding and must be performed from time to time due to the electrodes' drift. However, there are several challenging points that make difficult the design of a BSS method for this case. Firstly, the mixing model associated with an ISE array is nonlinear. Secondly, in a real application, the number of samples is usually reduced and the typical assumption of statistical independence between the sources may be not realistic in some scenarios.

---

[*] L. T. Duarte is grateful to the CNPq (Brazil) for funding his PhD research. The authors are grateful to Pierre Temple-Boyer, Jérôme Launay and Ahmed Benyahia (LAAS-CNRS) for the support in the acquisition of the dataset used in this work.
[1] Due to the usual scale indeterminacy of BSS methods, at least one calibration point is necessary.

In this work, we propose a Bayesian nonlinear BSS algorithm in order to overcome the above-mentioned difficulties. Our motivation is twofold. Firstly, by relying on a Bayesian BSS framework, we can easily exploit prior information other than the statistical independence [4,5]. For instance, the sources in our problem are always non-negative. Secondly, elaborated sampling methods [6,7], like the Gibbs' sampler, allow an efficient implementation of a Bayesian inference scheme even in high-dimensional and complex models as the one treated in this work. Concerning the organization of the paper, we start, in Section 2, with a description of the mixing model associated with an ISE array. In Section 3, we describe the proposed Bayesian BSS method. In Section 4, experiments are carried out. Finally, in Section 5, we state our conclusions.

2   Mixing Model

Let us consider the analysis of a solution containing n_s different kinds of ions via an array of n_c ISEs. Due to the interference problem, the response y_{it} of the i-th ISE in the array at the instant t is dependent not only on the activity of its target ion s_{it} but also on the activities of the other ions in the solution, which are represented by s_{jt}. The Nicolsky-Eisenman (NE) equation [1] states that

$$y_{it} = e_i + d_i \log \left(s_{it} + \sum_{j=1, j \neq i}^{n_s} a_{ij} s_{jt}^{z_i/z_j} \right), \tag{1}$$

where e_i is a constant, z_j is the valence of the j-th ion and a_{ij} denotes the selectivity coefficients. Finally, $d_i = RT/z_i F$, where R is the gas constant, T the temperature in Kelvin, and F the Faraday constant. In a matrix representation, the data provided by the ISE array can be described as follows

$$\mathbf{Y} = \mathbf{e} \cdot \mathbf{1}_{1 \times n_d} + diag(\mathbf{d}) \log \left(\mathbf{A} \otimes^{\mathbf{z}} \mathbf{S} \right), \tag{2}$$

where n_d denotes the number of samples, $\mathbf{Y} \in \mathbb{R}^{n_c \times n_d}$, $\mathbf{e} = [e_1, \ldots, e_{n_c}]^T$, $\mathbf{d} = [d_1, \ldots, d_{n_c}]^T$. Matrix $\mathbf{A} \in \mathbb{R}_+^{n_c \times n_s}$ contains the selectivity coefficients a_{ij}. The j-th row of the matrix $\mathbf{S} \in \mathbb{R}_+^{n_s \times n_d}$ corresponds to the temporal evolution of the activity of the j-th ion. Finally, $\mathbf{z} = [z_1, \ldots, z_{n_s}]^T$, and the operator $\otimes^{\mathbf{z}}$ describes the nonlinear transformation inside the log term in the NE model (see (1)). If the valences z_i are equal, then $\otimes^{\mathbf{z}}$ results in a simple matrix multiplication and, thus, (2) becomes a particular case of the class of post-nonlinear (PNL) models [8].

In view of a possible model inaccuracy and/or of the errors introduced by the measurement system, a more realistic description of the ISE array is given by $\mathbf{X} = \mathbf{Y} + \mathbf{N}$, where $\mathbf{N} \in \mathbb{R}^{n_c \times n_d}$ represents the noise terms. We assume a zero mean white Gaussian noise with covariance matrix $\mathbf{C} = diag([\sigma_1^2, \ldots, \sigma_{n_c}^2])$. Finally, SNR_i denotes the resulting signal-to-noise ratio at the ISE i.

3   Bayesian Source Separation Method

Given the mixing model description, we can now formulate the source separation problem treated in this work: estimate the elements of $\mathbf{S}$ by using the ISE array response $\mathbf{X}$ and by assuming that the vector of valences $\mathbf{z}$ is known. Since we envisage a blind method, all the parameters related to the mixing model (except $\mathbf{z}$) and the noise variance at each electrode are unknown and, thus, should be estimated. Furthermore, as it will become clear later, there are other unknown parameters, denoted by ϕ, which are related to the prior distributions assigned to the sources. Henceforth, all these parameters will be represented by the vector $\boldsymbol{\theta} = [\mathbf{S}, \mathbf{A}, \mathbf{d}, \mathbf{e}, \boldsymbol{\sigma}, \phi]$ and we will adopt the following notation: $\boldsymbol{\theta}_{-\theta_q}$ represents the vector containing all elements of $\boldsymbol{\theta}$ except θ_q.

A first step in the development of a Bayesian method concerns the definition of prior distributions for each element of $\boldsymbol{\theta}$. Then, by relying on the model likelihood and on the Bayes' rule, one can obtain the posterior distribution of $\boldsymbol{\theta}$ given $\mathbf{X}$. Finally, we use Markov Chain Monte Carlo methods to generate samples from the posterior distribution. This permits us to approximate the Bayesian minimum mean square error (MMSE) estimator [9] for $\boldsymbol{\theta}$ in a straightforward manner. In the sequel, we shall detail each of these steps.

3.1   Priors Definition

Ionic activities: The sources are assumed i.i.d. and mutually independent. Also, since the sources represent ionic activities, it is natural to consider a non-negative prior distribution. In this context, [4] has shown that a modeling based on Gamma distributions provides a flexible solution, since it can model from sparse to almost uniform sources. However, in this work, we adopt a lognormal distribution[2] for each source, i.e., $p(s_{jt}) = Log\mathcal{N}(\mu_{s_j}, \sigma_{s_j}^2)$ since one can find a conjugate prior[3] for the estimation of the unknown parameters $\phi_j = [\mu_{s_j}\ \sigma_{s_j}^2]$. Indeed, this can be done by setting the priors $p(\mu_{s_j}) = \mathcal{N}(\tilde{\mu}_{s_j}, \tilde{\sigma}_{s_j}^2)$ and $p(1/\sigma_{s_j}^2) = \mathcal{G}(\alpha_{\sigma_{s_j}}, \beta_{\sigma_{s_j}})$, where $\tilde{\mu}_{s_j}$, $\tilde{\sigma}_{s_j}$, $\alpha_{\sigma_{s_j}}$, $\beta_{\sigma_{s_j}}$ are hyperparameters. Also, there is a practical argument behind the choice of a log-normal distribution. Ionic activities are expected to have a small variation in the logarithmic scale. This particularity can be taken into account by the log-normal distribution, since such a distribution is nothing but a Gaussian distribution in the logarithmic scale.

Selectivity coefficients: The parameters a_{ij} are also non-negative [1,2]. Moreover, it is rare to find a sensor whose response depends more on the interfering ion than on the target one. Given that, we assume that a_{ij} is uniformly distributed in $[0, 1]$, i.e., $p(a_{ij}) = \mathcal{U}(0, 1)$.

Non-linear model parameters: Since the parameters d_i are related to physical parameters, *a priori*, they could be known beforehand. For instance, at room temperature, one has $d_i = 0.059/z_i$, and the ISEs with such sensibility are said to have a Nernstian response. However, in a practical scenario, due to the sensor

---

[2] The following notation is used. $\mathcal{N}(\mu, \sigma^2)$, $Log\mathcal{N}(\mu, \sigma^2)$, $\mathcal{G}(\alpha, \beta)$ and $\mathcal{U}(a, b)$ represent the Gaussian, Lognormal, Gamma and Uniform distributions, respectively.

[3] From the Bayes' rule $p(X/Y) \propto p(Y/X)p(X)$. Then $p(X)$ is a conjugate prior with respect to the likelihood $p(Y/X)$ when $p(X/Y)$ and $p(X)$ belong to the same family. Conjugate priors ease the simulations conducted in a MCMC algorithm.

fabrication process and to aging, an important deviation from this theoretical value can be observed. In order to take into account this deviation, we apply a Gaussian prior $p(d_i) = \mathcal{N}(\mu_{d_i} = 0.059/z_i, \sigma_{d_i}^2)$, where σ_{d_i} is an hyperparameter. We assume that the elements of the vector $\mathbf{d}$ are statistically independent.

Offset parameters: In contrast to the parameters d_i, there is no theoretical value for e_i. However, this parameter usually [10] lies on the interval $[.05, 0.35]$. Hence, we can set $p(e_i) = \mathcal{N}(\mu_{e_i} = 0.20, \sigma_{e_i}^2)$ where the variance $\sigma_{e_i}^2$ must be defined so that the resulting prior goes toward a non-informative prior in this interval (for instance, see [6] for a discussion on non-informative priors.). The elements of the vector $\mathbf{e}$ are assumed mutually independent.

Noise variances: Finally, we assign inverse Gamma distributions for the noise variances, i.e., $p(1/\sigma_i^2) = \mathcal{G}(\alpha_{\sigma_i}, \beta_{\sigma_i})$. By proceeding this way, one obtains a conjugate pair, which eases the sampling step. Moreover, it is possible to set the hyperparameters α_{σ_i} and β_{σ_i} to obtain a non-informative prior [4].

3.2 The Posterior Distribution

A first step to obtain the posterior of $\boldsymbol{\theta}$ is to find the likelihood $p(\mathbf{X}|\boldsymbol{\theta})$ associated with the mixing model. From the assumption of i.i.d. Gaussian noise which is also spatially uncorrelated, it asserts that

$$p(\mathbf{X}|\boldsymbol{\theta}) = \prod_{t=1}^{n_d} \prod_{i=1}^{n_c} \mathcal{N}_{x_{it}} \left(e_i + d_i \log \left(\sum_{j=1}^{n_s} a_{ij} s_{jt}^{z_i/z_j} \right), \sigma_i^2 \right), \tag{3}$$

where $\mathcal{N}_{x_{ik}}(\mu, \sigma^2)$ is a Gaussian distribution in x_{ik} with parameters μ and σ^2.

Having defined the likelihood $p(\mathbf{X}|\boldsymbol{\theta})$ and the prior distribution $p(\boldsymbol{\theta})$, we can use the Bayes' rule to write the posterior distribution as $p(\boldsymbol{\theta}|\mathbf{X}) \propto p(\mathbf{X}|\boldsymbol{\theta})p(\boldsymbol{\theta})$. Since the unknown variables of our problem are, by assumption, mutually independent (except $\mathbf{S}$ and $\boldsymbol{\phi}$), we can factorize $p(\boldsymbol{\theta}|\mathbf{X})$ in the following manner

$$p(\boldsymbol{\theta}|\mathbf{X}) \propto p(\mathbf{X}|\boldsymbol{\theta}) \cdot p(\mathbf{S}|\boldsymbol{\phi}) \cdot p(\boldsymbol{\phi}) \cdot p(\mathbf{A}) \cdot p(\mathbf{e}) \cdot p(\mathbf{d}) \cdot p(\boldsymbol{\sigma}). \tag{4}$$

With the posterior distribution in hand we can set an inference scheme. In this work, we consider the Bayesian minimum mean square error (MMSE) estimator [9] which is defined as $\boldsymbol{\theta}_{MMSE} = \int \boldsymbol{\theta} p(\boldsymbol{\theta}|\mathbf{X}) d\boldsymbol{\theta}$. The problem here is the analytical resolution of this integral, which culminates in a complex task when one deals with a high-dimensional problem with non-standard distributions [6].

A possible approach to overcome the calculation of the integral related to $\boldsymbol{\theta}_{MMSE}$ is to generate samples from the posterior distribution $p(\boldsymbol{\theta}|\mathbf{X})$ and then approximate the MMSE estimator using these samples. If, for instance, the generated samples are represented by $\boldsymbol{\theta}^1, \boldsymbol{\theta}^2, \ldots, \boldsymbol{\theta}^M$, then the MMSE estimator can be approximated by $\widetilde{\boldsymbol{\theta}}_{MMSE} = \frac{1}{M} \sum_{i=1}^{M} \boldsymbol{\theta}^i$. According to the law of large numbers, $\widetilde{\boldsymbol{\theta}}_{MMSE} = \boldsymbol{\theta}_{MMSE}$ as $M \to +\infty$. This important result gives the theoretical foundation for the above described methodology, which is referred as Monte Carlo integration [6]. In the next section, we apply this methodology to the estimation problem treated in this work.

3.3   Bayesian Inference through the Gibbs Sampler

The generation of samples from $p(\boldsymbol{\theta}|\mathbf{X})$ is performed by the Gibbs sampler. Given each conditional distribution $p(\theta_i|\boldsymbol{\theta}_{-\theta_i}, \mathbf{X})$, the Gibbs sampler uses the procedure described in Table 1 to generate a Markov Chain having $p(\boldsymbol{\theta}|\mathbf{X})$ as stationary distribution. Therefore, after a burn-in period, which is necessary for the generated Markov Chain to reach its stationary distribution, the algorithm given Table 1 provides samples from $p(\boldsymbol{\theta}|\mathbf{X})$. Then, as discussed before, one can use these samples to obtain the MMSE estimator of $\boldsymbol{\theta}$.

Table 1. Simulation of $p(\boldsymbol{\theta}|\mathbf{X})$ through the Gibbs' sampler

1. Initialize the actual samples $\boldsymbol{\theta}^0 = [\theta_1^0, \theta_2^0, \ldots, \theta_N^0]$;
2. For $p = 1$ to P do

$$\theta_1^p \sim p(\theta_1|\boldsymbol{\theta}_{-\theta_1}^p, \mathbf{X})$$
$$\theta_2^p \sim p(\theta_2|\boldsymbol{\theta}_{-\theta_2}^p, \mathbf{X})$$

$$\vdots$$

$$\theta_N^p \sim p(\theta_N|\boldsymbol{\theta}_{-\theta_N}^p, \mathbf{X})$$

end

The conditional densities $p(\theta_q|\boldsymbol{\theta}_{-\theta_q}^p, \mathbf{X})$ are summarized in Tab 2. Due to the reduced space, the derivation of these expressions is omitted here. It can be noted that, because of the selected priors, we obtained conjugate pairs for almost all parameters. As a consequence, the sampling in these cases becomes straightforward. However, for $p(s_{jt}|\boldsymbol{\theta}_{-s_{jt}}, \mathbf{X})$ and $p(a_{ij}|\boldsymbol{\theta}_{-a_{ij}}, \mathbf{X})$, we obtained the following non-standard distributions:

$$p(s_{jt}|\boldsymbol{\theta}_{-s_{jt}}, \mathbf{X}) \propto \exp\left[\sum_{i=1}^{n_c} -\frac{1}{2\sigma_i^2}\left(x_{it} - e_i - \right.\right.$$

$$\left.\left. d_i \log\left(a_{ij} s_{jt}^{z_i/z_j} + \sum_{\ell=1, \ell\neq j}^{n_s} a_{i\ell} s_{\ell t}^{z_i/z_\ell} \right) \right)^2 - \frac{(\log(s_{jt}) - \mu_j)^2}{2\sigma_j^2} \right] \frac{1}{s_{jt}} \mathbb{1}_{[0,+\infty[}, \quad (5)$$

$$p(a_{ij}|\boldsymbol{\theta}_{-a_{ij}}, \mathbf{X}) \propto \exp\left[-\frac{1}{2\sigma_i^2} \sum_{t=1}^{n_t}\left(x_{it} - e_i - \right.\right.$$

$$\left.\left. d_i \log\left(a_{ij} s_{jt}^{z_i/z_j} + \sum_{\ell=1, \ell\neq j}^{n_s} a_{i\ell} s_{\ell t}^{z_i/z_\ell} \right) \right)^2 \right] \mathbb{1}_{[0,1]}, \quad (6)$$

where $\mathbb{1}$ denotes the indicator function. The sampling from these distributions is conducted through the Metropolis-Hasting (MH) algorithm [6]. A truncated Gaussian distribution was adopted as instrumental distribution.

Table 2. Prior and conditional distributions of the Bayesian model parameters

θ_q	$p(\theta_q \mid \boldsymbol{\theta}_{-\theta_q}, \mathbf{X})$	Auxiliary parameters	Prior $p(\theta_q)$
s_{jk}	see Eq. (5)	-	$Log\mathcal{N}(\mu_{s_j}, \sigma^2_{s_j})$
a_{ij}	see Eq. (6)	-	$\mathcal{U}(0,1)$
μ_{s_j}	$\mathcal{N}\left(\dfrac{\mu_{L_j}\tilde{\sigma}^2_{s_j} + \tilde{\mu}_{s_j}\sigma^2_{L_j}}{\sigma^2_{L_j} + \tilde{\sigma}^2_{s_j}}, \dfrac{\sigma^2_{L_j}\tilde{\sigma}^2_{s_j}}{\sigma^2_{L_j} + \tilde{\sigma}^2_{s_j}}\right)$	$\mu_{L_j} = \dfrac{\sum_{t=1}^{n_d}\log(s_{jt})}{n_d}$ $\sigma^2_{L_j} = \dfrac{\tilde{\sigma}^2_{s_j}}{n_d}$	$\mathcal{N}(\tilde{\mu}_{s_j}, \tilde{\sigma}^2_{s_j})$
$1/\sigma^2_{s_j}$	$\mathcal{G}(\alpha_{P_j}, \beta_{P_j})$	$\alpha_{P_j} = \alpha_{\sigma_{s_j}} + n_d/2$ $\beta_{P_j}^{-1} = 0.5\sum_{t=1}^{n_d}\left(\log(s_{jt}) - \mu_{s_j}\right)^2 - \beta_{\sigma_{s_j}}^{-1}$	$\mathcal{G}(\alpha_{\sigma_{s_j}}, \beta_{\sigma_{s_j}})$
d_i	$\mathcal{N}\left(\dfrac{\mu_{L_{d_i}}\sigma^2_{d_i} + \mu_{d_i}\sigma^2_{L_{d_i}}}{\sigma^2_{L_{d_i}} + \sigma^2_{d_i}}, \dfrac{\sigma^2_{L_{d_i}}\sigma^2_{d_i}}{\sigma^2_{L_{d_i}} + \sigma^2_{d_i}}\right)$	$\mu_{L_{d_i}} = \dfrac{\left(\sum_{t=1}^{n_d}(x_{it} - e_i)\right)\log\left(\sum_{\ell=1}^{n_s} a_{i\ell} s_{\ell t}^{\frac{z_i}{z_\ell}}\right)}{\left(\log\left(\sum_{\ell=1}^{n_s} a_{i\ell} s_{\ell t}^{\frac{z_i}{z_\ell}}\right)\right)^2}$ $\sigma^2_{L_{d_i}} = \sigma^2_{d_i}\left(\log\left(\sum_{\ell=1}^{n_s} a_{i\ell} s_{\ell t}^{\frac{z_i}{z_\ell}}\right)\right)^{-2}$	$\mathcal{N}(\mu_{d_i}, \sigma^2_{d_i})$
e_i	$\mathcal{N}\left(\dfrac{\mu_{L_{e_i}}\sigma^2_{e_i} + \mu_{e_i}\sigma^2_{L_{e_i}}}{\sigma^2_{L_{e_i}} + \sigma^2_{e_i}}, \dfrac{\sigma^2_{L_{e_i}}\sigma^2_{e_i}}{\sigma^2_{L_{e_i}} + \sigma^2_{e_i}}\right)$	$\mu_{L_{e_i}} = \dfrac{\sum_{t=1}^{n_d}\left(x_{it} - d_i\log\left(\sum_{\ell=1}^{n_s} a_{i\ell} s_{\ell t}^{\frac{z_i}{z_\ell}}\right)\right)}{n_d}$ $\sigma^2_{L_{e_i}} = \sigma^2_{e_i} n_d^{-1}$	$\mathcal{N}(\mu_{e_i}, \sigma^2_{e_i})$
$1/\sigma^2_i$	$\mathcal{G}(\alpha_{P_i}, \beta_{P_i})$	$\alpha_{P_i} = \alpha_{\sigma_i} + \dfrac{n_d}{2}, \ \beta_{P_i}^{-1} = \Psi - \beta_{\sigma_i}^{-1}$ $\Psi = 0.5\sum_{t=1}^{n_d}\left(x_{it} - e_i - d_i\log\left(\sum_{\ell=1}^{n_s} a_{i\ell} s_{\ell t}^{\frac{z_i}{z_\ell}}\right)\right)^2$	$\mathcal{G}(\alpha_{\sigma_i}, \beta_{\sigma_i})$

4   Experimental Results

We test our algorithm using artificial data and also in a real situation involving an array constituted of electrodes of potassium (K^+) and of ammonium (NH_4^+). The quality related to the estimation $\hat{s}_i(t)$ (after scale normalization) of the source $s_i(t)$ is assessed using the following index: $SIR_i = 10\log\left(\frac{E\{s_i(t)^2\}}{E\{(s_i(t) - \hat{s}_i(t))^2\}}\right)$.

Experiments with artificial data: Aiming to define testing scenarios that are as close as possible to real situations, we make use of the selectivity coefficients database presented in [2]. Concerning the sources, we considered a set of $n_d = 900$ samples endowed with a temporal structure, each sample being generated by a lognormal distribution. In a first situation, we simulate an array of two electrodes ($n_c = 2$) (each one having a different ion as target) for estimating the activities of NH_4^+ and K^+ ($n_s = 2$). As discussed in Section 2, one has a PNL model in this situation given that the valences are equal ($z_1 = z_2 = 1$). The parameters of the mixing system for this case were $a_{12} = 0.3$, $a_{21} = 0.4$, $d_1 = 0.056$, $d_2 = 0.056$, $e_1 = 0.090$, $e_2 = 0.105$ and $SNR_1 = SNR_2 = 18\,\text{dB}$. Concerning the Gibbs sampler, 3000 iterations with a burn-in period of 1400 were conducted. The burn-in value was fixed empirically after a visual inspection of the chains and no convergence monitoring strategy was applied. Our method was able to provide a good estimate of the sources in this situation. The performance indexes for this case were $SIR_1 = 19.84$ dB, $SIR_2 = 18.75$ dB and $SIR = 19.29$ dB.

We also analyze the more complicate case of estimating the activities of calcium (Ca^{2+}) and sodium (Na^+) using an array of two ISEs $(n_c = 2$ and $n_s = 2)$, each one having a different ion as target. Since the valences are different now, the mixing system is composed of a nonlinear mapping followed by component-wise logarithm functions (see Eq. (1)). We consider the same source waveforms of the last case and the mixing system parameters were $a_{12} = 0.3$, $a_{21} = 0.4$, $d_1 = 0.026$, $d_2 = 0.046$, $e_1 = 0.090$, $e_2 = 0.105$ and $SNR_1 = SNR_2 = 18$ dB. The performance indexes in this case were $SIR_1 = 14.96$ dB, $SIR_2 = 17.37$ dB and $SIR = 16.17$ dB. Despite the performance deterioration with respect to the first case, our method provided fair estimations for the sources. The number of iterations for the Gibbs sampler was 5000 with a burn-in period of 2000.

Experiments with real data: We consider the analysis of a solution containing K^+ and NH_4^+ through an array composed of one K^+-ISE and one NH_4^+-ISE. This situation is typical in water quality monitoring. The actual sources and the data acquired $(n_d = 169)$ by the array are shown in the left and in the right side, respectively, of Fig. 1. For the Gibbs' sampler, we considered 3000 iterations and a burn-in period of 1000. In Fig. 1, we plot the retrieved sources. The performance indexes in this case were $SIR_1 = 25.10$ dB, $SIR_2 = 23.69$ dB, $SIR = 24.39$ dB. Despite a small residual interference, mainly for the K^+ activity, our method was able to provide a good estimation even in this difficult scenario in which the sources are dependent and only a reduced number of samples is available. Conversely, this situation poses a problem to ICA-based methods. Indeed, the performance of the PNL-ICA method proposed in [11] was poor in this case $(SIR_1 = 7.67$ dB, $SIR_2 = -0.33$ dB, $SIR = 3.67$ dB).

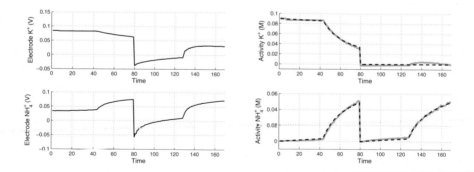

Fig. 1. Left: mixtures. Right: actual sources (dashed black) and estimations (after scale normalization) provided by the Bayesian method (gray)

5   Conclusion

In this work, we developed a MCMC-based Bayesian nonlinear BSS method for processing the outputs of an array of ion-selective electrodes. The Bayesian framework allowed us to consider prior information such as the non-negativity of the sources and of the selectivity coefficients. Experiments with artificial and real data attested the viability of our proposal. Moreover, given that in the Bayesian formulation the independence hypothesis is not as important as in ICA methods, our method could provide good estimations even in a scenario where the sources were dependent and where only a reduced number of samples was available.

A first perspective of this work is related to the sources modeling. Indeed, we assume i.i.d. sources which means that we are not taking advantage of the fact that chemical sources do have a time-structure (they are usually slowly varying signals). Also, there are other two points that deserve further investigation: 1) the case where neither the number of ions in the solution nor their valences are available. 2) Although widely used, the NE equation may be not precise in some cases [12]. Therefore, by considering more precise models, one could obtain a better separation and then work even when a very high precision is needed.

References

1. Fabry, P., Fouletier, J. (eds.): Microcapteurs chimiques et biologiques. Lavoisier (2003) (in French)
2. Umezawa, Y., Bühlmann, P., Umezawa, K., Tohda, K., Amemiya, S.: Potentiometric selectivity coefficients of ion-selective electrodes. Pure and Applied Chemistry 72, 1851–2082 (2000)
3. Duarte, L.T., Jutten, C.: Blind source separation of a class of nonlinear mixtures. In: Davies, M.E., James, C.J., Abdallah, S.A., Plumbley, M.D. (eds.) ICA 2007. LNCS, vol. 4666, pp. 41–48. Springer, Heidelberg (2007)
4. Moussaoui, S., Brie, D., Mohammad-Djafari, A., Carteret, C.: Separation of non-negative mixture of non-negative sources using a bayesian approach and mcmc sampling. IEEE Trans. on Signal Processing 54(11), 4133–4145 (2006)
5. Févotte, C., Godsill, S.J.: A Bayesian approach for blind separation of sparse sources. IEEE Trans. on Audio, Speech and Language Processing 14, 2174–2188 (2006)
6. Robert, C.P.: The Bayesian Choice. Springer, Heidelberg (2007)
7. Gilks, W.R., Richardson, S., Spiegelhalter, D. (eds.): Markov chain Monte Carlo in practice. CRC Press, Boca Raton (1995)
8. Taleb, A., Jutten, C.: Source separation in post-nonlinear mixtures. IEEE Trans. on Signal Processing 47(10), 2807–2820 (1999)
9. Kay, S.M.: Fundamentals of statistical signal processing: estimation theory. Prentice-Hall, Englewood Cliffs (1993)
10. Gründler, P.: Chemical sensors: an introduction for scientists and engineers. Springer, Heidelberg (2007)
11. Duarte, L.T., Suyama, R., de Faissol Attux, R.R., Von Zuben, F.J., Romano, J.M.T.. Blind source separation of post-nonlinear mixtures using evolutionary computation and order statistics. In: Rosca, J.P., Erdogmus, D., Príncipe, J.C., Haykin, S. (eds.) ICA 2006. LNCS, vol. 3889, pp. 66–73. Springer, Heidelberg (2006)
12. Bakker, E., Pretsch, E., Bühlmann, P.: Selectivity of poentiometric ion sensors. Analytical Chemistry 72, 1127–1133 (2000)

A Robust and Precise Solution to Permutation Indeterminacy and Complex Scaling Ambiguity in BSS-Based Blind MIMO-OFDM Receiver

Mahdi Khosravy[1], Mohammad Reza Alsharif[1,*], Bin Guo[2], Hai Lin[2], and Katsumi Yamashita[2]

[1] Department of Information Engineering, Faculty of Engineering, University of the Ryukyus, 1 Senbaru, Nishihara, Okinawa 903-0213, Japan
k078661@eve.u-ryukyu.ac.jp, asharif@ie.u-ryukyu.ac.jp
[2] Graduate School of Engineering, Osaka Prefecture University, 1-1 Gakuen-cho, Sakai, Osaka, Japan
yamashita@eis.osakafu-u.ac.jp

Abstract. The multiuser separation problem in a multi-input multi-output (MIMO) orthogonal frequency division multiplexing (OFDM) system can be represented as an instantaneous blind source separation (BSS) problem in complex domain. But reconstruction of multiuser signals suffers from permutation indeterminacy, amplitude scaling ambiguity and phase distortion which are inherent to complex BSS. This paper presents a robust and precise solution to these problems in order to reconstruct multiuser OFDM signals. It is based on the characteristics introduced to transmitted OFDM symbols by convolutional encoding at the transmitter. The accuracy of these characteristics is obtained by optimizing the convolutional code. Despite the previous BSS-based MIMO-OFDM multiuser signal reconstruction method, the proposed method is independent of channel characteristics. Experimental results show that even over channels with necessities required for efficiency of the previous method, the proposed method concludes significantly better performance.

Keywords: OFDM, MIMO systems, Blind source separation.

1 Introduction

Multiple-input multiple-output orthogonal frequency division multiplexing (MIMO -OFDM) has recently drawn research and technology attention, and it has been considered as a strong candidate for the next generation wireless communication systems [1]. MIMO-OFDM systems have the advantage of increasing reliability of wireless transmission because of their OFDM side, and they have high data rate and capacity because of their MIMO side. Their spectral efficiency is also increased more by deploying blind channel estimation.

---

* Also, Adjunct Professor at School of Electrical and Computer Engineering, University of Tehran.

T. Adali et al. (Eds.): ICA 2009, LNCS 5441, pp. 670–677, 2009.
© Springer-Verlag Berlin Heidelberg 2009

Blind Source Separation (BSS) [2,3] is separation of sources from mixed observed data without any information about sources and mixing process, except some assumptions that should be fulfilled [4]. The blind channel estimation and multiuser detection in MIMO-OFDM systems using BSS has been proposed by [5]. It has shown how the channel estimation in a MIMO-OFDM system can be transferred into a set of standard BSS problems with complex mixing matrices, that each problem is associated with one of orthogonal subcarriers. Although the complex BSS [6] successfully separates different user signals at frequency bin (FB) level, the recomposition of them suffers from permutation indeterminacy, amplitude scaling ambiguity and phase distortion which theses challenging problems are inherent to complex BSS.

It is assumed in [5] that adjacent MIMO channels are approximately correlated, and the correlation between adjacent FB tracks and their adjacent separating matrices are used to solve the permutation and scaling problems respectively. However, this assumption is not always true, and the method is not robust enough to perfectly reconstruct multiuser signals. Moreover, since the approximation error sequentially propagates across subcarriers, it is not precise.

This paper presents a channel independent method for reconstruction multiuser signals in a BSS-based blind MIMO-OFDM system. The proposed method possesses not only robustness but preciseness because of deployment an optimized convolutional encoder at the transmitter.

The organization of the paper is as follows. Section 2 briefly explains the BSS-based multiuser separation in a MIMO-OFDM system, and Section 3 presents the proposed method for reconstruction of multiuser data after separation by BSS. Section 4 describes optimization of the convolutional code. Section 5 provides the simulation results, and finally, section 6 concludes the paper.

2   BSS-Based Multiuser Separation

Consider a MIMO-OFDM system with M_T transmit and M_R receive antennas. $\boldsymbol{S}_i^{(k)}$ the N-length data symbol block at the ith user and time k is modulated by a N-point IDFT, where $\boldsymbol{S}_i^{(k)} = [S_i(0,k), S_i(1,k), \cdots, S_i(N-1,k)]^T$ and $i = 1, 2, \cdots, M_T$. After adding the cyclic prefix (CP) to avoid inter-symbol interference (ISI), the modulated signals are transmitted. The transmitted signals pass through different propagation channels and are received by No.j antenna at the receiver. After removal of the CP and demodulation by N-point DFT, the received N-length data symbol block by the jth antenna at time k is

$$\boldsymbol{X}_j^{(k)} = \sum_{i=1}^{M_T} \boldsymbol{H}_{ji} \boldsymbol{S}_i^{(k)} + \boldsymbol{Z}_j^{(k)} \tag{1}$$

where $\boldsymbol{Z}_j^{(k)}$ represents zero-mean white Gaussian noise. $\boldsymbol{H}_{ji}$ is the diagonal $N \times N$ matrix of channel gains between i^{th} transmit antenna and j^{th} receive antenna. The received signal of m-th subcarrier at the j-th antenna is

$$X_j(m,k) = \sum_{i=1}^{M_T} \boldsymbol{H}_{ji}(m,m)S_i(m,k) + Z_j(m,k), \tag{2}$$

where $m = 0, \cdots, N-1$. For all receive antennas at time k in vector notation

$$\boldsymbol{X}(m) = \mathbb{H}(m)\boldsymbol{S}(m) + \boldsymbol{Z}(m) \tag{3}$$

where
$$\begin{aligned}
\boldsymbol{X}(m) &= [X_1(m,k), X_2(m,k), \cdots, X_{M_R}(m,k)]^T \\
\boldsymbol{S}(m) &= [S_1(m,k), S_2(m,k), \cdots, S_{M_T}(m,k)]^T \\
\boldsymbol{Z}(m) &= [Z_1(m,k), Z_2(m,k), \cdots, Z_{M_R}(m,k)]^T
\end{aligned}$$
$$\mathbb{H}(m) = \begin{pmatrix} \boldsymbol{H}_{11}(m,m) & \boldsymbol{H}_{12}(m,m) & \cdots & \boldsymbol{H}_{1M_T}(m,m) \\ \boldsymbol{H}_{21}(m,m) & \boldsymbol{H}_{22}(m,m) & \cdots & \boldsymbol{H}_{2M_T}(m,m) \\ \vdots & \vdots & \ddots & \vdots \\ \boldsymbol{H}_{M_R1}(m,m) & \boldsymbol{H}_{M_R2}(m,m) & \cdots & \boldsymbol{H}_{M_RM_T}(m,m) \end{pmatrix}.$$

The multiuser separation has been split into N BSS problems related to N subcarriers. But even after successful separation at each subcarrier by complex BSS, the permutation indeterminacy, amplitude scaling ambiguity and phase distortion are problems remain to be solved by multiuser data reconstruction.

3   The Proposed Multiuser Signal Reconstruction Method

In the proposed method, the symbols of each user are convolutional encoded before transmission. The encoded m^{th} carrier of kth symbol of the ith user is

$$S_i(m,k) = \sum_{l=0}^{L_c-1} c(l)D_i(m-l,k) \tag{4}$$

where $D_i(m,k)$ are symbols of the ith user, and $S_i(m,k)$ is encoded signal. L_c is the length of the convolutional code and $c(.)$ denotes its coefficients. The auto-correlation of No.m FB track of ith encoded user frame can be obtained as:

$$R_{S_{im}}(0) = E\left[|S_i(m,k)|^2\right] = c^2(0) + \cdots + c^2(L_c - 1) \tag{5}$$

where $E[.]$ is the expectation with respect to k. Similarly, $R_{S_{im}}(1)$ the correlation between m^{th} and $m+1^{st}$ FB tracks of the same i^{th} user encoded symbol block can be obtained as

$$R_{S_{im}}(1) = E\left[S_i(m,k)S_i^*(m+1,k)\right] = c(0)c(1) + \cdots + c(L_c-2)c(L_c-1) \tag{6}$$

where $E[.]$ is the expectation with respect to k. Note that $R_{S_{im}}(0)$ and $R_{S_{im}}(1)$ both are independent of input user signals, and they depend only on the coefficients of the convolutional code. We respectively denote them as R_0 and R_1 hereafter. By applying the the proposed encoding, each FB track has a known auto-correlation as well as there is a known nonzero cross-correlation value between the adjacent FB tracks of the same user. These two characteristics will be respectively used for solving the amplitude scaling ambiguity and permutation indeterminacy at the receiver. The equal average phase of contiguous FB tracks is also used to solve the phase distortion problem at the receiver. The accuracy of the above characteristics is fulfilled by optimizing the encoder code as it has been explained in section 4.

The MIMO-OFDM system described in section 2 is applied over encoded multiuser symbols $S_i(m, k)$, and after solving the N complex BSS problems related to N subcarriers in Eq.(3) the separated FB tacks are obtained as

$$\boldsymbol{Y}(m) = \boldsymbol{W}(m)\boldsymbol{X}(m) \tag{7}$$

where $\boldsymbol{W}(m)$ is the un-mixing matrix related to m^{th} subcarrier.

3.1 Permutation Alignment

Since, different user signals are mutually independent, the correlation of FB tracks of different users equals zero. So, for adjacent p^{th} and $p + 1^{st}$ FB tracks of transmitted multiuser symbol blocks,

$$E[S_i(p, k)S_l^*(p + 1, k)] = \begin{cases} 0 & \text{if } i \neq l, \\ R_1 & \text{if } i = l, \end{cases} \tag{8}$$

where R_1 is a nonzero known value from Eq.(6), and $E[.]$ is the expectation with respect to k. Two adjacent FB tracks recovered by complex BSS are as follows

$$\boldsymbol{Y}_{ip} = A_{ip}e^{j\theta_{ip}}[S_i(p, 1), S_i(p, 2), \ldots, S_i(p, K)] \tag{9}$$

$$\boldsymbol{Y}_{l,p+1} = A_{l,p+1}e^{j\theta_{l,p+1}}[S_l(p + 1, 1), S_l(p + 1, 2), \ldots, S_l(p + 1, K)] \tag{10}$$

where i and l are unknown user ownership indices, A_{ip} and $A_{l,p+1}$ are unknown scaling amplitudes , θ_{ip} and $\theta_{l,p+1}$ are unknown phase distortions and K is the frame length. Using the Eq.(8), the cross-correlation of the above FB tracks will be as follows

$$E[\boldsymbol{Y}_{ip}\boldsymbol{Y}_{l,p+1}^H] = \begin{cases} 0 & \text{if } i \neq l, \\ R_1 A_{ip}A_{l,p+1}e^{(\theta_{ip}-\theta_{l,p+1})} \neq 0 & \text{if } i = l. \end{cases} \tag{11}$$

As it is seen, the cross-correlation between adjacent FB tracks of the same user is a nonzero value, while it is zero for adjacent FB tracks of different users. So, by doing a correlation-based grouping in the sequence from No.1 FB to No.$(N - 1)$ FB, the permutation corrected multiuser symbols $\dot{Y}_i(m, k)$ will be obtained.

3.2 Amplitude Scaling Correction

After permutation alignment, by having the prior knowledge about the auto-correlation of each FB track R_0 from Eq.(5), we can resolve the amplitude scaling ambiguity. Consider the No.p FB track of ith user symbol block after permutation alignment will be

$$\dot{\boldsymbol{Y}}_{ip} = A_{ip}e^{j\theta_{ip}}[S_i(p, 1), S_i(p, 2), \ldots, S_i(p, K)], \tag{12}$$

where A_{ip} and θ_{ip} are respectively unknown amplitude scaling and phase distortion, and K is the frame length. The auto-correlation of the above complex sequence is obtained as

$$\phi_{ip}^{ip} = E[\dot{\boldsymbol{Y}}_{ip}\dot{\boldsymbol{Y}}_{ip}^{H}] = A_{ip}^2 R_0 \tag{13}$$

So, the amplitude scaling of $\dot{\boldsymbol{Y}}_{ip}$ can be corrected by being multiplied by $\frac{1}{A_{ip}} = \sqrt{\frac{R_0}{\phi_{ip}^{ip}}}$. Thus the permutation and amplitude scaling corrected user symbol $\ddot{Y}_i(m,k)$ will be obtained as follows

$$\ddot{Y}_i(m,k) = \sqrt{\frac{R_0}{\phi_{im}^{im}}}\dot{Y}_i(m,k). \tag{14}$$

3.3   Phase Distortion Removal

Next, we deal with unknown phase distortion of FB tracks. Consider two adjacent p^{th} and $p+1^{st}$ FB tracks of the same user after permutation alignment and complex scaling amplitude correction as

$$\ddot{\boldsymbol{Y}}_{ip} = e^{j\theta_{ip}}[S_i(p,1), S_i(p,2), \ldots, S_i(p,K)] \tag{15}$$

$$\ddot{\boldsymbol{Y}}_{i,p+1} = e^{j\theta_{i,p+1}}[S_i(p+1,1), S_i(p+1,2), \ldots, S_i(p+1,K)], \tag{16}$$

where θ_{ip} and $\theta_{i,p+1}$ are unknown phase distortions, and K is the frame length. Fortunately, since the code of the convolutional encoder is optimized to conclude a zero phase for R_1, the phase deviation of each FB track with respect to its following one can be obtained as follows

$$\phi_{i,p+1}^{ip} = E[\ddot{\boldsymbol{Y}}_{ip}\ddot{\boldsymbol{Y}}_{i,p+1}^{H}] = e^{(\theta_{ip}-\theta_{i,p+1})}E[\boldsymbol{S}_{ip}\boldsymbol{S}_{i,p+1}^{H}] = e^{(\theta_{ip}-\theta_{i,p+1})}R_1. \tag{17}$$

Since, the phase of R_1 is zero,

$$e^{j(\theta_{ip}-\theta_{i,p+1})} = e^{j\angle\phi_{i,p+1}^{ip}} = \frac{\phi_{i,p+1}^{ip}}{|\phi_{i,p+1}^{ip}|}. \tag{18}$$

Therefore by multiplying $\ddot{\boldsymbol{Y}}_{i,p+1}$ by $\frac{\phi_{i,p+1}^{ip}}{|\phi_{i,p+1}^{ip}|}$, its phase deviation will be the same as $\ddot{\boldsymbol{Y}}_{ip}$. By doing this operation from No.1 FB to No.$N-1$ FB, each FB track is adjusted to its preceding FB track which was adjusted to its prior one. So, the phase distortion of all FB tracks become θ_{i1}, that is a case similar to uncertain carrier phase in single carrier system. The same unknown resultant phase deviation for all symbols of the user can be eliminated by noncoherent detection [7]. At this point all OFDM user signals have been reconstructed, and they are ready to be transferred to user identification unit. Fig. 1 shows the complete structure of the BSS-based MIMO-OFDM system which deploys the proposed method of multiuser signal reconstruction.

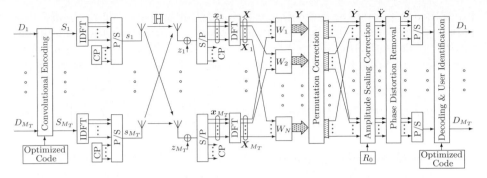

Fig. 1. The complete structure of BSS-based MIMO-OFDM system with the proposed method of multiuser signal reconstruction

4   The Optimum Convolutional Code

The phase distortion removal is the most sensitive part of the method to the convolutional code. Consequently the code is optimized to result in minimum error in estimation of the average phase of R_1. R_1 can be accurately written as:

$$R_1 = \sum_{l=0}^{L_c-2} c(l)c(l+1) + E\left[\sum_{l_i=0}^{L_c-1} \sum_{l_j=0}^{L_c-1\, l_j\neq l_i+1} c(l_i)c(l_j)e^{j(\alpha_{l_i,k}-\alpha_{l_j,k})}\right] \quad (19)$$

$$= \sum_{l=0}^{L_c-2} c(l)c(l+1) + \sum_{l_i=0}^{L_c-1} \sum_{l_j=0}^{L_c-1\, l_j\neq l_i+1} c(l_i)c(l_j)E\left[\cos(\alpha_{ij,k})\right]$$

$$+j\sum_{l_i=0}^{L_c-1} \sum_{l_j=0}^{L_c-1\, l_j\neq l_i+1} c(l_i)c(l_j)E\left[\sin(\alpha_{ij,k})\right] \quad (20)$$

where $E[.]$ is the expectation with respect to k, and $\alpha_{ij,k} = (\alpha_{l_i,k} - \alpha_{l_j,k})$. $c(.)$ are coefficients of the code, and L_c is the code length. $k = 1, 2, \cdots, K$, where K is the frame length. It can be shown that $\alpha_{ij,k}$ is the phase of a point in the same constellation. Since, $\alpha_{ij,k}$ is the phase of one of the points of a symmetric constellation with equal probability independent form l_i and l_j, $E\left[\cos(\alpha_{ij,k})\right]$ and $E\left[\sin(\alpha_{ij,k})\right]$ is taken as the same random variable x which tends to zero. While the code is constant, the phase of R_1 can be a function of x as follows

$$\varphi = \angle R_1(x) = Arctan\left(\frac{x\sum_{l_i=0}^{L_c-1} \sum_{l_j=0}^{L_c-1\, l_j\neq l_i+1} c(l_i)c(l_j)}{\sum_{l=0}^{L_c-2} c(l)c(l+1) + x\sum_{l_i=0}^{L_c-1} \sum_{l_j=0}^{L_c-1\, l_j\neq l_i+1} c(l_i)c(l_j)}\right). \quad (21)$$

It can be shown that

$$\lim_{x\to 0} d\varphi = \frac{\sum_{l_i=0}^{L_c-1} \sum_{l_j=0}^{L_c-1\, l_j\neq l_i+1} c(l_i)c(l_j)}{\sum_{l=0}^{L_c-2} c(l)c(l+1)}dx. \quad (22)$$

Eq.(22) conveys that the estimation error of $\angle R_1$ is proportional to the fraction part of the equation. Therefore, minimization of this fraction leads to minimum error of phase estimation as well as optimum code coefficients. After simplification of the fraction, the optimum convolutional code can be expressed as follows

$$[c(0), c(1), \cdots, c(L_c - 1)] = argmin_{c(.)} \left(\frac{\left[\sum_{l=0}^{L_c-1} c(l) \right]^2}{\sum_{l=0}^{L_c-1} c(l)c(l+1)} \right) \qquad (23)$$

Solving the above optimization problem lead us to a set of optimized codes with an equivalent result. Note that the convolutional encoding of the symbols changes their constellation map, and new constellation depends on the convolutional code. To reduce the error probability in detection of demodulated symbols at the receiver, among optimized codes we choose a code which concludes a symmetrical constellation with as much as possible equiprobable points.

5   Simulation Results

The proposed method has been evaluated over random frequency selective fading MIMO channels as well as realistic Rayleigh ones with exponential power delay profiles. The additive noise is taken zero mean white Gaussian. The employed configuration is a 4×4 MIMO system with monopolized transmit antennas assigned for each user. All user signals are modulated by 4QPSK scheme. The optimized code $[\frac{1}{3} \; \frac{1}{2} \; \frac{1}{6}]$ is used for convolutional encoding at the transmitter. The carrier frequency and system bandwidth are respectively 5 GHz and 20 MHz, and the resulting RMS delay is 50 ns. The number of subcarriers is set to $N = 64$, and the cyclic prefix is 16. The synchronization is assumed perfect.

The bit error rate (BER) vs. signal to noise ratio (SNR) results have been averaged over 1000 transmissions by using OFDM frames with the length of 2000 symbols per subcarrier. User signals are randomly generated for each experiment.

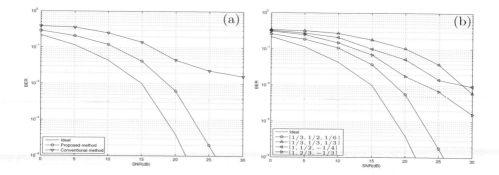

Fig. 2. (a) Performances of the proposed method and the conventional one over Rayleigh exponential fading channels. (b) Performance of the proposed method when it uses the optimized convolutional code and when it uses some other codes.

The BER vs. SNR result of the BSS based MIMO-OFDM system which deploys the proposed data reconstruction has been compared with the result of one which uses the method proposed in [5]. This comparison is over realistic Rayleigh channels with exponential fading. The BSS method that is used for both methods is ICA based on information maximization approach[8]. As it is seen in Fig. 2(a) the efficiency of the proposed data reconstruction leads to higher performance of MIMO-OFDM system. Fig. 2(b) compares the effect of optimized encoding code with some other codes on the system performance. This comparison is over independent random channels which do not fulfill the requirements of the conventional method. Fig.2(b) conveys two points. One point is the independence of the proposed method from channel, and the other one is the important role of the optimized encoding code in near perfect efficiency of the proposed method. In both Figures. 2(a) and 2(b), the curve without marks is for ideal reconstruction of BSS separated signals, wherein their source signals are used.

6   Conclusion

A robust solution to permutation indeterminacy and complex scaling ambiguity inherent to complex BSS for reconstruction of multiuser data in BSS based MIMO-OFDM system has been proposed. The proposed method not only significantly outperforms the previous method which requires correlated MIMO channels, but also it makes the MIMO-OFDM system independent of channels characteristics. By using the proposed method the MIMO-OFDM system is capable of recovering data even over random independent channels. The method is robust because it depends only on the convolutional code, and since the code is accurately optimized, it is precise too.

References

1. Sampath, H., Talwar, S., Tellado, J., Erceg, V., Paulraj, A.: A fourth-generation MIMO-OFDM broadband wireless system: design, performance, and field trial results. IEEE Commun. Mag. 40(9), 143–149 (2002)
2. Haykin, S. (ed.): Unsupervised Adaptive Filtering. Blind Source Separation, vol. 1. John Wiley & Sons, New York (2000)
3. Cichocki, A., Amari, S.: Adaptive Blind Signal and Image Processing. John Wiley & Sons, New York (2000)
4. Cardoso, J.F.: Blind signal separation: Statistical principles. Proc. IEEE 86(10), 2009–2025 (1998)
5. Wong, C.S., Obradovic, D., Madhu, N.: Independent component analysis (ICA) for blind equalization of frequency selective channels. In: Proc. 13th IEEE Workshop Neural Networks Signal Processing, pp. 419–428 (2003)
6. Smaragdis, P.: Blind separation of convoluted mixtures in frequency domain. J. Neurocomputing 22(1-3), 21 34 (1998)
7. Nee, R.V., Prasad, R.: OFDM for Wireless Multimedia Communications. Artech House (2000)
8. Bell, A.J., Sejnowski, T.J.: An information-maximization approach to blind separation and blind deconvolution. J. Neural Computation 7, 1129–1159 (1995)

Underdetermined BSS of MISO OSTBC Signals

Ali Mansour[1], Joe Youssef[2], and Koffi-Clement Yao[2]

[1] Depart. of ECE, Curtin University, Perth, WA 6152, Australia
[2] LEST UMR 6165, UBO, Brest, France
mansour@ieee.org, koffi-clement.yao@univ-brest.fr

Abstract. To improve the bit rate, the effectiveness of wireless transmission systems and limit the effects of fading transmission channel, an increased attention has recently been paid to MIMO systems. In fact, Alamouti's space-time block code is introduced in various wireless standards and systems. This manuscript deals with the problem of presence identification as well as blind separation of Orthogonal Space-Time Block Code (OSTBC) in the context of non data aided.

Keywords: MIMO channel, OSTBC, ICA, Alamouti's code, BSS of underdetermined mixture.

1 Introduction

In the last two decades, wireless communication systems become attractive economical sectors and the most challenging technology issues facing worldwide engineering institutions. In modern society, wireless communication systems and gadgets are very important in our every day life. Wireless systems are the most up-to-date technology and they can be found in various applications such as: Remote control toys, Global Positioning System (GPS), Mobile phone, radio, TV, satellite, robotics, *etc.*

The actual mobile phone standards such as, GSM (Global System for Mobile communications), GPRS (General Packet Radio Service) and UMTS (Universal Mobile Telecommunications System) can't support a very high data rate. In addition, the transmission system reliability and the transmission quality depend much on channel conditions. To improve the overall transmission performance using the diversity of transmission channel, Multiple-Input Multiple Output (MIMO) transmission systems have recently been introduced by many researchers and engineers. In fact, these systems have shown their abilities to tackle fading effects in multipath channels [13,4,8]. MIMO systems use Space-Time Block Coding (STBC) designs. Various STBC systems can be designed. In the literature, STBC which have orthogonal coding matrices are the most used ones and they are called the Orthogonal STBC (OSTBC). Recently, OSTBC have been introduced in recent wireless standards such as Worldwide Interoperability for Microwave Access (WiMax), and IEEE Standard 802.16e for 2.5GHZ bands.

One possible solution to mitigate the multipath fading problem is based on time or frequency diversity both at the transmitter and the receiver. The transmit diversity approach previously proposed by Alamouti [1] seems to be the most realistic scheme which is commonly used in MIMO systems.

T. Adali et al. (Eds.): ICA 2009, LNCS 5441, pp. 678–685, 2009.

When we are dealing with intercepted signals, problems become more challenging. This scenario can be found in various applications such as: Control of civilian authorities over the radio-band frequency, the control of communication quality, warfare, design of universal receiver, *etc.* For various reasons, this subject is not well addressed in the literature [2,9]. In this manuscript, the identification and the separation of OSTBC MIMO transmitted signals, in non data aided context, are considered.

2   Mathematical Model and Background

In this manuscript, the two widely used OSTBC codes are considered:

- Alamouti's code is an OSTBC2 (i.e. $n_T = 2$). Recently, this code has been introduced in many wireless communication standards and systems [14] (WiFi, IEEE 802.11n, 4G, etc). Alamouti's code can be implemented using the following coding matrix:

$$
\mathbf{C} = \overbrace{\begin{bmatrix} s_1 & -s_2^* \\ s_2 & s_1^* \end{bmatrix}}^{Time} \begin{matrix} \leftrightarrow A_1 \\ \leftrightarrow A_2 \end{matrix} \tag{1}
$$

- When $n_T = 3$, the orthogonal space-time block code is denoted by OSTBC3. Various OSTBC3 can be found in the literature [14,13]. In our study, one of the most efficient codes is considered (i.e. it has the maximum ratio of symbol number $N_s = 3$ to the transmission period $L = 4$):

$$
\mathbf{C} = \overbrace{\begin{bmatrix} s_1 & 0 & s_2 & -s_3 \\ 0 & s_1 & s_3^* & s_2^* \\ -s_2^* & -s_3 & s_1^* & 0 \end{bmatrix}}^{Time} \begin{matrix} \leftrightarrow A_1 \\ \leftrightarrow A_2 \\ \leftrightarrow A_3 \end{matrix} \tag{2}
$$

Let us denote by $\mathbf{x}(t)$, a $n_R \times 1$ complex vector, the signals received by an antenna array and by $\mathbf{s} = (s_j(t))$ the signals emitted by n_T transmitter antennas. The relationship among the received signals and the emitted signals can be represented by the following equation:

$$
x_i(t) = \sum_{j=1}^{n_T} h_{ij}(t,\tau) \otimes s_j(t) + b_i(t) \tag{3}
$$

Here $\otimes$ stands for the convolutive product, h_{ij} denotes the channel impulsive response between the ith emitter and the jth transmitter, and $\mathbf{b} = (b_i(t))$ represents the noise vector. Hereinafter, many realistic assumptions are made:

- The transmitted signals are mutually independent signals.
- The noises and the signals are independent of each other.
- The noise $\mathbf{b} = (b_i(t))$ is a zero-mean complex additive white gaussian noise.

- The channel parameters are unknown. However in many wireless applications, the channel can be considered as a quasi-stationary flat fading channel. In this case, one can neglect the Inter-Symbol Interference (ISI) and simplify the previous model (3) as following [4]:

$$x_i = \sum_{j=1}^{n_T} h_{ij} s_i + b_i \implies \mathbf{x} = \mathbf{Hs} + \mathbf{b} \tag{4}$$

- The receiver number n_R can be higher or lower than the emitter number n_S.
- The Symbol Rate (SR) and Carrier Wave Frequency (CWF) are unknown but they can be estimated [5,10]. This subject is beyond the scope of this manuscript, for further details please see [5,10] and their references.
- Using the previous assumption, the received signals can be over-sampled, that means the number of samples per symbol is greater than one. In addition a Carrier Frequency estimation error Δf_0 is introduced in our simulations.
- The emitted and received signals are not synchronize and a random demodulation phase $\phi \in [0, 2\pi]$ is considered.
- An asynchronous reception scheme is assumed. In this case, the sampling process is not necessarily synchronized with the symbol sequence.

3   Presence Detection of OSTBC Signals

The main idea of this manuscript consists in blindly separating OSTBC signals. In order to reach our goal, the OSTBC coding matrices are taken into consideration, this point is addressed in the next section. The latest statement means that prior information about the transmitted OSTBC signals should be available. The priori information can be considered as a strong assumption. In order to relax that assumption, we briefly describe here our new and simple approach to detect the presence of an OSTBC in the mixed and observed signals.

In order to clarify the main idea, we assume that at most one Alamouti-coded signal could be transmitted. The proposed approach can be generalized to deal with similar OSTBC signals. For non-noisy Alamouti-coded signal the eigenvectors of the observed signal covariance matrix $\mathbf{R}_{N \times N} = \mathbf{XX^H} = \mathbf{U\Delta U}^H$ are given by the following relationship, see [12]:

$$\mathbf{u}_1 = \mathcal{X}(\mathbf{s}_{(1)}, \mathbf{s}_{(2)}) = (a\mathbf{s}_{(1)} + be^{-j\theta}\mathbf{s}_{(2)})e^{i\psi_1}/\sqrt{\lambda_1} \tag{5}$$

$$\mathbf{u}_2 = \mathcal{Y}(\mathbf{s}_{(1)}, \mathbf{s}_{(2)}) = (b\mathbf{s}_{(1)} - ae^{-j\theta}\mathbf{s}_{(2)})e^{i\psi_2}/\sqrt{\lambda_2} \tag{6}$$

where $\mathbf{s}_{(1)} = \begin{bmatrix} s_1 & -s_2^* & \cdots & s_{N-1} & s_N^* \end{bmatrix}$ and $\mathbf{s}_{(2)} = \begin{bmatrix} s_2 & s_1^* & \cdots & s_N & s_{N-1}^* \end{bmatrix}$ stand for the vectors of the transmitted symbols, a, b, θ, ψ_1 and ψ_2 could be any number satisfying the following constraint $a^2 + b^2 = 1$.

Fig. 1 shows the main scheme of our approach. In this scheme, a modified cross-correlation matrix of the shifted eigenvectors is used:

$$\Gamma_\tau = \frac{1}{N} \sum_{i=0}^{\left[\frac{N}{2\tau}\right]-1} \sum_{k=0}^{\tau-1} \mathcal{X}_{2i\tau+k}\mathcal{Y}_{2i\tau+\tau+k} + \mathcal{Y}_{2i\tau+k}\mathcal{X}_{2i\tau+\tau+k} \tag{7}$$

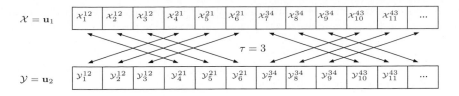

Fig. 1. The cross-correlation of an Alamouti's sequence using 3 samples/symbol, synchronization error ($\tau = 3$). The maximum cross correlation value can only be obtained if the two sequence are synchronized. Here, $\mathcal{X}_k^{mn} = \mathcal{X}(s_m, s_n)$, $\mathcal{X}_k^{nm} = \mathcal{X}(-s_n^*, s_m^*)$, $\mathcal{Y}_k^{mn} = \mathcal{Y}(s_m, s_n)$, $\mathcal{Y}_k^{nm} = \mathcal{Y}(-s_n^*, s_m^*)$ and $m < n$.

where $\left[\frac{N}{2\tau}\right]$ is the integer part of $\frac{N}{2\tau}$. Using the previous definition, the cross-correlation obviously becomes a function of the selected delay between the two shifted eigenvectors, τ and the synchronization error represented by $i = t_0$ $\ldots \left[\frac{N}{2\tau}\right] - 1$, i.e. the number of missed samples. This 3D function is shown in Fig. 2. The function maxima stands for the beginning of a symbol sequence. In this case, the cross-correlation function can be used to reduce the synchronization problem as well as to estimate the over-sampling ratio. Using that figure, it can be shown that:

- Two consecutive maximum are separate by 8 samples. One can conclude that the over-sampling ratio is 4 (i.e. 4 samples/symbol).
- The first maximum in Fig. 2 occurs at 5, that means we missed the beginning of the symbol sequence by 5 samples.

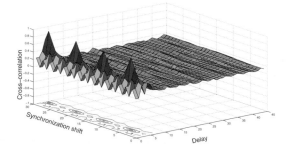

Fig. 2. Eigenvectors cross-correlation modified matrix defined by equation (7)

4 BSS of an Underdetermined Mixture of OSTBC Signals

In this section, the separation of underdetermined mixtures of OSTBC signals is addressed. The separation is achieved using the Multi User Kurtosis (MUK) algorithm and the specific structure of OSTBC coding matrices.

In [7], the authors proposed a BSS algorithm called Multi-User Kurtosis (MUK). MUK achieve the separation by maximizing a contrast function based on the kurtosis. Since the ninetieth of the last century, similar contrast functions have been proposed separately by other authors [15,3], further details can be founded in [6]. Similar to previous algorithms, MUK consists of two major steps:

- Orthogonalization and Whiteness.
- Separation which is achieved by estimating a rotation matrix to maximize the proposed contrast function.

In [7], the orthogonalization process is done using Gram-Schmidt algorithm. By taken into consideration the structure of OSTBC signals, the orthogonalization procedure can be improved. In our simulations at any iteration, the most modified weight vector is considered and used to establish the weight orthogonal matrix. Fig. 3 shows the effectiveness of the modified version. The comparison shown in the previous figure is obtained by averaging 10000 randomly initialized separations of an Alamouti's coded signals by using only one observed signal. As it was mentioned before, only two interesting OSTBC signals are considered in this manuscript. The actual study can be straightforward generalized to consider different OSTBC signals. Hereinafter, the transmission channel is assumed to be a Multiple-Input-Single-Output channel (MISO).

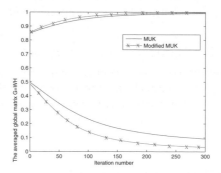

Fig. 3. Comparison between the original and the modified version of MUK

4.1   Extraction of an Alamouti's Coded Signal

To transmit an Alamouti's coded signal, two transmission antennas are needed. In noise free channels, the output of the MISO transmission channel is given by the following relationship:

$$\mathbf{x} = \begin{bmatrix} h_1 & h_2 \end{bmatrix} \begin{bmatrix} s_1 & -s_2^* & s_3 & -s_4^* & \cdots & s_{2N-1} & s_{2N}^* \\ s_2 & s_1^* & s_4 & s_3^* & \cdots & s_{2N} & s_{2N-1}^* \end{bmatrix} = \begin{bmatrix} h_1 s_1 + h_2 s_2 \\ -h_1 s_2^* + h_2 s_1^* \\ \vdots \\ h_1 s_{2N} + h_2 s_{2N-1} \\ -h_1 s_{2N}^* + h_2 s_{2N-1}^* \end{bmatrix}^T$$

Using the odd and the even components, the observed vector $\mathbf{x}$ can be split into the following two vectors:

$$\mathbf{x}' = \begin{bmatrix} x_1 & x_3 & \cdots & x_{2N-1} \\ x_2^* & x_4^* & \cdots & x_{2N}^* \end{bmatrix} = \begin{bmatrix} h_1 & h_2 \\ h_2^* & -h_1^* \end{bmatrix} \begin{bmatrix} s_1 & s_3 & \cdots & s_{2N-1} \\ s_2 & s_4 & \cdots & s_{2N} \end{bmatrix} \tag{8}$$

Without loss of generality, one can assume that transmitted symbols are Independent and Identically Distributed (iid) [11]. It is obvious that the previous transmission model is similar to an instantaneous complex mixture model.

Many simulations were conducted. Our experimental results show the effectiveness of the new separation scheme. Fig. 4 shows the separation of a QAM16 transmitted signal using an Alamouti's coding matrix.

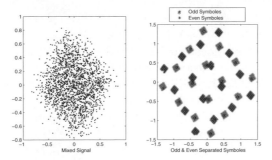

Fig. 4. Extraction of an Alamouti's QAM-16 coded signals using one receiver

4.2   Extraction of an OSTBC3 Signal

The previous separation scheme can be easily modified to take into consideration another OSTBC coding matrix. In fact, let us consider an OSTBC signal generated using the coding matrix of equation (2). The main key of our separation scheme is to find the MISO channel and the Equivalent Virtual Instantaneous mixture (EVI). Using equation (2), one can easily prove that an observed signal of a MISO channel should satisfy the following relationship:

$$\mathbf{x} = \begin{bmatrix} h_1 & h_2 & h_3 \end{bmatrix} \begin{bmatrix} s_1 & 0 & s_2 & -s_3 & \dots & s_{3N-2} & 0 & s_{3N-1} & -s_{3N} \\ 0 & s_1^* & s_3^* & s_2^* & \dots & 0 & s_{3N-2} & s_{3N}^* & s_{3N-1}^* \\ -s_2^* & -s_3 & s_1^* & 0 & \dots & -s_{3N-1}^* & -s_{3N} & s_{3N-2}^* & 0 \end{bmatrix}$$

$$\mathbf{x}^T = \begin{bmatrix} x_1 \\ x_2 \\ x_3 \\ x_4 \\ \vdots \\ x_{4M-3} \\ x_{4M-2} \\ x_{4M-1} \\ x_{4M} \end{bmatrix} = \begin{bmatrix} h_1 s_1 - h_3 s_2^* \\ h_2 s_1 + h_3 s_3 \\ h_1 s_2 + h_2 s_3^* + h_3 s_1^* \\ -h_1 s_3 + h_2 s_2^* \\ \vdots \\ h_1 s_{3N-2} + h_3 s_{3N-1} \\ h_2 s_{3N-2} + h_3 s_{3N} \\ h_1 s_{3N-1} + h_2 s_{3N}^* + h_3 s_{3N-2}^* \\ -h_1 s_{3N} + h_2 s_{3N-1}^* \end{bmatrix}$$

In this case, the observed vector could be divided into four sub-vectors using a cyclic selection of its components;

$$\begin{bmatrix} \mathbf{x}_{(1)} \\ \mathbf{x}_{(2)} \\ \mathbf{x}_{(3)}^* \\ \mathbf{x}_{(4)} \end{bmatrix} = \begin{bmatrix} x_1 & x_5 & \dots & x_{4M-3} \\ x_2 & x_6 & \dots & x_{4M-2} \\ x_3^* & x_7^* & \dots & x_{4M-1}^* \\ x_4 & x_8 & \dots & x_{4M} \end{bmatrix} = \begin{bmatrix} h_1 & -h_3 & 0 \\ h_2 & 0 & -h_3 \\ h_3^* & h_1^* & h_2^* \\ 0 & h_2 & -h_1 \end{bmatrix} \begin{bmatrix} s_1 & s_4 & \dots & s_{3N-2} \\ s_2^* & s_5^* & \dots & s_{3N-1}^* \\ s_3 & s_6 & \dots & s_{3N} \end{bmatrix} = \mathbf{HS}$$

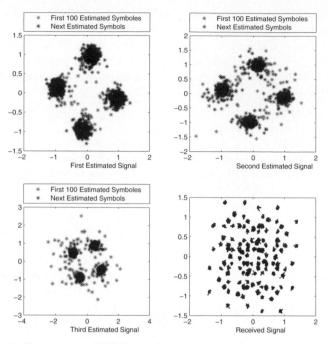

Fig. 5. Extraction of PSK4 symbols transmitted using an OSTBC3

Using the result of the previous section, one can make the sampling ratio $\rho = 1$ sample/symbol. If transmitted symbols are iid, then the previous model becomes an EVI model. The separation is done using the modified version of MUK. Fig. 5 shows the separation result of a PSK4 modulated using OSTBC3. In order to conduct that simulation, 2000 symbols, 4 transmission antennas and one receiver are used. We should mention here, that the previous separation scheme can still be applied to extract an OSTBC from a mixed modulated signals if its baud rate is different than the other ones. The extraction of other signals can be conducted using a deflation approach. With a high signal to noise ratio, good experimental results are obtained. Due to the lack of space, the details of this part of study is omitted. These points will be the goal of future studies.

5   Conclusion

In this manuscript, the identification, features extraction approach and an information retrieval method of an OSTBC signals are addressed. In fact, under realistic assumptions, an identification approach based on the cross-correlation of covariance matrix eigenvalues is proposed. In addition, a modified version of Multi-User Kurtosis (MUK) is briefly discussed. Finally, an OSTBC extraction scheme is considered. The new scheme uses the structure of the OSTBC coding matrix and a modified version of MUK.

References

1. Alamouti, S.: A simple transmit diversity technique for wireless communication. IEEE Journal on Selected Areas in Communications 16(8), 1451–1458 (1998)
2. Azzouz, E.E., Nandi, A.K.: Automatic modulation recognition of communication signals. Kluwer Academic Publishers, Dordrecht (1996)
3. Delfosse, N., Loubaton, P.: Adaptive blind separation of independent sources: A deflation approach. Signal Processing 45(1), 59–83 (1995)
4. Larsson, E.G., Stoica, P.: Space-time block coding for wireless communications. The Press Syndicate of the University of Cambridge (2003)
5. Le Guen, D., Mansour, A.: Automatic recognition algorithm for digitally modulated signals. In: 6th Baiona workshop on signal processing in communications, Baiona, Spain, 25-28 June (2003)
6. Mansour, A., Kardec Barros, A., Ohnishi, N.: Blind separation of sources: Methods, assumptions and applications. IEICE Transactions on Fundamentals of Electronics, Communications and Computer Sciences E83-A(8), 1498–1512 (2000)
7. Papadias, C.: Globally convergent blind source separation based on a multiuser kurtosis maximization criterion. IEEE Trans. on Signal Processing 48, 3508–3519 (2000)
8. Paulraj, A., Nabar, R., Gore, D.: Introduction to Space-Time Wireless. Cambridge University Press, New York (2003)
9. Pedzisz, M., Mansour, A.: Automatic modulation recognition of MPSK signals using constellation rotation and its 4-th order cumulant. Digital Signal Processing 15(3), 295–304 (2005)
10. Pedzisz, M., Mansour, A.: Hos-based multi-component frequency estimation. In: 13th European conference on Signal Processing, Theories and Applications (EU-SIPCO 2005). Elsevier, Turkey (2005)
11. Proakis, J.: Digital Communications. McGraw-Hill, New York (1983)
12. Riediger, M.L., Ho, P.K.M.: An eigen-assisted noncoherent receiver for alamouti type space-time modulation. IEEE Journal on Selected Areas in Communications 33(9), 1811–1820 (2005)
13. Shahbazpanah, S., Gershman, A.B., Manto, J.: Closed-form blind mimo channel estimation for orthogonal space-time block codes. IEEE Trans. on Signal Processing 53(12), 4506–4516 (2005)
14. Tarokh, V., Jafarkhani, H., Calderbank, A.R.: Space-time codes high data rate wireless communication: performance crirterion and construction. IEEE Trans. on Information Theory 44(2), 744–765 (1998)
15. Tarokh, V., Jafarkhani, H., Calderbank, A.R.: Space-time block codes from orthogonal designs. IEEE Trans. on Information Theory 45(5), 1456–1467 (1999)
16. Weinstein, E., Feder, M., Oppenheim, A.V.: Multi-channel signal separation by decorrelation. IEEE Trans. on Speech, and Audio Processing 1(4), 405–414 (1993)

Narrow-Band Short-Time Frequency-Domain Blind Signal Separation of Passive Sonar Signals

Natanael N. de Moura[1], Eduardo F. Simas Filho[1,2], and José M. de Seixas[1,*]

[1] Signal Processing Laboratory, COPPE/Poli/UFRJ, Rio de Janeiro, Brazil
[2] Federal Center for Technological Education, Bahia, Brazil
{natmoura,esimas,seixas}@lps.ufrj.br

Abstract. Sonar systems are very important for several military and civil navy applications. Passive sonar signals are susceptible to cross-interference from underwater acoustic sources (targets) present at different directions. In this work, a frequency-domain blind source separation procedure is proposed aiming at reducing cross-interferences, which may arise from adjacent signal source directions. As a consequence, target detection and classification may both be performed on cleaner data and one can expect an overall sonar efficiency improvement. As the underwater acoustic environment is time-varying, time-frequency transformation is performed using short-time windows. Original free of interference sources are estimated using ICA algorithms over narrow-band frequency-domain signals. It is shown that the proposed passive sonar signal processing approach attenuates in more than 10 dB the interference signals measured from two nearby directions and reduces the common background noise level in 7 dB.

Keywords: Passive Sonar, Spectral Analysis, BSS, Convolutive Mixtures, Interference Removal.

1 Introduction

If in a given operational condition exists more than one target to be detected, acoustic signals measured at adjacent directions (bearings) may be corrupted by cross-channel interference, contaminated by a background noise (from underwater acoustic environment) and the self-noise (acoustic signals generated by the own submarine in which the passive sonar system is located). Boths, target detection and classification tasks, may suffer from this interference, which may even provoke errors in important SO (Sonar Operator) decisions.

A narrow-band analysis (DEMON - Demodulation of Envelope Modulation On Noise) [1] is used as a pre-processing step selecting only the frequency band of interest for target characterization. DEMON is performed through Short-time Fast Fourier Transform to deal with the non-stationarity of the passive sonar signals.

* The authors would like to express their gratitude for the Brazilian Navy for providing the experimental data used in this work and for CNPq and FAPERJ for the financial support.

T. Adali et al. (Eds.): ICA 2009, LNCS 5441, pp. 686–693, 2009.

Envisaging cross-channel interference removal, a frequency-domain signal separation procedure is proposed in this paper for a passive sonar system. Original signal source estimation is then performed in frequency-domain using ICA (Independent Component Analysis) over preprocessed signals. Thus, independent components are extracted for each target.

2   Passive Sonar Systems

Passive sonar systems listen to sound radiated by a target using a hydrophone array [2] and detect signals against the background noise, composed by the ambient noise of the sea and the self-noise of the ship. After detection, the target must be identified based on its radiated noise [1]. The system present aural and visual information to the sonar operator, who will use this information to derive his decision, in terms of target identification.

2.1   DEMON Analysis

The DEMON analysis (Demodulation of Envelope Modulation On Noise) is often applied to obtain information about the propulsion of the target [7]. By demodulating the noise produced by cavitation propellers, it is possible to obtain shaft rotation, along with the number of blades and even the number of shafts of the target ship. This extracted information is extremely useful for the identification task [1].

Figure 1 displays the bearing time information obtained from the passive sonar system used in this work. The purpose of the bearing time analysis (beamforming) is to detect the acoustic signals directions of arrival. For that, a Cylindrical Hydrophone Array (CHA) is used in this work, allowing the system to perform omnidirectional surveillance [3].

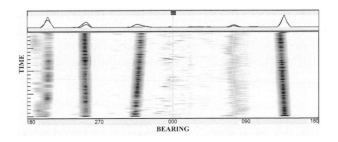

Fig. 1. Bearing time

The broadband noise from a propeller may be amplitude modulated at the blade rate frequency and its harmonics. Typically, target identification is supported by a DEMON analysis [1]. This is a narrow-band analysis that is applied on bearing information and helps the identification of the number of shafts, shaft rotation frequency and the number of blades [4,5]. As they provide a detailed

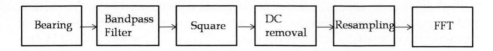

Fig. 2. Blocks diagram of DEMON analysis

knowledge of the threat's radiated noise, narrow-band sonar systems usually show good detection and classification capabilities [1,7].

Figure 2 shows a block diagram for classical DEMON analysis. Acquired time signals are filtered by a bandpass filter, typically between 1 to 10 kHz, which is the frequency band where cavitation is more evident. In sequence, signal is squared as in a traditional demodulation scheme and a TPSW (Two Pass Split Window) algorithm is used to estimate the mean of the background noise [1].

Using TPSW, it is possible to emphasize target signal peaks. Resampling is then performed to reach the band frequency of interest, 0 to 25 Hz, that corresponds to 0 to 1500 rpm. Finally, a Fast Fourier Transform (FFT) [6] is applied for each acquisition, which is 160 second long in this case. As the acoustic signals are time-variant due to modifications on the underwater acoustic scenario, here the FFT algorithm is applied to short lengths of time, which are selected from a moving Hanning-window (approximate length: 500 ms) applied to raw data.

As it has been already mentioned, contamination and interference may occur in neighbour bearings. Interference produces inaccurate peak detection and harmonics poorly definition. The self noise from the submarine may also produce interference in target bearing making it even more difficult the detection procedure.

3   The Proposed Signal Separation Procedure

In this work is proposed a frequency-domain ICA (FD-ICA) method for interference removal in passive sonar systems. As illustrated in Figure 3, DEMON analysis is initially performed over raw-data and frequency information from the three directions are used as inputs for an ICA algorithm, producing the independent (frequency-domain) components. Most of noise and non-relevant signals are eliminated by DEMON, allowing more accurate estimation of the independent components.

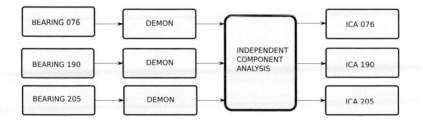

Fig. 3. Interference Removal in frequency domain

A particular characteristic is that DEMON analysis is usually performed over finite time-windows and the frequency components are estimated within these windows. Aiming at reducing the random noise generated in time-frequency transformation, an average spectrum is computed using frequency information from these time-slots.

Frequency-domain ICA (FD-ICA) is closely related to the convolutive mixing model [8], which assumes that the observed signals are generated by delayed versions of the sources. In the frequency-domain, convolutions are reduced to multiplications. A characteristic of FD-ICA is that the mixing matrix is frequency-dependent. In this work, as we are dealing with a narrow frequency band (0 to 25 Hz), for simplification, the mixing matrix is considered invariant in this band.

In ICA algorithms, the order and the amplitude of the estimated components are random parameters and thus different initializations may lead to different scaling factors and ordering [8]. In the proposed approach, ICA algorithms are executed after DEMON estimation at each time window, independent components from a certain direction may appear in different ordering at adjacent time-windows in this sequential procedure. Considering this, the short-time independent components must be reordered and normalized in amplitude. The normalization is performed by converting signal amplitude into dB scale. The reordering procedure is executed by computing the correlation between independent components estimated from adjacent time slots. High correlation indicates that these components are related to the same direction.

In the Section 4, experimental results obtained through the proposed approach are detailed. To estimate the independent components, JADE [10] and the Newton-like [12] algorithms were used. A performance comparison between these algorithms is presented.

3.1   Joint Approximate Diagonalization of Eigenmatrices

JADE (Joint Approximate Diagonalization of Eigenmatrices) [10] refers to one principle of solving the problem of equal eigenvalues of cumulant tensor achieving a diagonalization of the tensor through the eigenvalue decomposition. The tensorial method is used to carry through the independent components analysis. Tensors may be considered as a linear generalization of matrices or operators. The second and fourth order cumulant tensors are used to search for signal independence.

3.2   Multiplicative Newton-Like Algorithm for ICA

A multiplicative ICA algorithm was proposed by Akuzawa and Murata in [12]. Using the kurtosis as cost function, this method applies second-order optimization (through a Newton-like algorithm) in the search for independent components (instead of first-order gradient iterations used in most of ICA algorithms).

This algorithm does not require pre-whitening and thus operates directly over the data. Some experimental results obtained in [13] indicate that Akuzawa's algorithm outperforms classical ICA algorithms such as FastICA [8] and JADE in the presence of Gaussian noise.

4   Experimental Results

Experimental data used in this work comprises signals acquired from a CHA (sample frequency of 31,250 Hz). Initially, DEMON analysis is applied to raw data as a pre-processing. Figure 4 shows typical DEMON displays from three bearings. The horizontal axis represents the rotation scale (in rpm) while the vertical axis corresponds to signal amplitude (in dB). The largest peak amplitude reveals the speed of shaft rotation, while the subsequent harmonics indicate the number of blades.

Two targets are present in this experimental run at directions $190°$ and $205°$. As illustrated in Figure 4, the frequency components after DEMON analysis (here DEMON spectrums are composed by 512 frequency bins, spanning from zero up to 1400 RPM) at $190°$ target (FA=148 RPM and its multiples) are mixed together with information from the $205°$ direction (FB=119 RPM). The same problem exists in the signal measured at bearing $205°$. It was also observed that both signals ($190°$ and $205°$) are contaminated by a third component (FC=305 RPM), that is the main frequency present at direction $076°$. It is known from the experimental setup that the last bearing ($076°$) contains information from the noise radiated by the submarine where the hydrophones array is allocated. It can also be verified that, signal measured at direction $076°$ presents interference from target at $205°$ (FB).

The proposed narrow-band frequency-domain signal separation procedure was applied to the underwater acoustic signals measured at directions $076°$, $190°$ and $205°$. A frequency-time display (demongram) for direction $190°$ is provided in Figure 5. It can be depicted that an interference frequency component (FB) is present during the observed interval (from 65 to 85 seconds) in raw data at direction 190^0 (Figure 5-a). Figure 5-b illustrates the frequency-time display for the independent component related to the same direction and it can be observed

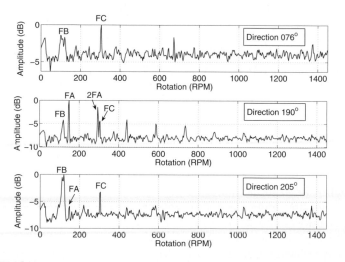

Fig. 4. DEMON analysis for direction $076°$ (top), $190°$ (middle) and $205°$ (bottom)

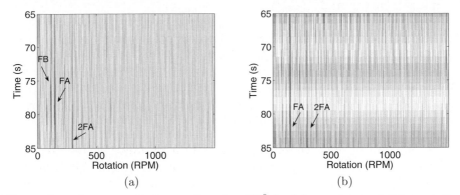

Fig. 5. Time-frequency display at direction 190^0 for (a) raw data and (b) short-time frequency-domain independent component

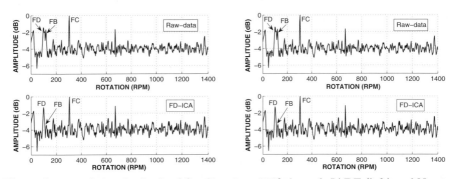

Fig. 6. Separated signals obtained for directions 076° through JADE (left) and Newton-like (right) algorithms

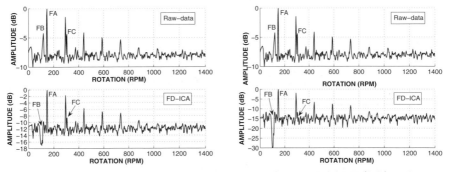

Fig. 7. Separated signals obtained for directions 190° through JADE (left) and Newton-like (right) algorithms

that the interference is removed and the second harmonic component highlighted as a result of FD-ICA. In this plot, JADE algorithm was applied.

In the classical DEMON processing, the mean amplitude of the frequency components obtained at each short-length time windows is calculated and a

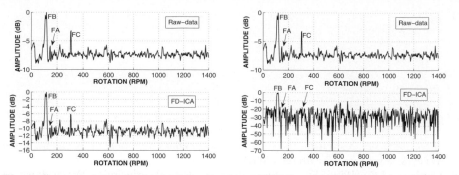

Fig. 8. Separated signals obtained for directions 205° through JADE (left) and Newton-like (right) algorithms

mean frequency-amplitude plot is generated. The results (considering the mean spectrums) obtained through JADE and the Newton-like algorithms are illustrated in Figures 6, 7 and 8 respectively for directions 076°, 190° and 205°. It can be observed from these Figures that in the independent signals, obtained through both algorithms, the cross-interference was significantly reduced.

For the Newton-like algorithm, it was also observed that the background noise level (which can be estimated from the mean amplitude of the non-relevant frequency components) was more significantly reduced when compared to JADE. For direction 190° the estimated background noise levels are approximately -8dB (raw-data), -12dB (FD-ICA / JADE) and -15dB (FD-ICA / Newton-like). Considering now direction 205°, the noise level was reduced from -7dB (raw-data) up to -11dB (FD-ICA / JADE) and -30dB (FD-ICA / Newton-like). The cross-interference peaks were also more severely attenuated at the independent components estimated through the Newton-like algorithm. These results indicates that the independent components estimation through the Newton-like algorithm is more suitable for this application.

5   Conclusions

On passive sonar signals, target identification relies very much on narrow-band frequency-domain information obtained through DEMON analysis. Signals acquired at adjacent directions (bearings) may be corrupted by cross-channel interference from multiple targets. In this work, Frequency-Domain Independent Component Analysis (FD-ICA) was used to reduce this interference. The performance obtained through two ICA algorithms, JADE and Newton-like, were compared and it was observed that the Newton-like method presented better results.

References

1. Nielsen, R.O.: Sonar Signal Processing. Artech House Inc., Nortwood (1991)
2. Van Veen, B.D., Buckley, K.M.: Beamforming: A versatile approach to spatial filtering. In: IEEE ASSP Magazine, pp. 4–24 (April 1988)

3. Krim, H., Viberg, M.: Two decades of array signal processing research: The parametric approach. IEEE Signal Processing Magazine 13(4), 67–94 (1996)
4. Van Trees, H.L.: Detection, Estimation and Modulation Theory - Pat I. John Wiley & Sons, New York (2001)
5. Nielsen, R.O.: Cramer-Rao lower bound for sonar broadband modulation parameters. IEEE Journal of Ocean Engineering 24(3), 285–290 (1999)
6. Diniz, P.S.R., Netto, S.L., da Silva, E.A.B.: Digital Signal Processing:System Analysis and Design. Cambridge University Press, Cambridge (2001)
7. Waite, A.D.: Sonar for practicing Engineers, New York (2003)
8. Hyvarinen, A., Karhunen, J., Oja, E.: Independent Component Analysis. John Wiley & Sons Inc., Chichester (2001)
9. Cardoso, J.F.: Blind Signal Separation: Statistical principles. Proceedings of IEEE 86(10), 2009–2025 (1998)
10. Cardoso, J.F., Souloumiac, A.: Blind beamforming for non-gaussian signals. IEE Proceedings 140, 362–370 (1993)
11. Yan Li, D.P., Peach, J.: Comparison of Blind Source Separation Algoritms. In: Advances in Neural Networks and Applications, pp. 18–21 (2000)
12. Akuzawa, T., Murata, N.: Multiplicative Nonholonomic Newton-like Algorithm. Chaos, Solutions and Fractals 12(4), 785–793 (2001)
13. Akuzawa, T.: Extended Quasi-Newton Method for the ICA. In: Int. Workshop Independent Component Anal. Blind Signal Separation, pp. 521–525 (2000)

Blind Channel Identification in (2 × 1) Alamouti Coded Systems Based on Maximizing the Eigenvalue Spread of Cumulant Matrices

Héctor J. Pérez-Iglesias[1], Daniel Iglesia[1], Adriana Dapena[1], and Vicente Zarzoso[2]

[1] Departamento de Electrónica y Sistemas, Universidade da Coruña, Facultad de Informática, Campus de Elviña, 5, 15071 A Coruña, Spain
{hperez,dani,adriana}@udc.es
[2] Laboratoire I3S, Université de Nice-Sophia Antipolis, Les Algorithmes, Euclide-B, BP 121, 06903 Sophia Antipolis, France
zarzoso@i3s.unice.fr

Abstract. Channel estimation in the (2 × 1) Alamouti space-time block coded systems can be performed blindly from the eigendecomposition (or diagonalization) of matrices composed of the receive antenna output 4th-order cumulants. In order to estimate the channel, we will propose to choose the cumulant matrix with maximum eigenvalue spread of cumulant-matrix. This matrix is determined in closed form. Simulation results show that the novel blind channel identification technique presents a satisfactory performance and low complexity.

1 Introduction

A large number of Space Time Coding (STC) techniques have been proposed in the literature to exploit spatial diversity in systems with multiple elements at both transmission and reception (see, for instance, [1] and references therein). The Orthogonal Space Time Block Coding (OSTBC) is remarkable in that it is able to provide full diversity gain with linear decoding complexity [2,3]. The basic premise of OSTBC is the encoding of the transmitting symbols into an unitary matrix to spatially decouple their Maximum Likelihood (ML) detection, which can be seen as a matched filter followed by a symbol-by-symbol detector.

The OSTBC scheme for MIMO systems with two transmit antennas is known as the Alamouti code [2] and it is the only OSTBC capable of achieving full spatial rate for complex constellations. The (2 × 1) Alamouti coded systems are attractive in wireless communications due to their simplicity and their ability to provide maximum diversity gain while preserving the channel capacity. Because of these advantages, the Alamouti code has been incorporated in the IEEE 802.11 and IEEE 802.16 standards [4].

Coherent detection in (2 × 1) Alamouti coded systems requires the identification of a (2 × 2) unitary channel matrix. The transmission of data known to the receiver, known as pilot or training symbols, is often used to perform

T. Adali et al. (Eds.): ICA 2009, LNCS 5441, pp. 694–701, 2009.
© Springer-Verlag Berlin Heidelberg 2009

channel estimation required for a coherent detection of OSTBCs. However, training symbols reduce the throughput and such schemes are inadequate when the bandwidth is scarce. Strategies that avoid this limitation include the so-called Differential STBC (DSTBC) [5] which is a signalling technique that generalizes differential modulations to the transmission over MIMO channels. DSTBCs can be incoherently decoded without the aid of channel estimates but they incur in a 3 dB performance penalty when compared to coherent detection.

Training sequences can also be avoided by the use of blind channel identification methods. In particular, this contribution focuses on blind channel identification in (2×1) Alamouti coded systems using higher-order eigen-based approaches. Under the assumption of independent symbol substreams, the channel can be estimated from the eigendecomposition of matrices composed of 2nd- or higher-order statistics (cumulants) of the received signal. The so-called joint approximate diagonalization of eigenmatrices (JADE) method for blind source separation via independent component analysis is optimal in that it tries to simultaneously diagonalize a full set of 4th-order cumulant matrices. In order to reduce the computational complexity, we propose to diagonalize a linear combination of cumulant matrices, which is judiciously chosen by maximizing its expected eigenvalue spread.

2   The (2×1) Alamouti Coding Scheme

Figure 1 shows the baseband representation of Alamouti OSTBC with two antennas at the transmitter and one antenna at the receiver. Each pair of symbols $\{s_1, s_2\}$ is transmitted in two adjacent periods using a simple strategy: in the first period s_1 and s_2 are transmitted from the first and the second antenna, respectively, and in the second period $-s_2^*$ is transmitted from the first antenna and s_1^* from the second one, symbol $(\cdot)^*$ denoting complex conjugation. In the sequel, we assume that the symbol substreams are complex-valued, zero-mean, stationary, non-Gaussian distributed and statistically independent; their exact probability density functions are otherwise unknown.

The transmitted symbols (sources) arrive at the receiving antenna through fading paths h_1 and h_2 from the first and second transmit antenna, respectively. Hence, the signal received during the first and the second symbol period have the

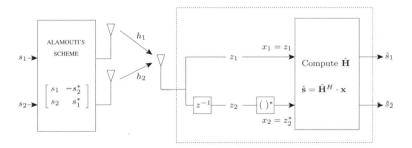

Fig. 1. Alamouti Coding Scheme

form $z_1 = s_1 h_1 + s_2 h_2 + n_1$ and $z_2 = s_1^* h_2 - s_2^* h_1 + n_2$, respectively. The term n_i denotes the additive white Gaussian noise at symbol period i. By defining the observation vector as $\mathbf{x} = [x_1, x_2]^T = [z_1, z_2^*]^T$, symbol $(\cdot)^T$ standing for the transpose operator, the relationship between the observation vector $\mathbf{x}$ and the source vector $\mathbf{s} = [s_1, s_2]^T$ is given by

$$\mathbf{x} = \mathbf{H s} + \mathbf{n} \tag{1}$$

where $\mathbf{n} = [n_1, n_2^*]^T$ is the noise vector and $\mathbf{H}$ represents the (2×2) channel matrix

$$\mathbf{H} = \begin{bmatrix} \mathbf{h}_1 | \mathbf{h}_2 \end{bmatrix} = \begin{bmatrix} h_1 & h_2 \\ h_2^* & -h_1^* \end{bmatrix} \tag{2}$$

It is interesting to note that matrix $\mathbf{H}$ is unitary up to a scalar factor, i.e.,

$$\mathbf{H} \mathbf{H}^H = \mathbf{H}^H \mathbf{H} = \|\mathbf{h}\|^2 \mathbf{I}_2 \tag{3}$$

where $\|\mathbf{h}\|^2 = |h_1|^2 + |h_2|^2$ is the squared Euclidean norm of the channel vector, $\mathbf{I}_2$ is the (2×2) identity matrix and $(\cdot)^H$ is the Hermitian operator. It follows that the transmitted symbols can be recovered, up to scale, as $\hat{\mathbf{s}} = \hat{\mathbf{H}}^H \mathbf{x}$, where $\hat{\mathbf{H}}$ is a suitable estimate of the channel matrix. As a result, this scheme supports ML detection based only on linear processing at the receiver. Consequently, the correct detection of the transmitted symbols $\mathbf{s}$ requires the accurate estimation of the channel matrix $\mathbf{H}$ from the received data $\mathbf{x}$.

3   Blind Channel Estimation Based on Eigenvalue Spread

In this section, we will propose a novel higher-order eigen-based approach to estimate the channel matrix in the (2×1) Alamouti system. For the sake of simplicity, we restrict the exposition to zero-mean distributions and circular statistics. Given a random vector $\mathbf{x} = [x_1, x_2] \in \mathbb{C}^2$, its 2nd-order cumulants are simply defined as $\mathrm{cum}(x_i, x_j^*) = \mathrm{E}[x_i x_j^*]$, and the 4th-order cumulants as

$$\mathrm{cum}(x_i, x_j^*, x_k, x_\ell^*) =$$

$$= \mathrm{E}[x_i, x_j^*, x_k, x_\ell^*] - \mathrm{E}[x_i x_j^*]\mathrm{E}[x_k x_\ell^*] - \mathrm{E}[x_i x_\ell^*]\mathrm{E}[x_j x_k^*] - \mathrm{E}[x_i x_k]\mathrm{E}[x_j^* x_\ell^*] \tag{4}$$

Given a matrix $\mathbf{M} \in \mathbb{C}^{2 \times 2}$

$$\mathbf{M} = \begin{bmatrix} m_{11} & m_{12} \\ m_{21} & m_{22} \end{bmatrix} \tag{5}$$

the 4th-order cumulant matrix $\mathbf{Q}^{(4)}(\mathbf{M})$ is defined as the (2×2) matrix with components [6]

$$[\mathbf{Q}^{(4)}(\mathbf{M})]_{ij} = \sum_{k,\ell=1}^{2} \mathrm{cum}(x_i, x_j^*, x_k, x_\ell^*) m_{\ell k} \tag{6}$$

Under a linear model like (1) with statistically independent sources and unitary mixing matrix, the cumulant matrix takes the form $\mathbf{Q}^{(4)}(\mathbf{M}) = \mathbf{H}\boldsymbol{\Delta}(\mathbf{M})\mathbf{H}^{\mathrm{H}}$. Matrix $\boldsymbol{\Delta}(\mathbf{M})$ being diagonal with

$$[\boldsymbol{\Delta}(\mathbf{M})]_{ii} = \rho_i \mathbf{h}_i^{\mathrm{H}} \mathbf{M} \mathbf{h}_i \tag{7}$$

where $\rho_i = \mathrm{cum}(s_i, s_i^*, s_i, s_i^*)$ is the marginal 4th-order cumulant (kurtosis) of the ith source, and $\mathbf{h}_i$ represents the ith column of $\mathbf{H}$. Therefore, the separating matrix $\mathbf{H}$ diagonalizes $\mathbf{Q}^{(4)}(\mathbf{M})$ for any $\mathbf{M}$. Hence, the eigendecomposition of (6) allows the identification of the remaining unitary part of $\mathbf{H}$ if the eigenvalues of $\mathbf{Q}^{(4)}(\mathbf{M})$ are different, i.e., if matrix $\boldsymbol{\Delta}(\mathbf{M})$ contains different entries: $\rho_i \mathbf{h}_i^{\mathrm{H}} \mathbf{M} \mathbf{h}_i \neq \rho_j \mathbf{h}_j^{\mathrm{H}} \mathbf{M} \mathbf{h}_j, \forall i \neq j$. To increase robustness to eigenspectrum degeneracy, a set $\{\mathbf{Q}^{(4)}(\mathbf{M}_k)\}_{k=1}^m$, may be (approximately) jointly diagonalized. The full set comprises $m = 2^2 = 4$ linearly independent (e.g., orthonormal) matrices $\{\mathbf{M}_k\}_{k=1}^m$. A simplified version of the algorithm is obtained by considering the set of matrices verifying $\mathbf{Q}^{(4)}(\mathbf{M}_k) = \lambda_k \mathbf{M}_k$. As there are only 2 such eigenmatrices, this version is, in theory, computationally more efficient. However, the eigenmatrices depend on matrix $\mathbf{H}$ itself and they must also be estimated from the data. JADE admits an efficient implementation in terms of the Jacobi technique for matrix diagonalization. In Alamouti Coding Scheme, the channel matrix is unitary, and can then be identified by this procedure.

3.1 Maximizing the Eigenvalue Spread

The performance of eigendecomposition-based methods depend on the eigenvalue spread of the matrix to diagonalize because the eigenvectors associated with equal eigenvalues can only be determined up to a unitary transformation [7]. For the (2×1) Alamouti OSTBC, with the same kurtosis for s_1 and s_2, $\rho = \rho_1 = \rho_2$, from equation (7) the eigenvalue spread of cumulant matrix $\mathbf{Q}^{(4)}(\mathbf{M})$ is

$$L(\mathbf{M}) = |\rho| \left| \mathbf{h}_1^{\mathrm{H}} \mathbf{M} \mathbf{h}_1 - \mathbf{h}_2^{\mathrm{H}} \mathbf{M} \mathbf{h}_2 \right| \tag{8}$$

In order to obtain the matrix $\mathbf{M}_{opt}$ that maximizes $L(\mathbf{M})$ we will introduce the following notation,

$$\tilde{\mathbf{h}} = \begin{bmatrix} |h_1|^2 - |h_2|^2 \\ 2h_1^* h_2^* \\ 2h_1 h_2 \\ |h_2|^2 - |h_1|^2 \end{bmatrix}, \quad \mathbf{m} = \begin{bmatrix} m_{11} \\ m_{21} \\ m_{12} \\ m_{22} \end{bmatrix} \tag{9}$$

Substituting in (8), the eigenvalue spread takes the form

$$L(\mathbf{M}) = |\rho| \left| \mathbf{h}_1^H \mathbf{M} \mathbf{h}_1 - \mathbf{h}_2^H \mathbf{M} \mathbf{h}_2 \right| =$$

$$|\rho| \left| (|h_1|^2 - |h_2|^2)(m_{11} - m_{22}) + 2(h_1 h_2 m_{21} + h_1^* h_2^* m_{12}) \right| = |\rho| \left| \tilde{\mathbf{h}}^H \mathbf{m} \right| \tag{10}$$

Our objective is to find the vector $\mathbf{m}$ that maximizes the eigenvalue spread, i.e.,

$$\mathbf{m}_{opt} = \begin{array}{c} arg\ max \\ ||\mathbf{m}|| = 1 \end{array} \left| \tilde{\mathbf{h}}^H \mathbf{m} \right| \tag{11}$$

The scalar product between $\mathbf{h}$ and $\mathbf{m}$ is maximum when $\mathbf{m}$ has the direction and sense of $\mathbf{h}$. Therefore, the optimum value is

$$\mathbf{m}_{opt} = \frac{\tilde{\mathbf{h}}}{||\tilde{\mathbf{h}}||} \tag{12}$$

As a result, the optimum matrix is given by

$$\mathbf{M}_{opt} = \frac{1}{\sqrt{2 + 2|\gamma|^2}} \begin{bmatrix} 1 & \gamma \\ \gamma^* & -1 \end{bmatrix}, \text{where } \gamma = \frac{2h_1 h_2}{|h_1|^2 - |h_2|^2} \tag{13}$$

Since this optimum matrix depends on the actual values of unknown channel coefficients h_1, h_2, we propose to estimate parameter γ using 4th-order cross-cumulants of the observations,

$$\gamma = \frac{\text{cum}(x_1, x_2^*, x_1, x_2^*)}{\text{cum}(x_1, x_2^*, x_1, x_1^*)} = \frac{\rho 2 h_1^2 h_2^2}{\rho(|h_1|^2 - |h_2|^2)h_1 h_2} = \frac{2h_1 h_2}{|h_1|^2 - |h_2|^2} \tag{14}$$

Note that the parameter γ is used to weighted 4th-order cross-cumulants in equation (6) and, therefore, errors in its estimation can degrade the performance of the method proposed to identify the channel matrix $\mathbf{H}$.

3.2   Blind Channel Estimation Based on Eigenvalue Spread (BCEES)

To evaluate the the 4th-order cumulant function (6) in the optimum matrix $\mathbf{M}$ given in equation (13) requires to compute sixteen 4th-order cross-cumulant. According to the symmetric property of the cumulants, it can be reduced to compute six fourth-order cross-cumulants. In this subsection, we will propose a simplified approach which only consider to fourth-order cross cumulant matrices:

$$\mathbf{M}_1 = \begin{bmatrix} 1 & 0 \\ 0 & 0 \end{bmatrix}, \quad \mathbf{M}_2 = \begin{bmatrix} 0 & 0 \\ 1 & 0 \end{bmatrix} \tag{15}$$

For these two matrices, the cumulant sum given in equation (6) is reduced to the computation of only one 4th-order cumulant matrix. Particularizing equation (10) to these matrices, we can determine which of the two matrices provides the maximum eigenvalue spread using to the following decision criterion

$$|\gamma| = \frac{L(\mathbf{M}_2)}{L(\mathbf{M}_1)} = \frac{2|h_1||h_2|}{||h_1|^2 - |h_2|^2|} = \frac{2|h_1||h_2|}{||h_1|^2 - |h_2|^2|} \overset{\mathbf{M}_1}{\underset{\mathbf{M}_2}{\lessgtr}} 1 \tag{16}$$

In the practice, γ can be estimated using (14).

4   Experimental Results

This section reports several numerical experiments carried out to evaluate and compare the performance of the blind channel estimation algorithms proposed

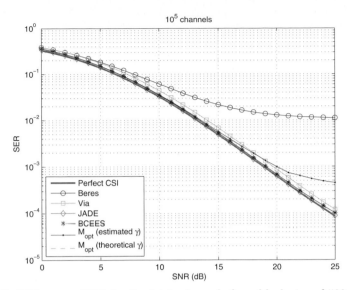

Fig. 2. SER versus SNR for Rayleigh channels for a block size of 500 symbols

in this paper. The experiments have been performed on QPSK source symbols coded with the Alamouti Scheme. The channel is assumed to remain constant during the transmission of a block of K symbols and it has a Rayleigh distribution. The cumulants are calculated by sample averaging over the symbols of a block. Performance has been measured in terms of the Symbol Error Rate (SER) of the source symbols estimated after channel identification. SER figures are obtained by averaging over 10^5 independent blocks of symbols and channel realizations.

Over this simulated scenario, the following techniques are compared:

- The method proposed by Beres *et al.* [8] corresponding to use $m_{11} = 1, m_{12} = m_{21} = m_{22} = 0$ or $m_{11} = 0, m_{12} = m_{21} = m_{22} = 2$.
- The SOS-based approach proposed by Via *et al.* [9].
- The JADE algorithm proposed by Cardoso and Souloumiac [6].
- The novel method based on maximizing the eigenvalue spread proposed in Subsection 3.1 using the theoretical γ.
- The novel method based on maximizing the eigenvalue spread proposed in Subsection 3.1 using the estimated γ in equation (14).
- The BCEES technique proposed in Subsection 3.2, using the estimated γ in equation (14).

As a bound of performance, we also present the SER obtained with Perfect Channel Side Information (CSI).

Figure 2 shows the performance obtained with these methods for different values of SNR. The 4th-order cumulants have been estimated using 500 symbols of each observed signals. Note that the results obtained with the novel approaches and JADE are very close to the Perfect CSI. Note, however, that for high SNRs

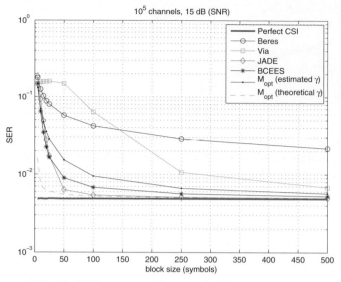

Fig. 3. SER versus packet size for Rayleigh channels

the novel method with the estimated γ presents a flooring effect due to errors in the estimation of γ.

Figure 3 shows the performance for different packet sizes at an SNR of 15 dB. The good performance of JADE, the novel approach with theoretical γ and BCEES is apparent.

The decoding complexity of methods based on cumulant-matrix diagonalization depends on two parameters: the number of cumulant matrices to be computed and the size of the matrix to be diagonalized. Note that the load of diagonalizing 2×2 matrices is very low because it can be used closed expressions [10]. Table 1 summarizes these parameters for the approaches considered in our simulations. Considering the computational load and the results presented in Figure 2 and Figure 3, we conclude that BCEES presents a satisfactory performance and low complexity.

Table 1. Computational load

Approach	Number of cumulant matrices	Size of the matrix to diagonalize
Beres et al., and Via et al.	1	2×2
JADE	4	3×3
$\mathbf{M}_{opt}$	4	2×2
BCEES	2	2×2

5   Conclusion

This paper addresses the problem of blind channel identification in (2×1) Alamouti coded system. In order to estimate the channel matrix, we have proposed to diagonalize a linear combination of cumulant matrices, which is judiciously chosen by maximizing its expected eigenvalue spread. We have also proposed a simplified approach, called BCEES, which presents a satisfactory performance and low complexity since it needs to diagonalize a single cumulant matrix.

References

1. Paulraj, A.J., Papadias, C.B.: Space-time processing for wireless communications. IEEE Signal Processing Magazine 14(6), 49–83 (1997)
2. Alamouti, S.M.: A simple transmit diversity technique for wireless communications. IEEE Journal Select. Areas Communications 16, 1451–1458 (1998)
3. Tarokh, V., Jafarkhani, H., Calderbank, A.R.: Space-Time Block Codes from Orthogonal Designs. IEEE Transactions on Information Theory 45(5), 1456–1467 (1999)
4. Andrews, J.G.: Fundamentals of WiMAX: Understanding Broadband Wireless Networking. Prentice Hall Communications Engineering and Emerging Technologies Series (2007)
5. Hughes, B.L.: Differential space-time modulation. IEEE Transactions on Information Theory 46(7), 2567–2578 (2000)
6. Cardoso, J.-F., Souloumiac, A.: Blind beamforming for non-Gaussian signals. IEE Proceedings F 140(46), 362–370 (1993)
7. Golub, G.H., Van Loan, C.F.: Matrix Computations, 3rd edn. Johns Hopkins Studies in Mathematical Sciences (1996)
8. Beres, E., Adve, R.: Blind Channel Estimation for Orthogonal STBC in MISO Systems. In: Proc. of Global Telecommunications Conference, vol. 4, pp. 2323–2328 (November 2004)
9. Vía, J., Santamaría, I., Pérez, J., Ramírez, D.: Blind Decoding of MISO-OSTBC Systems based on Principal Component Analysis. In: Proc. of International Conference on Acoustic, Speech and Signal Processing, vol. IV, pp. 545–549 (May 2006)
10. Blinn, J.: Consider the lowly 2x2 matrix. Computer Graphics and Applications 16(2), 82–88 (1996)

ICA Mixture Modeling for the Classification of Materials in Impact-Echo Testing

Addisson Salazar, Arturo Serrano, Raúl Llinares, Luis Vergara, and Jorge Igual

Universidad Politecnica de Valencia, Departamento de Comunicaciones, Camino de Vera s/n, 46022, Valencia, Spain
asalazar@dcom.upv.es

Abstract. This paper presents a novel ICA mixture model applied to the classification of different kinds of defective materials evaluated by impact-echo testing. The approach considers different geometries of defects build from point flaws inside the material. The defects change the wave propagation between the impact and the sensors producing particular spectrum elements which are considered as the sources of the underlying ICA model. These sources and their corresponding transfer functions to the sensors make a signature of the resonance modes for different conditions of the material. We demonstrate the model with several finite element simulations and real experiments.

Keywords: ICA, ICA mixtures, Non-destructive evaluation, Impact-echo testing.

1 Introduction

This paper introduces the application of the independent component analysis mixture modeling (ICAMM) in non-destructive testing (NDT). Particularly, we apply ICAMM to NDT by impact-echo. In impact-echo testing, the response of a material (vibrations) excited by an impact is measured by a set of sensors located in its surface [1][2][3]. The approach is formulated from the linear system theory dividing the wave path propagation in two parts: impact to point flaws and point flaws to sensors. It is assumed that the set of point flaws build defective areas with different geometries, such as, cracks (small parallelepipeds), holes (cylinders), and multiple defects (combination of cracks and holes). Depending on the kind of defective area, the spectrum measured by the sensors changes and this allow the kind of defect condition of the material to be discerned.

The ICAMM [4][5] has recently emerged as a flexible approach to model arbitrary data densities using mixtures of multiple ICA models [6][7][8] with non-Gaussian distributions for the independent components (i.e., relaxing the restriction of modelling every component by a multivariate Gaussian density). ICAMM is a kind of nonlinear ICA technique that extends the linear ICA method by learning multiple ICA models and weighting them in a probabilistic manner [4]. In the NDT area, there are there are a few references of applying ICA algorithms [9][10][11], but there are not references of application of ICAMM.

Recently, we classified impact-echo signals from defective materials using temporal and frequency features, and neural networks as classifiers [12], obtaining the best

T. Adali et al. (Eds.): ICA 2009, LNCS 5441, pp. 702–709, 2009.

results with the frequency features. In addition, we demonstrated in [13][14] that the ICA can be used to separate information of the defects in impact-echo testing. This work is focused in analyzing the signal spectra in order to exploit the resonant modes generated in the impact-echo experiment. A PCA procedure was applied to reduce dimensionality of the estimated spectrum for the multichannel configuration of the experiment. There is evidence that the first components of PCA retain essentially all of the useful information and this compression optimally removes noise and can be used to identify and classify unusual spectra [15][16].

We apply an ICAMM-based classifier [17], showing the impact-echo data fit into the developed ICAMM model. Results are included for different kinds of defective simulated models and laboratory specimens and other classifiers as LDA (Linear Discriminant Analysis) and MLP (Multi-Layer Perceptron) [18][19].

2 ICA Mixture Statement of the Problem

Suitability of mixtures of ICA for a given problem of data analysis and classification can be approached from different perspectives. On one end, we have the least physical explanation: the ICA mixture learning underlies an estimation/modeling of the probability density of multivariate data [5]. The degrees of freedom afforded by mixtures of ICA suggest a good candidate for a broad class of problems. In the middle we have the interpretation of ICA as a way of learning some appropriate bases (usually called activation functions) which are more or less connected to actual behaviors implicit in the physical phenomena under analysis [20]. In the other end, there is the most physical explanation which tries to indentify where sources are originated and how they mix before arriving to the sensors, i.e. it tries to provide a physical interpretation of the linear mixture model.

In this section we have made some effort from this later approach to model the data coming from the impact-echo inspection. It is clear the convenience to have as much knowledge as possible about the underlying physical phenomena, with the aim of making better interpretation of the results and the general performance of the method.

Recently, we proposed an ICA model for the impact-echo problem [13][14]. This model considered the transfer functions between the impact location and point defects spread in a material bulk as "sources" for blind source separation. In this section, we formulate a new ICA model considering the resonance phenomenon involved in the impact-echo method. We extend this model to defects with different shapes, such as cracks or holes, and formulate the quality condition determination of homogeneous and defective materials as an ICA mixture problem.

The signals measured by the sensors can be considered as a convolutive mixture of the input signal and the defects inside the material as shown in Fig. 1.

In Fig. 1, there is one attack point that generates the wave $r_0(n) = p(n)$; F internal focuses (point flaws) that generate the waves $f_j(n)$ $j = 1,...,F$; and N sensors that measure the waves $v_i(n)$ $i = 1,...,N$. To simplify, we note the impact as another focus; thus, $f_j(n)$ $j = 0,...,F$, being $f_0(n) = r_0(n)$. Linear propagation is assumed.

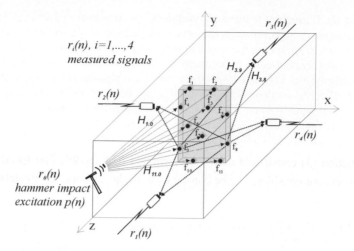

Fig. 1. Scheme of a material inspection by impact-echo using 4 sensors. Material with 11 internal focuses due to point flaws that build a crack-shape-like defect that is oriented in the plane xy (the path between the point flaws and the sensors is depicted only for a few focuses).

Let us call $\mathbf{R}_{ij}$ to the contribution to the spectrum of the signal recorded at sensor i, due to the focus j. We consider $\mathbf{R}_{ij}$ to be the observation vectors for an ICA model. This spectrum is obtained by the DFT computed on an interval of the signal $r_i(n)$. Thus,

$$\mathbf{R}_{ij} = diag\left[\mathbf{H}_{ij}\right] \cdot \mathbf{F}_j \ , \tag{1}$$

where

$\mathbf{H}_{ij}$:   transfer function vector between the internal focuses $j = 1, \ldots F$ and the sensors $i = 1, \ldots N$

$\mathbf{F}_j$:   vector of spectra elements from the focuses f_j, $j = 1, \ldots F$

The dimensionality of $\mathbf{R}_{ij}$ depends on the number of spectral frequencies used to calculate the DFT. A proper number for the frequency resolution in this application was 1024. This resolution allows the spectra differences in the resonance modes be captured for different kinds of defective materials. However, there is redundant information in the spectra that we reduce with PCA, obtaining only the significant components [21]

$$\mathbf{R}_{ij}^{(PCA)} = \mathbf{P} \cdot diag\left[\mathbf{H}_{ij}\right] \cdot \mathbf{F}_j = \mathbf{M}_{ij} \cdot \mathbf{F}_j \tag{2}$$

where $\mathbf{P}$ is a unitary transformation provided by PCA that allows the observation vectors to be ordered by their powers. PCA finds a rotated orthogonal system such that the elements of the original vectors in the new coordinates become uncorrelated, so the redundancy induced by correlation is removed [7]. We assume the data are centred.

The equation (2) represents an ICA model for the sensor i under the assumption of independence in the spectra components caused by the internal defective focuses [13][14]. In order to obtain a complete model of the multichannel impact-echo setup, we can write

$$\begin{bmatrix} \mathbf{R}_{1j}^{(PCA)} \\ \vdots \\ \mathbf{R}_{Nj}^{(PCA)} \end{bmatrix} = \begin{bmatrix} \mathbf{M}_{1j} \\ \vdots \\ \mathbf{M}_{Nj} \end{bmatrix} \cdot \mathbf{F}_j \tag{3}$$

The equation (3) consists of an over-determined ICA model. The spectral components (sources) are considered to be constant for all the sensors. $\mathbf{F}_j$ models the effect produced in the sensors by a certain focus j. In addition, values of $\mathbf{F}_j$ can be considered constant for close defect points and therefore they build a localized defective zone with a particular geometry.

We apply again projection by PCA in order to reduce the ICA model of equation (3) to the square case [21], obtaining the following expression,

$$\mathbf{R}_j^{((PCA))} = \mathbf{P}' \cdot \mathbf{M}_j \cdot \mathbf{F}_j = \mathbf{H}_j \cdot \mathbf{F}_j \tag{4}$$

where the recovered sources $\mathbf{F}_j' = \mathbf{H}_j^{-1} \cdot \mathbf{R}_j^{((PCA))}$ will be estimates of $\mathbf{F}_j$ up to scaling and permutation of the elements.

Adding the contributions of all the focuses $j = 1, \ldots F$, we obtain a model $\mathbf{R}$ for the inspected material as

$$\mathbf{R} = \sum_{j=1}^{F} \mathbf{R}_j^{((PCA))} = \sum_{j=1}^{F} \mathbf{H}_j \cdot \mathbf{F}_j \tag{5}$$

An underlying ICA model can be derived from equation (5), considering an amount of dependence between the spectra components for all the defective focuses. We can define $\mathbf{F}_j = \mathbf{s} + \mathbf{a}_j$ where $\mathbf{s}$ and $\mathbf{a}_j$ are vectors of independent and dependent random latent variables, respectively. Thus, we can write,

$$\mathbf{R} = \sum_{j=1}^{F} \mathbf{H}_j \cdot \left(\mathbf{s} + \mathbf{a}_j \right) = \sum_{j=1}^{F} \mathbf{H}_j \cdot \mathbf{s} + \sum_{j=1}^{F} \mathbf{H}_j \cdot \mathbf{a}_j = \mathbf{H} \cdot \mathbf{s} + \mathbf{b} \tag{6}$$

The spectra obtained from materials with different kinds of defects inspected by impact-echo should correspond to different models following the equation (6), depending on the shape of the defective area. Consequently, we can formulate the problem of classification of materials with different quality condition inspected by impact-echo in the ICAMM framework. From (6), the spectra (observation vectors) $\mathbf{R}^{(k)}$ sensed in a multichannel impact-echo testing corresponding to a given material class C_k ($k = 1 \ldots K$) are the result of applying a linear transformation $\mathbf{H}^{(k)}$ to a source vector $\mathbf{s}^{(k)}$, whose elements are independent random variables, plus a bias vector $\mathbf{b}^{(k)}$. Thus we can write the corresponding ICAMM expression,

$$\mathbf{R}^{(k)} = \mathbf{H}^{(k)} \cdot \mathbf{s}^{(k)} + \mathbf{b}^{(k)} \quad k = 1, \ldots, K \;, \tag{7}$$

where

$\mathbf{R}^{(k)}$:   compressed spectra of the multichannel impact-echo setup for the defective material class k

$\mathbf{H}^{(k)}$:   mixture matrix for the defective material class k

$\mathbf{s}^{(k)}$:   compressed spectra from the focuses f_j, $j = 1, \ldots F$ for the defective material class k

The set of classes in this application, for instance, can be: homogeneous, one-defect, and multiple-defect materials.

3   Results and Discussion

The set of impact-echo signals were extracted from 3D finite element models and lab specimens of aluminium alloy series 2000. These models and specimens belonged to four classes: homogeneous, holes, cracks, and multiple-defects. The holes were cylinders of $\phi = 1$ cm. oriented in the axis x or y, and the cracks were 1 mm. wide oriented in the planes xy, zy, or zx. The multiple-defect materials contained cracks and holes.

The set of simulated signals came from the full transient dynamic analysis of 100 3D finite element models. These models simulated parallelepiped-shape material of 0.07x0.05x0.22 m. (width, height and length), which was supported at one third and two thirds of the block length (direction z). Simulated finite models corresponded to one class of homogeneous and eleven classes of inhomogeneous models. The dynamic response of the material structure (time-varying displacements in the structure) under the action of a transient load was estimated from the transient analysis. The impact-echo real experiments were performed on lab specimens of he same dimensions of the models using the following equipment: i.) Instrumented impact hammer 084A14 PCB, ii.) 7 accelerometers 353B17 PCB located surrounding the faces of the material, iii.) ICP signal conditioner F482A18, iv.) Data acquisition module 6067E, and v.) Notebook. The total of experiments were: 200 (homogeneous), 341 (holes), 1044 (cracks), 276 (multiple-defects).

After the acquisition stage the spectrum features were extracted with the following procedure: (1) DFT 1024 points per channel, (2) PCA on the spectrum per channel, (3) Selection of 20 components per channel for a total of 140 components (explained variance >=95%), (4) PCA on the 140 spectra components, (5) Selection of 50 components (explained variance>=92%), (6) Classification with LDA using different number of components, (7) Selection of the number of components for the best classification with LDA, (8) Classification by MLP, and the ICAMM algorithm. Several Montecarlo simulations were made using this procedure dividing 70% of the data for training and 30% for testing.

We applied the ICAMM algorithm in [17] using JADE [22] and a non-parametric kernel-based algorithm for the steps of ICA parameter updating. Fig. 2 shows a set of ICAMM parameters obtained during training for the impact-echo experiments using 10 spectra components.

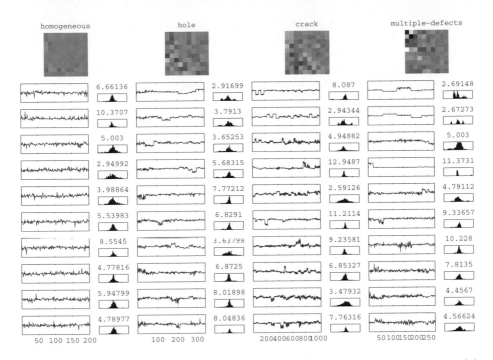

Fig. 2. Mixing matrixes and sources estimated for four different kinds of defective materials tested in impact-echo experiments

Estimated mixing matrixes (represented in grey scale), and sources with their kurtosis values for the four classes of materials are depicted in Fig. 2. The differences in these parameters between the classes are clear showing the suitability of the ICAMM model to classify different kinds of defective materials inspected by impact-echo.

The estimated sources represent linear combinations of the spectrum elements produced by the defects that activate different resonant modes of the material. In general, the pattern of the defects was detected independently of their orientation and dimension. These patterns seem related with the number of the point flaws that build the defects and the spatial relationship between the flaws. General results of classification accuracy were: ICAMM -nonparametric (82.7%), ICAMM -Jade (84%), MLP (80.3%), and LDA (79%).

4 Conclusions

A new ICA mixture model has been proposed for non destructive testing using impact-echo. The material under evaluation is modeled as a linear system which describes the resonance modes involved in the wave propagation phenomenon of the impact-echo. A compressed and representative pattern of the spectra for the multichannel setup of the impact-echo has been obtained using ICAMM. This modeling allowed the spectra differences in the resonance modes were discerned for different kinds of defective materials in several simulations and lab experiments.

There are a range of future directions for this research, such as, to attempt the application in industrial contexts and to obtain higher insights into the sources and mixing matrix of the model in order to exploit all the information collected by the sensors. From these insights, an accurate localization of the defects and a 3D reconstruction of the internal structure of the material can be attempted.

Acknowledgments. This work has been supported by Spanish Administration and the FEDER Programme of the European Community under grants TEC 2005-01820, TEC 2008-02975/TEC, and *Universidad Politécnica de Valencia* under the Programme of Support to Research and Development.

References

1. Sansalone, M., Street, W.: Impact-echo: Non-destructive evaluation of concrete and masonry. Bullbrier Press, New York (1997)
2. Carino, N.J.: The impact-echo method: an overview. In: Chang, P.C. (ed.) Structures Congress and Exposition 2001, pp. 1–18. American Society of Civil Engineers (2001)
3. Sansalone, M., Carino, N.J., Hsu, N.N.: Transient stress waves interaction with planar flaws. Batim-Int-Build-Res-Pract. 16, 18–24 (1998)
4. Lee, T.W., Lewicki, M.S., Sejnowski, T.J.: ICA mixture models for unsupervised classification of non-Gaussian classes and automatic context switching in blind signal separation. IEEE Transactions on Pattern Analysis and Machine Intelligence 22(10), 1078–1089 (2000)
5. Choudrey, R., Roberts, S.: Variational Mixture of Bayesian Independent Component Analysers. Neural Computation 15(1), 213–252 (2002)
6. Common, P.: Independent component analysis—a new concept? Signal Processing 36(3), 287–314 (1994)
7. Hyvärinen, A., Karhunen, J., Oja, E.: Independent Component Analysis. John Wiley & Sons, Chichester (2001)
8. Cichocki, A., Amari, S.: Adaptive Blind Signal and Image Processing: Learning algorithms and applications. Wiley, John & Sons (2001)
9. Morabito, C.F.: Independent Component Analysis and Extraction Techniques for NDT Data. Materials Evaluation 58(1), 85–92 (2000)
10. Igual, J., Camacho, A., Vergara, L.: Blind Source Separation Technique for Extracting Sinusoidal Interferences in Ultrasonic Non-Destructive Testing. Journal of VLSI Signal Processing 38, 25–34 (2004)
11. Salazar, A., Gosalbez, J., Igual, J., Llinares, R., Vergara, L.: Two Applications of Independent Component Analysis for Non-destructive Evaluation by Ultrasounds. In: Rosca, J.P., Erdogmus, D., Príncipe, J.C., Haykin, S. (eds.) ICA 2006. LNCS, vol. 3889, pp. 406–413. Springer, Heidelberg (2006)
12. Salazar, A., Unió, J., Serrano, A., Gosalbez, J.: Neural networks for defect detection in non-destructive evaluation by sonic signals. In: Sandoval, F., Prieto, A.G., Cabestany, J., Graña, M. (eds.) IWANN 2007. LNCS, vol. 4507, pp. 638–645. Springer, Heidelberg (2007)
13. Salazar, A., Vergara, L., Igual, J., Gosalbez, J.: Blind source separation for classification and detection of flaws in impact-echo testing. Mechanical Systems and Signal Processing 19, 1312–1325 (2005)

14. Salazar, A., Vergara, L., Igual, J., Gosálbez, J., Miralles, R.: ICA model applied to multichannel non-destructive evaluation by impact-echo. In: Puntonet, C.G., Prieto, A.G. (eds.) ICA 2004. LNCS, vol. 3195, pp. 470–477. Springer, Heidelberg (2004)
15. Bailer-Jones, C., Irwin, M., Hippel, T.: Automated classification of stellar spectra - II. Two-dimensional classification with neural networks and principal components analysis. Monthly Notices of the Royal Astronomical Society 298, 361–377 (1998)
16. Xu, R., Nguyen, H., Sobol, P., Wang, S.L., Wu, A., Johnson, K.E.: Application of Principal Component Analysis to the FTIR Spectra of Disk Lubricant to Study Lube–Carbon Interactions. IEEE Transactions on Magnetics 40, 3186–3189 (2004)
17. Vergara, L., Salazar, A., Igual, J., Serrano, A.: Data clustering methods based on mixture of independent component analyzers. In: 2006 ICA Research Network International Workshop, Liverpool, England, pp. 127–130 (2006)
18. Duda, R., Hart, P.E., Stork, D.G.: Pattern Classification, 2nd edn. Wiley Interscience, Hoboken (2000)
19. Bishop, C.M.: Neural Networks for Pattern Recognition. Oxford University Press, Oxford (2004)
20. Karklin, Y., Lewicki, M.S.: Leaning higher-order steructures in natural images. Network: Comput. Neural Syst. 14, 483–499 (2003)
21. Joho, M., Mathis, H., Lambert, R.H.: Overdetermined blind source separation: using more sensors than source signals in a noisy mixture. In: 2nd International Workshop on Independent Component Analysis and Blind Signal Separation, Helsinki, Finland, pp. 81–86 (2000)
22. Cardoso, J.F., Souloumiac, A.: Blind beamforming for non Gaussian signals. IEE Proceedings-F 140(6), 362–370 (1993)

Binary Nonnegative Matrix Factorization Applied to Semi-conductor Wafer Test Sets

Reinhard Schachtner[1,2], Gerhard Pöppel[1], and Elmar W. Lang[2]

[1] Infineon Technologies AG, 93049 Regensburg, Germany
[2] CIMLG / Biophysics, University of Regensburg, 93040 Regensburg, Germany
reinhard.schachtner@infineon.com

Abstract. We introduce a probabilistic extension of non-negative matrix factorization (NMF) by considering binary coded images as a probabilistic superposition of underlying continuous-valued elementary patterns. We provide an appropriate algorithm to solve the related optimization problem with non-negativity constraints which represents an extension of the well-known NMF-algorithm to binary-valued data sets. We demonstrate the performance of our method by applying it to the detection and characterization of hidden causes for failures during semi-conductor wafer processing. We decompose binary coded (pass/fail) wafer test data into underlying elementary failure patterns and study their influence on the performance of single wafers during testing.

1 Introduction

Manufacturing a microchip requires hundreds of processing steps, depending on the complexity of the final microchip. Lifetime, performance speed and other quality aspects afford a set of specifications to be tailored to the intended application field. The overall functionality of the completed chip is measured in a test series after the last processing step. A chip is labeled *pass*, if it satisfies all investigated test features, and *fail* otherwise. A disordered or miscalibrated production line can cause the failure of a quality check of a whole series of chips. The identification and characterization of such systematic errors is an acute but nontrivial problem in semi-conductor manufacturing. While several individual root causes can be the responsible trigger for a dropped-out device, only the overall *pass / fail* - information for the chip is available at any case. In this paper we model the systematic part of errors by a superposition of individual failure causes. A preliminary version of this technique has been presented recently in a classification context [7].

Notation. Measurement data from N wafers constitute a binary $N \times M$ data matrix $\mathbf{X}$, each row of which contains all M chips of one wafer aligned. Matrix entry X_{ij} contains the information whether chip j on wafer i has passed all functionality tests (0), or failed at at least one of them (1). In the following, we use $\mathbf{X}_{i*}$ to denote the i-th row and $\mathbf{X}_{*j}$ for the j-th column of $\mathbf{X}$. Thus the rows reflect the test signatures of each wafer across all chips and the columns contain the test profiles of a specific chip across all wafers.

T. Adali et al. (Eds.): ICA 2009, LNCS 5441, pp. 710–717, 2009.

2   Nonnegative Matrix Factorization

Nonnegative matrix factorization (NMF) is a very popular technique for the analysis of real-valued multivariate data sets which more than often show non-negative entries only. In the context of Blind Source Separation (BSS), NMF is intended to explain a data generating process as a strictly additive superposition of nonnegative source modes. A nonnegative $N \times M$ data matrix $\mathbf{X}$ is approximately decomposed into a $N \times K$ weight matrix $\mathbf{W}$ and a $K \times M$ matrix $\mathbf{H}$ of underlying modes such that

$$\mathbf{X} \approx \mathbf{WH} \qquad (\mathbf{W}, \mathbf{H} \geq 0). \tag{1}$$

The number of modes K is usually chosen such that $(N + M) \cdot K \leq N \cdot M$. The product $\mathbf{WH}$ can thus be regarded as a compressed version of the original data $\mathbf{X}$ [1]. Technically, the task of a NMF can be formulated as an optimization problem by minimizing a suitable cost function, such as the Euclidean distance

$$f(\mathbf{W}, \mathbf{H}) := \frac{1}{NM} \sum_{i=1}^{N} \sum_{j=1}^{M} (X_{ij} - [\mathbf{WH}]_{ij})^2 \tag{2}$$

with respect to the non-negativity constraints $\mathbf{W}, \mathbf{H} \geq 0$. However, this plain cost function does not lead to unique solutions, hence additional constraints need to be added. Constrained cost functions considering Kullback-Leibler- [1], Bregman- [5] or Csiszár's [3] divergences have been proposed in the literature. Additional sparseness, smoothness or volume constraints to enforce solutions with desired characteristics, as well as a variety of optimization techniques to achieve the desired matrix decomposition have been discussed (see e.g. [2],[4] for a survey).

Alternating Least Squares Algorithm for NMF. A very popular method to minimize the squared Euclidean distance $f(\mathbf{W}, \mathbf{H})$ (eq. 2) is called Alternating Least Squares procedure which iterates the following equations:

$$\text{Solve for } \mathbf{H} \text{ in matrix equation } \frac{\partial f}{\partial \mathbf{H}} = 0. \tag{3}$$

$$\text{Set all negative elements in } \mathbf{H} \text{ to } 0. \tag{4}$$

$$\text{Solve for } \mathbf{W} \text{ in matrix equation } \frac{\partial f}{\partial \mathbf{W}} = 0. \tag{5}$$

$$\text{Set all negative elements in } \mathbf{W} \text{ to } 0. \tag{6}$$

after random initialization of $\mathbf{W}$(see e.g. [2]). ALS-procedures which properly enforce non-negativity of $\mathbf{W}$ and $\mathbf{H}$ can be proven to converge towards a local minimum of the cost function [2]. Unfortunately, the brute force projection onto the nonnegative orthant after every optimization step can cause convergence problems. In case of convergence, however, the projected ALS algorithm is extremely fast. Computing several runs using different random initializations thus still outperforms other methods like gradient descent and multiplicative update rules with respect to the required computational time and is very attractive for NMF applications (see [4]).

3   NMF for Binary Datasets

3.1   Defect Generating Model

Assuming a Poisson-distributed random variable with parameter λ, the probability of n events is given by $e^{-\lambda}\frac{\lambda^k}{k!}$. We interpret a $N \times M$ data matrix $\mathbf{X}$ as a collection of $N \cdot M$ realizations of such Poisson-distributed random variables, each related with an individual parameter λ_{ij}. Therefore, we distinguish between the two cases whether there was no event at all or at least one event:

$$X_{ij} \begin{cases} = 0 & (\text{'pass'}, n = 0) \\ = 1 & (\text{'fail'}, n \geq 1) \end{cases} \quad \text{with probability} \begin{cases} e^{-\lambda_{ij}} \\ 1 - e^{-\lambda_{ij}} \end{cases} . \tag{7}$$

The assumption that the data was generated by several individual failure causes is reflected in the relations between the parameters λ_{ij}. We refer to an elementary mode $\mathbf{H}_{k*} = (H_{k1}, \ldots, H_{kM}) \geq 0$ as a M-dimensional row vector whose j-th component H_{kj} expresses the probability that observation X_{ij} is either 0 or 1. Suppose that mode k is the only event that causes a 1 in $\mathbf{X}$. $H_{kl} < H_{km}$ then implies that it is more probable to get a 1 on position m than on l. We further assume that the kth mode can vary its probability of occurrence. Therefore, we add weighting factors $W_{ik}(\geq 0)$ such that $W_{rk} < W_{sk}$ implies that the mode H_{k*} exerts a stronger influence on observation X_{s*} than on X_{r*}. The final expression for the probability of observing a zero on position (i,j) is

$$P(X_{ij} = 0 | H_{kj}, W_{ik}) = e^{-W_{ik} \cdot H_{kj}}, \tag{8}$$

if mode k is the only possible elementary event.

Next, we assume several individual modes $\mathbf{H}_{k*}$, $k = 1, \ldots, K$ to contribute. The contribution of mode k to observation i is denoted W_{ik}. The overall probability in case of K causes is then

$$P(X_{ij} = 0 | H_{1j}, \ldots, H_{Kj}, W_{i1}, \ldots, W_{iK}) = \prod_{k=1}^{K} e^{-W_{ik} \cdot H_{kj}}. \tag{9}$$

Summarizing, the conditional probabilities, given the hidden causes, read

$$P(X_{ij} = 0 | \mathbf{H}, \mathbf{W}) = e^{-[\mathbf{WH}]_{ij}}, \tag{10}$$

$$P(X_{ij} = 1 | \mathbf{H}, \mathbf{W}) = 1 - e^{-[\mathbf{WH}]_{ij}}. \tag{11}$$

Note that this model is not symmetric in the two cases $X_{ij} = 0/1$. The asymmetry reflects the nature of the defect generating process. If there are several causes for a defect, the probability for an object to be overall *pass* is the product of the probabilities being *pass* given each single cause. On the other hand if one source causes an event *fail*, the other sources cannot alleviate this state. Both matrices $\mathbf{W}$ and $\mathbf{H}$ are nonnegative and are related to the binary matrix $\mathbf{X}$ as described above. The challenge of finding these matrices can thus be viewed as an extension of NMF for this kind of binary data sets.

3.2   Bernoulli Likelihood

The Bernoulli likelihood is a natural choice for modeling binary data. Denoting p_{ij} the probability that $X_{ij} = 1$, the Bernoulli likelihood of one entry is

$$P(X_{ij}|\mathbf{H}, \mathbf{W}) = p_{ij}^{X_{ij}} (1 - p_{ij})^{1 - X_{ij}} \qquad (12)$$

leading to an overall (log-)likelihood

$$LL := \sum_{i=1}^{N} \sum_{j=1}^{M} X_{ij} \ln(1 - e^{-[\mathbf{WH}]_{ij}}) - [\mathbf{WH}]_{ij} + X_{ij}[\mathbf{WH}]_{ij}. \qquad (13)$$

In [6], a symmetric linear model is considered to approximate the Bernoulli parameter of a similar problem. The authors use an *Expectation Maximization*(EM) - type approach to maximize a lower bound for the log-likelihood. Here, we propose a completely different strategy for the optimization. We combine an Alternating Gradient Ascent (AGA) algorithm in the variables $\mathbf{W}$ and $\mathbf{H}$ together with a preceding search for appropriate initial values in order to reduce the risk of getting stuck in a local maximum.

3.3   Optimizing the Log-Likelihood

Alternating Gradient Ascent Algorithm. After an appropriate initialization of the parameter matrices $\mathbf{W}$ and $\mathbf{H}$, an iterative gradient ascent scheme for the log-likelihood (13) is given by

$$W_{ik} \leftarrow W_{ik} + \eta_W \frac{\partial LL}{\partial W_{ik}} \qquad (14)$$

$$H_{kj} \leftarrow H_{kj} + \eta_H \frac{\partial LL}{\partial H_{kj}}. \qquad (15)$$

While one of the matrices is updated, the other one is kept fixed. Due to the non-negativity constraints to all entries W_{ik}, H_{kj}, the step size parameters η_W and η_H have to be controlled carefully. Especially for small step sizes convergence can be unduly slow. Even in the unconstrained case, gradient ascent algorithms can only be guaranteed to find a local maximum for sufficiently small η_W, η_H. Particularly the logarithm in eq. (13) can cause serious global convergence problems by inducing local maxima to the log-likelihood function. Single entries $X_{ij} = 1$ with a small probability $1 - e^{-[\mathbf{WH}]_{ij}}$ may pin the optimization algorithm. In the following, we derive a strategy how to find a "good" starting point for the Alternating Gradient Ascent algorithm.

Alternating Least Squares on a Simplified Problem. In order to obtain suitable initial matrices $\mathbf{W}$ and $\mathbf{H}$ we preprocess a simplified version of the true optimization task with a standard NMF algorithm. We introduce an auxiliary variable $\alpha \in]0, 1[$ by setting

$$\begin{array}{lll} P(X_{ij} = 1) = 0, & \text{if } X_{ij} = 0 & \\ P(X_{ij} = 1) = \alpha, & \text{if } X_{ij} = 1 & \end{array} \quad \text{for all } i, j. \qquad (16)$$

The variable α can be regarded an average probability $P(X_{ij} = 1)$ given the observed realization was $X_{ij} = 1$. For all (i, j) this can be summarized by

$$\alpha X_{ij} = 1 - e^{-[\mathbf{WH}]_{ij}} \Leftrightarrow -\ln(1 - \alpha X_{ij}) = [\mathbf{WH}]_{ij} \tag{17}$$

Since the left hand side of the last equation is always nonnegative we recover a standard NMF problem $\mathbf{X}' \approx \mathbf{WH}$ when substituting $X'_{ij} =: -\ln(1 - \alpha X_{ij})$.

In the following we choose the squared Euclidean distance as a cost function

$$E(\alpha, \mathbf{W}, \mathbf{H}) = \sum_{i=1}^{N} \sum_{j=1}^{M} \left(\ln(1 - \alpha X_{ij}) + [\mathbf{WH}]_{ij} \right)^2 \tag{18}$$

and apply the Alternating Least Squares Algorithm as described in section 2 in order to minimize (18) with respect to $\mathbf{W} \geq 0$ and $\mathbf{H} \geq 0$. The ALS-updates are then given by

$$H_{rs} \leftarrow \max\{\epsilon, -\sum_{i=1}^{N} [(\mathbf{W}^T \mathbf{W})^{-1} \mathbf{W}^T]_{ri} \ln(1 - \alpha X_{is})\} \tag{19}$$

$$W_{lm} \leftarrow \max\{\epsilon, -\sum_{j=1}^{M} \ln(1 - \alpha X_{lj}) [\mathbf{H}^T (\mathbf{H}\mathbf{H}^T)^{-1}]_{jm}\}. \tag{20}$$

To avoid local minima of the cost function (18), the procedure is repeated using different random initializations of $\mathbf{H}$ and $\mathbf{W}$ and only the solution with the smallest Euclidean distance is retained.

Determining the Parameter α. Note that the global minimum of eq.(18) as a function of $\alpha, \mathbf{W}$ and $\mathbf{H}$ is given by $E = 0$ when $\alpha \to 0$, $\mathbf{W}, \mathbf{H} = 0$ independently from the data $\mathbf{X}$. Thus, we determine the optimal α by the log-likelihood of the estimated $\mathbf{W}(\alpha), \mathbf{H}(\alpha)$. If the parameter α is chosen too small, the probabilities $P(X_{ij} = 1)$ are consistently estimated too small and the related log-likelihood will be small. On the other hand, a large $\alpha \approx 1$ leads to very large values $[\mathbf{WH}]_{ij}$ for any $X_{ij} = 1$. Due to the matrix product this implies an increase of the whole column $\mathbf{H}_{*j}$ and/or row $\mathbf{W}_{i*}$ at the expense of the reconstruction accuracy for zeros in the same row and column ($X_{is} = 0, X_{rj} = 0$).

From simulations on toy data sets (see Section 4.1 for details), we observed that the best obtained log-likelihood $LL(\mathbf{X}, \mathbf{W}(\alpha), \mathbf{H}(\alpha))$ among several randomly initialized runs resembles a concave function of α (see Figure 1). Thus, a Golden Section Search procedure can be applied to obtain the optimal α in a reasonable amount of trials and computational time.

4 Results

4.1 Toydata Example

First, we present the performance of the above algorithm on a constructed toy data example.

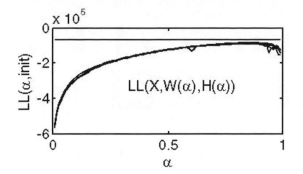

Fig. 1. Log-likelihood of the approximations computed by the ALS-method as a function of α for 10 random initializations. The best value is obtained for $\alpha = 0.87$ in this example. The horizontal line denotes the true log-likelihood.

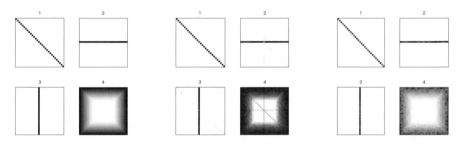

Fig. 2. Left: 30×30 source patterns $\mathbf{H}_{k*}$ valued in $[0,1]$ (white:0, black:1). Center: Reconstructions gained via ALS, Right: Maximum likelihood solutions.

A 1000×900 binary data matrix $\mathbf{X}$ was created by setting the (i,j)-th entry to 1 with probability $p_{ij} = 1 - e^{[\mathbf{WH}]_{ij}}$ (see Figure 3 for examples) from $K = 4$ fixed failure patterns $\mathbf{H}_{1*}, \dots, \mathbf{H}_{4*}$ and a randomly generated 1000×4 coefficient matrix $\mathbf{W}$. We use three binary patterns (white: 0, black: 1) and one pattern of values graded from zero in the center to one on the edges (see Figure 2,left hand side).

The simplified ALS-method yields quite good approximations of the original source patterns in this example. After 1000 iterations refinement by Alternating Gradient Ascent, nearly perfect reconstruction of the original patterns is achieved (see Figure 2). Note that in the images $\mathbf{W}$ and $\mathbf{H}$ are re-scaled such that the maximum value in each pattern $\mathbf{H}_{k*}$ is given by one. The top row of Figure 3 contains examples for randomly generated coefficients W_{ik}, the second row shows the corresponding binary images $\mathbf{X}_{i*}$. The last two rows contain the estimated coefficients by the ALS-method and after refinement by Alternating Gradient Ascent respectively.

4.2 Real World Example

The real world example shows a decomposition of $M = 3043$ wafers, each containing $M = 500$ chips into $K = 6$ source patterns (see Figure 4). The data stems from measurements which aim to identify latent structures and detect

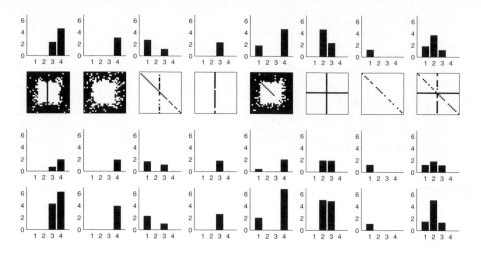

Fig. 3. Toydata examples. Top: Original coefficents $\mathbf{W}_{i*}$, Second row: Binary realizations $\mathbf{X}_{i*}$, Third row: Coefficients gained by ALS, Bottom: Coefficients after refinement by Gradient Ascent.

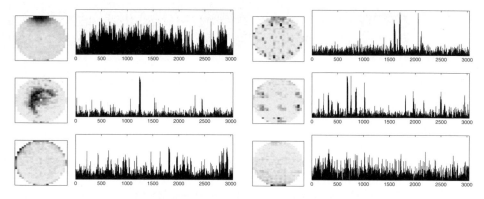

Fig. 4. Elementary modes $\mathbf{H}_{1*}, \ldots, \mathbf{H}_{6*}$ and contribution coefficients $\mathbf{W}_{*1}, \ldots, \mathbf{W}_{*6}$, estimated from a real dateset comprising 3043 wafers and 500 chips

potential failure causes in an early stadion of the processing chain. A different decomposition of the same dataset is shown in [7].

The estimated $K = 6$ source patterns $\mathbf{H}_{1*}, \ldots, \mathbf{H}_{6*}$ have clearly different characteristics: One pattern of higher fail-probability on the upper side of the wafer, a bead centered on the wafer, a ring of fails on the edge zone, two different repeated structures, and defects concentrated on the bottom of the wafer were detected. The related weight coefficients $\mathbf{W}_{*1}, \ldots, \mathbf{W}_{*6}$ store the activity of each of the 6 putative modes on each wafer separately. This new representation of the data contrasts wafers affected by the detected sources with untouched ones and is intended to support the detection of potential error causes.

5   Conclusion

We introduced a probabilistic framework to model systematic failure causes in the microchip production. A new methodology was presented which utilizes an extension of nonnegative matrix factorization to this kind of binary data sets. An optimization technique was presented which maximizes a log-likelihood function using a fast alternating least squares algorithm followed by gradient ascent refinement. The performance of the overall procedure was demonstrated on an artificial and a real world dataset.

References

1. Lee, D.D., Seung, H.S.: Learning the Parts of Objects by Non-negative Matrix Factorization. Nature 401, 788–791 (1999)
2. Berry, M.W., Browne, M., Langville, A.N., Pauca, V.P., Plemmons, R.J.: Algorithms and applications for approximate nonnegative matrix factorization. Computational Statistics & Data Analysis 52(1), 155–173 (2007)
3. Cichocki, A., Zdunek, R., Amari, S.-i.: Csiszár's divergences for non-negative matrix factorization: Family of new algorithms. In: Rosca, J.P., Erdogmus, D., Príncipe, J.C., Haykin, S. (eds.) ICA 2006. LNCS, vol. 3889, pp. 32–39. Springer, Heidelberg (2006)
4. Cichocki, A., Zdunek, R., Amari, S.: Nonnegative Matrix and Tensor Factorization. IEEE Signal Processing Magazine, 142–145 (January 2008)
5. Dhillon, I., Sra, S.: Generalized Nonnegative Matrix Approximations with Bregman Divergences. In: Weiss, Y., Schölkopf, B., Platt, J. (eds.) Advances in Neural Information Processing Systems 18, pp. 283–290. MIT Press, Cambridge (2006)
6. Kabán, A., Bingham, E., Hirsimäki, T.: Learning to Read Between the Lines: The Aspect Bernoulli Model. In: Proceedings of the 4th SIAM International Conference on Data Mining, Lake Buena Vista, Florida, April 22-24, pp. 462–466 (2004)
7. Schachtner, R., Pöppel, G., Lang, E.W.: Nonnegative Matrix Factorization for Binary Data to Extract Elementary Failure Maps from Wafer Test Images. In: Proc. 32th Annual Conference of the Gesellschaft für Klassifikation, Helmut Schmidt University Hamburg, July 16-18, 2008. Springer, Heidelberg (2008)

High-Energy Particles Online Discriminators Based on Nonlinear Independent Components

Eduardo F. Simas Filho[1,2], José M. de Seixas[1], and Luiz P. Calôba[1]

[1] Signal Processing Laboratory, COPPE/Poli/UFRJ, Rio de Janeiro, Brazil
[2] Federal Center for Technological Education, Bahia, Brazil
{esimas,seixas,caloba}@lps.ufrj.br

Abstract. High-energy detectors operating in particle collider experiments typically require efficient online filtering to guarantee that most of the background noise will be rejected and valuable information will not be lost. Among these types of detectors, calorimeters play an important role as they measure the energy of the incoming particles. In practical designs, calorimeter exhibit some sort of nonlinear behavior. In this paper, nonlinear independent component analysis (NLICA) methods are applied to extract relevant features from calorimeter data and produce high-efficient neural particle discriminators for online filtering operation. The study is performed for ATLAS experiment, one of the main detectors of the Large Hadron Collider (LHC), which is a last generation particle collider currently under operational tests. A performance comparison between different NLICA algorithms (PNL, SOM and Local ICA) is presented and it is shown that all outperform the baseline discriminator, that is based on classical statistical approach.

Keywords: NLICA, High-Energy Physics, Online Filtering, Feature Extraction, Calorimeters.

1 Introduction

Modern high-energy physics experiments are large scale facilities in which numerous physicists, engineers and technicians join together searching for the fundamental nature of mater. Particle colliders produce huge amounts of information and thus, are often used to look for new physics channels. The Large Hadron Collider (LHC) started its operational tests at CERN (European Center for Nuclear Research) and, when operating at full capacity (high luminosity), will be colliding bunches of protons at every 25 ns for a wide research program. Although a very high event rate is produced (approximately 10^6 interactions per second), the interesting channels will rarely occur [1]. Considering this, a high-efficient filtering (triggering) system is required to guarantee that most of the background noise will be rejected and valuable information will not be lost. In order to search for the Higgs particle and other new phenomena, a new energy range will be explored by LHC.

Placed at one of the LHC collision points, the ATLAS experiment [2] is responsible for the selection of relevant signatures, which are immerse in huge

T. Adali et al. (Eds.): ICA 2009, LNCS 5441, pp. 718–725, 2009.

background noise. Due to LHC bunch crossing rate and high segmentation of the detector, the total information generated at ATLAS will be near 60 TB/s (1.5 MB/colision). Considering this amount of data, the filtering procedure must be performed online within short latency time.

ATLAS filtering (trigger) system [2] comprises three sequential decision levels (see Figure 1-a), in which the amount of non-relevant signals retained in temporary memories is gradually reduced until final storage/rejection decision is made. The first-level is implemented in hardware as very fast decision (within 2.5 ms) is required. The second-level (LVL2) and the event filter (EF) are both implemented in software and processed in parallel by thousand of PC-like processors [2]. The latencies for LVL2 and EF are respectively 40 ms and 4 s. The particle discrimination procedure at LVL2 is split into feature extraction, where relevant information is extracted from the measured signals, and hypothesis testing, where particle discrimination is performed.

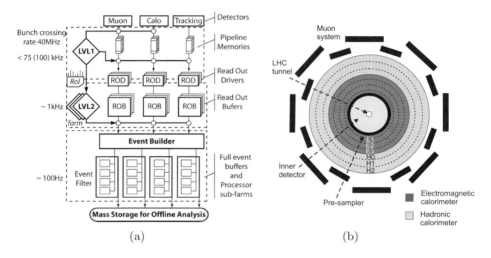

Fig. 1. ATLAS detector (a) triggering system architecture and (b) sub-detectors

For LHC, interesting signatures can be found through decays that produce electrons as final-state particles. Important information that guides electron identification process is the energy deposition profile measured by the calorimeter, one of the ATLAS subdetectors (see Figure 1-b). The calorimeter system is segmented into four electromagnetic (PS, E1, E2 and E3) and three hadronic (H0, H1 and H2) layers, producing more than 100,000 readout channels. Hadronic jets present energy deposition profiles similar to electrons (highly concentrated in the electromagnetic sections and almost no energy left in the hadronic layers), forming a huge background noise for the experiment.

In practical calorimeter design, nonlinearities may arise [3]. Considering this, the nonlinear independent component analysis (NLICA) model [4,5] is applied in this work to extract, from calorimeter signals, relevant features for particle discrimination. Identifying independent sources that form the energy deposition profiles may help revealing subtle differences between signal (electrons) and background noise (jets).

Calorimeter layers present different physical characteristics and cell granularity, considering this, feature extraction is performed in a segmented way (at layer level). The calorimeter signals from layer L ($\mathbf{x}^{(L)} = [x_1, ..., x_K]^T$) are considered to be generated through: $\mathbf{x} = \mathcal{F}(\mathbf{s})$, where $\mathbf{s} = [s_1, ..., s_N]^T$ are the independent sources and $\mathcal{F}(.)$ the nonlinear mixing mapping.

In this paper, different algorithms are used to extract features from calorimeter data envisaging optimum electron identification. The estimated segmented nonlinear independent components are used to feed the input nodes of a supervised neural classifier, which produces electron/jet decision.

2   Segmented NLICA-Based Discriminators

Included in the feature extraction phase of the ATLAS online triggering system, firstly calorimeter raw data is pre-processed and formatted into concentric energy rings (see Figure 2). In the following, nonlinear independent components are extracted from the ring signals in a segmented way (at layer level). For the hypothesis testing phase, a supervised MLP neural network [6], fed from the independent components, is used.

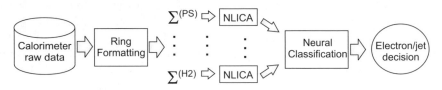

Fig. 2. Proposed Segmented NLICA-based classifiers

In this work, three NLICA algorithms were used to extract relevant features from ring-formatted calorimeter data. More details concerning these algorithms and the pre-processing procedure will be provided in the following.

2.1   Data Pre-processing

Electron/jet discrimination at ATLAS second-level trigger (LVL2) is based on the energy deposition profiles measured by the calorimeters. The first-level provides for LVL2 information on the Regions of Interest (RoI), which are the detector regions where relevant interactions may have occurred. A complete RoI is described by approximately 1,000 calorimeter cells.

As proposed in [7], here the calorimeter information is formatted into concentric rings through the following procedure: 1. In each calorimeter layer the most energetic cell is considered as the first ring; 2. The next rings are sequentially formed around the first one; 3. Ring signals are obtained by summing the energy of the cells belonging to a given ring. The ring energy is normalized within each layer.

This procedure makes signal description independent from the impact point in the detector, compacts RoI information by a factor of 10 (from 1000 cells up to 100 rings) and preserves the energy deposition profile. Ring signals from typical electron and jet are illustrated in Figure 3. One can see that the signals present similar patterns and their discrimination is not a trivial task.

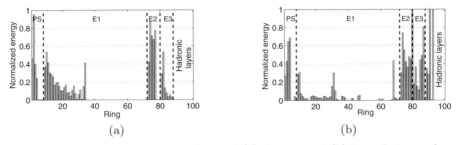

(a)   (b)

Fig. 3. Ring-formatted signatures of typical (a) electron and (b) jet, calorimeter layers are limited by vertical dotted lines

2.2   Post-nonlinear ICA Algorithm

The post-nonlinear (PNL) model [8] assumes that the observed signals are generated through a linear mixing followed by component-wise nonlinearities (cross-channel nonlinearities are not allowed). The observed signals can be expressed as [4]: $\mathbf{x} = F(\mathbf{As})$, where the nonlinear mapping is $F(.) = [f_1(.), f_2(.), ..., f_N(.)]^T$ (the number of sources N and sensors K is assumed to be the same). This model is usually applied when the source signals propagate through a linear channel and the nonlinearities are present on the sensors (which seems to be the case for calorimeters). The nonlinear functions are usually estimated through neural networks and the de-mixing matrix by a linear ICA algorithm [8].

In order to deal with high-dimensional data, here, as proposed in [9], a modified PNL model for the overdetermined case (when there exists more sensors than sources) was used. As illustrated in Figure 4, a linear block $\mathbf{B}$ is added to the standard PLN mixing model, allowing N<K.

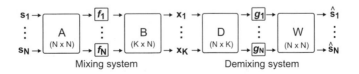

Fig. 4. Modified Post-Nonlinear mixing/de-mixing model

As the number of source signals may be unknown, in some cases, the estimation of matrix $\mathbf{D}$ can only be solved approximately. In this work, Principal Component Analysis (PCA) [5,6] is applied to estimate the number of nonlinear independent components to be computed.

2.3 Self-Organizing Map Based NLICA

One of the first attempts to perform nonlinear ICA was made through a Self Organizing Map (SOM) [10]. The SOM [11] is an unsupervised trained neural network, which learns a nonlinear mapping from the data and provides a topological organization of the input data set. SOM transforms a k-dimensional input space into a discrete (generally bi-dimensional) characteristic map.

It can be proved that, the coordinates of the winner neuron in the map are independent and roughly uniformly distributed [10]. To perform nonlinear ICA, SOM is trained using as inputs the observed signals and the coordinates of the winner vector correspond to the estimated independent components, which are assumed to have uniform probability distribution function (pdf).

2.4 Local ICA

Local ICA [4,12] can be viewed as a compromise between linear and nonlinear ICA. The purpose is to obtain better data representation when compared to linear ICA (by exploring local characteristics of the dataset), while avoiding the high computational cost of the nonlinear models.

In Local ICA model, a k-dimensional input space $Q \subset \mathbb{R}^k$ is divided into a finite number of subsets Q_k, $k = 1, ..., K$, which satisfy: $Q_1 \cup Q_2 \cup ... \cup Q_K = Q$. Clustering is responsible for an overall nonlinear representation. Linear ICA models are applied to data belonging to each cluster ($\mathbf{x}^{(k)}$) in order to estimate the local independent components $\mathbf{s}^{(k)} = \mathbf{B}^{(k)} \mathbf{x}^{(k)}$, where $\mathbf{B}^{(k)}$ is a local de-mixing matrix.

3 Experimental Results

The database used in this work was obtained through a Monte Carlo simulator for proton-proton collisions that considers both detector characteristics (including sensor and electronic noise) and first-level trigger effects [2]. The available dataset, which comprises approximately 500,000 electron and 500,000 jet signatures with energy spanning from 7 to 80 GeV, was equally split into development and testing sets.

Discrimination performance was evaluated through both ROC (Receiver Operating Characteristic) curve [13] and SP product. The SP is defined as [7]: $SP = [(Ef_e + Ef_j) \times \sqrt{(Ef_e \times Ef_j)}]/2$, where Ef_e and Ef_j are the detection efficiencies respectively for electrons and jets. The threshold value that maximizes SP provides both high probability of detection ($P_D = Ef_e$) and low false alarm ($P_F = 1 - Ef_j$).

Considering the Local ICA implementation, several energy retention levels through PCA were tested. Better discrimination performance was obtained while retaining 75% of signal energy, and thus, dimensionality was reduced from 100 rings to 41 segmented principal components (SPCA). The same number of SPCA was computed in the modified PNL model. To estimate the local independent components, FastICA algorithm was applied, in a segmented way, to data clustered into four groups (using more groups implies on a sparse representation). As proposed in a previous work [14], SOM was used for the clustering procedure. Figure 5-a illustrates the probabilities of occurrence for electrons and jets in these clusters. It can be observed from clusters composition that: group 1 concentrates most of typical electrons and group 4 typical jets; groups 2 and 3 are probably composed by signatures that are quite different from expected patterns for both classes.

Bi-dimensional (square) maps were used in SOM-based NLICA. The number of neurons (NH) in each map was chosen to minimize the mean error obtained by approximating the input signals by the SOM weights. NH vary from 16 in the hadronic layers up to 49 in the electromagnetic front layer.

A performance comparison between the proposed methods and algorithms already implemented in the detector software platform (T2Calo and Neural Ringer) is provided in Table 1. The baseline algorithm for electron/jet discrimination used at ATLAS second-level trigger (T2Calo) [2] is based on classical statistical analysis applied to parameters extracted from the energy deposition profiles. The Neural Ringer [7] consists basically on a neural classifier operating directly over the ring signals. One can see, from Table 1, that (Nonlinear and Linear) ICA-based classifiers outperforms both T2Calo and Neural Ringer.

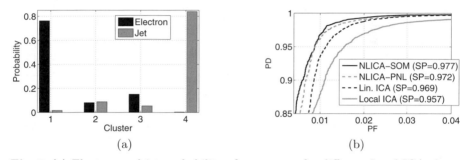

(a)   (b)

Fig. 5. (a) Electron and jet probability of occurrence for different Local ICA clusters and (b) ROC curves for different discriminators

Discrimination efficiency of NLICA and ICA based classifiers are also compared in Figure 5-b. Analyzing the ROC curves, it was observed that, SOM and modified PNL based feature extraction produced higher discrimination efficiency (SOM performs slightly better). It was also verified that linear ICA outperforms local ICA, it is important to note that, in both implementations the same ICA algorithm (FastICA) and PCA compaction level were used. From this comparative analysis it is clear that NLICA feature extraction (SOM and PNL) increases discrimination efficiency when compared to the linear model. It was also observed that through Local ICA, poorer results (among nonlinear and linear ICA implementations) were obtained. This indicates that, may be this model does not describe properly the data set.

Table 1. SP products for different discriminators

SOM	Mod. PNL	Local ICA	Linear ICA	NeuralRinger	T2Calo
0.977	0.972	0.957	0.969	0.870	0.852

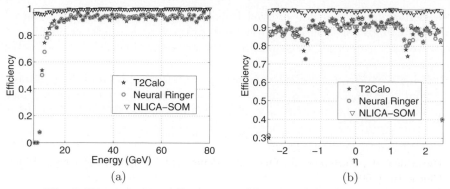

Fig. 6. Discrimination efficiencies in (a) energy (b) pseudorapidity (η)

For physics analysis, it is important to verify the proposed discriminators performance for variations on parameters such as the energy level and the position where the interaction occurred, which is expressed in terms of η and ϕ, axis of ATLAS coordinates system [2]. It can be depicted from Figure 6-a that, through NLICA feature extraction, high electron discrimination efficiency was achieved for the full energy range, contrasting with T2Calo and Neural Ringer, which produced lower performance (particularly for energy below 20 GeV). Considering the pseudorapidity (η), it can be observed in Figure 6-b that both T2Calo and Neural Ringer presented poor performance around $\eta = 1.5$. In this region exists a gap in calorimeter sensing elements. Through this gap pass maintenance and communication cables to the inner detector. The proposed discriminator is proved to be less influenced by the gap, producing high efficiency at this region. The detector is symmetric in ϕ and thus, no significant performance variation was observed in this axis. In these comparisons, results from the NLICA-SOM method were used, as this approach presented highest discrimination performance among NLICA implementations.

4   Conclusions

A novel signal processing procedure was proposed in this work for the ATLAS detector second-level trigger (electron/jet channel). High-dimensional calorimeter data was initially formatted into concentric energy rings, reducing signal dimensionality by a factor of 10. Nonlinear independent components were extracted from ring signals and used as inputs for a neural filter. Electron discrimination efficiency of 99.1% was obtained for 1.6% jet misclassification, outperforming the baseline discriminator.

Acknowledgments

The authors are thankful for the support provided by CNPq, FINEP and FAPERJ (Brazil), CERN (Switzerland), and for ATLAS Trigger/DAQ collaboration for providing the simulation data and for fruitful discussions concerning this work.

References

1. Evans, L., Bryant, P.: LHC Machine. Journal of Instrumentation, 3, S08001 (2008)
2. ATLAS Colaboration: ATLAS Experiment at CERN Large Hadron Collider. Journal of Instrumentation, 3, S08003 (2008)
3. Wigmans, R.: Calorimetry: Energy Measurement in Particle Physics. Clarendon Press (2000)
4. Jutten, C., Karhunen, J.: Advances in Nonlinear Blind Source Separation. In: 4th Int. Symp. on Independent Component Analysis and Blind Signal Separation, Nara, Japan, pp. 245–256 (2003)
5. Hyvarinen, A., Karhunen, J., Oja, E.: Independent Component Analysis. Wiley, Chichester (2001)
6. Haykin, S.: Neural Networks, Principles and Practice. Bookman (2001)
7. Anjos, A., et al.: Neural Triggering System Operating on High Resolution Calorimetry Information. Nuclear Instruments and Methods in Physics Research 559, 134–138 (2006)
8. Taleb, A., Jutten, C.: Source Separation in Post-nonlinear Mixtures. IEEE Trans. on Signal Processing 47, 2807–2820 (1999)
9. Simas Filho, E.F., Seixas, J.M., Caloba, L.P.: Segmented Overdetermined Nonlinear Independent Component Analysis for Online Neural Filtering. In: 10th Brazilian Symposium on Neural Networks, pp. 159–164 (2008)
10. Pajunen, P., Hyvarinen, A., Karhunen, J.: Nonlinear Blind Source Separation by Self-Organizing Maps. In: Int. Conf. on Neural Information Processing, Hong Kong, pp. 1207–1210 (1996)
11. Kohonen, T.: Self Organizing Maps, 3rd edn. Springer, Heidelberg (2001)
12. Karhunen, J., Malaroiu, S., Ilmoniemi, M.: Local Linear Independent Component Analysis based on Clustering. Int. Journal of Neural Systems 10, 439–451 (2000)
13. Trees, H.L.V.: Detection, Estimation, and Modulation Theory. Pt. I. Wiley, Chichester (2001)
14. Simas Filho, E.F., Seixas, J.M., Caloba, L.P.: Local Independent Component Analysis Applied to Highly Segmented Detectors. In: IEEE Int. Symp. on Circuits and Systems, Seatle, pp. 3005–3008 (2008)

Network Traffic Flow Separation and Control through a Hybrid ICA-Fuzzy Adaptive Algorithm

Flávio Henrique Teles Vieira[1], Lígia Maria Carvalho Sousa[2], George E. Bozinis[1],
Wesley F. de Miranda[1], and Charles Casimiro Cavalcante[2]

[1] Escola de Engenharia Elétrica e de Computação (EEEC)
Universidade Federal de Goiás (UFG)
flavio@eeec.ufg.br, bozinis@gmail.com, wesleymiranda@gmail.com
[2] Departamento de Engenharia de Teleinformática (DETI)
Universidade Federal do Ceará (UFC)
ligia@gtel.ufc.br, charles@gtel.ufc.br

Abstract. In this paper we present a hybrid ICA-fuzzy adaptive algorithm for traffic flow separation and control in contemporary computer networks. Our approach is composed by an ICA Block corresponding to the gradient algorithm proposed by Bell and Sejnowski for the information maximization at the output of a neural network as well as a Fuzzy Control System Block. The ICA algorithm is used to separate the controllable to the non-controllable network traffic sources. Additionally, we developed a predictive fuzzy controller following the Takagi and Sugeno fuzzy modeling. The combination of blind separation and control algorithm is applied to real network traffic traces. Finally, we verify that the proposed ICA-fuzzy adaptive control algorithm yields prominent control performances for single buffer server network environments.

Keywords: ICA, Traffic Separation, Network Traffic, Fuzzy Control, Prediction, Bell-Sejnowski algorithm, Independent Component Analysis.

1 Introduction

Nowadays, computer networks and the Internet have been facing an impacting transformation due to rapid growing of the number of real-time applications that require QoS (quality of service) guarantees. In order to attain the service demands of the network user QoS requirements, many strategies have been designed, capable of taking into account some particular statistical characteristics of the network traffic flows [1, 2]. These requirements and the additional flexibility of accommodating different traffic sources may cause serious congestion problems. Due to the fluctuation and burstiness of traffic flows within multimedia networks, traffic congestion can frequently occur, causing severe buffer overflow and QoS degradation. Therefore, an efficient congestion control mechanism is mandatory in order to overcome this problem. For this aim, the buffer length should be adequately controlled around a reference level [3].

T. Adali et al. (Eds.): ICA 2009, LNCS 5441, pp. 726–733, 2009.

Precise network traffic modeling is an important issue regarding traffic congestion control. The more realistic is the traffic source model, the more accurate is the estimate of network behavior and so, more appropriate will be the service offered to the user. On the other hand, if the traffic model does not accurately represent the actual network traffic, one may overestimate or underestimate network performance.

Many studies in fuzzy modeling have been carried out since the fuzzy theory was initially developed [4, 5]. The reason for these researches is due to the advantages of the fuzzy models of describing a given system over linear models. Besides a precise traffic modeling, it would be interesting if the network could separate the controllable (best-effort service) to the non-controllable network traffic flows. As a result, the QoS parameters of priority traffic flows could be guaranteed by decreasing the transmission rate of the best-effort service applications.

In this work, we mixture two different traffic sources (controllable and non-controllable) in order to study the capability of blind separation of the ICA algorithm proposed by Bell and Sejnowski in [6]. More specifically, we intend to evaluate the impact of such separation mechanism to the network performance as well as to an effective network traffic control mechanism. To this end, we introduce a novel TSK type fuzzy modeling in order to model the buffer queue size and the controllable source behavior, taking into account the nonlinear time-varying characteristics of network traffic flows. Our fuzzy model runs an adaptive clustering algorithm that divides the traffic data into several linear clusters by fuzzy interpolation, each one described by a TSK fuzzy model. In addition, a steepest descent algorithm is used to refine the obtained model and to improve the modeling accuracy. We propose an adaptive recursive fuzzy predictor, which predicts the future system behavior in order to overcome the long wait due to feedback delay and also, to solve the large bandwidth-delay product caused by congestion problems. Further, an optimal regulation rate is calculated via the parameters of the fuzzy predictor for traffic control.

Our traffic control strategy consist of: first, to separate the controllable and non-controllable traffic flows through an ICA algorithm, second, to derive an optimal prediction buffer length estimate and next, to control the controllable input traffic rates so that the buffer length is maintained at a desired buffer level and minimization of buffer length variance occurs.

2 Blind Network Traffic Separation

We aim to verify the possibility of separating the controllable and non-controllable traffic sources of a mixture traffic trace, assuming they are mutually independent processes. This kind of mixture frequently occurs in computer networks and an exact separation is a difficult task [7, 8]. Moreover, we intend to evaluate the impact of the ICA algorithm of Bell and Sejnowski to the network performance.

2.1 The Bell-Sejnowski Algorithm

Independent component analysis (ICA) is a linear transformation in which the desired representation is the one that minimizes the statistical dependence of its components.

The Bell-Sejnowski (or Infomax) [6] algorithm finds such components by maximizing the network entropy.

Let us assume that we have a given neural network of the type:

$$y = g(W x) \tag{1}$$

where $x = [x_1, \ldots, x_M]^T$ is a conveniently preprocessed input vector, W is an $M \times M$ matrix containing weights, g is an invertible nonlinear monotonic function and $y = [y_1, \ldots, y_M]^T$ is an output vector. The output entropy $H(y)$ can be calculated:

$$H(y) = H(x) + E\left[ln \left| (det\ W) \prod_i y_i' \right| \right] \tag{2}$$

where $H(x)$ is the input entropy, E is the expectancy operator and y_i' is the derivative of a given output i of y. This may be rewritten as:

$$H(y) = H(x) + ln|det\ W| + \sum_i E[ln\ y_i'] \tag{3}$$

The derivative of this expression is calculated as:

$$\Delta W \propto \frac{\partial H(y)}{\partial W} = [W^T]^{-1} + \frac{\partial}{\partial W}\left(\sum_i E[ln\ y_i'] \right) \tag{4}$$

It was proved that if the function g is well chosen then this framework enables the estimation of the ICA model [9] and is equivalent to the maximum likelihood approach [10]. The selection of g has been called a "black art" in [6] but according to [9, 11] $g = tanh$ works well for super-Gaussian independent components. The gradient of the output of $y = tanh(W x)$ regarding the coefficients of W may be calculated yielding the entropy maximization learning rule given by:

$$\Delta W \propto [W^T]^{-1} - 2\ y\ x^T \tag{5}$$

or alternatively:

$$\Delta w_{ij} \propto \frac{cof\ w_{ij}}{det\ W} - 2\ x_j\ y_i \tag{6}$$

3 ICA-Fuzzy Adaptive Predictive Flow Control

The proposed Adaptive Fuzzy Regression Clustering with Covariance Resetting (AFRCCR) algorithm simultaneously defines the fuzzy subspace and find the parameters for the consequent parts of the TSK rules.

3.1 AFRCCR TSK Fuzzy Modeling and Control

The input traffic arrived at the multiplexer buffer belongs to two distinct traffic groups. One consists of the controllable traffic $\mu(k)$, which can adjust its rate according to the network condition status. The other group consists of the uncontrollable traffic $v(k)$, which is delay-sensitive and has a higher priority than controllable type. Therefore, the controllable sources can only share the remaining link bandwidth left by the higher priority sources.

Let $\{(\vec{x}(1), y(1)), (\vec{x}(2), y(2)), \dots, (\vec{x}(N), y(N))\}$, be a set of observations with $\vec{x}(i) \in \mathbb{R}^n$ and $y(k) \in \mathbb{R}$, where N is the number of training data, $\vec{x}(k) = [x_1(k), x_2(k), \dots, x_n(k)]$ the k-th input vetor, and $y(k)$ the desired output for the input $\vec{x}(k)$, with $1 \leq k \leq N$. In this work, as mentioned previously, the input vector consist of the past and current value of the buffer length and the regulation rate, and the output will be the d-step ahead of the buffer length value, e.g., $\vec{x}(k) = [b(k - p), b(k - p + 1), \dots, b(k), \mu(k - q), \mu(k - q + 1), \dots, \mu(k)]$ and $y(k) = b(k + d)$.

Typically, a TSK fuzzy model consists of IF-THEN rules that have the following form:

$$\mathbb{R}^i : If \ b(k - p) \ is \ A_1^i(\vec{\theta}_1^i) \ \dots \ and \ b(k) \ is \ A_{p+1}^i(\vec{\theta}_{p+1}^i) \ and \ \mu(k$$
$$- q) \ is \ A_{p+2}^i(\vec{\theta}_{p+2}^i) \ \dots \ and \ \mu(k) \ is \ A_{p+q+2}^i(\vec{\theta}_{p+q+2}^i) \tag{7}$$

$$Then \ y^i(k) = b^i(k + d)$$

$$= \alpha_0^i + \alpha_1^i b(k - p) + \cdots + \alpha_{p+1}^i b(k) \tag{8}$$

$$+ \beta_{p+2}^i \mu(k - q) + \cdots + \beta_{p+q+2}^i \mu(k)$$

for $i = 1, 2, \dots, C$, where C is the number of IF-THEN rules, $p + 1$ is the number of input buffer length values, $q + 1$ is the number of input regulation rates ($n = p + q + 2$ is the number of total inputs), $A_j^i(\vec{\theta}_j^i)$ is the fuzzy set of the i-th rule for $x_j(k)$ with the adjustable parameter set with $1 \leq j \leq n$, and $\vec{\alpha}^i(k) = (\alpha_0^i, \alpha_1^i, \dots, \alpha_{p+1}^i, \beta_{p+2}^i, \beta_{p+3}^i, \dots, \beta_{p+q+2}^i)$ is the parameter set in the consequente part for the k-th output.

The prediction output of the fuzzy model is inferred as:

$$\hat{y}(k) = \hat{b}(k + d) = \frac{\sum_{i=1}^C b^i(k + d) w^i(k)}{\sum_{i=1}^C w^i(k)} \tag{9}$$

where $b^i(k + d)$ is the output of the i-th rule, $w^i(k) = \min_{j=1\dots n} A_j^i\left(\vec{\theta}_j^i; x_j(k)\right)$ is the i-th rule's firing strength, which is obtained as the minimum of the fuzzy membership degrees of all fuzzy variables. In (7) and (8), both the parameters of the premise parts (e.g., $\vec{\theta}_j^i$) and of the consequent parts (i.e., $\vec{\alpha}^i$) of the TSK fuzzy model are required to be identified.

The FRCM (Fuzzy C-Regression Model) clustering algorithm proposed in [12] does not consider the spatial distribution of training data. We propose the AFRCCR algorithm which takes into account the training data distribution and consider both the regression error and the distance between the input data and the clusters. Hence, the cost function of the AFRCCR algorithm is defined as:

$$J = \sum_{i=1}^{C} \sum_{k=1}^{N} u_{ik}^2 (r_{ik} d_{ik})^2 \tag{10}$$

subject to

$$\sum_{i=1}^{C} u_{ik} = 1, for\ 1 \le k \le N \tag{11}$$

where u_{ik} is the firing strength of the i-th rule for the k-th training pattern, C is the number of fuzzy rules and N is the number of the training data. In (10), r_{ik} is the error between the k-th desired output of the modeled system and the output of the i-th rule with the k-th input data, i.e., $r_{ik} = y(k) - f_i \left(\vec{x}(k); \vec{a}^i(k) \right)$. In addition, d_{ik} is the distance between the k-th input data and the center of i-th cluster (β_i), i.e., $d_{ik} = \vec{x}(k) - \beta_i$, where $i = 1, 2, ..., C$.

To minimize J in (10) subject to (11), the Lagrange multiplier method is applied, the following equations are obtained:

$$\vec{a}^i(k+1) = \vec{a}^i(k) + \left(P_i(k+1)x(k+1)\omega(k+1) \right) \times \left(y(k+1) - x(k+1)^T \vec{a}^i(k) \right) \tag{12}$$

and,

$$P(k+1) = P(k) - \frac{\omega(k+1)P_i(k)x(k+1)x(k+1)^T P_i(k)}{1 + \omega(k+1)x(k+1)^T P_i(k)x(k+1)} \tag{13}$$

where $x(k+1)$ is the $(k+1)$-th row of matrix X and $w(k+1)$ is the $(k+1)$-th diagonal element of matrix $Q_i(k+1)$.

$$u_{ik} = \frac{1/(2r_{ik}^2 d_{ik}^2)}{\sum_{i=1}^{C} 1/(2r_{ik}^2 d_{ik}^2)} \tag{14}$$

And the center of the i-th cluster (β_i) is given by the following equation:

$$\beta_i(k) = \frac{\sum_{z=1}^{N} r_{ik}^2 d_{ik}^2 \vec{x}(z)}{\sum_{z=1}^{N} r_{ik}^2 d_{ik}^2} \tag{15}$$

In practice, this recursive method of parameters $\vec{a}^i(k+1)$ estimation presents a high initial convergence rate. However, the algorithm gain is reduced when the elements of covariance matrix P become small latter than some iterations later. To avoid this problem, the covariance matrix is reseted periodically every 10 time units, so that is, the algorithm gain will not be reduced and a high convergence rate can be obtained. Furthermore, we apply a steepest descent algorithm to refine the obtained model and to improve the modeling accuracy according to [3].

The optimal flow control rate $\mu^o(k)$ is a function of the input vetor $\vec{x}(k)$, the parameters of the consequent parts $\vec{a}^i(k)$ and the reference buffer level b^τ by minimizing the following d-step ahead quadratic cost function:

$$J(k+d) = E\left[\frac{1}{2}(b(k+d) - b^\tau)^2 + \frac{\lambda}{2}\mu^2(k)|\Im_k\right] \tag{16}$$

where λ is a weighting factor and b^τ is the reference buffer level. Explicitly, the rate $\mu^o(k)$ is given by:

$$\mu^o(k) = \beta_0(k)b^\tau - \left\{\sum_{i=1}^{C} h_i(k)\left[\sum_{j=0}^{p} a_{ij}b(k-j) + \sum_{j=1}^{q} \beta_{ij}u(k-j)\right]\right\} \tag{17}$$

where

$$\beta_0(k) \triangleq \frac{\sum_{i=1}^{C} h_i(k)\beta_{i0}}{(\sum_{i=1}^{C} h_i(k)\beta_{i0})^2 + \lambda} \tag{18}$$

and

$$h_i(k) = \frac{w^i(k)}{\sum_{i=1}^{C} w^i(k)}; \ w^i(k) = \min_{j=1...n} A_j^i\left(\vec{\theta}_j^i; x_j(k)\right) \tag{19}$$

4 Simulation Results

In order to describe our simulation procedures of validating the proposed separating and control approaches, it is necessary to specify the uncontrollable and controllable sources. Two TCP/IP wide network traffic traces are used for our experimental investigation (dec-pkt-1-512.tcp and dec-pkt-2-512.tcp of Digital Equipment Corporation) [14].

In the simulations, the sampling time interval T is chosen as $512ms$, the link capacity as $190\ kbytes/s$, the maximum buffer size B_{max} as 1.5×10^6 and d as $512ms$. We considered two relevant past terms in equation (7) and (8), i.e., $p = 2$ and $q = 2$.

For the estimation of the optimal traffic control rate $\mu^o(k)$, the desired level b^τ was set to 30% of B_{max}, i.e., b^τ, and the design parameter λ was defined as 1. Table 1 compares some network performance parameters resulting from the application of the proposed network traffic control procedure and without control, as well as, the queueing behavior considering the flow separation of a trace composed by different linear combinations of traffic flows represented by the mixing matrix A. That is, when A is given by:

$$A = \begin{bmatrix} 0.8 & 0.2 \\ 0.2 & 0.8 \end{bmatrix} \tag{20}$$

the input x of the system in function of the original signals s is:

$$x = As \Longrightarrow \begin{cases} x_1 = 0.8\ s_1 + 0.2\ s_2 \\ x_2 = 0.2\ s_1 + 0.8\ s_2 \end{cases} \tag{21}$$

where the traffic trace s is composed of s_1 which is a controllable traffic flow and s_2 which is a non-controllable traffic flow.

Table 1. Results

	Original Traces	Recovered Traces for: $A = \begin{bmatrix} 0.8 & 0.2 \\ 0.2 & 0.8 \end{bmatrix}$	Recovered Traces for $A = \begin{bmatrix} 0.7 & 0.3 \\ 0.3 & 0.7 \end{bmatrix}$	Recovered Traces for $A = \begin{bmatrix} 0.6 & 0.4 \\ 0.4 & 0.6 \end{bmatrix}$
Buffer Mean (without control)	1.10E+06	1.10E+06	1.25E+06	1.30E+06
Buffer Mean (with control)	4.79E+03	2.58E+03	1.32E+03	2.36E+03
Buffer Variance (without control)	3.63E+11	3.63E+11	2.74E+11	2.31E+11
Buffer variance (with control)	3.21E+08	1.10E+08	6.67E+07	1.26E+08
Utilization (without control)	99.9022	99.9021	99.95	99.95
Utilization (with control)	90.6696	95.1729	94.1407	99.6133
Packet Loss Rate (without control)	9.0304	8.9788	15.7405	22.9086
Packet Loss Rate (with control)	0	0	0	0

Analyzing Table 1, we note that with the control approach the buffer length (buffer mean) is maintained below the desired level, which is 4.5×10^5 *bytes*. In addition, the variance of the buffer length is decreased as well as the PLR (Packet Loss Rate) is reduced at the price of slightly decreasing network utilization. It can also be seen that the better the traffic flow separation the closer are the results using the recovered traces to those of using the original traffic sources. The traffic separation performance was verified by analyzing the correlation coefficient between the original and recovered traffic traces for different A values. In the simulations, we obtained the best traffic separation for $A = \begin{bmatrix} 0.8 & 0.2 \\ 0.2 & 0.8 \end{bmatrix}$, resulting in a network performance similar to that with the original traffic traces when no control is applied. In this same case, by applying the proposed control, the lowest PLR is achieved.

5 Conclusion

A separation and control procedure based on ICA and fuzzy modeling was proposed. The major advantages of this approach are as follows. The proposed control algorithm provides a blind separation of controllable (Best-effort) and non-controllable (e.g. real-time applications) traffic flows as well as an efficient control rate control of the controllable traffic sources. Particularly, we applied our control approach to a network link showing that we could control the traffic flow rates in order to maintain the desired buffer level. We also verified that ICA algorithm can be used to separate traffic flows for specific values of the matrix A. Accordingly, the simulation results showed that the variance of the queue size process and the byte loss rate were decreased when compared to the case using the same server but without fuzzy control. Therefore, the proposed fuzzy model based adaptive fuzzy control algorithm can be seen as an important step towards a reliable, adaptive and QoS based control system for computer networks.

References

1. Frost, V.S., Melamed, B.: Traffic modeling for telecommunications networks. Communications Magazine 32, 70–81 (1994)
2. Adas, A.: Traffic models in broadband networks. Communications Magazine 35, 82–89 (1997)
3. Chen, B.S., Yang, Y.S., Lee, B.K., Lee, T.H.: Fuzzy adaptive predictive flow control of ATM network traffic. IEEE Transactions on Fuzzy Systems 11, 568–581 (2003)
4. Takagi, T., Sugeno, M.: Fuzzy identification of systems and its applications to modeling and control. IEEE transactions on systems, man, and cybernetics 15, 116–132 (1985)
5. Sugeno, M., Yasukawa, T.: A fuzzy-logic-based approach to qualitative modeling. IEEE Transactions on Fuzzy Systems 1 (1993)
6. Bell, A., Sejnowski, T.: An Information-Maximization Approach to Blind Separation and Blind Deconvolution. Neural Computation 7, 1129–1159 (1995)
7. Jiang, J.: Detection of network anomalies and novel attacks in the Internet via statistical network traffic separation and normality prediction. New Jersey Institute of Technology (2005)
8. Jiang, J., Papavassiliou, S.: Enhancing network traffic prediction and anomaly detection via statistical network traffic separation and combination strategies. Computer Communications 29, 1627–1638 (2006)
9. Roberts, S., Everson, R.: Independent Component Analysis: Principles and Practice. Cambridge University Press, New York (2001)
10. Hyvarinen, A., Karhunen, J., Oja, E.: Independent component analysis. John Wiley & Sons, Chichester (2001)
11. Hyvarinen, A.: Survey on independent component analysis. Neural Computing Surveys 2, 94–128 (1999)
12. Hathaway, R., Bezdek, J.: Switching regression models and fuzzy clustering. IEEE Transactions on Fuzzy Systems 1, 195–204 (1993)
13. Haykin, S.: Adaptive filter theory. Prentice-Hall, Inc., Upper Saddle River (1991)
14. Traces In The Internet Traffic Archive. The Internet Traffic Archive, Lawrence Berkeley National Laboratory, http://ita.ee.lbl.gov/html/traces.html

The 2008 Signal Separation Evaluation Campaign: A Community-Based Approach to Large-Scale Evaluation

Emmanuel Vincent[1], Shoko Araki[2], and Pau Bofill[3]

[1] METISS Group, IRISA-INRIA
Campus de Beaulieu, 35042 Rennes Cedex, France
emmanuel.vincent@irisa.fr
[2] Signal Processing Research Group, NTT Communication Science Labs
2-4, Hikaridai, Seika-cho, Soraku-gun, Kyoto 619-0237, Japan
[3] Departament d'Arquitectura de Computadors, Universitat Politècnica de Catalunya
Campus Nord Mòdul D6, Jordi Girona 1-3, 08034 Barcelona, Spain

Abstract. This paper introduces the first community-based Signal Separation Evaluation Campaign (SiSEC 2008), coordinated by the authors. This initiative aims to evaluate source separation systems following specifications agreed between the entrants. Four speech and music datasets were contributed, including synthetic mixtures as well as microphone recordings and professional mixtures. The source separation problem was split into four tasks, each evaluated via different objective performance criteria. We provide an overview of these datasets, tasks and criteria, summarize the results achieved by the submitted systems and discuss organization strategies for future campaigns.

1 Introduction

Large-scale evaluations are a key ingredient to scientific and technological maturation by revealing the effects of different system designs, promoting common test specifications and attracting the interest of industries and funding bodies. Recent evaluations of source separation systems include the 2006 Speech Separation Challenge[1] and the 2007 Stereo Audio Source Separation Evaluation Campaign [1]. The subsequent panel discussion held at the 7th International Conference on Independent Component Analysis and Signal Separation (ICA 2007) resulted in a set of recommendations regarding future evaluations, in particular:

- splitting the overall problem into several successive or alternative tasks,
- providing reference software and evaluation criteria for each task,
- considering toy data as well as real-world data of interest to companies,
- letting entrants specify all aspects of the evaluation collaboratively.

These general principles aim to facilitate the entrance of researchers addressing different tasks and to enable detailed diagnosis of the submitted systems.

---

[1] http://www.dcs.shef.ac.uk/~martin/SpeechSeparationChallenge.htm

T. Adali et al. (Eds.): ICA 2009, LNCS 5441, pp. 734–741, 2009.

This article introduces the 2008 community-based Signal Separation Evaluation Campaign (SiSEC) as a tentative implementation of these principles. Due to the variety of the submitted systems, we focus on the general outcomes of the campaign and let readers refer to the website at `http://sisec.wiki.irisa.fr/` for the details and results of individual systems. We describe the chosen datasets, tasks and evaluation criteria in Section 2. We summarize the results and provide bibliographical references to the submitted systems in Section 3. We conclude and discuss organization strategies for future campaigns in Section 4.

2   Specifications

The datasets, tasks and evaluation criteria considered in the campaign were specified in a collaborative fashion. A few initial specifications were first suggested by the organizers. Potential entrants were then invited to give their feedback and contribute additional specifications using collaborative software tools (wiki, mailing list). Although few people eventually took advantage of this opportunity, those who did contributed a large proportion of the evaluation materials. All materials, including data and code, are available at `http://sisec.wiki.irisa.fr/`.

2.1   Datasets

The data consisted of audio signals spanning a range of mixing conditions. The channels $x_i(t)$ $(1 \leq i \leq I)$ of each mixture signal were generally obtained as

$$x_i(t) = \sum_{j=1}^{J} s_{ij}^{\text{img}}(t) \tag{1}$$

where $s_{ij}^{\text{img}}(t)$ is the *spatial image* of source j $(1 \leq j \leq J)$ on channel i, that is the contribution of this source to the mixture in this channel. *Instantaneous* mixtures are generated via $s_{ij}^{\text{img}}(t) = a_{ij}s_j(t)$, where $s_j(t)$ are single-channel source signals and a_{ij} positive mixing gains. *Convolutive* mixtures are obtained similarly from mixing filters $a_{ij}(\tau)$ via $s_{ij}^{\text{img}}(t) = \sum_{\tau} a_{ij}(\tau)s_j(t - \tau)$. *Recorded* mixtures are acquired by playing each source at a time on a loudspeaker and recording it over a set of microphones. Four distinct datasets were provided:

D1. Under-determined speech and music mixtures
This dataset consists of 36 instantaneous, convolutive and recorded stereo mixtures of three to four audio sources of 10 s duration, sampled at 16 kHz. Recorded mixtures were acquired in a chamber with cushion walls, using the loudspeaker and microphone arrangement depicted in [1], while convolutive mixtures were obtained with artificial room impulse responses simulating the same arrangement. The distance between microphones was set to either 5 cm or 1 m and the room reverberation time (RT) to 130 ms or 250 ms. The source signals include unrelated female or male speech and synchronized percussive or non-percussive music.

D2. Determined and over-determined speech and music mixtures
This dataset includes 21 four-channel recordings of two to four unrelated speech sources of 10 s duration, sampled at 16 kHz, acquired in four different rooms:

two chambers with cushion walls, an office room and a conference room. Some mixtures were directly recorded instead of computed via (1). The microphones were placed either at a height of 1.25 m or at different heights, near the walls or near the center and at a distance of about 5 cm or 1 m. The sources were placed either randomly or at 1 m distance from the microphones. The dataset also includes a 2-channel mixture of 2 speech sources recorded via cardioid microphones placed on either side of a dummy head.

D3. Head-geometry mixtures of two speech sources in real environments
This dataset consists of 648 two-channel convolutive mixtures of two unrelated speech sources of about 10 s, sampled at 16 kHz. The mixing filters were real-world impulse responses from two rooms, an anechoic chamber and an office room, measured by hearing aid microphones mounted on a dummy head. The sources were placed in the horizontal plane at fixed distance from the head. In the anechoic chamber, the distance was set to 3 m and the direction of arrival (DOA) varied over $360°$ in $20°$ steps. In the office room, the distance was set to 1 m and the DOA varied over the front $180°$ hemisphere in $10°$ steps. All possible combinations of two different DOAs were generated.

D4. Professionally produced music recordings
This dataset consists of two stereo music signals sampled at 44.1 kHz involving two and ten synchronized sources of 13 and 14 s duration, respectively. The stereo spatial image of each source was generated by a combination of professional recording and mixing techniques. Special effects applied to individual sources include chorus, distortion pads, vocoder, delays, parametic equalization and dynamic multi-band compression.

All datasets except D2 include both test and development data generated in a similar fashion, but from different source signals and source positions. The true source signals and source positions underlying the test data were hidden to the entrants, except for D3 where the source positions were provided as prior information. The true number of sources was always available.

2.2 Tasks

The source separation problem was split into four tasks:

T1. Source counting
T2. Mixing system estimation
T3. Source signal estimation
T4. Source spatial image estimation

These tasks consists of finding, respectively: (T1) the number of sources J, (T2) the mixing gains a_{ij} or the discrete Fourier transform $a_{ij}(\nu)$ of the mixing filters, (T3) the single-channel source signals $s_j(t)$ and (T4) the spatial images $s_{ij}^{\text{img}}(t)$ of the sources over all channels i. Entrants were asked to submit the results of their system to T3 and/or T4 and on an optional basis to T1 and/or T2.

Reference software was provided to address tasks T1 and T2 over instantaneous mixtures [2] (R1) and tasks T3 and T4 either via binary masking (R2) or via l_p-norm minimization [3] (R3). This software aims to facilitate entrance and to provide baseline results for benchmarking purposes. Two oracle systems were also considered for the benchmarking of task T4: ideal binary masking over a short-time Fourier transform (STFT) [4] (O1) or over a cochleagram [5] (O2). These systems require knowledge of the true source spatial images and provide theoretical upper performance bounds for binary masking-based systems.

2.3   Evaluation Criteria

Although standard evaluation criteria exist for task T2 when the number of sources is smaller than the number of sensors, there are no such criteria in a more general setting so far. For instantaneous mixtures, the vector $\widehat{\mathbf{a}}_j$ of estimated mixing gains for a given source j was decomposed as

$$\widehat{\mathbf{a}}_j = \mathbf{a}_j^{\mathrm{coll}} + \mathbf{a}_j^{\mathrm{orth}} \tag{2}$$

where $\mathbf{a}_j^{\mathrm{coll}}$ and $\mathbf{a}_j^{\mathrm{orth}}$ are respectively collinear and orthogonal to the true vector of mixing gains $\mathbf{a}_j$ and are computed by least squares projection. Accuracy was then assessed via the mixing error ratio (MER) in decibels (dB)

$$\mathrm{MER}_j = 10 \log_{10} \frac{\|\mathbf{a}_j^{\mathrm{coll}}\|^2}{\|\mathbf{a}_j^{\mathrm{orth}}\|^2} \tag{3}$$

where $\|.\|$ is the Euclidean norm. This criterion allows arbitrary scaling of the gains for each source. It is equal to $+\infty$ for an exact estimate, 0 when the estimate forms a 45° angle with the ground truth and $-\infty$ when it is orthogonal. For convolutive mixtures, the accuracy of estimated mixing filters for source j was similarly assessed by computing the MER in each frequency bin ν between $\widehat{\mathbf{a}}_j(\nu)$ and $\mathbf{a}_j(\nu)$ and averaging it over frequency. Since the sources can be characterized only up to an arbitrary permutation, all possible permutations were tested and the one maximizing the average MER was selected.

Tasks T3 and T4 were evaluated via the criteria in [6] and [1], respectively, termed signal to distortion ratio (SDR), source image to spatial distortion ratio (ISR), signal to interference ratio (SIR) and signal to artifacts ratio (SAR). These criteria can be computed for any separation system and do not necessitate knowledge of the unmixing filters or masks. The SDR for task T3 allows arbitrary filtering of the target source, while that for T4 allows no scaling or filtering distortion, which is separately measured by the ISR. The signals were permuted so as to maximize the average SIR. The resulting permutations were found to be relevant and identical to that estimated from the MER, except in cases involving much interference. For dataset D4, performance was also measured via a magnitude Signal-to-Error Ratio (mSER) between the true and estimated magnitude STFTs of the source spatial images over each channel.

3   Results

The details and the results of the thirty submitted systems are available for viewing and listening at http://sisec.wiki.irisa.fr/. The systems [7], [8],

Table 1. Average MER for task T2 over the instantaneous mixtures of dataset D1

System	[11]	[7]	[8]	R1
MER	80.9	81.7	42.4	49.0

Table 2. Average performance for tasks T3 or T4 over the instantaneous mixtures of dataset D1. Figures relate to T4 when the ISR is reported and to T3 otherwise.

System	[12]	[13]	[3]	[14]	[15]	[11][2]	[16]	[17]	[9][2]	[8]	R2	R3	O1	O2
SDR	9.8	14.0	11.7	11.3	7.8	10.7	12.6	6.8	6.2^3	5.5	8.8	11.1	10.7	8.1
ISR	18.2	24.1	22.9	20.1	16.4	21.9				11.4	17.5	22.7	20.0	14.4
SIR	14.9	20.6	18.5	16.9	17.3	18.6	18.5	16.8	11.6^3	12.9	18.6	18.4	21.6	17.4
SAR	11.4	15.4	13.2	12.8	8.8	11.9	13.2	7.7	9.0^3	8.2	9.6	12.3	11.5	9.1

Table 3. Average performance for tasks T3 or T4 over the convolutive/recorded mixtures of dataset D1. Figures relate to T4 when the ISR is reported and to T3 otherwise.

System	RT=130ms				RT=250ms			
	SDR	ISR	SIR	SAR	SDR	ISR	SIR	SAR
[18]	2.0^3	6.3^3	5.8^3	5.5^3	1.0^3	5.2^3	2.9^3	5.2^3
[19]	0.8	5.9	2.7	5.7	0.3	5.1	0.6	6.1
[20]	1.6	5.8	3.2	7.1	1.1	5.0	1.6	7.9
[10]	2.9^3	6.5^3	7.1^3	8.6^3	3.7^3	6.5^3	5.0^3	8.8^3
[21]	-1.1^3		6.8^3	1.3^3	-1.1^3		6.6^3	1.5^3
[8]	3.3	6.7	4.3	7.9	3.1	6.2	3.9	8.4
O1	9.7	18.3	19.9	10.2	8.7	16.2	19.4	10.4
O2	6.9	12.1	16.5	7.6	6.6	11.5	16.0	7.9

Table 4. Average SIR for task T3 over dataset D2

System	Cushioned rooms			Office/lab rooms			Conference room		
	$J=2$	$J=3$	$J=4$	$J=2$	$J=3$	$J=4$	$J=2$	$J=3$	$J=4$
[22]	14.4	16.3	8.9	14.1	5.7	-0.3	8.2	1.5	-2.3
[23]	5.3	12.8	9.0	19.6^3					
[24]	3.9	4.3	0.9	6.9	2.6	-2.9	7.1	2.2	-0.6
I. Takashi	11.3^3	7.4^3	5.6^3	2.8^3					
[25]	10.6	9.2	4.1	4.2^3	-0.4	-3.7	2.3	-1.2	-3.5
[26]	8.7	6.8	2.5	3.2^3	-1.3	-4.0	2.2	-1.3	-5.1

R1, [9] and [10] addressed task T1 without error. Summary performance figures for other tasks are provided in Tables 1 to 6 and in Figure 1 after averaging over all sources then over several mixtures. An analysis of each table is beyond the scope of this paper, due to the wide variety of prior knowledge and computation resources used by different systems. We observe that the mixing matrix estimation task is now solved for instantaneous mixtures, that the source

---

[2] Variant or extension of the system presented in the bibliographical reference.

[3] Figure computed by averaging over an incomplete set of mixtures or sources.

Table 5. Average performance for task T4 over dataset D3

System	Anechoic chamber				Office room			
	SDR	ISR	SIR	SAR	SDR	ISR	SIR	SAR
[19]	4.4	6.9	19.2	13.6	1.9	7.9	5.8	10.7
[19]²	4.7	7.0	19.3	14.3	2.2	8.4	5.9	11.2
[27]					1.4	6.7	6.1	10.7
O1	13.7	24.5	24.7	14.1	13.0	23.7	23.9	13.4
O2	11.0	20.1	19.9	11.7	10.6	19.8	19.6	11.3

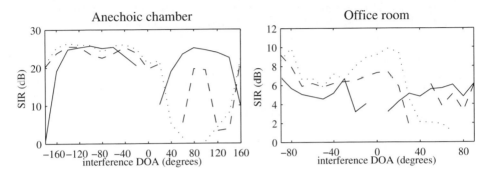

Fig. 1. Average SIR achieved by system [19] for task T4 over dataset D3 as a function of the interference DOA for three target DOAs (plain: 0°, dashed:40°, dotted:80°)

Table 6. Average performance for task T4 over dataset D4. The SIR quantifies interference from the target sources only, while the SAR includes that from other sources.

System	Tamy ($J = 2$)					Bearlin ($J = 10$)				
	mSER	SDR	ISR	SIR	SAR	mSER	SDR	ISR	SIR	SAR
[13]	6.0	4.5	10.0	8.9	8.6	4.2	-0.4	7.8	6.9	1.6
[16]²	6.8	5.9	10.2	8.8	10.7	4.9	3.8	9.7	8.3	4.8
[28]	9.5	8.6	17.3	16.4	9.5					
[29]	8.4	7.7	16.5	15.4	8.4					
[15]²	8.5	7.2	16.5	15.7	8.3	3.3	2.6	8.6	12.9	1.6
[30]	3.5	4.5	7.4	18.5	4.6	3.4	3.2	8.4	12.0	1.8
[31]²	9.3	8.3	15.1	23.5	8.0					
[31]²	10.0	9.1	15.1	24.1	9.1					
[31]²						5.7³	5.4³	9.8³	17.7³	5.2³
O1	12.8	11.0	21.4	21.1	11.4	9.1	7.5	14.7	18.0	7.9
O2	9.0	8.0	14.5	15.1	8.9	-0.3	2.0	8.0	10.7	0.4

signal estimation task can now be addressed with a mean SIR around 20 dB for instantaneous or anechoic mixtures and that the separation of monophonic instruments from professional music recordings can also be achieved with a SIR above 15 dB. Nevertheless, the separation of reverberant mixtures remains a challenge for any number of sources and channels despite continued progress, as illustrated by an average SIR around 6 dB for office recordings of three sources.

4   Conclusion

We summarized the specifications and outcomes of the first community-based Signal Separation Evaluation Campaign. We hope that this campaign fosters interest for evaluation in the source separation community, so that more entrants contribute feedback, datasets or code in the future. With thirty submissions but three organizers only, the current organization scheme has reached its goal of attracting many entrants, but failed to provide detailed analysis of the results. We believe that increased participation from the community is key to maximizing the benefits of future campaigns. We advocate the creation of a larger organization committee with members dedicated to the evaluation of a particular dataset or task and invite all willing researchers to become part of it.

Acknowledgments

We would like to thank all the entrants, as well as H. Sawada, M. Dyrholm, L.C. Parra, K.E. Hild II, H. Kayser, J. Anemüller, M. Vinyes Raso and J. Woodruff for sharing their datasets and evaluation code. The collection of dataset D3 was supported by the European Commission under IP DIRAC.

References

1. Vincent, E., Sawada, H., Bofill, P., Makino, S., Rosca, J.: First stereo audio source separation evaluation campaign: Data, algorithms and results. In: Davies, M.E., James, C.J., Abdallah, S.A., Plumbley, M.D. (eds.) ICA 2007. LNCS, vol. 4666, pp. 552–559. Springer, Heidelberg (2007)
2. Bofill, P.: Identifying single source data for mixing matrix estimation in instantaneous blind source separation. In: Koutník, J., Kůrková, V., Neruda, R. (eds.) ICANN 2008, Part I. LNCS, vol. 5163. Springer, Heidelberg (2008)
3. Vincent, E.: Complex nonconvex l_p norm minimization for underdetermined source separation. In: Davies, M.E., James, C.J., Abdallah, S.A., Plumbley, M.D. (eds.) ICA 2007. LNCS, vol. 4666, pp. 430–437. Springer, Heidelberg (2007)
4. Vincent, E., Gribonval, R., Plumbley, M.D.: Oracle estimators for the benchmarking of source separation algorithms. Signal Processing 87, 1933–1950 (2007)
5. Wang, D.L.: On ideal binary mask as the computational goal of auditory scene analysis. In: Speech Separation by Humans and Machines. Springer, New York (2005)
6. Vincent, E., Gribonval, R., Févotte, C.: Performance measurement in blind audio source separation. IEEE Trans. on Audio, Speech and Language Processing 14, 1462–1469 (2006)
7. Arberet, S., Gribonval, R., Bimbot, F.: A robust method to count and locate audio sources in a stereophonic linear instantaneous mixture. In: Rosca, J.P., Erdogmus, D., Príncipe, J.C., Haykin, S. (eds.) ICA 2006. LNCS, vol. 3889, pp. 536–543. Springer, Heidelberg (2006)
8. El Chami, Z., Pham, A.D.T., Servière, C., Guerin, A.: A new model based underdetermined source separation. In: Proc. IWAENC (2008)
9. Gowreesunker, B.V., Tewfik, A.H.: Blind source separation using monochannel overcomplete dictionaries. In: Proc. ICASSP, pp. 33–36 (2008)
10. Araki, S., Nakatani, T., Sawada, H., Makino, S.: Stereo source separation and source counting with MAP estimation with Dirichlet prior considering spatial aliasing problem. In: Adali, T., Jutten, C., Romano, J.M.T., Barros, A.K. (eds.) ICA 2009. LNCS, vol. 5441, pp. 742–750. Springer, Heidelberg (2009)

11. Deville, Y., Puigt, M., Albouy, B.: Time-frequency blind signal separation: extended methods, performance evaluation for speech sources. In: Proc. IJCNN, pp. 255–260 (2004)
12. Nesbit, A., Vincent, E., Plumbley, M.D.: Extension of sparse, adaptive signal decompositions to semi-blind audio source separation. In: Adali, T., Jutten, C., Romano, J.M.T., Barros, A.K. (eds.) ICA 2009. LNCS, vol. 5441, pp. 605–612. Springer, Heidelberg (2009)
13. Ozerov, A., Févotte, C.: Multichannel nonnegative matrix factorization in convolutive mixtures. With application to blind audio source separation. In: Proc. ICASSP (in press, 2009)
14. Vincent, E., Arberet, S., Gribonval, R.: Underdetermined instantaneous audio source separation via local Gaussian modeling. In: Adali, T., Jutten, C., Romano, J.M.T., Barros, A.K. (eds.) ICA 2009. LNCS, vol. 5441, pp. 775–782. Springer, Heidelberg (2009)
15. Cobos, M., López, J.J.: Stereo audio source separation based on time-frequency masking and multilevel thresholding. Digital Signal Processing 18, 960–976 (2008)
16. Arberet, S., Ozerov, A., Gribonval, R., Bimbot, F.: Blind spectral-GMM estimation for underdetermined instantaneous audio source separation. In: Adali, T., Jutten, C., Romano, J.M.T., Barros, A.K. (eds.) ICA 2009. LNCS, vol. 5441, pp. 751–758. Springer, Heidelberg (2009)
17. Gowreesunker, B.V., Tewfik, A.H.: Two improved sparse decomposition methods for blind source separation. In: Davies, M.E., James, C.J., Abdallah, S.A., Plumbley, M.D. (eds.) ICA 2007. LNCS, vol. 4666, pp. 365–372. Springer, Heidelberg (2007)
18. Cobos, M., López, J.J.: Blind separation of underdetermined speech mixtures based on DOA segmentation. IEEE Trans. on Signal Processing (submitted, 2009)
19. Mandel, M.I., Ellis, D.P.W.: EM localization and separation using interaural level and phase cues. In: Proc. WASPAA, pp. 275–278 (2007)
20. Mandel, M.I., Ellis, D.P.W., Jebara, T.: An EM algorithm for localizing multiple sound sources in reverberant environments. In: Advances in Neural Information Processing Systems (NIPS 19) (2007)
21. Izumi, Y., Ono, N., Sagayama, S.: Sparseness-based 2ch BSS using the EM algorithm in reverberant environment. In: Proc. WASPAA, pp. 147–150 (2007)
22. Nesta, F., Omologo, M., Svaizer, P.: Multiple TDOA estimation by using a state coherence transform for solving the permutation problem in frequency-domain BSS. In: Proc. MLSP, pp. 43–48 (2008)
23. Lee, I.: Permutation correction in blind source separation using sliding subband likelihood function. In: Proc. ICA (2009)
24. Lee, I., Kim, T., Lee, T.W.: Independent vector analysis for blind speech separation. In: Blind speech separation. Springer, Dordrecht (2007)
25. Gupta, M., Douglas, S.C.: Scaled natural gradient algorithms for instantaneous and convolutive blind source separation. In: Proc. ICASSP, pp. II–637–II–640 (2007)
26. Douglas, S.C., Gupta, M., Sawada, H., Makino, S.: Spatio-temporal FastICA algorithms for the blind separation of convolutive mixtures. IEEE Trans. on Audio, Speech and Language Processing 15, 1511–1520 (2007)
27. Weiss, R.J., Ellis, D.P.W.: Speech separation using speaker-adapted eigenvoice speech models. Computer Speech and Language (in press, 2008)
28. Durrieu, J.L., Richard, G., David, B.: Singer melody extraction in polyphonic signals using source separation methods. In: Proc. ICASSP, pp. 109–172 (2008)
29. Durrieu, J.L., Richard, G., David, B.: An iterative approach to monaural musical mixture de-soloing. In: Proc. ICASSP (2009)
30. Bonada, J., Loscos, A., Vinyes Raso, M.: Demixing commercial music productions via human-assisted time-frequency masking. In: Proc. AES 120th Convention (2006)
31. Cancela, P.: Tracking melody in polyphonic audio. In: Proc. MIREX (2008)

Stereo Source Separation and Source Counting with MAP Estimation with Dirichlet Prior Considering Spatial Aliasing Problem

Shoko Araki, Tomohiro Nakatani, Hiroshi Sawada, and Shoji Makino

NTT Communication Science Laboratories, NTT Corporation
2-4 Hikaridai, Seika-cho, Soraku-gun, Kyoto 619-0237, Japan
{shoko,nak,sawada,maki}@cslab.kecl.ntt.co.jp

Abstract. In this paper, we propose a novel sparse source separation method that can estimate the number of sources and time-frequency masks simultaneously, even when the spatial aliasing problem exists. Recently, many sparse source separation approaches with time-frequency masks have been proposed. However, most of these approaches require information on the number of sources in advance. In our proposed method, we model the phase difference of arrival (PDOA) between microphones with a Gaussian mixture model (GMM) with a Dirichlet prior. Then we estimate the model parameters by using the maximum a posteriori (MAP) estimation based on the EM algorithm. In order to avoid one cluster being modeled by two or more Gaussians, we utilize a sparse distribution modeled by the Dirichlet distributions as the prior of the GMM mixture weight. Moreover, to handle wide microphone spacing cases where the spatial aliasing problem occurs, the indeterminacy of modulus $2\pi k$ in the phase is also included in our model. Experimental results show good performance of our proposed method.

Keywords: Dirichlet distribution, prior, number of sources, blind source separation, sparse, spatial aliasing problem.

1 Introduction

Blind source separation (BSS) is an approach for estimating source signals that uses only the mixed signal information observed at each microphone. The BSS technique for speech dealt with in this paper has many applications, including the hands-free teleconference systems and preprocessing for an automatic speech recognizer.

Let us formulate the task. Suppose that $N_s \geq 2$ speech sources $s_1, \ldots, s_{N_s}$ are convolutively mixed and observed at N_m microphones,

$$x_j(t) = \sum_{i=1}^{N_s} \sum_l h_{ji}(l)\, s_i(t-l), \quad j=1,\ldots,N_m, \tag{1}$$

where $h_{ji}(l)$ represents the impulse response from source i to microphone j. Our goal is to obtain estimates y_i of each source signal s_i from the microphone observations x_j without information about the number of sources N_s, the speech sources s_i or the mixing process h_{ji}.

T. Adali et al. (Eds.): ICA 2009, LNCS 5441, pp. 742–750, 2009.

Two approaches have been widely studied and employed to solve the BSS problem: one is based on independent component analysis (ICA) (e.g., [1]) and the other relies on the sparseness of source signals (e.g., [2]). In this paper, we focus on the latter approach, more specifically, the time-frequency mask approach [2,3]. With the time-frequency mask approach, we classify the phase difference of arrival (PDOA) between microphone observations, and separate each signal by collecting the observation signal at time-frequency points in each cluster.

In previous work [2,3], to automatically find clusters, the number of sources N_s is assumed to be known. However, in real situations we usually cannot obtain information on the number of sources N_s in advance. Especially for an under-determined case ($N_s > N_m$), the source counting is difficult, and few papers have dealt with this problem. Moreover, when the microphone spacing is large, the spatial aliasing problem occurs. This problem makes it difficult to classify the PDOA because the phase has the indeterminacy of modulus $2\pi k$ in high frequencies. [4] considered the spatial aliasing problem in a time-frequency mask approach, however, the number of sources N_s should be known.

In this paper, we propose a novel sparse source separation method that can estimate the number of sources and time-frequency masks simultaneously, even when spatial aliasing occurs. We model the PDOA distribution with a Gaussian mixture model (GMM) with a Dirichlet prior [5], and estimate the model parameters by using the EM algorithm. In order to avoid one cluster being modeled by two or more Gaussians, thus making it possible to estimate the number of sources correctly, we propose utilizing a sparse distribution modeled by the Dirichlet distribution as the prior of the GMM mixture weight. The authors of [6,7] also derived the EM algorithm, however, they still needed to know the number of sources N_s in advance. On the other hand, our proposed algorithm does not require information on the source number, thanks to the weight prior. Because the indeterminacy of $2\pi k$ in phase is modeled in our GMM, we can also overcome the difficulty in the PDOA clustering even in a spatial aliasing case.

The experimental results with a wide microphone spacing (20 cm) show that our proposed method can estimate the number of sources and can separate signals by time-frequency masks obtained by the posterior probability for each cluster.

2   Mixing and Separation Processes

This paper employs a time-frequency domain approach. With an F-point short-time Fourier transform (STFT), (1) is converted into:

$$x_j(n, f) = \sum_{i=1}^{N_s} h_{ji}(f)s_i(n, f), \tag{2}$$

where $h_{ji}(f)$ is the frequency response from source i to microphone j, $s_i(n, f)$ is the STFT of a source s_i. $f \in \{0, \frac{1}{F}f_s, \ldots, \frac{F-1}{F}f_s\}$ is a frequency (f_s is the sampling frequency) and $n(= 0, \cdots, N - 1)$ is a time-frame index.

In this paper, we assume the sparseness of the sources [2]:

$$x_j(n, f) \approx h_{ji}(f)s_i(n, f), \tag{3}$$

where $s_i(n, f)$ is a dominant source at the time-frequency slot (n, f). This is approximately true for speech signals in the time-frequency domain [2,3].

2.1   Separation Method

In this paper, we assume that $h_{ji}(f)$ in (2) is modeled by an anechoic model (e.g., eq. (13) of [3]), that is, the PDOA between microphones is given as:

$$\arg\left[\frac{x_1(n, f)}{x_2(n, f)}\right] = 2\pi f\tau(n, f) = 2\pi f\frac{d\cos\varphi(n, f)}{v}, \tag{4}$$

where $\tau(n, f) = d\cos\varphi(n, f)/v$ is the time difference of arrival (TDOA), $\varphi(n, f)$ is the dominant source direction at the time-frequency (n, f), and d and v denote the microphone spacing and the sound speed.

First, by assuming the source sparseness, we calculate the PDOA at each time-frequency slot by the left-side of (4). Then, by considering the frequency dependence in the PDOA, we classify the PDOA values in some way. For example, if there is no spatial aliasing problem and we know the number of sources, the k-means clustering algorithm can be applied to TDOA $\tau(n, f) = \frac{1}{2\pi f}\arg[x_1(n, f)/x_2(n, f)]$. Our method, which considers the aliasing problem and the unknown source number, is introduced in the following section.

Finally, we estimate the separated signals $y_i(n, f)$ with time-frequency masks $M_i(n, f)$, which extract time-frequency points of members in the i-th cluster:

$$y_i(n, f) = x_1(n, f)M_i(n, f). \tag{5}$$

3   Proposed Method

3.1   Problems in PDOA Clustering

The first problem for the PDOA clustering is the spatial aliasing problem. As can be seen in (4), when the frequency f or the microphone spacing d are large, $\arg[x_1(n, f)/x_2(n, f)] = 2\pi fd\cos\varphi(n, f)/v$ exceeds $\pm\pi$. However, since the arg operation has the indefiniteness of modulus $2\pi k$, (4) should be:

$$2\pi f\tau(n, f) = \arg\left[\frac{x_1(n, f)}{x_2(n, f)}\right] + 2\pi k = o(n, f) + 2\pi k, \tag{6}$$

where $o(n, f) = \arg[x_1(n, f)/x_2(n, f)]$, $-\pi \leq o(n, f) < \pi$, and k is an integer. Note that we can observe just $o(n, f)$, and k is unknown when the source direction φ is unknown. This is the spatial aliasing problem. Figure 1 gives an example of the observed PDOA $o(n, f)$ for a wide microphone spacing of 20 cm. In the next subsection, k in (6) is considered as a hidden variable.

The second problem occurs when we apply a GMM fitting method for an unknown number of mixtures. Figure 2 shows an example. Here we have two clusters in the histogram (Fig. 2(a)). Figure 2(b) shows the fitting result of GMM of eight Gaussians, to Fig. 2 (a). From Fig. 2(b), we can see that multiple Gaussians are fit to each cluster. However, we expect just one Gaussian for each peak, in order to estimate the number of sources by counting the number of dominant Gaussians. In this paper, in order to avoid the case where one cluster is modeled by two or more Gaussians, we propose utilizing a sparse distribution for the prior of the GMM mixture weight parameter in the next subsection.

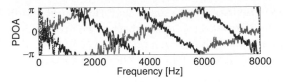

Fig. 1. Example PDOA for a microphone spacing of 20 cm and a sampling rate of 16 kHz. The PDOA for a source at 70° and a source at 150° are drawn individually for illustrative purposes.

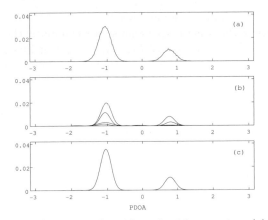

Fig. 2. Example GMM fitting result with and without prior. (a) Histogram of two Gaussians, (b) estimated Gaussians of GMM without prior ($\phi = 1.0$), (c) estimated Gaussians of GMM with prior ($\phi = 0.9$).

3.2   Probabilistic Model

To begin with, let us consider that we observe one source from one direction. Hereafter, a notation $o_{nf} = o(n, f)$ is utilized. Because the spatial aliasing issue in (6) can be considered as a phase wrapping problem, we can model the PDOA with a Gaussian distribution by considering the unwrapped data $o_{nf} + 2\pi k$. In other words, the phase wrapping process can be modeled by summing the Gaussians at intervals of 2π. That is, we assume that the PDOA follows a wrapped Gaussian distribution [8],

$$p(o_{nf}; \mu, \sigma) = \sum_{k=-K_f}^{K_f} p(o_{nf}, k; \mu, \sigma) = \sum_{k=-K_f}^{K_f} \frac{1}{\sqrt{2\pi\sigma^2}} \exp\left(\frac{-(o_{nf} + 2\pi k - 2\pi f \mu)^2}{2\sigma^2} \right),$$

$$(7)$$

where $-\pi \le o_{nf} < \pi$, μ gives us the expectation value of the TDOA τ of the source, o^2 is the variance of the PDOA, and k is an integer to handle the spatial aliasing (6). The value K_f is a frequency dependent integer, and it can be determined if we know the microphone spacing d and the frequency f. If we do not know d, we can set a sufficiently large value for K_f for all frequencies. This model is inspired by a wrapped Gaussian model [8].

In our observed mixture, we assume that there are a sufficient number of source signals from different directions, where some are dominant and others are much less dominant. Each source is modeled by (7). We also assume that the PDOA for an observed mixture follows a Gaussian mixture model (GMM):

$$p(o_{nf}; \mu_m, \sigma_m) = \sum_{m=1}^{M} \sum_{k=-K_f}^{K_f} \frac{\alpha_m}{\sqrt{2\pi\sigma_m^2}} \exp\left(\frac{-(o_{nf} + 2\pi k - 2\pi f \mu_m)^2}{2\sigma_m^2}\right). \tag{8}$$

We prepare a sufficient number M of Gaussians for our GMM model and estimate the mean μ_m, variance σ_m^2 and weight α_m for each Gaussian m.

In order to solve the second problem mentioned in Section 3.1, that is, in order to model the observed PDOA data by allocating one Gaussian to each source, we assume *the sparseness of the source directions*, where each direction is dominated by at most one source. For this purpose, as the prior of the mixture weight, we employ the Dirichlet distribution:

$$p(\alpha) = \frac{1}{B(\phi)} \prod_{m}^{M} \alpha_m^{\phi-1}, \tag{9}$$

where $\alpha = \{\alpha_1, \cdots, \alpha_m, \cdots, \alpha_M\}$, $\sum_m^M \alpha_m = 1$, $0 \leq \alpha_m \leq 1$, and $B(\phi)$ is the beta distribution (regularization term). When we set small hyper parameter ϕ ($\phi < 1$), the prior takes a larger value as the number of mixture weights whose values are close to zero increases, which is desirable for representing the sparseness of the source direction [5]. In addition, the Dirichlet distribution is known to be a conjugate prior of the mixture weight [5], and it can be incorporated into the GMM fitting approach in a computationally efficient manner.

Figure 2 (c) shows a GMM fitting result with prior ((9) with $\phi = 0.9$) for the distribution in Fig. 2 (a). In spite of utilizing eight Gaussians, we can see that just two Gaussians are dominant in Fig. 2 (c). That is, using the prior, more correct GMM fitting can be performed.

3.3   Cost Function Based on GMM

Let $\theta = \{\alpha_m, \mu_m, \sigma_m\}$ be a model parameter set. The observations are $o = \{o_{11}, o_{12}, \cdots, o_{nf}, \cdots, o_{NF}\}$ and power values $a = \{a_{11}, a_{12}, \cdots, a_{nf}, \cdots, a_{NF}\}$, where $a_{nf} = a(n, f) = |x_1(n, f)|^2$. In the following, Gaussian indices m and k in the PDOA model (8) are assumed not to be observed, and therefore dealt with as hidden variables.

The cost function of the maximum a posteriori (MAP) estimation is defined based on a log of a joint probability density function (pdf) as

$$\mathcal{L}(\theta) = \log p(o, \theta) = \log p(o|\theta) + \log p(\alpha) + \text{const.} \tag{10}$$

$$= \sum_{n}^{N} \sum_{f}^{F} f(a_{nf}) \log p(o_{nf}|\theta) + \log p(\alpha) + \text{const.} \tag{11}$$

$$= \sum_{n}^{N} \sum_{f}^{F} f(a_{nf}) \log \left(\sum_{m}^{M} \sum_{k=-K_f}^{K_f} p(m, k, o_{nf}|\theta)\right) + \log p(\alpha) + \text{const.}, \tag{12}$$

where

$$p(m, k, o_{nf}|\theta) = \frac{\alpha_m}{\sqrt{2\pi\sigma_m^2}} \exp\left(\frac{-(o_{nf} + 2\pi k - 2\pi f \mu_m)^2}{2\sigma_m^2}\right). \tag{13}$$

We disregarded the priors of the model parameters except for α in (10), and

$$f(a_{nf}) = ca_{nf}/\sum_n \sum_f a_{nf} \tag{14}$$

gives a power weight ($a_{nf} = |x_1(n, f)|^2$ in this paper) and controls the importance of the observation relative to the prior term (2nd term of (12)), where c is a control parameter.

In (12), the mixture weight α follows the Dirichlet distribution (9) and $\sum_m^M \alpha_m = 1$, $0 \leq \alpha_m \leq 1$ holds. For the sparse representation of the GMM, $\phi < 1$ is preferred for the Dirichlet distribution (9). Note that $\phi = 1$ is equivalent to the case without a prior for the mixture weight.

3.4   EM Algorithm

Here we derive an algorithm for estimating parameter θ by the EM algorithm.

The auxiliary function Q is given as

$$Q(\theta|\theta^t) = \mathrm{E}\left[\log p(o_{nf}; \theta)|o_{nf}; \theta^t\right] \tag{15}$$

$$= \sum_n \sum_f \sum_m \sum_k \left[p(m, k|o_{nf}, \theta^t)f(a_{nf}) \log p(m, k, o_{nf}|\theta)\right] + \log p(\alpha), \tag{16}$$

where θ^t is the estimate of the parameters after the t-th iteration, and

$$p(m, k|o_{nf}, \theta^t) = \frac{p(m, k, o_{nf}|\theta^t)}{\sum_m \sum_k p(m, k, o_{nf}|\theta^t)}. \tag{17}$$

By setting $\frac{\partial Q(\theta|\theta^t)}{\partial \mu_m} = 0$ and $\frac{\partial Q(\theta|\theta^t)}{\partial \sigma_m^2} = 0$, we obtain

$$\mu_m^{t+1} = \frac{\sum_n \sum_f \sum_k p(m, k|o_{nf}, \theta^t)f(a_{nf})(o_{nf} + 2\pi k)}{\sum_n \sum_f \sum_k 2\pi f p(m, k|o_{nf}, \theta^t)f(a_{nf})} \tag{18}$$

$$(\sigma_m^2)^{t+1} = \frac{\sum_n \sum_f \sum_k p(m, k|o_{nf}, \theta^t)f(a_{nf})(o_{nf} + 2\pi k - 2\pi f \mu_m)^2}{\sum_n \sum_f \sum_k p(m, k|o_{nf}, \theta^t)f(a_{nf})}. \tag{19}$$

Moreover, by using the Lagrange multiplier method, $\sum_m^M \alpha_m = 1$ and (14), the mixture weight is obtained as follows:

$$\alpha_m^{t+1} = \frac{1}{c + M(\phi - 1)} \left\{\sum_n \sum_f \sum_k p(m, k|o_{nf}, \theta^t)f(a_{nf}) + (\phi - 1)\right\}. \tag{20}$$

Since $\alpha_m > 0$, $c > M(1 - \phi)$ must hold from (20).

In the E-step we calculate (17), then in the M-step the parameters θ are calculated by using (18), (19) and (20). Sometimes $\alpha_m < 0$ occurs. In such a case, we can factor out the corresponding Gaussian (by setting $\alpha_m = \epsilon$, where ϵ is a very small value) and recalculate the parameters.

3.5   Source Counting

Thanks to the Dirichlet prior (9), most of the mixture weight parameter α_m becomes very close to zero and some have dominant values. Sometimes, some weight parameters α_m do not come to zero sufficiently because of very large variance σ_m. So, we can determine the number of sources N_s by counting the number of Gaussians whose parameters meet conditions $\alpha_m \geq \epsilon$ and $\sigma_m \leq th$, where ϵ is a sufficiently small threshold value and th is an appropriate threshold value. $\epsilon = 0.2$ and $th = \pi/3$ degrees are used in this paper.

3.6   Source Separation

The time-frequency mask $M_m(n, f)$ for the m-th separated source (see (5)) is obtained by marginalizing the estimated pdf (17) with respect to k,

$$M_m(n, f) = p(m|o_{nf}, \theta) = \sum_{k=-K_f}^{K_f} p(m, k|o_{nf}, \theta). \tag{21}$$

The separated signal is obtained by

$$y_m(n, f) = x_1(n, f)M_m(n, f) = x_1(n, f)p(m|o_{nf}, \theta). \tag{22}$$

4   Experiments

4.1   Experimental Setup

We performed experiments with measured impulse responses h_{ji} in a room whose reverberation time was 130 ms (see Fig. 9's setup A of [4]). We utilized two microphones whose spacing was 20 cm. The numbers of sources N_s were two and three. Mixtures were made by convolving the measured room impulse responses and 5-second English speech signals sampled at 16 kHz. The frame size F for STFT was 1024 (64 ms), and the frame shift was 256 (16 ms).

In the EM algorithm, we utilized $M = 8$ Gaussians. From the microphone spacing and sampling rate, the aliasing problem occured above 850 Hz. In our implementation, $K_f = K = 5$ was utilized for all frequencies f. For the comparison with an aliasing-unconsidered case, $K_f = K = 0$ for all frequencies f was also tested. As the hyper parameter for (9), we utilized $\phi = 0.9$ for our proposed method and $\phi = 1.0$ for a conventional EM algorithm that corresponds to the case without any prior for the mixture weights. The number of iterations was 10, and the control parameter c for (14) was 5.

We evaluated the signal-to-interference ratio (SIR) as a separation performance measure, and the signal-to-distortion ratio (SDR) as a sound quality measure. Their definitions can be found in [3]. We calculated SIR and SDR values for the separated sources that are counted as the sources by the method in Section 3.5. We conducted 20 trials with different speech source combinations and location combinations, and then averaged the results.

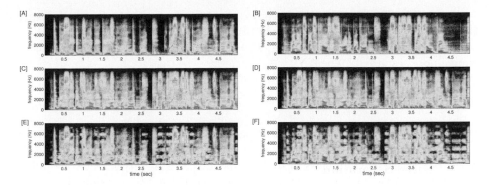

Fig. 3. Example spectra of (A)(B) sources, (C)(D) observations, (E)(F) separated signals. $N_s = 2$, $\phi = 0.9$ and $K = 5$.

Table 1. Experimental results. Input SIR was 0.0 [dB] ($N_s = 2$), and -3.1 [dB] ($N_s = 3$).

N_s	ϕ	K	$\hat{N}_s$:1	2	3	4	5	6	7	8	Output SIR	SDR
2	0.9	5		100							11.5	12.4
	0.9	0		35	60	5					11.4	7.5
	1.0	5		0	5	40	35	20			8.8	9.7
3	0.9	5		15	75	10					7.5	8.2
	0.9	0	5	25	50	20					2.0	8.7
	1.0	5			0	90	10				6.9	7.5

The header spans: "Accuracy of $\hat{N}_s$ estimation [%]" over columns $\hat{N}_s$:1 through 8, and "Performance [dB]" over Output SIR and SDR.

4.2 Results

Figure 3 shows the example spectra of sources, observations, and separated sources for a two source case. The source directions were 70° and 150°, whose example PDOA is shown in Fig. 1. By comparing the source spectra Fig. 3 (A)(B) and the separation spectra Fig. 3 (E)(F), it can be seen that the spatial aliasing problem does not occur in most frequencies. However, it is also seen that at the frequencies where the PDOA of two sources lap over each other, say around 1500, 3000, 4500, 6000, 7500 Hz (see Fig. 1), the signals are not separated well. Such phenomena can be seen in the separated spectra Fig. 3 (E)(F).

Table 1 reports the experimental results. In the table, $\phi = 0.9$ means the results with sparse prior (9) and $\phi = 1.0$ indicates the results without a prior. $K = 5$ and $K = 0$ mean the spatial aliasing is considered and unconsidered, respectively. The percentage values are shown where the method estimates the number of sources as $\hat{N}_s$ within 20 trials. The average separation performance results, SIR and SDR in dB, are also reported.

From Table 1, we can see that with the prior ($\phi = 0.9$) by considering the aliasing ($K = 5$), the number of sources is almost perfectly estimated. On the other hand, without the prior, the number of sources is overestimated, and the accuracy rate was quite low.

As for the separation performance, we obtained better performance by using the prior ($\phi = 0.9$) than without the prior ($\phi = 1.0$) when $K = 5$. When we did not consider the spatial aliasing, $K = 0$, the separation performance was of course poor, especially when $N_s = 3$.

5 Conclusion

We proposed a speech source separation method that can estimate both the number of sources and separation masks. We model the PDOA with a GMM, where the phase indefiniteness in spatial aliasing cases is considered. We employ the Dirichlet distribution as the prior of the GMM mixture weight to model each cluster by a single Gaussian. Our experimental results show that the proposed method can estimate the number of sources correctly. We also confirmed that the proposed method gives good separation performance in a room with reverberation time of 130 ms.

References

1. Hyvärinen, A., Karhunen, J., Oja, E.: Independent Component Analysis. John Wiley & Sons, Chichester (2001)
2. Yilmaz, O., Rickard, S.: Blind separation of speech mixtures via time-frequency masking. IEEE Trans. Signal Processing 52(7), 1830–1847 (2004)
3. Araki, S., Sawada, H., Mukai, R., Makino, S.: Underdetermined blind sparse source separation for arbitrarily arranged multiple sensors. Signal Processing 77(8), 1833–1847 (2007)
4. Sawada, H., Araki, S., Mukai, R., Makino, S.: Grouping separated frequency components by estimating propagation model parameters in frequency-domain blind source separation. IEEE Trans. Audio, Speech and Language Processing 15(5), 1592–1604 (2007)
5. Bishop, C.M.: Pattern recognition and machine learning. Springer, Heidelberg (2008)
6. Mandel, M., Ellis, D., Jebara, T.: An EM algorithm for localizing multiple sound sources in reverberant environments. In: Proc. Neural Info. Proc. Sys. (2006)
7. O'Grady, P., Pearlmutter, B.: Soft-LOST: EM on a mixture of oriented lines. In: Puntonet, C.G., Prieto, A.G. (eds.) ICA 2004. LNCS, vol. 3195, pp. 430–436. Springer, Heidelberg (2004)
8. Smaragdis, P., Boufounos, P.: Learning source trajectories using wrapped-phase hidden markov models. In: Proc. of WASPAA 2005, pp. 114–117 (2005)

Blind Spectral-GMM Estimation for Underdetermined Instantaneous Audio Source Separation

Simon Arberet[1], Alexey Ozerov[2,*], Rémi Gribonval[1], and Frédéric Bimbot[1]

[1] METISS Group, IRISA-INRIA
Campus de Beaulieu, 35042 Rennes Cedex, France
{simon.arberet,remi.gribonval,frederic.bimbot}@irisa.fr
[2] LTCI (TELECOM ParisTech & CNRS) - TSI Department
37-39 rue Dareau, 75014 Paris, France
alexey.ozerov@telecom-paristech.fr

Abstract. The underdetermined blind audio source separation problem is often addressed in the time-frequency domain by assuming that each time-frequency point is an independently distributed random variable. Other approaches which are not blind assume a more structured model, like the Spectral Gaussian Mixture Models (Spectral-GMMs), thus exploiting statistical diversity of audio sources in the separation process. However, in this last approach, Spectral-GMMs are supposed to be learned from some training signals. In this paper, we propose a new approach for learning Spectral-GMMs of the sources without the need of using training signals. The proposed blind method significantly outperforms state-of-the-art approaches on stereophonic instantaneous music mixtures.

1 Introduction

The problem of underdetermined Blind Source Separation (BSS) is to recover single-channel source signals $s_n(\tau)$, $1 \leq n \leq N$, from a multichannel mixture signal $x_m(\tau)$, $1 \leq m \leq M$, with $M < N$. Taking the Short Time Fourier Transform (STFT) $X_m(t, f)$ of each channel $x_m(t)$ of the mixture, the instantaneous mixing process is modeled in the time-frequency domain as:

$$\mathbf{X}(t, f) = \mathbf{A}\mathbf{S}(t, f) \tag{1}$$

where $\mathbf{X}(t, f)$ and $\mathbf{S}(t, f)$ denote respectively the column vectors $[X_m(t, f)]_{m=1}^{M}$ and $[S_n(t, f)]_{n=1}^{N}$, and $\mathbf{A}$ is the $M \times N$ real-valued mixing matrix.

The underdetermined BSS problem is often addressed in a two step approach where: first the mixing matrix is estimated, and then the source coefficients are estimated with the Maximum A Posteriori (MAP) criterion given a sparse source prior and the mixing matrix. Sources are then recovered using the inverse

---

* A part of this work was done while A. Ozerov was with KTH, Stockholm, Sweden.

T. Adali et al. (Eds.): ICA 2009, LNCS 5441, pp. 751–758, 2009.
© Springer-Verlag Berlin Heidelberg 2009

STFT. In the audio domain, sparse prior distributions are usually a Laplacian [1], a generalized Gaussian [2], a Student-t [3], or a mixture of two Gaussians [4]. This approach however suffers from the following issues:

1. In each time-frequency point, the maximum number of nonzero sources is usually assumed to be limited to the number M of channels [1,2].
2. The assumed nonzero sources are always the M neighboring directions which points toward the direction of the observed mixture [5].
3. Each time-frequency coefficient is estimated independently of the others without taking into account the structure of each source in the time-frequency domain. In other words, the signal redundancy and structure are not fully exploited.

In this paper we assume that $\mathbf{A}$ is known or has already been estimated [6], and the columns are pairwise linearly independent. Issues two and three have been addressed by the Statistically Sparse Decomposition Principle (SSDP)[5], which exploit the correlation between the mixture channels and more recently, the three issues have been addressed by the Local Gaussian Modeling (LGM) [7], where time-frequency coefficients are modeled via Gaussian priors with free variances. The third issue has been indeed partially addressed by SSDP and LGM, which exploit the neighborhood of the time-frequency points, in order to estimate the source distribution of the coefficients.

A more globally structured approach (to address these three issues) consists in assuming a spectral model of the sources via Spectral Gaussian Mixture Models (Spectral-GMMs) [8,9]. This approach has been successfully used to separate sources in the monophonic case ($M = 1$) [8,9], when sparse methods are unsuitable. However this approach is not blind because the models need to be learned from some training sources which should have characteristics similar to those of the sources to be separated. An EM algorithm could be used to learn GMMs directly from the mixture [10,11], but this approach suffers from two big issues. First, the number of Gaussians in the observation density grows exponentially with the number of sources, which often leads to an intractable algorithm. Second, the algorithm can be very slow and converges to a local maximum depending on the initial values.

In this paper, we propose a framework to blindly learn Spectral-GMMs with a linear complexity, provided that we have for each n-th source and for each time-frequency point (t, f) the following two estimates:

1. an estimate $\tilde{S}_n(t, f)$ of the source coefficient $S_n(t, f)$;
2. an estimate $\tilde{\sigma}^2_{n,t}(f)$ of the coefficient estimation squared error:

$$e^2_{n,t}(f) \triangleq \left| \tilde{S}_n(t, f) - S_n(t, f) \right|^2. \tag{2}$$

The paper is organized as follows. In section 2, we describe the Spectral-GMM source estimation method assuming known models. In section 3, we recall the LGM source estimation method and show that, with this approach, we can also provide the two above-mentioned estimates required by the proposed framework.

In section 4, we describe our approach to blindly estimate Spectral-GMMs of the sources, with a linear complexity EM algorithm. Finally, we evaluate the performances of our approach on musical data in section 5.

2   Spectral-GMM Source Estimation

In this section, we briefly describe the principles of the Spectral-GMM source estimation methods [9], that we extend to the multichannel case. The short time Fourier spectrum $S(t) = [S(t, f)]_f^1$ of source S at time t is modeled by a multidimensional zero-mean complex valued $\mathcal{K}$-states GMM with probability density function (pdf) given by:

$$P(S(t)|\lambda) = \sum_{k=1}^{\mathcal{K}} \pi_k \, N_c(S(t); \bar{0}, \Sigma_k) \tag{3}$$

where $N_c(S(t); \bar{0}, \Sigma_k) \triangleq \prod_f \frac{1}{\pi \sigma_k^2(f)} \exp\left[-\frac{|S(t,f)|^2}{\sigma_k^2(f)}\right]$, $\lambda \triangleq \{\pi_k, \Sigma_k\}_k$ is a Spectral-GMM of source S, π_k being a weight of Gaussian k of GMM λ, and $\Sigma_k \triangleq \mathrm{diag}([\sigma_k^2(f)]_f)$ is a diagonal covariance matrix of Gaussian k of GMM λ.

Provided that we know the Spectral-GMMs $\boldsymbol{\Lambda} = [\lambda_n]_{n=1}^N$ of the sources, the separation is performed in the STFT domain with the Minimum Mean Square Error (MMSE) estimator, which can be viewed as a form of adaptive Wiener filtering:

$$\widehat{\mathbf{S}}(t, f) = \sum_{\mathbf{k}} \gamma_{\mathbf{k}}(t) \mathbf{W}_{\mathbf{k}}(f) \mathbf{X}(t, f) \tag{4}$$

where $\mathbf{k} \triangleq [k_n]_{n=1}^N$, and $\gamma_{\mathbf{k}}(t)$ is the state probability at frame t ($\sum_{\mathbf{k}} \gamma_{\mathbf{k}}(t) = 1$):

$$\gamma_{\mathbf{k}}(t) \triangleq P(\mathbf{k}|\mathbf{X}(t); \mathbf{A}, \boldsymbol{\Lambda}) \propto \pi_{\mathbf{k}} \prod_f N_c\left(\mathbf{X}(t, f); \bar{0}, \mathbf{A}\Sigma_{\mathbf{k}}(f)\mathbf{A}^T\right) \tag{5}$$

with $\mathbf{X}(t) \triangleq [\mathbf{X}(t, f)]_f$, $\Sigma_{\mathbf{k}}(f) \triangleq \mathrm{diag}([\sigma_{n,k_n}^2(f)]_{n=1}^N)$, and the Wiener filter is given by:

$$\mathbf{W}_{\mathbf{k}}(f) \triangleq \Sigma_{\mathbf{k}}(f)\mathbf{A}^T \left(\mathbf{A}\,\Sigma_{\mathbf{k}}(f)\mathbf{A}^T\right)^{-1} \tag{6}$$

Thus, at each frame t, the source estimation is done in two steps:

1. *decoding step*, where the state probabilities $\gamma_{\mathbf{k}}(t)$ are calculated with equation (5);
2. *filtering step*, where the source coefficients are estimated by the weighted Wiener filtering of equation (4).

In such an approach the models λ_n are usually learned separately [8] by maximization of the likelihoods $P(\bar{S}_n|\lambda_n)$, where $\bar{S}_n$ is the STFT of the training signal for source s_n. This maximization is achieved via the Expectation Maximization (EM) algorithm [12] initialized by some clustering algorithm (e.g., K-means). As we will see in section 5, the performances of this method can be very good. However, it suffers from two big issues:

---

[1] The notation $[S(t, f)]_f$ means a column vector composed of elements $S(t, f), \forall f$.

- The approach requires availability of training signals, that are difficult to obtain in most realistic situations [9].
- As the mixture state $\mathbf{k}$ is a combination of all the source states $k_n, 1 \leq n \leq N$, the decoding step of equation (5) is of complexity $O(\mathcal{K}^N)$.

3   LGM Source Estimation

The LGM [7] is a method which consists in estimating local source variances $\mathbf{v}(t, f) = [\sigma_n^2(t, f)]_{n=1}^N$ in each time-frequency point and then estimating the source coefficients with the MMSE estimator given by the Wiener filter:

$$\tilde{\mathbf{S}}(t, f) = \mathbf{W}(t, f)\mathbf{X}(t, f) \tag{7}$$

where $\mathbf{W}(t, f) = \widehat{\boldsymbol{\Sigma}}(t, f)\mathbf{A}^T \left(\mathbf{A}\widehat{\boldsymbol{\Sigma}}(t, f)\mathbf{A}^T\right)^{-1}$, and $\widehat{\boldsymbol{\Sigma}}(t, f)$ is a diagonal matrix whose entries are the estimated source variances: $\widehat{\boldsymbol{\Sigma}}(t, f) = \text{diag}\left(\widehat{\mathbf{v}}(t, f)\right)$.

The LGM is based on the empirical local covariance matrix in the time-frequency domain, which has already been used by mixing matrix estimation methods [6,13] so as to select time-frequency regions where only one source is supposed active, and which is defined by:

$$\widehat{\mathbf{R}}_x(t, f) = \sum_{t', f'} w(t - t', f - f')\mathbf{X}(t', f')\mathbf{X}^H(t', f') \tag{8}$$

where w is a bi-dimensional normalized window function which defines the neighborhood shape, and H denotes the conjugate transpose of a matrix. If we assume that each complex source coefficient $S_n(t, f)$ in a time-frequency neighborhood follows an independent (over time t and frequency f) zero-mean Gaussian distribution with variance $\sigma_n^2(t, f)$, then the mixture coefficients in that neighborhood follow a zero-mean Gaussian distribution with covariance matrix:

$$\mathbf{R}_x(t, f) = \mathbf{A}\boldsymbol{\Sigma}(t, f)\mathbf{A}^T. \tag{9}$$

The Maximum Likelihood (ML) estimate of source variances σ_n^2 (we drop the (t, f) index for simplicity) is obtained by minimization of the Kullback-Leibler (KL) divergence between the empirical and the mixture model covariances [14]:

$$\widehat{\boldsymbol{\Sigma}} = \underset{\boldsymbol{\Sigma}=\text{diag}(\mathbf{v}),\mathbf{v}\geq 0}{\arg\min} \quad KL(\widehat{\mathbf{R}}_x|\mathbf{R}_x), \tag{10}$$

where $KL(\widehat{\mathbf{R}}_x|\mathbf{R}_x)$ is defined as:

$$KL(\widehat{\mathbf{R}}_x|\mathbf{R}_x) = \frac{1}{2}\left(\text{tr}\left(\widehat{\mathbf{R}}_x\mathbf{R}_x^{-1}\right) - \log\det\left(\widehat{\mathbf{R}}_x\mathbf{R}_x^{-1}\right) - M\right). \tag{11}$$

The LGM method [7] uses a global non-iterative optimization algorithm to solve the problem of equation (10), which roughly consists of estimating variances $\mathbf{v}$ by solving the linear system $\widehat{\mathbf{R}}_x \approx \mathbf{R}_x$ with a sparsity assumption on the variance vector $\mathbf{v}$.

The MMSE estimate of $(\tilde{\mathbf{S}} - \mathbf{S})(\tilde{\mathbf{S}} - \mathbf{S})^H$ is given by the covariance matrix of $\mathbf{S}$ given $\mathbf{X}$:

$$\mathbf{C} \triangleq \mathbb{E}\left[(\tilde{\mathbf{S}} - \mathbf{S})(\tilde{\mathbf{S}} - \mathbf{S})^H \Big| \mathbf{X} \right] = (\mathbf{I} - \mathbf{W}\mathbf{A})\hat{\boldsymbol{\Sigma}}, \tag{12}$$

where $\mathbf{W}$ is defined just after equation (7).

So the MMSE estimate of $e_{n,t}^2(f)$ defined by equation (2) is given by the corresponding diagonal element of matrix $\mathbf{C}(t, f)$:

$$\tilde{\sigma}_{n,t}^2(f) = \mathbb{E}\left[\left| \tilde{S}_n(t, f) - S_n(t, f) \right|^2 \Big| \mathbf{X} \right] = \mathbf{C}(t, f)_{n,n}. \tag{13}$$

4   Spectral-GMM Blind Learning Framework

The aim of the proposed framework is to learn the Spectral-GMM λ_n for each source s_n, provided that at each time-frequency point (t, f), we have an estimate $\tilde{S}_n(t, f)$ of the source coefficient $S_n(t, f)$ together with an estimate $\tilde{\sigma}_{n,t}^2(f)$ of the squared error $e_{n,t}^2(f)$ defined by equation (2).

The learning step is done for each source independently, so in the following we drop the source's index n for simplicity. Let us denote the error of source estimation as $\tilde{E}(t, f) \triangleq \tilde{S}(t, f) - S(t, f)$. Now we assume that $\tilde{E}(t) = [\tilde{E}(t, f)]_f$ is a realization of a Gaussian complex vector with zero mean and a diagonal covariance matrix $\tilde{\Sigma}_t = \mathrm{diag}([\tilde{\sigma}_t^2(f)]_f)$, i.e., $P(\tilde{E}(t)|\tilde{\Sigma}_t) = N_c(\tilde{E}(t); \bar{0}, \tilde{\Sigma}_t)$. The relation:

$$\tilde{S}(t, f) = S(t, f) + \tilde{E}(t, f) \tag{14}$$

can be interpreted as a single sensor source separation problem with mixture $\tilde{S}$ and sources S and $\tilde{E}$, where source $\tilde{E}$ is modeled by $\hat{\boldsymbol{\Sigma}} = [\tilde{\Sigma}_t]_{t=1}^T$, which is fixed, and source S is modeled by GMM $\lambda = \{\pi_k, \Sigma_k\}_{k=1}^K$ that we want to estimate in the ML sense, given the observed mixture $\tilde{S}$ and fixed model $\hat{\boldsymbol{\Sigma}}$. Thus, we are looking for λ optimizing the following ML criterion:

$$\lambda = \arg\max_{\lambda'} p(\tilde{S}|\lambda', \hat{\boldsymbol{\Sigma}}). \tag{15}$$

Algorithm 1 summarizes an Expectation-Maximization (EM) algorithm for optimization of criterion (15) (see [9] for derivation). Initialization is done by applying K-means clustering algorithm to the source estimate $\tilde{S}$.

Once we have learned the source models $\boldsymbol{\Lambda} = [\lambda_n]_{n=1}^N$, we could estimate the sources with the procedure of section 2, but in that case the decoding step at each frame t will still be of complexity $O(\mathcal{K}^N)$. In order to have a linear complexity method, we do not calculate all the $\mathcal{K}^N$ mixture state probabilities, but only the $\mathcal{K}$ state probabilities of each source using $\gamma_{k_n}^{(L+1)}(t)$, where $\gamma_{k_n}^{(L)}(t)$ (defined by equation (16) in algorithm 1) are the state probabilities of source S_n calculated during the last iteration of algorithm 1. The source coefficients are then estimated with the Wiener filter $\mathbf{W}_{\mathbf{k}^*(t)}(f)$ of equation (6), with $\mathbf{k}^*(t) = [k_n^*(t)]_{n=1}^N$ and where $k_n^*(t) = \arg\max_k P(q(t) = k|\tilde{S}_n, \lambda_n^{(L+1)}, \hat{\Sigma}_n)$ is the most likely state of model $\lambda_n^{(L+1)}$ at frame t, given $\tilde{S}_n(t, f)$ and $\hat{\Sigma}_n$.

Algorithm 1. EM Algorithm for source Spectral-GMM estimation in ML sense (index (l) in power denotes the parameters estimated at the lth iteration of the algorithm)

1. Compute the weights $\gamma_k^{(l)}(t)$ satisfying $\sum_k \gamma_k^{(l)}(t) = 1$ and

$$\gamma_k^{(l)}(t) \triangleq P(q(t) = k|\tilde{S}, \lambda^{(l)}, \tilde{\Sigma}) \propto \pi_k^{(l)} N_c(\tilde{S}(t); \bar{0}, \Sigma_k^{(l)} + \tilde{\Sigma}_t) \qquad (16)$$

   where $q(t)$ is the current state of GMM λ at frame t.
2. Compute the expected Power Spectral Density (PSD) for state k

$$\langle |S(t,f)|^2 \rangle_k^{(l)} \triangleq \mathbb{E}_S \left[|S(t,f)|^2 \,\middle|\, q(t) = k, \tilde{S}, \lambda^{(l)}, \tilde{\Sigma} \right] =$$

$$= \frac{\sigma_k^{2,(l)}(f)\tilde{\sigma}_t^2(f)}{\sigma_k^{2,(l)}(f) + \tilde{\sigma}_t^2(f)} + \left| \frac{\sigma_k^{2,(l)}(f)}{\sigma_k^{2,(l)}(f) + \tilde{\sigma}_t^2(f)} \tilde{S}(t,f) \right|^2 \qquad (17)$$

3. Re-estimate Gaussian weights

$$\pi_k^{(l+1)} = \frac{1}{T} \sum_t \gamma_k^{(l)}(t) \qquad (18)$$

4. Re-estimate covariance matrices

$$\sigma_k^{2,(l+1)}(f) = \frac{\sum_t \langle |S(t,f)|^2 \rangle_k^{(l)} \gamma_k^{(l)}(t)}{\sum_t \gamma_k^{(l)}(t)} \qquad (19)$$

5   Experimental Results

We evaluate our method[2] over music mixtures, with the number of sources N varying from 3 to 6. For each N a mixing matrix was computed as described in [15], given an angle of $50 - 5N$ degrees between successive sources, and ten instantaneous mixtures were generated from different source signals of duration 10 s, sampled at 22.05 kHz. The STFT was computed with a sine window of length 2048 (93 ms). The performance measure used is the Signal-to-Distortion Ratio (SDR) defined in [16]. The bi-dimensional window w defining time-frequency neighborhoods of the LGM method was the outer product of two Hanning windows with length 3 as in [7]. The Spectral-GMMs were learned with 30 iterations of algorithm 1 using LGM parameters given by equations (7) and (13) as entries, and the number $\mathcal{K}$ of states per GMM was chosen equal to 8, because it yielded the best results in SDR. Figure 1 compares the average SDR achieved by the proposed Spectral-GMM method, the LGM method presented in Section 3 and the classical DUET [17]. The proposed algorithm outperforms DUET by more than 5 dB in the 3 sources case and LGM by at least 2 dB whatever the number of sources. We also plotted the performance of the (oracle) Spectral-GMM separation when models, with the same number $\mathcal{K}$ of states, are learned and

---

[2] This method was also submitted to the 2008 Signal Separation Evaluation Campaign.

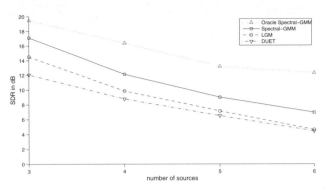

Fig. 1. Source separation performance over stereo instantaneous musical mixtures. STFT window length is 2048 (93 ms) and $\mathcal{K} = 8$.

decoded using the original sources. We can notice that the performance of the proposed method remains between 2 dB and 5 dB below the oracle performance, and that this gap increases with the number of sources, showing the difficulty to blindly learn Spectral-GMM when the number of sources is high. As for the computational load, the MATLAB implementation of the proposed algorithm on a 3.4 GHz CPU runs in 133 s in the 4 sources case, while it runs in 120 s for LGM and in 2 s for DUET.

6   Conclusion

In this paper, we proposed a new framework for the blind audio source separation problem in the multichannel instantaneous mixture case. In this framework Spectral-GMM models of sources were blindly learned, i.e. without using any other informations than the mixture and the mixing matrix, with an EM algorithm having a linear $O(N\mathcal{K})$ complexity, in contrast to some related state-of-the-art methods having an exponential $O(\mathcal{K}^N)$ complexity. As opposed to the other blind audio source separation methods, the proposed method exploits the structure of each source in the time-frequency domain. The proposed method outperforms the state-of-the-art methods tested by between 2 dB and 5 dB in SDR. Further work include an extension of the method to the anechoic and convolutive cases, evaluation of the robustness of the method by using mixing matrices which are not perfectly estimated, and improvement of the method to fill the gap between the blindly learned models and the oracle ones.

References

1. Zibulevsky, M., Pearlmutter, B.A., Bofill, P., Kisilev, P.: Blind source separation by sparse decomposition in a signal dictionary. In: Independent Component Analysis: Principles and Practice, pp. 181–208. Cambridge Press (2001)
2. Vincent, E.: Complex nonconvex l_p norm minimization for underdetermined source separation. In: Davies, M.E., James, C.J., Abdallah, S.A., Plumbley, M.D. (eds.) ICA 2007. LNCS, vol. 4666, pp. 430–437. Springer, Heidelberg (2007)

3. Fevotte, C., Godsill, S.: A bayesian approach for blind separation of sparse sources. IEEE Transactions on Audio, Speech, and Language Processing 14(6), 2174–2188 (2006)
4. Davies, M.E., Mitianoudis, N.: Simple mixture model for sparse overcomplete ICA. IEE Proceedings on Vision, Image and Signal Processing 151(1), 35–43 (2004)
5. Xiao, M., Xie, S., Fu, Y.: A statistically sparse decomposition principle for underdetermined blind source separation. In: Proc. Int. Symp. on Intelligent Signal Processing and Communication Systems (ISPACS), pp. 165–168 (2005)
6. Arberet, S., Gribonval, R., Bimbot, F.: A robust method to count and locate audio sources in a stereophonic linear instantaneous mixture. In: Rosca, J.P., Erdogmus, D., Príncipe, J.C., Haykin, S. (eds.) ICA 2006. LNCS, vol. 3889, pp. 536–543. Springer, Heidelberg (2006)
7. Vincent, E., Arberet, S., Gribonval, R.: Underdetermined audio source separation via gaussian local modeling. In: Proc. Int. Conf. on Independent Component Analysis and Blind Source Separation (ICA) (2009)
8. Benaroya, L., Bimbot, F.: Wiener based source separation with HMM/GMM using a single sensor. In: Proc. ICA, pp. 957–961 (2003)
9. Ozerov, A., Philippe, P., Bimbot, F., Gribonval, R.: Adaptation of bayesian models for single-channel source separation and its application to voice/music separation in popular songs. IEEE Transactions on Audio, Speech and Language Processing 15(5), 1564–1578 (2007); see also: IEEE Transactions on Speech and Audio Processing
10. Moulines, E., Cardoso, J.F., Gassiat, E.: Maximum likelihood for blind separation and deconvolution of noisy signals using mixture models. In: IEEE International Conference on Acoustics, Speech, and Signal Processing, vol. 5, pp. 3617–3620 (April 1997)
11. Attias, H.: Independent factor analysis. Neural Comput. 11(4), 803–851 (1999)
12. Dempster, A.P., Laird, N.M., Rubin, D.B.: Maximum likelihood from incomplete data via the em algorithm
13. Arberet, S., Gribonval, R., Bimbot, F.: A robust method to count and locate audio sources in a stereophonic linear anechoic mixture. In: ICASSP 2007, vol. 3, pp. 745–748 (April 2007)
14. Pham, D.T., Cardoso, J.F.: Blind separation of instantaneous mixtures of non stationary sources. IEEE Trans. on Signal Processing 49(9), 1837–1848 (2001)
15. Pulkki, V., Karjalainen, M.: Localization of amplitude-panned virtual sources I: stereophonic panning. Journal of the Audio Engineering Society 49(9), 739–752 (2001)
16. Vincent, E., Gribonval, R., Févotte, C.: Performance measurement in blind audio source separation. IEEE Trans. on Audio, Speech and Language Processing 14(4), 1462–1469 (2006)
17. Yılmaz, O., Rickard, S.T.: Blind separation of speech mixtures via time-frequency masking. IEEE Trans. on Signal Processing 52(7), 1830–1847 (2004)

Application of Two Time-Domain Convolutive Blind Source Separation Algorithms to the 2008 Signal Separation Evaluation Campaign (SiSEC) Data

Scott C. Douglas

Department of Electrical Engineering
Southern Methodist University
Dallas, Texas 75275 USA
douglas@lyle.smu.edu

Abstract. This paper describes the application of two time-domain convolutive blind source separation algorithms – the scaled natural gradient algorithm [1] and the spatio-temporal FastICA algorithm with symmetric orthogonality constraints [2] – to a portion of the determined and overdetermined acoustic data sets created for the 2008 Signal Separation Evaluation Campaign (SiSEC). As the 2008 SiSEC competition provides no ground truth data and thus no *a priori* method for numerical performance calculation, our approach to determining overall performance is a *decoding* of the contents of the recorded sources used to create the data through the two algorithms used. Information about the sources themselves, such as the instrumentation and structure of the musical selections chosen, the qualities of the voices and written transcripts of what is spoken, and additional information about the signals extracted, are provided without our ever having heard the sources in isolation. A qualitative performance comparison of the two approaches is also provided.

Keywords: Convolutive blind source separation, natural gradient, FastICA, speech separation.

1 Introduction

Convolutive blind source separation (BSS) is an interesting and promising approach for enhancing audio signals that have been linearly-mixed by physical processes such as acoustic propagation. When the number of sources being mixed is no greater than the number of sensors used to collect the individual mixtures, the separation problem is linear, such that multichannel filters can be used to extract estimates of each of the source signals in the mixtures. The convolutive BSS problem thus reduces to a multichannel adaptive filtering task, in which the coefficients of the adaptive filter are adapted according to the statistics of the extracted output signals. Numerous algorithms have been developed for the convolutive BSS task for these types of mixtures, such that it is nearly impossible to list all methods or types in a conference paper let alone carry out a fair comparison between competing algorithms. As a result, it can be extremely challenging to judge the performance of any one convolutive BSS

T. Adali et al. (Eds.): ICA 2009, LNCS 5441, pp. 759–766, 2009.

algorithm published in the scientific literature, particularly since the data used to illustrate algorithm performance are often collected by the developers of the algorithms being tested. Such situations can lead to a positive bias in the reported results, even if such a bias is unintended.

The 2008 Signal Separation Evaulation Campaign (SiSEC) attempts to overcome the limitations of past author-driven evaluations of convolutive BSS algorithms by establishing several candidate data sets for the scientific community to explore. For details on the SiSEC competition, the reader is encouraged to visit the SiSEC website [3]. The unique feature of the SiSEC campaign that makes it interesting to the BSS community is simple: *The source data is initially made unavailable.* Thus, although the mixtures and any separated results provided by the larger BSS community can be well-characterized by the Campaign developers, algorithm designers in the BSS community cannot rely on source data to tune their algorithms for best performance. This "blind" approach to algorithm evaluation means that members of the BSS community must find good BSS approaches that work independently of the data sets on which they might be applied. Unfortunately, the approach also provides little quantitative data to help algorithm developers describe the results of their separation methods outside of simply making the audio files available for listening by others,

In this paper, we report on the results of applying two different algorithms – the scaled natural gradient algorithm [1] and the spatio-temporal FastICA algorithm with symmetric orthogonality constraints [2] – to a portion of the determined and overdetermined acoustic data sets created for the 2008 Signal Separation Evaluation Campaign (SiSEC). To judge overall performance, we provide a careful description of the contents of the data sets as illuminated by the separation algorithms, which is the only quantitative data one can provide outside of the separated audio signals and the separation system coefficients themselves in this situation. We also describe qualitative differences between the two approaches as applied to these data sets and point out specific features of the algorithms that allow them to function properly in the tested environments.

2 Algorithm Settings

We now describe the algorithm settings used in the evaluation process. Because of the "blind" nature of the evaluation process, we decided to not attempt to carefully tune the algorithms chosen for testing to each of the data sets. Instead, we chose specific parameter values for each of the two algorithms under test and kept these parameters the same for every data set. Since every data set used for testing was a four-microphone recording, we further chose *not to tune* the algorithms to the number of sources being extracted. Thus, each algorithm attempted to extract four sources from every data set, even in situations where there were only two or three underlying sources contained within the sensor data. Such a scenario is practical and realistic, as it is unlikely that one would know *a priori* how many sound sources are in an acoustic environment.

The scaled natural gradient algorithm in [1] is a modified version of the truncated natural gradient algorithm in [3], in which a clever step size strategy is employed to allow scale-invariant adaptation of the algorithm parameters. The step size scaling used in the algorithm is based on the sums of the absolute values of the gradient entries in the block-based algorithm. Further details about the algorithm can be found in [2]. For this algorithm, the following settings were used:

- The signals were prewhitened using two stages of multichannel least-squares linear prediction in a fashion identical to the decorrelation processing used in [2]. The lengths of the linear predictors in each stage was $M = 400$ taps.
- The scaled natural gradient algorithms was then applied with the following parameters: $L = 512$ filter taps, mu = 0.45 as the step size, $f(y) = $ sgn(y) as the algorithm nonlinearity, and "center-spike" initialization, in which the only non-zero taps are in the center-lag of the FIR filter at tap position $L/2$. A One hundred iterations of the algorithm were used in each case.

The spatio-temporal FastICA algorithm in [2] is a temporal extension of the well-known FastICA algorithm of Hyvarinen and Oja, in which spatio temporal prewhitening using multichannel least-squares is first performed. The algorithm employs a unique iterative method for maintaining paraunitary constraints on the associated multichannel adaptive filter coefficients of the separation system. Two version of the algorithm are provided in [2]; one with asymmetric deflationary-type constraints, and one with symmetric paraunitary constraints. For this algorithm, the following settings were used:

- The signals were prewhitened using two stages of multichannel least-squares linear prediction. The lengths of the linear predictors in each stage was $M = 400$ taps.
- The spatio-temporal FastICA iteration with symmetric paraunitary constraints was then applied with the following parameters: $L = 300$ filter taps, $f(y) = \tanh(20y)$ as the algorithm nonlinearity, $df(y)/dy = 20\text{sech}^2(20y)$ as the algorithm nonlinearity derivative, and "center-spike" initialization. Note that the tanh(.) nonlinearity is differentiable, a requirement for the FastICA algorithm. One hundred iterations of the iteration were used in each case.

Both algorithms employed prewhitening as a processing stage, and they do not attempt to restore a natural spectral characteristic to the extracted sources. In addition, ambient noise in the recordings can become emphasized due to the prewhitening process. To mitigate these effects, two stages of linear filtering were applied to each of the extracted signals as a post-processing step to improve their spectral characteristics.

1. A 201-tap FIR linear-phase bandpass filter designed using MATLAB's $fir1$ command was applied to each extracted source. The cutoff frequencies for this filter were chosen to be 50Hz and 7500Hz, in which the original sampling rate of all signals was 16kHz.
2. Two single-pole IIR filters were applied to the data one causal, and one anti-causal. The transfer function of the combined filter was $H(z) = 0.2775 (1 - 0.85z^{-1})^{-1}(1-0.85z)^{-1}$. The resulting filter has a zero-phase response.

3 Separation Results: Source Content Discovery

For our evaluations, we focused on the *parra.zip* data set within the 2008 SiSEC competition. This 23MB data set contains four-microphone recordings of two-, three-, and four-source mixtures as recorded in seven different rooms with different microphone arrangements and source positions. Additional information about this data set can be found in [3], although "ground-truth" source signals have not been made available for this testing. The seven rooms shall be called *Rooms 1, 2, 3, 4, 5, O,* and *C,* in deference to the naming convention on the 2008 SiSEC website.

After application of both the scaled natural gradient and the spatio-temporal FastICA with symmetric constraints to this data, we performed post-processing of each extracted signal as described previously. We then listened carefully to the resulting output signals. What follows is the information we obtained by carefully listening to the separated results from both algorithms.

3.1 Rooms 1, 2 and 3

Rooms 1 through 3 of the *parra.zip* data set were found to contain mixtures of recorded music. Each room contained the same recorded sources organized in the following way. The two-source mixtures contained Sources A and B; the three-source mixtures contained Sources A, B, and C; and the four-source mixtures contained Sources A through D. A careful description of each source now follows.

Source A: The musical selection is that of a jazz piece, perhaps near the end of the song. The selection begins with a four-second segment of eight notes played in unison by a xylophone or vibraphone together with flute, along with string bass and drums accompaniment. This portion of the musical excerpt is played with a slow jazz swing feel. This portion is followed by six seconds of a flourish of notes from the flute, string bass, and drums, the latter played using brushes on what sounds like a snare drum. There is no regular time signature to the final six seconds of the excerpt.

Source B: The musical selection is of a different jazz piece with four-piece instrumentation: alto saxophone, marimba or vibraphone, string bass, and a jazz drum set. The musical melody being played on alto saxophone is easily transcribable by someone trained in the musical arts, as is the accompaniment played on string bass. This musical selection has a time signature of 3/4, corresponding to a jazz waltz, and the tempo is 112 bpm. The snare drum is being played with brushes, and the timing of the snare's beats is easily determined from plots of the signal. The very last note played by the saxophonist in the musical excerpt is joined by a chord played by the mallet instrument.

Source C: This musical selection begins with approximately 0.8 seconds of silence, indicating that it may be at the beginning of a song. After this silence period, a fast jazz swing song is played, with a flurry of notes from an alto saxophone, string bass, marimba, and jazz drum set. The time signature of the song is 4/4 and the tempo is approximately 146 bpm. The notes being played on the alto saxophone are recognizable and could be transcribed with enough effort. The timing of the marimba chords can also be determined readily. Exactly 6.8 seconds into the audio selection, the alto saxophonist quotes the first three measures of the well-known song, "In Walked Bud," by Thelonious Monk. During this portion of the song, the drummer is heard playing the snare drum with drumsticks in a fast eighth-note syncopated pattern.

Source D: This musical selection is the hardest to determine, as it can only be found in the four-source mixtures in Rooms 1, 2, and 3. The song is again a jazz piece, with an alto saxophone as the lead instrument. This instrument is the only one that is clearly identified in the selection, and it may be a solo alto saxophone excerpt. The meter of the musical selection is not clear, indicating that it might be a solo jazz cadenza. Towards the end of the musical selection, the saxophonist plays several slurred notes in time at about 96 bpm. The notes being played on alto saxophone are recognizable and could be transcribed by a trained musician.

3.2 Rooms 4 and 5

Rooms 4 and 5 of the *parra.zip* data set were found to contain mixtures of recorded voices speaking in the English language. Each room contained the same recorded voices organized in the following way. The two-source mixtures contained Voices A and B; the three-source mixtures contained Voices A, B, and C; and the four-source mixtures contained Voices A through D. A careful description of each source now follows.

Voice A: This male talker has a voice with a lower register. The words spoken are as follows:
> "The historical film moved her deeply.
> Canada was established only in 1867.
> There was another woman in the same group.
> You are elected an honorary member—"

Voice B: The female voice of this talker has a slightly soft character, and she is speaking somewhat quickly. The words spoken are as follows:

> "On the 200th day it was reopened.
> Standing 30 meters below ground level.
> Inverted pyramid in the shopping mall.
> Two hundred have not been displayed properly.
> At—"

Voice C: This talker has a male voice with a soft character, and he speaks very quickly with almost no pause between sentences. The words spoken are as follows:

> "She went on turning it around and around.
> They will be happy until it's bedtime.
> Maria didn't mind all the attention.
> You must be ready to listen to them.
> Matthew smiled and started to cheer up.
> The chi—"

Voice D: This female talker speaks a bit slowly with a high degree of enunciation. The words spoken are as follows:

> "Debussey enjoyed the occasional hand bridge.
> Christmas decorations are going up this year.
> Executives should be good judges of character.
> Competitors should stay—"

3.3 Rooms C and O

Rooms C and O of the *parra.zip* data set were found to contain mixtures of recorded sounds of various types. Each room contained the same sounds organized in the following way. The two-source mixtures contained Sources I and II; the three-source mixtures contained Sources I, II, and II; and the four-source mixtures contained Sources I through IV. A careful description of each source now follows.

Source I: This source is that of a male talker speaking very slowly. The slowness of the speech suggests that the audio may be artificially slowed-down, *e.g.* because of a resampling error. The phrase spoken is as follows:

> "Sea velkun mouwenston. Soulen aih, ul, ow-"

A quick web search suggests that the language being spoken could be Dutch. Using the Babelfish language translator, the above phrase with portions in English becomes

> "(Sea) is possible sleeve barrel. Soul music (aih, ul, ow-)"

Source II: This source is a musical work, with alto saxophone, marimba, string bass, and drums. The time signature is 4/4, with a tempo of 56 bpm. The song begins with a three-note phrase that is played in similar time by all instruments and syncopated faster than the 4/4 time signature, followed by four beats of near silence. The musical phrase ends with three saxophone notes in an ascending motif, with a single bass tone and four notes played by the marimba.

Source III: This source is a solo stringed instrument with high-tension strings, such as a banjo or mandolin, playing three collections of notes with pauses between them, in which the second collection closely matches that of the first. The notes are clearly delineated to the point of being transcribable. The highest pitch played is in the third collection of notes. There does not appear to be a time signature to this musical piece.

Source IV: This source is the sound of a person typing on a computer keyboard or keypad. The timing of each tap is readily heard, and the equalized signal produced by the separation algorithm provides sharp transients with which timing information of key strikes could be determined numerically.

4 Qualitative Comparison of Algorithm Performance

In this section, we compare qualitatively the performance of the scaled natural gradient algorithm and the spatio-temporal FastICA algorithm with symmetric constraints in separating the mixtures contained within the *parra.zip* data set. Again, since no ground truth information about this data is available, we can only describe what we hear through careful listening of the signal outputs. Issues such as the amount of distortion in the separated outputs is unknown at this time, because no "ground-truth" signals are available for numerical evaluation or listener training.

From this listening, several features of the algorithms are apparent:

1. *Both algorithms are successful at extracting fewer sources from overdetermined mixtures.* In such cases, the remaining output signals largely contain

broadband noise or distant-sounding mixtures of the sources. For example, when applying these algorithms to four-microphone recordings of mixtures of two sources, they produce four outputs, with two outputs containing each of the two sources and the other two outputs containing broadband noise or distant-sounding mixtures of the sources. It was quite easy to determine through listening to the output signals which of the outputs contained good estimates of the sources in these cases.

2. *Both algorithms produce unordered signal outputs.* In other words, the extracted sources were not produced at *a priori* known output positions. This attribute is due to the symmetric nature in which these algorithms treat the separated outputs.

3. *The same source type was typically extracted at the same output channel for different-order mixtures.* In other words, if Voice A was extracted on Output Channel 2 for the two-source mixtures in Room 3, then Voice A would be extracted on Output Channel 2 for the three- and four source mixtures. This attribute strongly suggests that the physical positions of the sources did not change in the two-source, three-source, and four-source mixtures within any one room when the data was collected.

4. *Both algorithms performed best in separating speech signal mixtures.* This fact is largely due to the choice of nonlinearity for each algorithm, which has been optimized for heavy-tailed source distributions such as speech.

5. *The scaled natural gradient algorithm appeared to out-perform the spatio-temporal FastICA algorithm overall.* This result is at best an educated guess, ashis result is a bit surprising, as the scaled natural gradient algorithm is designed to separate only those sources for which the algorithm nonlinearity guarantees local stability, and the choice $f(y) = \text{sgn}(y)$ is designed for impulsive sources. We expected the spatio-temporal FastICA algorithm to perform better in the case of music mixtures, as the statistics of music is not necessarily impulsive and FastICA should work for arbitrary source types. Perhaps the good performance of scaled natural gradient in these cases is due to the choice of music, as jazz has a sparse musical form.

5   Conclusions

In this paper, we have described numerical experiments on convolutive blind source separation performed on acoustic data taken from the 2008 Signal Separation Evaluation Campaign. Two algorithms have been applied to this data – the scaled natural gradient algorithm [1] and the spatio-temporal FastICA algorithm with symmetric constraints [2]. Using these algorithms, we have carefully described what we believe the sources to be within these data. Not having access to the underlying ground truth recordings, we cannot be sure of the correctness of our conclusions, but we are highly-confident that we have gotten these source descriptions right. Moreover, we describe the attributes of the algorithms' performances in these scenarios, indicating that the scaled natural gradient algorithm is preferable overall in this task. We look forward to the numerical evaluations promised by the SiSEC organizers as well as the opportunity to interact with others who competed in this data challenge.

Acknowledgments. The author would like to thank Dr. Malay Gupta for assistance with coding of the software tools used in this evaluation and Michael Danser for assistance with the interpretation of the audio signals produced in the separation experiments.

References

1. Gupta, M., Douglas, S.C.: Scaled Natural Gradient Algorithms for Instantaneous and Convolutive Blind Source Separation. In: Proc. IEEE International Conference on Acoustics, Speech, and Signal Processing, Honolulu, HI, vol. 2, pp. 637–640. IEEE Press, Los Alamitos (2007)
2. Douglas, S.C., Gupta, M., Sawada, H., Makino, S.: Spatio-Temporal FastICA Algorithms for the Blind Separation of Convolutive Mixtures. IEEE Trans. Audio, Speech, and Language Processing 15(5), 1511–1520 (2007)
3. 2008 Signal Separation Evaluation Campaign, http://sisec.wiki.irisa.fr

Permutation Correction in Blind Source Separation Using Sliding Subband Likelihood Function

Intae Lee

Institute for Neural Computation, University of California, San Diego
9500 Gilman Drive, La Jolla, CA 92093
intelli@ucsd.edu

Abstract. Proportional variance dependency among the frequency components is characteristic of natural signals and has been utilized in frequency-domain blind source separation to solve the permutation problem. In order to increase robustness in such methods, overall measures have been preferred to the measures between directly neighboring frequency components. The overall variance dependency pattern in the fullband, however, can vary by signals and is difficult to be modeled, whereas in smaller subbands the proportional variance dependency is more definite. Here, a novel permutation correction method that utilizes the proportional variance dependency in small subbands is proposed. A windowed likelihood function that uses source priors with internal variance dependency is employed as the measure of permutation correction. This method not only shows robust separation performance but also is computation-wise very efficient.

1 Introduction

Blind source separation (BSS), which is also known as blind signal separation, refers to a set of problems that aim to separate individual source signals from their mixtures where the mixing process and the original source signals are unknown and only the observed signal mixtures are available. Generally, the unknown mixing process can be expressed as the following:

$$x_{jt} = \sum_i a_{jit} * s_{it} + n_{jt} \qquad (1)$$

where s_{it} denotes the i-th source signal, a_{jit} denotes the filter from s_{it} to the j-th microphone, $*$ denotes convolution, n_{jt} denotes the noise signal which is either sensor noise or ambient noise, and x_{jt} denotes the mixed signal that is observed in the j-th sensor.

Independent component analysis (ICA) is one of the most popular algorithmic methods that have been very successful in the field of BSS [1, 2]. A major assumption in ICA is that the source signals are statistically independent and thus the original signals can be recovered by exploiting the independence among

T. Adali et al. (Eds.): ICA 2009, LNCS 5441, pp. 767–774, 2009.
© Springer-Verlag Berlin Heidelberg 2009

them. In the simplest form of ICA, the mixing process is assumed instantaneous, the noise signals are neglected, and the number of source signals is at most the number of sensor signals. In this setting, given that there are at most one Gaussian-distributed source signal, ICA can separate the original source signals from their mixtures up to some scaling and permutation.

In most practical situations, however, where there are time delay and reverberation, the process of source mixing is not instantaneous but convolutive. Solving such convolutive BSS problem in the time domain has been very challenging and thus many researchers have tackled the problem in the time-frequency (T-F) domain. The advantage of the T-F domain approach is that, given that the frame length of the short time Fourier transform (STFT) is long enough to cover the reverberation time, the convolutive BSS problem decomposes into a set of smaller (frequency) bin-wise separation problems where the bin-wise mixing process can be approximated to an instantaneous one. And thus the complex-valued ICA algorithms designed for instantaneous mixtures can be applied yielding excellent bin-wise separation.

In bin-wise separation approaches, however, the inherent permutation indeterminacy of ICA gives rise to permutation disorder along the frequency bins and thus additional algorithmic methods follow. Inter-frequency covariance of the absolute values [3], direction of arrival (DOA) estimation [4], or dominance measure [5] have been used to solve the permutation problem. Meanwhile, in a way of avoiding the permutation problem, a multidimensional approach termed independent vector analysis (IVA) has been proposed [6, 7]. IVA employs multivariate source priors and thus applies a multivariate objective function for the separation such that the frequency components automatically align. In this draft, a novel method is proposed for the correction of the permutation disorder. A sliding multivariate likelihood function that is closely related to IVA is applied as the objective function. Simulation results demonstrate the strength and the effectiveness of the approach.

2   Notations

In this section, the notations are defined. It should be noted that the signals and the filters being defined here are the T-F-domain ones. Scalars (including complex-valued ones), vectors, and matrices will be denoted with normal lowercase, bold lowercase, and bold uppercase letters, respectively.

- $(\cdot)^{\mathrm{T}}$: the transpose operator.
- $(\cdot)^{\mathrm{H}}$: the Hermitian transpose operator.
- $(\cdot)^{-1}$: the matrix inverse operator.
- $|\cdot|$: either the absolute value of a scalar or the cardinality of a set.
- $\|\cdot\|_p$: L^p norm of a vector, i.e. $\|\mathbf{z}\|^p = \left(|z_1|^p + |z_2|^p + \cdots\right)^{\frac{1}{p}}$.
- N: the number of frames in the short-time Fourier domain.
- F: the number of frequency bins.
- $\widetilde{\mathrm{E}}[\cdot]$: the expectation with respect to the empirical distribution, i.e. $\frac{1}{N}\sum_{n=1}^{N}\cdot$.
- $s_i^f[n]$: the STFT-ed i-th source signal in the f-th bin.

- $\mathbf{s}^f[n]$: the column vector whose i-th component is $s_i^f[n]$.
- $x_j^f[n]$: the STFT-ed j-th microphone signal in the f-th bin.
- $\mathbf{x}^f[n]$: the column vector whose j-th component is $x_j^f[n]$.
- $\mathbf{A}^f$: the mixing matrix in the f-th bin. Here we assume that the impulse responses are time-invariant.
- $\mathbf{Q}^f$: the pre-whitening matrix in the f-th bin.

Hence, their relation is

$$\mathbf{x}^f[n] = \mathbf{A}^f \mathbf{s}^f[n], \tag{2}$$

The signals and filters that have to be learned are denoted as the following.

- $\mathbf{W}^f$: the estimate of the inverse of $\mathbf{A}^f$.
- $\mathbf{w}_i^f$: the i-th row of $\mathbf{W}^f$.
- $\hat{s}_i^f[n]$: the estimate of $s_i^f[n]$ with respect to $\mathbf{w}_i^f$.

Hence,

$$\hat{s}_i^f[n] = \mathbf{w}_i^f \mathbf{x}^f[n]. \tag{3}$$

3   Prior Art and Motivations

The IVA approach is illustrated in Fig. 1 in a 2×2 mixing case. In the figure, the unknown bin-wise mixing process which includes the source signals and the mixing matrices is bracketed and the frequency components of each estimated source signal are grouped together representing a multivariate signal. Instead of applying ICA algorithms bin-wise and correcting the permutation disorder afterwards, IVA deals with the frequency components together as a multivariate signal where there is internal dependency among the components. Thus, using probabilistic approach, multivariate probability density functions (PDFs) are employed as the source priors instead of uni-variate PDFs.

As many ICA algorithms employ fixed representative source priors, e.g. Laplace distribution for time-domain speech and other super-Gaussian-distributed signals, IVA algorithms have employed certain super-Gaussian-distributed multivariate PDFs that show proportional variance dependency of the complex-valued components [8]. However, it is usually difficult to choose a proper source prior for IVA because the complicated dependency among the components has to be captured efficiently in the source prior. And that, since the fast Fourier transform (FFT) frame size needs to be large enough to cover the lengths of the impulse responses, it results in high dimensionality of the source signals. In Fig. 2, the normalized covariance matrix of the magnitudes of the frequency components are shown for two pieces of speech and a piece of music. As it can be seen, the dependency types are complicated and they even differ among speeches. Hence, when the source signal is unknown, errors in the fixed source priors are inevitable and such errors can degrade the separation performance of IVA.

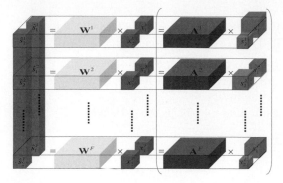

Fig. 1. The approach of independent vector analysis (IVA) for solving the permutation problem

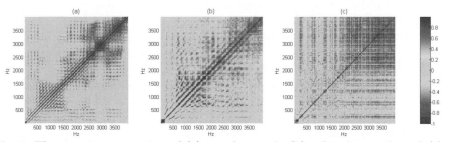

Fig. 2. The covariance matrices of (a) a male speech, (b) a female speech, and (c) a piece of music where the acoustic signals were 12 seconds long

Instead of using fixed source priors, we can also use ones that are able to adapt to various types of probability densities as the mixture of multivariate Gaussians [9]. In this case, however, huge number of data points is essential for learning the unmixing matrices and the parameters of the source priors.

The permutation correction methods also provided reasonable solutions to the permutation problem. However, they also suffer from drawbacks. The DOA method depends much on the robustness of the DOA estimator and its performance easily degrades when the source signals are located in close directions or when the recording environment is reverberant such that the reflections hinder the DOA estimation. In the magnitude covariance or dominance measure methods, the base measure needs pair-wise computation. When using this measure for adjoining pairs only, the permutations can easily flip especially in such bins where the signals are ill separated or where the signal data is statistically ill behaved. Once flipped in the middle, wrong permutations are likely to follow. Thus, as an overall measure, the pair-wise measures are often evaluated and combined for all possible pairs of frequency signals. However, the overall computation is heavy and inefficient. In addition and more importantly, not all pairs of frequency components show variance dependency or similar dominance patterns. In Fig. 3, which is a closer look of Fig. 2, it can be seen that significant number of pairs of frequency components do not show proportional variance dependency. A number of pairs even show inversely proportional variance dependency.

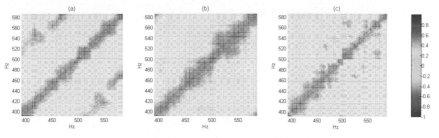

Fig. 3. A closer look of Fig. 2

4   Proposed Permutation Correction Method

Usually, in acoustic signals, there are several types of inter-frequency dependency patterns to be seen. Among those, proportional variance dependency is the key factor that has been modeled in the source priors of the previously proposed IVA methods. With respect to proportional variance dependency there are roughly two types. One is the dependency among harmonic frequency components, and the other is the one among neighboring frequency components.

Since the number of harmonic patterns can be infinitely large, it becomes intractable to employ a representative model that captures the harmonic structure of an unknown acoustic signal. On the meanwhile, the dependency pattern in small subbands (as opposed to fullband or large subbands) is relatively less random. In Fig. 3, it can be seen that most of the normalized covariance components that are sufficiently close to the diagonal have positive values for all three acoustic signals. This implies, in sufficiently small subbands, the proportional inter-frequency variance-dependency is more definite and stronger.

By taking this into account in the permutation problem, it will be advantageous to apply likelihood functions over sliding subbands. The following are the fullband source priors (of normalized signals) that have been proposed in IVA framework [6, 8, 9]:

$$\mathrm{f}(\mathbf{z}) = \frac{1}{Z}\mathrm{e}^{-\frac{1}{\sigma}\|\mathbf{z}\|_p^{\frac{1}{m}}}, \tag{4}$$

$$\mathrm{f}(\mathbf{z}) = \frac{1}{Z}\mathrm{e}^{-\left(\sum_k \frac{1}{\sigma_k}\|\mathbf{z}_{b_k}^{e_k}\|_2\right)}, \tag{5}$$

$$\mathrm{f}(\mathbf{z}) = \sum_k \pi_k \frac{1}{Z_k}\mathrm{e}^{-\mathbf{z}^{\mathrm{H}}\mathbf{D}_k\mathbf{z}}. \tag{6}$$

where $\mathbf{z} = [z_1, z_2, \cdots, z_F]^{\mathrm{T}}$ is a multivariate dummy variable denoting the frequency components of a source, Z and Z_k are normalization factors, σ and σ_k are the coefficients that control the variance in the PDFs, $\mathbf{D}_k$ is the inverse of the k-th covariance matrix in a multivariate Gaussian mixture model, and π_k is the state probability of the k-th multivariate distribution. As mentioned earlier, these source priors are characterized by capturing the proportional variance dependency among the frequency components, and they can be employed as the subband source priors, too. Here in this paper, the results by the spherically symmetric Laplace distribution ((4) when $p = 2$ and $m = 1$) will be shown.

5   Algorithm Description

In large, the algorithm is composed of three parts.

5.1   Bin-Wise ICA Separation

For the bin-wise ICA separation, the FastICA algorithm for complex-valued signals is applied [10]. FastICA algorithm spatially pre-whitens the signals and constrains the unmixing matrices $\mathbf{W}^f$'s to be orthogonal such that the output signals are uncorrelated. Assuming $\mathbf{x}^f$'s are spatially pre-whitened by the following equation

$$\mathbf{x}^f = \mathbf{Q}^f \mathbf{x}^f_{\text{old}}, \tag{7}$$

the algorithm is written as

$$\mathbf{w}^f_i \leftarrow \widetilde{\mathrm{E}}\Big[\mathrm{G}'\big(|\hat{s}^f_i|^2\big) + |\hat{s}^f_i|^2 \mathrm{G}''\big(|\hat{s}^f_i|^2\big)\Big]\mathbf{w}^f - \mu\widetilde{\mathrm{E}}\Big[\mathbf{x}^f(\hat{s}^f_i)^* \mathrm{G}'\big(|\hat{s}^f_i|^2\big)\Big] \tag{8}$$

with the symmetric decorrelation of

$$\mathbf{W}^f \leftarrow \big(\mathbf{W}^f(\mathbf{W}^f)^{\mathrm{H}}\big)^{-1}\mathbf{W}^f. \tag{9}$$

For the nonlinearity function $\mathrm{G}(\cdot)$, the following was chosen:

$$\mathrm{G}(z) = \sqrt{z}. \tag{10}$$

5.2   Permutation Correction

As from the previous discussion, the (normalized) subband log-likelihood function with the source prior in (4) is employed as the measure of permutation correction:

$$\mathcal{L}_{\mathcal{S}} = -\frac{1}{\sigma}\sum_i \widetilde{\mathrm{E}}\left[\sqrt{\sum_{f \in \mathcal{S}}(\hat{s}^f_i)^2}\right] + \sum_{f \in \mathcal{S}}\log|\det(\mathbf{W}^f)| \tag{11}$$

where $\mathcal{S}$ denotes the set of the bins in a sliding subband. Note that the second term in the right-hand-side of the likelihood function can be ignored since the FastICA algorithm that is used for the bin-wise separation constrains the unmixing matrices to be orthogonal. The size of the subband that is chosen in this draft is fixed around 40 Hz. The permutation in the f-th bin is determined such that it maximizes the measure $\mathcal{L}_{\mathcal{S}}$ in (11) where $\mathcal{S} = \{f - |\mathcal{S}| + 1, \cdots, f\}$.

5.3   Scaling Correction

In frequency-domain BSS, other than the permutation problem, the scaling problem which includes the phase ambiguity arises because of the scaling indeterminacy of ICA. Here, the minimal distortion principle [11] is applied to fix the scaling problem. After all unmixing matrices $\mathbf{W}^f$'s are learned, the diagonal of $(\mathbf{W}^f\mathbf{Q}^f)^{-1}$ is multiplied to the left of $\mathbf{W}^f\mathbf{Q}^f$ [8]:

$$\mathbf{W}^f_{\text{final}} \leftarrow \mathrm{diag}\big((\mathbf{W}^f\mathbf{Q}^f)^{-1}\big)\mathbf{W}^f\mathbf{Q}^f. \tag{12}$$

6   Experiments

The BSS algorithm has been applied to 2×2 speech separation problems. The mixed speech signals were synthetic signals generated in a simulated room environment. In generating synthetic data, 12-second-long clean speech signals were used. Also, 4096-point FFT and a 4096-tab long Hanning window with the shift size of 512 samples were chosen.

     The geometric configuration of the simulated room environment is depicted in Fig. 4(a). We set the room size to 7 m $\times$ 5 m $\times$ 2.75 m and set all heights of the microphone and source locations to 1.5 m. A reverberation time of 200 ms was chosen and the corresponding reflection coefficients were set to 0.57 for every wall, floor, and ceiling. Clean speech signals were convolved with room impulse responses that were obtained by an image method [12] using this room configuration. Various 2×2 case simulations (Fig. 4(b)) were carried out. The separation performance was measured by the signal to interference ratio (SIR) in dB. The separation performance of the proposed algorithm was compared with the separation performance of an IVA algorithm that uses the same kind of source prior except that the source prior is overall. Thus the objective function of the compared method is equal to the measure in (11) where $S = \{1, \cdots, F\}$. In order to have similar conditions in the algorithms, the FastIVA algorithm [8] that also keeps the output data uncorrelated and uses approximated Newton update optimization method is employed for comparison. The results are shown in Fig. 4(c).

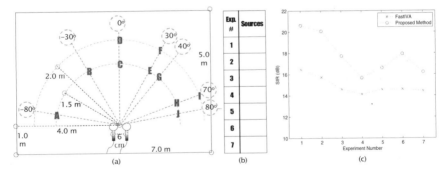

Fig. 4. The experimental results are shown in this figure

7   Discussion

The idea of using a sliding subband likelihood function is advantageous in that it is computation-wise very efficient. Also, it can be adopted in the IVA framework to yield subband-wise IVA algorithms. Furthermore, in the sliding subbands, not only the maximum likelihood approach but also other machine learning or decision making techniques can be employed. More extensive work will be shown in a longer version of this draft.

References

1. Herault, J., Jutten, C.: Space or time processing by neural network models. In: Proc. AIP Conf.: Neural Networks for Computing, vol. 151, pp. 206–211 (1986)
2. Comon, P.: Independent component analysis-a new concept? Signal Processing 36, 287–314 (1994)
3. Anemueller, J., Kollmeier, B.: Amplitude modulation decorrelation for convolutive blind source separation. In: Proc. Int. Conf. on Independent Component Analysis and Blind Source Separation, pp. 215–220 (2000)
4. Ikram, M.Z., Morgan, D.R.: A beamforming approach to permutation alignment for multichannel frequency-domain blind speech separation. In: Proc. IEEE Int. Conf. on Acoustics, Speech, and Signal Processing, pp. 881–884 (2002)
5. Sawada, H., Araki, S., Makino, S.: Measuring dependence of bin-wise separated signals for permutation alignment in frequency-domain BSS. In: Proc. IEEE Int. Symp. on Circuits and Systems, pp. 3247–3250 (2007)
6. Kim, T., Attias, H.T., Lee, S.Y., Lee, T.W.: Blind source separation exploiting higher-order frequency dependencies. IEEE Trans. on Speech and Audio Processing 15(1), 70–79 (2007)
7. Lee, I., Kim, T., Lee, T.W.: Complex FastIVA: a robust maximum likelihood approach of MICA for convolutive BSS. LNCS, pp. 625–632 (2006)
8. Lee, I., Kim, T., Lee, T.W.: 6. In: Independent Vector Analysis for Blind Speech Separation, pp. 167–190. Springer, Heidelberg (2007)
9. Lee, I., Hao, J., Lee, T.W.: Adaptive independent vector analysis for the separation of convoluted mixtures using EM algorithm. In: Proc. IEEE Int. Conf. on Acoustics, Speech, and Signal Processing, pp. 145–148 (2008)
10. Bingham, E., Hyvärinen, A.: A fast fixed-point algorithm for independent componenet analysis of complex-valued signals. Int. J. of Neural Systems 10(1), 1–8 (2000)
11. Matsuoka, K., Nakashima, S.: Minimal distortion principle for blind source separation. In: Proc. Int. Conf. on Independent Component Analysis and Blind Source Separation, pp. 722–727 (2001)
12. Allen, J.B., Berkley, D.A.: Image method for efficiently simulating small room acoustics. J. Acoust. Soc. Amer. 65, 943–950 (1979)

Underdetermined Instantaneous Audio Source Separation via Local Gaussian Modeling

Emmanuel Vincent, Simon Arberet, and Rémi Gribonval

METISS Group, IRISA-INRIA
Campus de Beaulieu, 35042 Rennes Cedex, France
{emmanuel.vincent,simon.arberet,remi.gribonval}@irisa.fr

Abstract. Underdetermined source separation is often carried out by modeling time-frequency source coefficients via a fixed sparse prior. This approach fails when the number of active sources in one time-frequency bin is larger than the number of channels or when active sources lie on both sides of an inactive source. In this article, we partially address these issues by modeling time-frequency source coefficients via Gaussian priors with free variances. We study the resulting maximum likelihood criterion and derive a fast non-iterative optimization algorithm that finds the global minimum. We show that this algorithm outperforms state-of-the-art approaches over stereo instantaneous speech mixtures.

1 Introduction

Underdetermined source separation is the problem of recovering the single-channel source signals $s_j(t)$, $1 \leq j \leq J$, underlying a multichannel mixture signal $x_i(t)$, $1 \leq i \leq I$, with $I < J$. The mixing process can be modeled in the time-frequency domain via the Short-Term Fourier Transform (STFT) as

$$\mathbf{x}(n, f) = \mathbf{A}(f)\mathbf{s}(n, f) \tag{1}$$

where $\mathbf{s}(n, f)$ is the vector of source STFT coefficients in time-frequency bin (n, f), $\mathbf{x}(n, f)$ is the vector of mixture STFT coefficients in that bin, and $\mathbf{A}(f)$ is a complex mixing matrix. This problem can be addressed by first estimating the mixing matrices then computing the Maximum *A Posteriori* (MAP) source coefficients given some prior distribution and inverting the STFT. For audio data, a common sparse prior such as the Laplacian [1], a mixture of Gaussians [2] or a generalized Gaussian [3], is usually assumed for all source coefficients. This model suffers from two issues. Firstly, a maximum number of I nonzero coefficients can often be recovered in each time-frequency bin, with the $J - I$ remaining coefficients being estimated as zero [1,3]. Secondly, the corresponding columns of the mixing matrix must point towards the closest directions to the observed mixture direction. A proof is given in [4] for a Laplacian prior.

In this paper, we aim to overcome both issues in the popular setting of stereo ($I = 2$) instantaneous mixtures, where the mixing matrices $\mathbf{A}(f)$ are equal to the same real-valued matrix $\mathbf{A}$ for all f. We assume that $\mathbf{A}$ is known and that its

T. Adali et al. (Eds.): ICA 2009, LNCS 5441, pp. 775–782, 2009.

columns are pairwise linearly independent, *i.e.* the sources have different directions. We build upon the Statistically Sparse Decomposition Principle (SSDP) presented in [4], which addresses the second issue using the correlation between the mixture channels but provides poor separation performance due to time-domain modeling and constraining of the number of nonzero sources per bin.

The structure of the paper is as follows. We apply the SSDP to the time-frequency domain in Section 2 and prove that it implicitly assumes a local Gaussian source model in Section 3. In Section 4, we extend this model to a larger number of nonzero sources and derive a new source separation algorithm. We evaluate its performance on speech data in Section 5 and conclude in Section 6.

2   Time-Frequency Statistically Sparse Decomposition

The SSDP is based on the empirical multichannel covariance matrix of the mixture over short time frames. In the time-frequency domain, we define this quantity over the neighborhood of each time-frequency bin (n, f) instead as

$$\widehat{\mathbf{R}}_{\mathbf{xx}}(n, f) = \frac{1}{\sum_{n',f'} w(n' - n, f' - f)} \sum_{n',f'} w(n' - n, f' - f)\mathbf{x}(n', f')\mathbf{x}(n', f')^H \tag{2}$$

where w is a bi-dimensional window specifying the shape of the neighborhood and ^{H} denotes the conjugate transpose of a matrix. In the rest of the paper, bin indexes (n, f) are dropped for the sake of legibility. The quantity (2) has long been exploited by mixing matrix estimation methods, *e.g.* [5,6], to obtain accurate direction estimates by selecting the bins where a single source is active. These bins are characterized by the fact that the cross-correlation between the mixture channels, also termed interchannel coherence, is high.

More generally, the cross-correlation is higher when the active sources have close directions. This fact can be exploited for source separation as follows. Let us assume that the number of active sources in each time-frequency bin is equal to two. For each pair of source indexes (j_1, j_2), the empirical covariance matrix of these sources may be defined as [4]

$$\widehat{\mathbf{R}}_{\mathbf{s}_{j_1 j_2} \mathbf{s}_{j_1 j_2}} = \mathbf{A}_{j_1 j_2}^{-1} \, \widehat{\mathbf{R}}_{\mathbf{xx}} \, (\mathbf{A}_{j_1 j_2}^{-1})^T \tag{3}$$

where $\mathbf{A}_{j_1 j_2}$ is the 2×2 matrix composed of the columns $\mathbf{A}_j$ of $\mathbf{A}$ indexed by $j \in \{j_1, j_2\}$ and ^{T} denotes transposition. The best pair of active sources may be selected via the SSDP [4]

$$(\widehat{j}_1, \widehat{j}_2) = \arg\min_{j_1, j_2} \frac{|\widehat{R}_{s_{j_1} s_{j_2}}|}{\sqrt{\widehat{R}_{s_{j_1} s_{j_1}} \, \widehat{R}_{s_{j_2} s_{j_2}}}} \tag{4}$$

with $\widehat{R}_{s_{j_k} s_{j_l}}$ denoting the (k, l)-th entry of $\widehat{\mathbf{R}}_{\mathbf{s}_{j_1 j_2} \mathbf{s}_{j_1 j_2}}$. The source STFT coefficients are then estimated by local mixing inversion as

$$\begin{cases} \widehat{\mathbf{s}}_{\widehat{j}_1 \widehat{j}_2}(n, f) = \mathbf{A}_{\widehat{j}_1 \widehat{j}_2}^{-1} \, \mathbf{x}(n, f) \\ \widehat{s}_j(n, f) = 0 \qquad\qquad \text{for all } j \notin \{\widehat{j}_1, \widehat{j}_2\}. \end{cases} \tag{5}$$

3   Interpretation as Constrained Local Gaussian Modeling

This algorithm admits the following probabilistic interpretation. Let us assume that the source coefficients follow independent zero-mean Gaussian priors over the neighborhood of each time-frequency bin (n, f) whose variances v_j depend on that bin. This assumption appears well suited to audio signals, which are typically non-sparse over small time-frequency regions but non-stationary hence sparse over larger regions. Given this model, the mixture coefficients in a given neighborhood follow a zero-mean Gaussian prior with covariance matrix

$$\mathbf{R_{xx}} = \mathbf{A} \operatorname{Diag}(\mathbf{v}) \mathbf{A}^T. \tag{6}$$

where the operator $\mathbf{Diag}$ applied to a vector denotes the diagonal matrix whose entries are those of the vector. The log-likelihood of the source variances is equal up to a constant to minus the Kullback-Leibler (KL) divergence $KL(\widehat{\mathbf{R}}_{\mathbf{xx}}|\mathbf{R_{xx}})$ between the empirical and expected mixture covariances [7][1] with

$$KL(\widehat{\mathbf{R}}|\mathbf{R}) = \frac{1}{2}[\operatorname{tr}(\mathbf{R}^{-1}\widehat{\mathbf{R}}) - \log\det(\mathbf{R}^{-1}\widehat{\mathbf{R}})] - 1. \tag{7}$$

Assuming that at most two sources have nonzero variance, their indexes and variances may be estimated in the Maximum Likelihood (ML) sense by

$$(\widehat{j}_1, \widehat{j}_2, \widehat{\mathbf{v}}_{\widehat{j}_1\widehat{j}_2}) = \arg\min_{j_1, j_2, \mathbf{v}_{j_1 j_2} \geq \mathbf{0}} KL(\widehat{\mathbf{R}}_{\mathbf{xx}}|\mathbf{R_{xx}}). \tag{8}$$

The KL divergence is invariant under invertible linear transforms. When applied to $\mathbf{A}_{j_1 j_2}^{-1}$, this property yields

$$KL(\widehat{\mathbf{R}}_{\mathbf{xx}}|\mathbf{R_{xx}}) = KL(\widehat{\mathbf{R}}_{\mathbf{s}_{j_1 j_2}\mathbf{s}_{j_1 j_2}}|\mathbf{Diag}(\mathbf{v}_{j_1 j_2})) \tag{9}$$

$$= \frac{1}{2}\left[\frac{\widehat{R}_{s_{j_1}s_{j_1}}}{v_{j_1}} + \frac{\widehat{R}_{s_{j_2}s_{j_2}}}{v_{j_2}} - \log\frac{\widehat{R}_{s_{j_1}s_{j_1}}\widehat{R}_{s_{j_2}s_{j_2}} - |\widehat{R}_{s_{j_1}s_{j_2}}|^2}{v_{j_1}v_{j_2}}\right] - 1. \tag{10}$$

By finding the zeroes of the partial derivatives of this expression with respect to v_{j_1} and v_{j_2}, we get

$$\begin{cases} \widehat{v}_{j_1} = \widehat{R}_{s_{j_1}s_{j_1}} \quad\text{and}\quad \widehat{v}_{j_2} = \widehat{R}_{s_{j_2}s_{j_2}} \\ (\widehat{j}_1, \widehat{j}_2) = \arg\min_{j_1, j_2} -\frac{1}{2}\log\left(1 - \frac{|\widehat{R}_{s_{j_1}s_{j_2}}|^2}{\widehat{R}_{s_{j_1}s_{j_1}}\widehat{R}_{s_{j_2}s_{j_2}}}\right). \end{cases} \tag{11}$$

This criterion is equivalent to (4), hence the SSDP does estimate the two sources with nonzero variance in the ML sense. In addition, the ML variances of these sources are equal to the diagonal entries of the empirical source covariance matrix. It can also be shown that the MAP source coefficients given the ML source variances are obtained via (5).

---

[1] This relation holds provided that $\widehat{\mathbf{R}}_{\mathbf{xx}}$ has full rank. We consider the KL divergence because of its well-known invariance and nonnegativity properties. However it can be shown from the expression of the log-likelihood that the following derivations remain true otherwise.

4   Minimally Constrained Local Gaussian Modeling

While the SSDP allows the separation of sources not pointing to close directions, the number of nonzero source coefficients that can be estimated in each time-frequency bin remains constrained to two. The above probabilistic interpretation provides a natural way of relaxing this constraint by assuming that the source coefficients follow independent zero-mean Gaussian priors over the neighborhood of each time-frequency bin, whose variances v_j are free. This model has been exploited in the context of determined mixtures, albeit with the different goal of estimating the mixing matrix given estimates of the source variances [7]. In the under-determined context, ML variance estimates are obtained by

$$\widehat{\mathbf{v}} = \arg\min_{\mathbf{v} \geq 0} KL(\widehat{\mathbf{R}}_{\mathbf{xx}}|\mathbf{R}_{\mathbf{xx}}) \tag{12}$$

and MAP source coefficients are classically derived via the Wiener filter

$$\widehat{\mathbf{s}}(n, f) = \mathbf{Diag}(\widehat{\mathbf{v}})\,\mathbf{A}^T\,(\mathbf{A}\,\mathbf{Diag}(\widehat{\mathbf{v}})\,\mathbf{A}^T)^{-1}\,\mathbf{x}(n, f). \tag{13}$$

The above vector minimization problem may be solved via standard iterative optimization techniques based on the gradient. However these methods are computationally intensive and the result may be a local minimum or one of several possible global minima. We avoid these issues by characterizing the minima. We show below that global minima with three or more nonzero entries satisfy the equality $\mathbf{R}_{\mathbf{xx}} = \Re(\widehat{\mathbf{R}}_{\mathbf{xx}})$, where $\Re$ denotes the real part of a complex matrix. If no vector satisfies this equality, the global minima consequently have two nonzero entries and can be obtained via the SSDP as shown in Section 3. This suggests an efficient way of computing the global minima: find the vectors $\mathbf{v} \geq \mathbf{0}$ such that $\mathbf{R}_{\mathbf{xx}} = \Re(\widehat{\mathbf{R}}_{\mathbf{xx}})$ and, if none exists, apply the SSDP instead. We also study below the cases where several solutions arise and propose minimal constraints to select a single solution. The reader is advised to skip the proofs of the following lemmas on first reading and to proceed directly with the details of Algorithm 1 at the end of this section.

Lemma 1. *The KL divergence criterion is always larger than $KL(\widehat{\mathbf{R}}_{\mathbf{xx}}|\Re(\widehat{\mathbf{R}}_{\mathbf{xx}}))$ and equal to that value if and only if $\mathbf{R}_{\mathbf{xx}} = \Re(\widehat{\mathbf{R}}_{\mathbf{xx}})$.*

Proof. Since the mixing matrix $\mathbf{A}$ is real-valued, $\mathbf{R}_{\mathbf{xx}}$ is real-valued and admits a real-valued square root $\mathbf{R}_{\mathbf{xx}}^{1/2}$. The matrix $\mathbf{R}_{\mathbf{xx}}^{-1/2}\widehat{\mathbf{R}}_{\mathbf{xx}}\mathbf{R}_{\mathbf{xx}}^{-1/2}$ is Hermitian, hence using the commutativity of the trace $\mathrm{tr}(\mathbf{R}_{\mathbf{xx}}^{-1}\widehat{\mathbf{R}}_{\mathbf{xx}}) = \mathrm{tr}(\mathbf{R}_{\mathbf{xx}}^{-1/2}\widehat{\mathbf{R}}_{\mathbf{xx}}\mathbf{R}_{\mathbf{xx}}^{-1/2}) = \mathrm{tr}(\mathbf{R}_{\mathbf{xx}}^{-1/2}\Re(\widehat{\mathbf{R}}_{\mathbf{xx}})\mathbf{R}_{\mathbf{xx}}^{-1/2}) = \mathrm{tr}(\mathbf{R}_{\mathbf{xx}}^{-1}\Re(\widehat{\mathbf{R}}_{\mathbf{xx}}))$. By combining this equality with (7), we get $KL(\widehat{\mathbf{R}}_{\mathbf{xx}}|\mathbf{R}_{\mathbf{xx}}) = KL(\Re(\widehat{\mathbf{R}}_{\mathbf{xx}})|\mathbf{R}_{\mathbf{xx}}) + \log\det(\widehat{\mathbf{R}}_{\mathbf{xx}}^{-1}\Re(\widehat{\mathbf{R}}_{\mathbf{xx}}))$. The second term of this equation does not depend on $\mathbf{v}$, while the first term is nonnegative and equal to zero if and only if $\mathbf{R}_{\mathbf{xx}} = \Re(\widehat{\mathbf{R}}_{\mathbf{xx}})$ by property of the KL divergence. $\qquad\square$

Lemma 2. *If $\mathbf{v}$ is a local minimum of the criterion with $K \geq 3$ nonzero entries $v_{j_1}, \ldots, v_{j_K}$, then $\mathbf{v}$ is a global minimum and satisfies $\mathbf{R}_{\mathbf{xx}} = \Re(\widehat{\mathbf{R}}_{\mathbf{xx}})$.*

Proof. The gradient of the criterion is given by

$$\frac{\partial KL(\widehat{\mathbf{R}}_{\mathbf{xx}}|\mathbf{R}_{\mathbf{xx}})}{\partial v_j} = \langle \mathbf{E}, \mathbf{A}_j \mathbf{A}_j^T \rangle \quad \text{where } \mathbf{E} = \frac{1}{2}[\mathbf{R}_{\mathbf{xx}}^{-1}(\mathbf{R}_{\mathbf{xx}} - \Re(\widehat{\mathbf{R}}_{\mathbf{xx}}))\mathbf{R}_{\mathbf{xx}}^{-1}] \quad (14)$$

and $\langle .,. \rangle$ is the Euclidean dot product over the space $S_2(\mathbb{R})$ of real-valued symmetric 2×2 matrices. If $\mathbf{v}$ is a local extremum, the gradient is zero for all entries v_j that are not boundaries of the optimization domain. Hence $\mathbf{E}$ is orthogonal to the matrices $\mathbf{A}_j \mathbf{A}_j^T$, $j \in \{j_1, j_2, j_3\}$.

Let us consider the 3×3 matrix $\mathbf{B}_{j_1 j_2 j_3}$ whose columns consist of the upper triangular entries of the latter matrices:

$$\mathbf{B}_{j_1 j_2 j_3} = \begin{pmatrix} A_{1_{j_1}}^2 & A_{1_{j_2}}^2 & A_{1_{j_3}}^2 \\ A_{2_{j_1}}^2 & A_{2_{j_2}}^2 & A_{2_{j_3}}^2 \\ A_{1_{j_1}} A_{2_{j_1}} & A_{1_{j_2}} A_{2_{j_2}} & A_{1_{j_3}} A_{2_{j_3}} \end{pmatrix}. \quad (15)$$

By computing and factoring the determinant of $\mathbf{B}_{j_1 j_2 j_3}$, we get

$$\det \mathbf{B}_{j_1 j_2 j_3} = \det \mathbf{A}_{j_1 j_2} \det \mathbf{A}_{j_2 j_3} \det \mathbf{A}_{j_3 j_1}. \quad (16)$$

Since the columns $\mathbf{A}_j$ of $\mathbf{A}$ are pairwise linearly independent, all the terms of this equation are nonzero and the columns of $\mathbf{B}_{j_1 j_2 j_3}$ form a basis of $\mathbb{R}^3$. This is equivalent to $\mathbf{A}_j \mathbf{A}_j^T$, $j \in \{j_1, j_2, j_3\}$, being a basis of $S_2(\mathbb{R})$.

We deduce from the above results that $\mathbf{E} = \mathbf{0}$ hence $\mathbf{R}_{\mathbf{xx}} = \Re(\widehat{\mathbf{R}}_{\mathbf{xx}})$. Therefore $\mathbf{v}$ is a global minimum of the criterion according to lemma 1. □

Lemma 3. *The matrix equality* $\mathbf{R}_{\mathbf{xx}} = \Re(\widehat{\mathbf{R}}_{\mathbf{xx}})$ *can be equivalently rewritten as*

$$\mathbf{B}_{j_1 \ldots j_K} \mathbf{v}_{j_1 \ldots j_K} = \widehat{\mathbf{w}} \quad (17)$$

where $\mathbf{v}_{j_1 \ldots j_K}$ *is the vector of nonzero entries of* $\mathbf{v}$,

$$\mathbf{B}_{j_1 \ldots j_K} = \begin{pmatrix} A_{1_{j_1}}^2 & \cdots & A_{1_{j_K}}^2 \\ A_{2_{j_1}}^2 & \cdots & A_{2_{j_K}}^2 \\ A_{1_{j_1}} A_{2_{j_1}} & \cdots & A_{1_{j_K}} A_{2_{j_K}} \end{pmatrix} \quad and \quad \widehat{\mathbf{w}} = \begin{pmatrix} \widehat{R}_{x_1 x_1} \\ \widehat{R}_{x_2 x_2} \\ \Re(\widehat{R}_{x_1 x_2}) \end{pmatrix}. \quad (18)$$

Proof. From (6), $\mathbf{R}_{\mathbf{xx}} = \Re(\widehat{\mathbf{R}}_{\mathbf{xx}})$ is equivalent to $\sum_{k=1}^{K} v_{j_k} \mathbf{A}_{j_k} \mathbf{A}_{j_k}^T = \Re(\widehat{\mathbf{R}}_{\mathbf{xx}})$. By rearranging the upper triangular terms of this matrix equality into vectors, this is in turn equivalent to (17). □

Lemma 4. *With $J \geq 4$ sources, if the criterion admits a global minimum with $K \geq 3$ nonzero entries, then there exists a global minimum with $K \leq 3$ nonzero entries. Moreover, if $\mathbf{A}$ is nonnegative and there is a global minimum with $K = 3$ nonzero entries, then there are several global minima with $K \leq 3$ nonzero entries.*

Proof. Let $\mathbf{v}$ be a global minimum of the criterion with $K \geq 4$ nonzero entries. According to lemma 2 and its proof, $\mathbf{v}$ satisfies (17) and $\mathbf{B}_{j_1 \ldots j_K}$ has rank 3. The null space of $\mathbf{B}_{j_1 \ldots j_K}$ is therefore of dimension $K - 3 > 0$. Let $\mathbf{z}$ be a vector such that $\mathbf{z}_{j_1...j_K} \neq \mathbf{0}$ belongs to that null space and $z_j = 0$ for all $j \notin \{j_1,\ldots,j_K\}$. We define the vector $\mathbf{v}'$ as

$$\mathbf{v}' = \mathbf{v} - \frac{v_{j_l}}{z_{j_l}}\mathbf{z} \quad \text{with } l = \arg\min_{k,z_{j_k}\neq 0} \frac{v_{j_k}}{|z_{j_k}|}. \tag{19}$$

The entries of this vector are given by $v'_{j_k} = v_{j_k} - v_{j_l}z_{j_k}/z_{j_l}$. Clearly, $v'_j = 0$ for all $j \notin \{j_1,\ldots,j_K\}$ and $v'_{j_l} = 0$. If z_{j_k} and z_{j_l} have different signs, then $z_{j_k}/z_{j_l} \leq 0$ and $v'_{j_k} \geq v_{j_k} \geq 0$. If z_{j_k} and z_{j_l} have the same sign, then $v'_{j_k} = v_{j_k} - v_{j_l}|z_{j_k}|/|z_{j_l}| \geq 0$ given (19). Hence $\mathbf{v}'$ has nonnegative entries and at most $K-1$ positive entries. In addition, $\mathbf{B}_{j_1...j_{l-1}j_{l+1}...j_K}\mathbf{v}'_{j_1...j_{l-1}j_{l+1}...j_K} = \mathbf{B}_{j_1...j_K}\mathbf{v}'_{j_1...j_K} = \mathbf{B}_{j_1...j_K}\mathbf{v}_{j_1...j_K} - v_{j_l}/z_{j_l}\mathbf{B}_{j_1...j_K}\mathbf{z}_{j_1...j_K} = \widehat{\mathbf{w}} - v_{j_l}/z_{j_l}\mathbf{0} = \widehat{\mathbf{w}}$. This shows that $\mathbf{v}'$ is a global minimum of the criterion with $K' \leq K-1$ nonzero entries. By recurrently applying the above construction, we find a global minimum $\mathbf{v}''$ with $K'' \leq 3$ nonzero entries.

Let us now assume that $\mathbf{A}$ is nonnegative and $K'' = 3$. We denote by j_1, j_2, j_3 the nonzero entries of $\mathbf{v}''$ and by j_4 any other index. Since the matrix $\mathbf{B}_{j_1j_2j_3j_4}$ is nonnegative and all its 3×3 submatrices have rank 3, the non-null vectors of its null space have no zero entry and both positive and negative entries. Let $\mathbf{z}'$ be a vector such that $\mathbf{z}'_{j_1j_2j_3j_4} \neq \mathbf{0}$ belongs to that null space, $z'_{j_4} < 0$ and $z'_j = 0$ for all $j \notin \{j_1,j_2,j_3,j_4\}$. We define the vector $\mathbf{v}'''$ as

$$\mathbf{v}''' = \mathbf{v}'' - \frac{v''_{j_l}}{z'_{j_l}}\mathbf{z}' \quad \text{with } l = \arg\min_{k,z'_{j_k}>0} \frac{v''_{j_k}}{z'_{j_k}}. \tag{20}$$

Similarly to above, it can be proved that $\mathbf{v}'''$ is a global minimum of the criterion with $K''' \leq 3$ nonzero entries indexed by some $j \in \{j_1,j_2,j_3,j_4\}$ and $j \neq j_l$. □

Lemma 4 shows that ML estimation of the source variances is an ill-posed problem with $J \geq 4$ sources. Appropriate constraints must be set over the source variances in order to obtain a unique solution. While probabilistic hyperpriors may model flexible constraints, the resulting MAP solution may not match any of the ML solutions, so that the benefit of characterizing ML solutions is lost. Instead, we select the sparsest ML solution: we restrict the optimization domain to vectors with $K \leq 3$ nonzero entries and select the ML solution with minimum l_p norm $\|\widehat{\mathbf{v}}\|_p$ [3] in case several ML solutions can be found in this domain.

Given these constraints and the characterization of ML source variances in Section 3 and lemma 3, we perform source separation in each time-frequency bin (n, f) via the following fast global optimization algorithm.

Algorithm 1

1. *Compute the empirical mixture covariance $\widehat{\mathbf{R}}_{\mathbf{xx}}$ in (2) and derive the vector $\widehat{\mathbf{w}}$ in (18).*
2. *Compute the candidate source variances $\mathbf{v}_{j_1j_2j_3} = \mathbf{B}_{j_1j_2j_3}^{-1}\widehat{\mathbf{w}}$ for all triplets of source indexes $\{j_1, j_2, j_3\}$, with $\mathbf{B}_{j_1j_2j_3}$ defined in (15).*
3. *If some candidates have positive entries only, then they are solutions of the ML estimation problem. Select the one with minimum l_p norm among these and derive the MAP source coefficients via (13).*

4. *Otherwise, compute the empirical source covariance matrices* $\widehat{\mathbf{R}}_{\mathbf{s}_{j_1 j_2} \mathbf{s}_{j_1 j_2}} = \mathbf{A}_{j_1 j_2}^{-1} \widehat{\mathbf{R}}_{\mathbf{xx}} (\mathbf{A}_{j_1 j_2}^{-1})^T$ *for all pairs of source indexes* $\{j_1, j_2\}$. *Select the ML pair via* (4) *and estimate the MAP source coefficients via* (5).

5 Experimental Results

We evaluated this algorithm over the speech data in [8]. The number of sources J was varied from 3 to 6. For each J, a nonnegative mixing matrix was computed from [9], given an angle of $50 - 5J$ degrees between successive sources, and ten instantaneous mixtures were generated from different source signals resampled at 8 kHz. The STFT was computed with a sine window of length 512 (64 ms). The bi-dimensional window w defining time-frequency neighborhoods was chosen as the outer product of two rectangular or Hanning windows with variable length. The l_p norm exponent was set to $p \to 0$ [3]. The results were evaluated via the Signal-to-Distortion Ratio (SDR) defined in [10]. The best results were achieved for w chosen as the outer product of two Hanning windows of length 3. The computation time was then between 1.8 and 3.7 times the mixture duration depending on J, using Matlab on a 1.2 GHz dual core CPU.

Figure 1 compares the average SDR achieved by the proposed algorithm, the time-frequency domain SSDP in Section 2 and two state-of-the-art algorithms: l_p norm minimization [3] and DUET [11]. The proposed algorithm outperforms all other algorithms whatever the number of sources. Nevertheless, it should be noted that its performance remains about 10 dB below the theoretical upper bound obtained by local mixing inversion (5) given the best pair of active sources [8] and 11 dB below the theoretical upper bound obtained by Wiener filtering (13) given the true sources variances.

This algorithm was submitted to the 2008 Signal Separation Evaluation Campaign with the same parameters, except a STFT window length of 1024 and step size of 256. Mixing matrices were estimated via the software in [6].

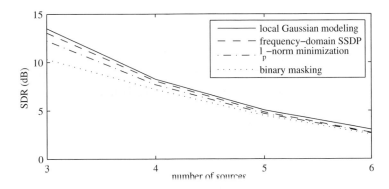

Fig. 1. Source separation performance over stereo instantaneous speech mixtures

6   Conclusion

In this paper, we proposed a new source separation algorithm for stereo instantaneous mixtures based on the modeling of source STFT coefficients via local Gaussian priors with minimally constrained variances. This algorithm can estimate up to three nonzero source coefficients in each bin, as opposed to two for state-of-the-art methods, and provides improved separation performance. This suggests that local mixture covariance can be successfully exploited for underdetermined source separation in addition to mixing matrix estimation. Further work includes the generalization of this algorithm to convolutive mixtures with $I \geq 2$ channels. A larger improvement is expected, since up to $I(I+1)/2$ nonzero source coefficients could be estimated in each time-frequency bin. Local nongaussian source priors could also be investigated.

References

1. Zibulevsky, M., Pearlmutter, B.A., Bofill, P., Kisilev, P.: Blind source separation by sparse decomposition in a signal dictionary. In: Independent Component Analysis: Principles and Practice, pp. 181–208. Cambridge Press (2001)
2. Davies, M.E., Mitianoudis, N.: Simple mixture model for sparse overcomplete ICA. IEE Proceedings on Vision, Image and Signal Processing 151(1), 35–43 (2004)
3. Vincent, E.: Complex nonconvex l_p norm minimization for underdetermined source separation. In: Davies, M.E., James, C.J., Abdallah, S.A., Plumbley, M.D. (eds.) ICA 2007. LNCS, vol. 4666, pp. 430–437. Springer, Heidelberg (2007)
4. Xiao, M., Xie, S., Fu, Y.: A statistically sparse decomposition principle for underdetermined blind source separation. In: Proc. Int. Symp. on Intelligent Signal Processing and Communication Systems (ISPACS), pp. 165–168 (2005)
5. Belouchrani, A., Amin, M.G., Abed-Meraïm, K.: Blind source separation based on time-frequency signal representations. IEEE Trans. on Signal Processing 46(11), 2888–2897 (1998)
6. Arberet, S., Gribonval, R., Bimbot, F.: A robust method to count and locate audio sources in a stereophonic linear instantaneous mixture. In: Rosca, J.P., Erdogmus, D., Príncipe, J.C., Haykin, S. (eds.) ICA 2006. LNCS, vol. 3889, pp. 536–543. Springer, Heidelberg (2006)
7. Pham, D.T., Cardoso, J.F.: Blind separation of instantaneous mixtures of non stationary sources. IEEE Trans. on Signal Processing 49(9), 1837–1848 (2001)
8. Vincent, E., Gribonval, R., Plumbley, M.D.: Oracle estimators for the benchmarking of source separation algorithms. Signal Processing 87(8), 1933–1950 (2007)
9. Pulkki, V., Karjalainen, M.: Localization of amplitude-panned virtual sources I: stereophonic panning. Journal of the Audio Engineering Society 49(9), 739–752 (2001)
10. Vincent, E., Gribonval, R., Févotte, C.: Performance measurement in blind audio source separation. IEEE Trans. on Audio, Speech and Language Processing 14(4), 1462–1469 (2006)
11. Yılmaz, O., Rickard, S.T.: Blind separation of speech mixtures via time-frequency masking. IEEE Trans. on Signal Processing 52(7), 1830–1847 (2004)

Author Index

Printing: Mercedes-Druck, Berlin
Binding: Stein+Lehmann, Berlin